GW00726845

THE JURASSIC SYSTEM
IN GREAT BRITAIN

Oxford University Press, Ely House, London W. 1

GLASGOW   NEW YORK   TORONTO   MELBOURNE   WELLINGTON
CAPE TOWN   SALISBURY   IBADAN   NAIROBI   DAR ES SALAAM   LUSAKA   ADDIS ABABA
BOMBAY   CALCUTTA   MADRAS   KARACHI   LAHORE   DACCA
KUALA LUMPUR   SINGAPORE   HONG KONG   TOKYO

Portland Beds, West Weare Cliff, Isle of Portland.

Cliff of Portland Sand and the Cherty Series of the Portland Stone, from the top of which the Freestone Series has been quarried. The beginning of Chesil Beach is seen on the left, with Portland Harbour beyond.

THE
JURASSIC SYSTEM
IN GREAT BRITAIN

BY

W. J. ARKELL

OXFORD

AT THE CLARENDON PRESS

'If the elucidation of earth history and the origin and evolution of life on the globe are not of prime importance as ends in themselves, if the whence and the why and the whither are not supreme, then indeed has our lot fallen among evil days.'

E. W. BERRY, 1919
Address before the Geological Society of Washington
Amer. Journ. Sci., *vol.* xlviii, *p.* 11.

FIRST PUBLISHED 1933
REPRINTED LITHOGRAPHICALLY IN GREAT BRITAIN
AT THE UNIVERSITY PRESS, OXFORD
BY VIVIAN RIDLER
PRINTER TO THE UNIVERSITY
1970

PREFACE

IN England, in the sphere of Jurassic geology, we are wardens of a classic area, for our cliffs and quarries are the standards of comparison for the whole world. A German authority, Dr. Hans Salfeld, remarked after a brief study in 1914: 'Research on the faunas and their succession shows that the English Upper Jurassic can be taken as the type of that of North-West Europe, in the most complete development anywhere yet known.' He had studied only the Upper Jurassic, but the same could with equal truth be said of the Lower.

This is no mean heritage. In our Jurassic rocks all the principles of stratigraphy are illustrated perhaps more clearly than in any other part of the geological record. Palaeontologically, too, the system contains an unequalled wealth of materials; and for the evolutionist, the ecologist and the palaeogeographer no more favourable field exists.

The aim of this book is, first and foremost, to provide a general description of the Jurassic rocks of the British Isles, indicating what work has been done and where the information is to be obtained, and to illustrate some of the magnificent type-sections. It has been found that to give even an outline account of the rocks and the various changes that they undergo as they are traced across the country, pointing out the presence and significance of discordances, and arranging the facts in sequence as data for the elucidation of earth-history, is a matter of considerable difficulty within the limits of one volume.

Although such gigantic compilations as *The Jurassic Rocks of Britain* (in 5 volumes, although Scotland and Ireland are not included) have long remained invaluable sources of facts for those who knew how to interpret the information they contain in the light of subsequent discoveries, it is hardly likely that they will be repeated or brought up to date. They labour under a fundamental disability, for by striving after an illusory completeness they become more and more encumbered with detail as new discoveries are made —a negation of progress. Unless the endless description of sections and listing of fossils be regarded as a worthy aim in itself, it will be agreed that the discovery of new exposures, the filling in of gaps, should lead to the elimination of detail and enable generalizations to be made.

I have not attempted, therefore, to compile a summary of everything known to date concerning the Jurassic System, or to write a handbook suitable for taking into the field, with descriptions of quarries for the illustration of local geology. It is assumed that any one who wishes to ascertain the exact succession displayed at any particular exposure will consult one of the many excellent Sheet or District Memoirs of the Geological Survey, or the detailed papers cited in the bibliography and footnotes.

The difficulties experienced, especially by foreigners, in extracting any coherent story from the existing literature of the Jurassic System are illustrated by several authoritative treatises on stratigraphy that have been published on the Continent in the last twenty or thirty years. Haug's *Traité de Géologie* and the equally comprehensive work with the same title by De Lapparent contain perhaps the most complete reviews of the Jurassic System yet written; and although both are twenty years out of date and familiarize the reader with classifications and terminologies that have been superseded, they still remain the best introductions to the subject extant. But when the British Isles are considered, the statements are not twenty but forty years out of date. The reader might think that the type-localities of the Lias, the Bathonian, the Oxfordian, the Kimeridgian, the Portlandian and the Purbeckian—names used the world over—were in a little-known land unfrequented by geologists.[1] For this we, or at least our predecessors, are largely to blame. It is unreasonable to expect a foreigner to search through scores of back numbers of the journals of country field-clubs, unknown to him and virtually inaccessible; or to read the exhausting reports of innumerable excursions, in which, among accounts of the weather and the tea, so many conclusions are interred. And if the specialist, as has been proved, will not undertake this work even though he have a definite object in view, how much less will the student do so, when he can make his choice from many attractive sciences ?

I think, therefore, that the need for some co-ordinating work will hardly be questioned; but any such attempt as this is so largely dependent on the personal factor of selection that it is too much to hope that the result will find favour with all. Nevertheless, I take heart from the thought that if a score of geologists were to make their own selections of relevant facts, there would be as many differences of opinion.

One or two points need explanation. I have striven hard to avoid unevenness of treatment in the Stratigraphical Part, but am aware of having failed to eliminate it altogether. A good deal of unevenness, however, is independent of the writer and reflects inconsistencies in our knowledge. Some formations are much better known than others and are likely long to remain so; yet, paradoxically, it is not these that require the most space in description: rather it is the formations about which there is still considerable uncertainty in some parts of the country, and in which correlations are not always firmly established. The Great Oolite Series and the Corallian Beds, for instance, cannot be so readily summed up as the Inferior Oolite. Thanks to the work of the late S. S. Buckman and Mr. L. Richardson, the Inferior Oolite Series from Dorset to Northamptonshire is so thoroughly known that almost perfect generalization is possible: a detailed section of the beds can be drawn right across the country. Where there have been no recent revisions, on the other hand, it is necessary to give more details and also to document the statements more fully. Quite apart from this, it is obvious that the oolites, with their numerous transformations of facies, occupy more space in treatment than relatively uniform deposits like the Lias.

A different form of unevenness that is difficult to avoid entirely is geographical. The emphasis laid on Dorset, however, is intentional. The description

---

[1] Haug was able to write so recently as 1911: 'Malgré que les coupes de l'île de Portland et des environs de Swanage et de Kimeridge soient classiques, les nombreuses divisions locales établies par les géologues anglais ne permettent guère de parallélismes précis avec les autres régions, car les Céphalopodes sont rares et la plupart des espèces que l'on a citées sont basées sur des figures très médiocres et auraient besoin d'une sérieuse revision' (*Traité*, p. 1081). That the cephalopods were not the only matters about which he was misinformed is shown by a statement on the next page, to the effect that in Dorset: 'Le Purbeckien repose en discordance sur le Portland Stone et supporte lui-même en discordance le Wealdien.' Both assertions are untrue. Similar instances may be multiplied almost indefinitely when foreign literature is consulted.

of every formation has been started from there, using its cliffs as a standard, because they undoubtedly show the full marine sequence of the Jurassic System incomparably better developed and better exposed than any other sections in the British Isles.

I am conscious of having perhaps dwelt in the first chapter too long for some tastes on the stratigraphical and chronological terms in modern use and on the refinements of meaning attached to them. But I hope that if these terms are unmasked at the outset they will have ceased to trouble the reader when he comes across them in the ensuing chapters, and afterwards in other works. The terms have their uses in clarifying our conceptions. Moreover, it is essential to realize from the beginning that, formidable though they seem at first, and fraught with weighty possibilities, they all lead to nothing. Although we have built up an imaginary time-scale as a pure abstraction, entirely separate from any known sequence of strata at any one place, yet it advances us no further towards the measurement of earth history by an absolute chronology; for the units are still of unknown and unequal duration, whatever scale we employ. Whether we speak of hemeræ, of biochrons or of zone-moments, is a matter of personal choice or local expediency, but whichever we choose we are only, as it were, playing a game, in which we decide beforehand what are to be the rules. As if to compensate themselves for their ignorance of the real values of their units, geologists have made much of the theoretical differences between them and have invented wonderful words to distinguish them.

One day, perhaps, it may become possible to attach our present chronology to an absolute time-scale, measured in fixed, definite units. Either we may gain some conception of the average duration in years of ammonite species, or we may discover the frequency of the pulse of earth-movements controlling the cycles of sedimentation (and so, indirectly, the life-succession). The realization of this ideal seems as far off as ever, but meanwhile it is something to perfect our shadow-chronology and have it ready, awaiting the time when the key will be supplied to interpret its relative conceptions in absolute terms.

Any value this book may have is largely due to the help I have received from a number of kind friends. First and foremost I am indebted to Prof. W. J. Sollas, F.R.S., in whose department the work has been done, and on whose students much of the matter has been tried out in lectures. He and Mr. L. Richardson, F.R.S.E., Prof. A. E. Trueman, Prof. A. Morley Davies, Dr. R. H. Rastall, Dr. J. A. Douglas, Dr. F. L. Kitchin, F.R.S., and Dr. J. Pringle, have very kindly read through one or more chapters in manuscript or typescript and have made valuable comments and corrections. How much the stratigraphical part owes to conversations and correspondence with Mr. Richardson, I cannot attempt to assess. I will only say that without his interest and collaboration in certain researches in the field it would have been impossible to write the chapters on the Inferior Oolite and Great Oolite in a form presentable for publication. He has also given me reprints of many papers and put at my disposal his collections and the manuscript of his recently published revision of the Inferior Oolite of the Sherborne district. To Dr. J. Pringle I am also indebted for his help in replying to inquiries on various points and for kind gifts of rare papers; and to Dr. L. F. Spath for much information imparted in the course of discussions on the ammonites.

Professor P. F. Kendall, F.R.S., generously lent me some of his note-books; Dr. R. H. Rastall put at my disposal the results of his work on the heavy minerals of the Yorkshire Estuarine Series; Dr. Vernon Wilson kindly helped with the Yorkshire Corallian, and Mr. Malcolm MacGregor with the rocks of Northern Trotternish.

To the skilful collaboration of Mr. J. W. Tutcher is largely due the series of plates at the end of the volume, illustrating the zonal species of ammonites. It was my own experience to be familiar for years with the names of some of the zonal ammonites, which appear in every paper and memoir, before being able to discover to what the names actually applied. The lack of figures of even the most commonly quoted species is astonishing. It was therefore my intention to figure all the principal species, from either the type-specimens or properly authenticated material, or, failing both, from the protographs; and to this end Mr. Tutcher readily consented to collaborate. Since the zonal indices are frequently foreign and are by no means always abundant in this country, even when the proper zones are developed, absolute authenticity cannot be claimed for every species figured; but it is believed that most of the identifications are close enough to fall within the range of individual variation. Mr. Tutcher has been almost entirely responsible for selecting the specimens from the Lias and Inferior Oolite, and nearly all of them are from his own collection; for the higher formations I have been mainly responsible. Where the exact identification of a species is notoriously a matter of controversy (as with the Perisphinctids of the Corallian) I have thought it better to refigure the type-specimens, even though they may provide indifferent material, and thus to make the characters of the types more familiar, rather than to risk adding to the confusion by publishing yet another mis-identification. It is to be regretted that the figures of some of the species are of necessity greatly reduced in size, but it was considered better to give a reduced figure of the larger ammonites than none at all. The amount of the reduction and the dimensions of the original are always stated. For the loan of specimens or the gift of casts, photographs and advice, I am indebted to Dr. F. L. Kitchin, Dr. W. D. Lang, Dr. L. F. Spath, and Prof. A. Morley Davies.

For permission to reproduce or use drawings or photographs from their works throughout the book, my thanks are due to Prof. A. E. Trueman, Mr. J. W. Tutcher, Dr. R. H. Rastall, Mr. L. Richardson, Prof. P. F. Kendall, Dr. F. A. Bather, Dr. R. Brinkmann, Dr. E. Neaverson, Mr. M. Black, Prof. S. H. Reynolds and Prof. A. Morley Davies (who has also kindly supplied a revised map of the outcrop of the Arngrove Stone); for further permission to make use of figures I am indebted to Prof. A. Gilligan (for allowing me to select some photographs from the Godfrey Bingley Collection), to Mrs. Buckman, and to the Councils of the Royal Society, the Geological Society of London, the Geologists' Association, the Palaeontographical Society, and the Yorkshire Geological Society, the Director of the Geological Survey, and the Controller of H.M. Stationery Office.

Finally, the unfailing courtesy and patience shown by all at the Clarendon Press have made the production of this book a pleasure.

W. J. A.

OXFORD, *January* 1933.

CONTENTS

PART I
CLASSIFICATION AND CHRONOLOGY

PART II
DISTRIBUTION AND TECTONICS

PART III
STRATIGRAPHY
Lower Jurassic or Lias

Middle Jurassic: Lower Oolites or Dogger

Upper Jurassic: Middle and Upper Oolites or Malm

PART IV
PALAEOGEOGRAPHY

LIST OF PLATES

Figures 15, 30, 35 are folders facing pages 84, 165, 197 respectively.

CLASSIFICATION AND CHRONOLOGY

CHAPTER I

THE CLASSIFICATION OF THE ENGLISH JURASSIC ROCKS, AND THE PARTITION OF JURASSIC TIME

'England, by reason of the simple stratigraphical relations of its beds, the usually marked lithological peculiarities of each individual member of the series, and the extraordinary wealth of its fossil remains, became the classic ground of the Jurassic System. Thus the classification of the British Jurassic strata, laid down by William Smith and but little modified by Conybeare, Phillips, Buckland and de la Beche, supplied the names under which geologists have endeavoured to range the contemporaneous formations all over the world.'

K. von Zittel, 1899, *Geschichte der Geologie und Paläontologie*, p. 659.

IN the early stages of the inception of this book it was proposed to translate all our haphazard terms, such as Inferior Oolite and Corallian Beds, into the properly systematized stage-names used on the Continent—Bajocian, Bathonian, Argovian and the rest. But the difficulties soon began to pile up to such an extent that they formed a mountain of polemical matter barring further progress: the objects of the book began to be in danger of being lost sight of amid the monstrous and sterile problems of stratigraphical nomenclature. These problems have obstructed geologists for a century and they can only be solved by an international committee with dictatorial powers, similar to that which has in some measure introduced order into the nomenclature of zoology. My object never was to attempt a solution of these problems single-handed, but to give descriptions and correlations of the Jurassic rocks of the British Isles, properly arranged as documents of history, and to draw as many historical conclusions from them as possible. The system of classification is therefore a secondary consideration, and all that is required of it is efficiency.

Although many of the systematized stage-names used on the Continent are taken from English localities (for example Oxfordian, Kimeridgian) they do not correspond even approximately with the strata which English geologists have always associated with those names. Directly we begin to apply them to the English rocks we are faced with fundamental difficulties. Is such a term as Portlandian to be used as if it were synonymous with Portland Beds, or as its author, d'Orbigny, used it, to embrace about half of our Kimeridge Clay? Is Oxfordian to be applied only to the Corallian Beds and uppermost Oxford Clay because it was so used by d'Orbigny, although he was oblivious of the fact that he was excluding the bulk of the Oxford Clay from the Oxfordian? Or, alternatively, which of the numerous subsequent attempts at adjustment are to be accepted? Once we deviate from the original meanings of the terms and abandon the principle of priority, we lose our hold on the only lifeline that can save us from the slough of conflicting opinions.[1]

If we are to describe our rocks under names defined according to the

---

[1] It must be remembered that a rigid application of the rule of priority would invalidate nearly all d'Orbigny's stage-names, for most of them had been previously used as stratigraphical terms with different meanings. There seems no reason why the addition of *-ien* or *-ian* to a name should justify an author in overriding all previous usages (see p. 13).

development of the strata on the Continent, a close study and revision of the foreign rocks must precede any attempt to classify our own. Even were such an undertaking possible, or the necessary postponement of work on our own rocks desirable, we should often find that when a French or German term had at last been thoroughly understood and defined, the development of the strata in the place of its origin would be so different from anything we know in England as to render the term quite unsuitable and indefinite in our own country. The tribulations that beset the English stratigrapher, attempting to use d'Orbigny's stage-names conscientiously, resemble those of a Chinese historian struggling to marshal his dynasties into such vague and exotic periods as The Renaissance, The Perpendicular, The Hyksos, or The Pre-Raphaelite.

When geologists of every nation have studied, described and redescribed the succession of rocks, and above all of the fossil faunas, in their own countries, bringing their correlations up to date with the advances of the sole international medium, palaeontology, only then will the time be ripe for standardizing stratigraphical nomenclature. Meanwhile each country must use its own terminology, evolved to suit its own particular circumstances. The multiplication of local names is no evil, for no one need learn them all. On the other hand, the more detailed they are, the more accurately will they enable foreigners to comprehend the palaeontological succession, on which alone correlations can be made, and which must eventually form the basis of any international classification.

We need not, therefore, despise our own indigenous terminology in speaking or writing of the English rocks. Apart from the advantage of their perfectly definite meaning, the names by which the Jurassic formations in England have been known for so long are worthy to be preserved because they are the earliest terms ever used in stratigraphical geology. They embody the first beginnings of the science, and most of them are a direct heritage from its originator, William Smith. Sir Archibald Geikie wrote in 1897:

'The growth of stratigraphical nomenclature is eminently characteristic of the early rise and progress of the study of stratigraphy in Europe. Precisians decry this inartificial and haphazard language, and would like to introduce a brand new harmonious and systematised terminology. But the present arrangement has its historical interest and value, and so long as it is convenient and intelligible, I do not see that any advantage to science would accrue from its abolition. The method of naming formations or groups of strata after the districts where they are typically developed has long been in use and has many advantages, but it has not supplanted all the original names, and I for my part hope that it never will.'[1]

I. WILLIAM SMITH AND THE STRATAL TERMS

As the history of the familiar names Cornbrash, Great Oolite and the rest is closely bound up with the history of William Smith and the development of his fundamental theory of the constant order of superposition of the strata and the possibility of recognizing them by means of their organic contents, it is of no small interest to trace the origin of what Adam Sedgwick called 'those arbitrary and somewhat uncouth terms which we derive from him as our master'.[2]

---

[1] A. Geikie, 1897, *The Founders of Geology*, p. 244.
[2] Adam Sedgwick, Address to the Geological Society at the presentation of the first Wollaston Medal to William Smith, Feb. 18th, 1831.

A manuscript note of William Smith's, dated 'Scarborough, May 17, 1839', throws interesting light on their origins.[1]

'For several years after the foundation of the earth's history was securely laid', he wrote in reminiscence, 'we had no words for the science, no language in which we could convey our ideas; its present comprehensive name of Geology remained un-noticed in dictionaries and unuttered in England, and usage had scarcely settled whether the word strata should not have an *s* appended; but how numerous are now the words from the dead languages which geology has revived and brought into common use all over the world!

'Much doubt remained for a long time whether the science, like chemistry, should not have a language of its own; and I, so very incompetent to the task, thought much about a new nomenclature, and have been at different times strongly urged to it by deep-learned men; but having dictated, off-hand, in the plain language of the country, a tabular view of the science to my two first pupils, the Rev. Benjamin Richardson and the Rev. Joseph Townsend, that crude manuscript, without any revision whatever, was faithfully transcribed from one to another, and soon des-patched to remote parts of the world.

'The new cultivators of the science found, as I had done, the necessity of accom-modating their language to those in the country from whom they had to collect the facts; and so, in transmitting by the press the knowledge acquired, some old Saxon and British words have been brought into use. . . .'[2]

The crude manuscript referred to was the first stratigraphical table, a 'Table of the Order of Strata and their embedded Organic Remains in the vicinity of Bath, examined and proved prior to 1799'. It was designed to accompany the first geological map ever made, showing the country within a radius of five miles round Bath. Both were presented to the Geological Society of London, by whom they are still preserved in Burlington House[3] (see p. 4, col. 1).

The story of how this table came to be drawn up has been interestingly told by William Smith's nephew, John Phillips, who wrote the *Memoirs* of his illustrious uncle.

During the years 1794–9, when Smith was living near Bath as resident engineer of a branch of the Kennet and Avon Canal, and was compiling this first geological map, he made the acquaintance of an enthusiastic collector of fossils, the Rev. Benjamin Richardson of Farleigh Hungerford.

'The result of a meeting between two such reciprocally adjusted minds was an electric combination; the fossils which the one possessed were marshalled in the order of strata by the other, until all found their appropriate places, and the arrange-ment of the cabinet became a true copy of nature.

[1] Printed in Phillips's *Memoirs of William Smith*, 1844, pp. 72–3.

[2] For instance: marl, clunch, lias, brash. Martin Lister used the words clunch, marle and Fuller's Earth as early as 1683, but only as types of clay in general.

[3] This can be truly said to be the first *geological* map, in which colours were used to denote strata as distinct from mere soils, and to show their order of superposition. Claims have been made for several others: Zittel asserts that the first coloured geological maps were those of Gottlieb Gläser (Leipzig, 1775) and Wilhelm von Charpentier (Chur-Saxony, 1778), (Zittel, *History of Geology and Palaeontology*, English ed., 1901, p. 38), but these are only soil-maps or agriculture maps, of the kind suggested by Martin Lister as early as 1683 ('An Ingenious Proposal for a new Sort of Maps . . .', *Phil. Trans. Royal Soc.*, 1684, vol. xiv, No. 164, pp. 739–46). Of the same kind are the county survey maps published by the Board of Agriculture in 1794, 1796 and subsequent years, of which Conybeare and Phillips uphold the priority in their *Geology of England and Wales* (1822, p. xlv). Mr. T. Sheppard remarks that, although these may have suggested to William Smith his system of colouring, they give no indication of stratigraphical structure ('William Smith: his Maps and Memoirs', 1915, *Proc. Yorks. Geol. Soc.*, N.S., vol. xix, pp. 100–1).

TABLE I.—COMPARISON OF THE EARLIEST TABLES OF STRATA (JURASSIC PORTIONS)

Smith's MS. Table 1799 (unpublished).	Smith's MS. Map Table, 1812 (unpublished).	Townsend's Table 1813.	Smith's Improved Table 1815–16.	Buckland's Table 1818.	Divisions now used.
	Purbeck Stone, Kentish Rag			Purbeck Beds	PURBECK BEDS
	Limestones of the Vales of (Pickering and) Aylesbury	? Superior Oolite	9. Portland Rock / 10. Sand	Portland Stone	PORTLAND BEDS: Limestone / Sand
	(Iron Sand and Carstone) [misplaced; Cretaceous]		11. Oaktree Clay	Kimmeridge Clay (from Webster, 1816)	KIMERIDGE CLAY
		Calcareous Grit / Coral Rag	12. Coral Rag and Pisolite / 13. Sand	Upper or Oxford Oolite, Coral Rag Calcareous Grit	CORALLIAN BEDS: Upper Calc. Grit / Oolites / Lower Calc. Grit
3. Clay	Dark blue shale		14. Clunch Clay and Shale	Oxford, Forest or Fen Clay	OXFORD CLAY
4. Sand and Stone			15. Kelloway's Stone	Kelloway Rock	KELLAWAYS BEDS: Sandstone / Clay
	Cornbrash		16. Cornbrash	Corn Brash	CORNBRASH
5. Clay / 6. Forest Marble / 7. Freestone	Forest Marble Rock / Great Oolite Rock	Forest Marble / Great Oolite or Bath Stone	17. Sand and Sandstone [Hinton Sand] / 18. Forest Marble / 19. Clay over Upper Oolite [Bradford Clay] / 20. Upper Oolite	Forest Marble / Stonesfield Slate / Great Oolite	GREAT OOLITE SERIES: Forest Marble / Limestones / Stonesfield Sl.
8, 9. Blue and yellow clay / 10. Fuller's Earth / 11. Bastard Fuller's Earth and sundries	Clay	Clay	21. Fuller's Earth and Rock	Fuller's Earth	FULLER'S EARTH: Clay / Rock / Clay
12. Freestone	Under Oolite	Inferior Oolite	22. Under Oolite	Inferior or Bastard Oolite	INFERIOR OOLITE SERIES: Upper / Middle / Lower
13. Sand	Sand	Sand	23. Sand	Sand of Inferior Oolite	
14. Marl, blue	Blue Marl	Blue Clay	24. Marlstone / 25. Blue Marl	Lias	UPPER LIAS
					MIDDLE LIAS
15. Blue Lias	Blue Lias	Lyas	26. Blue Lias	Lias	LOWER LIAS
16. White Lias / 17. Marlstone, indigo and black marls	White Lias		27. White Lias		RHÆTIC BEDS

'That such fossils had been found in such rocks was immediately acknowledged by Mr. Richardson to be true, though the connection had not before presented itself to his mind; but when Mr. Smith added the assurance that everywhere throughout this district, and to considerable distances around, it was a general law that the "same strata were found always in the same order of superposition and contained the same peculiar fossils", his friend was both astonished and incredulous.'

A number of excursions were undertaken, and Smith soon convinced his friend of the reality of his discoveries. His first triumph consisted in correctly forecasting both the nature of the rock and the contained fossils that would be found on the Inferior Oolite outlier of Dundry Hill.

About this time he also became associated with the Rev. Joseph Townsend, a man of letters, and later Rector of Pewsey, Wilts., who accompanied them on their geological excursions in the neighbourhood of Bath.

'One day, after dining together at the house of the Rev. Joseph Townsend,' Phillips narrates, 'it was proposed, by one of this triumvirate, that a tabular view of the main features of the subject, as it had been expounded by Mr. Smith, and verified and enriched by their joint labours, should be drawn up in writing. Richardson held the pen and wrote down, from Smith's dictation, the different strata according to their order of succession in descending order, commencing with the chalk, and numbered, in continuous series, down to the coal, below which the strata were not sufficiently determined.

'To this description of the strata was added, in the proper places, a list of the most remarkable fossils which had been gathered in the several layers of rock. The names of these fossils were principally supplied by Mr. Richardson, and are such as were then, and for a long time afterwards, familiarly employed in the many collections near Bath. Of the document thus jointly arranged each person present took a copy, under no stipulation as to the use which should be made of it, and accordingly it was extensively distributed, and remained for a long period the type and authority for the descriptions and order of superposition of the strata near Bath.' [1]

In consequence of this liberality with which Smith shared his ideas with his friends, he was anticipated in the publication of all his stratal terms. The Rev. Joseph Townsend printed the first list in 1813, at the same time paying a handsome tribute to his teacher.

'To a strong understanding, a retentive memory, indefatigable ardour, and more than common sagacity, this extraordinary man unites a perfect contempt for money, when compared with science. Had he kept his discoveries to himself, he might have accumulated wealth: but with unparalleled disinterestedness of mind, he scorned concealment, and made known his discoveries to every one who wished for information.' [2]

In the same year, 1813, the list of strata quoted by Townsend was copied in a text-book.[3]

This first published list (see p. 4, col. 3) is in some respects more complete than the manuscript table of 1799, and partly anticipates the revisions published by William Smith with his map of England and Wales in 1815 and in the *Stratigraphical System of Organised Fossils* (1817), but there are serious omissions and errors. It will be seen that the manuscript table of 1799 is incomplete, for it has a large lacuna between the Clunch or Oxford Clay and

[1] J. Phillips, 1844, *Memoirs of William Smith*, pp. 28–9.
[2] J. Townsend, *The Character of Moses Established for Veracity as an Historian, recording events from the Creation to the Deluge*, 1813, vol. i, preface. Tables of Strata, pp. 100 et seq.
[3] Bakewell's *Introduction to Geology*, 1813, p. 259.

the Sand beneath the Chalk (Upper Greensand). William Smith was perfectly well aware of the existence of the Corallian Beds, Kimeridge Clay and Portland Beds, but his table referred to the country on the road from Bath to Warminster, and along this line the strata which he failed to show are wanting, for the Gault and Greensand overstep on to the Oxford Clay. The Cornbrash, which he added to his revised tables, was probably included in Bed 4, the 'stone' of the 'sand and stone' above the Forest Marble clay, the sand certainly referring to the Hinton Sand (a part of the 'Forest Marble' Series). Townsend attempted to make the table more generally applicable by inserting the 'Superior Oolite' (Portland Stone, see p. 167) and the Coral Rag and [Lower] Calcareous Grit, but he placed the last two in inverted relation to one another and omitted the Oxford and Kimeridge Clays. Thus, while William Smith's table is accurate for the district of which it treats, although incomplete when a wider area is considered, Townsend's version, copied by Bakewell, is inaccurate and misleading.

A new point in Townsend's table is the first mention of the term Inferior Oolite, in place of the Under Oolite consistently used by Smith. In it also appeared for the first time the names Great Oolite or Bath Stone, not figuring in the manuscript table of 1799.

In the improved tables published by William Smith in 1815 and 1816 (p. 4, col. 4) the missing strata above the Forest Marble are filled in, the Cornbrash, Kellaways Rock (spelled Kelloway's Stone) and other beds being enumerated, up to the Portland and Purbeck Stones. Just as some of Smith's other words, such as clunch, marl, and brash, were borrowed from agriculture, so these last two terms, Portland Stone and Purbeck Stone, were simply adopted from the current language of quarrying, building and architectural circles. That they were in use at least as early as 1668 is proved from their appearance in the *Discourses* of Dr. Robert Hooke,[1] and they had already been adopted in 1811–12 by Thomas Webster.[2]

Webster was a gifted artist and geologist, who came into prominence in 1814 with a treatise on the Isle of Wight Tertiaries read before the Geological Society. In 1811–12 he was sent on a mission to Purbeck by Sir Henry Englefield, to study the strata and to compare them with those in the Isle of Wight. His series of letters were published in 1816 in Englefield's *Description of the Principal Picturesque Beauties &c. of the Isle of Wight*, accompanied by an excellent geological map and accurate drawings. One of the drawings, showing the quarrying of the stone at Tilly Whim, where the industry has long since died out, is reproduced here (fig. 81, p. 484). In addition to adopting the terms Purbeck Beds and Portland Oolite, Webster seems to have been the first to use the expression Kimeridge Clay.[3] The Portland Sand, although

---

[1] Posthumous Works of Robert Hooke, 1705: *Lectures and Discourses of Earthquakes and Subterraneous Eruptions* (delivered 1668), pp. 289, 320. Burford-stone and Northamptonshire-stone are also mentioned. This advanced thinker and observer, unlike his contemporaries, Robert Plot and Martin Lister, perceived the true nature of fossils, and he figured a number of Jurassic and Carboniferous species on admirable plates. From the surmise that ''tis not improbable, but that many Inland Parts of this Island, if not all, may have been heretofore all cover'd with the Sea, and have had Fishes swimming over it' he goes on by a comparison between the fossils and their living counterparts to suggest that there have been changes of climate produced by alteration in the earth's axis. He was also the first discoverer of vegetable cells.

[2] Who was wrongly credited with the first use of them by H. B. Woodward in his *Jurassic Rocks of Britain*.   [3] In Englefield's *Description . . . of the Isle of Wight*, 1816, Plate 47.

entered as 'sand' in its proper place in William Smith's table, was not named, but received its name later (in 1827) from Fitton.

It is noticeable that in Smith's revised tables the term Marlstone is for the first time used in its present sense, having been previously applied to some part of the Rhætic Beds or Tea Green Marls, between the White Lias and the Red (Keuper) Marls.

The next and last great addition to the table of the Jurassic strata was made by Dean Buckland, in 1818, the year before he was elected to the newly founded chair of Geology at Oxford, where he already held the Readership in Mineralogy. His table[1] (p. 4, col. 5) shows that the scene had now shifted from Bath to Oxford, for we have Upper or Oxford Oolite, Oxford, Forest or Fen Clay, and Stonesfield Slate. Oxford Clay has survived, while its synonyms, descriptive of the kind of country to which it gives rise, have, like Smith's earlier term, Clunch Clay, been forgotten. On the other hand Oxford Oolite, which is in many ways preferable to Corallian or Coralline Oolite, has in spite of its priority and legitimate form been discarded, although Buckland mentioned as typical localities Headington (Oxon.), Calne (Wilts.) and New Malton (Yorks.). He described it as 'perishable freestone, composed of oolitic concretions and shelly fragments, united by a calcareous cement, with a waterworn surface at Headington'; clearly, therefore, he meant it to apply to the detrital or 'Wheatley Limestones' facies of the Osmington Oolite Series (see p. 405). We now know that the Coral Rag around Oxford, which he placed below the Oxford Oolite, is in reality contemporaneous with it, and Oxford Oolite therefore has prior claim to be the name used for the Osmington Oolite Series (so called by Blake and Hudleston in 1877). But the Oxford Oolite of Headington and elsewhere round Oxford is only a detrital deposit made up of coral- and shell-fragments, and is as abnormal a facies as the Coral Rag. There seems, therefore, to be some justification for retaining Blake and Hudleston's name for the series as a whole, embodying as it does the name of a locality where the oolite is largely a true oolite and is far more thickly developed. (The typical sections at Osmington are shown in Plate XVI, p. 380.)

Perhaps the chief objection to reviving the term Oxford Oolite in its original sense (more important than the essentially non-oolitic structure of the great bulk of the Wheatley Limestones) is the variety of meanings which the name suggests. Phillips in 1871 used it for the whole Corallian plus Oxford Clay, and Buckman and some foreign geologists have used it as equivalent to Corallian Beds. The term Oxfordian, moreover, has been employed in England and abroad in so many senses that any name including the word Oxford is liable to be misinterpreted.

The uncertainty prevailing at that time as to the position of the Stonesfield Slate is indicated by Buckland's placing it above the Great Oolite and correlating it on the one hand with the sandy beds at Hinton Charterhouse, near Bath, which we now know to be part of the Forest Marble, and on the other with the Collyweston Slate, since proved to be Inferior Oolite. This last mistake survived until nearly fifty years later.

Of the other terms, the Calcareous Grit was transferred to its proper position below the Coral Rag, where it already figured in Smith's table as 'sand'.

[1] Appended to William Phillips's *Selection of Facts . . . to form an Outline of the Geology of England and Wales*, 1818.

The prefix Lower was added by Phillips in 1829, in describing Yorkshire; and at the same time he recognized its companion, the Upper Calcareous Grit. Webster's term Kimeridge (or Kimmeridge) Clay was adopted by Buckland to replace Smith's Oak Tree Clay, and has been in use ever since.[1]

II. D'ORBIGNY AND THE STAGES; A SYSTEMATIZED CLASSIFICATION BASED ON A COMBINATION OF PALAEONTOLOGY AND STRATIGRAPHY

William Smith had no claims to rank as a palaeontologist according to the modern meaning of the word, and yet he was certainly a practical stratigraphical palaeontologist of no mean attainments. Since the early days of his boyhood in the Oxfordshire village of Churchill, where he played at marbles with the Terebratulids from the *Clypeus* Grit and noticed that the large *Clypei* were used as 'poundstones' or weights by the cottagers, he was a close observer and collector of fossils. He was the first to recognize their value in the determination of strata, and we are told how he rarely returned from his numerous rides and walks without his pockets bulging with 'identifying fossils', to guide him in entering the boundary-lines upon his map. In his stratigraphical arrangement of the Rev. Richardson's collection, which caused, as we have just seen, such incredulity and later, when the principles were demonstrated, such unbounded admiration, he showed himself to have mastered thoroughly the successive assemblages of fossils contained in the oolitic formations. That he could turn them to such useful account although he had no names for most of them is an object-lesson which many modern workers might with advantage take to heart.

The axiom that 'the same strata are found always in the same order of superposition and contain the same peculiar fossils' was quickly accepted and the correlation of the strata in other parts of England, partly by means of lithology and partly by means of their fossils, was rapidly pushed forward by Smith's successors. Before 1830 there had already appeared the two monographs which remain undisputedly the prototypes of all regional stratigraphical treatises—Fitton's great memoir on *The Strata between the Chalk and the Kimeridge Clay in the South-East of England* (1827) and John Phillips's *Illustrations of the Geology of Yorkshire, or a Description of the Strata and Organic Remains of the Yorkshire Coast* (1829). As new areas came to be studied it was soon realized that the lithological characters of the formations changed from place to place, but that, even if they lost their identity entirely, as often happened in Yorkshire, they could still be recognized by their contained fossils.

This conception of the independence of fossils from facies was perfectly familiar to Phillips by 1829 and can scarcely have been foreign to William Smith ten years earlier, though we seldom find it formulated. It is only occasionally that a gleam of light reveals the inner working of men's minds about this time, for the output of a great mass of important descriptive matter was engaging most of their attention.

Some of the first definite attempts to formulate ideas of detailed correlation solely by means of fossils emanated from an obscure geologist, Louis Hunton, whose collecting ('zonal' in the full modern sense of the word) among the

[1] For note on the spelling of this see footnote on p. 441.

Middle and Upper Lias of the Yorkshire Coast, at the same time as W. C. Williamson's, will be referred to later on.

If these ideas were slow to take shape in England, they soon afterwards made rapid progress in France, led by a man of vision, Alcide d'Orbigny. Realizing the all-important factor in the correlation of strata in different localities to be the fauna, and impatient at the indefinite multiplication of local names for beds of differing lithological development but of the same age, he sought to sweep aside lithology and give to the beds containing each successive assemblage a single name. The groups of strata indicated by these names he called **stages**. 'It will be seen from the nomenclature adopted in naming these stages', he explained, 'that I have endeavoured . . . to choose names drawn from the localities where a stage is best developed,[1] in order to put an end to this jumbled nomenclature, based on the local lithology, which varies so greatly according to the place, and on the fossils which happen to be predominant at one locality, but may be wanting elsewhere.'[2]

Ten of these Stages were to be recognized in the Jurassic System, and no ambiguity was left as to the basis on which they were to be established. 'Geologists in their classifications', wrote d'Orbigny, 'allow themselves to be influenced by the lithology of the beds, while I take for my starting-point . . . the annihilation of an assemblage of life-forms and its replacement by another. I proceed solely according to the identity in the composition of the faunas, or the extinction of genera or families.'[3]

These words are momentous ones. D'Orbigny's stage-names have been so widely adopted that it is important that there should be no misunderstanding of their true nature, and it seems advisable that all who use them should have some acquaintance with d'Orbigny's methods and aims expressed as nearly as possible in his own words. At the end of the first volume of the *Paléontologie française; terrains jurassiques* he gives the following as part of a *Résumé géologique*:

'*Division of the Jurassic Rocks in Stages.*'

'Many subdivisions have already been proposed for the Jurassic rocks, some based on the lithological characters of the beds, others on the dominance of some fossil or another. I do not intend to discuss the worth of the successions established by these methods; all, when they are the result of direct observation and not of theorizing, set forth local or general facts of great interest; but when one comes to co-ordinate them, obstacles at once appear. How are groupings based only on lithological characters to be dealt with, when they have been seen to be misleading? On the other hand, how can one rely on the nomenclature of the fossils recorded in a succession of beds, when one sees these fossils identified by authors with such irresponsibility that it is often necessary to ignore half of the identifications? . . . Confronted with these insurmountable difficulties I have found only one solution possible, and that was to consult nature herself. Since my first observations on the rocks of France, I have realized that in crossing the successive beds from the older to the younger, I have everywhere met with the same sequence of fossil faunas, restricted within the same vertical limits in the geological succession,

---

[1] His system of naming the strata with terminations in -*ien* was started by Brongniart, who published a remarkable *Tableau des terrains qui composent l'écorce du globe* in 1829, in which he adapted some of the old names (Oxfordien, Portlandien, Purbeckien) and coined some new (Hâvrien); see table on p. 11.

[2] A. d'Orbigny, 1842–9, *Paléontologie française; terrains jurassiques*, vol. i, p. 604.

[3] A. d'Orbigny, loc. cit., p. 9.

whatever might be the lithological composition of the beds containing them.[1] I have also recognised that the lithology has done nothing but mislead observers, by making them see imaginary likenesses, such as between the ferruginous beds on the opposite sides of France, which contain totally distinct faunas (nevertheless confounded) although the same faunas are found on corresponding horizons in beds of different lithology and often remote from one another. I therefore set myself the task of following up the palaeontological horizons in the beds of which I was making a preliminary study, wherever they were to be found, in order to satisfy myself whether they signalized a definite period or only a local facies, governed by bathymetrical considerations. After having obtained in all parts of France—north, south, east and west, in Provence as in Normandy, in the Ardennes as well as in the Vendée,—the same results, and having found nothing but one confirmation after another over a period of fifteen years, without encountering a single contradictory fact, I at last became convinced that the Jurassic rocks were divisible into ten zones or stages, demarcated as well by the different faunas they contain as by stratigraphical boundaries [2] which reappeared again and again at every point. I have followed them one after another around the basins in France and outside; I have ascertained that in no single locality do they become confused, and that they represent as many distinct geological epochs succeeding one another in a constant and regular order. This fact gained, it remained to assure myself whether these different stages established in our country were the result of local circumstances peculiar to France, or if they were due to general causes operative simultaneously all over the world. Happily I had the means of solving this last problem.

'The researches carried out in Russia by Messrs. Murchison, de Verneuil, von Keyserling and Hommaire de Hell placed within my reach the knowledge, derived from a comparative study of the faunas in the Jurassic rocks of that country with those of France, that from the Crimea to Moscow, and thence as far as the north of the Urals, all the several members of the Jurassic formation observed corresponded, in their faunas, with our French stages. The same may be said of England and Germany, whither extend the same ancient basins as those in France. The Jurassic fossils of Central America, collected by Messrs. Darwin and Domeiko, have led me to the same conclusions, just as have those of the Province of Cutch in India. In fact these isolated localities, immensely remote from one another, have in common with our French stages not only a general similarity in their faunas, but frequently also several identical species, which prove their perfect contemporaneity. These distant confirmations of my own observations on the Jurassic rocks assured me at the same time that the causes of the separation of the stages were general. . . . I was therefore bound to adopt them for the double reason that there is nothing arbitrary about them and that they are, on the contrary, the expression of the divisions which nature has delineated with bold strokes across the whole earth.' [3]

These are the words of an enthusiast carried away by his subject and unfettered by too much knowledge. The idea that inspired them was magnificent, and if d'Orbigny, without being anticipated by any one with less breadth of outlook, had lived and launched his system half a century later, he would probably have rendered stratigraphy the greatest service standing to the credit of any man. By then he could no longer have entertained such an idea of universal uniformity and simplicity, and his stages would have been far more numerous. Like the many other aspects of nature, which have appeared to some to offer scope for grand generalizations, this turned out to be of almost infinite complexity.

---

[1] So far he is merely following William Smith.   [2] 'lignes de démarcation stratigraphique'.
[3] Alcide d'Orbigny, 1842–9, loc. cit., pp. 600–3.

TABLE II. *D'Orbigny's stages and zones, and the stratigraphical terms employed by Brongniart.*

BRONGNIART'S TERMS, 1829.	D'ORBIGNY, 1842–9.	
	Étages	*Zones de*
Calcaire PURBECKIEN	—	—
Calcaire PORTLANDIEN	PORTLANDIEN	*Ammonites giganteus* *Ammonites irius* *Trigonia gibbosa*
Marne argileuse HÂVRIENNE	KIMMÉRIDGIEN	*Ammonites lallieri* *Ostrea deltoidea* *Ostrea virgula*
Calcaire CORALLIQUE	CORALLIEN	*Ammonites altenensis* *Iceras arietina*
Marne OXFORDIENNE	OXFORDIEN	*Ammonites cordatus* *Plicatula tubifera* *Trigonia clavellata*
—	CALLOVIEN	*Ammonites Jason* *Ammonites refractus* *Ostrea dilatata*
Oolithe Miliaire	BATHONIEN	*Ammonites bullatus* *Terebratula digona* *Ostrea acuminata*
—	BAJOCIEN	*Ammonites interruptus* Brug. (*parkinsoni* Sow.) *Trigonia costata*
—	TOARCIEN	*Ammonites bifrons* *Lima gigantea*
—	LIASIEN	*Ammonites margaritatus* *Ostrea cymbium*
—	SINÉMURIEN	*Ammonites bisulcatus* *Ostrea arcuata*

Few of the geologists working in other countries at first adopted d'Orbigny's stages, for with their more detailed local knowledge they were unable to recognize any such ten divisions which might have been 'delineated by nature with bold strokes across the whole earth'. Quenstedt, who published his principal work *Der Jura* a decade later, but had already completed a detailed study of the Jurassic rocks of Württemberg, and was undoubtedly the greatest authority on the subject at that time, was withering in his criticism of the superficiality and inaccuracy of d'Orbigny's work. His ire was in large measure aroused by the fact that d'Orbigny had ignored the priority of his work and that of his compatriots. Nevertheless his eight pages of criticisms are extremely telling.[1]

Other workers (for the most part French and Swiss) took up d'Orbigny's idea but made more stages to suit their own districts. By 1860 there had already been added VESULIEN, ARGOVIEN and SEQUANIEN (Marcou, 1848), PTÉROCERIEN, ASTARTIEN and VIRGULIEN (Thurmann, 1852), DICERATIEN

[1] F. A. Quenstedt, 1858, *Der Jura* (Tübingen), pp. 17–24.

and SPONGITIEN (Étallon, 1857), PLIENSBACHIEN (Oppel, 1858), BRADFORDIEN and DUBISIEN (Desor, 1859), and MANDUBIEN (Marcou, 1860). The greatest manufacturer of stages was Mayer-Eymar, who brought out four works in 1864, 1881, 1884 and 1888, in which he introduced no less than 30 new names, all taken from places but differing slightly from the standard set by Brongniart in having terminations in -*on* or -*in*.[1] The game has not ceased since Mayer-Eymar, and there are now at least 100 stage-names to be fitted into the Jurassic System, defined with varying degrees of accuracy according to the capacity or the outlook of the author (see Appendix).

Thus by their ardour the disciples of d'Orbigny have entirely defeated his ends. They have 'improved' his system until nothing is left of it but a meaningless complex of overlapping stages of differing values, not in the smallest degree more efficient than the old nomenclature which it was designed to replace.

Yet by this painful process a truer picture of nature has been achieved. There was a cataclysmic bias which gave an inherent improbability to d'Orbigny's theory of ten successive faunas, occupying the whole earth for a span and then being swept away everywhere simultaneously. The wider and the more detailed that our knowledge grows, the more apparent does it become that the changes took place in a continuous stream. Earth movements, like evolution, never ceased, only varying in degree of slowness at different times, in different parts of the globe. By their agency the environment of the forms of life inhabiting the sea was constantly changing, and thus giving rise to incessant migration and to the extinction of some species and the evolution of new. A sentence written by Sir Andrew Ramsay in 1864 in reference to the Lias might well apply in some region or other to the whole sedimentary series:

'I incline to the idea that, considering the frequent large percentages of passage [of species from one part of the Lias to the next] (ranging as high as 50 per cent.), we are justified in supposing that migration of what were old species here into new areas elsewhere, and of certain older species from other areas into ours, may account for the very incomplete breaks in the succession of Liassic life in England, more especially if there were occasional pauses in the deposition of the strata.'[2]

In other respects too, there is much to criticize in d'Orbigny's scheme of Stages. In the first place, while decrying 'this jumbled nomenclature, based on local lithology . . . and on the fossils which happen to be predominant at one locality but may be wanting elsewhere', he not only adopts Thurmann's term Corallien, derived from William Smith's Coral Rag, but also coins an altogether new and horrible hybrid, 'Liasien'.[3] That he should have left these

---

[1] These terminations were designed to show that they were only to rank as substages, and the principle was accepted at the Third International Geological Congress at Berlin in 1883 (*Compte Rendu*, published in 1888, pp. 323 et seq.). The substage in -*in* was defined as equivalent in value to the French term 'assise', which has no exact equivalent in English. The higher of two substages was to end in -*in*, the lower in -*on*, in every stage subdivided.

[2] A. Ramsay, 1864, Address to the Geological Society, *Q.J.G.S.*, vol. xx, p. li.

[3] William Smith's term 'Lias', in use to this day among the quarrymen of the West of England, is often supposed to be a corruption of the word 'layers'; Buckman, however, considered it to be an ancient Gaelic word *leac*, meaning a flat stone, which is embodied in such place-names as Llechryd, Lechlade, Leckhampton, and survives in Brittany as *liach*, *leach*, a stone. In old French it occurs as *liais*, meaning a hard freestone from which steps and tombstones were made. (*Type Ammonites*, vol. i, 1910, p. xv.) Buckman is probably right, for in the West of England dialects 'layers' would never become 'lias' as it would in the mouths of cockneys.

inconsistencies in his otherwise uniform table of adjectival place-names is strange. Many of his followers have been equally inconsistent, and from any finally revised scheme of classification by stages that may be adopted in the future will have to be expunged the terms ASTARTIEN, PTEROCERIEN, STROM-BIEN, VIRGULIEN, CORALLIEN, CORALLINIEN, ZOANTHARIEN, SPONGITIEN, CYMBIEN, DICERATIEN, GLYPTICIEN, OPALINIEN, and PHOLADOMYEN. Besides these there are a few terms that stand in a class apart, for they are not founded on place-names and so are not strictly conformable to the scheme, although they may be useful. Among these are Woodward's FULLONIAN (1894), which is anyhow antedated by CADOMIN (Mayer-Eymar, 1881, after Caen); and Oppel's TITHONIAN,[1] which is akin to such terms as Eocene and Miocene. Liasien, which is equivalent to the Middle and part of the Lower Lias, has been superseded by DOMERIAN[2] (Bonarelli, 1894) and CHARMOUTHIAN (Mayer-Eymar, 1864). Oppel first suggested PLIENSBACHIEN as a substitute, but as it included part of the Lower Lias it has fallen into disuse as being too crude.[3] Charmouthian is open to the same objection, and in 1913 Dr. Lang suggested replacing the name for the lower part of the Charmouthian, below the Domerian, by the form CARIXIAN (from Carixa = Charmouth).[4]

A far more serious criticism of d'Orbigny's method is that he adopted names of places in foreign countries, particularly in England, which had already been associated for many years with a particular series of strata, without ascertaining exactly what those names meant in the place of their origin. Just as any name of a place conferred on a fossil, if confusion is to be avoided, should be that of the locality from which the type-specimen was derived, so when a stage is introduced into stratigraphy it should be defined according to the development of the strata in the locality from which it takes its name. By adding -ien to names with such old-established geological associations as Kimeridge, Portland, Oxford, and Bath, d'Orbigny sought to change their meanings entirely, adapting them to suit the succession where he knew it best, in the North of France. The result is that many of his names refer to now well-known but then only vaguely-known successions in England, while his palaeontological definitions refer to the succession in France. There is strong justification for the movement to redefine his stages so that they shall correspond with the strata developed at the localities after which they are named, and so with the strata bearing the same names previously.[5]

All this points to the fact that d'Orbigny's scheme of classification was premature. The ideal would have been to wait until a tolerably complete record of the succession of Jurassic faunas had been elucidated, by piecing together the results obtained by workers in different parts of Europe, and then, bearing in mind the original scope of the old terms, to divide up the column so that each division should carry the name of the locality or region where its particular

[1] Tithon was the spouse of Eos, Goddess of the Dawn.
[2] From Monte Domero, in the Lombardy Alps.
[3] A. Oppel, 1856–8, *Die Juraformation*, p. 815. Pliensbach is a village in Swabia.
[4] See, however, the terms VIRTONIAN (from Virton, Belgium) and LOTHARINGIAN from Lothringen = Lorraine), both already applied to all or part of the Pliensbachian. (See Appendix at end of this book.)
[5] Various opinions on the necessity for consistency in the names used in stratigraphy were offered at the Second and Third International Geological Congresses at Bologna in 1880 and Berlin in 1883, but no satisfactory conclusions were arrived at and, as no rules were formulated, the results of the debates have been little heeded.

fauna or group of faunas was best developed. For this the time was not yet ripe, and much of the confusion in international nomenclature and correlation has been caused by d'Orbigny's attempt to take a short cut, omitting an indispensable stage in the evolution of the science.

III. QUENSTEDT AND OPPEL TO BUCKMAN; ZONES AND THE SEARCH FOR AN INDEPENDENT TIME-SCALE

The laborious course which stratigraphy had still to take, and of which it has not yet reached the end (the following chapters are a humble contribution to its advance) was tersely enunciated by Quenstedt in his criticism of d'Orbigny's method.

'The next task which we have to fulfil in connection with the Jurassic is to draw up sections as faithfully as possible in the different districts. . . . The accurate comparison of two successive beds three inches thick by means of their actual contents can contribute more fruitfully to the development of the science than the cataloguing of stages from the farthest corners of the earth, when it has immediately to be admitted, as a matter of course, that they are not correct. . . . Of what avail is it if a man has seen the whole world, and he does not understand aright the things which lie in front of his own doors?'

And again:

'Let us not weary of searching our strata; let each one of us collect as much as he can in his own neighbourhood, labelling the specimens exactly with their localities, and compare them with the material collected by others; then at least the first goal of all geological research should not remain far from our reach—*a true table of the succession of the strata.*' [1]

This detailed and truly scientific palaeontological method, of which Quenstedt was at once the leading champion and the foremost exponent, had been pursued quietly in various parts of Europe for at least twenty years. In England it was earliest cultivated by a band of palaeontologists, far ahead of the thought of their time, in the intellectual centres of Whitby, Scarborough and York. A short paper published in 1836, by one of the first of them, Louis Hunton, is a truly remarkable document. This pioneer had already perceived that 'Of all organic remains, the Ammonites afford the most beautiful illustrations of the subdivision of the strata, for they appear to have been the least able . . . to conform to a change of external circumstances.' He applied his discovery to a minute study of the Middle and Upper Lias in the Whitby cliffs and alum works, with results which show the surprising progress already being made, at this early date, in the observation and selection of short-ranged fossils of zonal value. Next to his paper was printed a more ambitious project of the same type by W. C. Williamson, but this was more a catalogue of fossils arranged according to their horizons.[2]

After describing the Jet Rock, and remarking that 'on inspection of the section, it will be perceived, that the Jet Rock has its peculiar suite of Ammonites', Hunton passes on to an accurate analysis of the fauna of the Marlstone of the Middle Lias:

'The species of Ammonites, though few in number, are, however, highly characteristic; thus we find *A. vittatus* about the centre of the series, confined to a very

[1] F. A. Quenstedt, 1858, *Der Jura*, pp. 23, 823.
[2] W. C. Williamson, 1836, *Trans. Geol. Soc.* [2], vol. v, pp. 223–42, and vol. vi, 1841, pp. 143–52.

small range, associated in nodules with the *Cardium multicostatum*, *Turbo undulatus*, and *Pecten planus*; but the two latter occur in other parts of the formation. The *A. maculatus* is constantly found at the junction of the marlstone with the lower lias, which here pass so gradually into each other, that it is impossible to determine where the sandstones end and the blue shale begins. I have long sought for *A. maculatus* in the upper and central portions of the *marlstone*, but have never found it many feet above the junction beds; and though this and other Ammonites from unequal geographical distribution, may be more abundant in one place than in another (*A. maculatus* is in greater number at Staithes, *A. Hawskerensis* at Hawsker-bottoms), yet they constantly maintain an invariable relative position. . . .

'The above description', he concludes, 'may not, in some instances, exactly accord with previous statements, but one great source of error has hitherto been the collecting of specimens from the debris of the whole formation, accumulated at the foot of cliffs or other similar situations, where they have long laid, and the inferring of their position from the nature of the matrix.'[1]

The publication of Hunton's paper preceded by one year the appointment of Martin Simpson to be Curator of the Museum of the Whitby Literary and Philosophical Society in succession to John Bird. The new curator on taking up his position found in the museum large collections of ammonites from the Yorkshire Lias (among them perhaps Hunton's specimens), and he set himself to describe them. In 1843 he published descriptions of more than 100 species,[2] followed in 1868 by a detailed section of the Whitby cliffs. Simpson was a close follower of Hunton's methods. 'Being convinced by observation', he wrote, 'that few species of the Lias fossils had existed during the deposition of any great thickness of strata, but, on the contrary, were often confined to thin seams, I measured carefully, with a two foot rule, all the beds and seams of Lias both to the south and north of Whitby, and at the same time collected the fossils from each stratum.'[3]

These examples serve to illustrate the kind of work that was proceeding in various parts of Europe, by means of which alone, as Quenstedt perceived, real progress was to be made. The names of many other Englishmen could be added to those of Simpson and Hunton; those of Leckenby, Bean and Williamson especially should not be omitted.

The man, however, who was to place the whole science of stratigraphical geology on a new footing and to breathe new life into it was Albert Oppel. It was appropriate that one who accomplished so much in this sphere should have been a pupil of Quenstedt's at Tübingen. A young man, gifted with more than ordinary powers of observation, generalization and exposition, he acquired world-wide fame at the age of twenty-five by his studies in the Jurassic rocks. It was therefore the more tragic that in 1865 he fell a victim to typhoid fever and died, at the early age of thirty-four.

After acquainting himself thoroughly with the strata at home, in Swabia and Württemberg, so lucidly and minutely described by his master, Oppel set out in 1854 on a tour of the Jurassic regions of Switzerland, France and England, with the object of correlating them in greater detail by means of

[1] Louis Hunton, 1836, *Trans. Geol. Soc.* [2], vol. v, pp. 216–18. For calling my attention to the remarkable qualities of this forgotten paper I am indebted to Prof. Davies's 'Geological Life-Work of S. S. Buckman' and H. B. Woodward's *History Geol. Soc.*, p. 121.

[2] Many of these have been figured in Buckman's *Yorkshire Type Ammonites*, which was originally undertaken expressly for that purpose.

[3] Published in 1868 in a *Guide to the Geology of the Yorkshire Coast*, ed. 4; reprinted in *Fossils of the Yorkshire Lias*, ed. 2, 1884, pp. ix–xxiii.

fossils especially chosen for their wide horizontal and small vertical range. He stated the problem before him as follows:

'Comparison has often been made between whole groups of beds, but it has not been shown that *each horizon*, identifiable in any place by a number of peculiar and constant species, is to be recognized with the same degree of certainty in distant regions. This task is admittedly a hard one, but it is only by carrying it out that an accurate correlation of a whole system can be assured. It necessarily involves exploring the vertical range of each separate species in the most diverse localities, while ignoring the lithological development of the beds; by this means will be brought into prominence those **zones** which, through the constant and exclusive occurrence of certain species, mark themselves off from their neighbours as distinct horizons. In this way is obtained an ideal profile, of which the component parts of the same age in the various districts are characterized always by the same species.' [1]

This passage represents one of the most important landmarks in the progress of stratigraphical geology, for it contains the first germ of the conception of a detailed time-scale, abstracted from all local considerations, either lithological or palaeontological. Before it geological history had been as confused as the history of Assyria and Babylonia at the time of the city-kingdoms, each with its individual local chronology, overlapping those of its neighbours. Since Oppel, historians have been provided with an orderly system of dynasties, subdivided into reigns, and even in countries as distant as the Himalayas it has been possible to discern marks appropriate to the periods when the more important of the dynasties held sway, although the influence of the individual reigns was not always felt outside North-Western and Central Europe.

It is remarkable that Oppel nowhere defined what he meant by a zone. He is frequently credited with the first use of the word, but it had in fact been employed by several French geologists before him, and a definite meaning was already attached to it. Oppel adopted the term and accepted its meaning and no doubt it seemed to him in consequence unnecessary to give a definition.

The first occurrences seem to be in the writings of d'Orbigny (1842–52) and Hébert (1857). D'Orbigny employed it in exactly the same way as the term Stage, except that he reserved it for palaeontological use; thus he speaks of the Bajocian Stage as constituting the zone of a named group of fossils.[2]

An example will make the matter clearer:

7ᵉ Étage: SINEMURIEN, d'Orb.

Première apparition, de l'ordre des Insectes diptères . . . (&c.).

Règne des genres *Cardinia, Spiriferina, Octocœnia.*

Première pèriode. De la faune spéciale aux terrains jurassiques.

Zone du *Belemnites acutus,* des *Ammonites bisulcatus* et *catenatus,* du *Cardinia hybrida,* de *l'Unicardium cardioides,* de *l'Ostrea arcuata,* et du *Spiriferina walcotii.*

Hébert, on the other hand, chose index-fossils and referred to the zone of a certain named species; e.g. 'la zone (or assise) à *Am. primordialis*'.[3]

From these earlier uses of the term zone Oppel did not deviate. If he had

---

[1] A. Oppel, 1856–8, *Die Juraformation Englands, Frankreichs und des Südwestlichen Deutschlands,* Stuttgart, p. 3.

[2] A. d'Orbigny, 1849–52, *Cours élémentaire de Paléontologie et de Géologie stratigraphique,* Paris, pp. 433 et seq. (The Palaeozoic 'stages' were treated in the same way on earlier pages. The same use was made of the word 'zone' in the *Paléontologie française,* but not in tabular form; I append a tabulation in Table II, p. 11.)

[3] E. Hébert, 1857, *Les mers anciennes et leurs rivages dans le bassin de Paris,* 1ᵉ *Partie, Terrains jurass.* Paris, pp. 18, 23 (assise, p. 22).

given a definition of what he understood by the term it would have been in fact superfluous, for his meaning is apparent on almost every page of his book. He even says that although he has elected to name his zones after fossils, they could equally well have been named after places.

To Oppel, therefore, is due, not the credit for the inception of the zonal idea, but for a very great refinement in its use, and, most important of all, for emancipating the zones from the thralls both of local facies, lithological and palaeontological, and of cataclysmic annihilations, thus giving them an enormous extension and transferring them from mere local records of succession to correlation-planes of much wider (theoretically universal) application.

Oppel's first zonal table for the Jurassic, here reproduced (Table III, p. 18), is a document of considerable interest. He evidently appreciated, as had Louis Hunton in England twenty years earlier, that 'of all organic remains the Ammonites afford the most beautiful illustrations of the subdivision of the strata', for of his 33 Jurassic zones 22 were assigned ammonites as index-fossils. Of the remainder, 5 were named after lamellibranchs, 2 after brachiopods, 2 after echinoderms and 1 after a gastropod, while 1 was left to be named later. All the zones to which Oppel originally assigned indices other than ammonites have since had ammonites found for them.

Since the time of Oppel nearly every geologist who has pursued local studies of the Jurassic System has found it necessary to increase the numbers of the zones in order to bring his stratigraphical scheme into closer relationship with nature. Oppel himself wrote: 'The difficulty [of constructing an adequate zonal table] arises chiefly from the insufficient number of well described species. The more accurately the species are defined, the more exactly can the beds be subdivided.'[1] Thirty-five years later Buckman was still echoing the same complaint: 'Anyone unacquainted with the Dorset-Somerset Inferior Oolite, and the richness of its deposits, could scarcely credit the difficulty experienced in these investigations from want of names for the ammonites. Species which have been perfectly well known for years as indicators of certain horizons are altogether devoid of any specific name. They could not, therefore, be recorded with any precision.'[2] Buckman in order to make good this defect devoted much of his time in the later part of his life to figuring and naming more and more ammonites in his work, *Yorkshire Type Ammonites*, and its continuation, *Type Ammonites*. At the time of his death he had figured in this series no less than 797 species, illustrated in 1,052 plates.

It need not concern us here to trace the stages by which Oppel's first essay in a zonal table has become elaborated by a host of stratigraphers until the present formidable tables of zones have been evolved. Suffice it to say that, as with the designation of beds and stages, complete independence of action is allowed the individual author, and he is at liberty to cumber the terminology with as many new zones as may seem to him desirable, no matter how local his requirements, his knowledge, or his outlook.

IV. BUCKMAN AND THE MODERN CHRONOLOGY

From the examples quoted of the earliest uses of the word zone, as adopted, elaborated and refined by Oppel, it will be evident that it is purely a stratigraphical term—a *bed* or group of *beds*, identified by palaeontological criteria

[1] A. Oppel, 1858, *Die Juraformation*, p. 3.   [2] S. S. Buckman, 1893, *Q.J.G.S.* vol. xlix, p. 482.

TABLE III. *Oppel's Table of Zones.*

Forma-tionsab-theilun-gen.	Etagen *oder* Zonengruppen.	Zonen (*Lager oder Stufen, d. h. paläontol. bestimmbare Schichtencomplexe*).	Conybeare & Phillips. 1822. England.	Dufrénoy & Élie de Beaumont. 1848. Frankreich.
Oberer Jura oder Malm.	*Kimmeridge-gruppe.*	Zone der Trigonia gibbosa.	Upper Division of Oolites.	Ét. supér. du système oolithique.
		Zone der Pterocera Oceani.		
		Zone d. Astarte supracorallina.		
	=========?{	Zone der Diceras arietina.	}?======	
	Oxford-gruppe.	Zone des Cidaris florigemma.		Étage moyen du système oolithique.
		Low. calc. grit & Scyphienkalke.	Middle Division of Oolites.	
		Zone des Amm. biarmatus.		
	Kelloway-gruppe.	Zone des Amm. athleta.		
		Zone des Amm. anceps.		
		Zone des Amm. macrocephalus.		
Mittlerer Jura oder Dogger.	*Bathgruppe.*	Zone der Terebr. lagenalis.		Étage inférieur du système oolithique.
		Zone der Terebr. digona.		
	Bayeux-gruppe.	Zone des Amm. Parkinsoni.	Lower Division of Oolites.	
		Zone d. Amm. Humphriesianus.		
		Zone des Amm. Sauzei.		
		Zone des Amm. Murchisonae.		
		Zone der Trigonia navis.		
		Zone des Amm. torulosus.		
Unterer Jura oder Lias.	*Thouars-gruppe.*	Zone des Amm. jurensis.		
		Zone der Posidonia Bronni.		
	Pliensbach-gruppe. (Liasien d'Orb.)	Zone des Amm. spinatus.	Lias.	
		Obere Z. d. A. margaritatus.		
		Untere Z. d. A. margaritatus.		
		Zone des Amm. Davöi.		
		Zone des Amm. ibex.		
		Zone des Amm. Jamesoni.		
	Semur-gruppe.	Zone des Amm. raricostatus.		Calcaire à Gryphées arquées ou Lias.
		Zone des Amm. oxynotus.		
		Zone des Amm. obtusus.		
		Zone des Pentacr. tuberculatus.		
		Zone des Amm. Bucklandi.		
		Zone des Amm. angulatus.		
		Zone des Amm. planorbis.		

(by a fossil or an assemblage of fossils). The chosen fauna is the constant term; the lithic characters and their concomitant facies-faunas may be subject to indefinite variation. It is important that this conception should be perfectly clear. It is perhaps best stated in the form of a definition; and one of the most concise is due to Prof. Marr: 'Zones are belts of strata, each of which is characterized by an assemblage of organic remains, of which one abundant and characteristic form is chosen as index.'[1] An official definition was enunciated at the third session of the International Geological Congress in 1883: 'A zone is a group of beds, of an inferior status, characterized by one or several special fossils which serve as indices.'[2] From time to time erroneous interpretations are put forward, due to confused thinking. For instance, H. B. Woodward, in a special article on zones, defined them as 'assemblages of organic remains'.[3] As Buckman remarked, the fossils in a museum might be called a zone according to this definition.

It will be noticed that in all these definitions and uses of the word zone there is no mention of *time*. It is, of course, implicit in Oppel's and all other authors' tables of zones, stages or strata, that they occupied a certain time in forming, and that the subdivisions probably occupied a lesser time than the major divisions. Before 1893, however, we find no attempt to formulate any strictly chronological ideas, or to construct a time-scale independent of strata, whereby might be compared the relative time taken to deposit strata in different localities. No vocabulary for any such conceptions existed.

The year 1893 saw the passing of another landmark along the path of stratigraphical geology. This time the advance was due to the work of an Englishman, and again it resulted from studies in the Jurassic System. Sydney S. Buckman was born in 1860 at Cirencester, Gloucestershire, where his father was Professor of Geology at the Agricultural College, and, when three years old, he was moved to Bradford Abbas, near Sherborne, Dorset, whither his father retired to take up farming.[4] He was thus brought up amid one of the most richly fossiliferous districts in the whole world, where the ammonites are packed in incredible numbers in the thin beds of the Inferior Oolite formation. His surroundings had a powerful effect on him. At the age of eighteen he was already the author of a paper on the Astartes of the Inferior Oolite, which he followed up at the age of twenty-one with another paper on the ammonites. These studies were only the prelude to activities of body and of mind which were to revolutionize much of the accepted geological thought of his time. After making himself master of his own district, the rich and intricate Sherborne country, he set forth to apply his restless mind to ever-expanding problems, until nothing less than the whole British Jurassic System, to the far north of Scotland, was his undisputed province.

The work with which we are here concerned was the first of Buckman's stratigraphical papers, published at the age of thirty-three, and setting forth the mature results of his researches in the Sherborne district.[5] The

[1] J. E. Marr, 1898, *Principles of Stratigraphical Geology*, p. 68.
[2] *Compte Rendu, 3e Session, Congrès géol. int.*, Berlin, 1883, p. 323.
[3] H. B. Woodward, 1892, *P.G.A.*, vol. xii, p. 298.
[4] A. Morley Davies, 1930, 'The Geological Life-Work of S. S. Buckman', *P.G.A.*, vol. xli, pp. 221–40.
[5] S. S. Buckman, 1893, 'The Bajocian of the Sherborne District', *Q.J.G.S.*, vol. xlix, pp. 479–522.

stratigraphical portion of this paper will be noticed in the appropriate place (Chapter VIII); the part of greatest importance is that dealing with certain theoretical matters. Impatient with the inadequacy of the existing terminology to deal with any but spatial conceptions, he introduced a new term **hemera** (ἡμέρα).

'Its meaning is "day", or "time"; and I wish to use it as the chronological indicator of the faunal sequence. Successive "hemeræ" should mark the smallest consecutive divisions which the sequence of different species enables us to separate in the maximum developments of strata. In attenuated strata the deposits belonging to successive hemeræ may not be absolutely distinguishable, yet the presence of successive hemeræ may be recognized by their index-species, or some known contemporary; and reference to the maximum developments of strata will explain that the hemeræ were not contemporaneous but consecutive.

'The term "hemera" is intended to mark the acme of development of one or more species. It is designed as a chronological division and will not replace the term "zone" or be a subdivision of it, for that term is strictly a stratigraphical one.'[1]

And again he insists:

'It must be particularly understood that it is used in a chronological sense as a subdivision of an "age".'[2]

It would be difficult to imagine anything stated much more clearly than this. Nevertheless, from the moment of its birth, Buckman's new term began to be misunderstood and misrepresented. In the discussion which followed the reading of the paper at the Geological Society all three speakers misunderstood its import and criticized it on irrelevant grounds (perhaps because Buckman did not himself read the communication and make the important passages clear). The conception of a time-scale entirely independent of deposit, so that it could be said that each hemera must have passed everywhere, whether any deposit was formed or not, seems to have been too much for many of the geologists of the time. The advantages of the system went unnoticed in the general disapproval.

Nine years later, finding that there were still some who considered that a hemera was simply a subdivision of a zone, Buckman published a fresh explanation,[3] and this is an interesting document, both because it introduces some new theoretical considerations, and because it throws new light on the changes and development of Buckman's own ideas. He now gives two definitions of a hemera. One is simply a restatement in different words and an amplification of the first—'the subdivision of an "age", it indicates the period of time from the rise of one dominant species to the rise of the next'; the other is entirely different—'The hemera is the time during which a certain piece of work, namely, the deposition of what is called "the zone", was done.'

This second definition is irreconcilable with the first and must be rejected, because it is based on a misconception of the meaning of 'zone'. The same word cannot be used with two different meanings, and we must keep to the original definitions, namely that the hemera is based on 'the acme of development of one or more species' (1893), that it is 'the time during which a particular species—generally an ammonite—had dominant existence' (1898),[4]

---

[1] S. S. Buckman, 1893, loc. cit., pp. 481–2.   [2] Ibid., p. 518.
[3] S. S. Buckman, 1902, 'The Term "Hemera",' *Geol. Mag.* [4], vol. ix, pp. 554–7.
[4] S. S. Buckman, 1898, *Q.J.G.S.*, vol. liv, p. 443.

or that it is the time during which one or more species gradually reached their acme, passed it, and declined (1902).

The time taken to deposit a zone is another conception altogether. We have just seen that a zone is not a strictly biological unit, but a *bed* or group of *beds* characterized by an *assemblage* of organisms, one of which is chosen as index-species, but need not necessarily be either confined to its zone or found throughout every part of the zone. It will be obvious that such a period of indefinite length as may be required for the formation of a zone may embrace several hemeræ. Buckman indicated this when he first introduced the word hemera, for he placed several beside every zone. Jukes-Browne was misled by this, as many others have probably been, into supposing that a hemera was merely a subdivision of a zone, and so a synonym of a subzone. Against this confusion of spatial and chronological conceptions Buckman rightly protested, but there is a germ of truth in Jukes-Browne's criticism. He would have been very near the truth if he had said that a hemera, according to Buckman's first and only acceptable definition, is a subdivision *of the time taken to deposit a zone*. This broader time-unit he believed to be without a name, and he coined for it the term **secule** (from *seculum*).[1] Prof. Diener, however, rejects the term in favour of an earlier, having prior claim: the word **moment** was proposed with precisely this meaning, to be the time-equivalent of a zone, by the Swiss Committee for the Unification of Nomenclature, at the International Geological Congress at Bologna in 1881.[2] Strictly speaking, therefore, 'moment' should take priority over secule, and Diener adopts it, with the variations time-moment and zone-moment.[3] On the other hand, it may with justice be objected that the word 'moment' was already long ago preoccupied in ordinary parlance for an indefinite but certainly small fraction of time, and to misapply it to a period of many thousands of years duration is absurd.

Thus we have, as the unit of time corresponding to the zone, the secule, the time-moment or the zone-moment. What, then, is the stratigraphical unit corresponding to the hemera?

Several hemeræ may be contained in one secule or zone-moment. On the other hand, a hemera cannot exactly be said to be a subdivision of a secule, for the two expressions involve different conceptions: the secule is based on the duration or acme of the assemblage, the hemera on the acme of a chosen species. The two are units on different scales, just as are centimetres and feet, and each must have a separate spatial equivalent. To fill this want Prof. Trueman in 1923 introduced the term **epibole**, 'as a stratigraphical term to cover deposits accumulated during a hemera'.[4] The word is a useful one, but it has not received the recognition that it deserves, for reasons which we shall analyse shortly.

At this point we must examine two other terms introduced by Buckman in 1902,[5] the **biozone** and the **faunizone**.

The biozone was introduced as a time-term 'to signify the range of organisms in time as indicated by their entombment in the strata. Thus we might

---

[1] A. J. Jukes-Browne, 1903, 'The Term "Hemera",' *Geol. Mag.* [4], vol. x, pp. 36-8.
[2] *Compte Rendu, 2nd session,* Bologna, 1881 (1883), p. 542.
[3] C. Diener, 1918, *Neues Jahrb. für Min., &c.,* Beilage Bd. xlii, p. 91; and 1925 *Grundzüge der Biostratigraphie,* Vienna and Leipzig, pp. 217-18.
[4] A. E. Trueman, 1923, *P.G.A.,* vol. xxxiv, p. 200.
[5] S. S. Buckman, 1902, loc. cit., pp. 556-7.

have the biozone of a species, of a genus, of a family, or of a larger group. The biozone of Ammonites would be equal to Mesozoic time; . . . the biozone of an Ammonite genus, *Coroniceras*, would be through two or three hemeræ; the biozone of an Ammonite species would be about equal to a hemera.' The distinction between the biozone of a given species and its hemera may be expressed as the difference between its absolute duration (as indicated by its total range) and its acme; but since in practice neither can be very accurately defined, owing to the imperfections of our collecting ('collection-failure'), the two units become virtually synonymous.

With hemeræ and secules by which to measure time, there would seem to be little necessity for a third unit which is elastic and must be qualified whenever used. It is as easy to talk of the 'duration of *Coroniceras*' as of the 'biozone of *Coroniceras*'. In Germany, however, the term has been adopted in a spatial or stratigraphical sense, to denote the deposit equivalent to the time-interval for which it was originally coined. The Germans have not adopted the hemera and consequently they have felt no want for the corresponding stratigraphical term, the epibole. Instead of thinking in terms of acme they think in terms of total range, using the biozone. There is a difference of theory and of words, but in practice the result is much the same.

In this somewhat changed sense the term biozone is useful, for we have no other word to denote the deposit formed during the total existence of a species, as indicated by its absolute range. In Buckman's original sense, however, it is objectionable, for it contains the root 'zone', which is a spatial term, and employs it in a chronological sense. Such misapplication may lead to still further distortions of the meaning of 'zone'; its insidious influence may perhaps be detected in such definitions as that of Wedekind, who thinks a zone is 'a time-interval based on the absolute duration of a species'.[1] This is indeed precisely what Buckman meant by a biozone.

Happily the biozone in this sense was anticipated by one year by a much better term, the American biochron, introduced by H. S. Williams in 1901.[2]

'It is essential', Williams wrote, 'to distinguish the **geochron** (expressed in terms of feet thickness of stratified sediments of uniform lithologic constitution) from the **biochron** (expressed in terms of presence in the sediments of fossils of the same species, genus, or family). Thus the time-value of the Hamilton formation would be spoken of as the Hamilton geochron; while the time-value of the species *Tropidoleptus carinatus* would be the *Tropidoleptus [carinatus]* biochron.'

Again,

'Palaeontologists are familiar with the very long range of the species *Atrypa reticularis*; *Rhynchonella cuboides*, on the other hand, has a very short range. In the nomenclature proposed (so long as both are considered to be species), the fact would be expressed by saying that the *Atrypa reticularis* biochron is longer than the *Rhynchonella cuboides* biochron.'

It is clear that Williams intended biochrons to apply to the absolute duration of a species, genus, family, or any larger taxonomic group; or, as he put it, that they were to be 'units whose measure is the endurance of organic

---

[1] R. Wedekind, 1918, *Centralblatt für Min.*, &c., p. 270: and see also his book, *Über die Grundlagen und Methoden der Biostratigraphie*, 1916.

[2] H. S. Williams, 1901, *Journ. Geol.*, Chicago, vol. ix, pp. 579–80.

characters'.[1] In summarizing at the end of his paper, however, he tabulated the word under a looser definition as 'the time-equivalent of a fauna or flora',[2] which would make it the equivalent of a secule or zone-moment. Diener accepted this latter meaning and so dropped the word in favour of 'moment', saying 'The duration of a species I call, in agreement with Buckman, a bio-zone'.[3] I think it will be agreed, however, that Williams's earlier definitions in the text of his paper must stand, and from them there can be no possible doubt that in introducing the biochron he had the same idea in mind as Buckman expressed a year later in the biozone.[4]

Buckman's second term, **faunizone**, was defined as follows: 'faunizones are, to paraphrase Mr. Marr, "belts of strata, each of which is characterized by an assemblage of organic remains", with this provision, that faunizones may vary horizontally or vertically, or the strata may not vary and yet may show several successive faunas. So faunizones are the successive faunal facies exhibited in strata.'[5] The term therefore expresses almost exactly what was originally meant by a zone. It has been adopted in Germany in the form Faunenzone, while the simple term zone by itself has come more and more to be dropped altogether, as not being sufficiently precise or involving ambiguity. Biozones and faunizones were seldom, if ever, afterwards mentioned by Buckman. He always kept to his earlier unit, the hemera.

The next step was the grouping of his hemeræ into larger divisions or **Ages**. In 1896 he published a paper on the Inferior Oolite of Dundry Hill, near Bristol (where William Smith had demonstrated his law of superposition to Joseph Townsend and Benjamin Richardson about a hundred years before) and in it he gave his Ages the same names as the stratigraphical stages of d'Orbigny—Bajocian, Bathonian, &c. In 1898 he submitted another paper to the Geological Society, proposing to carry on this usage. The Council pointed out, however, that he was using stratigraphical terms in a chronological sense, a practice which would lead to confusion, and should be discontinued. It was at the behest of the Council of the Geological Society, therefore, that Buckman first introduced the dual terminology now in use.

Freed from the toils of stratigraphical nomenclature, he immediately threw himself with zest into the elaboration of what must then have seemed a very daring enterprise, the foundation of a time-scale based purely on zoology.

'If the ammonites are recognized as the best indicators of the faunal sequence', he wrote, 'and since the chronological subdivisions depend upon this sequence, then the further grouping of the chronological subdivisions must be controlled by the zoological affinities of the ammonites. For instance, the shortest geological time-division is a hemera: that is, the time during which a particular species—generally in Mesozoic chronology, of an ammonite—had dominant existence. A longer space of time contains so many hemeræ; it is at present designated by the very faulty title of "an Age"; but it is obvious that, as a hemera depends on the ammonite-species, an "Age" must depend on the duration of allied series of ammonite-species . . . As a family has its periods of rise, of maturity, and of decline (or, in scientific language, its epacme, acme, and paracme), so the duration of the Age would be principally

---

[1] H. S. Williams, 1901, loc. cit., p. 578.   [2] p. 583.
[3] C. Diener, 1918, loc. cit., p. 93; and 1925, *Grundzüge der Biostratigraphie*, p. 230.
[4] Williams gave, incidentally, an entirely erroneous definition of a hemera (loc. cit., p. 583).
[5] S. S. Buckman, 1902, loc. cit., p. 557.

TABLE IV. *The Ages of Jurassic Time.*

Mainly after S. S. Buckman, *T.A.*, vol. iv, 1922, pp. 6–13, and vol. v, 1925, pp. 71–8, and vol. vii, 1930 (Index) with emendations by E. Neaverson, L. F. Spath, and the author.

(For the upward continuation of this table see p. 546).

PORTLAND BEDS	42 Titanitan 41 Behemothan	INFERIOR OOLITE	22 Parkinsonian 21 Stepheoceratan 20 Sonninian 19 Ludwigian	
KIMERIDGE CLAY	40 Pavlovian 39 Pectinatitan 38 Allovirgatitan 37 Gravesian 36 Physodoceratan 35 Rasenian 34 ?Prionodoceratan	UPPER LIAS	18 Canavarinan 17 Dumortierian 16 Grammoceratan 15 Haugian 14 Hildoceratan 13 Harpoceratan	
CORALLIAN BEDS	33 Ringsteadian 32 Perisphinctean	MIDDLE LIAS	12 Amalthean	
OXFORD CLAY	31 Cardioceratan 30 Quenstedtoceratan 29 Kosmoceratan 28 Proplanulitan	LOWER LIAS	11 Liparoceratan 10 Polymorphitan 9 Deroceratan 8 Oxynoticeratan 7 Asteroceratan 6 Microderoceratan	
CORNBRASH AND GREAT OOLITE SERIES	27 Macrocephalitan 26 Clydoniceratan 25 ?Oxyceritan 24 Tulitan 23 Zigzagiceratan		5 Agassiceratan 4 Coroniceratan 3 Vermiceratan 2 Schlotheimian 1 Psiloceratan	

NOTES

1–6. The first 6 Ages were introduced in 1925, *T.A.*, vol. v, pp. 71–8, to take the place of the two Ages Caloceratan and Coroniceratan. They are accepted by Neaverson (1928, *Stratigraphical Palaeontology*) except that he uses Arietitan in place of Microderoceratan.

24. Below this Buckman placed a Gracilisphinctean Age, based on the rare genus *Gracilisphinctes*, known only from a single species found in the Stonesfield Slate; but the Stonesfield Slate was deposited in the middle of the Tulitan Age, as it occurs between the Fuller's Earth Rock and the base of the Great Oolite Limestones, which both yield Tulitidae (see Chapter X).

25. The existence of an Oxyceritan Age distinct from the Tulitan has not been proved, but is inserted on the opinion of Buckman.

28. Synonymous (or rather contemporaneous) with this is the Reineckeian Age, always placed by Buckman after it. See Spath, *The Naturalist*, 1926, pp. 324–5.

30. Synonymous with this is Buckman's Vertumniceratan Age. *Vertumniceras* being a rare genus almost confined to Yorkshire, it is more useful to retain Quenstedtoceratan, as Spath and Neaverson have done. See Spath, *The Naturalist*, 1926, p. 324.

34. The separate existence of a Prionodoceratan Age is far from surely established, as the genus *Prionodoceras* was coexistent with *Ringsteadia* at least in England it appears before it (in the Sandsfoot Clay). It is inserted here only because so far no other term has been found for the time between the Ringsteadian and the Rasenian. Pictonian might be better; but if the *Pictonia* zone be regarded as belonging to the Rasenian, then Prionodoceratan is redundant.

36 = the hemeræ of *Aulacostephanus yo* and *A. pseudomutabilis*.

37. Above this Buckman supposed to be the position of the 'Mazapalitan' fauna of Mexico, but there is no evidence (see *T.A.*, vol. iv, 1922, pp. 6–13).

38–40. The positions of these are according to Neaverson, 1928, *Strat. Pal.*, pp. 373–6. As will be explained on pp. 465–6, Buckman misunderstood the stratigraphy of the Upper Kimeridge Clay. Pavlovian = Holcosphinctean of Neaverson and Buckman (*Holcosphinctes* being synonymous with *Pavlovia*; see p. 440).

42 = Gigantitan of Buckman, *Gigantites* Buckman being, according to Spath, a synonym of the earlier *Titanites* Buckman.

governed by the period of acme, and would be less concerned with the epacmastic and paracmastic periods of the same family. . . . Following on this again as a logical conclusion is the recognition that any time-division larger than an Age must depend upon the duration in time of allied ammonite-families.'

The larger time-divisions he called Epochs.[1]

In the first table (published in 1898) seven Ages and two Epochs were introduced to cover the Lias and Inferior Oolite deposits; thus:

STAGES.	AGES.	EPOCHS.
Bathonian	Parkinsonian	Stepheoceratan
Bajocian	Sonninian	
Aalenian	Ludwigian	Arietidan
Toarcian	Harpoceratan	
Pliensbachian	Deroceratan	
Sinemurian	Asteroceratan	
Hettangian	Caloceratan	
Rhétian	—	

At this time Buckman advocated relegating the Hettangian as well as the Rhætic stages to the Trias and beginning Jurassic time with the Arietidan Epoch. However logical this may seem from a purely laboratory study of the ammonites, it overrides too many geological considerations to have proved acceptable to any subsequent authors.

The elaboration of the new chronological classification was proceeded with apace and extended to embrace the Upper Jurassic, but it was not until 1922 that a complete table of the whole of Jurassic time was ventured upon.[2] Buckman now proposed 43 Ages, of which 18 occupied Caloceratan-Parkinsonian time, for which he had suggested only 7 Ages in 1898, and for which 22 are accepted in more recent works (see Table IV, p. 24). The new additions resulted from a much finer discrimination in Ammonite identifications and a hemeral table growing by leaps and bounds every year. At the same time Buckman worked backwards from palaeontology to stratigraphy and, having recognized 7 Ages in the Lower Lias, he proceeded to coin new stage-names for the deposits to which they were supposed to correspond—Lymian, Mercian, Deiran, Raasayan, Wessexian, Hwiccian.[3] Unfortunately he overlooked the work of Mayer-Eymar, who had already assigned a series of substage names to the Lower Lias (Filderin, Balingin, Rottorfin, Mendin), some of them with exactly the same meanings as Buckman's.[4] As there is, moreover, difference of opinion regarding the reality of some of the separate Ages upon which Buckman's Lower Liassic stages were founded, the new names are not adopted here.

V. THE LIMITATIONS OF BUCKMAN'S CHRONOLOGY: PRACTICAL DIFFICULTIES AND THEORETICAL CRITICISMS

Detailed collecting from the Lias and Inferior Oolite, by Dr. W. D. Lang on the Dorset Coast, by Prof. A. E. Trueman and Mr. J. W. Tutcher in the

[1] S. S. Buckman, 1898, 'The Grouping of some Divisions of so-called "Jurassic" Time', Q.J.G.S., vol. liv, p. 443.   [2] S. S. Buckman, 1922, T.A., vol. iv, pp. 6–13.
[3] S. S. Buckman, 1917, 'Jurassic Chronology, I, Lias', Q.J.G.S., vol. lxx, pp. 259 et seq.
[4] K. Mayer-Eymar, 1881, Classification internationale, &c., Zurich, 4°; also 1884 'Die Filation der Belemnites acuti, Vierteljahrsschrift der Zürcher Naturforsch. Gesellsch. (transcribed, Q.J.G.S., 1917, vol. lxxiii, p. 283); and repeated in 1888, Tabl. des Terrains, Zurich 4°.

Radstock district and South Wales, by Buckman and Mr. L. Richardson in the Inferior Oolite from Dorset to the Cotswolds, and by the late Dr. Lee in the Hebrides, led to so great an elaboration of the hemeral table that many lost patience with Jurassic chronology, and the principles on which it is based are frequently discredited. It may, therefore, be apposite to pass in brief review the steps by which the long lists were built up and to attempt to analyse their meaning.

For many years Dr. Lang has studied the Lower Lias of the Dorset Coast, collecting the ammonites inch by inch through the long and perfect sections exposed in the cliffs on either side of Lyme Regis and Charmouth. He has proved that, as Hunton and Simpson perceived in Yorkshire nearly a century ago, the successive species have a restricted vertical range, and that, although their biochrons may have overlapped to a greater or a lesser extent, the successive acmes marking the hemeræ of the species can be recognized and separated by detailed collecting. Thus Dr. Lang has been able to divide up the Lower Lias of Dorset into 38 epiboles, each representing the hemera of a certain ammonite (see p. 28). These epiboles average between 9 and 10 ft. in thickness, but some are much thinner. In the Belemnite Marls, for instance, there are 13 epiboles, with an average thickness of 5–6 ft. Some of them may be mere layers no thicker than the ammonites. The boundaries have to be fixed arbitrarily, for the fossils often occur only in thin seams, separated by barren clays, which cannot be assigned with certainty to any particular epibole.

Correlation by means of such minute subdivisions can only be attempted where the exposures are exceptionally favourable. But it so happens that there are several other localities in Britain where equally detailed work is possible and has been carried out with success. The first of these was the Radstock district of Somerset, where Mr. Tutcher and Prof. Trueman were able to put the so-called 'polyhemeral system' to the test. By carefully collecting from numerous quarries they were able to show an almost equally detailed hemeral succession, but the details were in many instances different. Whole blocks of epiboles appeared which were unrepresented in Dr. Lang's sequence, while others recognized by Dr. Lang could not be found. The same occurred in the Inner Hebrides, where the Geological Survey collected large numbers of ammonites from the Lower Lias in the cliffs of Pabba and Skye and submitted them to Buckman. At the same time there was enough general similarity to render broad correlations, by means of some dozen zones, an easy matter.

The supporters of the polyhemeral system of correlation, led by Buckman, interpreted these facts as indicating that the existing epiboles present at any one locality represent only a fraction of the total number of hemeræ that elapsed during the Lower Lias period. Each locality, they suppose, shows epiboles that have been removed from all the other districts by contemporaneous erosion, and consequently the geological record is far more incomplete than was imagined. If this is true, we can only hope to build up anything like a complete succession by piecing together the fragments that have come down to us in separate localities—a process involving many dangers, since it depends largely on preconceived notions of ammonite evolution.

This principle was often summed up by Buckman in favourite axiomatic form; for instance 'Different contiguous exposures show faunas of partly or wholly unlike facies: therefore their faunas are partly or wholly hetero-chronoeous';[1] or 'as a proved sequence shows the meaning of dissimilar faunas, so dissimilar faunas give reason for expecting heterochroneous de-position'.[2] One version was printed on the fly-leaf of *Type Ammonites* (vol. iii) for all to ponder who opened the cover: 'Additions to fauna decrease the imperfection of the zoological, but increase that of any local geological record: the gaps caused by destruction stand revealed more plainly.' It was, in fact, the very key-note of Buckman's teaching. Let us, therefore, examine the foundations on which it is based.

When Buckman states that, if two areas are on the same latitude and a fauna is well developed in one and absent in the other, the deposits are not con-temporaneous (synchronous),[3] he is assuming that the fauna in question (usually by 'fauna' he means an assemblage of ammonites or even only one ammonite species) was ubiquitous during the period of its acme; further that it attained its acme for all practical purposes simultaneously in every part of the area of its distribution.

The validity of this assumption has been challenged on various grounds, and when the facts of animal ecology at the present day are taken into con-sideration it seems to have little justification. An elaborate attack has been directed by Prof. L. D. Stamp, who has marshalled all the facts he can find to show how the local distribution of the faunas at present inhabiting the sea-bed is governed by the depth, salinity and temperature of the water, the nature of the bottom, the direction of the currents, the availability of food, and all the other familiar ecological conditions making up a suitable or an unsuitable environment.[4] Without recapitulating all his arguments, it can be said that Prof. Stamp's thesis amounts to this: that observed dissimilarity of fossil faunas can be explained by reference to the partial distribution of marine organisms at the present day.

This claim makes a ready appeal to the biologist and more especially to the beginner in geology, who may have come to the subject fresh from a study of biology. It is, however, an impertinence to suppose that such obvious con-siderations as Prof. Stamp enumerates have not received attention from the field-palaeontologists. As Dr. Lang, in restrained language, replied in their defence, the palaeontologists have always kept the possibility of reconstructing the conditions under which the fossils lived in the forefront of their investiga-tions.[5] How otherwise could they hope to detect a facies fauna and to select the fossils of value in making their correlations?

In the course of an argument with Buckman over his insertion of an exces-sive number of hemerae into the time-table of the Corallian rocks, I attempted to justify my view (which I still hold) that many of his hemeral indices lived side by side on the same sea-bed, by reminding him of Lyell's principle that the present is a key to the past and instancing the sporadic distribution of many modern sea-shells around our coasts. He caused considerable provoca-tion by remarking with a smile 'Ah! So you have been reading Lyell: a most misleading book'. At the time there seemed no more to be said, so outrageous was the heresy. But the remark often recurred to me and I realized that in

[1] S. S. Buckman, 1917, loc. cit., p. 259.   [2] Ibid.   [3] Ibid., p. 278.
[4] L. D. Stamp, 1925, *P.G.A.*, vol. xxxvi, pp. 11–25.
[5] W. D. Lang, 1925, in discussion of Stamp's paper.

TABLE V. *Ammonite Epiboles Determined in the Lower Lias.*

FAUNIZONES.	EPIBOLES DORSET W. D. LANG.	EPIBOLES RADSTOCK J. W. TUTCHER and A. E. TRUEMAN.	EPIBOLES VALE OF EVESHAM A. E. TRUEMAN and D. M. WILLIAMS.
DAVOEI	*Oistoceras spp.* *striatum* *latæcosta*	*?brevilobatum* *latæcosta* *Beaniceras sp.* *cheltiense* *sparsicosta*	*Oistoceras spp.* *brevilobatum* *latæcosta* *Beaniceras sp.* *cheltiense* *sparsicosta*
IBEX	*centaurus* *actæon* *maugenesti* and *valdani*	*actæon* *maugenesti* *ibex*	*centaurus* *actæon* *maugenesti* (*valdani*) *ibex*
JAMESONI	*masseanum* *pettos* *jamesoni* *obsoleta* *brevispina* *polymorphus* *Tetraspidoceras spp.* *peregrinum* *taylori*	*masseanum* *jamesoni* *obsoleta* *brevispina* *polymorphus* *Phricodoceras*	*masseanum* *jamesoni* *brevispina* *Phricodoceras*
RARICOSTATUM	*leckenbyi* *exhæredatum* *raricostatoides* *æneum* *obesum* and *armatum* *bispinigerum* *densinodulum*	*leckenbyi* *lorioli* *aplanatum* (derived) *macdonnellii* (derived) *raricostatoides* (derived) *zieteni* *subplanicosta*	*nodoblongum* *subplanicosta* *planum* *armatum* *bispinigerum* *densinodulum*
OXYNOTUM	*lymense*	*Gleviceras* (derived) *Oxynoticeras* (derived)	*polyophyllum* *oxynotum* *biferum*
OBTUSUM	*stellare* *landrioti* *planicosta* *obtusum* *capricornoides* *turneri* *birchi*	*stellare* *planicosta* *obtusum* *turneri*	*lacunata* *subpolita* *denotatus* *stellare* *sagittarium* *obtusum* *turneri* *birchi*
SEMICOSTATUM	*hartmanni* *brooki* *sulcifer* *Arnioceras spp.* *alcinoe* *Agassiceras sp.*	*hartmanni* *alcinoe*	 *nodulosum* *alcinoe*
BUCKLANDI	*striaries* *Arnioceras sp.* *pseudokridion* *scipionianum* *gmuendense* and *bucklandi* *rotiforme* *conybeari*	*sauzeanum* *scipionianum* (derived) *gmuendense* (derived) *?vercingetorix* (derived) *meridionalis* (derived) *Coroniceras sp.* (derived)	*sauzeanum* *scipionianum* *gmuendense* *rotator* *conybeari*
ANGULATUM	*marmorea* *liasicus* *laqueus* *hagenowi* *portlocki*	*liasicus*	*angulatum* (*marmorea*) *liasicus* *megastoma*
PLANORBIS	*johnstoni* *planorbis*	*johnstoni* *planorbis*	*johnstoni* *planorbis*

NOTE: the boundaries between the faunizones are fixed arbitrarily.

this particular connexion there was at least a germ of truth in it. It showed me that Buckman was not oblivious of ecological arguments such as those adduced by Prof. Stamp, but that he did not consider them valid.

The whole crux of the matter upon which the two schools of opinion differ is the factor of *time*. Ecological studies at the present day take no account of time as the geologist knows it. They presuppose a stability of animal distribution which the palaeontologist cannot tolerate. Thus it is interesting to notice that those who invoke partial distribution to explain faunal dissimilarity in the rocks picture the past ages in which the geologist is accustomed to soar without restriction as a series of kaleidoscoped presents, rather like a reel of motion-picture films, each hemera small and rigid. Prof. Stamp complains that an ammonite or a brachiopod would not have time to attain wide distribution in 'such a short space of time as a hemera';[1] later he again refers to a hemera as 'a brief space of time'.[2]

The views of the other school are strongly contrasted with this. Prof. Trueman believes that 'a hemera must represent some thousands of years',[3] or again 'must represent a very long period of time, certainly many thousands of years'.[4] Buckman goes farther and says 'What is there to prevent giving to a hemera a length of time like a million years'.[5]

Given sufficient time, a floating and free-swimming species like an ammonite would disseminate its shells, borne along by the wind and the tides after its death, over vast areas of the earth's surface. Moreover, the geographical barriers, which are the principal factor in the separation of local faunas, undergo, in the course of time, removal by erosion and earth-movements, thus giving free play to migration.

Every serious student of palaeontology and geology comes to take this extreme slowness of the accumulation of the sedimentary series for granted. Buckman put the case characteristically in 1922, three years before Stamp's criticisms were made:

'A hemera, though taken as the chronological unit, must be regarded as a very lengthy stretch of time. Migration of Ammonites would be a slow process; but . . . the rate of Ammonite migration to that of deposition was like the flight of an aeroplane to the progress of brick-laying.

'Present-day phenomena of deposition or of faunal dispersal are very unsafe guides. Geological strata are made by the net result of a constant battle of addition versus subtraction, in which are seen, locally, the small, slow victories of addition, after many vicissitudes. The same applies to modern faunal irregularities—they cannot be true criteria of what the ultimate geological record in the rocks will be: they are only records of temporary local phenomena, observed during a length of time quite negligible in comparison with the length of a hemera.'[6]

Ammonites are not the only shells capable of wide dispersal in the course of a hemera. Prof. Morley Davies has shown by simple arithmetic that even the sessile brachiopods are capable of migrating a mile in twenty years during their free-swimming larval stages.[7] In this way is explained the extraordinarily widespread occurrence of certain forms in thin zones; for instance of *Homœorhynchia cynocephala* in the *Scissum* Beds, *Acanthothyris spinosa* in the

[1] L. D. Stamp, 1925, loc. cit., p. 19.   [2] Ibid., p. 23.
[3] A. E. Trueman, 1923, *P.G.A.*, vol. xxxiv, p. 196.   [4] Ibid., p. 206.
[5] S. S. Buckman, 1925, *T.A.*, vol. v, p. 70.
[6] Ibid., vol. iv, p. 24.   [7] A. M. Davies, 1930, loc. cit., p. 235.

garantiana zone, or *Cererithyris intermedia* at the base of the Cornbrash. Contrasted with these, however, are a number of brachiopods of such restricted colonial habits that they are worthless for all but local correlation: for instance *Plectothyris fimbria* and a number of others in the Inferior Oolite of the Cotswolds, a whole group of localized Cornbrash species (*Ornithella classis*, *O. rugosa*, *O. foxleyensis*, *Kutchirhynchia idonea*, &c.), and certain notable forms in the Upper Jurassic, for example *Rhynchonella sutherlandiæ*. From such instances it is obvious that brachiopods cannot be used indiscriminately for zonal purposes, but that each species has to be considered on its own merits. If a widespread species is absent from any locality, its absence may lead to the detection of stratal failure; but the absence of the many restricted colonial forms means nothing.

The gist of the argument in favour of the utility of dissimilar faunas for the detection of stratal failure may be summed up as follows: A certain species is found in remotely separated localities, such as in the Lower Inferior Oolite of Dorset and the Hebrides, and it is thus proved to have been of widespread distribution and not merely a local form. It is contended that the period of its dominance was so long that the species would have had time to penetrate to all parts of the sea-bed between the places where it is found, and over a wide area beyond, during at least some part of its hemera. If it is absent from any area where the lithological facies is the same and there are no obvious impediments to its migration (such as coral reefs), then it is contended that the strata which would have contained the missing species have been removed by erosion.

The most serious objections to correlating by the polyhemeral system arise, not out of the theoretical considerations advanced by what might be termed the ecology school, but from the practical results obtained by the field-palaeontologists themselves, in the course of their collecting in distant regions. As, with the aid of the newly acquired refinement in discrimination of species, detailed work is extended beyond the classic areas, an increasing body of evidence is being collected tending to show that the succession of ammonite species is not everywhere repeated in the same order.

One of the most difficult anomalies to explain was found by the Survey when they collected from the *raricostatum* zone of the Lower Lias in the Island of Raasay, in the Hebrides. The beds are here developed much more thickly than in England, and they were found to contain a remarkable alternation of Echiocerates and Derocerates. At the base was a horizon of *Echioceras*, followed by one of *Deroceras*, and then another of *Echioceras*, and finally a second horizon of *Deroceras* at the top. According to Buckman's interpretation these horizons represent four distinct hemeræ, and he endeavoured to show that each could be correlated with the so-called *raricostatum* or *armatum* zones of some locality in England, and that therefore these zones were of different dates in different places.[1] With this view Prof. Trueman and Miss Williams are in agreement.[2] An alternative explanation, however, which Prof. Morley Davies advances, is that the Echiocerates and Derocerates merely migrated to and fro; or as he puts it 'that dispersal was slow in relation to species-duration—probably

[1] S. S. Buckman, 1914, *T.A.*, vol. ii, p. 96 c, and in Lee, 1920, 'Mes. Rocks of Applecross, Raasay and N. E. Skye', *Mem. Geol. Surv.*, pp. 82–3.
[2] A. E. Trueman and D. M. Williams, 1925, *Trans. Roy. Soc. Edinburgh*, vol. liii, pp. 732–6.

owing to climatic checks of a temporary or recurrent kind'.[1] Some such explanation is also preferred by Dr. Spath.[2]

Dr. Spath has drawn attention to far more serious anomalies resulting from the work of Bovier in the Sinemurian (Lower Lias) near Champfromier, in the Department of Ain, and by Burckhardt in the Lower Lias of Mexico. If the identifications of these observers are to be relied upon, the epiboles at those places are arranged in entirely different order. Dr. Spath gives the following table for comparison of the ammonite sequence at Champfromier with that established in England; for the purpose of comparison five species common to the two areas are selected:[3]

(a) Buckman's 12 hemeræ of the Oxynoticeratan Age.	(b) Champfromier, Beds 2–13. (E. Bovier).
l.	13. . . . . . .
k.	12. *simpsoni*
j.	11. *bifer*
i.	10. . . . . . .
h.	9. . . . . . .
g.	8. . . . . . .
f. *oxynotum*	7. *lacunata*
e. *bifer*	6. . . . . . .
d. *simpsoni*	5. *gagateum*
c. *gagateum*	4. . . . . . .
b. *lacunata*	3. *oxynotum*
a.	2. . . . . . .

The species in the Champfromier column are placed in the order of their acmes—representing the hemeræ of the species—but the total ranges (bio-zones) are much greater. Thus *O. oxynotum* ranges up into Bed 13.

If, as Dr. Spath believes, these results are soundly established, they prove (what might have been expected) that the ascertained acme of a given species was a local event depending on favourable local conditions. Three species may have attained their acmes in one basin of deposition in the order A–B–C, in another in the order C–B–A, and another B–A–C, in another C–A–B, in another B–C–A, in another A–C–B. Such possibilities cut at the roots of polyhemeral chronology, for if the hemera, founded on the acme of a species, is likely to have occurred at different times in different places, it is clearly useless as a time-unit. Epiboles therefore lose their value in correlation over all but short distances.

The results obtained at Champfromier are not surprising when we come to examine the idea of the acme of a species, upon which the hemera is based. There is no evidence that the *visible* acme in any district corresponds even approximately with the complete acme of the species, in the total area over which it migrated. Still less does it bear any constant relation to the total range, which is the only concrete and sometimes determinable quantity. Almost invariably the species perforce chosen as hemeral indices are unrelated to those either above or below. They appear upon the scene without antecedents and disappear as suddenly, leaving no immediate descendants behind

[1] A. M. Davies, 1930, *P.G.A.*, vol. xli, p. 239.
[2] L. F. Spath, 1931, *Geol. Mag.*, vol. lxviii, p. 184; and 1924, *P.G.A.*, vol. xxxv, p. 190.
[3] L. F. Spath, 1931, loc. cit., p. 184, where full references are given.

them in the succeeding strata. Now, unless all these species were specially created and extinguished (if evolution has been continuous), they must have had antecedents and descendants somewhere. It is unlikely that every species entered our district immediately after it came into existence or left it at the moment of its modification to form a new species elsewhere: it is improbable, in other words, that every lineage remained in one particular district during exactly the time of existence of a single species, no more and no less. Since they did not remain longer, we must assume they did not remain so long. Therefore all that we can see is a part, and an unknown proportion, of any biozone.

When the epiboles are thus analysed, it is not surprising to find that many of the species, which appear to have a restricted range in certain localities, prove to have a much longer range—become 'zone-breakers'—when the area of observation is extended.[1] Rather, it seems inevitable that they should. It follows from the fact that the local acme (epibole) of such a species is only a fraction of the complete biozone, that any fraction may be represented in any place—the epibole in different places may be on different horizons within the much larger biozone. Put in different words, the hemera during which the local epibole was formed may represent different parts of the biochron in different places.

Now, in view of this, the fundamental fallacy in the polyhemeral system of correlation seems to be, that it takes no account of the probability of several, or even any number, of species belonging to unrelated stocks having thrived and migrated simultaneously in the same basin. Yet if the Jurassic seas were anything like those of the present day, this postulate should be at the forefront of any speculations. The paths of the migrating species and stocks would intersect in criss-cross fashion as shown in Fig. 1, and the order of their appearance in any given district would be entirely fortuitous.

Behind this fallacy there lies confusion of thought in the definition of a hemera. The original definitions make no distinction between 'local hemeræ' and 'ideal hemeræ'. From what has just been said it will be evident that the two are by no means the same thing. Yet Buckman, on recognizing a thin epibole, assumed that the 'hemera' during which it was formed coincided with the 'hemera' or time of acme of the species over the whole area of its occurrence. As we have seen, such an assumption is quite unjustifiable. The two conceptions are so entirely different that there should be two distinct time-terms to express them.

The Germans overcome this difficulty, for they have two different words for these distinct concepts. They do not speak of hemeræ and epiboles, however, but prefer to think in terms of total range, using biozones (of which the equivalent time-units, as we have seen, are biochrons). In practice, since neither acme nor total range can generally be accurately defined, as has been said, the result is the same. The ideal biochron is as elusive as the ideal hemera. The visible range at any given places bears no constant relation to the complete or absolute total range of the species, considered over the whole area of its occurrence. Consequently the Germans do not use the word biozone

[1] 'Such species of ammonites, that have shown themselves to indicate a distinct zone in certain districts, but transgress the boundaries of that zone in other districts of the same zoological province, I call zone-breaking species (zonenbrechende Spezies).'—C. Deiner, 1918, loc. cit, p. 113.

for the visible fragments or parts of biozones with which they usually have to deal. For these they have another, an excellent, word—Teilzones.[1]

A Teilzone is a part, and an unknown proportion, of the theoretically complete biozone. All so-called 'zones' founded on species unrelated to those above or below (that is, on species whose antecedents and descendants are unknown) are Teilzones. The term is therefore precisely what we want in order to clear up the existing confusion in the use of 'hemeræ'. The corresponding time-unit would be most logically expressed by some such term as 'Teilchron'.

At present almost the whole of our zonal table is built up of these fragmentary units or Teilzones, and it is all-important not to lose sight of their

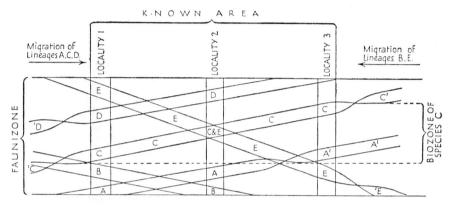

Fig. 1. Diagram illustrating the relations between faunizones, biozones and epiboles. At locality 1, within a faunizone is a sequence of 5 local epiboles, marking the local acmes of five unrelated species, A, B, C, D, E. At localities 2 and 3 the sequence is found to be different. The diagram illustrates the suggested explanation of this, the species being represented by long pencils.

The lineage to which A belongs is the only one that gives rise to a new mutation or transient (A′) within the area covered by the three localities. All the other species both came into existence and were modified to form descendant species or transients outside the known area. Within the known area all the species are therefore represented only by teilzones.

(The evolution of the several species is incorrectly represented as occurring in jerks for the sake of diagrammatic clearness.)

limitations. Only very rarely can we dispense with them—only when we are able to deal with lineages.

When we can detect lineages running up through a stratified series and so can get to know the antecedents and descendants of a species, then only can we define its range and be satisfied that the visible range is also the absolute range. For obviously the biozone of a species cannot embrace either antecedent or descendant species of the same lineage, and so bounds are set to the biozone. Unfortunately, opportunities for making use of lineages in zonal work are extremely rare. Brinkmann has been able to employ them in the Kosmocerates of the Oxford Clay,[2] and Lange has shown that they exist in the genera *Schlotheimia*, *Scamnoceras* and *Saxoceras* of the *angulatum* and *planorbis* zones of the Lower Lias in North Germany.[3] Famous instances

[1] See especially H. Frebold, 1924, *Centralblatt für Min.*, &c., pp. 313–20. I am unable to suggest an adequate and unugly English translation.
[2] R. Brinkmann, 1929, see p. 353 below.
[3] W. Lange, 1931, *Centralblatt für Min.*, &c., pp. 349–72.

have also been worked out in other formations, using corals, echinoderms, gastropods, &c. Lineages are the exception, however, and usually we have to do without them in our correlations. In the shallow epeiric seas of North-West Europe in Jurrassic times, conditions of life seem to have been normally too unstable for any stocks to remain long enough in one place to evolve *in situ*. They continually moved to pastures new. All that we can do is to compare the successions of unrelated forms that passed across our different districts in the course of their migrations; and let us be under no delusion as to the incompleteness and the inadequacy of the record with which we have to deal.

Before we pass on to other considerations, it may be useful to summarize, by means of the following table, the various conceptions and terms which we have been discussing. The term 'zone' by itself has now become a kind of family term, which may be very ambiguous unless qualified.

ZONES BASED ON ASSEMBLAGES

Basis.	Stratal Term.	Chronological Term.
Acme or Duration	FAUNIZONE (German Faunenzone)	SECULE or MOMENT (Zeitmoment or Zonenmoment)

ZONES BASED ON SINGLE SPECIES

Basis.	Stratal Terms.	Chronological Terms.
Acme	EPIBOLE [1]	HEMERA [1] (Blütezeit einer Art)
Absolute Duration	BIOZONE	SPECIES-BIOCHRON (Absolute Lebensdauer einer Art)
Local Duration	TEILZONE	TEILCHRON (Locale Existenzdauer einer Art)

There can be no doubt that in practical stratigraphy the old zones or faunizones, broad enough to contain a number of epiboles or teilzones, have many advantages. In the first place, they accord with the subdivisions universally employed in the other systems; in the second place they are frequently all that it is possible to recognize or map, and they usually suffice for elucidating the structure of a district. For this reason they should be known by all geologists, who need not trouble about (and could scarcely memorize) some hundreds of local epiboles. Lastly, they do give a tolerably true picture of the palaeontological sequence. They will therefore be adhered to wherever possible in the stratigraphical part of this book.

A faunizone, however, as Lange remarks, is nothing more than the sum of the biozones of a number of species, and its use is therefore subject to all the same reservations and limitations as a biozone. Most of the faunizones with which the stratigrapher has to deal are fragmentary—they are only 'faunizone-teilzones', made up of bundles of teilzones (see fig. 1).

It is noteworthy that the distinction between faunizones and the smaller local units or epiboles, founded on single species, although obvious in a

---

[1] Since neither definitions nor usage indicate which of the two alternative meanings should be attached to the hemera and the epibole, I restrict their use to the only ascertainable values, those based on the local or visible acme. It was from these empirical units that Buckman built up his 'polyhemeral' tables, and they remain unaffected even though they were assumed to be of universal application. It seems more logical to restrict the terms to what Buckman actually worked with than to what he thought he was working with.

comparatively uniform clay-formation such as the Lower Lias, breaks down when we come to deal with strata of more varied lithology. For instance, in the Inferior Oolite Series, a varied limestone formation, the smallest subdivision that can be made by means of the ammonites (the epiboles) are only 11 or 12 in number (see pp. 189 and 230). But these subdivisions can be, and have been, followed all along the outcrop in the South of England, often without finding any specimen of the index-fossil, simply by the general assemblage of fossils, with the aid, locally, of the lithology. Such stratal divisions answer in every way to the definition of true zones or faunizones. Here, therefore, the faunizones and the epiboles are in practice the same thing.

In the Great Oolite Series ammonites are so rare that they are of little account in field-work, but it appears that certain species of *Tulites* range through a thick series of varied deposits. The faunizones, on the other hand, are numerous and thin—considerably thinner than the epiboles of the few ammonites. The same is true also of the Corallian Beds, where certain Perisphinctids range through almost as great a thickness of rock.

In the Upper Jurassic clays, as in the Lias, the ammonite epiboles are again considerably thinner and more numerous than the faunizones. Moreover, as in the Lias, lineages can sometimes be recognized.

From this a general law may be formulated, that in a slowly-deposited clay formation the faunizones are much larger than the epiboles, while in varied and presumably more rapidly-formed deposits, the epiboles may equal or exceed the faunizones in thickness.

This suggests a means of answering the question: Which changed more regularly, the ammonite epiboles or the faunizones, and so which are the more satisfactory units for measuring the periods of earth-history? The greater thickness of the ammonite epiboles and the greater thinness of the faunizones in the more rapidly-formed deposits indicates that migration of the faunal assemblages was accelerated, *pari passu* with the acceleration of deposition, relative to ammonite migration. Conversely, the concentration of the ammonite epiboles and the spreading out of the faunizones in the clay-formations indicates a slowing down of both deposition and general faunal dispersal relative to ammonite migration.

Further, the fact that lineages sometimes occur in clay-formations in such a way that several complete successive ammonite biozones (as opposed to mere teilzones) appear to be telescoped into a single faunizone, suggests that in the clay-periods there was an actual slowing down of general faunal dispersal relative to ammonite evolution as well as to ammonite migration.

It is therefore probably true to say of ammonites that in a very general way the frequency in the change of dominant species is some measure of the passage of time, even though it can only be gauged by local teilzones. Their migrations seem to have been determined by forces largely independent of local conditions, and therefore, when they are available, they provide by far the best zonal indices.

VI. OTHER CRITICISMS

The tables of ammonite sequences which we have been considering, based as they are on careful field-work, represent definite and valuable additions to knowledge; for even if the restricted vertical distribution of ammonites observed in one locality be found to have changed in another, the mere discovery of this fact is knowledge gained. All such detailed field-investigations carry out the sound precept of Quenstedt, that we must get to know our successions minutely, as the very first step in stratigraphical geology. It is absolutely essential that this work should be undertaken, for until the detailed knowledge of the rocks and faunas it provides is available for large areas, generalizations are impossible. The task of establishing the palaeontological succession in monotonous clay-formations is a thankless one. Those who delight in proclaiming that all the results are worthless would therefore do well to keep their triumphant tones for more deserving objects.

Tributes to the value of the results accruing from this detailed work come from all parts of the world, and it is clear that without Buckman's chronology even general correlation between distant continents would often be impossible. The foremost authority on the American Jurassic writes: 'The rocks of the American Jurassic are too broken by diastrophism, too discontinuous, too poor in fossils, to serve for the working out of the succession except by correlation of their parts by comparison with a standard chronology. Buckman's biologic chronology provides such a standard.'[1] The work of McLearn in Canada points to the same conclusion.

Hitherto, however, we have confined our attention to hemeral tables based on actual field-work. Unfortunately other hemeral tables have been propounded without the necessary basis in fact, and for such work scorn need not be spared. It is one of the greatest misfortunes for Jurassic geology that when increasing age and frailty prevented Buckman from continuing active field-work, he lost sight of the distinction between results obtained with hammer, collecting bag and field notebook, and those arrived at by speculation and deduction from matrices at home. The extent of the calamity is magnified by the fact that this loss of grasp of the importance of keeping surmise distinct from fact came when his reputation was at its highest, and the two kinds of results are almost inextricably interwoven in his later published works. Only those with intimate local knowledge of the English Upper Jurassic rocks can hope to distinguish the two. Foreign geologists have been all too ready to adopt Buckman's conclusions indiscriminately and an appalling amount of error has already been propagated in the world's literature; but they cannot be blamed for this. If the following chapters help in some small degree to guide future workers to distinguish between fact and fable, this book will not have been written in vain.

That Buckman, who had tramped the Cotswolds and the Sherborne country from end to end and knew every quarry intimately, whose earlier work was built up solely on sound field-work, could also be the author of his last paper on 'Some Faunal Horizons in Cornbrash'[2] and of some of the later parts of the fifth and sixth volumes of *Type Ammonites*, is difficult to believe. Without any practical knowledge of the Cornbrash, without describing so much as a single section, he proceeded to divide it up into 11 brachiopod zones and coined for it 5 new stage-names. Neither zones nor stages have any foundation in fact—one stage ('Hintonian') even brought together Upper Cornbrash and Forest Marble (Hinton Sands). The zonal indices represent

---

[1] C. H. Crickmay, 1931, 'Jurassic History of North America: its bearing on the Development of Continental Structure', *Proc. Amer. Phil. Soc.*, vol. lxx, p. 73.

[2] S. S. Buckman, 1927, *Q.J.G.S.*, vol. lxxxiii.

no more than a list of the brachiopod fauna of the Cornbrash, enumerated in no special order, while the definitions of all the stage-names, without exception, are fanciful.[1]

On Buckman's privately-printed work there was not even the check of an appointed referee (however lenient) or a publication committee. The immense hemeral tables drawn up for the Corallian Beds in volume v of *Type Ammonites* are entirely fictitious. Almost every ammonite hitherto named from the Oxford district is supposed to be a separate hemeral index and there is no observational backing for the order of succession in which most of the species are placed. The method of analysing Blake and Hudleston's paper on the Yorkshire Corallian Beds, explained by Buckman, speaks for itself;[2] and some of his work on the Portland Beds and Kimeridge Clay was little better.[3] The more important matters will receive attention in their proper places in the ensuing chapters.

[1] For a more detailed examination see J. A. Douglas and W. J. Arkell, 'The Stratigraphical Distribution of the Cornbrash, Part I', 1928, *Q.J.G.S.*, vol. lxxxiv, and Part II, 1932, vol. lxxxviii.

[2] S. S. Buckman, 1924, *T.A.*, vol. v, p. 35. For criticisms of Buckman's work on the Corallian Beds see Arkell, 1927, *Phil. Trans.*, vol. ccxvi B, and 1929, 'Mon. Corall. Lamell.' *Pal. Soc.*; I have pointed out the absurd results arrived at for the Yorkshire sequence by Buckman's juggling with numbered slips of paper, loc. cit., 1929, p. 7.

[3] See L. R. Cox, 1925, *Proc. Dorset N.F.C.*, vol. xlvi, and F. L. Kitchin, 1926, *Ann. Mag. Nat. Hist.* [9], vol. xviii, pp. 499-54.

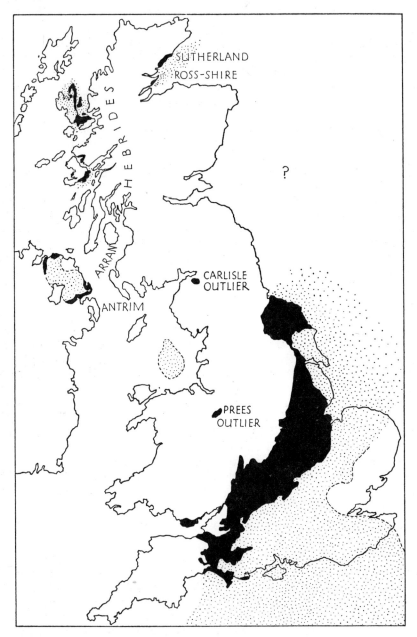

FIG. 2. Sketch-map showing the present distribution of the Jurassic rocks.
Black—Jurassic outcropping at the surface or covered only by Pleistocene accumu-
lations.
Dotted—Jurassic concealed by later deposits or under the sea.

PART II

THE TROUGHS OF SEDIMENTATION AND THEIR TECTONIC HISTORY

CHAPTER II

THE GENERAL STRUCTURE OF THE TROUGHS

THE distribution of the Jurassic rocks in the British Isles is shown in fig. 2, p. 38. It will be seen that the main outcrop crosses England in a broad band from the coast of Dorset to the coast of Yorkshire, while a former extension far to the north and west is proved by outlying patches in Shropshire, near Carlisle, around the basalt plateau of Antrim, in the Inner Hebrides and on the east coasts of Sutherland and Ross. South-eastward the system passes underground, sweeping round the London Basin into Kent and reappearing on the other side of the English Channel in the Boulonnais and Normandy, whence it is prolonged southward as a continuous sheet under the Paris Basin.

To what extent this distribution reflects the geography of the Jurassic period is a problem upon which many have speculated. The basis of any inquiry must be investigation of the following subjects: (1) the nature and configuration of the Palaeozoic platform upon which the Mesozoic rocks rest; (2) the extent and effects of any subsequent movements or denudations suffered by the platform or its covering; (3) the connexion between sedimentation and contemporaneous earth-movements, and therefore the nature of those movements; (4) the lithological and palaeontological facies of the sediments and the distribution of the fauna and flora, involving as accurate a correlation as possible of all the surviving occurrences of Jurassic rocks.

The first three problems will be dealt with briefly in this part, while the fourth will form the subject of the next or stratigraphical part of the book. The final synthesis, a palaeogeographical restoration, will then be briefly attempted in the concluding chapter.

I. THE PALAEOZOIC PLATFORM ON WHICH THE MESOZOIC ROCKS REST

In the year 1856 R. Godwin-Austen published a remarkable paper, setting forth reasoned speculations on the possible extension of coal-basins beneath the Secondary rocks of South-East England. He showed especially that the anticlinal axis of Artois, in the Chalk to the south-east of the Boulonnais, could be traced through the Boulonnais and the anticline of the Weald to the Vale of Pewsey and the Mendips. From the occurrence of coalfields to the north of this axis at both ends, where it emerges from beneath the Secondary covering (the Somerset coal-field in the west and the Franco-Belgian fields in the east) he concluded that Coal Measures might be expected to run north of the axis all the way across Southern England, and might be bored for with success along the Thames Valley. He also prophesied that they would be met with 'along and beneath some of the longitudinal folds of the Wealden denudation',[1] and he complained bitterly that there was not as yet one single boring in the South-East of England deep enough to give any certain information concerning the Palaeozoic rocks below the Secondary covering.

This paper aroused general curiosity, and as the result of a realization of the economic gain that would accrue if coal seams could be found at a workable distance below the Chalk, borings soon began to be undertaken. The first were the famous Sub-Wealden borings near Battle, in Sussex (1872–5), which, however, unexpectedly proved an enormous thickness of Jurassic rocks and were abandoned in Oxford Clay at a depth of 1,605 ft. below sea-level. As is now well known, numerous subsequent shafts farther north and east, in Kent, struck Coal Measures at workable depth and the economic working of the Kent coal-field is now an accomplished fact. The data obtained in the Kentish borings and published by the Geological Survey have been of incalculable value to theoretical geology and stratigraphy, and most of all to the stratigraphy of the Jurassic.

Immediately after the sinking of the Sub-Wealden shafts a deep boring was undertaken at Burford Signet, near Burford, Oxfordshire (1875–7), which encountered Coal Measures at a depth of 834 ft. below sea-level or 1,184 ft. below the surface of the Great Oolite. This was followed in the late 'seventies and early 'eighties by several borings for water in and round London—at Meux's Brewery in the Tottenham Court Road (1877), at Streatham (1882) and at Richmond (1882–4). These and a number of subsequent sinkings in the London area proved Old Red Sandstone and Devonian at depths of from 900 ft. to 1,100 and even 1,200 ft. below sea-level, without any Coal Measures. They thus provided an explanation of a problematic deep boring for water made at Kentish Town as early as 1853, which had passed from Gault into red sandstones, supposed by some to be an abnormal development of the Lower Greensand and by others regarded as Trias, but which were in fact Old Red Sandstone.[2]

Farther north, at Turnford and at Ware, the Palaeozoic platform was reached at shallower depths—879 ft. and 686 ft. below sea-level—and at Ware it consisted of Lower Palaeozoic strata (Silurian). In 1887 Charnian igneous rocks were struck at Bletchley, only 159 ft. below sea-level. This occasioned such surprise that the record was for many years misinterpreted; but early in the present century the matter was set at rest by two shafts sunk at Calvert, Bucks., only 12 miles from Bletchley, which encountered Shineton Shales (Cambrian) at almost exactly the same level (153 ft. below sea-level or 443 ft. below the surface of the Oxford Clay). Still more recently, Silurian has been struck again in Buckinghamshire, at Little Missenden, at a depth of −741 ft.[3]

In East Anglia the platform has been reached at Culford near Bury St. Edmunds, at three points at and near Harwich (see map, fig. 3, p. 44) and at Lowestoft. At all of these places it consists of Lower Palaeozoic rocks (Silurian or Ordovician or older), directly overlain by the Cretaceous.

In the South of England, therefore, there remain only two spaces beneath which we know nothing of the Palaeozoic platform. In the more northerly,

[1] R. A. C. Godwin-Austen, 1856, *Q.J.G.S.*, vol. xii, p. 73.
[2] J. Prestwich, 1856, *Q.J.G.S.*, vol. xii, pp. 6–14. The matter was complicated by Cretaceous fossils falling down the hole and being recorded from the lowest levels.
[3] A. Strahan, 1916, *Sum. Prog. Geol. Surv.* for 1915, pp. 43–6.

which runs from the estuary of the Thames north-westward between Ware and the Harwich district, the nature of the platform may in the absence of evidence be inferred from the nearest borings on either side of it. The other blank space concerning which we are ignorant is much larger, embracing all the country south-west of a line drawn along the edge of the North Downs from Brabourne to Slough and onward to Burford. Here the Palaeozoic floor is so deeply buried that, although deep borings have been undertaken, it has never been reached. A sinking at Southampton was stopped at 1,168 ft. below sea-level without reaching the bottom of the Chalk, while in Sussex the borings at Battle and Penshurst reached 1,605 ft. and 1,760 ft. below sea-level without passing through the whole of the Upper Jurassic. Even so near the western Palaeozoic outcrops as Lyme Regis, on the border of Dorset and Devon, a boring started near the base of the Lower Lias was sunk for 1,302 ft. but failed to reach the bottom of the Keuper Marl.[1]

The data that are available enable a general idea of the configuration of the Palaeozoic platform under the south-eastern counties to be obtained. In 1913, in a masterly presidential address to the Geological Society, Sir Aubrey Strahan described its relief and presented a contoured map of its surface (see p. 44).[2]

The information as to the composition of the platform is obviously too scattered for the insertion of geological boundaries, but it is possible to make the general statement that older rocks take part in its formation towards the north. All the borings proved Lower Palaeozoics north of a line running somewhere between Ware and Turnford and again at Cliffe and Chilham in the extreme north of Kent; a Devonian belt was encountered under London; while still farther south, under most of Kent, and again towards the west at Burford, Moreton in Marsh and Bradford on Avon, the Carboniferous System comes in. As has been pointed out by Dr. Rastall, this arrangement, with the older rocks to the east and north, the Devonian to the south and west, and the Carboniferous forming all the southern margin, is a mirror-image, as it were, of the Plateau of Brabant, around Brussels.[3]

There can be no doubt that the Plateau of Brabant and the London–East Anglian Plateau are parts of one and the same massif. To the south of both are important coal-basins—in Belgium that of Namur, in England those of Kent and, farther west, of Radstock. The Carboniferous rocks of both the Namur and the Radstock coal-fields are intensely folded and overthrust. Those of the Kent coal-field appear to have suffered less intense distortion, but their true disposition is not thoroughly known, although there are certain anomalies of dip and bedding which show at least some degree of disturbance.[4]

The junction of the coal-basins and the London–Brabant Plateau is one of the most important tectonic boundaries in Europe. It separates the old territory of the Caledonian fold-system, termed Palaeo-Europe, from the Armorican and Variscan chains thrust against it from the south. The denuded remnants of these chains form Meso-Europe, and against them in turn surged

[1] H. B. Woodward and W. A. E. Ussher, 1911, 'Geol. Country near Sidmouth and Lyme Regis', *Mem. Geol. Surv.*, p. 20.
[2] Sir A. Strahan, 1913, *Q.J.G.S.*, vol. lxix.
[3] R. H. Rastall, 1927, *Geol. Mag.*, vol. lxiv, p. 15.
[4] R. H. Rastall, 1927, loc. cit., p. 16. Published information concerning the Coal Measures in Kent is scant, owing to commercial secrecy.

the Alpine System, the boundaries of which enclose a tectonically still newer territory, Neo-Europe.[1] Thus, as Dr. Rastall has described, 'the plateau of Brabant acted as a horst, against which the advancing earth-waves of the Armorican-Variscan system broke and expressed themselves as a series of folds and overthrusts; the synclinal of Namur; the anticlinal of the Condroz; the synclinal of Dinant; . . . the folds of the Ardennes . . . These constituted a mountain chain of Alpine type'.[2]

The conception of a mountain mass in this position, arising as late as Permo-Carboniferous times, is clearly of profound importance for our reconstructions of Jurassic palaeogeography. It bears out in a remarkable way the forecasts of Godwin-Austen, who in his map of 1856 marked East Anglia north of the Thames Estuary, the Netherlands and most of the North Sea as land in the Oolitic period.[3] By the Upper Cretaceous period, however, the mountain region was completely levelled down and the Gault was everywhere deposited over it. How much of it was still above water in Jurassic times is a question which constantly recurs in studies on the English Jurassic, and will now be briefly considered.

II. THE ELIMINATION OF THE EFFECTS OF TERTIARY FOLDING: THE ORIGINAL RELATIONS OF THE JURASSIC ROCKS TO THE PLATFORM

Before we can hope to obtain any idea of the original configuration of the Jurassic sea-bed we have to take into account the changes which it has subsequently undergone. How great these changes have been, and how misleading can be their effects, was shown by Sir A. Strahan in his well-known presidential address, referred to on page 41.

The discovery of Charnian igneous rocks in the Bletchley boring at so shallow a depth as 159 ft. below sea-level showed that an elevated ridge exists there below the Jurassic covering, and it was assumed to be a south-easterly extension of the ancient complex of Charnwood Forest. Even if the igneous rock brought up from the bottom of the boring was only derived from boulders, these must have been too large to have travelled far, and since they cannot have moved upwards, still higher ground would have to be postulated in the immediate neighbourhood. Consequently it was assumed that, as Professor Kendall put it in 1905, 'the Oxford Clay at Bletchley was simply bedded round about and finally over a rocky islet of Charnian igneous rock'.[4] 'The great Charnian ridge which I have shown to extend far to the south-east of its exposure at Charnwood', he wrote, 'I regard as the dominant factor in determining the deposition of the whole series of rocks from the Carboniferous to the Cretaceous. . . . There are many considerations which make it probable that portions of this ancient ridge stood actually above water, in late Palaeozoic times and right through the secondary period up to the time of the deposition of the Chalk.'[5]

In spite of this, Prof. Kendall mentioned that he suspected the presence of lower portions of the Jurassic Series than might have been supposed from the

[1] H. Stille, 1924, *Grundfragen der vergleichenden Tektonik*, p. 232.
[2] R. H. Rastall, 1927, loc. cit., p. 13.
[3] R. A. C. Godwin-Austen, 1856, loc. cit., map facing p. 46*.
[4] P. F. Kendall, 1905, *Final Report to the Royal Commission on Coal Supplies*, p. 196.
[5] Ibid., p. 197.

accepted record of the Bletchley boring; indeed less than 5 miles away, two borings at Stony Stratford showed a succession of Lias and Oolites. Further, he confessed that careful search had not resulted in the discovery of any trace of pebbles of foreign rocks of Charnian type in any of the Liassic, Oolitic or Neocomian deposits—a very remarkable fact.

In 1913 Prof. A. Morley Davies put forward a fresh interpretation of the Bletchley boring, based on the new shafts at Calvert, 12 miles away in the direction of the strike of the Oxford Clay. He was able to show almost conclusively that both the Bletchley and the Calvert borings passed through Lower Oolites and Lias, and that the Charnian rocks at Bletchley formed part of the ancient floor beneath the Lower Lias, corresponding to the Cambrian shales in the same position at Calvert.[1] At the same time he published a map showing the contours at the base of the marine Jurassic (Rhætic and Lias) in this part of England.

Sir A. Strahan showed in the same year that the ridge at Bletchley was a comparatively recent feature, of little account in Jurassic palaeogeography.

After reducing all the depths at which the platform had been encountered in borings to their distance below sea-level, and from this compiling the contoured map shown in fig. 3 (p. 44), he proceeded to correct it for post-Cretaceous earth-movements. This was done by selecting a convenient stratigraphical plane in the Upper Cretaceous rocks, of widespread occurrence and uniform lithology, and ascertaining to what extent it has been distorted, assuming that at the time of deposition it was approximately horizontal.

The plane selected was the base of the Gault. Its altitude was determined at as many points as possible, at the outcrop and from numerous well-records, and from the data so collected was constructed a second map, shown in fig. 4. Seven contours were drawn at intervals of 500 ft., relative to present sea-level, ranging from +500 ft. (above sea) in several districts to −2,500 ft. (below sea) in the Hampshire Basin. All the contours below −1,000 ft. were estimated from the level of the top of the Chalk as found in borings, and by comparison with the thickness of the Upper Cretaceous rocks as developed at the nearest outcrops.

The map so produced represents in a striking way the long E.-W. flexures imposed on the Mesozoic rocks in the Tertiary (?Miocene) period, at the time of the Alpine orogeny. Their intensity is surprising, even when one is acquainted with the great thrust-fault of that date in the Isle of Purbeck. In the Hampshire Basin the base of the Gault is shown to have been forced down to a depth of more than 2,500 ft. below sea-level, while in the Wealden Anticline, before removal by denudation, it must have stood at not less than 1,500 ft. above sea-level—a total difference of 4,000 ft. In the London Basin the greatest depth attained is probably −1,500 ft. in a small area north of the Hog's Back.

By combining this ingenious map with that showing the present depth of the Palaeozoic platform, Sir A. Strahan then proceeded to eliminate all the folding subsequent to the deposition of the Gault. The second map showed the correction necessary at any point to bring the base of the Gault back to horizontality at present sea-level. By applying this correction to the depth

[1] A. M. Davies, in Davies and Pringle, 1913, *Q.J.G.S.*, vol. lxix, p. 333.

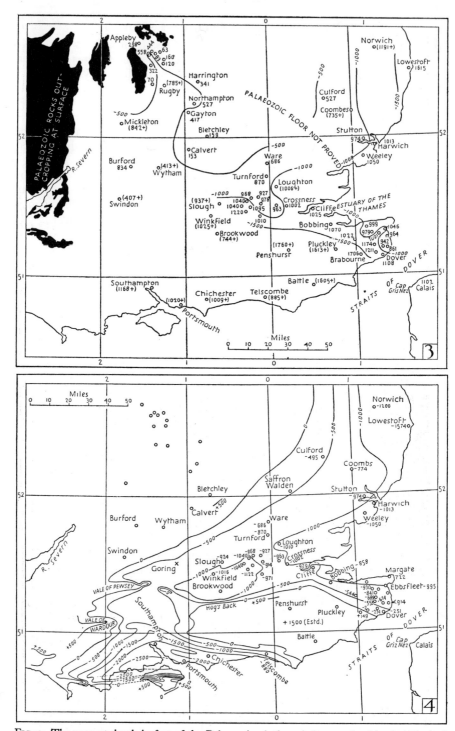

FIG. 3. The present depth in feet of the Palaeozoic platform below sea-level in the principal deep borings in South-East England, with contour lines drawn on the surface of the platform. After A. STRAHAN, 1913, *Q.J.G.S.*, vol. lxix.　　　　FIG. 4. The contour lines in the base of the Gault, and thus illustrating the deformation to which the Gault has been subjected. After A. STRAHAN, 1913.

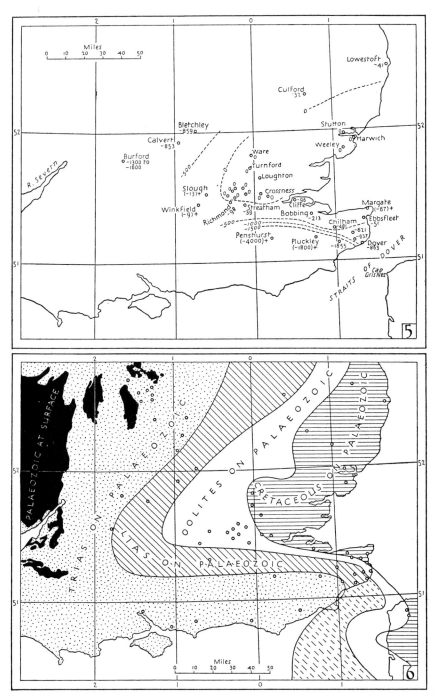

FIG. 5. The contour lines on the Palaeozoic platform at the beginning of Upper Cretaceous times, assuming that the platform has since undergone the same deformation as the Gault. After A. STRAHAN, 1913.   FIG. 6. The Mesozoic rocks which rest directly upon the Palaeozoic platform in the South-East of England, and illustrating the easterly overlap. After R. H. RASTALL, 1925, *Geol. Mag.*, vol. lxii, p. 211, fig. 2.

of the platform at the same spot, he determined the level at which the platform supposedly stood when the Gault was deposited. Thus at Richmond, where the base of the Gault lies 1,122 ft. below sea-level, the platform must have been depressed by that amount, to −1,220 ft. from −98 ft.; while at Bletchley, where the Gault no longer exists, but where its base is estimated to have stood at 700 ft. above sea-level, he argued that the platform must have been raised in Tertiary times by that amount, to −159 ft. from −859 ft. All points where the Gault rests directly upon the platform are assumed to have lain at sea-level (actually sea-bottom) at the time of the Gault, and are represented, when corrected, by the figure o.

From these data Sir Aubrey constructed the third map, fig. 5, showing the supposed configuration of the platform at the time of the Albian transgression. Some interesting results emerge, the most fundamental being

'to accentuate the importance of the elevated tract of Eastern England at the expense of the Palaeozoic areas of Western England and Wales; for, of course, the general tilt of the Upper Cretaceous rocks towards the east has been eliminated, as well as the synclines and anticlines.[1]

'The ridge apparently proved by the Calvert and Bletchley borings is no longer in evidence. Those places now appear as being situated on the north-western slopes of the elevated tract; nor is there any reason to doubt that the same slope extends south-westwards towards Burford. . . . Thus the existence of a ridge in the Palaeozoic platform is not in itself sufficient to prove the continuation of an ancient axis, such as that of Charnwood. On the other hand, the fact that Cambrian rocks constitute the platform at Calvert, and the probability that Charnian rocks were reached at Bletchley are highly significant of an old line of upheaval.'[2]

Thus Strahan's method presents us with an entirely new picture of the Palaeozoic platform beneath the Mesozoic covering of Eastern England at the time when the Gault was formed. Under London, the Thames Estuary, Essex and East Suffolk to about as far north as Southwold, stood an elevated tract of Palaeozoic and older rocks which appears to have been submerged for the first time by the Albian transgression. Around it on all sides except the east and north-east lay a trough filled with Jurassic and Lower Cretaceous sediments. The slope of the trough, as shown by the spacing of the contours, was steepest towards the south, in which direction the floor descended to much greater depths than elsewhere. Over the Southern Midlands it was comparatively shallow and shelved gently.

If now we wish to correct the picture still further, to show the configuration of the platform at the close of Portlandian times, when the marine conditions of the Jurassic came to an end, we can arrive at some approximation to the truth by stripping off in imagination the Lower Cretaceous and Purbeck strata. In late Jurassic and early Cretaceous times there were considerable earth-movements—part of the Saxonian movements, which have been more minutely studied in North Germany than in this country. Their effects are seen in the faulting and gentle folding of the Jurassic rocks under the Cretaceous covering in many parts of England (some of the most easily studied are the pre-Albian faults in the Upper Jurassic of Ringstead Bay, Dorset, which appear in Plate xxi, p. 451). It is impossible to allow for faulting, but where depression was marked by sedimentation it becomes measurable, and in

---

[1] A. Strahan, 1913, loc. cit., p. lxxviii.   [2] Ibid., p. lxxvii.

comparison with the depression that has taken place the effects of the faulting are probably so local as to be negligible. Southern England was a region of almost continuous sedimentation during the Upper Jurassic and Lower Cretaceous (as will be seen in Chapter XVII, where the matter will be discussed more fully) and the magnitude of this sedimentation is a measure of the depression which the region underwent.

The thicknesses of strata between the top of the Portland Limestone and the base of the Gault in the areas south of the London landmass and to the west and north-west of it make an interesting comparison:[1]

	Counties to the South.				Counties to the West and North-West.			
	Dorset	Hants	Sussex	Kent	Wilts.	Berks. and Oxon.	Bucks.	Beds.
L.G.S.	200	800	490	253	40	100	250	280
Wealden	2350	2000+	2100	2000+	30	50	45	—
Purbeck	400	500(?)	466	562	85	4	25	—
TOTAL	2950	3300+	3056	2815+	155	154	320	280

From these figures it appears that the portion of the floor of the trough which by Gault times was the deepest, namely that lying south of the London landmass, received its extra deepening after the cessation of normal marine sedimentation at the end of the Portlandian; in fact that nearly all the ascertained extra deepening is post-Jurassic. But we shall see later that the greater thickness of the Upper Jurassic rocks in the deepened area proves the movement to have begun actually long before the Cretaceous period. It may be significant that the area which suffered such great depression relatively to the rest corresponds with the portion of the platform which belongs to Meso-Europe, while the more rigid portion to the north is founded on the older Palaeo-Europe.

Six years after Sir A. Strahan's address to the Geological Society on 'The Form and Structure of the Palaeozoic Platform', the Society listened to another Presidential Address with a kindred theme by G. W. Lamplugh, entitled 'The Structure of the Weald and Analogous Tracts'.[2] In this address the president summed up the results of the long-continued investigations by him and his colleagues of the Survey, Dr. Kitchin and Dr. Pringle, into the Mesozoic strata penetrated in the Wealden borings. He showed that, as Topley had discovered many years ago, before the borings were started,[3] the Wealden anticline is only a superficial structure, imposed on a deeper-seated syncline, the syncline being the trough filled with Jurassic strata which we have been discussing. The borings are sufficiently numerous to show that each member of the Jurassic Series thickens towards the centre of the Weald from both the north and the south (see fig. 7).

Following Topley, but with a far greater body of evidence, Lamplugh then proceeded to show that over the trough in other parts of England also, wherever the Jurassic sediments are thickest, the Upper Cretaceous beds are arched up to form a compensating anticline. When allowance is made for the

[1] Compiled from 'Thicknesses of Strata', *Mem. Geol. Surv.*, 1916.
[2] G.W. Lamplugh, 1919, *Q.J.G.S.*, vol. lxxv. [3] W. Topley, 1875, *Q.J.G.S.*, vol. xxx, pp. 186-95.

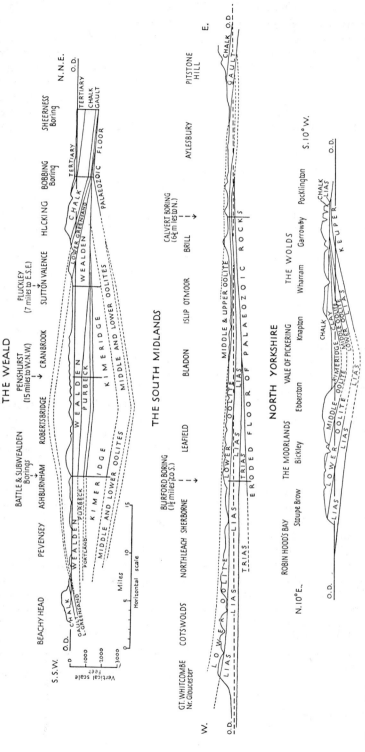

Fig. 7. Sections across the Jurassic rocks in three parts of England, showing the lenticular shape of the sediments filling the trough, and the superposition of an anticline in the Cretaceous rocks upon a syncline in the Jurassics. All three sections are on the same scale. The thick vertical lines represent borings.
After G. W. LAMPLUGH, 1919, *Q.J.G.S.*, vol. lxxv, figs. 2-4.

post-Cretaceous tilt already eliminated by Strahan, the general section across the Southern Midlands reveals a closely analogous, though flatter, structure, while farther north the same arrangement is again seen in the Yorkshire Basin. In the Midlands most of the Chalk arch has been removed by erosion, and the western side of the Jurassic lens is also missing, but when the lines that can be ascertained are carried on, as was done by Lamplugh in the sections reproduced in fig. 7, the analogy is unmistakable. In the Yorkshire Basin the Palaeozoic platform under the Jurassic area has never been reached by borings, but the manner in which the Jurassic formations dip inwards on both sides, while the Chalk rises over them from the south until it is truncated by erosion, leaves no doubt but that we have there also the relic of a similar structure.[1]

More recently the Mesozoic rocks near the centre of the Paris Basin have for the first time been pierced right through to the Palaeozoic platform by a boring at Ferrières, in the anticline of the Pays de Bray, and M. Pierre Pruvost has corrected the level of the Palaeozoic platform for post-Cretaceous movements by Strahan's method. He points out that the well-known anticline in the Cretaceous rocks at the surface, which provides an invaluable glimpse of the Upper Jurassics in a region where they would otherwise be unknown, coincides with a thickness of Jurassic rocks which is one-third in excess of the thickness at Rouen on the south-west and quadruple that in the Boulonnais on the north. Below the Pays de Bray the Palaeozoic platform, which was found to consist of ancient mica-schist, lies at a depth of 1,010 metres (3,367 ft.) below the surface, or 1,200 metres (4,000 ft.) below the base of the Cretaceous, a depth greater than at any other point known in the Paris Basin.[2] Here also, therefore, the anticline in the Cretaceous rocks is superimposed on a much deeper synclinal trough filled with Jurassic sediments.

Even without the additional evidence from the French boring, Lamplugh came to the conclusion that the same cause had operated all along the troughs of maximum deposition. He regarded the superficial anticlines as due to isostatic recovery of the base of the troughs after the prolonged subsidence which had proceeded all through Jurassic times. 'The infilling and sinking have gone on for a very long time in apparent association, as if due to compensative adjustment; but, in all cases, there has been a final recovery which was greatest where the deposits were thickest. This recovery implies a shallowing of the trough, and may have been due to a stretching of the floor with consequent reduction of the curvature.'[3]

The conception of a *stretching* of the floor seems difficult to reconcile with what is known of the Alpine movements—and post-Oligocene and pre-Pliocene the major part of the movements in England certainly were, as indicated by the Pliocene platform and the relative positions of the Pliocene Lenham Beds in Kent and the Oligocene strata in the Isle of Wight and Creech Barrow in Dorset. In Dorset there is incontrovertible evidence of lateral pressure in the Isle of Purbeck thrust-fault, which resulted from the fracture of a steep monoclinal fold. This fold, as Strahan showed, carried the Gault

[1] For an elaboration of this theme as applied to Yorkshire see Lamplugh, 1920, *Proc. Yorks. Geol. Soc.*, N.S., vol. xix, pp. 383–94.
[2] P. Pruvost, 1930, *Livre Jubilaire, Centenaire Soc. géol. France*, vol. ii, pp. 548–9, and see also P. Pruvost, 1928, 'Le sondage de Ferrières-en-Bray', *Ann. de l'Office nat. des Combustibles liquides*, 3ᵉ ann. 3ᵉ livr., pp. 429–57; also 1928, *R. Acad. Sci.*, vol. 186, pp. 242 and 386.
[3] G. W. Lamplugh, 1919, loc. cit., p. xc.

down under the Hampshire Basin to more than 2,500 ft. below sea-level. Compression, then, rather than tension is a force which it is much more easy to imagine in operation about this time, and it would seem to be equally capable of accounting for the anticlines and synclines.

The Chalk, Greensand, and Gault covering which spread evenly over troughs and landmass alike, when subjected to lateral pressure, would have to adjust itself to occupy a smaller area. So far from folding downwards into the troughs, as it would have done had ordinary subsidence continued, it would be forced upwards by the compression of the Jurassic sediments already filling them. By this process the formation of the anticlines is visualized as a bulging up from below relatively to the stable landmass under London, rather than a regular folding into anticlines and synclines; a conception which was already advocated by Prof. Kendall in his *Report*.

If this be the correct explanation of the anticlinal structures, there seems to be a fallacy in Strahan's method of correcting the depths of the Palaeozoic platform by eliminating the effects of Tertiary folding. The method depends on the assumption that the platform was folded in harmony with the Cretaceous base-line, the two maintaining a constant relationship throughout the movements. It would seem, however, that if the anticlines in the Chalk and Gault were formed as a result of compression of the troughs which already existed beneath, the distance between the surface of the platform and the Cretaceous base-line (the Gault) was increased in the process. Thus, when the base of the Gault was forced up to 1,500 ft. above present sea-level in the crest of the Wealden arch, the Palaeozoic platform below was not necessarily lifted by the same amount, but may have been actually depressed in the centre.[1] This consideration would very materially alter Strahan's figures, but the effect on his final map of the Palaeozoic platform (fig. 5) would be only to lessen the depth of the Jurassic trough relatively to the London landmass. His main theme, the existence of this landmass surrounded by troughs of deposition, remains unaffected.

Before leaving the question of the earth-movements to which the Mesozoic rocks were subjected in the Tertiary era, and as a corrective against minimizing them, it is as well to mention here the large throws of the faults bounding the Mesozoic formations in Scotland and the enormous outpourings of basalt by which they were covered, as well as the great volcanic activity of Tertiary times. In the Hebrides the preservation of such small relics of the Mesozoic formations as remain is generally acknowledged to be due to their having been lowered into favourable positions among the ancient Pre-Cambrian and Cambrian floor by faults with throws of 1,000 ft. or more.[2] The whole of the main mass of the Mesozoic rocks has been swept away by denudation, and all that is left is fragments that were faulted down into the ancient platform beneath. Similarly in Eastern Scotland the Jurassic rocks are faulted into contact with the Moine Schists, Helmsdale Granite or Old Red Sandstone, by the Great Boundary Fault, and all indications of their original position have been lost.

---

[1] Since writing this I have found that the same criticism seems to have occurred to the late Dr. J. W. Evans, who once remarked in the course of a discussion following the reading of an unpublished paper by Mr. H. A. Baker, 'The east-and-west foldings of the Wealden rocks were secondary structures resulting from the lateral compression and deepening of the basin of older rocks in which they lay' (*Abstr. Proc. Geol. Soc.*, 1917, p. 21).

[2] G. W. Lee, 1920, 'The Mesozoic Rocks of Applecros, Raasay and North-East Skye', *Mem. Geol. Surv.*, pp. 61-2.

CHAPTER III

CONTEMPORANEOUS TECTONICS AND THEIR EFFECTS ON SEDIMENTATION

I. EPEIROGENIC OSCILLATIONS AND SUBSIDENCE OF THE TROUGHS; CYCLIC SEDIMENTATION

HAVING now gained some idea of the configuration of the ancient platform below the principal areas of deposition and of the subsequent changes for which allowance has to be made, we are in a better position to examine the processes by which the troughs came into existence and were filled with sediment. The subject for our inquiry is the fact, as Lamplugh put it, that across England 'through all the Jurassic Period there existed a steadily deepening trough which swept in a bold curve from north to south, bulging towards the west, with land not far distant on both sides'.[1]

Before the discovery of the landmass under London Edward Hull wrote a long paper demonstrating the south-easterly attenuation of the Jurassic formations in Central England, and remarking that if denudation had caused the Jurassic escarpment to recede as far back as the present position of the Chalk escarpment, very little of the Jurassic strata would be left.[2] He attributed this attenuation to increasing remoteness from the assumed source of supply of all the sediment in the west and north-west. He little thought that as the strata thinned out towards London they were in reality approaching another shore-line; that, in fact, they were nearer to land on this side than the thick accumulations of the Cotswolds were to the known land on the other.

There is now no doubt that the south-easterly attenuation remarked by Hull, and so conspicuous in the records of the Bletchley and Calvert borings, is due primarily to proximity to the London–Ardennes landmass, and there are strong indications that sediments were contributed from both sides of the trough. Mr. J. G. A. Skerl has found evidence that the minerals of the Northampton Ironstone were derived directly from the east or south-east, and he even believes that in certain localities the outcrop is not far from the mouths of streams that brought the materials into the sea.[3] I have also shown that there seems no escape from the conclusion that the sands of the Lower Calcareous Grit in Oxfordshire, Berkshire and Wiltshire, with their small lydite pebbles, entered the sea somewhere to the south-east of Oxford.[4] A similar source is also probable, in view of their distribution, for the lydite pebbles and sands in the Upper Kimeridge Clay and Lower Portland Beds (see Chapter XV, pp. 514–15). The same land was still contributing sediment in the Wealden period, for Mr. Baker has been able to trace the mouths of streams which opened towards the south-west into the Wealden lake in East Kent.[5]

The gradual overlap of the successive members of the Jurassic Series and the manner in which they overstep on to the London landmass on the south

---

[1] G. W. Lamplugh, 1919, loc. cit., p. xc.
[2] E. Hull, 1860, 'The South-Easterly Attenuation', &c., *Q.J.G.S.*, vol. xvi, pp. 63–81.
[3] J. G. A. Skerl, 1927, *P.G.A.*, vol. xxxviii, pp. 388, 393.
[4] W. J. Arkell, 1927, *Phil. Trans. Roy. Soc.*, vol. ccxvi B, pp. 80–2.
[5] H. A. Baker, 1917, *Geol. Mag.* [6], vol. iv, pp. 547–9.

854371
E

side has been demonstrated by the Survey and is well seen in the section by Lamplugh (fig. 8). Dr. Rastall has shown the extent of the overlaps and of the total overstep for the whole landmass by means of a map (fig. 6). This is, in effect, a geological map of the under side of the Mesozoic covering, and it is a highly instructive object for study. Now that so many borings have revealed what lies beneath the South-East of England, it is evident that the protective action of the Cretaceous covering has preserved from erosion far more evidence of the relations of the Jurassic rocks to a contemporary landmass than can be obtained in any other part of Britain. Nowhere along the margins of the western land could sections like Lamplugh's or a map like Rastall's be drawn,

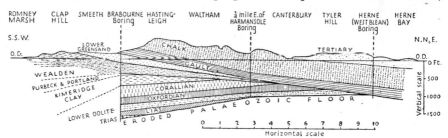

Fig. 8. Section across Kent from SSW. to NNE., illustrating the manner in which the Jurassic rocks overlap one another against the London landmass and are overstepped by the Cretaceous. After G.W. LAMPLUGH, 1919, *Q.J.G.S.*, vol. lxxv, fig. 1.

for the necessary data have been lost. The Cretaceous blanket, which hid everything from our forefathers, has since the days of deep borings shown itself to be our best friend.

Of all the many interesting facts which have come to light, that which most immediately concerns us here is the evidence of continual movements through-out the Jurassic period. As the formations approach the shore, not only do they become thinner and the higher beds overlap the lower, but they become increasingly interrupted by non-sequences or stratal lacunæ of varying extent. As has been pointed out by Mr. Baker, there is evidence that in Kent during some parts of the Jurassic period denudation was proceeding in the north-east near the shore at the same time as deposition in the south-west.[1]

It emerges from a study of the cores of the borings sunk into the deepest parts of the troughs, such as those at Battle, Penshurst and Pluckley in the Weald, and at Ferrières in the Pays de Bray, that the facies of the rocks at the bottom of the deepest troughs is the same as at the sides. M. Pierre Pruvost has shown that at the bottom of the Ferrières boring, below a covering of 4,000 ft. of Jurassic sediments, the Lower Lias begins with a deposit of marl containing vegetable fragments, interbedded with limestones yielding Hettangian mollusca, the whole having a littoral appearance.[2] Thence upward through the whole sequence of Jurassic formations the rocks are of the same shallow-water or even littoral facies as are met with near the sides of the trough in the Boulonnais and in Normandy. In the Bathonian, coral reefs

[1] H. A. Baker, 1917, loc. cit., p. 545. Compare with this the over-confident statement by Jukes-Browne in 1911 (and reprinted in 1922) that the absence of Upper Jurassic rocks under London was 'unquestionably due to erosion and planation' after Jurassic times and prior to the deposition of the Gault (*Building of the British Isles*, 3rd and 4th editions, p. 277).

[2] P. Pruvost, 1930, loc. cit., pp. 548, 550.

were found at a depth of over 2,300 ft. below the base of the Cretaceous—a depth at which, as M. Pruvost says, at the bottom of the sea 'clay deposits predominate, their thickness is reduced, neritic formations are unknown, coral reefs cannot live, and the fauna has special characters determined by the absence of light'. The coral reefs, which occur at two levels 1,000 ft. apart in the same boring, in the Bathonian and the Corallian, afford absolutely conclusive proof of subsidence during Jurassic times, for reef corals cannot grow at greater depths than 40 fathoms. The most recent conclusion, which agrees substantially with all previous ones, is that the ordinary depth at which coral reefs can commence to be built, except in the abnormally clear water of the open ocean, is 120–50 ft.[1] Since the whole of the rest of the circumstances are opposed to the idea that the water was abnormally clear or that the reefs grew in an open ocean, we may safely limit the depth of the sea in the centre of all the troughs where fossil coral reefs are found at 120–50 ft.

If, therefore, we wish to measure the extent of the subsidence in any period of Jurassic time, or to compare the relative subsidence over a given time in two localities, every bed that maintains a constant facies over a sufficiently wide area can be used as a datum-plane in the same way as Strahan used the base of the Gault. This principle was extended to various horizons in the Jurassic by Lamplugh, who enunciated what might be termed the Law of Planes of Horizontality:

'When a particular bed in a stratified sequence bears evidence of having been deposited at the same depth, and at about the same time, over a wide area, it marks a plane originally parallel to the sea-level of its period—i.e. a plane of horizontality. When several beds of this kind occur at intervals in the sequence and are not parallel to each other, we must infer that there has been relative change in the local plane of horizontality during the accumulation of the sequence, and we are enabled to determine the direction and amount of the change.'[2]

In Yorkshire, of which he was writing, he used the Rhætic, the Dogger, the Millepore Bed, the Grey Limestone, the Cornbrash, the Corallian limestone and the Red Chalk, and as many convenient planes can be found in almost any sequence.

If we restore each plane in turn to horizontality, we come at last to the Rhætic Beds, which contain several distinct planes within the series, and provide perhaps the most constant and the most widespread, and therefore the most trustworthy, guide of all. It is certain that at the beginning of the Jurassic period, when the sea first invaded the Triassic salt lakes and the surrounding shores, the Rhætic Beds were spread out over an almost perfectly level plain of Triassic marls and sandstones. The only undulations that have been traced in the underlying Keuper Marls are wide and low, and the crests (if so they can be called) were probably all covered before the end of Lower Rhætic times.[3] But the subsidence which brought in the sea and allowed it to penetrate across North Germany and England to the far north of Scotland was the first of the long-continued downward movements, which kept pace with sedimentation until the troughs had been filled to depths of 3,000–4,000 ft.

[1] J. Stanley Gardiner, 1931, *Coral Reefs and Atolls*, pp. 63–5.
[2] G. W. Lamplugh, 1920, *Proc. Yorks. Geol. Soc.*, vol. xix, p. 383.
[3] L. Richardson, 1904, *Q.J.G.S.*, vol. lx, pp. 356–7.

It is inconceivable that movements of such magnitude and so protracted did not exert a profound influence on, even control, sedimentation. In fact their importance has in recent years come to be increasingly recognized, and a substantial literature has already grown up on the subject.

So long ago as 1822 Conybeare and W. Phillips remarked on the rhythmic manner in which clay is followed by sand and sand by limestone through the Oolitic sequence, and on this account it early acquired the name 'Tripartite Series'. J. Phillips enlarged on the theme in 1871, pointing out five alternations of clay, sand and limestone from the Upper Lias to the Portland Stone.[1] Woodward remarked that when the rock-succession is examined in more detail 'the exceptions are almost as frequent as the rule',[2] and certainly the rule does not hold everywhere, for one or more steps are often missing. Nevertheless, in the South of England at least, where 'Estuarine' Series do not interfere, no one could fail to be struck with the remarkable rhythmic alternation of sediments to which Conybeare and W. Phillips and J. Phillips called attention. Indeed some of the apparent exceptions, when inquired into over a wide area, are found to furnish new instances conforming to the rule, their apparent nonconformity having been due to local failure of one of the stages. A good instance of this is the Upper Calcareous Grit, which Phillips mentioned as an anomalous sand found locally above the Corallian limestone. Where the sequence is fully developed the Upper Calcareous Grit is separated from the Corallian limestone by up to 40 ft. of clay (the Sandsfoot Clay) and is succeeded by an oolite (the Westbury Iron Ore and Ringstead Coral Bed). Closer examination of Corallian stratigraphy has revealed, besides this, another complete rotation of clay, sand and limestone, within what was formerly grouped as one limestone.[3] We can now draw up a fuller list of 'tripartite' rotations in the Jurassic Series of the South of England than was attempted by Phillips:

SEQUENCE	REMARKS
Portland Stone Portland Sand Kimeridge Clay	The Shotover Grit Sands of Wilts. and Oxon. in the Upper Kim. Clay may indicate another incomplete local rotation, the limestone having been removed before the deposition of the Lower Lydite Bed (see Ch. XIV).
Westbury Ironshot Oolite Sandsfoot Grit Sandsfoot Clay	The Westbury Iron Ore and Ringstead Coral Bed are more thickly developed in the Boulonnais as the Oolite d'Hesdin l'Abbé.
Osmington Oolite and *T. clavellata* Beds Bencliff and Highworth Grits Nothe and Highworth Clays	The sand and clay are only present in a few places.
Berkshire Oolite Limestones Lower Calcareous Grit Oxford Clay	

[1] J. Phillips, 1871, *Geology of Oxford*, pp. 393–4.
[2] H. B. Woodward, 1894, *J.R.B.*, vol. iv, p. 6.
[3] W. J. Arkell, 1927, 'Mon. Corallian Lamellibranchia', *Pal. Soc.*, p. 6.

SEQUENCE	REMARKS
Kellaways Rock Kellaways Sand Kellaways Clay	The sand and limestone are rarely separable, but seem to have been so at Kellaways.[1] Limestone is usually absent except as calcareous gritstone.
Cornbrash	
Main Forest Marble Limestone Hinton Sand and Forest Marble Bradford Clay and Forest Marble	This rotation is much interfered with by the development of the Forest Marble facies, but it is one of Phillips's original instances. He included the Cornbrash, which can scarcely belong.
Great Oolite Limestones Stonesfield Slate Beds Fuller's Earth (with argillaceous rock bands)	Woodward cites the Fuller's Earth Rock as breaking the series, but the 'rock' is only an argillaceous stone and occurs at various horizons throughout an essentially clay series. The intensely sandy nature of the Stonesfield Slate Beds is best appreciated in the Cotswolds (as at Hampen cutting, p. 276).
Inferior Oolite Series Upper Lias Sands and *Scissum* Beds Upper Lias Clay	The Inferior Oolite Series may contain two minor cycles in the Cotswolds (Naunton Clay, Harford Sand, Snowshill Clay, Tilestone, &c.).
Marlstone Middle Lias Sands Middle Lias Clay	
Lower Lias and Rhætic Beds	These may span as much time as several of the later cycles.

It is not suggested that sand, clay or limestone every time overspread the whole area of deposition simultaneously. Such a suggestion would be absurd. Moreover, it is known that sand often passes laterally into clay and vice versa; this is of especially frequent occurrence in the Upper Lias, Lower Calcareous Grit and Portland Sand, and is due to sand having begun to be deposited earlier in some places than in others. The conclusion is unavoidable, however, that for a large part of Jurassic time (not, apparently, at first, there being no sand or notable thicknesses of primary limestones in the Lower Lias except near the shoreline in the Mendip–S. Wales region) the sediment brought into the sea underwent periodic changes; and that in general a rotation from clay through sand to limestone is observable. In the relatively clear-water phases when limestones were formed, reef corals grew in many parts of the South of England on at least six occasions—Inferior Oolite (several times), Great Oolite, Berkshire Oolite, Osmington Oolite, Ringstead Coral Bed (representing a broken-up reef hidden from view) and possibly the Portland Oolite. Probably the time-values of the cycles are enormously different.

The explanation of these cyclic rotations (in other systems besides the Jurassic) was long ago 'furnished', as Phillips expressed it, 'on the simple and sure basis of interrupted depression of the sea-bed',[2] though it is only recently that the precise mechanism and meaning of the movements have been

[1] W. Lonsdale, 1832, *Trans. Geol. Soc.* [2], vol. iii, p. 260.
[2] J. Phillips, 1871, loc. cit., p. 393.

investigated. Their important bearing on the problems of stratigraphy seem to have been first recognized by Andrée [1] and elaborated by Klüpfel. [2]

Forms of cyclic sedimentation on a small scale are frequently discernible in the Lias and certain other formations consisting predominantly of clay. These have been closely investigated in Germany by W. Klüpfel, H. Frebold and others. [3] At variable intervals through the series are found bands of limestone with the upper surface water-worn and drilled by *Lithophaga* and other boring organisms, or encrusted with oysters. Above such eroded surfaces there is a sudden lithological change to clay, and afterwards a gradual passage through marl to limestone once more. The limestone bands, being the last deposits of each cycle, are called Dachbänke (roof-beds). There are numerous modifications of this scheme, and Frebold states that in Germany almost every small subdivision of the Lias has its own scheme. Some of the stages may be missing, while new kinds of sediment, foreign to the cycle, may come in. The changes, however, are seldom so important that the signs of rhythmic deposition cannot be recognized in some form or another, and the investigator has to accustom himself to the special manifestations peculiar to each area or group of strata.

A common variant, instead of the usual limestone Dachbank, is a thin bed crowded with some particular kind of shell, with or without signs of erosion. A characteristic form of this is familiar to collectors in the Lower Lias of Dorset, namely the beds strewn with thousands of belemnite guards, picturesquely called by the Germans 'belemnite battlefields' (*Belemnitenschlachtfelder*). Other beds are composed of ammonite shells, oysters or crinoids.

A point of fundamental importance, noticed by Klüpfel and confirmed by Frebold, is that the Dachbänke and the 'battlefield beds' often accurately delimit the faunizones or faunizone-teilzones. Frebold has been able to show that most of the ammonites of importance for zonal purposes, excepting certain long-ranged forms, do not transgress beyond the Dachbank next above them. Each cycle, in fact, *corresponds with a faunizone* (usually in the form of a teilzone).

From this it follows that the sedimentary cycles and the ecological changes which gave rise to the succession of faunizones probably had a common cause. And the cause was earth-movement.

'It seems justifiable to suppose', wrote Buckman in 1922, 'that crustal movements were like the waves of the sea, continuous, widespread and of variable magnitude, able to raise even deep-sea formations to within reach of denuding agencies.' [4] He was writing as a palaeontologist, seeking only to account for the non-sequences which his palaeontological researches had revealed to him, but the investigators into sedimentation in Germany have strangely re-echoed his views. How many times has Buckman, purely from

[1] K. Andrée, 1908, 'Über stetige und unterbrochene Meeressedimentation, ihre Ursachen, sowie über deren Bedeutung für die Stratigraphie', *Neues Jahrb. für Min.*, &c., Beilage Band xxv, pp. 366–42.

[2] W. Klüpfel, 1916, 'Über die Sedimente der Flachsee im Lothringer Jura', *Geol. Rundschau*, vol. vii, pp. 97–109.

[3] H. Frebold, 1924, 'Ammonitenzonen und Sedimentationszyklen und ihrer Beziehung zueinander', *Centrablatt für Min.*, &c., 1924, pp. 313–20; 1925, *Ueber cyclische Meeressedimentation*, Leipzig; and 1927, 'Die paläogeographische Analyse der epirogenen Bewegungen und ihre Bedeutung für die Stratigraphie', *Geol. Archiv.*, vol. iv, pp. 223–40.

[4] S. S. Buckman, 1922, *T.A.*, vol. iv, p. 24.

the standpoint of a palaeontologist, reiterated sentences which, if they were translated into German, would be indistinguishable from some of Frebold's; for example 'Gegenüber der Zeitdauer der Sedimentation nimmt die Zeitdauer der Sedimentationsunterbrechungen einen nicht geringen Umfang an'; [1] and again 'Wir erkennen, dass die häufigen Schichtlücken, die durch tektonische Bewegungen bedingt sind, nicht etwas Absonderliches, sondern etwas Natürliches sind.' [2]

That each Dachbank and 'battlefield-bed', even when no signs of erosion are visible, represents not only a shallowing of the sea but also a prolonged pause, seems justifiably deduced by Frebold from the difference between the faunas found above and below. He concludes that, during the interruptions of sedimentation which these beds mark, the faunas migrated elsewhere, and even that some of the more quickly-evolving ammonites became extinct— 'that their biochrons expired'—before the return of suitable conditions enabled them to migrate back again. In other words, the time occupied in the completion of each sedimentary cycle, including the interval during which there was no deposition, exceeded the biochrons of most of the ammonite species.

Interesting confirmation of this deduction was obtained by Brinkmann in his work on the Oxford Clay of Peterborough.[3] He distinguished there two kinds of interruptions in the sedimentation, which he called Dachbänke and Sohlbänke (he has kindly allowed his drawings of these to be reproduced here as fig. 9. The descriptions are appended below). The Dachbänke and Sohlbänke consist principally of masses of whole and broken shells, chiefly ammonites, but including many lamellibranchs. Veins of differently coloured clay pierce the undisturbed clay beneath the Sohlbänke like the burrows of boring organisms. In the clays fossils consist almost solely of plankton and nekton, showing that they were probably laid down under anaerobic conditions in which benthos could not live. With the shallowing of the water during the formation of the Dachbänke and Sohlbänke came increased current action, bringing with it improved aeration, and the bottom-dwelling lamellibranchs swarmed. Owing to his more refined method of zoning, it being possible to work with true biozones instead of faunizones or teilzones (as explained on p. 33), Brinkmann discovered that the periods of shallow water (marked by the Dachbänke and Sohlbänke) correspond with apparent leaps in the evolution of the several ammonite stocks. In the intervening clays evolution proceeded slowly and regularly. These apparent accelerations he surmised must indicate prolonged pauses in sedimentation, the pauses often exceeding in length the biochrons of several of his subspecies, which some other palaeontologists would call species.

Analysis of the movements involved in these cycles of sedimentation always shows that they were caused by gradual shallowing of the water (raising of the sea-bed) followed, after a pause, by relatively sudden deepening (subsidence of the sea-bed), after which shallowing began again. The same movements were repeated dozens of times and they occurred through the length and breadth of the sedimentary troughs of Europe. Frebold has traced the same individual cycles, their sharp boundaries marking off the same faunizones, in

[1] Hans Frebold, 1927, loc. cit., p. 240. [2] Idem., 1925, loc. cit., p. 55.
[3] R. Brinkmann, 1929, loc. cit.

the Middle and the Upper part of the Lower Lias of North-West Germany, Swabia, Lorraine and Aveyron in France.[1] They were therefore widespread and of epeirogenic origin.

An explanation of the crustal movements involved has been advanced by Stille in numerous papers and finally in his great work *Grundfragen der*

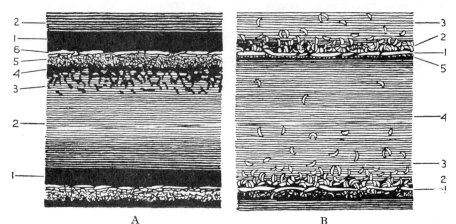

FIG. 9. Diagrammatic sections, illustrating the 'roofed' (Dachbank) (A) and 'floored' (Sohlbank) (B) types of sedimentary cycle in the Oxford Clay of Peterborough.

A	B
2. Brownish, sticky shaly clay, gradually passing down into 　1. Greenish clay.	3. Grey-brown shaly clay with *Nucula* shells, which become so common towards the base that they form a 　2. Shell bed. 　1. Pavement of ammonite shells.
Sharp boundary (denoting a considerable pause in sedimentation).	Sharp boundary (denoting a considerable pause in sedimentation).
6. Pavement of shells, overgrown with oysters. 　5. Comminuted shells, coarse above, becoming finer below and passing downwards into 　4. Greenish clay. 　3. Brownish clay, penetrated by veins of greenish clay, which fade downwards. 　2. Brownish sticky shaly clay, gradually passing down into 　1. Greenish clay.	5. Thin breccia of *Pseudomonotis*, passing rapidly into 　4. Grey-brown shaly clay, which passes down gradually into 　3. Similar clay with *Nucula* shells, which form a basal 　2. Shell-bed. 　1. Pavement of ammonite shells.
	Sharp boundary, as above
Sharp boundary, as above	*and so on*

and so on

After R. BRINKMANN, 1929, *Abh. Gesell. Wiss. Göttingen*, M.-P.-Klasse, N.S., vol. xiii, figs. 23, 24.

vergleichenden Tektonik. The earth's crust is divided into tectonically unstable and sinking regions or geosynclines and tectonically stable or rising regions or geanticlines. The whole of North-West Europe is, and was in the Jurassic period, a large geanticline (Meso- and Palaeo-Europe) upon which continental (epeiric) seas occupied shallow troughs. To the south lay the deep

[1] H. Frebold, 1927, loc. cit., p. 229.

geosyncline of Tethys, corresponding with what is now Neo-Europe—the Alps and the Mediterranean.

The general tendency of the geanticline was one of gradual elevation. But the shallow troughs containing the Jurassic seas acted as small unstable and sinking regions within the larger stable unit. These smaller sinking units Stille regards as diminutive geosynclines of a special kind, but Dacqué and Frebold consider that Haug's 'aires d'ennoyage' or 'drowned regions' is a better term, less likely to lead to confusion.[1] According to the theory, it was the interplay of oscillations set up by the upward-rising larger unit and the periodic down-sinkings of the drowned regions or troughs within it that gave rise to the cyclic sedimentation. Since it follows that the effects of the same causes were contemporaneous in different parts of the same trough, the cycles admit of correlation over wide areas. The chief point that remains doubtful is whether the sea bed actually rose between the major periodical down-sinkings, or whether it was merely built up by the accumulating sediment.[2]

II. DIFFERENTIAL MOVEMENTS WITHIN THE TROUGHS: 'AXES OF UPLIFT'

The general subsidence of the drowned areas did not everywhere proceed at the same rate. They were crossed at intervals by more stable ridges or axes, where sedimentation was slow and interrupted, just as near the margins of the landmasses. Relatively to the subsiding sea-bed on either side these ridges had an upward tendency, on account of which they have been commonly termed 'axes of uplift'. Their effect was to subdivide the troughs into a number of more or less separate basins of deposition, and their influence locally in inhibiting sedimentation in their vicinity was so marked that they early attracted attention. In the ensuing chapters they will be used as guiding lines in subdividing the outcrop for purposes of description. By this means more naturally defined districts are obtained than by the arbitrary methods previously employed (as in *The Jurassic Rocks of Britain*), when the distribution of towns or the modern physiographical features were made the basis of the descriptions. The major axes of uplift generally mark off regions distinguished by peculiarities of facies, both lithological and palaeontological.

Research has shown that the axes almost invariably coincide with older anticlines in the Palaeozoic rocks beneath, while they are as often indicated by subsequent foldlines in the Cretaceous and Tertiary covering above. Godwin-Austen already remarked on this in his astute paper on the possible underground extension of the Coal Measures, which in this, as in other ways, was many years ahead of its time. He wrote:

'The general law seems to be, that when any band of the earthy crust has been greatly folded or fractured, each subsequent disturbance follows the very same lines—and that, simply because they are the lines of least resistance. In this way, marked physical features in any region become unerring guides as to the character and extent of the earliest disturbance which took place there.'[3]

[1] E. Dacqué, 1915, *Grundlagen und Methoden der Paläogeographie*, pp. 132–3; see E. Haug, 1900, 'Les géosynclinaux et les aires continentales', *Bull. Soc. géol. France*, [3], vol. xxviii, pp. 617–711.
[2] As argued by K. Beurlen, 1927, *Neues Jahrb. für Min.*, &c., B. B. lvii, pp. 161–230.
[3] R. A. C. Godwin-Austen, 1856, *Q.J.G.S.*, vol. xii, p. 62.

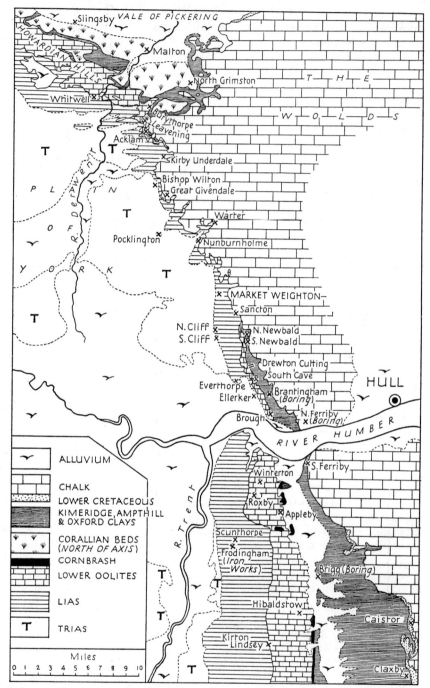

Fig. 10. Sketch-map of the region of the Market Weighton Axis, showing the westward overstep of the Chalk. The deflections in the Chalk and Lower Oolite escarpments attributed to the Caistor Axis are also shown. *Based on the Geol. Survey index map.*

Buckman restated the principle in 1901, in a form which has often been quoted. He came to the same conclusion as Godwin-Austen after studying the flexures in the Inferior Oolite of the Cotswolds.

'An anticlinal axis indicates a line of weakness. A line of weakness, once formed, tends to produce subsequent lines of weakness. Therefore the Jurassic lines of weakness may indicate former lines of weakness; hence former anticlines; hence denudation. There may, then, have been several elevations and several denudations on the same lines.' [1]

The English Jurassic area was crossed by a number of these axes, of which three rank as of primary importance—the Market Weighton, Vale of Moreton, and Mendip Axes—and they serve also to illustrate three different types of phenomena. The Market Weighton Axis ran WNW.–ESE. across South Yorkshire, effectually separating the Yorkshire Basin from the rest of the trough of deposition throughout the Jurassic period. Of all the axes this was the most persistent and the most important. It lies on a direct continuation of the Wharfe Anticline in the Carboniferous System of the Pennines.

The next in order of magnitude was the Mendip Axis. This was more acutely anticlinal than the Market Weighton ridge, but it dies out rapidly towards the east, and although it completely interrupts the continuity of the Lias and Lower and Middle Inferior Oolite, it has little or no effect on the Upper Jurassic formations outcropping farther east. The Mendip Axis is directly traceable to the well-known hogsback of steep periclines in the Carboniferous and Devonian rocks forming the Mendip Hills, with a curving strike from NW.–SE. to W.–E. (the individual periclines are directed east and west).

The Vale of Moreton Axis is a broader and more complex feature, the investigation of which is less straightforward owing to the greater part of the Mesozoic evidence having been destroyed by Tertiary erosion. It seems to have originally lain across the plain north of the present Oolitic escarpment, striking N.–S. from the Warwickshire coal-field down the Vales of Moreton and Bourton. All that we now see of it is the roots, as it were, close to the margin of the trough, where deposition was defective over a wide area from the eastern margin of the Cotswolds to the borders of Oxfordshire and Northants. Formerly, however, the N.–S. axis separated two great basins of deposition—those of the Cotswolds and Lincolnshire—where the deposits, especially of the Inferior Oolite period, are very different.

In a useful summary of English tectonics, Prof. Morley Davies has grouped together some of the leading flexures purely according to their directrix, after the classification of the late Professor Lapworth. [2]

Armoricanoid (or *Armorican*)—those with E.–W. directrix, or varying from NE.–SW. to NW.–SE.
Malvernoid (or *Malvernian*)—those with N.–S. directrix.
Charnoid (or *Charnian*)—those with NW.–SE. directrix.
Caledonoid (or *Caledonian*)—those with a NE.–SW. directrix.

We will now review rapidly the more important axes that cross the Jurassic areas, referring very briefly to the evidence for any intra-Jurassic and later

[1] S. S. Buckman, 1901, *Q.J.G.S.*, vol. lvii, pp. 147–8.
[2] A. M. Davies, 1929, in *Handb. Geol. Great Brit.*, p. 1.

movements along them. Details of the stratigraphical evidence, or at least summaries with references to the literature, will be found in the second part of the book. The importance of the role which these differential movements played in the sedimentation during Jurassic times can hardly be overestimated. To facilitate reference I have delineated the principal anticlinal flexures of South Central England on the accompanying sketch-map (fig. 15, facing p. 86), so as to show the connexion between the Jurassic and earlier and later folds. It will be shown also that to a certain extent classification is possible by means of the periods of activity.

A Specimen Axis: The Market Weighton Axis.

The Market Weighton Axis is the most regular example, as well as the most persistent, and serves as a model for understanding the rest. It was rendered classic by Prof. P. F. Kendall's masterly treatment in his *Report to the Royal Commission on Coal Supplies* in 1905.

'The line which I have termed the Wharfe [or Market Weighton] axis', wrote Prof. Kendall,[1] 'was a critical zone throughout the Jurassic and much of the Cretaceous periods. Most of the formations then deposited display pronounced abnormalities of thickness and character when that region is approached either from south or north, while for some it constitutes the line of transition or change from a northern to a southern type, and in regard to the whole series down to the Lower Lias there is unmistakable evidence of drastic and complete denudation immediately before the deposition of the Chalk.

'In face of coincidences so numerous and so complete it seems impossible to avoid the conclusion that we have here a line of repeated movement, an axis of unrest, which has operated continuously or with brief intermissions through the greater part of the secondary period, its position being in direct alignment with the anticlinal axis of the Wharfe.

'The Lower Lias shows a thinning on the Wharfe axis, such as might be explained by the maintenance during its deposition of shallow water conditions, which would prevent any great amount of sedimentation; or there might have been, in consequence of remoteness from a source of supply, some deficiency of sediment; or, again, if the area were undergoing oscillations of level, there might have been frequent recurrences of conditions of scour, by which the already formed sediments might be winnowed away.

'The first and last of these seem to me to have been the causes chiefly operative. The great prevalence of oyster beds in the neighbourhood of Cave and Market Weighton appears to indicate conditions of shallow water which I have not noted in the Lias to the same extent elsewhere in Yorkshire, and the thinning of the individual zones may not improbably be attributed to scouring.

'I should interpret the facts to imply a ridge of shallows here maintained by oscillations of small amplitude, while to the north over the Cleveland area and to the south in the Fen region depression was much more continuous and rapid.

'In Middle Lias times this tendency went even further, and . . . over the axis itself this formation is actually absent.

'The Upper Lias was continuously deposited over the ridge, but greatly reduced in thickness, so that we may, I think, safely regard the movement as still continuing in the same way as in Lower Lias times.

'The unconformable overlap of the Lower Oolites on to the Lower Lias at Givendale may mark an upward movement of that region in post-Liassic times.

[1] P. F. Kendall, 1905, loc. cit., pp. 199–200 (29–30).

'The Lower Oolites again show a most marked reduction when traced from the north and south. This reduction is largely, though not entirely, due to the failure of the great sediments of the Estuarine beds to reach so far from their source, but

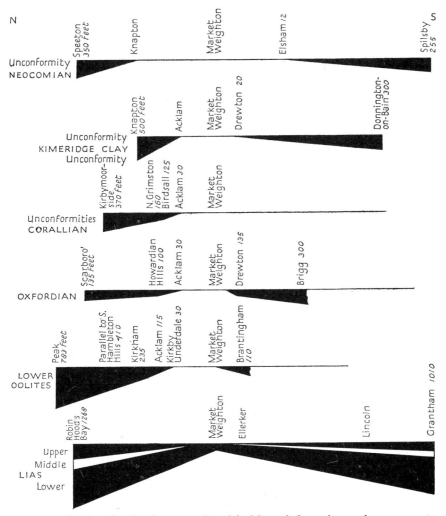

FIG. 11. Sections showing the attenuation of the Mesozoic formations as they pass over the Market Weighton Axis. After P. F. KENDALL, 1905, *Rept. Roy. Commission on Coal Supplies.*

(NOTE: at the time this figure was drawn rocks of Corallian age were thought to be absent south of the axis, but it is now certain that some are present: see text, p. 419.)

the marine beds are of less magnitude than in the Cleveland areas and in Lincoln-shire, and signs of unconformable overlap of the higher beds across the lower are traceable at several places in Lincolnshire.'

Prof. Kendall then proceeds to point out similar reductions in thickness in all the Upper Jurassic formations. Even a bed elsewhere so widespread as the Cornbrash is entirely absent over the axis and for a long distance on either side of it (in all for more than 40 miles). More recently additional contributions to our knowledge of the anomalies in the stratification of the Middle and Upper Jurassic rocks have been made by Messrs. Versey and Beilby and by Brinkmann, who extended hither his work on the Oxford Clay and compared the phenomena with the attenuation against the coastline in the Baltic region.[1] Mr. Beilby has paid special attention to the various components of the Inferior Oolite, and he finds that the principal movements took place after the end of each deltaic phase, before or with the marine incursions. Moreover, if the axes of the uplifts are considered to have lain midway between the points where any particular bed disappears when traced from both directions along the outcrop, then the successive uplifts did not always follow exactly on the same line. He has been able to determine the positions of maximum uplift along several subsidiary anticlines, during various parts of the Inferior Oolite period.[2]

Axes of Malvernian Trend.

(1) THE VALE OF MORETON AXIS

The literature bearing on the Vale of Moreton Axis is more scanty and more scattered, having suffered from the want of so able an historian as Prof. Kendall; therefore as much relevant stratigraphical evidence as possible has been brought together in the appropriate places in the following chapters.

The first to recognize the existence of the anticline seems to have been E. Hull, who stated in 1855 that

'The valley of Bourton on the Water appears to have originated in the existence of an anticlinal traversing its centre from north to south.' 'It appears', he added, 'that, though the strata of the Cotteswold Hills and the valleys by which they are surrounded have a general dip toward the south, yet the district is traversed by a series of gentle rolls, with north and south axes, and that the anticlines have produced lines of weakness, originating valleys; and the synclinals lines of strength, originating headlands.'[3]

He was speaking, however, only of the more recent folds superimposed on the Jurassic rocks, and he made no mention of intra-Jurassic movements. In his memoir on the Cotswold region [4] he pointed out the easterly thinning of the Lias and Inferior Oolite towards Oxfordshire, and implied that it was part of the general south-easterly attenuation due to remoteness from the source of sediment, to which he drew attention in the paper already quoted.

The first to recognize differential movements in the neighbourhood in Jurassic times appears to have been Judd, who approached the area from the opposite direction and found the evidence of attenuation just as striking far to the east of the Vale of Moreton, in Oxfordshire. In 1875, he wrote:

'During a considerable portion of the Jurassic period, the area now forming the county of Oxford underwent a far less amount of subsidence than that to the north-

[1] R. Brinkmann, 1925, 'Über sed. Abbild. epirog. Bewegungen', *Nachr. Gesell. Wiss. Göttingen*, M. Phys. Klasse, pp. 202–28.

[2] E. M. Beilby, 1930, 'The Market Weighton Axis in Middle Jurassic Times', *Trans. Leeds Geol. Soc.*, pt. xx, pp. 10–12; see also H. C. Versey, 1929, *Proc. Yorks. Geol. Soc.*, vol. xxi, p. 219.

[3] E. Hull, 1855, 'On the Physical Geog. of the Cotteswold Hills', *Q.J.G.S.*, vol. xi, p. 483.

[4] E. Hull, 1857, 'Geol. Cheltenham', *Mem. Geol. Surv.*, Plate 2.

east and south-west of it respectively. The consequences of this inequality of movement are manifested in the disappearance of some members of the series, the
attenuation of many others, and the littoral characters presented by nearly all of
them.' [1]

It was Buckman who saw in the modern anticline of the Vales of Moreton
and Bourton a revival of a Jurassic and still earlier anticline; and carried away
by Godwin-Austen's ideas of posthumous uplift along old-established lines
of weakness, and partly as the result of his own detection of the Birdlip Axis
in the Cotswolds, he tended to minimize the importance of the Oxfordshire
evidence adduced by Judd. In his North Cotswold paper of 1901[2] he contro-
verted Hull's view of a general south-easterly attenuation hereabouts, con-
tending that the Lower Lias was little affected in the neighbourhood of the
axis. The principal period of activity, Buckman pointed out, was certainly
from the Middle Lias to the end of Middle Inferior Oolite times. The Lower
and Middle divisions of the Inferior Oolite, which exceed 250 ft. in thickness
in the Cotswold Basin, thin out rapidly towards the Vales of Moreton and
Bourton and in these valleys and that of the Evenlode they are entirely absent,
Upper Inferior Oolite resting upon Upper Lias. To the north-east they
thicken again until in Lincolnshire they exceed 125 ft. Similarly the Upper
Lias, which is 300 ft. thick in the Cotswolds and nearly as thick in the
Northants-Lincolnshire Basin, is reduced to less than 20 ft. in the Evenlode Valley. The Fuller's Earth and Stonesfield Slate also disappear as they
approach the axis.

Now to a certain extent these appearances are deceptive, as will be seen by
reference to the map on p. 207, where I have indicated the southern boundary
of the *Scissum* Beds and so of the Lower and Middle Inferior Oolite in
Oxfordshire and Northants. It will be seen that these formations are not only
absent over the region of the Vale of Moreton, but also everywhere to the
south of a line which can be traced in a direction rather east of north-east for
nearly 50 miles, from Stow on the Wold by Chipping Norton and Towcester
to a point near Higham Ferrers, where the strike of the Oolites changes to
nearly north. This line seems to indicate approximately the boundary of the
area of deposition of the Lower and Middle Inferior Oolite—a supposition which receives support from Skerl's researches on the minerals of the
Northampton Ironstone. The gradual return of the missing Inferior Oolite as
the outcrop is followed northwards is undoubtedly due mainly to the change
in the strike. A horizontal section through Lincolnshire from Northampton
shows that after the change of strike the outcrop cuts obliquely across a trough
of deposition, of which it has been following the margin from the Vale of
Moreton.[3] In this respect, therefore, the Vale of Moreton area shows no
more signs of anticlinal uplift than does the Cherwell Valley, or than would
any valley excavated by denudation south of the line bounding the *Scissum*
Beds.

Moreover, when we come to examine the Upper Inferior Oolite we find
that, although the Upper *Trigonia* Grit (*garantiana* zone) stops short at the
Vale of Bourton, the *Clypeus* Grit (*truellei* and *schloenbachi* zones), which is

[1] J. W. Judd, 1875, 'Geol. Rutland', *Mem. Geol. Surv.*, p. 52.
[2] S. S. Buckman, 1901, *Q.J.G.S.*, vol. lvii, pp. 138–49.
[3] See fig. 47, p. 245.

much thicker, passes on across the 'axis' region and finally comes to an end *beyond* it, somewhere between the valleys of the Evenlode and the Cherwell (as shown in the map, p. 207). The Great Oolite Series, Cornbrash and Oxford Clay do not seem to show any special features in this region indicative of an axis of uplift, beyond the gradual disappearance of the Fuller's Earth and Stonesfield Slate, overlapped by the Taynton Stone. Again, however, this is only part of a much more general overlap towards the east, continued across Oxfordshire into Northamptonshire (fig. 47, p. 245).

The most convincing indications of anticlinal uplift are perhaps those displayed by the Upper Lias. As already stated, this formation is 300 ft. thick in the Cotswolds, but thins to 80 ft. along the western side of the Vale of Moreton, to 30–40 ft. at Chipping Norton, and 5–12 ft. in the Evenlode Valley near Charlbury. It soon thickens again to the east, being already 70 ft. thick at Bloxham, before reaching the valley of the Cherwell. But it is noticeable that the thinnest records are those farthest south. This is especially evident when the records of 5–12 ft. at Charlbury and 20 ft. at Milton under Wychwood are compared with the 30–40 ft. at Chipping Norton, some 5 miles due north. Bloxham, where 70 ft. is recorded, is still farther to the north, and no records of the thickness of the Upper Lias seem to be available farther south in the longitude of the Cherwell Valley. It may be hazarded that here no greater thickness will be proved than at Chipping Norton. There is, in fact, a record of a boring at Wytham, near Oxford, on the same longitude as Bloxham but much more to the south, and the thickness proved was rather less than at Charlbury.

Thus it would seem probable that two different types of phenomena have been confounded, having all been explained as being 'due to the Vale of Moreton Axis'. Certainly some anticlinal appearances in the Corallian near Oxford are unlikely to have any connexion with the attenuation of the Lias and Inferior Oolite, and I have kept the two phenomena separate by referring the Corallian uplift to the elevation of an Oxford Axis,[1] which will be discussed later.

Borings made since Buckman wrote in 1901 have proved that Hull was fully justified in his opinion that the Lower Lias also thins out from west to east into Oxfordshire, and that this thinning of all the Lower Jurassic formations is primarily a part of the general south-easterly attenuation to which he was the first to draw attention. A boring at Lower Lemington (Batsford) near Moreton in Marsh,[2] proved the Lower Lias to have a thickness of only 418 ft. in the centre of the Vale, or, making allowance for subaerial denudation, perhaps 500–50 ft. at most. It has thus dwindled to half its thickness at Mickleton, in the North Cotswold Syncline, only 7 miles away. Before reaching Calvert, in Buckinghamshire, it has halved a second time, until no more than 240 ft. remains.

I would venture to explain this thinning of the Lias and Inferior Oolite as due primarily to the relations of the present outcrop to the margin of the Jurassic trough of deposition. There seems to have been a shelf of shallows hereabouts, projecting from the London landmass, and the present outcrops curve south-eastward to meet it. This is cause and effect, for all the formations

---

[1] W. J. Arkell, 1927, loc. cit., pp. 118–22.
[2] A. Strahan, 1913, *Sum. Prog. Geol. Surv.* for 1912, pp. 90–1.

are thinner here than on either side and, in particular, the thick limestone mass of the Lower and Middle Inferior Oolite is absent. Much less resistance has therefore been offered to subaerial denudation and the retreat of the escarpments has been accelerated.

Nevertheless, the differences between the facies of the Inferior Oolite in the Lincolnshire and Northamptonshire part of the trough and that in the Cotswold part point to a barrier or 'axis of uplift', like that at Market Weighton, partially separating them during the time of deposition. Several facts are favourable to the location of this axis along the Vales of Moreton and Bourton. In the first place there is the sharp angle in the '*Scissum* Line' (map on p. 207) near Stow on the Wold and Bourton on the Water, which suggests that the shelf was suddenly terminated at this point. Secondly, there is the fact that the individual members of the Inferior Oolite Series in the North Cotswolds thin out everywhere towards the Vale of Moreton, their feather-edges striking almost due N.-S. (Two are inserted as examples in the map, p. 207.) Thirdly, as Richardson has pointed out, the Rhætic Beds show a number of peculiarities about Stratford on Avon.[1] All these facts are highly suggestive of uplift along a N.-S. axis passing down the centre of the Vales of Moreton and Bourton, perhaps along a prolongation of the main Pennine Axis of Derbyshire via the Warwickshire coal-field. Permian rocks, in fact, tongue southward to Warwick, only 21 miles north of Moreton in Marsh. What we now see on the line of the present outcrop of the Lower Oolites is only the roots of such an axis, near the region where it joined the land. Probably the part which was comparable with that still to be seen near Market Weighton was completely destroyed by erosion during the formation of the Vale of Worcester and Warwick.

No effects of any movements subsequent to the time of the Lower Oolites have so far been convincingly demonstrated along the southward prolongation of the Vale of Moreton Axis. The Upper Jurassic and Cretaceous rocks seem to be unaffected, unless some anomalies in the Corallian near Purton, to be discussed later, are partly due to this cause.[2]

(2 & 3) THE MALVERN AND WINCHOMBE AXES

Two other anticlinal flexures of parallel N.-S. strike call for notice before we pass on to others of different directrices. On the west side of the main promontory of the North Cotswold hill-mass the map shows another conspicuous tongue of Lias penetrating southward into the oolitic hills. This is the Vale of Winchcombe. It is complementary, as it were, to the Vale of Moreton, though much smaller, and its anticlinal structure was likewise pointed out by Hull in 1855.[3] That the folding has affected the Lower Lias is proved by the occurrence of strata so low as the *obtusum* zone near Toddington, on the Stow road,[4] but no intra-Jurassic movements have so far been shown to have taken place along it.

[1] L. Richardson, 1912, *Geol. Mag.* [5], vol. ix, p. 25.

[2] Cox and Trueman have suggested (*Geol. Mag.*, 1920, vol. lvii, p. 204) that the overstep of the Lower Greensand on to Corallian at Faringdon might be accounted for by posthumous movements along this axis, but Faringdon lies 5 or 6 miles east of the line, and the occurrence, as we shall see, is better explained in another way.

[3] E. Hull, 1855, loc. cit., p. 483, and 1857, 'Geol. Cheltenham', *Mem. Geol. Surv.*, p. 99, fig. 15.

[4] L. Richardson, 1929, 'Geol. Moreton in Marsh', *Mem. Geol. Surv.*, p. 8.

F

A much more important axis with the same strike is the line of the Malvern and Abberley Hills, where Pre-Cambrian and Lower Palaeozoic rocks are lifted high above the sea. This line corresponds approximately with the boundary between the Palaeozoic and Mesozoic terrains, but the junction between them is nearly always a faulted one. Although there are many gaps, the eye is carried on southward by the Lower Palaeozoic inliers of May Hill (about 7 miles from the southern end of the Malverns) and Tortworth, and eventually almost to Bath by the upturned eastern margin of the Bristol coal-field. Carboniferous Limestone crops out continuously as far south as Chipping Sodbury, and even beyond this Old Red Sandstone and probably still

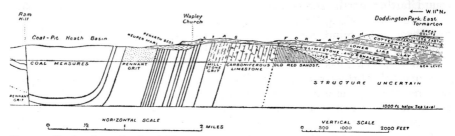

FIG. 12. Section from the South Cotswold escarpment to the Bristol coal-field, showing the Rhætic and Lias overlapping the Trias on to the Palaeozoic rocks over the continuation of the Malvern Axis. From L. RICHARDSON, 1930, 'Wells and Springs of Gloucestershire', *Mem. Geol. Surv.*, fig. 2

earlier rocks immediately underlie the Mesozoic covering (often directly under the Rhætics) a short distance to the east (fig. 12). The Lower Lias, moreover, here becomes very thin just where the Trias is missing. The unusually straight N.-S. alignment of the South Cotswold escarpment from Dursley to Doynton is also suggestive of an uplift along the line of strike. The same trend is carried on by the Inferior Oolite escarpment south of the Mendips to Milborne Port, on the Dorset border, but no actual line can be drawn on the map.

Axes of Armorican Trend in the South of England.

(1) THE MENDIP AXIS

The Mendip Hills present a third type of axis, differing somewhat from the others. The structure of the chain has been elucidated with great care and described by Dr. F. B. A. Welch and others.[1] It consists of four steep, elongated domes or periclines, their axes striking E.-W., but arranged *en échelon* at such an angle that a line joining their centres runs almost NW.-SE. along the central ridge of the hills. In the core of each pericline Devonian rocks are exposed, while the cover consists of Carboniferous Limestone. According to Dr. Welch the pressure that gave rise to the folding came from the south, and the resistance offered by the syncline of the Radstock coal-field, the axis of which ran N.-S. parallel to the Malvern Axis, caused thrusting and some over-folding of the Carboniferous strata.

Part of the Mendip Hills stood above water as an island or a chain of islands,

[1] F. B. A. Welch, 1929, 'The Structure of the Central Mendips', *Q.J.G.S.*, vol. lxxxv pp. 45–76; and 'of the Eastern Mendips', ibid., vol. lxxxviii, 1932.

certainly throughout Lower Lias times, and possibly until the end of the Bajocian (Middle Inferior Oolite) or even Vesulian periods. Before the deposition of the Vesulian the folded Palaeozoics suffered peneplanation. Subsequent downwarping of the Jurassic trough carried any more easterly periclines that there may be far down underground, where they are concealed beneath thick accumulations of Upper Jurassic rocks which do not seem to be affected at all in their passage over the prolongation of the axis.

Godwin-Austen, in his paper of 1856, joined up the eastern end of the Mendips with the Vale of Pewsey and so with the Wealden Anticline and the Axis of Artois. Buckman in 1901 took further liberties with it, swinging it round farther north, across the mouth of the Vale of Pewsey and northeastward almost along the edge of the Chalk Downs.[1] By 1927, however, he had changed his mind, for he depicted the eastern extension as straight and short, directed due W.–E., under the middle of Salisbury Plain.[2]

Now Dr. Welch has shown how on theoretical grounds it is probable that only one half of the structure is visible in the Mendip Hills, while the other half, resembling it as a mirror image, lies buried beneath the Mesozoic covering to the north-east. The resistance offered by the older Malvernian fold of the Radstock coal-field to the northward-advancing Armorican earth-wave, which gave rise to the formation of the series of periclines arranged *en échelon* from SE. to NW. on one side, would have produced a corresponding series on the east side of the obstacle, arranged from SW. to NE. or ENE. Thus the chances are that the general line of the Mendip fold, regarded as a whole, extends in a NE. or ENE. direction.

Theoretical considerations are here entirely borne out by the arrangement of the outcrops, as may be seen by a glance at the map (fig. 13, p. 70). Each formation, from the Carboniferous Limestone to the Cornbrash inclusive, tongues deeply into the next in an ENE. direction, and all the tongues point straight at the Vale of Pewsey. It seems logical to follow Godwin-Austen and to join up the Mendip fold with the Vale of Pewsey; and no sooner do we do so than we are committed to follow him still farther, for the Pewsey monocline with its periclinal inliers of Upper Greensand, the Vales of Shalbourne and Kingsclere, is continued eastward almost to Farnham, where it joins the Peasemarsh or Hog's Back Monocline, the most important structural line of the Weald (see fig. 15, facing p. 86).

The Hog's Back–Vale of Pewsey Monocline involves Eocene strata. It is presumably, therefore, due chiefly to an uplift in the Miocene period, as is the Purbeck–Isle of Wight Monocline, with which it is roughly parallel.[3] There are also definite indications of movement between the Cretaceous and Eocene periods.[4] From this it might be supposed that a more or less continuous history of uplift could be traced along the axis through the Jurassic period; but on the contrary there seems to have been complete absence of differential uplift from Forest Marble times until the Lower Cretaceous.

[1] S. S. Buckman, 1901, loc. cit., p. 148.

[2] S. S. Buckman, 1927, *Q.J.G.S.*, vol. lxxxiii, p. 16, fig. 2.

[3] Messrs. Dines and Edmunds have shown that, as in the Isle of Purbeck, there was also a certain amount of thrusting at the Hog's Back ('On the Tectonic Structure of the Hog's Back', 1927, *P.G.A.*, vol. xxxviii, pp. 395–401).

[4] For a valuable summary of the evidence bearing on this point see H. J. Osborne White, 1907, 'Geol. Hungerford and Newbury', *Mem. Geol. Surv.*, pp. 43–6.

The relations of the various Jurassic formations to the axis will be treated in greater detail in the ensuing chapters, but, owing principally to the wealth of exposures in the vicinity of the Radstock coal-field, more is known of the

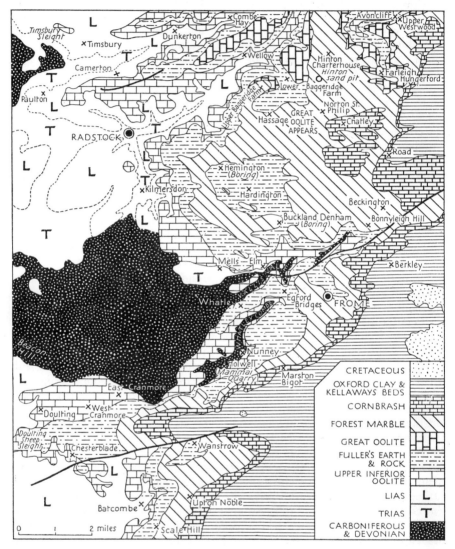

FIG. 13. Sketch-map of the region around Frome and Radstock, where the Jurassic outcrops cross the end of the Mendip Axis. (*Based on the 1 inch map of the Geol. Survey.*)

detailed movements that took place in Liassic times along several subsidiary axes farther north, about Radstock and Keynsham, than of those affecting the main anticline.[1] Small movements recurred restlessly along all of these axes throughout the deposition of the Lower and Middle Lias. The greatest uplift

[1] J. W. Tutcher and A. E. Trueman, 1925, *Q.J.G.S.*, vol. lxxxi; see below p. 131.

of all, however, took place at the end of Bajocian times, before the Vesulian transgression. This uplift was not confined to the Mendip Axis, but affected all the Armorican axes in the South-West of England, giving rise to what Buckman called the Bajocian Denudation. The last Jurassic uplift of the Mendip Axis apparently took place during the deposition of the Fuller's Earth (see pp. 261–2) and immediately before. The Fuller's Earth as a whole becomes very thin and the middle portion of the formation overlaps the lower; then finally, near Whatley it overlaps the Upper Inferior Oolite also and comes to rest on the Carboniferous Limestone (see fig. 13, opposite).

(2) THE NORTH DEVON AXIS

In his well-known memoir on the Inferior Oolite of Somerset between Doulting and Milborne Port, Richardson described a small syncline of Bajocian rocks centred at Cole near Bruton.[1] This he named the Cole Syncline. Subsequently he drew attention to the continuity of the syncline with the conspicuous basin-shaped outliers of Middle and Upper Lias forming the hills at the Pennards, Glastonbury Tor and Brent Knoll, the last no less than 22 miles west of the nearest point on the main outcrop, Lamyatt Hill.[2] The Cole Syncline (described in full on p. 195) preserves a small trough of Lower and Middle Inferior Oolite, of which the continuations on the north and south were uplifted and eroded away by the Bajocian Denudation, so that Upper Inferior Oolite (Vesulian) was deposited on Upper Lias. The Middle and Upper Lias of the Brent Knoll, Glastonbury and Pennard outliers were therefore let down into the low positions which led to their preservation, not by Tertiary movements, but between Bajocian and Vesulian times. As pointed out by Richardson, this syncline is approximately parallel with the Mendip Anticline, and its chief Jurassic period of activity was definitely contemporaneous with the principal movement along that anticline. The relics of a corresponding trough on the north of the Mendip Anticline may be discernible in the outlier of Dundry Hill, between the Mendips and the South Wales Anticline.

Since it is in the nature of a syncline to lie between two anticlines, the next step is to inquire whether another anticline bordered the Cole Syncline on the south. Evidence of such an anticline—'a well-marked line of weakness' as Richardson termed it,[3]—there undoubtedly is, and I propose to use for it the name North Devon Axis, following Boyd-Dawkins.[4]

For about 6 miles south of Castle Cary, as will be explained in Chapter IX, the Upper Inferior Oolite rests on Upper Lias, while farther south, beyond Corton Downs, the Lower and Middle Inferior Oolite return and extend through the Sherborne district. This anticlinal structure lies directly on the prolongation of Boyd-Dawkins's main axis of North Devon, which probably runs approximately along the coast at Minehead, where the lowest beds of

---

[1] L. Richardson, 1916, *Q.J.G.S.*, vol. lxxi, pp. 495–503.
[2] L. Richardson, 1926, *Proc. Somerset Arch. N. H. Soc.* [4], vol. lxxii, pp. 73–5. The synclinal structure of the Brent Knoll and Glastonbury Tor outliers was pointed out by H. B. Woodward in 1887, *Proc. Bath N. H. and A. F. C.*, vol. vi, p. 130.
[3] L. Richardson, 1916, loc. cit., p. 518.
[4] W. Boyd-Dawkins, 1894, 'The Probable Range of the Coal-Measures under Oxfordshire and the Adjoining Counties', *Geol. Mag.*, N.S. [4], vol. i, p. 459.

the Lower Devonian exposed on the North Devon–Somerset border come to the surface.

The line is indicated by a long elliptical inlier of Rhætic Beds at Sparkford and Camel, close to the Inferior Oolite escarpment; in fact Rhætics are here faulted up within less than a mile of Inferior Oolite, and not many hundred yards from Middle Lias (p. 122). West of this the long Lower Lias and Rhætic escarpment of the Polden Hills, running nearly straight for 30 miles, from Charlton Mackrell to near Watchet, at once claims attention. The dip here is gentle and the outcrops consequently broad. The nearest tongue of Trias, between Somerton and Charlton Mackrell, is probably deceptive, owing its existence apparently to the erosion of the River Cary. It is safer to continue the anticline through Langport straight for the nearest Devonian rocks, which tongue a long way into the Triassic tract as the eastward continuation of the Quantocks, until they reach immediately north of Taunton. Thence the line is followed easily north-westward through the centre of the Quantock Anticline, which probably stood above the Rhætic and Liassic sea as an island, like the Mendips.[1] No Jurassic rocks are seen to overlap the Trias on to the Devonian of the Quantocks, but the Rhætic Beds, which approach nearest, diminish greatly in thickness. In the outlier of Chedzoy, near Bridgwater, they are 30 ft. thinner than at Puriton, only 3 miles farther away.[2]

If we continue the curve of this axis eastward, keeping the same radius (slightly longer than that of the Mendip Axis but approximately equal to that of the Cole Syncline), it passes directly up the centre of the Vale of Wardour, parallel to the great fault which bounds the Vale on the north side. The Vale of Wardour Axis in turn loses itself in the Plain north of Salisbury, within a few miles of where the new Stockbridge and Winchester folds begin; and these carry the same line on into the Weald (see fig. 15). The small gap near Salisbury is bridged by the parallel anticline a few miles to the south, which is continued from Ports Down by further periclines, at Dean Hill and Bower Chalke, almost as far west as Shaftesbury.[3]

There can be little doubt, in view of these facts, that the two principal 'notches' in the western escarpment of the Chalk, the Vales of Wardour and of Pewsey, which owe their existence to denudation acting on approximately E.–W. anticlines formed during the Miocene orogeny, are based on much older Hercynian axes. A study of the Jurassic rocks shows that the uplifts had a curiously intermittent history, for the Vesulian and later formations (except the Fuller's Earth) are entirely unaffected, although a somewhat violent uplift took place along both axes immediately prior to the Vesulian. Throughout the whole Upper Jurassic they seem to have sunk steadily as part of the normal trough of active sedimentation, differential movements only recurring with the onset of tangential pressure, first in the Lower Cretaceous and again in the Miocene periods.

[1] The 'Quantock Isle' of Lloyd Morgan's map of the islands in the Keuper lake, reproduced in Jukes-Browne, *Building Brit. Isles*, 1911, ed. 3, p. 250.
[2] H. B. Woodward and W. A. E. Ussher, 1908, 'Geol. Quantock Hills', *Mem. Geol. Surv.*, p. 71.
[3] Most of these axes in the Chalk country have been worked out by Osborne White, in *Mems. Geol. Surv.*: see especially the Memoirs for Winchester, Andover, Basingstoke and Shaftesbury.

(3) THE WEYMOUTH, PURBECK AND ISLE OF WIGHT AXIS

A third great E.-W. monoclinal fold across the South of England forms the central axis of the Isles of Wight and Purbeck and the Weymouth district. This fold was the most intense of all the Miocene disturbances in England; the steep northern limb was first inverted and then fractured, resulting in the Isle of Purbeck and Ridgeway thrust-fault. The fact that the Hamstead Beds of the Middle Oligocene are involved in the folding in the Isle of Wight, as are also the Oligocene Beds of Creech Barrow in Dorset, proves that the greater part of the movement took place during or after the Miocene period. On the other hand, the detailed investigation and mapping of the Eocene Beds by Clement Reid showed that considerable upheaval and denudation of the Chalk was already in progress in the Eocene period, just as on the Pewsey Axis. This is proved by the composition of the Eocene gravels: the Reading Beds are largely composed of Chalk flint and Greensand chert, while the Bagshot Beds contain in addition chert of Purbeck age.[1]

This axis has been clearly described by Sir A. Strahan, whose intensive studies have revealed many features of interest.[2] In the first place, the mapping shows the presence of separate subsidiary axes on the north, such as the anticline of Chaldon, Poxwell and Ridgeway. All, though roughly in line, are not a continuous fold but a series of elongated domes or periclines, like those of Dean Hill, Ports Down, Kingsclere and Shalbourne. It is evident that this structure is characteristic everywhere, uplift not having operated evenly along the whole course of any axis. The Brixton and Sandown Anticlines on either side of the Isle of Wight are separate and *en échelon* (see fig. 15), and so are those of Weymouth and Purbeck, though there is no doubt that they formed part of one and the same major line of upheaval.

In the second place Sir A. Strahan was able to show conclusively that these axes had undergone a phase of elevation between the Lower and Upper Cretaceous periods—that is, before the Albian transgression. At Ringstead and Chaldon the evidence for this is so well displayed that it is possible to measure the dip of the Upper Jurassic and Wealden rocks and of the Albian (Upper Greensand) above them, and so to determine the degree of intra-Cretaceous uplift quantitatively; and it constitutes no inconsiderable proportion of the total movement.

From the arrangement of the outcrops around the mouths of the Vales of Wardour and Pewsey and the presence of a pebble-bed at the base of the Albian, which rests on much-attenuated Aptian sands, there is little doubt that movements took place immediately before the Albian along those axes also. Pre-Aptian movements are attested by the unconformable relations of the thin Aptian sands to the Wealden Beds.[3]

With regard to the question of intra-Jurassic movements along the Weymouth Axis we are almost entirely in the dark. The earliest Jurassic sediments

[1] C. Reid, 1896, *Q.J.G.S.*, vol. lii, p. 490.
[2] A. Strahan, 1898, 'Geol. Isle of Purbeck and Weymouth', *Mem. Geol. Surv.*, pp. 212–29; see also his earlier memoir on the Isle of Wight; a paper 'On Overthrusts of Tertiary Date in Dorset', *Q.J.G.S.*, vol. li, 1895, pp. 549–62; and 1906, 'Guide to the Geological Model of the Isle of Purbeck', *Mem. Geol. Surv.*, pp. 5–12 (new edition, 1932).
[3] H. B. Woodward and C. Reid, 1903, 'Geol. Salisbury', *Mem. Geol. Surv.*, pp. 34–6.

brought to the surface in the centre of the axis before it runs out to sea near Langton Herring belong to the Great Oolite Series (Upper Fuller's Earth). As might be expected by analogy with the Upper Jurassic rocks over the North Devon Axis, neither this nor most of the subsequent formations so well exposed around Weymouth show any signs of disturbance or abnormality of any kind. The sections are more complete than any others in England, and there can be no doubt that the Weymouth and Purbeck Anticline remained for all practical purposes quiescent from Bathonian times onward until the end of the Wealden. It is just possible, however, that over the centre of the axis the Portland Beds suffered erosion in Purbeck times, for there are pebbles of Portland Limestone in the Lower Purbeck Dirt Beds on Portland Island (see p. 530).

The nearest outcrops of Lower Oolitic rocks and Lias, on the coast between Bridport and Lyme Regis, are at least 5 miles north of the line of the axis. Nevertheless, the Lower and Middle Inferior Oolite are thinner and more incomplete on the coast than farther inland towards Beaminster, and they contain two pebble-beds, which disappear inland. Buckman attributed in addition certain phenomena of the Upper Lias Junction Bed to proximity to the axis, but the evidence for this seems rather doubtful, some of the data upon which he based his conclusions having been subsequently shown to be erroneous.[1]

Farther west the course of the anticline is uncertain, but it seems to have abutted against Dartmoor.

Although the inroads of the sea in Dead Men's Bay have obliterated all the Lower Jurassic portion of the Weymouth Anticline, there is still much to be learnt from the syncline which separates it from the North Devon Anticline. This is broader than the syncline between the North Devon and Mendip Anticlines in proportion as the anticline to the south is more acute. The first features noticed on consulting the map and running the eye westward along the continuation of the Tertiary trough of Hants and Dorset are the long tongues of Oolites and Lias which extend westward by Ilminster and Chard. The Lower Lias reaches to the longitude of Taunton, which is 4 miles farther west than the most easterly Devonian rocks of the Quantocks, on the anticline a short distance to the north. Next to claim attention is the still more extended outcrop of Upper Cretaceous rocks, which reach almost to the edge of the Trias.[2] More especially remarkable is the continuous, even curve of their northern margin, which accurately reflects the curve of the North Devon Axis. From Melcombe Horsey, in the angle of the Dorset Downs, the curve is continued westward with only the most trivial deviations, obviously due to subsequent denudation, to the end of the Black Down Hills; these in turn point straight into the more northerly of the two main troughs of the Mid-Devon Syncline, where the Permian rocks tongue conspicuously into the Culm Measure region past Tiverton. The detailed stratigraphy of the Inferior Oolite bears witness to the existence of a minor anticline in the Jurassics

---

[1] See below, pp. 168, footnote.

[2] The marked overstep of the Albian along this line does not, in my opinion, indicate any anticlinal pre-Albian uplift, since if Albian rocks were preserved anywhere else so far west they would likewise overstep the Jurassics towards the margin of the basin (as towards London); rather the preservation of the Albian along this line points to its having been lowered in a subsequent (Tertiary) syncline.

north of Crewkerne, but the evidence does not suffice for entering its course upon the map (for particulars see pp. 192–3). In the continuation of the synclinorium in the Culm Measures to westward there are many minor flexures.

(4) THE BIRDLIP AXIS

To be classed with the E.–W. axes which we have been discussing, rather than with those of Charnian trend, is the Birdlip Anticline in the Cotswolds, detected and described in detail by Buckman. This uplift divides the main Cotswold basin or trough into two minor synclines, those of Cleeve Hill and Painswick. It strikes approximately WNW.–ESE. from near Gloucester through Birdlip Hill, thence curving SE. along Ermine Street towards Cirencester, where it becomes lost beneath the Great Oolite and later rocks.

The principal uplift took place during the Bajocian (immediately prior to the Vesulian) but there were also movements during the Aalenian. In the Vesulian and subsequent epochs of the Jurassic the axis seems to have had no influence whatever.

Movements at earlier times are more easily traceable in the synclines on either side than along the axis itself. Thus Richardson has shown that the Rhætic Beds reach a greater thickness than anywhere else in Worcestershire along the continuation of the Cleeve Hill–Bredon Hill Syncline, at the end of the long tongue of Lias and Rhætics that extends to Droitwich, between the anticline of Birdlip and that of the Vale of Moreton.[1] Opposite the Painswick Syncline he has similarly found 'very striking evidence of a syncline' in the Rhætic Beds of Chaxhill, Westbury on Severn.[2] But although at Denny Hill, $2\frac{1}{4}$ miles north-east of Chaxhill, 'there is evidence of the proximity of an anticline', he was unable to determine its exact position. It may lie near Lassington, but the evidence is equivocal and long since obscured [3] (fig. 14). Consequently the axis cannot be localized more accurately than between Denny Hill and Wainlode Cliff. As may be seen on the map (fig. 15), there seems considerable probability that the Birdlip Axis is either a direct continuation of or a side-branch from the Malverns. The Malverns and May Hill are not in continuity, but seem to be separated by a synclinal area, where Trias and Permian rest on Devonian.

If we classify by means of the principal periods of activity, as it is here proposed to do, the Birdlip Axis falls into line with the E.–W. or Armorican group. The NW.–SE. direction seems at first an objection, but it should be noted that at least two of the other E.–W. axes, those of the Mendips and North Devon, swing NW. at their western extremities; in fact the visible portion of the Birdlip Anticline is almost exactly parallel with the western extremity of the Mendip Axis, repeating its curve faithfully. It is possible that it was deviated southward by the Vale of Moreton Axis.

When viewed in this new light, the failure of the Birdlip Axis to affect the Upper Jurassic rocks is explained. Instead of seeking such effects we look for oversteps at the base of the Aptian and the Albian, and for final uplift in the Miocene period. East of the Mendip Axis is the notch in the Chalk Downs forming the Vale of Pewsey; east of the North Devon Axis lies similarly the twin notch of the Vale of Wardour; can it be mere coincidence that the

[1] L. Richardson, 1904, Q.J.G.S., vol. lx, p. 352.   [2] Ibid., p. 356.   [3] Ibid., pp. 352–3.

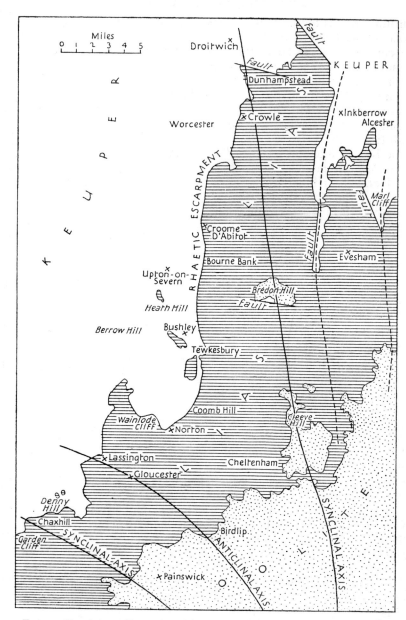

FIG. 14. Sketch-map illustrating the anticlinal and synclinal axes along which there is evidence of movements in Keuper-Rhætic and in Inferior Oolite times. Inferior Oolite and later beds dotted; Lias shaded; Rhætic and earlier beds white (boundaries approximate). From L. RICHARDSON, 1904, *Q.J.G.S.*, vol. lx, p. 350, fig. 1.

Birdlip Axis has to eastward of it the great angle of the Downs, the incipient 'Vale' of Wallingford, at the entrance of the Goring Gap? Let us continue the curve on the map below Cirencester, swinging it round thence eastward. First we come to the tongue of Lower Greensand resting on Corallian at Faringdon and Badbury Hill, then to the long overstep of the Gault on to Kimeridge Clay through the Vale of the White Horse—protracted perhaps because cutting our line at a very low angle—and so to Wallingford. On comparing the disposition of the Greensand and Chalk of the Wallingford Gap with that in the Vales of Pewsey and Wardour, it seems that the differences could be accounted for by supposing that the anticline intersects the strike of the rocks much more obliquely and is feebler. Finally, the southerly dip of the Marlborough and Berkshire Downs demands an E.–W. anticline north of them, and this want the Birdlip Axis as here visualized supplies.

Beyond this (perhaps even so far) it is rash to speculate, but to my knowledge no other axis of uplift passes anywhere near the sharp post-Eocene anticline superimposed on a pre-Eocene and post-Cretaceous forerunner at Windsor. It may be significant that this lies just where we should draw the continuation of the Birdlip Axis in order to make it roughly parallel with the adjoining portions of the Vale of Pewsey–Hog's Back fold; moreover, the strike of the Windsor Anticline is just in the right direction, E. 15° S.[1]

Other explanations of the entrance to the Goring Gap at Wallingford have been put forward (see below), but the one here offered applies to the Albian overstep in the Vale of the White Horse, the Aptian overstep at Faringdon, and the southerly dip of the Chalk Downs of Berkshire, and it has the advantage in addition of accounting for the analogous phenomena at the Vales of Wardour and Pewsey and at Wallingford by analogous causes.

(?5) THE MELTON MOWBRAY AXIS

Perhaps the first feature that strikes the eye on glancing at the geological index map of Lincolnshire, Leicestershire and Rutland is the long, narrow tongue of Middle and Upper Lias and Inferior Oolite which extends north of Melton Mowbray for about 12 miles westward of the general line of outcrop, to Wartnaby and Old Dalby. The general strike is NNE.–SSW., but at Grantham it changes to NE.–SW. and then, at the extremity of the tongue, swings round through 315° to due E.–W., to revert equally suddenly near Sproxton to its original direction of NNE.–SSW. This at once suggests a syncline striking ENE.–WSW., running down the centre of the tongue.

The connexion of the phenomena with geological structure was proved by Prof. A. H. Cox, who investigated the area with a view to tracing the relationship between geological structure and magnetic disturbances. He showed that an anticline runs south of the Wartnaby and Old Dalby tongue, passing along the valley through Melton Mowbray, that the age of the anticline is at least as late as post-Triassic, and that it probably 'follows and is founded upon the line of an older and more pronounced anticlinal uplift of pre-Permian date'.[2]

In a later paper Professors Cox and Trueman stated that the appearances

---

[1] C. N. Bromehead, 1915, 'Geol. Windsor', *Mem. Geol. Surv.*, pp. 14–15.
[2] A. H. Cox, 1919, *Phil. Trans. Roy. Soc.*, vol. ccxix A, pp. 73–135; and Cox and Trueman, 1920, *Geol. Mag.*, vol. lvii, p. 198.

'suggest that the movements which gave rise to the anticlinal structure had died down during or prior to the deposition of the Bathonian rocks',[1] and Dr. Rastall in discussing the anticline goes even farther, saying that 'it seems impossible to attribute the whole thing to Tertiary disturbance, since the Upper Jurassic strata are not affected'.[2]

Now, examination of the map shows that, although the rocks above the Inferior Oolite may show no disturbance over the anticline, they certainly are affected by the corresponding syncline. In this respect they agree with the Inferior Oolite and Lias, for by examination of the map alone it would be impossible to detect the presence of the Melton Mowbray Anticline. It is the companion syncline to the north of it that has the conspicuous effect on the strike, and here too the Great Oolite, Cornbrash and Oxford Clay tongue westward, repeating faithfully the behaviour of the lower formations. The Oxford Clay extends past Bassingthorpe at least 5 miles west of the general boundary of the outcrop, and it would extend much farther if it were not cut off by a small fault upthrowing west. There is, even in spite of the fault, a small outlier of Great Oolite and Cornbrash at Shillington, 4 miles still farther west, making the total length of the projection 9 miles.

It can hardly be supposed that the anticline and syncline, so close and so perfectly parallel, did not take part in the same movements; it is necessary to consider them as one. It can therefore be definitely said that the last movements along the Melton Mowbray Axis were later than the highest Jurassic rocks in the district—the Oxford Clay; and from this it follows that they were probably intra-Cretaceous or Tertiary or both. If the anticline really dies out eastwards as Messrs. Cox and Trueman consider is suggested by the map, this does not prove the movements to have been intra-Jurassic, any more than would the dying out in both directions of any of the periclines strung along the other axes (e.g. Kingsclere and Bower Chalke) were the Cretaceous strata removed from the visibly folded areas.

As Dr. Rastall has remarked, the Melton Mowbray Axis is difficult to correlate with any other system of folds, owing to its anomalous direction. Messrs. Cox and Trueman called its strike E.–W., but Dr. Rastall, apparently with reason, regards it as more nearly ENE.–WSW. It is, therefore, not parallel with the Market Weighton Axis. We have seen, however, how general the tendency is to curve and wander from any true line in the larger E.–W. axes farther south. It is probably legitimate to regard this as the most northerly of the great E.–W. folds, which suffered their maximum uplift in the Miocene and are most acute in the south, becoming fainter northward away from the source of pressure. Before the question can profitably be discussed further, detailed research on the stratigraphy of the Inferior Oolite is essential. If such research reveals an important phase of movement between the Bajocian and Vesulian, there will be strong grounds for correlation with the E.–W. group. Professors Cox and Trueman have already shown that there is evidence of movement between the Middle and Upper Lias, since the *acutum* zone appears to be absent over the anticline, while it is well developed on either side.[3]

---

[1] A. H. Cox and A. E. Trueman, 1920, loc. cit., p. 198.
[2] R. H. Rastall, 1927, loc. cit., p. 21.
[3] Cox and Trueman, 1920, loc. cit., p. 201.

(?6) THE MARKET HARBOROUGH AXIS

Professors Cox and Trueman and Dr. Rastall have discussed the appearances of an anticlinal axis running from near Market Harborough to Peterborough, and, although the evidence is as yet vague, it seems to be agreed that there is probably an axis along this line, parallel and contemporaneous with that at Melton Mowbray. As summarized by Dr. Rastall,

'the map shows a long projection of rocks up to Cornbrash as far as Peterborough, and the general lie of the strata indicates distinctly an anticlinal axis running from Market Harborough to Peterborough, parallel to the Melton Mowbray axis. It is probably significant that, as pointed out by Cox and Trueman in the first case, the Pre-Cambrian outcrops of Charnwood Forest and Nuneaton lie on the westward prolongations of these axes . . .' [1]

The only piece of stratigraphical evidence suggesting intra-Jurassic movements seems to be Prof. Trueman's observation that the *acutum* zone is absent, or at least has never been recorded, near Market Harborough.[2] Southwestward the Rhætic Beds show contemporaneous erosion near Rugby, but this is on the Nuneaton Axis also (see p. 110).

The Charnian Axes:

(1) CHARNWOOD, (2) NUNEATON, (3) SEDGLEY-LICKEY, (4) WOOLHOPE–MAY HILL, (?5) UPWARE AND OTHER POSSIBLE AXES.

In 1925 Dr. R. H. Rastall published an illuminating paper, in which he showed that the Charnian rocks in the Bletchley Boring were not, as had previously been supposed, part of a broad ridge continued southward from Charnwood Forest, but that they lay on a parallel axis striking NW.–SE. He also traced the courses of four axes of the same trend in the Southern Midlands, and concluded that

'the general structure of the Midlands is due to the superposition of a fanlike virgation of the Pennine axis on a pre-existing series of folds with a NW.–SE. (charnoid) trend, the whole being limited on the west by the outer margin of the Caledonian fold-system, and on the south by the outer margin of the Armorican system, while on the east the relations are unknown.' [3]

He likened the process to the bending of a sheet of corrugated iron by pressure applied obliquely to the corrugations. A third system of folds was imposed, as we have just seen, by pressure acting S.–N. This gave rise to the important axes of Armorican trend.

Dr. Rastall pointed out that the anticline in Charnwood Forest, as Prof. Watts had shown, strikes almost exactly NW.–SE., and moreover that Charnian rocks have been proved exactly on the continuation of this line in borings at Leicester and Orton (5 miles west of Kettering). Therefore it would only be by serious distortion that the line could be made to pass under

---

[1] R. H. Rastall, 1927, loc. cit., p. 20.

[2] Cox and Trueman, 1920, loc. cit., p. 201. These authors also point out that the especially deep retreats of the Lias and Oolite escarpments at Weedon and Banbury suggest axes, but no evidence for anticlines at these places seems to have been obtained.

[3] R. H. Rastall, 1925, 'The Tectonics of the Southern Midlands', *Geol. Mag.*, vol. lxii, p. 213.

Bletchley. On the other hand, about 17 miles south-west of the Charnwood Forest Axis, there is the parallel ridge of Pre-Cambrian and Cambrian rocks of Nuneaton and Atherstone, with the Caldecote Series of Pre-Cambrian volcanics. This line points straight for the Bletchley Boring, and the meagre specimens of volcanic rocks from the boring agree just as well with the Atherstone Series as with those in Charnwood Forest.[1]

Dr. Rastall showed also that over both of these axes the Lower Greensand is affected. On the continuation of the Charnwood Axis, south-west of Sandy, it becomes very thin, while 20 miles farther south-west, on the continuation of the Nuneaton Axis, it disappears altogether near Leighton Buzzard.

The Sedgley–Lickey Axis lies about the same distance to the south-west, where, as Lapworth showed, three of the Pre-Cambrian and Silurian inliers of the South Staffordshire coal-field (Sedgley, the Wren's Nest and Dudley) form a NNW.–SSE. line, pointing approximately towards the Lickey Hills.

Still farther to the south-west, and about the same distance beyond the Sedgley–Lickey Axis, a fourth line is suggested by the two dome-like inliers of Silurian rocks which rise through the Old Red Sandstone at Woolhope and May Hill. These are areas of superelevation due to the intersection of two anticlinal axes. The line joining them has again an approximately NW.–SE. strike and seems to belong to the Charnian fold system. It certainly forms with the others a remarkably regular series of axes of similar constitution, evenly spaced, and with approximately parallel strike. It might be expected that any movement would be felt along all of them simultaneously.

Detailed investigation of the Corallian Beds of the counties of Wilts., Berks. and Oxon. has revealed unmistakable signs of relative uplift in Corallian times in two areas, about Oxford and about Purton and Wootton Bassett. As I showed in 1927, in these two areas, each extending along 8 or 9 miles of the outcrop, the sea-bed was unstable and sedimentation was abnormally retarded throughout Corallian times.[2] Each of the subdivisions becomes thinner, overlaps the subdivisions below, or disappears altogether over the critical regions. These regions I therefore regarded as axes of uplift, and named them the Oxford and Purton (perhaps better Wootton Bassett) Axes. (For a more detailed account of the stratigraphical anomalies connected with them, see Chapter XIII, pp. 393, 403, et seq.)

I pointed out at the same time that the Oxford and Wootton Bassett Axes lie approximately along the continuations of the older Sedgley–Lickey and May Hill lines, and suggested that they were due to posthumous uplift along those lines. When the gentle curve of the line joining the centres of the Woolhope and May Hill domes is continued south-eastwards it passes up the Nailsworth Valley, under Charlton and Garsdon, near Malmesbury, towards Wootton Bassett or Tockenham—that is towards the centre of the Wootton Bassett Axis. The distance from May Hill is 32 miles.

The connexion with the Oxford Axis and the Sedgley–Lickey line is less convincing and the distance is greater—50 miles. The axis of the ancient anticline is here less clearly marked and it is correspondingly difficult to foretell where it would lie underground at so great a distance. If we join up the

---

[1] R. H. Rastall, 1925, loc. cit., pp. 199–202, where all references are given and the subject is treated in much greater detail.

[2] W. J. Arkell, 1927, *Phil. Trans. Roy. Soc.*, vol. ccxvi B, pp. 120–2.

three inliers of Sedgley, Wren's Nest and Dudley and project the line perfectly straight, it passes under Oxford City, which is on the centre of the Oxford Axis. This line does not lie along the Vale of Moreton, but passes rather east of Chipping Norton. On the other hand it also fails to pass through the Lickey Hills; to include them a slight detour of 2 miles to the SW. is necessary. The line may curve back again from the Lickey Hills and run towards Oxford, but if it continued to diverge at the same rate it might pass not merely through the Vale of Moreton but still farther west. However, as already shown, the phenomena in the Vale of Moreton are adequately explained by the Pennine Axis.

The other two axes, those of Charnwood and Nuneaton, lay beneath the Ampthill Clay area in Corallian times, where conditions of sedimentation were very different. Exposures in the clay district are so rare that we know nothing of the detailed stratigraphy from place to place along this part of the outcrop, and any changes there may be in the equivalent of the Corallian either near Leighton Buzzard or south-east of Bedford have still to be discovered. But some 25 miles north-east of the Charnwood Axis (that is, about as far beyond it as the average distance between the four known axes) lies the shallow-water reef and oolite shoal of Upware, completely surrounded by clay deposits. This isolated reef might possibly owe its existence to uplift of the sea-bed along a fifth axis of the Charnian group, but this is a matter of conjecture only.

Uplifts subsequent to those in the Corallian period have only been satisfactorily proved over the Charnwood and Nuneaton Axes, where, as Dr. Rastall showed, the Lower Greensand is affected. It may also be significant that the Nuneaton and Sedgley–Lickey Axes accurately delimit the area in which the Portland and Purbeck Beds occur in Oxon. and Bucks. —namely between Stewkley, near Leighton Buzzard, and Nuneham, near Oxford.

Earlier movements seem to be traceable with certainty only in the Rhætic Beds and Middle Lias. On the Charnwood Axis the Rhætic Beds at Leicester contain small pebbles of igneous rocks from Charnwood Forest, but this does not indicate special uplift. A short distance south of Leicester, however, the Langport Beds finally die out, not to reappear again (so far as is known) farther north (see p. 110). The Marlstone of the Middle Lias undergoes marked attenuation over the region of the axis, and between Hallaton and Keythorpe it disappears altogether (see p. 157).[1]

The Nuneaton Axis passes a mile or two south-west of Rugby, close to the village of Church Lawford, where Richardson has described 'abundant evidence for a non-sequence between the White Lias (Langport Beds) and the superincumbent Lower Lias'. There is a ferruginous deposit and the usual signs of erosion, while parts of the White Lias are pebbly, and the *planorbis* zone rests non-sequentially on its bored and pitted upper surface.[2]

The Sedgley–Lickey Axis too, wherever we draw its ultimate continuation, passes first of all through an area of disturbed and attenuated Rhætic Beds in

[1] Between Barrowden and Wakerley, east of Uppingham, the Northampton Sands disappear and Lincolnshire Limestone comes to rest on Upper Lias; but this is 10 miles from the nearest point on the line of the Charnwood Axis (see p. 213).
[2] L. Richardson, 1912, *Geol. Mag.* [5], vol. ix, p. 32.

the tongue of outcrop west of Stratford on Avon, about Binton, Grafton and Bickmarsh. This is the area where the Rhætics are capped by the 'Guinea Bed', a conglomeratic basement-bed of the Lower Lias, containing derived Rhætic fossils (see p. 135). This area lies directly across the North Cotswold Syncline, not many miles north of Mickleton, where the Lower Lias reaches its record thickness of about 1,000 ft. It seems, therefore, to prove conclusively that the syncline was crossed by a line of uplift on the continuation of the Sedgley–Lickey Axis. The whole of the area of disturbance is too far west to be due directly to the Vale of Moreton Axis, though it is about on the line of the Vale of Moreton Axis that the White Lias begins to make its reappearance, after being absent over most of Gloucestershire and Worcestershire. It is possible that this Sedgley–Lickey Axis may have been influential in limiting the Banbury iron-field (Marlstone ironstone) on the south-west, but there seems little definite evidence.

On the Woolhope–May Hill Axis the Rhætic Beds tell quite another story. Only 4 miles from the Silurian rocks of May Hill, and directly on the line of the anticlinal axis (i.e. SE. of May Hill) the Rhætics of Chaxhill, near Westbury on Severn, show 'very striking evidence of a syncline' [1] as noticed above. In fact the Woolhope–May Hill Anticline in the Lower Palaeozoic rocks seems to point nearly into the centre of the Painswick Syncline in the Inferior Oolite 10 miles away. Clearly here, as we pass farther south, the movements of Armorican trend were paramount and totally obscured those of any other directrix.

Evidence of movements along the Charnian axes during the time of the Lower Oolites is wholly inconclusive. The phenomena in the Evenlode Valley, which Dr. Rastall suggested might be due to the crossing of the Pennine Axis by the Sedgley–Lickey Axis near Kingham, in the Vale of Moreton,[2] are, as we have seen, capable of another more probable explanation. In Inferior Oolite times the Armorican and Malvernian axes seem to have controlled sedimentation without interference by movements of other trends, for the Malvern Axis seems to have determined the western boundary and the Pennine the eastern boundary of the Cotswold Syncline, while the Birdlip Anticline subdivided it transversely. Between the other Charnian axes and the phenomena displayed by the Inferior Oolite it seems impossible to establish any connexion. On the contrary there was steady overlap towards the east and south-east. The *Clypeus* Grit passes across the continuation of the Sedgley–Lickey Axis and comes to an end beyond it, between the valleys of the Evenlode and Cherwell; the Hook Norton Beds die out midway between this axis and the next, while the White Sands of the *fusca* zone seem to die out between the Nuneaton and Charnwood Axes (see pp. 207, 304).

In the principal periods of posthumous activity, therefore, there is a fundamental difference between the Charnian Axes and those of Armorican and Malvernian trends; for while two or three or possibly all of the Charnian Axes experienced activity in Corallian times but were quiescent in the Bajocian, the others displayed intense activity during the Bajocian Denudation but lapsed into tranquillity during the Upper Jurassic.

During the Great Oolite period differential movements of the sea-bed certainly occurred, but they have not yet been satisfactorily traced to any

---

[1] L. Richardson, 1904, loc. cit., p. 356.   [2] R. H. Rastall, 1925, loc. cit., p. 215.

particular system of axes. Such regions of uplift as are known in the Upper Estuarine Series of Northamptonshire [1] and in the Forest Marble of Oxfordshire [2] seem to have no connexion whatever with the known axes, either in regard to distribution or direction. It appears that the same may also be said of the local uplifts controlling the distribution of the Stonesfield Slate.

Stratigraphical anomalies are frequent along the outcrop of the Cornbrash, and a close study of this formation seemed at first more promising than almost any other for tracing possible connexions between sedimentation and ancient axes of upheaval. Only two of the anomalies, however, coincide with known axes, leaving many others unexplained. Those two are a boulder-bed in the Upper Cornbrash at Charlton and Garsdon, near Malmesbury (on the Woolhope–May Hill line), and extreme attenuation of both Upper and Lower Cornbrash in the Ouse Valley, from Bedford north-westward (approximately along the line of the Charnwood Axis). These two phenomena belong to different categories, however, for while the Lower Cornbrash is very thin in the Ouse Valley, it is exceptionally thick around Malmesbury; and the boulder-bed near Malmesbury forms part of the *siddingtonensis* zone, which is probably missing altogether in the Ouse Valley. All things considered, the evidence of movement along particular axes during the deposition of the Cornbrash is unsatisfactory.[3]

Before leaving the axes of Charnian trend, mention must be made of some suggestions put forward by Prof. Hawkins in 1918, before the appearance of Dr. Rastall's now well-known paper. From a consideration of the ideal resultants between dip and pitch along the southern margin of the London Basin as compared with the actual resultants shown by the strike of the rocks, he came to the conclusion that the western end of the basin was crossed by two very shallow cross-folds striking NNW.–SSE.[4] One of these anticlines was supposed to pass through White Horse Hill and the Shalbourne Pericline, the other from about Lockinge, near Wantage, through the Kingsclere Pericline, close to Kingsclere. By this means it was sought to explain the similarity in outline between the escarpments of the Berkshire Downs on the north and the Hampshire Downs on the south, as well as the origin of the two periclines. The further discussion of this problem does not concern us here, since no traces of any intra-Jurassic or even intra-Cretaceous movements have been recorded along the continuations of either of these lines. One point is of special interest, however: Prof. Hawkins speaks of the Goring Gap, through which the Thames flows at Goring, as a 'broad shallow syncline' [5] of Charnian trend. This is not at all consistent with Cox and Trueman's or Rastall's suggestions that the Goring Gap may have been determined by an *anticline* of this trend (for Cox and Trueman the anticline runs down the Cherwell Valley from Banbury; for Rastall it comes from the Lickey Hills via the Vale of Moreton). Prof. Hawkins has local knowledge, and *if the Thames runs*

[1] B. Thompson, 1930, *Q.J.G.S.*, vol. lxxxvi, pp. 447–9 and Pl. li. In spite of what Mr. Thompson says (p. 447) it may be surmised that his largest 'anticline' along the Nene Valley from Rushden to Wadenhoe is merely the marginal thinning of the Great Oolite Series south-eastwards, as Judd thought (see below, p. 310).

[2] W. J. Arkell, 1931, *Q.J.G.S.*, vol. lxxxvii, pp. 563–95.

[3] J. A. Douglas and W. J. Arkell, 1928–32, *Q.J.G.S.*, vols. lxxxiv and lxxxviii; also below, Chapter XI.

[4] H. L. Hawkins, 1918, *Proc. Hants F. C.*, vol. viii, part ii, pp. 16–21.

[5] Loc. cit., p. 20.

through a syncline at Goring, the line of uplift which caused the retreat of the escarpment between Wallingford and Blewbury must strike across it instead of along the gap from NW. to SE. as the other authors have supposed. This offers substantial support 'for the view of the underground course of the Birdlip Axis outlined above.

One more supposed anticline of Charnian trend must be mentioned, though it probably has no connexion whatever with the others. It affects the Chalk and underlying Jurassic near the sharp angle in the Dorset Downs, where, according to Osborne White, an ill-defined, depressed fold runs from Ibberton north-westward, towards Milborne Port.[1] It seems probable that the appearance of an anticline here is due to the margin of the Hants–Dorset E.–W. syncline intersecting the rocks dipping east, with a N.–S. strike, along the continuation of the Malvern Axis. The change of dip is too large and general a feature to be caused by the anticline, and it is more likely that the anticline is the effect rather than the cause.

Some Axes of Similar Trend but of a Different Type.

(1) THE CLEVELAND AXIS

So great an authority as Prof. Kendall believes that the Peak Fault resulted from the fracture of a monocline, and as there seems no doubt that the first fracture took place actually during the deposition of the Upper Lias, there is some reason for supposing that uplift may have started along the anticline which now forms the axis of the Cleveland Hills as early as Liassic times. The evidence will be discussed in more detail in Chapter VII, pp. 180–1. In general Cleveland, lying in the centre of the Yorkshire Basin, was certainly a region of subsidence and heavy sedimentation all through the Jurassic period. Any uplift at the Peak can only have been a local and relatively insignificant occurrence within the synclinorium.

(2) THE LOUTH–WILLOUGHBY AXIS

In 1905 Prof. Kendall drew attention to what he termed the Louth–Willoughby Axis, defining it as 'a long anticlinal fold [in the Cretaceous rocks] running through Lincolnshire from the Wash in a north-westerly direction as far as Louth'.[2] He remarked that the Jurassic rocks appear to be especially thick along this line, and he therefore considered it a post-Cretaceous anticline superimposed on a Jurassic syncline, analogous with the Yorkshire Basin (Cleveland). Its path is conspicuous on the map by reason of the long tongue of Lower Cretaceous rocks running from the Wash up into the Chalk area.

　If this line be swung round a few degrees to slightly west of north-west beyond Louth, it coincides with the line of uplift which Dr. Versey calls the Caistor Axis.[3] This passes off the Chalk outcrop a short distance north of Caistor, and Dr. Versey believes that its continuation is indicated by the deflections in the Lias and Lower Oolite scarps at Flixborough and Santon. It seems to be approximately parallel with the Market Weighton Axis, and

---

[1] H. J. Osborne White, 1923, 'Geol. Shaftesbury', *Mem. Geol. Surv.*, p. 5.
[2] P. F. Kendall, 1905, loc. cit., p. 201.
[3] H. C. Versey, 1931, *Proc. Yorks. Geol. Soc.*, vol. xxii, pp. 55–6.

Dr. Versey places the centre of the intervening syncline at Elsham, three miles north-north-east of Brigg (fig. 10, p. 60).

In Jurassic times this axis can have been in no way analogous with that at Market Weighton, but during the time of the Lower Cretaceous and immediately pre-Cretaceous movements it presented some analogies. Near Caistor the 'Carstone', believed to be locally of Lower and Middle Albian date, rests on Kimeridge Clay—a very considerable hiatus. About Willoughby, however, comparatively thick marine Neocomian strata are developed. It is evident, therefore, that uplift was less intense towards the south-east—that the axis died out in that direction. It is possible that if erosion or deep borings were to give any insight into what becomes of the Market Weighton Axis towards the south-east, it also would be found to fade away in the same direction.

The Willoughby–Caistor Axis is difficult to classify. In direction it is intermediate between the Market Weighton and the Charnian Axes; the strike at Caistor is more nearly parallel with the Market Weighton line,[1] but between Louth and Willoughby it is truly Charnian (due NW.–SE.). It agrees with both Charnian and Market Weighton Axes in giving rise to an overlap of the Lower Cretaceous by the Albian, but it is fundamentally different from both in having been a syncline through most of Jurassic times. Dr. Rastall describes it as 'an old Charnoid axis, probably synclinal in the Jurassic and anticlinal later, to which subsequent movements readily adapted themselves'.[2]

Definite evidence that the axis was synclinal at least in the Great Oolite period is provided by the Brigg boring (see p. 311).

More will be said of this type of axis in Chapter XVIII.

Axes of Caledonian Trend.

(1) THE ISLIP AXIS

North-east of Oxford an anticlinal axis running NE.–SW. is marked by a line of six Cornbrash inliers rising through the Oxford Clay lowlands near Ot Moor, at Islip, Oddington, Merton, Ambrosden, Blackthorn Hill and Marsh Gibbon. The largest of the inliers bring up Great Oolite in the centre. The axis seems to consist of a chain of small domes, but it can be traced for considerable distances in both directions. Towards the south-west its effects can be clearly seen in the Corallian rocks of Wytham Hill [3] (a distance of 16 miles from the farthest inlier at Marsh Gibbon), and the Corallian rocks are again arched up on the same line 12 miles farther on, at Shellingford.[4] Towards the north-east the last certain traces are seen in the brick pit at Calvert, where trouble has been caused by the shaly brick clays or 'knots' of the Lower Oxford Clay suddenly dipping underground and being replaced by soapy clays of higher zones, as the expansion of the works has caused the pit to extend across the southern limb of the fold. It can therefore be definitely said that the Islip Axis is at least 30 miles long and that its date is mainly post-Corallian.

The direction of strike is quite peculiar and cannot be reconciled with any

[1] Dr. Versey calls the strike of the Market Weighton Axis NW.–SE., but Dr. Rastall considers it only W. 5° N. (1927, *Geol. Mag.*, vol. lxiv, p. 24): apparently WNW.–ESE. would be a nearer approximation to the truth.

[2] R. H. Rastall, 1927, loc. cit., p. 18.

[3] T. I. Pocock, 1908, 'Geol. Oxford', *Mem. Geol. Surv.*, p. 20.

[4] W. J. Arkell, 1927, loc. cit., p. 103.

other known folds of comparable importance, but its agreement with the general strike of the rocks is noteworthy and may be significant. If the two are causally connected, a Tertiary age is suggested for the Islip Axis, but this does not exclude the possibility of earlier beginnings.

A few observations on the possible extension of this axis may not be entirely idle, since an axis that can be traced for 30 miles may well be much longer. In the first place, if the line be continued a further 29 miles south-west of Shellingford it is found to reach exactly to the spot where the Lower Greensand unaccountably oversteps the Kimeridge Clay and Corallian between Rowde and Bromham, near Devizes, coming to rest for about 1¾ miles on Oxford Clay. A few miles beyond this and almost on the same line are the Hilperton–Trowbridge inlier of Cornbrash and the Hardington inlier of Inferior Oolite. These two inliers indicate an anticline almost parallel with the Mendip–Vale of Pewsey Axis, but appearing to diverge from it at a low angle somewhere in the Mendip Hills. The chief objection to the idea that this anticline is really a continuation of that at Islip is the too-accurate straightness of the line—for, as we have seen, it is not in the nature of the other axes to pursue a straight course for very long. Nevertheless, the 30 miles that are definitely visible are almost mathematically straight, just as some of the Charnian Axes appear to be, and there is therefore no reason why it should not be the same for 75 miles.

If the axis be continued in the opposite direction some more remarkable coincidences appear. Three miles north-east of Calvert (where the fold is still intense) the general course of the River Ouse begins to coincide with the line and continues straight (notwithstanding minor meanders) for 25 miles. The last 8 miles of this reach, from Olney to Sharnbrook, are highly suggestive of some fundamental structural cause, a view that is confirmed by two Cornbrash inliers at Riseley, a few miles farther on, which continue the line after the river valley has bent south-eastwards at right angles. (This southeast part of the river, from Sharnbrook to Bedford, which is set so conspicuously at right-angles to the higher course, lies, as already mentioned, approximately along the Charnwood Axis.) The Riseley inliers are not quite mathematically in line with the Oxfordshire ones, the axis having apparently bent a few degrees northwards. If they are on one and the same line, they extend the Islip Axis by 33 miles—about as far as the Hilperton–Trowbridge inlier—making the total length possibly 100 miles. If this is all one fold it is evidently a highly important one. Hitherto, however, no signs of intra-Jurassic movements have been discovered along it.

(2) SOME MINOR YORKSHIRE AXES

Blake and Hudleston and Fox-Strangways remarked on 'a slight upthrow' or 'a slight roll' in the Lower Corallian strata at Appleton-le-Street in the Howardian Hills, which alters the strike for a short distance.[1] More recently Dr. Versey has re-examined the ground and has come to the conclusion that the anticline has a Caledonian (NE.–SW.) trend.[2] He has also detected three other small anticlines of similar strike in the vicinity, spaced fairly regularly at intervals of just under 7 miles, at Roulston Scar, Gilling Park, and on the

---

[1] Blake and Hudleston, 1877, *Q.J.G.S.*, vol. xxxiii, p. 362.
[2] H. C. Versey, 1929, *Proc. Yorks. Geol. Soc.*, vol. xxi, pp. 206–10, and fig. 8, p. 207.

other side of Appleton-le-Street, at Grimston Field House. There seems to have been poverty of sedimentation due to uplift at these localities during the formation of the Lower Calcareous Grit and perhaps also the Oxford Clay, but Dr. Vernon Wilson informs me that his researches show that any such movements had ceased before the deposition of the 'Upper Corallian' limestones. It is hoped that further minute stratigraphical work on the Jurassic rocks will throw more light on these and other small intra-Jurassic axes in Yorkshire.

TABLE VI. *The principal axes and their periods of activity.*

+ denotes proved uplift (relative to the surroundings).
− denoted proved subsidence (relative quiescence or downsinking).
? denotes that evidence is needed.

Periods / Names of Axes	Purbeck–Weymouth	North Devon–Wardour	Mendip–Pewsey	Birdlip–Wallingford	Market Harborough	Melton Mowbray	Market Weighton	(Willoughby–Louth.)	Charnwood	Nuneaton	Sedgley–Lickey	Woolhope–May Hill	Vale of Moreton	Malvern Range	Islip	Howardian Series
	ARMORICAN GROUP								CHARNIAN GROUP					MALVERNIAN GROUP		CALEDONIAN GROUP
TERTIARY	+	+	+	+	+?	+?	?	+	?	?	−?	−?	?	+?		?
PRE-ALBIAN	+	+	+	+	+?	+?	+	+	+	+	−?	−?	−	?	Post-Corallian Uplift	?
PRE-APTIAN	+	+	+	+	+?	+?	+	+	+	+	−?	−?	−	?		?
Kim. Portlandian	+?	−	−	−	+?	+?	+	+?	+?	+?	−	−	−	?		?
Corallian	−	−	−	−	?	?	+	−?	?	?	+	+	−	?	−	+
Oxfordian	−	−	−	−	?	?	+	−?	?	?	−	−	−	?	−	+
Bathonian	−	−	+	−?	?	?	+	−	−	−	−	−	−	?	−	−?
Pre-Vesulian	+	+	+	+	?	?	+	−	−	−	−	−	+	+?	−	−?
Pre-Bajocian	+	+	+	+	?	?	+	−	−	−	−	−	+	+?	−	−?
Upper Lias	+	?	+	+	?	?	+	−	−	−	−	−	+	?	−?	−?
Middle Lias	+?	?	+	+	+	+	+	−	+	−	−	−	+	?	−?	−?
Lower Lias	?	?	+	+	?	+	+	−	?	+	+	−	+	?	−?	−?
Rhætic	?	+	+	+	?	+?	?	−?	+?	+	+	−	+	?	−?	−?
PRE-JURASSIC	?	+	+	?	+?	+	+	?	+	+	+	+	+	+	?	?

III. THE DISCRIMINATION OF EPEIROGENIC AND OROGENIC PROCESSES AND THE CLASSIFICATION OF INTRA-JURASSIC TECTONICS

Although our knowledge is still very incomplete, the foregoing brief analysis of the differential movements along the so-called axes of uplift brings out several facts which help us to understand the nature of the processes in operation.

In the first place we find that movements tended to take place in different parts of the Jurassic period along axes of different directrix, as if answering to pressures exerted from different points of the compass.

In the second place it emerges that, although the movements were often protracted and gentle, they were not of the broad, regional type which Gilbert named epeirogenic; for they were restricted to certain tectonic lines, and these coincided with the buried axes of more ancient upheavals of the convulsive and episodic type, recognizable at once as orogenic. Moreover, at least all the axes of E.–W. (Armorican) direction were subsequently rekindled into orogenic activity during the Alpine folding of the Miocene period.

Thus these axes are really orogenic lines, and although through most of Mesozoic times they were probably almost stable, only appearing to rise in relation to the subsiding sea-bed around them, it would seem that they depend merely on a greater degree of pressure to make them rise in relation to sea-level and to take part in mountain-building movements which all would class as truly orogenic. Whether we regard the movements as the dying tremors of past orogenies or as the faint anticipations of others yet to come, they differ from more obvious orogenic processes only in date and degree, and it would be highly artificial to draw a sharp line of distinction between them by calling the Jurassic movements epeirogenic. We have only to go farther afield, into North-West Germany, to find that the time-distinction breaks down also, for there contemporaneous folding attained mountain-building proportions. Faced with the mountain-chains of the Teutoburger Wald, the Egge Kette, the Süntel and the Deister (to go no more than 60 miles from Hanover), all raised partly in Mesozoic times, we are hard pressed to find means of differentiating between the forces that gave rise to them and the much greater ones that built the Alps.

It is not surprising that the classification of these movements has evoked conflicting opinions. The differences depend ultimately upon the definitions of the terms orogenic and epeirogenic, a matter concerning which there is as little agreement as over the correct meaning of the expression geosyncline.

The leading authority on the subject is generally recognized to be Prof. Hans Stille of Göttingen, who has been investigating and publishing papers upon the Mesozoic movements in North-West Germany for the past thirty years. In 1924 he collected together the results in his great work *Grundfragen der vergleichenden Tektonik*, and from this must start any attempt to understand or correlate the English tectonic events.

As conceived by Stille:[1]

Orogenic processes define themselves
    1. as changes in the structure of the underlying platform,
    2. as episodic.
Epeirogenic processes, on the other hand,
    1. leave the tectonic structure of the platform intact,
    2. continue through long periods of geological time (are 'secular' processes),
    3. are widespread.
'And so the conception of orogeny partakes of the spirit of the convenient expression "mountain-building" . . .

. . . 'And when we treat the conception orogeny in the same way [as mountain-building used in its modern sense], we are hardly departing from Gilbert's original idea, for that was based on the contrast between widespread warping, the example of which is Lake Bonneville, and "real" tectonic events, such as have produced, for example, the separate mountain ridges of the type of the Basin Ranges.

[1] H. Stille, 1924, *Grundfragen*, p. 11.

'Epeirogenic events are the form assumed by tectonics in "anorogenic" times. They comprise protracted and more or less uniform movements, affecting wide tracts of the earth's crust—large-scale movements of bigger units than are involved in an orogeny.'

Again 'Orogeny produces anticlines and synclines (ridges and furrows); epeirogeny produces geanticlines and geosynclines (broad elevations, basins and troughs).'[1]

If we accept these definitions (and they are in harmony with the views of most modern writers) then the downsinking of the Jurassic troughs of sedimentation or 'aires d'ennoyage' is to be classed as epeirogenic, as well as the general uplift that is presumed to have affected the whole tectonically stable block by which the epeiric seas were supported (as explained on p. 59, in connexion with cyclic sedimentation). As orogenic events, on the other hand, appear the differential movements along the axes of uplift.

Earlier writers formulated different definitions. Haug,[2] for instance, basing himself on the work of James Hall, distinguished the kinds of tectonic activities largely according to the nature of the *milieu* in which they took place. According to the Haug school orogenic movements are confined to geosynclinal regions, epeirogenic to the continental areas or geanticlines. Thus Prof. Dacqué,[3] rejecting Stille's conception of the downsinking trough of North Germany as a small geosyncline, cannot accept the folding of the Teutoburger Wald and associated ranges within the trough as analogous with the Alpine movements on a small scale, because the Alpine folds arose out of a true geosyncline (the Tethys). He claims that the lesser folds of North-West Germany, which for him are of epeirogenic origin, are due to lateral pressure, while the great Alpine movements that arose out of the geosyncline of Tethys did not result from external pressure but from internal forces of a different kind, and therefore they alone should be called orogenic. Such a classification based on causes might be ideal were there any certainty of its correctness. But at present the whole question of causes is far too hypothetical for us to be able to take any cognizance of them, and we must classify phenomena *per se*. If this be agreed, then Stille's system has no equal.

It is to Stille that we owe our only rational ideas of the tectonics of the Mesozoic period. His pioneer researches in North-West Germany produced ordered knowledge where ignorance reigned before.

Finding all existing names ambiguous, owing to their having been used in a directional sense, he coined the new term Saxonian Folding for all the mountain-building movements, of any directrix whatever, that occurred in the extra-Alpine region of North-West Europe after the Variscan (or Hercynian, or Altaid, that is, Carbo-Permian) orogeny.[4] Analysing these movements, he found that they occurred in a number of phases spanning Mesozoic and Cainozoic time, and all those falling between the Trias and the Cretaceous he called Cimmerian (Kimmerisch), a term borrowed from the Crimean Peninsula (Krim). The Cimmerian Mountains were described by Suess[5] as 'The remains of a folded chain of Mesozoic [pre-Neocomian] age, which forms the Crimea and Dobrudscha (Dobrogea), embraces the mouths of the

---

[1] H. Stille, 1924, loc. cit., p. 15.
[2] E. Haug, 1907, *Traité de Géologie*, p. 160; also Kober, 1928, *Bau der Erde*, 2nd ed., p. 84.
[3] E. Dacqué, 1915, *Grundlagen und Methoden der Paläogeographie*, pp. 131–3.
[4] H. Stille, 1913, *Geol. Rundschau*, vol. iv, pp. 364, 366; and 1924, loc. cit., p. 131.
[5] E. Suess, *The Face of the Earth*, English ed., vol. iv, 1909, pp. 23 and 632.

Danube, and disappears beneath the projecting arc of the Carpathians'. He recognized two periods of folding: the first between the Trias and the Lias, which Stille calls Older Cimmerian, the second between the Jurassic and the Neocomian, which Stille calls Younger Cimmerian. These are nearly everywhere the two principal phases of the Cimmerian movements, but when the Jurassic system is traced into other parts of Europe a number of sub-phases are revealed. Three are recognizable in the North German Basin, where they have been given names by Dahlgrün: [1] the earliest, the Deister Phase, falls within the Lower Kimeridgian (before the *Gravesia* zones); the next, or Osterwald Phase, falls within the Purbeckian (before the Serpulite, or Middle Purbeckian); the last, or Hils Phase, within the Neocomian (between the freshwater Wealden and the marine Upper Valanginian or Hils Clay).

The criteria by which these tectonic phases may be recognized are the disturbance (folding and erosion) of the underlying strata and the overstepping of the superjacent stratum transgressively across their basset edges.

In England perhaps the most familiar discordance of this sort is the transgression of the Vesulian (*garantiana* and *truellei* zones) across all the underlying rocks down to the Carboniferous Limestone of the Mendips. If we regard such interruptions in the stratification purely from the point of view of tectonics and give them all special tectonic names as Dahlgrün has begun to do, we are faced with a completely new nomenclature, a chronology of tectonics bearing no relation to ordinary geological chronologies and requiring to be memorized independently. The giving of such names as Deister Phase would proceed almost indefinitely as knowledge became extended over new territories, and the prospect of the history of stratigraphical and chronological nomenclature repeating itself is highly disconcerting. Such a course seems wholly unnecessary. We already have two time-scales, and well may we ask what is the use of them if they cannot be brought into play for the purpose of dating tectonic as well as any other events.

Several methods of expressing the same ideas have already long been used by English geologists, always with the aid of the existing chronologies. Buckman spoke of the interruption between the Bajocian and the Vesulian as the Bajocian Denudation, and of the similar but somewhat lesser one between the Bajocian and the Aalenian as the Aalenian Denudation. Prof. Sollas refers to the same phenomena as the Bajocian Oscillation and the Aalenian Oscillation,[2] a change which is an improvement in so far as it directs attention to the cause rather than to the effect. But a still greater improvement would be to speak of the Vesulian Oscillation instead of the Bajocian, and the Bajocian Oscillation instead of the Aalenian, thereby emphasizing the transgression (which was caused by a widespread, epeirogenic, oscillation) instead of the previous uplift (which was local, orogenic). Prof. Sollas has, in fact, adopted this plan in speaking of other oscillations, not previously named by Buckman. Thus he speaks of the Forestian Oscillation, which is denoted by an eroded surface *below* (*at the base of*) the Forest Marble (Kemble Beds), of a *Ceteosaurus* Oscillation, which is *at the base of* the beds containing *Ceteosaurus* (the *Fimbriata-waltoni* Beds), and so on. The preceding disturbances would then be the Pre-Vesulian, &c.

[1] F. Dahlgrün, 1920, *Jahrb. Preuss. Geol. Landesanst.*, vol. xlii, p. 747.
[2] W. J. Sollas, 1926, *Nat. Hist. Oxford Dist.*, p. 36.

Such a system of naming oscillations (or whatever we choose to call them) after the transgressive stratum *above* the discordance has marked practical advantages. For instance, when we stand in Vallis Vale and look at the Vesulian (Upper Inferior Oolite) resting upon a smoothly-planed surface of the Carboniferous Limestone (as in plate XI, p. 238), the obvious way to describe what we see is to call it the Vesulian Discordance, resulting from the Vesulian Transgression. If we move to the escarpment in the South Cotswolds or Mid-Somerset we still find the discordance below the Vesulian, but there the transgressive bed rests upon various members of the Lias. Without previous knowledge we cannot describe this as the Bajocian Oscillation, for the Bajocian does not visibly enter into the matter. The only constant factor, the only criterion for dating, is always the Vesulian.[1] We do, in fact, know that where the sequence is as nearly complete as possible the Vesulian is immediately preceded by the Bajocian; that in England the Bajocian is involved in the pre-Vesulian folding; therefore that the 'denudation' or 'oscillation' is post-Bajocian. But in unexplored territory, where we had no knowledge of the ideal sequence, if we found a Jurassic stratum which could be dated to Vesulian resting discordantly upon Palaeozoic rocks, all that we could say would be that there had been a Vesulian transgression. The denudation or the oscillation might have taken place at any of the earlier periods unrepresented by sediments.

It is important to keep distinct in our minds the two entirely different processes involved in such a discordance as that in Vallis Vale. First there is the folding and uplift of the rocks now covered by the transgressive stratum—a purely orogenic process. Secondly there is the general downsinking which allowed the folded and peneplaned strata to be submerged uniformly with the unfolded strata elsewhere—an epeirogenic movement not restricted to the region of the axes.

We cannot always lightly use the word Transgression to signify nothing more than the transgressive overstep of one stratum by another, for the word has a special geological meaning. It signifies an extension of the sea over the land, a widening of the area of marine sedimentation, produced by epeirogenic movements; and of such an extension the mere occurrence of Vesulian rocks resting upon Palaeozoics on the axes of uplift is no evidence. Both in the Mendips and in Normandy there are often thin remanié beds between the Vesulian and the Palaeozoics, crowded with Liassic and Bajocian fossils, or Rhætic and other remains may have fallen down cracks in the Palaeozoic rocks (see p. 105), proving that the missing sediments once existed there but were removed as the result of the folding. The resumption of sedimentation in Vesulian times over the folded axes, therefore, does not necessarily imply any great extension of the sea; it does not of itself indicate a true transgression, although it did, in fact, coincide with one.

On the other hand a true transgression may occur without leaving any signs of discordance in the areas where the sea already existed and where no orogenic folding supervened. For instance the great Callovian Transgression, by which the sea for the first time in the Jurassic period overspread a great

---

[1] The lowest part of the transgressive stratum is usually slightly diachronic: for example, against the Mendips the *garantiana* zone is gradually overlapped by the *truellei* zone as the Carboniferous platform rises; but any diachronism through overlap is generally negligible in comparison with the differences in age of the underlying strata that are overstepped.

part of Russia, leaving Callovian sediments resting discordantly on Palaeozoics over hundreds of square miles,[1] is not recognizable by any conspicuous unconformity in Britain. But the sudden incoming of the Callovian fauna in the middle of the Cornbrash, the upper half of the Cornbrash containing the new Callovian and the lower half the old Bathonian fauna, while the two divisions are almost perfectly conformable, is probably due to the submergence of barriers and accelerated migration attendant on the Callovian Transgression.

The greatest transgression of Jurassic times in North-west Europe, the Rhætic, was accompanied by very little overstep or discordance. The Triassic salt-lakes were already at or below sea-level and the Rhætic sea merely flowed uniformly over them, distributing the same fauna from Germany to Dorset and the North of Scotland.

The type of phenomenon exemplified by the Rhætic Transgression may be distinguished as 'conformable transgression', the other as unconformable. The two different types are complementary manifestations of the same event, consequent on the previous history of the terrain invaded.

Just as transgressions should be distinguished from the effects of local orogenic movements, so too should regressions. Transgressive formations, such as the Callovian or Cretaceous, may rest non-sequentially or with slight unconformity upon much older Jurassic or Palaeozoic beds over hundreds of square miles, as does the Cretaceous in the northern half of the British Isles. In all the counties north of Bedford Cretaceous rocks repose on Kimeridge Clay or older beds, and no sign of Portland or Purbeck Beds is found. It is highly improbable that had these last ever been deposited they could have been so completely removed as to leave no vestige behind throughout the long outcrop from Cambridgeshire to Yorkshire or in Scotland. It seems much more likely that the whole of Northern Britain received an upward tilt about the end of the Kimeridge period, which brought sedimentation to a standstill. Detailed correlation of the Portland and Purbeck Beds of Oxon. and Bucks. with those of Dorset reveals that each subdivision of the formations thins out northward and would soon disappear beyond the present outcrops, without the aid of any subsequent erosion.[2] We seem, therefore, to be dealing with a regional upward warping, causing a regression of the sea—a truly epeirogenic event. Dr. Versey has attempted to correlate the movement with the Younger Cimmerian movements in North-west Germany, but he seems to me to be dealing with two different classes of phenomena.[3] In order to find the counterparts of the Younger Cimmerian movements we ought to study the tectonics of the South of England, where sedimentation was proceeding all the time they were in operation.

It may be useful to review very briefly the principal examples of unconformable transgressions, with their pebble-beds and allied phenomena, observed in the British Jurassic rocks, noticing which seem to have been preceded by orogenic activity.

---

[1] E. Haug, 1911, *Traité*, pp. 1002–3; and Suess, *Face of the Earth*, vol. ii, p. 273.

[2] More will be found on this subject at the beginning of Chapter XVII.

[3] H. C. Versey, 'Saxonian Movements in East Yorkshire', *Proc. Yorks. Geol. Soc.*, vol. xxii, p. 57. In interpreting the term 'Portlandian' it should be noted that, so far from the Osterwald and Deister Phases having taken place at the time of our Portland Beds (Portlandian as understood in England), one occurred in the Lower Kimeridge and the other in the Middle Purbeck (see below, Chap. XVII).

SUMMARY OF THE PRINCIPAL TRANSGRESSIONS RECOGNIZABLE IN THE
ENGLISH JURASSIC [1]

(Details will be found in the Stratigraphical Part of the Book.)

Rhætic Beds.

The great Rhætic Transgression has already been mentioned, and this seems to have coincided with the close of the Older Cimmerian orogenic phase, but folding in Britain was only slight. Minor disturbance continued, as we have seen, along several of the axes until the end of the period.

Lower Lias.

During the early part of Lower Lias times a series of small folds arose on the north of, and parallel with, the Mendip Axis, their crests undergoing repeated erosion. The *bucklandi* zone transgresses across all these until it comes to rest on Rhætic White Lias about Radstock. Similar movements doubtless occurred along the other axes, where the formation is much thinner than the normal, but lack of exposures prevents investigation (see p. 131).

In Scotland the *semicostatum* zone overlaps the earlier beds in Skye and comes to rest on Cambrian limestones on the shores of Loch Slapin.

In Kent and in Buckinghamshire (as proved by the Calvert Borings) there is an overlap, perhaps indicative of a minor transgression, of the *jamesoni* zone on to the Palaeozoic Platform.

Inferior Oolite.

The *scissum* zone is transgressive for some hundreds of miles across different members of the Whitbian, from Oxfordshire through Northants. and all Eastern England to Yorkshire. Chiefly on this account it has been selected here as the base of the Inferior Oolite, though the rarity of the zonal index fossil in Northants. and its gradual disappearance northwards suggests that there is some overlap, the basal member, with its pebble-bed, probably being slightly diachronic. In Yorkshire it may be of either *Ancolioceras* or *murchisonae* date. This transgression seems to be quite independent of the axes.

The *discites* zone (the basal zone of the Bajocian *sensu stricto*) is discordantly related to the Aalenian Stage in Dorset, Somerset and the Cotswolds, and in Dorset it has a conglomeratic bed at the base. There are indications that the same zone overlaps the Aalenian in North Lincolnshire; and in Yorkshire the transgression is marked as the most important of the marine interludes in the Estuarine Series by the Millepore Bed and Cave and Whitwell Oolites. The transgression was preceded by activity along the Birdlip Axis in the Cotswolds and probably also along several of the other axes.

The *garantiana* and *truellei* zones (Lower Vesulian) are highly unconformable with the underlying beds from Dorset to Oxfordshire. They mark one of the most important transgressions of Jurassic times; but in England the spectacular effects are greatly increased owing to the transgression having been preceded by a period of profound orogenic activity. All the axes of Armorican (E.–W.) directrix were uplifted and their crests eroded before the transgression. A sudden deepening of the sea seems to have taken place also

[1] It is not possible here to distinguish all the conformable transgressions.

in the Hebrides, where the *garantiana* zone coincides with an abrupt change of lithology, from several hundreds of feet of sandstones to black shale.

Great Oolite Series.

The Great Oolite Series is highly transgressive eastwards over the London landmass. The lowest divisions, the *zigzag* and *fusca* zones (equivalents of the Chipping Norton Limestone), overlap the Inferior Oolite in Oxfordshire before reaching the Cherwell Valley, and they are in turn overlapped by the middle portion of the series (including the Great Oolite limestone) all along the eastern outcrop from the neighbourhood of Northampton northwards. The middle and upper portions of the series transgress over the Palaeozoic platform as far as London.

There are local overlaps and oversteps within the series, such as those which cause the appearance and disappearance in short distances of the Stonesfield Slate Series and the Wychwood Beds in Gloucestershire and Oxfordshire. The Lower Cornbrash seems to overstep the Wychwood Beds towards the north and east, reducing them from perhaps 90 ft. to nothing between Wiltshire and East Oxfordshire. Farther north the Upper Cornbrash oversteps (or overlaps) the Lower and in Yorkshire it becomes a transgressive stratum resting on the Estuarine Series; over most of England, however, the Upper Cornbrash is perfectly conformable with the Lower, marking only a conformable transgression.

Oxford Clay and Corallian.

Considerable elevatory movements took place at the time of the Oxford Clay in the Hebridean area of Western Scotland, but the details have not been worked out (see pp. 370–2).

One of the most remarkable pebble-beds in the Jurassic System occurs in the Corallian of the Midlands, at the base of the Berkshire Oolite Series (*martelli* zone). It is continuous for more than 40 miles, from near Calne in Wiltshire to Oxford, and indicates a widespread epeirogenic movement like that which heralded the Inferior Oolite. Beyond Oxford the same horizon continues to be related non-sequentially, and with signs of erosion, to the Lower Calcareous Grit and Oxford Clay, which it oversteps eastward. North of Cambridge there is probably also an overlap of the Berkshire Oolite Series by higher parts of the Corallian Beds developed in clay facies.

At Oxford and around Wootton Bassett, Wiltshire, the Corallian Beds show unmistakable signs of local uplift and non-deposition or erosion, apparently unrelated to the more general movements and of orogenic type.

Kimeridge Clay and Portland Beds.

According to Salfeld's identification of *Gravesiæ* both in the *Gigas* Beds of Germany and in the clays of the lower part of Hen Cliff, Kimeridge, Dahlgrün's Deister Sub-Phase of the Younger Cimmerian movements took place during the deposition of the Lower Kimeridge Clay (not the 'Portlandian'). We should look for signs of it in England between the *Gravesia* and the *Aulacostephanus* zones, but so far no indications of discordance at that horizon have been recorded. The *Gravesia* zones often seem to be absent, and indeed their presence anywhere in England except Dorset is a matter of doubt; but this does not signify movements contemporaneous with the 'Deister Phase', for in North-west Germany it is the *Gravesia* zones that are transgressive and zones below and above that are wanting.

Two widespread pebble-beds, coinciding with considerable non-sequences, occur higher up in the series, from Wiltshire to Buckinghamshire. The first, known as the Lower Lydite Bed, appears at the base of the *rotunda* zone at Swindon and continues past Oxford into Bucks. It marks the disappearance of at least some of the 240 ft. of clays between the oil-shale and the *rotunda* zone on the Dorset coast. The second, the Upper Lydite Bed, first appears in the Vale of Wardour and also continues to Bucks. Its stratigraphical position is somewhere in the middle part of the Portland Sand of the South, and it too cuts down transgressively, so that almost as great a thickness of sands and clays is missing. Both of these pebble-beds denote epeirogenic movements and they seem to be connected with the proximity of the London landmass, whence their materials were probably derived (see pp. 514–15).

The regional warping that set in early in Kimeridge Clay times and by the end of the period put a stop to sedimentation in North Britain has already been mentioned. After the deposition of the Portland Stone a barrier was raised which cut off the South also from the open sea, and the peculiar conditions of the Purbeck period reigned in Southern England and North-west Germany, although subsidence continued in both areas.

Purbeck Beds.

In the Middle Purbeck the sea advanced simultaneously in Kent and Dorset and inland at least as far north as Mid-Wiltshire. This advance, a conformable transgression, seems to have been beyond much doubt contemporaneous with the unconformable transgression of the Serpulite in North-west Germany (the argument for this correlation is given in Chapter XVII, pp. 550–1). The causes of the discordant oversteps in Germany were the orogenic movements of Dahlgrün's 'Osterwald Phase', which do not seem to have operated in England. Here then, is a good illustration of the distinct effects produced by the same marine transgression, in regions which have and have not been first subjected to orogenic folding. Where the area invaded is already at sea-level, as in South England, there is perfect conformity in spite of a sharp change from fresh-water to marine fossils; where there has been orogenic activity locally, as in North-west Germany, the transgressive bed oversteps the folded strata and an unconformity results.

Post-Jurassic and Pre-Aptian Movements.

Earth-movements of both epeirogenic and orogenic types took place extensively all over the British Isles before both the Aptian and the Albian transgressions. The history of events during the transition from Jurassic to Cretaceous times will be traced in Chapter XVII, and so little need be said here, beyond pointing out that these movements were probably greater than any that occurred during the Jurassic period, and were moreover accompanied by extensive faulting. In Dorset, in a belt of country lying north of the Weymouth Axis, around Charmouth and Bridport, the Jurassic rocks are torn by numberless faults, of almost every strike, the key-lines however trending E.–W. In Watton Cliff one of these faults throws Forest Marble against

Middle Lias and at Burton Bradstock another throws Forest Marble against Upper Lias (Bridport Sands). North-eastward the system of faults passes under the overstepping Gault and Chalk without affecting them in the smallest degree. Under the Gault at Abbotsbury there is a fault of 600–700 ft., throwing down Kimeridge Clay against Forest Marble.[1] Still farther east is the fault in Ringstead Bay, already mentioned, where the Purbeck Beds are seen to be affected, but not the Gault. As the Wealden Beds are everywhere perfectly conformable with the Purbecks, these faults can be said to be post-Neocomian and pre-Albian.

Inland there is evidence that considerable faulting was not only pre-Albian but also pre-Aptian. This can be seen near Calne, between Bromham and Seend, where the Lower Greensand crosses a fault from Kimeridge Clay on to Lower Corallian Beds.[2]

Instances of unconformity between Jurassic and Aptian beds are too numerous to mention, for wherever we examine the relations of the two formations it is evident that great earth-movements supervened between them. Something will be said on the meaning of these movements in Chapter XVII. Meanwhile we must pass on to the next and most important part of our theme, the stratigraphy.

[1] Figured in Whitaker and Edwards, 1926, 'Water Supply of Dorset', p. 4, *Mem. Geol. Surv.*

[2] H. J. Osborne White, 1925, 'Geol. Marlborough', *Mem. Geol. Surv.*, p. 8.

PART III
STRATIGRAPHY

CHAPTER IV
RHÆTIC BEDS

Strata	Thickness	Principal Fossils (Plate XXIX)
WATCHET BEDS	0–8′	*Modiola langportensis* Rich. & Tutch.
LANGPORT BEDS[1] or WHITE LIAS (*sensu stricto*)	0–25′	*Ostrea liassica* Strickland *Dimyodon intus-striatus* (Emmr.)
COTHAM BEDS	0–19′	*Pseudomonotis fallax* (Pflücker) Ostracoda *Estheriæ* and *Lycopodites*
WESTBURY BEDS or CONTORTA SHALES	1–47′	*Pteria contorta* (Portlock) *Cardium cloacinum* Quenst. *Chlamys valoniensis* (Defr.) Vertebrata (esp. *Ceratodus*)
SULLY BEDS (fossiliferous top portion of the Grey Marls)	0–14′	*Ostrea bristovi* Etheridge

IT is now generally agreed that the Rhætic Beds are best classed as the basal member of the Jurassic System.[2] The old classification, which regarded them as part of the Trias, draws, at least in North-west Europe, an unnatural line of division between the Rhætic and the Lias, while ignoring a slight but widespread unconformity, coincident with an important and ubiquitous change of lithology, between the Keuper and the Rhætic. There is no datum so suitable for starting the Jurassic System as that marked by the Rhætic Transgression, the effects of which are marked all over North-west Europe.

Thanks principally to Mr. L. Richardson's detailed studies, spread over the last thirty years, a great deal is now known of the Rhætic rocks of Southern and Central England. These studies have tended to accentuate the importance of the break at the top of the Keuper. Richardson has shown that the Keuper Marls, in West and Central England, were arched up into a series of gentle flexures before the Tethyan sea advanced from the south over the desert plains and salt lakes and submerged them. The first Rhætic sediments were laid down in the hollows between the folds, and it was not until later that the crests sank also beneath the sea.

Wider knowledge of other parts of the Jurassic System, too, has thrown new light on the origin of the famous Bone Bed at or near the base of the Rhætic. As Strickland first suggested many years ago,[3] this stratum probably results,

---

[1] The term 'Langport Beds' was suggested by Richardson to replace 'White Lias' because the latter was used inconsistently—sometimes for true White Lias, sometimes for White Lias plus Cotham Beds, and at other times for Cotham Beds. It was originally introduced for the whole 'Upper Rhætic' by William Smith (see p. 4).

[2] Unless, as some have suggested, they be considered to form a diminutive system of their own.

[3] *Memoir of H. E. Strickland*, 1858, pp. 157–8; and see L. Richardson, 1904, *Q.J.G.S.* vol. lx, p. 354. The idea is also adopted by L. J. Wills, 1929, *Phys. Evol. Britain*, p. 132.

not from any sudden massacre of vertebrates or temporary abundance of vertebrate life as has been so often supposed, but from slowness of sedimentation. Comparable fossil-beds in other parts of the Jurassic Series, such as those composing much of the Inferior Oolite in the Dorset–Somerset area, or the Cephalopod Beds at the top of the Upper Lias of the Cotswolds, are now known to be natural segregations. They were accumulated over a long period of time, during which life flourished normally, but either little sediment was present to entomb the remains, or currents carried it continually away. It is now accepted, therefore, as indication of a prolonged pause in early Rhætic times.

I. THE DEVON–DORSET–W. SOMERSET AREA

(a) The Devon–Dorset Coast.[1]

A complete though discontinuous section of the Rhætic Beds is exposed in the cliffs of Culverhole Point, near the border between Devon and Dorset. The Upper Rhætic is again exposed in Charlton and Pinhay Bays to the east, between Culverhole Point and Lyme Regis (Plate I). The total thickness is about 47 ft.

The Keuper Red Marls pass up into about 75 ft. of grey-green marls known as the Tea Green Marls, which have in the past been called the Transition Beds, and variously assigned to the Keuper and to the Rhætic. They yield no fossils. At the top are widespread signs of erosion before the deposition of the Westbury Beds or *Contorta* Shales. Mainly on this evidence, Richardson assigns the Tea Green Marls to the Keuper and considers that there is a non-sequence at the base of the Rhætic, some fossiliferous grey marls (Sully Beds), which appear in the neighbourhood of the Bristol Channel, being unrepresented.

The top of the Rhætic too, consisting of the limestone of the White Lias, is bored and eroded, this erosion having destroyed the Watchet Beds of North Somerset.

SUMMARY OF THE SEQUENCE ON THE DORSET COAST

Langport Beds or White Lias (*sensu stricto*), 25 ft.:

Thin beds of white limestones, here and there false-bedded, the top layer massive and brownish, with the upper surface bored to a depth of 1 or 2 in., and some other irregular and apparently bored surfaces in the lower part. In the middle and upper portions *Ostrea liassica*, *Protocardia*, *Dimyodon intusstriatus*, and a few ill-preserved fragments of other fossils·occur. The upper part of this division and the junction with the Blue Lias may be studied to advantage on the east side of Pinhay Bay, in a long exposure at the foot of the cliff (Plate I).

Cotham Beds, 5 ft.:

The Cotham Marble is an impersistent, pale greenish-grey limestone, only from 1½ to 8 in. thick, at the top of the division. It is slightly pyritic and arborescent. The rest of the beds below are made up of dark shales with two

[1] Based on L. Richardson, 1906, 'On the Rhætic and Contiguous Deposits of Devon and Dorset', *P.G.A.*, vol. xix, pp. 401–9.

PLATE I

Photo. W. J. A.

Photo. W. J. A.

Cliffs of Blue Lias and White Lias, Pinhay Bay, Lyme Regis.

PLATE II

Photo.   *Sidney Pitcher.*

Photo.   *Sidney Pitcher.*

Garden Cliff, near Westbury on Severn, Glos.

Showing Rhœtic Beds with two prominent bands of *Pullastra* sandstone in the
lower part, resting on the Tea Green Marls of the Keuper. In the lower
photograph fallen blocks of the sandstone show ripple-marks. The cliff in the
distance is composed of red and green Keuper Marls capped by Tea Green
Marls.

thin limestone seams. The shales yield Ostracods (*Darwinula*), fish-scales, saurian coprolites, and *Cardium cloacinum*.

Westbury Beds or Contorta Shales, 17 ft.:

Black and brown fossiliferous shales with three thin limestone seams. The shales are laminated and contain much selenite, and yield the following fossils: *Pteria contorta* (very abundant), *Chlamys valoniensis, Myophoria emmrichi, 'Schizodus' ewaldi,*[1] *Placunopsis alpina, Protocardia rhætica, Pleurophorus elongatus*; also the fish *Acrodus minimus* and *Gyrolepis alberti*. At the base is the Bone Bed, 2 in. thick, consisting of indurated, gritty, black shale, full of small bones, scales and teeth of *Gyrolepis alberti, Acrodus minimus, Saurichthys acuminatus,* ?*Sargodon, Lepidotus* and *Hybodus*, with coprolites.

[Eroded surface of Tea Green Marls; Sully Beds and certain of the lower beds of the *Contorta* Shales missing.]

(b) West Somerset

The Devon and Dorset outcrop of the Rhætic runs north to the neighbourhood of Chard, where it is concealed for a short distance by the Cretaceous rocks. On emerging from beneath the Blackdown Hills it strikes in a low ridge north-westward to Langport and Somerton, thence circling round King's Sedgmoor north-eastwards, beneath the Blue Lias of the Polden Hills, to the coast at Watchet. Between the Polden Hills and the Mendips the Rhætic rises through the alluvium in several places near Wedmore.

This part of the outcrop contains several fine sections, notably those on the coast at Blue Anchor Point, St. Audries' Slip and Lilstock, all near Watchet, railway-cuttings at Puriton and Charlton Mackrell at either end of the Polden Hills, Langport railway-cutting, and a number of minor exposures and quarries. At Charlton Mackrell at the SE. end of the Polden Hills, the total thickness is 47 ft., as in Dorset; but towards the south the beds grow thinner, while towards the coast they become considerably thicker and the Watchet Beds appear above the White Lias and the Sully Beds below the *Contorta* Shales. The succession may be summarized as follows:

SUMMARY OF THE RHÆTIC SUCCESSION NEAR WATCHET [2]

Watchet Beds, 0–8 ft.:

Marls and Shales with inconspicuous layers of impure limestone, containing occasional specimens of *Ostrea liassica* and *Modiola langportensis*. These are best developed in the coast sections at Watchet, to which neighbourhood they are restricted. Over the rest of the area the paper-shale at the base of the Lower Lias lies non-sequentially on the Langport Beds.

Langport Beds or White Lias (*s.s.*)

The thickest development in the district is at Charlton Mackrell at the SE. end of the Polden Hills, where the White Lias is 20 ft. 8 in. thick. The mamillated and often bored upper surfaces of the limestone courses, their

---

[1] See note opposite Plate XXIX, fig. 10, at the end of the book.
[2] Based on L. Richardson, 1911, *Q.J.G.S.*, vol. lxvii, pp. 1–55; and 1914, *P.G.A.*, vol. xxv, pp. 97–102.

irregular under-surfaces, and irregular, rubbly and conchoidal fracture, indicate a slow rate of formation. Near the middle of the series at Charlton Mackrell is a coral limestone, 16 in. thick, full of *Thecosmilia? michelini* Terq. and Piette, associated with numerous *Ostrea liassica*. Towards the north-west the Langport Beds thin out. In the Dunball railway-cutting near Puriton, at the other end of the Polden Hills, they are reduced to 4 ft. 3 in., while on the coast they have still further dwindled, measuring less than 4 ft. at Lilstock and only 2 ft. 3 in. at Blue Anchor Point. Here all that is left is four thin bands of limestone.

Palaeontologically the Langport Beds of Somerset resemble those of Dorset. *Dimyodon intus-striatus* appears in great force, and *Modiola langportensis*, *Ostrea liassica*, *Lima valoniensis* and internal casts of *Cardinia* and *Protocardia* abound, but are of no zonal value.

Cotham Beds.

This division is much the same as in Dorset, usually from 5 ft. to 7 ft. thick, consisting of pale marls with at least three thin limestone bands, the Cotham Marble being sometimes represented at the top, rich in *Pseudomonotis fallax*. Ostracods and *Estheriæ* are common.

Westbury Beds or Contorta Shales.

This division thickens towards the coast at Watchet proportionately as the White Lias becomes thinner. At Langport and Charlton Mackrell it measures 22–3 ft., but at Lilstock a measurement of 32 ft. was obtained, and at Blue Anchor Point 46 ft. In lithic characters the *Contorta* Shales resemble those of Dorset, except that at Langport three thin bands of sandstone appear near the base. The general facies is sub-littoral and the mollusca are usually dwarfed.

In the upper part of the series mollusca abound, particularly *Pt. contorta*, *Chl. valoniensis* and *Dimyodon intus-striatus*. In the west there is often a *Pleurophorus* bed, and another characterized by *Cardium cloacinum* Quenst. In the lower part of the series vertebrate remains predominate, occurring at several horizons. But the chief bone bed, which contains *Ceratodus* at Blue Anchor Point, instead of forming the base, is found up to 22 ft. higher. Another bone bed, 2 in. thick, takes its place at the base, yielding *Acrodus minimus*, *Gyrolepis alberti*, *Saurichthys acuminatus* Ag., *Hybodus minor*, *Sargodon tomicus* Plieninger, and *Lepidotus*.

Sully Beds.

These beds are merely the uppermost 14 ft. of the Grey Marls, formerly classed with the Keuper, and always separated from the basal Bone Bed of the *Contorta* Shales by a surface of erosion. But in the neighbourhood of Watchet, where the *Contorta* Shales are thickest, higher horizons of the Grey Marls are preserved, and Rhætic fossils occur in them down to 14 ft. below the basal Bone Bed. These fossiliferous upper portions of the Grey Marls are best developed at St. Mary's Well Bay near Sully, on the opposite side of the Bristol Channel, after which Richardson has named them the Sully Beds. As long ago as 1864 the late Sir W. Boyd Dawkins insisted that these marls with Rhætic fossils, although they occurred below the Bone Bed, should be classed with the Rhætic.[1] In the Watchet coast-sections *Pt. contorta, Gervillia præcursor, Chl. valoniensis* and other fossils have been found down to 14 ft. below the Bone Bed, and at a depth of 10 ft. 6 in. Sir W. Boyd Dawkins found the earliest known mammal, the small *Hypsiprymnopsis rhæticus*,[2] in association with *Acrodus minimus, Sargodon tomicus*, scales of *Gyrolepis* spp., and a Pterodactylian bone.

(c) The Glamorgan Coast.[3]

The former continuation of the West-Somerset Rhætic Beds across what is now the Bristol Channel is proved by the exactly comparable sections in the cliffs between Cardiff and Barry. The finest sections are at Lavernock Point and the type locality of Sully, at Penarth, and in the cliff facing Barry Island, near Coldknap Farm, at which last place only the White Lias is well seen.

The basal Bone Bed of the Westbury Beds or *Contorta* Shales at Lavernock has long been known under the name of the 'Fish Bed', and besides the species recorded at Watchet, it yields *Saurichthys acuminatus* Ag., *Hybodus minor* Ag., and *H. cloacinus* Quenst. It rests on a conspicuously waterworn surface of the Sully Beds, filling hollows in them to a depth of up to 5 in., and sometimes containing pebbles of argillaceous stone. The Sully Beds consist of grey marl and bands of hard grey marlstone. The boundary between them and the Tea Green Marls is drawn arbitrarily at about 14 ft. below the Fish Bed, this being the lowest occurrence of Rhætic fossils. At a depth of about 6 ft. have been found remains of the Labyrinthodon *Mastodonsaurus*, a mandible and two teeth, believed to belong to *Palæosaurus*, bones of *Trematosaurus*, and numerous small teeth of *Sphærodus*.

As in the Watchet district, a second bone bed occurs in the *Contorta* Shales, containing vertebrae of *Plesiosaurus*. At Lavernock it is only 1 ft. above the base. A few inches higher is a layer with hundreds of teeth of *Acrodus minimus* and scales of *Gyrolepis alberti*. But the most persistent strata are two *Pecten*-beds in the upper part, full of *Chl. valoniensis* and the other mollusca common in Somerset.

The Cotham Beds and White Lias are thinner than at Watchet, measuring only about 3 ft. and 2 ft. 6 in., but palaeontologically and lithologically they are typical. The White Lias is especially fossiliferous at Coldknap, Barry.

At the top, overlain by the paper-shales of the Lower Lias, are the Watchet Beds, similar to those at the type locality, just over 6 ft. thick at Lavernock, 9 ft. thick at Barry.

II. THE LITTORAL OF THE SOUTH WALES COAL-FIELD

Inland towards Cowbridge, in the Vale of Glamorgan, the Rhætic Beds show signs of proximity to the shore-line of the South Wales highland. The edge of the Upper Carboniferous rocks forming the rim of the coal-field still rises as a wall high above the Vale of Glamorgan, but when the Rhætic waters

---

[1] W. B. Dawkins, 1864, *Q.J.G.S.*, vol. xx, p. 408.
[2] Formerly known as *Microlestes*. See W. B. Dawkins, 1864, loc. cit., pp. 409–12.
[3] L. Richardson, 1905, *Q.J.G.S.*, vol. lxi, pp. 385–425; F. F. Miskin, 1922, *Trans. Cardiff Nat. Soc.*, vol. lii, pp. 23–5, gives a different interpretation of the Rhætic base.

lapped against its foot, along a line now raised to 400 ft. above the sea, its elevation must have been many hundreds of feet greater (fig. 16).

In Triassic times an elongate island of Carboniferous Limestone, thrown up by an E.-W. anticline, ran off-shore from near the mouth of the present River Ogmore to north-east of Cowbridge, where it broke up into two tongues. Between the tongues, at their extremities, would have lain the city of Llandaff. During the Rhætic transgression the Keuper Marls were overlapped and much of the eastern end of the island became submerged. It was not yet

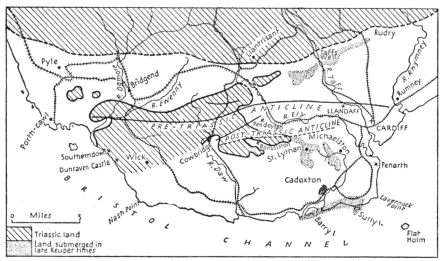

Fig. 16. Sketch-map showing the geography of the neighbourhood of Cowbridge Island at the end of the Triassic period. From Strahan and Cantrill, 'Geol. S. Wales Coal-field', part vi, p. 24, fig. 2. *Mem. Geol. Surv.*

completely covered, however, and it continued to form a barrier between the shore and the open sea to the south. In the channel between the north shore and the mainland were collected most of the sandy sediments brought down by the rivers from the Welsh mountains.

The first signs of shore-deposits in the Rhætic succession are met with in the neighbourhood of Cowbridge. At Bonvilston, on the southern edge of the old island, and at Pendoylen and other places to the east, normal black shales and shelly limestones still prevail. At Cowbridge, for the first time in passing westward, the *Contorta* Shales contain bands of grey sandstone and also an 8-inch band of hard conglomerate. It is an interesting indication of the prevailing currents that the constituents of these sandstones were drifted round the western end of the Carboniferous Limestone island.

The best sections in the Cowbridge district are in the Tregyff road- and railway-cuttings, where the Westbury Beds are highly fossiliferous. Richardson has identified a limestone and marlstone below, yielding *Acrodus* and *Sargodon*, which he correlates with the Sully Beds. The total thickness of the Rhætics from the top of the Tea Green Marls to the paper-shale at the base of the Lias is 19 ft.[1]

[1] L. Richardson, 1905, *Q.J.G.S.*, vol. lxi, pp. 400–2.

In the channel between the north side of the island and the mainland, the Rhætic Beds present an entirely different appearance. The upper parts of the formation are represented by red-streaked greenish marls and shales, greatly reduced in thickness, while the lower parts pass entirely into sandstones, which at Hendre, near Pencoed, and other places, are as much as 30 ft. thick. At Quarella Quarry, Bridgend, the Rhætic sandstones are quarried for building, and are seen in an unbroken succession to a thickness of 24 ft. For the most part they are white, fine-grained and massive below, and more thin-bedded above. It is probable that the whole of the Rhætic Series is represented in these sections. The uplands of Stormy Down and St. Mary Hill Down, which are strewn with large masses of the sandstone, resemble moorlands of Millstone Grit.

In the sandstone at Stormy Down have been found *Hybodus*, '*Schizodus*', *Myophoria* and a few other mollusca, and a jaw of a large megalosaurian reptile, *Zanclodon* (?*Avalonia*) *cambrensis* Newton. The associated marl beds yield *Pt. contorta*.[1]

The final stage of the littoral deposits is seen in an actual beach of Rhætic age on Cae Tor, near Pyle, capped by the Oyster Beds at the base of the Lias. The beach material was described by the late Sir Aubrey Strahan as a 'coarse calcareous grit with a few pebbles, in close association with a shelly and detrital limestone containing fragments of corals, molluscs and echinoderms, together with small fragments of an older granular limestone and grains of quartz'.[2] It rests upon or close above Triassic conglomerate and thins out within a few yards to the south, against the Cowbridge Island.

III. THE MENDIP ARCHIPELAGO[3]

The Cowbridge Island was one of an archipelago of Carboniferous Limestone islands which lay off the shore of the South Wales mountains in Rhætic and Liassic times. They dotted the sea on both sides of what is now the Bristol Channel, and although they have since been completely submerged and were covered up with later Jurassic and Cretaceous sediments, Tertiary and Quaternary denudation has stripped them bare once more, so that most of them now rise above the Triassic and Liassic plain as prominent hills.

The principal limestone masses were the ridge of the Mendips and the other Carboniferous inliers of Kingswood, Clifton, Clevedon, Wrington and Cowbridge (fig. 17). Lesser islets can be recognized in Worle Hill, Brean Down, and the Steep Holm and Flat Holm rocks, between Weston and Barry. These last have reverted to their original condition of islands, being now surrounded by the waters of the Bristol Channel.

While the mainland provided a catchment area for rivers and streams, which, as we have seen, brought down large quantities of sand and detritus from the Millstone Grit and Coal Measures of South Wales, to pour them into the Rhætic sea, the islands had little effect on the sedimentation. The Rhætic Beds deposited in the channels among the archipelago are thinner

---

[1] L. Richardson, 1905, loc. cit., pp. 406–10; E. T. Newton, 1899, *Q.J.G.S.*, vol. lv, pp. 89–96.
[2] A. Strahan, 1904, 'Geol. S. Wales Coal-field, part vi, Bridgend', *Mem. Geol. Surv.*, p. 54.
[3] L. Richardson, 1903, 'The Mesozoic Geography of the Mendip Archipelago', *Proc. Cots. N.F.C.*, vol. xiv, pp. 59–72.

than elsewhere, but their facies is much the same as that of the beds formed in more open water to the south and west. The most noticeable change of facies is from shales to limestones, the lime being supplied by solution and redeposition from the limestone cliffs and screes.

The normal clays and shales can sometimes be seen passing surprisingly close to the old shore-lines, limestones and conglomerates taking their place

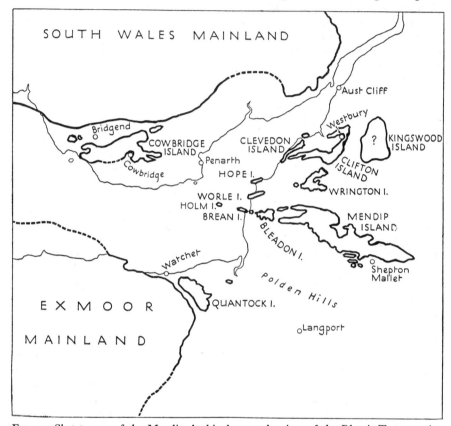

FIG. 17. Sketch-map of the Mendip Archipelago at the time of the Rhætic Transgression. Thick lines represent the presumed pre-Rhætic coastline; thin lines show the present coastline of the Bristol Channel. Mainly after Lloyd Morgan, Strahan and Cantrill.

only where the beds actually abut on the Palaeozoic surface. Perhaps the best idea of this is to be obtained near Shepton Mallet. In the railway-cutting by the Three-Arch Bridge, south of the town, the Rhætic Beds are 28 ft. thick and contain the usual fossils. The Westbury Beds at the base consist as usual of $11\frac{1}{2}$ ft. of black shales, and the Langport Beds or White Lias at the top comprise 12 ft. of the normal cream-coloured limestones. Only the Cotham Beds in the centre, though mainly shales, contain a thin band of conglomerate, 2 in. thick, formed of pebbles of Carboniferous Limestone.[1]

Less than a mile to the north, near the viaduct south of Downside, the

[1] L. Richardson, 1911, *Q.J.G.S.*, vol. lxvii, p. 60; and to the end of this section, based on op. cit., pp. 55–72.

Rhætic is represented only by a breccio-conglomerate, resting on the planed surface of the Carboniferous Limestone, and overlain by a massive sparry limestone representing the base of the Lower Lias. The constituents of the conglomerate are limestone and chert derived from the Carboniferous, embedded in a grey matrix, which yields *Pt. contorta*, vertebrae of *Plesiosaurus*, teeth of *Saurichthys*, *Sargodon*, *Lepidotus*, *Acrodus* and *Gyrolepis*, coprolites, small quartz pebbles, and reptilian bones.

The same rapid change of facies may be seen at various places round the eastern end and northern flank of the Mendips. Vallis Vale, cutting from the Frome Valley north of Frome westwards into the end of the Mendips, provides fine sections of attenuated Rhætic, Liassic and Inferior Oolite strata thinning out against the rising surface of the Carboniferous Limestone (map, fig. 13, p. 70). About 2 miles north-west of Frome the Rhætic, with the Bone Bed at the base, is normal. At Hapsford Mills it is 14½ ft. thick, consisting of alternate bands of conglomerate and clay in the lower part, and of conglomerate and limestone in the upper, with fish teeth at the base. A very short distance nearer the shore-line it is 4 ft. thick, consisting mainly of conglomerate, with thin seams of clay and some limestone and nodules, the whole richly fossiliferous. Many of the limestone pebbles are bored by *Polydora*,[1] and in the shales Charles Moore found specimens of *Eolepas* and *Chiton*. At both these places the base is cemented firmly on to an eroded surface of Carboniferous Limestone, while the top is in turn eroded and well bored and directly overlain by the Upper Inferior Oolite. In a similar section at Holwell the Rhætic is only 1 ft. thick, but contains numerous characteristic fossils. Charles Moore made Holwell famous by the discovery of about 70,000 teeth and small bones of vertebrates, among which were 29 teeth of mammals, now assigned to the genera *Microcleptes* and *Thomasia*. The remains were collected in fissures in the Carboniferous Limestone.[2]

Similar patches of Rhætic shore-deposits, which may or may not be covered by attenuated representatives of the Lias, are seen resting in hollows of the Carboniferous Limestone floor at Vobster, Harptree and elsewhere. At Harptree they are arenaceous, owing to the proximity of Old Red Sandstone. Vertebrate remains are often numerous in fissures in the underlying Carboniferous Limestone.

At Nempnett some small outliers provide a glimpse of the beds laid down in one of the channels between the islands: to the south lie the Mendips, to the north Wrington Island, the channel between them being about 3 miles wide. In the middle of this channel lie the Nempnett outliers of Rhætic Beds resting on the Tea Green Marls. From the base upwards, massive conglomerates replace the greater part of the Westbury Beds, Cotham Beds and Langport Beds. They are composed of pebbles of Carboniferous Limestone embedded in a matrix of shell debris and calcite, through which are sparsely distributed fish-scales and teeth, coprolites, and fragments of bone. At intervals normal shale and limestone are interbedded and yield the characteristic Rhætic fossils. At one place there is a *Chl. valoniensis* limestone. No Watchet Beds

[1] F. A. Bather, 1909, *Geol. Mag.* [5], vol. vi, p. 109.
[2] For a revision of the stratigraphy, a description of the circumstances of the find and references to all previous literature see L. Richardson, 1911, loc. cit., pp. 62–4; also G. G. Simpson, 1928, *Catal. Mesozoic Mammalia in Geol. Dept. B.M.*, where the mammal-remains are fully described and compared with those from the German Rhætic.

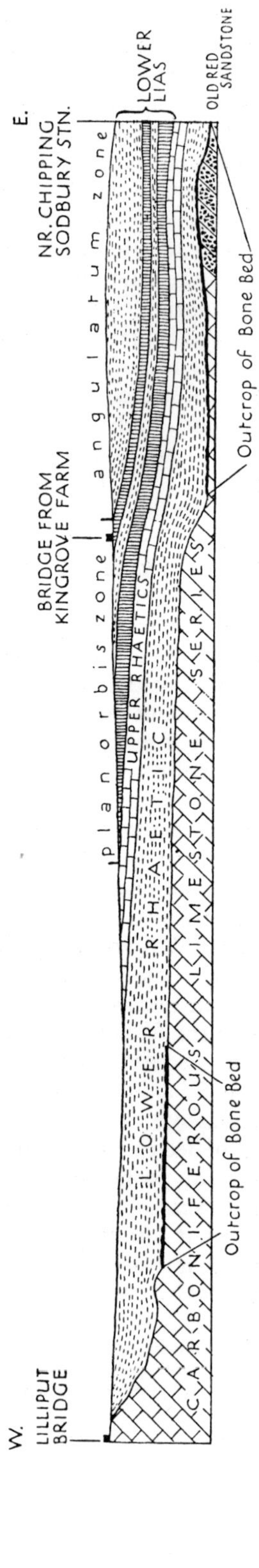

[Figure labels, left column: E. | NR. CHIPPING SODBURY STN. | LOWER LIAS | OLD RED SANDSTONE | angulatum zone | BRIDGE FROM KINGROVE FARM | planorbis zone | UPPER RHAETIC | WHITE RHAETIC LIMESTONE SERIES | Outcrop of Bone Bed | LOWER RHAETIC | CARBONIFEROUS LIMESTONE SERIES | Outcrop of Bone Bed | LILLIPUT BRIDGE | W.]

are present, and the top bed of the White Lias can be recognized, overlain by the Lower Lias limestones.

Hereabouts and over a great part of Somerset and the adjoining counties the top bed of the White Lias is a white, fine-grained limestone, having a curiously marked upper surface resembling dried mud cracked by the sun. On account of this feature it early received the name SUN BED, but the origin of the cracks is by no means certainly known. The wide distribution of the bed introduces serious difficulties in the way of accepting the popular explanation. The presence in it of worm-tubes and the wings of insects, however, shows that it was certainly formed in very shallow water.

IV. THE SEVERN VALLEY NORTH OF BRISTOL

When we pass out of the northern fringe of the Mendip Archipelago we see no more of the conglomeratic and beach deposits of Rhætic age, so well shown on the flanks of the southern islands. During the construction of the South-Wales-Direct railway, the cuttings between Chipping Sodbury and Stoke Gifford exposed Rhætic Beds lapping on to the Carboniferous rocks of Kingswood Island; but there were no conglomerates (fig. 18). The normal shaly facies of the Rhætic was seen contour-bedded over the hummocky surface of the Carboniferous Limestone, forming 'bedding-anticlines'. The thickness varied considerably according to the contours of the bottom, and at the base was an intensely hard, but impersistent, Bone Bed. In the railway-cutting at Lilliput the Bone Bed was seen resting on the Old Red Sandstone and was packed with vertebrate remains. Considerable lateral variation in the beds was also apparent, lenticular hard bands, sometimes gritty, but generally of limestone, appearing at various levels in the *Contorta* Shales.[1]

On the whole, over the region between Bristol and Tewkesbury the Rhætic succession shows remarkable correspondence in all the exposures. It is thinnest at Redland, Bristol, close to the shore of the

[1] S. H. Reynolds and A. Vaughan, 1904, *Q.J.G.S.*, vol. lx, pp. 194–8.

Fig. 18. Section in the railway-cutting between Chipping Sodbury Station and Lilliput Bridge, Glos., showing the Rhætic Beds, without littoral facies, contour-bedded over Carboniferous Limestone and Old Red Sandstone. From S. H. Reynolds and A. Vaughan, 1904, *Quart. Journ. Geol. Soc.*, vol. lx, p. 198, fig. 2. (Horizontal scale: 1 inch = 300 ft.; vertical scale: 1 inch = 75 ft.)

Clifton Island, where it measures only 13 ft. in thickness, although maintaining its normal characters. The more usual thickness is from 20 ft. to 25 ft., increasing to 30 ft. towards Tewkesbury.

The region is classic ground in the history of Rhætic geology. The most important sections are Aust Cliff and Sedbury Cliff on the left and right banks of the Severn opposite Chepstow; Goldcliff, in the Rhætic and Lias patch east of the mouth of the Usk, opposite Newport; the Stoke Gifford and Chipping Sodbury railway-cuttings, already mentioned, the former 5 miles north of the centre of Bristol; and a road-cutting at Redland, Bristol. Farther north are the fine cliff-sections on the banks of the Severn: Wainlode Cliff on the left bank between Gloucester and Tewkesbury, and a neighbouring road-cutting known as Coomb Hill, and Garden Cliff, Westbury on Severn, on the right bank 8 miles west-south-west of Gloucester (Plate II).

These sections may be summarized in the following sequence:

SUMMARY OF THE RHÆTIC BEDS OF THE SEVERN VALLEY [1]

Langport Beds or White Lias, 0–1½ ft.:

The White Lias is extremely attenuated and disappears altogether north of Bristol, not to reappear before Stratford on Avon is reached. At Redland, Bristol, and in various exposures around Bath, it consists only of the Sun Bed, the usual 1 ft.–1½ ft. band of hard, white or cream-coloured, fine-grained limestone, having a conchoidal fracture, and its upper surface marked with cracks. It contains *Ostrea liassica*, *Modiola langportensis*, insect wings and worm-tubes.

Cotham Beds.

The COTHAM or 'LANDSCAPE' MARBLE (from Cotham near Bristol) is best developed in Gloucester and Somerset. It is a peculiar white stone up to 8 in. in thickness, with arborescent markings which are supposed to have been caused by the escape of contained gases generated by decomposition in the mud. In the northern sections, at Wainlode and Garden Cliffs, it passes into a few inches of limestone crowded with the miniature *Pseudomonotis fallax* and noted for its insect fauna. The insect remains consist chiefly of elytra of beetles, but there are some wings allied to a Neuropteron, *Chauliodes*. They are associated with leaves of a fern, *Otopteris obtusa*, and small fragments of carbonized wood, with marine shells. At Sedbury Cliff the bed was removed by erosion at the beginning of Liassic times, the base of the Lower Lias being composed of a conglomerate of Cotham Marble fragments.

Below the Cotham Marble is a band of clay, varying in thickness but always identifiable, separating it from an *Estheria* and *Naiadites* limestone below. *Estheria minuta* var. *brodieana* and the plant *Naiadites lanceolata*, sometimes associated with *Pleurophorus* and '*Schizodus*' *ewaldi*, form a widespread and constant horizon in this area, usually in a bed of limestone. In the southern sections, around Bristol, the fossils occur in the top of from 2 ft. to 4 ft. of limestones, either massive or fissile; the lower parts are barren. In the

---

[1] See A. R. Short, 1904, *Q.J.G.S.*, vol. lx, pp. 170–93; L. Richardson, 1903, ibid., vol. lix, pp. 390–5; 1905, ibid., vol. lxi, pp. 374–84; 1903, *Proc. Cots. N.F.C.*, vol. xiv, pp. 127–74, 251–7.

northern sections (Sedbury, Wainlode, Westbury, &c., and at Goldcliff) the lower parts become argillaceous, *Estheria* and *Naiadites* occurring in a thin limestone or nodule-bed overlying barren shales. The transition is seen at Aust Cliff, where the *Estheria* Bed overlies yellow, thinly-bedded, very argillaceous limestone, often crumbly. *Naiadites* is of special interest because it is probably the earliest known fossil plant belonging to the Lycopodiaceæ.[1]

Westbury Beds or Contorta Shales.

This division is thinnest (6 ft. 10 in.) at Redland, Bristol, near the Clifton Island, and thickest in the north, at Westbury on Severn (19 ft. 8 in.). The mass of the shales may be divided into two palaeontological subzones, corresponding to the local acmes of *Chl. valoniensis* (above) and of *Pt. contorta* (below), with a vertebrate zone at the base. The top is often formed by a *Pecten*-bed limestone, and there is sometimes a lower *Pecten*-bed as much as 8 ft. lower (e.g. at Aust Cliff). Shales with *Chlamys* may occur above the upper *Pecten*-bed (e.g. at Sedbury Cliff and the northern sections). Mollusca are abundant in this upper part, as in Somerset and Dorset, and they belong to the same species. At Deerhurst Richardson found a specimen of *Heterastræa*, the only compound coral recorded from the British *Contorta* Beds.[2] At Wainlode and Westbury a few thin seams of sandstone occur in the shales, as at Langport.

The Bone Bed, with the usual vertebrates, which are often extraordinarily abundant, is impersistent; it usually takes the form of a sandstone, resting on a markedly eroded surface of the underlying Tea Green Marls, or (at Chipping Sodbury) Carboniferous Limestone. At Wainlode and Westbury, in the north, however, 2 ft. of black shales come in between the Bone Bed and the eroded surface, as at Watchet, in Somerset. This irregular distribution of the beds below the Bone Bed is explained by Richardson as due to the Tea Green Marls having been gently folded towards the end of Keuper times, with the result that sedimentation was first renewed in the hollows, each successive bed having an increasingly wide extent (see fig. 14).

The occurrence of thin seams of sandstone in the lower part of the Westbury Beds at Westbury and Wainlode may perhaps be attributed to proximity to the Palaeozoic rocks of the Welsh Border, where the coast-line presumably lay. The tendency to a sandy facies becomes considerably more marked farther north, towards the Malverns. An outlier on Berrow Hill, 7 miles west of Tewkesbury, which lies only 2 miles from the present outcrop of the Malvern Palaeozoic rocks, was especially excavated by Richardson in the hope of finding signs of a littoral facies of the Rhætics. Five thin seams of sandstone and sandy limestone were found interbedded with the *Contorta* Shales (the total thickness of which was reduced to about 9 ft.) and they were found to rest on a lower horizon in the Tea Green Marls.[3]

Similar sections of the Westbury Beds have been recorded by Richardson near the extremity of the long tongue of Lower Lias and Rhætic Beds that extends north of the general outcrop towards Worcester and Droitwich. At Crowle, 4 miles east-north-east of Worcester, and in a railway-cutting at

---

[1] I. B. J. Sollas, 1901, *Q.J.G.S.*, vol. lvii, p. 311.
[2] R. F. Tomes, 1903, *Q.J.G.S.*, vol. lix, pp. 403–7.
[3] L. Richardson, 1905, *Q.J.G.S.*, vol. lxi, pp. 425–30.

Dunhampstead near Droitwich, the lowest 8 ft. of the Rhætic consists largely
of numerous thin sandstone layers interbedded with shales. The Bone Bed
is represented by a 1 ft. sandstone layer 4 ft. from the base, containing fish
teeth and scales and vertebrae, while two other sandstone layers contain
'Schizodus' casts, and many show ripple-marks and tracks of annelid worms.
The junction with the Tea Green Marls, which are well exposed below, is
abrupt and non-sequential.[1]

V. FROM THE SEVERN TO THE HUMBER

From the Severn Valley the outcrop of the Rhætic Beds turns north-eastward across the Midland Plain of England, passing through the counties
of Worcester, Warwick, Leicester, Nottingham, and Lincoln. Between the
Severn and the Stratford Avon long tongues stretch northward, usually overlain by Lower Lias, to Droitwich and Wootton Wawen, near Henley in Arden,
while the farthest outlier is as distant as Knowle, midway between Birmingham
and Warwick. The outcrop sometimes forms a recognizable surface-feature,
though for long distances its position can only be inferred where the limestones
at the base of the Lower Lias form a subdued escarpment overlooking the
Triassic plain, and the Rhætic Beds crop out along it or at its foot. Exposures
are few and far between, but enough have been seen from time to time to
prove that the Rhætic formation maintains its essential characteristics all
across England.

The most interesting feature of the Midland area is the return of the true
White Lias or Langport Beds, which, after being absent from the greater part
of Gloucestershire, reappear near Stratford on Avon and continue nearly to
Wigston in Leicestershire, where they disappear for good. The Cotham Beds
and the Westbury Beds or *Contorta* Shales continue unchanged, while the
junction of the latter with the Tea Green Marls remains, as usual, abrupt and
non-sequential.

Apart from the exposures mentioned at the end of the last section, on the
west side of the Tewkesbury–Droitwich tongue, adjoining the Severn Valley,
few others of interest have been recorded in the area where the Langport
Beds are absent, west of the Stratford on Avon district. Perhaps the best was
described by Richardson in the bank of a sunk bridle-path at Woodnorton,
near Evesham, and this was said to be the only section in Worcestershire
showing the upper beds. Beneath the *Ostrea* Beds of the Lower Lias were
$19\frac{1}{2}$ ft. of grey, greenish and yellow shales, with the thin *Estheria* Bed of the
Cotham Marble Series only 8 ft. from the top, containing *Estheria minuta*
var. *brodieana* and *Naiadites lanceolata*. These rested on normal dark shales
belonging to the Westbury Beds, with the usual fossils, but the full thickness
could not be seen.[2] About Binton, Grafton, Wilmcote, and Bickmarsh, only
a short distance west of Stratford on Avon, the White Lias is still absent, and
over this area the Lower Lias has at its base a peculiar hard, shelly limestone
called the Guinea Bed, which is frequently conglomeratic. It is no longer
exposed, but old records show that it contains a typical Lower Lias assemblage
of shells, mixed with others apparently derived from the Rhætic. It may be

---

[1] L. Richardson, 1903, *Geol. Mag.* [4], vol. x, p. 80.
[2] L. Richardson, 1903, loc. cit., p. 82.

significant that this area lies across the continuation of the Sedgley–Lickey Axis.

The White Lias first appears as field rubble about half a mile south-east of Sweet Knowle, south of Stratford on Avon, and is to be seen in some abandoned workings on Wimpstone Field and again at Newbold Limeworks, all about on the line of the Vale of Moreton Axis.

'Thence, right through to Rugby', according to Richardson, 'it is well and persistently developed (maximum about 10 ft.), of very much the same appearance as the Somerset White Lias, contains specimens of *Dimyodon intus-striatus* (Emmr.) abundantly in places, and not infrequently corals. Certain of its beds are well bored by annelids, and at Church Lawford portions are pebbly; while in the same neighbourhood it has a ferruginous deposit on top, with which are associated ample indications of erosion by water: in other words there is abundant evidence for a non-sequence between the White Lias and the superincumbent Lower Lias deposits.'[1]

Church Lawford and Rugby, as has been noted, lie upon the prolongation of the Nuneaton Axis.

The Cotham Beds, consisting of the usual pale greenish-grey marls, continue with little change throughout Warwickshire and the adjacent counties farther north and east. The *Estheria* Bed has been proved at Wilmcote, Wootton Wawen, Summer Hill railway-cutting between Stratford on Avon and Alcester, and at other places, but no 'landscape marble' has been found.

The Westbury Beds or *Contorta* Shales are also normal and persistent, though their full thickness cannot be measured until Leicester is reached. The junction between them and the Cotham Beds is a gradual transition, but their lower limit is abrupt. Judging from old records and from specimens that have been obtained from Summer Hill, fossils abound at certain horizons, the commonest being *Pteria contorta* and '*Schizodus*' sp. Hard beds do not figure prominently, but there are a number of thin sandstone layers in the lower part of the shales in the Summer Hill railway-cutting, and Brodie obtained highly fossiliferous *Pecten*-limestone from temporary excavations at Brown's Wood, 3½ miles west of Bearley Junction. The Bone Bed is poorly developed, being nowhere very rich in vertebrate remains.[2]

The finest section of the Rhætic Beds in the Midlands is at Glen Parva Brickworks, near South Wigston, 4 miles south of Leicester. The total thickness is 40 ft.—Cotham Beds 10½ ft., Westbury Beds 20 ft. 9 in.[3] In the old Spinney Hill sections on the eastern outskirts of Leicester, minutely described by W. J. Harrison,[4] but now overgrown, there was another section of Westbury Beds visible to a thickness of 10 ft., with an *Estheria* Bed at the top.[5] The Westbury Beds in these exposures have proved rich in the usual fossils, and in addition a starfish, *Ophiolepis damesii*, was said by Harrison to occur in large numbers in a certain seam among the dark, finely-laminated shales. A noticeable feature is the almost entire absence of hard bands. At the base a conglomeratic Bone Bed (½ in.–2 in. thick), consisting of a pyritous impure

[1] L. Richardson, 1912, *Geol. Mag.* [5], vol. ix, pp. 24–33, for Warwickshire.
[2] Ibid., p. 26.
[3] Measurements and grouping according to the latest revision by L. Richardson, 1909, *Geol. Mag.* [5], vol. vi, pp. 366–70.
[4] W. J. Harrison, 1876, *Q.J.G.S.*, vol. xxxii, pp. 212–18.
[5] There are unconfirmed records of *Estheria minuta* at two levels in the *Contorta* Shales of the Glen Parva Brick Pit.

sandstone, is full of pebbles of Triassic sandstone, slaty and quartzose rocks from Charnwood Forest, phosphatic nodules, coprolites, and fragmentary remains of saurians and fish. From this have been obtained bones or teeth or scales of *Ichthyosaurus, Plesiosaurus, Saurichthys acuminatus, Acrodus minimus, Sargodon tomicus, Hybodus cloacinus, Nemacanthus monilifer*.

Similar sections have been described at various points along the outcrop farther north, but nearly all were railway-cuttings which have become overgrown, or old clay pits which have fallen into disuse and are now obscure. In a railway-cutting at Stanton on the Wolds,[1] between Nottingham and Melton Mowbray, a few reptilian bones occurred at the base of the *Contorta* Shales, but the main Bone Bed, yielding all the vertebrate species found at Leicester, with coprolites and quartz pebbles, appeared 2 ft. 8 in. higher up, in the form of a 1-in. seam of soft white sand. At Barnston, in a cutting on the line south-east of Bingham, also on the border of the Vale of Belvoir, the Bone Bed was hard and pyritic once more, as at Leicester, but its position was still 1½ ft. from the base. In neighbouring railway-cuttings at Cotham and Kilvington, a few miles south of Newark,[2] the whole Rhætic formation was cut through, but, although 34 ft. thick, it proved exceptionally unfossiliferous. The nodular *Estheria* Bed was present near the top, but all sign of the Bone Bed was lacking, both here and at Newark.

In the Lincoln district the outcrop is much concealed by superficial deposits, but the Rhætic Beds were exposed during the making of cuttings for the Great Northern line between Gainsborough and the adjoining village of Lea.[3] Here there was not only one bone bed, but three, as in parts of the West Country: one was at the base, consisting of 1 ft. of loose micaceous sandstone, and the other two were 8 ft. and 8 ft. 5 in. higher, comprising mere seams of coprolites, worn bones, fin-spines, teeth, fish-scales and small pebbles. A vertebrate fauna even richer than that at Leicester was found, and *Chl. valoniensis* abounded again near the top. In all 26 ft. of black shales were exposed, with occasional thin sandstone seams, and *Pteria contorta* and '*Schizodus*' *ewaldi* were abundant throughout, showing that the whole series belonged to the Westbury Beds. At other places in Lincolnshire the Upper Rhætic has been occasionally seen, and at the top a typical Sun Bed has been observed, covered with cracks, as in Somerset.

In all these Midland exposures, wherever the base of the Rhætic has come under observation, it rests upon a markedly eroded surface of the underlying Tea Green Marls, which in turn pass down without any visible break into the typical red marls of the Keuper.[4]

VI. THE MARKET WEIGHTON AXIS

Much detailed work still remains to be done before it can be ascertained what changes the Rhætic Beds undergo as they cross the Market Weighton

[1] E. Wilson, 1882, *Q.J.G.S.*, vol. xxxviii, p. 454.

[2] Ibid., pp. 453–4. (This Cotham is not to be confused with the one near Bristol.)

[3] M. F. Burton, 1867, *Q.J.G.S.*, vol. xxiii, pp. 315–22; and W. A. E. Ussher, 1888, 'Geol. Lincoln', *Mem. Geol. Surv.*, p. 12.

[4] There seems every probability that the Tea Green Marls were once also red and non-calcareous, but that in course of time they have been bleached and rendered partly calcareous owing to the downward infiltration of carbonate of lime and some deoxidizing chemical agent, possibly derived from the decomposition of the organic remains in the overlying Rhætics.

Axis. Fox-Strangways[1] followed the outcrop across the critical belt and obtained a number of indications of green and black shales with here and there thin bands of sandstone, of Rhætic aspect, at North Cliff, Market Weighton, Londesborough Park, Warter Priory, and Pocklington. There can thus be no doubt that the outcrop is continuous. A well at Market Weighton 14 ft. deep was said to be principally in black Rhætic shale with minute shells, while a road-cutting at North Cliff showed 18 ft. or more of shale with sandy bands below the *Pteromya crowcombeia* Bed at the base of the Lower Lias. Here there is no indication of White Lias, but north of Market Weighton, at Warter Priory near Pocklington, field rubble of White Lias was said to be abundant, and it is a feature of the top of the Rhætics of the Yorkshire Basin. It probably represents the Cotham Beds rather than the true White Lias or Langport Beds.[2]

VII. THE YORKSHIRE BASIN

Rhætic Beds are not exposed on the coast of Yorkshire, for the outcrop is covered by the alluvium of the River Tees, but they have been met with at numerous inland localities. The best sections are at Northallerton, where the Rhætic is 27 ft. thick. The sequence, which may be considered typical of Yorkshire by reason of its central position, is as follows:[3]

Cotham Beds (? and Langport Beds): 10 ft. of white shale, with a 3-in. band of white argillaceous limestone at the top (? Sun Bed) have been assigned, on the ground of lithological resemblance, to the White Lias *sensu lato*. No fossils have been found, but it is probable that only the Cotham Beds are represented.

Westbury Beds or Contorta Shales, 17 ft. The *Contorta* Shales consist mainly of black, crisp shales, with *Pt. contorta* throughout. They are interrupted by two hard bands of light-coloured, close-grained, siliceous rock full of '*Schizodus' ewaldi*. The upper band (about the centre) is 3 in. thick and yields also fragmentary crinoid stems. The lower band (1½ ft. from the base) is 6 in. thick, and in the shale immediately above it occur numerous fish-scales. This is thought to represent the Bone Bed.

Unfossiliferous Tea Green Marls appear below.

VIII. NORTHERN IRELAND: THE ANTRIM PLATEAU

Outliers of Lower Lias, which will be described in the following chapter, exist far from the main outcrop in the Shropshire and Cheshire Plain and also in the Plain of Carlisle. Their stratigraphy is imperfectly known owing to coverings of Drift, and although it is likely that Rhætic Beds exist beneath the Lias, as yet no definite evidence of their presence has been obtained.

About 120 miles due west of the Carlisle outlier further proof of the extension of the Liassic sea is found in Antrim, especially near Belfast, and here thick beds of fossiliferous Rhætics are exposed beneath. The deposits owe their preservation directly to the protective covering of the Tertiary basalt plateau, beneath which they probably have a considerable extension, since they crop out at many favourable points around the edge.

---

[1] C. Fox-Strangways, 1886, 'Geol. Country between York and Hull', *Mem. Geol. Surv.*, p. 10.   [2] L. Richardson, *in lit.*
[3] C. Fox-Strangways, 1886, 'Geol. Northallerton and Thirsk', *Mem. Geol. Surv.*, p. 13.

The beds were described by Tate in 1864,[1] in the days when all the strata above the *Contorta* Shales were classed together as White Lias. A detailed interpretation of Tate's records is difficult to make.

The best section was at Collin Glen, near Belfast, where the total thickness of beds assigned to the Rhætic was about 46 ft., but it is considerably faulted and now much obscured. At the top, underlying *Ostrea liassica* Beds forming the base of the Lower Lias, were described 16 ft. of grey and reddish marls, resting on 10 ft. of arenaceous shales. Both of these divisions contained *Protocardia rhætica*, and so can be classed at least as 'Upper Rhætic'. Beneath are the *Contorta* Shales, typically developed and 20 ft. thick. They consist of black shales with thin bands of blue argillaceous limestone and in the lower part seams of compact sandstone, with a 2-in. Fish Bed or Bone Bed 6 ft. 7 in. from the bottom. Tate recorded *Pt. contorta, Chl. valoniensis, Protocardia rhætica* and *Cardium cloacinum* throughout. A rich collection of fish-remains has been made from the Bone Bed, which has been likened to that in the West of England.[2]

At Cave Hill, 3 miles north of Belfast, the Upper Rhætic is only about 14 ft. thick. At Larne it contains pseudo-oolitic and pisolitic concretions disseminated through the grey marls. The *Contorta* Shales are still over 19 ft. thick at Whitehead railway-cutting, 14 miles north of Belfast, but only 16 ft. thick at Cave Hill.

Tate remarked on the abrupt change from the red and grey marls of the Keuper to the black *Contorta* Shales in Ireland. Nowhere is there any sign of gradation from one to the other.

IX. THE HEBRIDEAN AREA

South of Alt na Teangaidh, near Gribun, in Western Mull, about 30–40 ft. of Rhætic Beds follow conformably on the Triassic sandstone. They consist principally of dark sandy limestone or calcareous sandstone with somewhat wavy bedding, and subordinate yellow sandstone. The junction with the Lias is nowhere seen. Lamellibranchs are common and are the only fossils, excepting some fish-scales; the following species have been recorded: *Pt. contorta, Gervillia præcursor, Myophoria postera* (?) (Bronn), *Chl. valoniensis, Protocardia rhætica, Pleurophorus elongatus* and *Modiola* (?) *sp.* These species occur throughout the section at Gribun and suggest that the whole belongs to the Westbury Beds.

Later beds may be represented in the Wilderness, where Bailey described a little cliff showing 5 ft. of shales with cementstones, of rather doubtful relations, overlying 20 ft. of obviously Triassic sandstone with conglomerate (cornstone).[3]

In Morven, half a mile south of the Rannoch River, 10 ft. of fine-grained white sandstone, sometimes pebbly, with layers of ill-preserved lamellibranchs near the top, and underlain in one place by 6 in. of grey marl, are seen resting on the Triassic cornstone. The lamellibranchs, so far as they can be

---

[1] R. Tate, 1864, *Q.J.G.S.*, vol. xx, pp. 103–11.
[2] W. F. Hume, 1897, *Q.J.G.S.*, vol. liii, p. 549.
[3] E. B. Bailey, 1925, 'Pre-Tert. Geol. Mull, Loch Aline and Oban', *Mem. Geol. Surv.*, pp. 72–3.

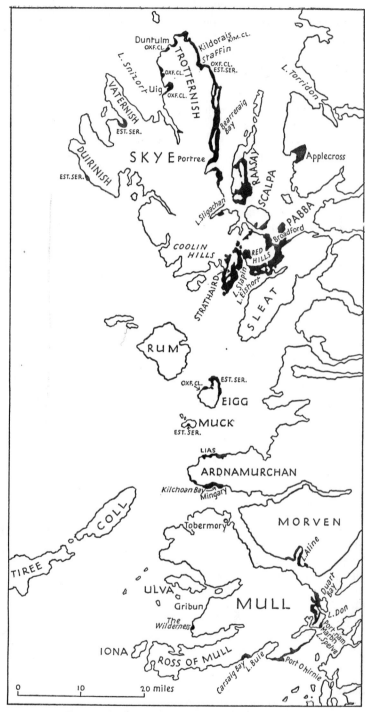

FIG. 19. Sketch-map of the Inner Hebrides, showing
the Jurassic outcrops (in black).

determined, suggest a Rhaetic age: *Pteromya simplex* Moore, *P. crowcombeia* (?) Moore, *Pleurophorus elongatus* (?) Moore, and some others not identified specifically. Lower Lias limestone crops out 20 ft. higher up, but the intervening strata cannot be seen.[1]

The only other place in Scotland where a definite Rhætic fauna has been found is the Isle of Arran. It was described here by Messrs. Peach, Gunn, and Newton, and lies in a mass of typical Rhætic shale tumbled down the throat of a Tertiary volcano, but the beds no longer occur *in situ*, having been otherwise completely removed by erosion. The fossils procured from the volcano are identified as *Pt. contorta*, *Chl. valoniensis*, '*Schizodus*' [*ewaldi*], *Protocardia rhætica* (?), *Modiola minima* (?), *Estheria minuta* (?), and *Gyrolepis alberti* (?), a thoroughly conclusive assemblage.[2]

As the Isle of Arran is some 40 miles from the edge of the Antrim Plateau and 50 miles from Mull, this occurrence provides a valuable link between the Rhætic areas of Ireland and the Hebrides, greatly facilitating palaeogeographical restorations.

In Skye, near Sconser, Loch Sligachan, 9 ft. of green sandstone and blue sandy limestone with shale partings are seen below blue Lower Lias limestone, and are called by the Survey 'Passage Beds (= Rhætic?)'. One band yields *Chl. valoniensis* and another is said to be full of a small *Ostrea*. A few feet of similar beds between the Trias and the Lower Lias limestones north-east of Heast, Skye, may also be of Rhætic age, but the evidence is inconclusive, since they have yielded only a fish scale, *Ostrea sp.* and plant remains.[3]

On the other side of Skye, and in Raasay, at Applecross and elsewhere, there is no recognizable trace of Rhætic Beds, the Lower Lias having apparently overlapped on to the older rocks. On the shores of Loch Slapin in southeast Skye, strata as high as the *semicostatum* zone can be clearly seen resting directly on Cambrian limestones.

Commenting on the Hebridean Rhætic sections in 1929, Richardson remarks:[4]

'The descriptions of the Rhætic sandstone in Mull given by Lee and Bailey call to mind the Rhætic sandstone of the Bridgend district of Glamorganshire; and the account by Bailey of "greenish-grey shale with cream-weathering cementstones", which appeared to him to succeed the sandstone, inclines one to suggest that the Cotham Beds are represented.'

The localized distribution of the Scottish Rhætic Beds and their tendency to be overlapped or overstepped by the Lower Lias are both unusual features.

X. EASTERN SCOTLAND

On the coast of Sutherlandshire, at the south end of the narrow coastal strip of Jurassic rocks (of which a great deal more will be said in ensuing chapters), the base of the Lias crops out in the foreshore below the pier of Dunrobin Castle, Golspie. The exposure is a poor one, only visible at low

---

[1] G. W. Lee, 1925, 'Pre-Tert. Geol. Mull, Loch Aline and Oban', *Mem. Geol. Surv.*, pp. 73–4.
[2] 1901, *Q.J.G.S.*, vol. lvii, pp. 226–43.
[3] H. B. Woodward, 1910, 'Geol. Glenelg, Lochalsh and S.-E. Skye', *Mem. Geol. Surv.*, pp. 94–7.
[4] 1929, *Handb. Geol. Great Britain*, p. 348.

tide, but it shows about 6 feet of unfossiliferous grits and conglomerates at the base of the Lias, resting upon Triassic cherty rock similar to some at Elgin, on the other side of the Firth. The conglomerate contains pebbles of chert, sandstone, and limestone derived from the Trias, and in masses of similar conglomerate found in the Boulder Clay on the south side of the Moray Firth have been found fossils similar to those in the Rhætic of Scania.

Although there is no definite evidence, the Dunrobin conglomerate is therefore generally supposed to be of Rhætic age.[1]

XI. KENT

In the Kent borings no Rhætic fossils were encountered.

[1] G. W. Lee, 1925, 'Geol. Golspie', *Mem. Geol. Surv.*, p. 68.

CHAPTER V

LOWER LIAS [1]

Stages.	Zones (Faunizones). (Plates XXX, XXXI).	Stratigraphical Divisions in Dorset.
CARIXIAN or LOWER PLIENSBACHIAN	*Prodactylioceras davœi* [2]	WEAR CLIFF or GREEN AMMONITE BEDS 30–105 ft.
	Tragophylloceras ibex	Belemnite Stone STONEBARROW BEDS or
	Uptonia jamesoni	BELEMNITE MARLS 75 ft.
SINEMURIAN	*Echioceras raricostatum* [3]	Hummocky Limestone
	Oxynoticeras oxynotum	BLACK VEN MARLS 140–50 ft.
	Asteroceras obtusum	*Birchi* Tabular, *Birchi* Nodular
	Arnioceras semicostatum [4]	SHALES WITH BEEF 70 ft.
	Coroniceras bucklandi	Table Ledge
HETTANGIAN	*Scamnoceras angulatum* [5]	LYME REGIS BEDS or BLUE LIAS LIMESTONES 85 ft.
	Psiloceras planorbis	
	Ostrea liassica	
	Pleuromya tatei	PRE-PLANORBIS BEDS 12 ft.

ALMOST the entire succession of the Lower Lias in its typical development across England from the Dorset coast consists of clays and shales, in the lower parts of which numerous bands of calcareous mudstone or clay-limestone have been formed by secondary chemical processes.

A rich and varied succession of ammonite faunas is preserved, and the lower zones are especially characterized by beds of *Ostrea*, *Gryphœa* and some other lamellibrachs; apart from the ammonites and Gryphæas, however, the molluscan fauna is relatively poor. The most conspicuous remains are the skeletons of the great marine reptiles. The presence of these, and also of fossil insects, points to deposition, not in deep water as frequently deduced from the clayey sediment, but in a comparatively shallow continental sea. The rivers flowing into the Liassic sea had reached maturity, for instead of bringing down detrital material from the surrounding land, they carried only fine mud. Mixed with the mud in suspension came a high proportion of iron, which, in the form of minutely disseminated pyrites, gives the usual dark grey to black colour to the Lias shales.

[1] For the origin of the word Lias see footnote 3 on p. 12.

[2] Called by Wright the *capricornu* zone (1863) and the *henleyi* zone (1878), but Dr. Lang sees no good reason for altering Oppel's name. The genus *Prodactylioceras* was proposed for *Cœloceras davœi* auct. by Dr. Spath, 1923, *Geol. Mag.*, vol. lx, p. 10.

[3] This includes the old *armatum* zone, which was above the *raricostatum* horizon in Yorkshire but was proved by Dr. Lang to occur below it in Dorset, and therefore to have no value. For full discussion see Spath, 1925, *Naturalist*, pp. 167–72.

[4] Oppel's *Pentacrinus tuberculatus* zone and approximately Wright's *turneri* zone; renamed the *semicostatum* zone by Judd and adopted as such by the Survey and subsequent writers.

[5] Formerly *Schlotheimia angulata*; for generic revision see Spath, 1925, *Naturalist*, pp. 201–6. Included in the *bucklandi* zone by Wright (1860 and 1863).

I. THE DORSET–WEST SOMERSET AREA

(a) The Dorset Coast

The Lias is cut by the Dorset coast in such a way that a complete section is laid bare across the extension of the trough of deposition that stretched from the Mendip Hills and South Wales in the north to the massif of the Cotentin and Normandy in the south. One after another its zones rise from the beach to build the line of magnificent cliffs extending from Bridport in Dorset westward past Charmouth and Lyme Regis to Seaton in Devon.

From the early days of geology attention has been drawn to the Lower Lias of these cliffs by the profusion of ammonities and by the discoveries of giant reptile skeletons made from time to time near Lyme Regis. A great deal of new light has been thrown on the succession by the careful studies and collecting of Dr. W. D. Lang, from whose papers the following account is principally derived. The remarkable sequence of ammonite epiboles established by him has been shown in Table V, p. 28, where some general remarks on the subject have already been made.

SUMMARY OF THE SUCCESSION ON THE DORSET COAST [1]

(Total thickness 430 ft.)

Davœi Zone (WEAR CLIFF OR GREEN AMMONITE BEDS), 100 ft.

This occupies a longer stretch of cliff than any other division of the Lower Lias on the Dorset coast. Faulted up from beneath the beach at Seatown, it forms the lowest precipice of Golden Cap, and then rises gradually westward to the highest precipice of Stonebarrow Cliff, thinning out beneath the Upper Greensand. Its upper limit is clearly defined by the prominent basal bed of the Middle Lias, called the Three Tiers, its lower limit by a hard band, the Belemnite Stone (Plates III and V).

The series consists of marly clays with ferruginous bands, somewhat sandy towards the top. It derives its name from the occurrence of the ammonite *Androgynoceras latæcosta* embedded in limestone nodules, with the gas-chambers filled with green calc-spar. Specimens used to be cut longitudinally, polished, and sold as ornaments.

The monotony of the clays is broken by three more or less continuous limestone bands, which mark off four approximately equal divisions, each from 20 ft. to 30 ft. thick under Golden Cap. The most conspicuous and constant is the central limestone, known as the Red Band. When fresh it is hard and firm, but being only 6 in. to 1 ft. thick, it has generally been weathered nearly to the core, so that it is crumbly and brown with a pinkish-red surface. The commonest fossils are species of *Liparoceras*, but *Tragophylloceras loscombi* and the zone fossil, *Prodactylioceras davœi*, also occur. *T. loscombi* ranges through the whole of the Green Ammonite Beds and passes up into the Middle Lias.

The other two limestone beds are of very different appearance from the Red Band, being hard, nodular and impersistent, while fossils are rarer (as

---

[1] W. D. Lang, 1913–28; for list of papers, see the Bibliography, at end. For a general account, giving the stratal succession, see H. B. Woodward, 1911, 'Geol. Sidmouth and Lyme Regis', 2nd ed., *Mem. Geol. Surv.*, pp. 21–40.

they are also in the intervening clays).
Dr. Lang has determined, however,
that ammonites of the genus *Oistoceras*
characterize the upper limestone band
and the lower part of the clays above
it, while *Androgynoceras latæcosta* is
found in and about the lower limestone
band.

The Green Ammonite Beds thin out
considerably from east to west. Only
the lower division, below the Red Band,
is fully exposed west of Golden Cap,
but in two miles this dwindles from
48 ft. one mile west of Golden Cap to
14 ft. at Stonebarrow. If the upper
division thinned at the same rate before
the Cretaceous denudation, the total
thickness of the Green Ammonite Beds
on Black Ven was only 30 ft.

Ibex and **Jamesoni Zones** (STONE-
BARROW BEDS OR BELEMNITE MARLS),
75 ft.

Underneath the Green Ammonite
Beds, rising westward along the face
of Stonebarrow Cliff, is a series of pale
marly clays known as the Belemnite
Marls, which build the third or upper
precipice of Black Ven, between Charmouth and Lyme Regis. Almost the
only abundant and well-preserved
fossils are belemnites.

The top is marked by the BELEMNITE
STONE, a 6-inch band of nodular but
persistent limestone, which weathers
creamy white. It is full of fossils, the
belemnites being *Passaloteuthis apici-
curvata* Lang, *Hastites spadix-ari* Lang,
H. microstylus, and *H. stonebarrowensis*
Lang, and the ammonites indicating
the hemera of *Beaniceras centaurus*. In
addition *Lytoceras fimbriatum* (Sow.)
occurs, together with species of *Tragophylloceras*, and numerous lamellibranchs, gastropods and brachiopods.

FIG. 20. Section along the cliffs from Pinhay
Bay past Lyme Regis and Charmouth to near
Seatown. Total distance 6 miles. After H. B.
Woodward, 'Jurassic Rocks of Britain', vol. iii,
1893, p. 53, *Mem. Geol. Surv.*

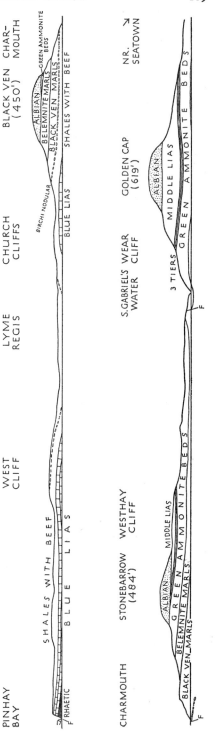

The rich belemnite fauna of the underlying marls, described by Dr. Lang, includes 5 genera and 26 species. The genus *Passaloteuthis* ranges through the whole series; *Angeloteuthis* and *Hastites* are confined to the uppermost 20 ft.; *Clastoteuthis* has a restricted range about the middle; *Pseudohastites* is confined to the lower half. A varied ammonite sequence has also been made out, as shown on p. 28.

Raricostatum, Oxynotum and Obtusum Zones (BLACK VEN MARLS), 140–50 ft.

At the base of the Belemnite Marls is a band of nodular limestone with *Epideroceras exhæredatum*, called the Hummocky Limestone, which rises from the beach under Westhay Cliff, the eastern end of Stonebarrow. Palaeontologically it marks the disappearance of the genus *Crucilobiceras* and may be taken as the topmost horizon of the Sinemurian. The limestone serves to define the upward limit of a thick series of much darker clays and marls, called the Black Ven Marls from their building the main and middle precipice of Black Ven and the cliffs on either side of Charmouth.

The uniformity of the series is broken by several layers of nodules and lenticular limestone bands, often enclosing ammonites. The nodules have a flattened ellipsoidal shape, suggestive of a secondary origin, and many are septarian.

The most fossiliferous horizon is a lenticular limestone 14 ft. from the top, called the AMMONITE MARBLE. This is full of small ammonites of the genera *Promicroceras*, *Echioceras*, *Euechioceras*, *Pleurechioceras*, and *Crucilobiceras*, and it is on about the horizon of a similar limestone in North Dorset, which was formerly cut and polished under the name of Ammonite or Marston Marble. An interesting feature of this horizon is the abrupt reappearance of *Gryphæa arcuata* in abundance, well incurved and indistinguishable from forms in the Blue Lias, of which it is primarily characteristic. In the marls immediately above the Ammonite Marble is a Pentacrinite bed yielding *Extracrinus briareus*.

About 48 ft. from the top is a layer of large nodules, up to 18 in. in diameter, enclosing specimens of *Asteroceras stellare*, while at 78–85 ft. down (about the middle) are three layers of 'flatstones', or large lenticular nodules, containing *Asteroceras obtusum*. The most conspicuous bands are twin layers at the base, 1 ft. apart, enclosing large specimens of *Microderoceras birchi*: the upper layer is called the *Birchi* Tabular, and the lower, which is merely a row of large concretions, the *Birchi* Nodular.

The lamellibranchs *Lima gigantea*, *Oxytoma inæquivalvis* and *Inoceramus faberi* are fairly abundant throughout.

Semicostatum Zone (SHALES WITH BEEF OR LOWER BLACK VEN BEDS), 70 ft.

The lowest precipice of Black Ven Cliff, below the row of *Birchi* Nodules, is formed of 70 ft. of dark shales, paper-shales and marls, with occasional indurated bands of limestone and nodule-beds. They differ in appearance from the Black Ven Marls above by the presence of numerous interbedded seams of fibrous calcite with a superficial resemblance to beef.

Palaeontologically the Shales with Beef may be divided, according to Dr. Lang, into three main divisions, an upper, corresponding with the range

PLATE III

Photo. *Purch.*

Black Ven (left), Stonebarrow Cliff (centre) and Golden Cap (right) from above the Cobb, Lyme Regis.

Photo. *Purch.*

Charmouth and Stonebarrow Cliff from the foot of Church Cliffs, Lyme Regis.

The pale band is the Belemnite Marls; Black Ven Marls below, and Green Ammonite Beds above. The cliff is capped with Upper Greensand.

PLATE IV

Photo.   *W. J. A.*

Frodingham Ironstone Workings, near Scunthorpe, Lincs.
Pleistocene sands are seen above the ironstone face.

Photo.   *W. J. A.*

Blue Lias cliffs west of Lyme Regis.
The cliff shows the characteristic banding produced by alternate beds
of limestone and shale.

of *Microderoceras* and including the *Birchi* Beds; a middle division characterized by *Arietites*, *Sulciferites*, and *Arnioceras*, but no *Agassiceras*; and a
lower division with both *Arnioceras* and *Agassiceras*.

Bucklandi, Angulatum and **Planorbis Zones** (LYME REGIS BEDS OR BLUE
LIAS LIMESTONES), 85 ft.

At the foot of Black Ven Cliff, below the Shales with Beef, a 3 ft. 6 in. stone
band rises from the beach, called the Table Ledge. This is the uppermost
band of the Blue Lias, a thick series of blue-grey limestones separated by
numerous shaly partings, which build the Church Cliffs to the east of Lyme
Regis, where Monmouth drew up his army after landing in 1685, and the sea
cliffs, rocks, and ledges westward to beyond Pinhay Bay (Plate IV).

It is now generally considered that the series was originally deposited as a
calcareous clay and that the formation of the limestones is a product of some
process of rhythmic deposition of calcium carbonate, analogous with that
depositing silica to produce bands of flint in the Chalk. The limestones here
(in contrast with those near the Mendip Axis, around Radstock) seem to have
been formed subsequently to any penecontemporaneous erosion that may
have taken place.

Palaeontologically, the most conspicuous features of the Blue Lias are an
almost unbelievable multitude of large Coronicerates and other ammonites,
with nests of *Gryphœa arcuata*, *Lima gigantea* and more rarely several other
lamellibranchs. The geologist who walks from Lyme Cobb round the foot of
the cliffs to Pinhay Bay beholds probably the greatest profusion of large
ammonites to be seen anywhere in the British Isles. Some of the ledges are
completely covered, so that one can hardly cross without stepping on ammonites, and fallen blocks may show several specimens over 2 ft. in diameter.[1]
The Crinoid, *Isocrinus* (*Pentacrinus*) *tuberculatus* (Miller), occurs at several
horizons.

The most famous products of Lyme Regis, the skeletons of the great marine
reptiles and the Pterodactyls, were obtained principally from the Blue Lias.
In particular, *Ichthyosaurus communis* and *I. platyodon* were found in the
scipionianum epibole, near the top of the series. Many of the fishes, too, such
as *Acrodus*, *Eugnathus*, *Hybodus*, *Pholidophorus* and *Dapedius*, came from the
Blue Lias, although others, such as *Chondrosteus* and some species of *Dapedius*,
are said to have been found in the *obtusum* zone in the Black Ven Marls.

Interesting evidence of the southerly and westerly extension of the Lower
Lias was the dredging up of *Psiloceras planorbis* or an allied species from the
bottom of the sea below 80 metres of water, 30 miles south of the Eddystone
Lighthouse.[2]

THE PRE-PLANORBIS BEDS, 12 FT.

The basal beds of the Lower Lias continue to Culverhole Point, but they
are much slipped and tumbled, and the most westerly good sections are to be
seen in Pinhay Bay. Here the Pre-*planorbis* Beds consist of about 11 ft. of

---

[1] For a study of the ammonites of the Blue Lias see Spath, 1924, *P.G.A.*, vol. xxxv,
pp. 186–211.
[2] Kilian and Blanchet, 1923, *C.R. Acad. Sci.*, Paris, vol. clxxvi, p. 156; quoted by Spath,
1924, *P.G.A.*, vol. xxxv, p. 190.

shelly limestone in twelve bands, separated by shale partings, with 1 ft. of dark paper shale at the base.

Ostrea liassica Strickland is fairly abundant almost throughout, together with allied species of oysters, and more rarely, *Lima gigantea* and *Modiola minima*. In inland sections nearby, *Pleuromya tatei* has been found, a fossil that is highly characteristic in other parts of England. One bed near the base is locally crowded with Echinoderm spines, belonging to the species *Cidaris edwardsi*, *Pseudodiadema lobata*, *Hemipedina bechei*, and *H. bowerbanki*.

(b) West Somerset

Inland through Dorset and Somerset the Lower Lias outcrop forms broad tracts of valley country, flooring the Vale of Marshwood, the levels between Ilminster and Ilchester, the Glastonbury levels, and that larger portion of Sedgmoor lying north of the Polden Hills (King's Sedgmoor is formed by the Keuper Marls). Like the Rhætic, the Lower Lias is overstepped and largely concealed by the Cretaceous rocks in the neighbourhood of Chard.

Throughout this tract most of the quarries and cuttings expose only the Blue Lias limestones, which form a low ridge along the western margin of the outcrop, culminating in the line of the Polden Hills. The clays of the upper portion of the Lower Lias, dipping under the marshland and largely covered by alluvium, yield nothing of commercial value and never necessitate deep cuttings for railways. In consequence little is known of the development of all but the lowest two zones and the Pre-*planorbis* Beds.

The limestones of the Lyme Regis district diminish rapidly northward, being largely represented by clays with only subordinate bands of stone in North Dorset and the Polden Hills. The most complete section was exposed in a railway-cutting across the inlier at Sparkford Hill, Queen Camel, Somerset, described by Charles Moore, and the lower parts of the section are duplicated and still accessible in neighbouring quarries at Camel Hill.[1]

Alternate bands of limestone and clay to the great thickness of 97 ft., with *Psiloceras planorbis* and *Caloceras johnstoni*, represent the *planorbis* zone, and Richardson has shown that here, as in the Dorset cliffs, the epibole of the ribbed *C. johnstoni* occurs above that of *P. planorbis* in the quarry.[2] Above this over 100 ft. of clay and marls, with some limestone bands, were formerly exposed in the cutting, and assigned to the *angulatum* and *bucklandi* zones. The Pre-*planorbis* Beds consist of about 11 ft. of shale and limestone bands, but *Ostrea liassica* and *Modiola minima* do not occur in the basal 4 ft., which instead yield remains of insects and the unique Crustacean, *Coleia wilmcotensis* (Woodward). As this locality lies directly over the North Devon Axis, it is evident that at least this part of the axis was quiescent during Rhætic and early Liassic times.

At Street, south of Glastonbury,[3] the lower part of the Blue Lias and the Pre-*planorbis* Beds have been quarried for many years for building-stone and paving-slabs, and have yielded quantities of Saurian remains. The quarries show about 20 ft. of blue limestone bands, alternating with shale, with

---

[1] L. Richardson, 1911, *Q.J.G.S.*, vol. lxvii, pp. 45–9.
[2] Though in the cutting Moore recorded *P. planorbis* from the top bed.
[3] H. B. Woodward, 1893, *J.R.B.*, pp. 79–81.

Psiloceras planorbis in the highest 8 ft. Most of the Saurian remains have been obtained from the Pre-*planorbis* Beds, but they also occur in the topmost bed of the quarries. The Pre-*planorbis* Beds contain *Ostrea liassica* and *Modiola minima* in every bed, and have also yielded *Rhynchonella calcicosta*, *Cardinia crassiuscula* and *Lima punctata*; while the last, together with *L. gigantea*, *Gryphæa arcuata* and *Heterastræa latimæandroidea*, have been obtained from the *planorbis* zone.

The vertebrates that have made Street famous are *Ichthyosaurus intermedius* and *I. tenuirostris*, *Plesiosaurus etheridgei* (= *P. hawkinsi*), *P. megacephalus* and *P. macrocephalus*, and the fish, *Amblyurus*, *Dapedius*, *Leptolepis*, and *Pholidophorus*.

The only indication of the higher beds of the Lower Lias of special interest in this district is at Marston Magna, north-east of Yeovil, where there is a lenticular layer of stone, composed almost entirely of a mass of small ammonites, with the white nacreous layer well preserved, the commonest being *Promicroceras marstonense* Spath. The stone was formerly polished and sold as Ammonite Marble or MARSTON MARBLE, but it is not visible *in situ* at the present day. Supplies are thought to have been obtained from an old marl pit in the eighteenth century, and a large mass was once discovered by the sinking of a well.[1] Other species of ammonites in the Marston Marble are *Præderoceras ziphus* (Ziet.), *P. trinodum* (Dunk.), *Asteroceras smithi*, and *A. marstonense* Spath, an assemblage which Dr. Spath considers to indicate perhaps a somewhat higher horizon than the supposed equivalent on the Dorset coast.[2]

The coast-sections about Watchet[3] are discontinuous and nowhere expose beds above the zone of *Arnioceras semicostatum*, or the top of the Shales-with-Beef of the Dorset cliffs. The total thickness of the Pre-*planorbis* Beds, Blue Lias, and Shales-with-Beef-equivalent is about the same as in Dorset (130–50 ft.). The main limestone development, however, has moved upward from its position at Lyme Regis into the zones of *bucklandi* and *semicostatum*, while the *angulatum* zone consists of grey shale and marl with only occasional bands of limestone. The *planorbis* and Pre-*planorbis* Beds have together shrunk to less than 20 ft. The absence of any conglomeratic or littoral facies so close to the Devonian rocks is remarkable, and may indicate that the present proximity to those rocks is in some measure due to faulting. That the Quantock Island still stood above water during the formation of the Lower Lias seems highly probable, however, and the lack of any littoral facies close to the shore is paralleled on the southern flank of the Mendip Island, where it is especially well displayed in the Rhætic Beds of Shepton Mallet (see above, p. 104). It is probably attributable to the small size of the islands and lack of streams bearing sediments into the surrounding sea.

(c) The Continuation of the Dorset–West Somerset Area on the Coast of Glamorgan

The zonal equivalents of the Dorset and West Somerset Blue Lias and Pre-*planorbis* Beds are continued on the opposite side of the Bristol Channel in

[1] H. B. Woodward, 1893, *J.R.B.*, p. 84.
[2] L. F. Spath, 1925, *The Naturalist*, pp. 305–6.
[3] H. B. Woodward, 1893, *J.R.B.*, pp. 91–7. There are some good photographs in L. Richardson, 1914, *P.G.A.*, vol. xxv, pp. 97–102.

the Vale of Glamorgan. They are exposed in the cliffs at Penarth and Lavernock, and between Barry and Southerndown, the last a section about 14 miles in length (fig. 21). The most easterly occurrences are in the form of outliers at Penarth, Lavernock, Leckwith, and St. Fagans, where the best exposure is afforded by the cliffs at Lavernock, that at Penarth being inaccessible. The highest beds preserved in these outliers belong to the *angulatum* zone. West of Barry, on the main outcrop, the *bucklandi* zone is also present, and probably that of *semicostatum*. The Lower Lias of Glamorgan has been studied in detail by Prof. A. E. Trueman, from whose accounts the following particulars are derived.

The facies of the rock in the eastern part of Glamorgan is normal and resembles that on the Somerset coast, except that there is a great thickening of the zones up to that of *semicostatum* to 300 ft., or double their thickness in Dorset and West Somerset. There is also a noticeably greater proportion, amounting to a preponderance, of limestones.

SUMMARY OF THE LOWER LIAS OF EASTERN GLAMORGAN [1]

Semicostatum Zone (= SHALES WITH BEEF OF THE DORSET COAST).

About 50 ft. of limestones and shales with nodules are assigned tentatively to this zone, but the only ammonite they have yielded is *Arnioceras bodleyi* (J. Buckman). They form the highest part of the cliff west of the mouth of the River Daw.

Bucklandi, Angulatum and Planorbis Zones, *c.* 265 ft. (= BLUE LIAS OF DORSET).

The Blue Lias forms most of the long stretch of cliffs from Barry to Nash Point and Dunraven, but west of Nash Point towards Dunraven a littoral facies is developed, which will be dealt with in the next section.

At the top is an uncertain thickness (? 60 ft.) of shale, in the lower part of which *Gryphæa* aff. *incurva* is so abundant that the shells make up fully one-half of the rock. These shales yield *Paracoroniceras* cf. *gmuendense*. Below come about 80 ft. of limestone bands with *Coroniceras rotiforme* (2–30 ft.), *Metophioceras* cf. *conybeari* (40 ft.), and *M.* cf. *rougemonti* (15 ft.). The base of the limestones, which corresponds with the base of the *bucklandi* zone, forms a conspicuous ledge projecting from the cliff, and in it occurs the coral *Montlivaltia haimei* (abundant near Sea Mouth), together with *Lima gigantea* and *Gryphæa* aff. *obliqua* grown to a large size.

The *angulatum* zone below (about 100 ft.) is more argillaceous than the *bucklandi* zone, consisting chiefly of shales with bands of small nodules. At Lavernock, however, where the lower 40 ft. are known as the Lavernock Shales, the upper 50 ft. contain a considerable percentage of limestone bands. The ammonite succession determined by Prof. Trueman is as follows:

Scamnoceras angulatum and other spp.	*c.* 50 ft.
S. spp. and *Alsatites* spp.	*c.* 30 ft.
S. spp. and *Caloceras* spp.	*c.* ?20 ft.
Wæhnoceras spp. and *Caloceras* spp.	6ft.

The *planorbis* zone is well seen only at Lavernock. It consists of alternate

---

[1] A. E. Trueman, 1920–30; for list of papers see Bibliography.

bands of shale and hard, blue, nodular limestone, with *Ostrea liassica*, &c. *P. planorbis* is found only in the lowest 5 ft., where it is abundant, while the remaining 20 ft. are characterized by *Caloceras johnstoni* and allied species.

PRE-PLANORBIS BEDS, *c.* 20 ft.

These consist of well-bedded but thin limestone bands, with shale partings, the whole abounding in *Ostrea liassica*. In addition *Lima gigantea, Modiola, Pleuromya* and other molluscs, with Saurian bones, are recorded.

II. THE LITTORAL FACIES OF GLAMORGAN

From Dunraven Castle westward in the cliffs to Southerndown and Sutton a remarkable littoral facies is developed. The lower zones may be seen lapping in turn against a rising floor of eroded Carboniferous Limestone, and as they approach within half a mile of the old shore they pass into blue and white limestone and conglomerate. In these deposits there is little arenaceous material derived from the Upper Carboniferous rocks of the mainland, such as was poured into the sea about Bridgend in Rhætic times. It is clear that the sediment is almost entirely the waste from the Carboniferous Limestone of Cowbridge Island.

With the close of the Rhætic period a further submergence had taken place, causing the Liassic sea to flood the greater part of the island. Across the centre a channel was opened up from north to

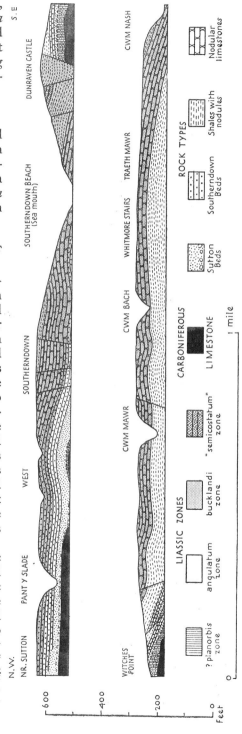

FIG. 21. Section of the Lower Lias in the cliffs from near Sutton to Nash Point, Glamorgan, showing the distribution of the zones and rock-types. From A. E. Trueman, 1922, *Proc. Geol. Assoc.*, vol. xxxiii, plate 10.

south, and Prof. Trueman considers that the remainder was separated into four islets, upon which the sea continued to encroach during Liassic times. On the line of section cut by the modern cliffs of Southerndown and Sutton, a portion of the island still remained above the surface during the *planorbis* secule, but the water passed over it about the middle of the *angulatum* secule, and all through late *angulatum* and *bucklandi* times it was submerged beneath the sea.

The littoral deposits are of two kinds: above, are dark blue limestones, weathering dull brown, irregularly bedded, sometimes feebly nodular, with seams of shale and bands of small Carboniferous Limestone pebbles; this type is known as the SOUTHERNDOWN SERIES. It merges downwards into the second type, a white or pale cream, massive limestone, frequently conglomeratic at the base, called the SUTTON STONE.

The origin and correlation of these two types of deposits have given rise to a great deal of discussion in the past, but their significance and relations have now been elucidated by Prof. Trueman.[1] By more detailed zonal work, he has shown that each zone in turn passes laterally first into 'Southerndown Beds' and then into 'Sutton Stone' as it approaches its own shore-line. Consequently, at any given point, the Sutton Stone always underlies the Southerndown Series, which in turn grades up into normal limestones and shales (fig. 22). Thus from Dunraven to Pant-y-Slade and Sutton, the *planorbis* zone is seen only in the form of Sutton Stone, and thins out against the Carboniferous floor before Pant-y-Slade is reached. Where it is banked against steep-sided Carboniferous Limestone, the conglomerates are thick and some of the boulders are several feet in diameter. The *angulatum* zone resting on it consists of normal shales-with-nodules in the east, passing into Southerndown Beds farther west, and finally into Sutton Stone at Pant-y-Slade. The *bucklandi* zone overlapped so much farther west that in the upper part the normal limestones and shales extend almost to Pant-y-Slade, but the greater part of the zone is represented there by Southerndown Beds. The Sutton Stone facies of the *bucklandi* zone, which presumably lies underground farther inland, is not seen.

Petrologically, the Southerndown Series consists of angular and subangular fragments and pebbles, chiefly of *Caninia* Oolite and chert in a hardened calcareous mud. The Sutton Stone matrix varies from hard crystalline calcite to limestone formed of ground and rounded shell-fragments. In places it is shelly and may contain abundant corals, chiefly broken.

In the channel between the Cowbridge Island and the mainland of the South Wales Coal-field, the Lower Lias is chiefly of normal shale and limestone facies. About Bridgend this condition stands in marked contrast with the sandstone facies of the Rhætic. In Liassic times sediment seems to have been derived no longer from rainfall on the mainland, but almost entirely from solution and erosion of the Carboniferous Limestone islands and comminution of the shells and corals on their shores.

Around these islands the water was sufficiently free from the prevalent mud-bearing currents to allow a rich coral colony to thrive. In the limited number of exposures no evidence has been seen that the corals formed actual reefs on the shores of the islands, but in a quarry at Brocastle, now overgrown,

[1] A. E. Trueman, 1922, *P.G.A.*, vol. xxxiii, pp. 245–84.

Charles Moore described the corals as growing upon boulders and encrusting cracks in the Carboniferous Limestone sea-bed.

This quarry showed Sutton Stone facies of the upper part of the *angulatum* zone or possibly the *bucklandi* zone, almost a solid mass of organic remains, resting in hollows in the Carboniferous Limestone. It was vividly described by Duncan, to whom we owe our knowledge of the corals:

'Some of the fossils are perfect even in their most delicate ornamentation; others are worn, having been rolled; and myriads are in fragments. The smaller fossils

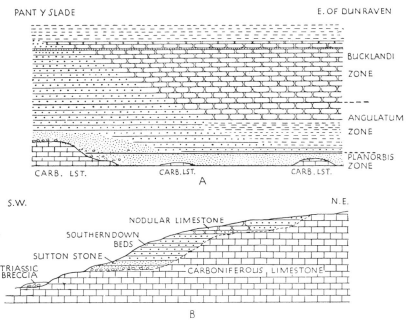

FIG. 22. Diagrams showing the relations of the littoral deposits in the Lower Lias near Sutton and Dunraven and at Pant-y-Slade to each other and to the Carboniferous Limestone. From A. E. Trueman, 1922, *Proc. Geol. Assoc.*, vol. xxxiii, figs. 60, 61. (The Conventional shadings for the different facies of the Lias are explained on fig. 21.)

stud the blocks, and consist for the most part of *Madreporaria* and *Pentacrinites*, of *Cidaris* spines and plates, and of fragments of large and small Lamellibranchiata; and with these are mixed the shells of tiny Gastropoda. The larger fossils consist of perfect spheres of *Isastræa globosa* Dunc. (and fragments of them), of blocks more or less gibbous of another *Isastræa*, of flat or dendroid pieces of *Astrocænia*, and of more or less fragmentary *Thecosmiliæ* and *Montlivaltiæ*; and amidst these are more or less perfect *Cerithia*, *Turritellæ*, large *Pleurotomariæ*, *Straparolli*, *Neritopsides*, many rugged *Ostreæ*, and more or less perfect *Limæ*. There are also Polyzoa and *Serpulæ*. All are mingled together, and here and there are some remanié species of *Syringopora*, *Amplexus*, *Cyathophyllum*, and *Lithostrotion*.'[1]

Long lists of mollusca were published by the early geologists from the Sutton Stone of Glamorgan, but since many of the identifications are doubtful

[1] P. M. Duncan, 1867, *Q.J.G.S.*, vol. xxiii, p. 14.

and the nomenclature is in need of revision, little purpose would be served by repeating them. The subject is one which would repay attention by a specialist. The corals are extraordinarily abundant for the Lias, in which corals are known but rarely in other parts of England.

III. THE MENDIP AXIS

The neighbourhood of the Mendips remained a region of intermittent disturbance after the original anticlinal uplift in Armorican times, until well into the Mesozoic. Not only the Rhætic Beds, as we have seen in the last chapter, but also the Lias and Inferior Oolite thin out and pass into semi-littoral deposits, overlapping on to the Triassic marls and even the Carboniferous and Devonian rocks as they approach the flanks of the hills.

A curious semi-littoral facies is seen in the neighbourhood of Chewton Mendip, on Harptree and Egar Hills, and near Emborrow and Binegar.[1] Here the basal zone of the Lias with *P. planorbis* and *C. johnstoni* and other fossils appears as a brown, grey or white chert, resting on ochreous sand with seams of clay, from which ochre was formerly obtained. The total thickness is nearly 30 ft., but the lower parts of the series, at least in places, are of Rhætic age. The siliceous condition of the rock is thought to have been brought about by the infiltration of hot silica-bearing water from some subterranean intrusion, for nearby, the Lower Lias maintains its normal characters of argillaceous limestones and clays, with an oyster-bed of *Ostrea liassica* in a sandy limestone at the base.

On Broadfield Down also, the lower zones of the Lias overlap the Rhætic and rest on the Carboniferous Limestone, where they take on a Sutton Stone facies as in Glamorgan.

At a slightly greater distance from the old shore-lines, about Radstock, Paulton and Timsbury, the Lower Lias is of more normal appearance, but all the zones except the uppermost are much condensed, while some contain beds of phosphatic nodules together with fossils derived from lower zones; others are absent altogether.

The long-continued researches of Mr. J. W. Tutcher of Bristol and Prof. A. E. Trueman have shown that, in contrast to the southern side of the Mendips, where deposition continued steadily throughout Liassic times, the northern flanks were an area of continued submarine disturbance, of erosion and redeposition. The shallow seas were particularly favourable to ammonite growth, and the wealth of their entombed shells has enabled the history of events to be made out in considerable detail.

The downward limit of the Lias is easily defined hereabouts by the Sun Bed at the top of the Langport Beds. The upward limit is usually the base of the Inferior Oolite or, in places, up to 9 ft. of sands, marl and ironshot limestone belonging to the Upper Lias.[2]

The total thickness of the Lower Lias in the Radstock district is very variable. The maximum may be 190 ft., but of this 120 ft. belong to the *davœi* zone, represented by normal clays, while the whole of the remaining zones are compressed into from 20 to 25 ft. of strata.

---

[1] H. B. Woodward, 1893, *J.R.B.*, p. 123.
[2] The Middle Lias seems to be absent altogether.

SUMMARY OF THE LOWER LIAS OF THE RADSTOCK DISTRICT [1]

Davœi Zone (= GREEN AMMONITE BEDS OF DORSET).

This zone is exposed at only a few places, notably Clandown Colliery Quarry, Radstock, and Huish Colliery Quarry, where 8 ft. of grey-green shaly clay, with rare specimens of *Liparoceras*, are exposed, immediately overlain by the Inferior Oolite. At other places, however, where exposures are lacking, there were synclines in which the clays were spared by the Bajocian Denudation, and their thickness is estimated at 120 ft.

Ibex and **Jamesoni Zones** (= BELEMNITE MARLS OF DORSET).

The Belemnite Marls of Dorset are represented by from 6 ft. to 10 ft. of limestone, in which belemnites are much less abundant, although still common. At the top there is occasionally a hard, splintery bed with *Tragophylloceras ibex* and *Acanthopleuroceras valdani*. Below this is the main *Jamesoni* Limestone; and at the base an '*Armatum*' Bed, ironshot and rubbly, and containing numerous derived ammonities and phosphatic nodules in the lower part. The JAMESONI LIMESTONE is uniform over a wide area and shows no signs of having undergone erosion. North of Timsbury Sleight and Paulton it passes into normal clays. The '*Armatum*' Bed, however, is absent from Timsbury to the neighbourhood of Dundry, having been apparently removed from the crest of an anticlinal uplift.

Raricostatum, Oxynotum and **Obtusum Zones** (= BLACK VEN MARLS OF DORSET).

This division is the most attenuated of all, the *raricostatum* zone consisting of from an inch to a foot of clay, the *obtusum* zone of a line of derived fossils and phosphatic nodules, and the *oxynotum* zone being absent altogether. The *Obtusum* Nodule Bed rests on various lower zones in different places and contains fossils derived from their denudation; it is, in fact, a typical remanié bed. The denudation marked by the Nodule Bed seems to have followed general uplift along the Mendip Axis. Movements continued during the *raricostatum* secule, causing great restriction of deposition: the fossils were exhumed and reinterred repeatedly, so that the true sequence is rarely discernible.

Semicostatum, Bucklandi, Angulatum and **Planorbis Zones** (= SHALES WITH BEEF AND BLUE LIAS OF DORSET).

The upper part of the *angulatum* zone and lower part of the *bucklandi* zone are absent, but the rest of the Blue Lias is rather less attenuated than some of the higher divisions. The Pre-*Planorbis* Beds and the *planorbis* zone are normally developed, consisting of from 2 to 30 ft. of limestone bands with shale partings, crowded with *Ostrea liassica*. Similar conditions seem to have continued into the earlier part of the *angulatum* secule. Deposition then ceased in the middle of the *angulatum* secule and a series of E.–W. folds, with a slight eastward pitch, came into being to the north of the Mendips and parallel with the main axis, gradually decreasing in intensity away from the hills. The crests of the uplifts suffered penecontemporaneous erosion, until,

[1] J. W. Tutcher, 1917, *Q.J.G.S.*, vol. lxxiii, pp. 278–81; and J. W. Tutcher and A. E. Trueman, 1925, ibid., vol. lxxxi, pp. 595–666.

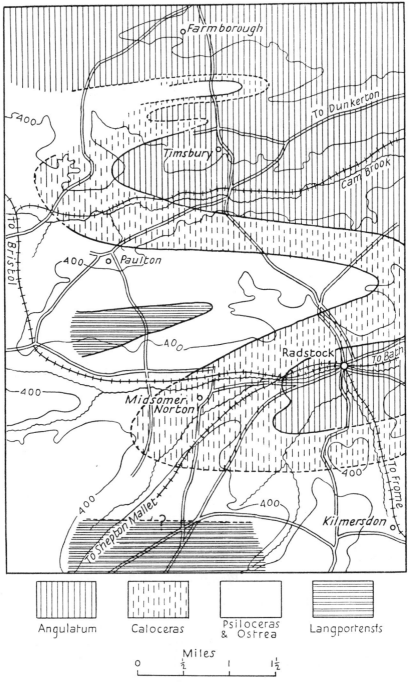

Angulatum   **Caloceras**   **Psiloceras & Ostrea**   **Langportensis**

Miles

0   ½   1   1½

FIG. 23. Sketch-map showing approximately the age of the strata upon which the *Bucklandi* Bed was deposited in the Radstock district, after the Sinemurian Denudation. (From Tutcher and Trueman, 1925, loc. cit., fig. 11 combined with fig. 2.) (Compare fig. 36, p. 199.)

towards the end of the *bucklandi* secule (in the hemera of the genus *Euagassi-ceras*), a general subsidence brought about the renewal of deposition. The erosion of the anticlinal crests has been termed by Tutcher and Trueman the Sinemurian Denudation (see figs. 23, 24).

The first stratum to be deposited across the eroded crests of the anticlines and in the synclines was a thin remanié-bed from a few inches to a foot thick, known as the *Bucklandi* Bed from its containing abundant Coroniceratids; over this was laid a layer of the large brachiopod, *Spiriferina walcotti*. The brachiopods are in fine preservation and are so abundant that several of the

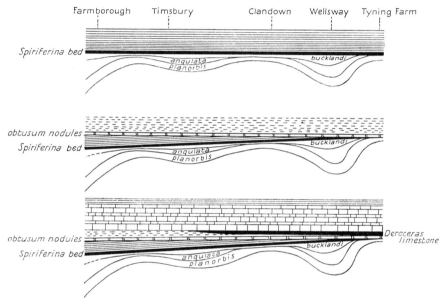

Fig. 24. Diagrams illustrating successive stages in the deposition of the Lower Lias near Radstock. (Compare fig. 26.) After Tutcher and Trueman, 1925, loc. cit., fig. 10.

exposures are famous for the products of the *Spiriferina* Bed. Finally, the *Spiriferina* Bed was sealed by up to 5 ft. of clay (the *Turneri* Clay). Signs that erosion still continued during the deposition of this clay are evident from the inclusion of partly fossilized and broken shells.

The Lower Lias history of the shallow and unstable region to the north of the Mendips may be summarized as follows: while deposition of the earlier zones was in progress the sea-bed was thrown into a series of E.–W. folds, decreasing in intensity northward, away from the main axis of the Mendip Hills. Repeated partial denudation of the crests of the folds resulted in the overlapping by the *Bucklandi* Bed and *Spiriferina* Bed of all the underlying zones down to the White Lias about Radstock, the Sinemurian deposits coming to be best preserved in a shallow syncline around Keynsham. Later, a general uplift along the Mendip Axis caused renewed denudation of the *Turneri* Clay in the south and the formation of the *Obtusum* Nodule Bed; after which the later zones were laid down conformably upon a sea-bed that had at last reached temporary stability.

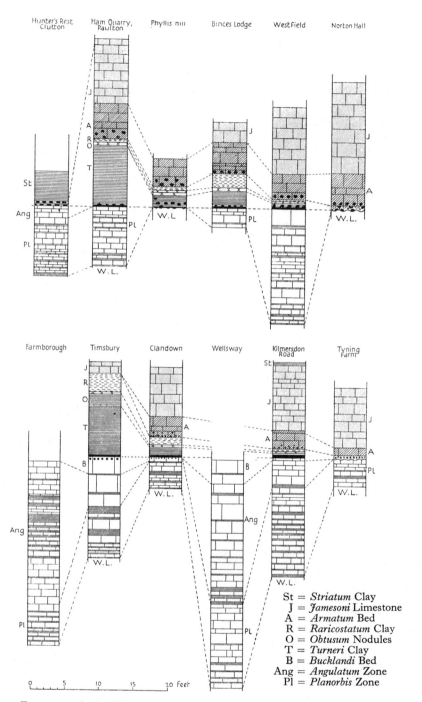

Hunter's Rest, Clutton   Ham Quarry, Paulton   Phyllis Hill   Binces Lodge   Westfield   Norton Hall

Farmborough   Timsbury   Clandown   Wellsway   Kilmersdon Road   Tyning Farm

St = *Striatum* Clay
J = *Jamesoni* Limestone
A = *Armatum* Bed
R = *Raricostatum* Clay
O = *Obtusum* Nodules
T = *Turneri* Clay
B = *Bucklandi* Bed
Ang = *Angulatum* Zone
Pl = *Planorbis* Zone

0   5   10   15   20 feet

FIGS. 25 and 26. Diagrams representing exposures of Lower Lias in the Radstock district, along two lines running nearly due north and south; the upper line west of the lower. From J. W. Tutcher and A. E. Trueman, 1925, *Quart. Journ. Geol. Soc.*, vol. lxxxi, pp. 600, 612.

IV. THE MIDLAND PLAINS
(a) From the Severn to the Witham at Lincoln

North of Bath the outcrop grows narrower, until between Wotton under Edge and Berkeley it occupies a strip less than a mile wide between the Cotswold escarpment and the Lower Palaeozoic rocks of the Tortworth inlier. Little is known of the stratigraphy in this region, but the Lower Lias seems to be considerably diminished in thickness over the buried continuation of the Malvern Axis (see fig. 12, p. 68).

Then in the Vale of Berkeley below Stroud, where the Lower Oolites turn north-eastward and the Palaeozoic outcrop continues due north, across the Severn to May Hill and the Malverns, the Lower Lias and Trias tracts widen rapidly.

The Lower Lias forms the broad and fertile Vale of Gloucester, merging northward into the fruit district of the Vale of Evesham and Worcestershire. Long tongues reach northward as far as Droitwich and Henley in Arden, and are continued in a small outlier at Knowle, almost on a line between Birmingham and Coventry, 24 miles north of the farthest projection of the North Cotswolds.

As might be expected, such widening of the outcrop is commensurate with great expansion in thickness. A deep boring at Mickleton in the Vale of Evesham proved a thickness of 960 ft. of Lower Lias; but this is probably exceptional. Another boring at Batsford (Lower Lemington) only 7 miles away, in the Vale of Moreton, proved only some 500 ft. and there is attenuation southward and eastward, though something approaching the thickness at Mickleton may be maintained near the northern margin of the outcrop in Warwickshire and Worcestershire.

Beyond, the formation continues to form wide clay vales and spreads over much of the Midland Plain, by way of Rugby, Leicester, Melton Mowbray, and the Vale of Belvoir to the Witham Valley in Lincolnshire, where the outcrop is much obscured by alluvium. A general attenuation sets in north of Rugby, until of 750 ft.–650 ft. in Leicestershire and the borders of Lincolnshire, only 300 ft. remains at the Humber.

South and east of the main outcrop, narrow feelers of Lower Lias invade the Oolite hill-masses along the valleys of the Rivers Evenlode, Cherwell, Nene, and Welland. Still farther down the dip-slope, under the covering of oolitic rocks, the Lower Lias has been pierced in deep borings and found to thin out to 450 ft. at Burford Signett, 460 ft. under parts of Northamptonshire, and 228 ft. at Claydon (Calvert), on the Oxford Clay outcrop south of Buckingham. This thinning was proved to be accompanied by overlapping of the higher zones over the lower, for in the Calvert boring the *jamesoni* zone was found to be resting directly on the Palaeozoic floor of Shineton Shales.[1]

Throughout the long tract from the Severn to the Witham there are few features meriting special attention. The great thickening in Gloucestershire and Worcestershire is mainly due to the expansion of the clays, but farther north the Blue Lias limestones come in again in force. These last, known in this part of the country as the Hydraulic Limestones, are exposed in numerous

---

[1] A. M. Davies and J. Pringle, 1913, *Q.J.G.S.*, vol. lxix, pp. 311–14.

lime and cement works, but the overlying clays are proportionately as rarely exposed as in West Somerset, and for the same reasons.

The full succession of zones has been elucidated in several districts, notably about Cheltenham, Gloucester and the Vale of Evesham, in Northampton-shire, and in South Lincolnshire. The epiboles detected in the western district are appended in the table on page 28, compiled from the works of Prof. Trueman and Miss D. M. Williams, largely from collections made by Mr. L. Richardson. A comparison of the table with that portraying the succession in the Radstock district shows that, although the *raricostatum* zone is represented in both districts, the faunas contained in it are largely different.

Probably the best section of the clays of the upper zones is to be seen in the Battledown Brickworks, Cheltenham, where the *ibex* and the lower half of the *davœi* zones can be studied. The section is continued into the upper half of the *davœi* zone by a brickworks at Aston Magna.[1] Knowledge of the great mass of the central parts of the Lower Lias in this region, however, is derived from small temporary openings and old railway-cuttings now overgrown.

In Warwickshire, Northamptonshire and South Lincolnshire all the zones have been recognized in various brickyards, cuttings and boreholes, but the epiboles still remain to be worked out in detail. The most remarkable addition to the palaeontological succession was a blue clay abounding in Rhynchonel-lids and the ammonite *Cœloceras pettos*, encountered in a cutting on the rail-way near Flecknoe.[2] The position of this bed is between the *jamesoni* and *ibex* zones, and Mr. Beeby Thompson has suggested raising it to the rank of a new zone. The ammonite has been found also in the Dorset cliffs, where its position has been fixed by Dr. Lang in his table of epiboles in the *jamesoni* zone.

Another interesting feature, formerly seen in a railway-cutting 1 mile south of Rugby, was an exposure, 25 ft. deep, of the *oxynotum* zone, yielding over a hundred species of fossils, especially great numbers of *Hippopodium pon-derosum* and *Dentalium etalense*.[3] The true *Oxynoticeras oxynotum*, which does not occur on the Dorset coast, was here found in hundreds, associated with *O. biferum*.

Occasionally large corals of the genus *Montlivaltia* occur at various levels— *M. rugosa* forms a thin coral bed in the *raricostatum* zone of Marle Hill Brick-yard, Cheltenham; *M. victoriæ* occurs in a hard band in the *davœi* zone, formerly polished and sold as Banbury Marble at Banbury; *M. mucronata* has been found in the *jamesoni* zone near Wolfhamcote, Northants; *M. mucronata* and *M. haimei* were found in the tunnel between Old Dalby and Saxelby on the railway from Melton Mowbray to Nottingham, which passed through the *oxynotum, raricostatum, jamesoni, ibex,* and possibly *davœi* zones.[4]

The lower zones of the Lower Lias are well known owing to the extensive exploitation of the Hydraulic Limestones for lime and cement.

The best inland exposure in England is at Victoria Quarry, Rugby,[5] where the limestones are 70 ft. thick (about the same thickness as in Dorset), con-

[1] L. Richardson, 1929, 'Geol. Moreton in Marsh', *Mem. Geol. Surv.*, pp. 12–15.
[2] B. Thompson, 1910, 'Geol. in the Field', p. 456; and 1899, *Q.J.G.S.*, vol. lv, pp. 65–88.
[3] B. Thompson, 1910, loc. cit., p. 455.
[4] For list of fossils collected on the tip heaps thrown out in the making of the tunnel see Trueman, *Geol. Mag.*, 1918, pp. 101–2.
[5] H. B. Woodward, 1893, *J.R.B.*, p. 163.

sisting of upwards of 35 bands of limestone with shale partings. The ammonites shew that the *planorbis* (*johnstoni* subzone), *angulatum* and *bucklandi* zones are represented, and as usual Saurian and fish remains are present (fragments of *Ichthyosaurus*, *Plesiosaurus*, &c.).

In Nottinghamshire and South Lincolnshire the Hydraulic or Blue Lias Limestones are about 25 ft. thick, and seem to include only the Pre-*planorbis*, *planorbis*, and *angulatum* zones. The Pre-*planorbis* deposits consist largely of fine-grained calcitic mudstones with bands of Ostracods and Foraminifera, suggestive of formation in lagoons.[1]

In the part of Warwickshire lying over the Sedgley–Lickey Axis, about Binton, Grafton, Wilmcote, and Bickmarsh, the Lower Lias has cut down through the Langport Beds of the Rhætic, for it has at the base a conglomeratic layer known as the Guinea Bed, which contains a Lower Lias fauna with an admixture of derived Rhætic fossils, and over the area of its occurrence the Langport Beds are absent. In the region of the Nuneaton Axis also, a mile or two south and west of Rugby (at Church Lawford) the Pre-*planorbis* Beds are absent and the *planorbis* zone rests non-sequentially on an eroded surface of the Langport Beds.

(b) The Shropshire and Cheshire Outlier[2]

The former continuation of the Lias far to the north of its present outcrop in the Midlands and West of England is proved by a basin-shaped outlier of Lower and Middle Lias in North Shropshire and South Cheshire. The outlier measures some 10 miles in length and about 3 to 4½ miles in breadth and, although largely Drift-covered, it rises into prominent hills between Whitchurch and Ightfield and at Prees, the highest part being capped with Middle Lias Marlstone.

The Lower Lias is only exposed meagrely in the sides of small brooks, and the information is mainly derived from borings. Both lithologically and palaeontologically, it is evidently a continuation of the main outcrop of Gloucestershire and Worcestershire, consisting for the most part of finely laminated shale, with occasional hard calcareous mudstone bands and layers of cementstone nodules. The total thickness is probably more than 400 ft. A typical Lower Lias fauna has been obtained from the borings, and the ammonites prove the presence of the *planorbis*, *angulatum*, *bucklandi*, *semicostatum* and *jamesoni* zones. Near Burley Dam the shales are hard enough to have been used in the past for roofing-slates.

The presence of Rhætic Beds is strongly suggested by a record of *Protocardia rhætica*, but no other evidence has been obtained.

(c) Lincoln to the Humber: the Frodingham Ironstone District

On the main outcrop north of Lincoln the Lower Lias continues to form low-lying ground, largely covered by Boulder Clay, marshland and alluvium, as far as the village of Scotter. North of this, for the last 16 miles to the Humber, the width of the outcrop averages 4 miles, and the limestones of the *angulatum* and *bucklandi* zones form an increasingly prominent ridge running

---

[1] A. E. Trueman, 1915, *Geol. Mag.* [6], vol. ii, pp. 150–2; and 1918, ibid., vol. v, pp. 64–73, 101–3.
[2] H. B. Woodward, 1893, *J.R.B.*, pp. 180–3.

S.–N. between the Oolite escarpment on the east and the Trent Valley on the west.

The tract east of this ridge, separating it from the Oolites, is of exceptional interest owing to the development of a ferruginous facies of the upper part of the *bucklandi* and perhaps also part of the *semicostatum* zones. The well-known Lincolnshire or FRODINGHAM IRONSTONE, which has given rise to the modern industrial town of Scunthorpe, has been quarried from large areas, leaving vast shallow pits, now cultivated or full of water; and active quarrying by modern methods directly into railway trucks is still carried on along numerous working faces ramifying out for several miles both to north and south of the smelting furnaces (Plate IV).

The total thickness of Lower Lias in the ironfield is just over 400 ft., and it is usually divided into three groups:[1]

> Upper Clays (maximum 200 ft.).
> Frodingham Ironstone (25–30 ft.).
> Lower Clays and Limestones (200 ft.).

Davœi-Obtusum Zones (THE UPPER CLAYS).

The uppermost 66 ft. of clays contain *Androgynoceras capricornum* throughout and therefore belong to the *davœi* zone. Below this is a 4 ft. band of ironstone known as the *Pecten* Bed from the abundance of *P. æquivalvis* and *P. lunularis*, together with other shells such as *Cardinia listeri*, *C. hybrida*, *Pseudotrapezium intermedium*, *Gryphæa* spp., *Lima hermanni*, *Modiola scalprum*, and brachiopods.

Between the *Pecten* Bed and the Frodingham Ironstone, the clays were proved by borings to thin from south to north from 140 ft. south-west of Kirton Lindsey Railway Station to 90 ft. south of Appleby Railway Station.

The records of ammonites from the *Pecten* Bed and the clay below it date from a time when identifications were made with little refinement, and they need revision by careful collecting. Cross[2] (who is followed by the Survey) stated that the *Pecten* Bed was full of *Ammonites striatus* and *A. henleyi* (both *davœi* zone) and also *A. armatus*. Either the last is a misidentification, or the records from the underlying clay are faulty, since they include *A. henleyi*, *A. [Androgynoceras] latæcosta*, *A. capricornum*, *A. [Tragophylloceras] loscombi* (all *davœi* zone), *A. [Phricodoceras] taylori* (base of *jamesoni* zone) and *A. [Echioceras] raricostatum*. From these records it would seem that there is a great expansion of the *davœi* zone, which includes the *Pecten* Bed, all of the clays above it and part of those below it. An *Oxynoticeras*, however, occurs in the clays not far above the main ironstone.

Bucklandi Zone (Pars) and Semicostatum Zone (THE FRODINGHAM IRONSTONE).

The ironstone is best described as a ferruginous oolitic limestone. The iron is originally present in the form of carbonate, but towards the surface it has been weathered to hydrated oxide. A quantity of iron silicate is also present and colours the Cardinias and other shells a bright green. Weathering is

---

[1] Based mainly on W. A. E. Ussher, &c., 1890, 'Geol. N. Lincs.', *Mem. Geol. Surv.*, pp. 12–50.
[2] 1875, *Q.J.G.S.*, vol. xxxi, pp. 118–20.

accompanied by obliteration of the oolitic structure, until on the removal of all the carbonate the rock becomes soft, incoherent, and purplish-brown in colour. The average iron-content is only 21·8–22·7 per cent. and the lime-content as high as 18·2 per cent. on account of the prevalence of bands of shells.[1] The most conspicuous are *Cardinia* of several species, *Gryphæa* cf. *cymbium, Lima gigantea* and *L. hermanni,* together with *Pectens, Pholadomya ambigua* and *Spiriferina walcotti,* which all occur in perfect preservation.

Large specimens of a *Coroniceras,* probably *Paracoroniceras gmuendense,* occur in the lowest 5 ft. of the ironstone, while from the higher parts are obtained an ammonite which has usually been known as *A. semicostatus.* Various species of *Arnioceras,* however, are common at several horizons in the *bucklandi* zone, and Dr. Spath avers that the species recorded as *A. semicostatus* has been misidentified, the true *Arnioceras semicostatum* (Y. and B.) belonging to a higher level; and consequently that the true horizon of the Frodingham Ironstone is about *gmuendense-Ætomoceras-Agassiceras* epiboles[2] (i.e. *bucklandi* zone). Support is given to this view by the old record from the ironstone of *Metophioceras conybeari,* which may be seen by reference to Dr. Lang's sequence on the Dorset coast (p. 28) to occur below *Paracoroniceras gmuendense,* at the very base of the *bucklandi* zone.

Bucklandi, Angulatum, Planorbis and Pre-Planorbis Zones (THE LOWER CLAYS AND LIMESTONES).

Two hundred feet of grey and blue, compact, often nodular, limestones, interstratified with shales, build the escarpment west of the ironstone workings, overlooking the Valley of the Trent. Certain bands contain thousands of *Gryphæa arcuata,* while *Lima gigantea, L. hermanni* and *Cardinia listeri* are conspicuous. The ammonites recorded are *Coroniceras bucklandi, Metophioceras conybeari* (*bucklandi* zone), *Scamnoceras angulatum* (*angulatum* zone) and *Caloceras johnstoni* (*planorbis* zone); but *Psiloceras planorbis* is, according to Cross, entirely absent. The lowest 20 ft. contain the fauna of the Pre-*planorbis* Beds—*Ostrea liassica, Pleuromya crowcombeia, Modiola minima* and *Ichthyosaurus.*

V. THE MARKET WEIGHTON AXIS[3]

Towards the Humber the Lower Lias escarpment diminishes in importance, and the ferruginous facies which caused so many sections to be opened in the higher beds is not seen again. After disappearing under the Humber and the alluvial tracts on either side, the Lower Lias reappears at North Cave, and thence to Market Weighton the limestones and clays of the Pre-*planorbis* and *Planorbis* Beds form a small but definite escarpment, rising abruptly above the Triassic plain.

Several well-known sections formerly existed along this escarpment, but they are now all more or less obscured. The best was at the village of North Cliff,[4] where a lane cutting and a pit for marling the sandy land to the west showed 55 ft. of *Planorbis* and Pre-*planorbis* Beds. The zonal ammonite (flattened) was found in a blue clay 18–27 ft. from the base, while the 28 ft.

[1] C. B. Wedd, 1920, Spec. Repts. Min. Resources, vol. xii, pp. 71–105, *Mem. Geol. Surv.*
[2] L. F. Spath, 1922, *Geol. Mag.,* vol. lix, p. 171.
[3] C. Fox-Strangways, 1892, *J.R.B.,* pp. 67–70.
[4] Detailed descriptions in J. F. Blake, *Q.J.G.S.,* vol. xxviii, pp. 132–46.

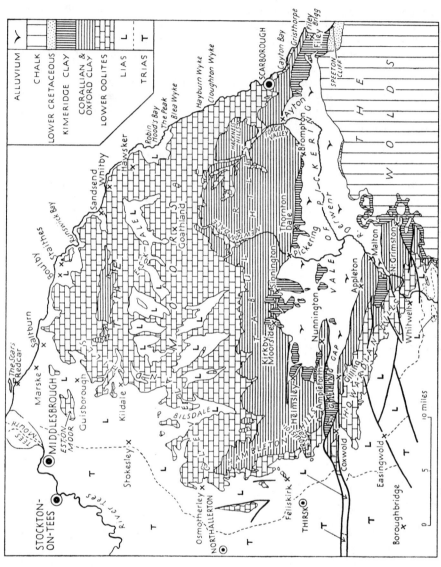

FIG. 27. Sketch-map of the Yorkshire Basin, to show the distribution of the Jurassic rocks. The thick lines are faults.

of alternating clay and limestone bands above yielded *Caloceras johnstoni* at several horizons, with *Lima gigantea, Protocardia philippiana, Modiola minima, Ostrea liassica* and other fossils. The Pre-*planorbis* Beds were also typical and a continuation of those south of the Humber.

At Market Weighton the Upper Cretaceous comes to rest for some miles directly on the Lower Lias. Then for 10 miles northward the Lower Lias forms a zone of slips along the base of the Chalk Wolds, and is much obscured by fallen debris. Unfortunately most of the scanty exposures are in the lower beds, so that it is not known how many of the higher zones are missing over the axis, or whether they are represented in attenuated form. The total thickness diminishes to 150 ft. and in places even to 100 ft.

VI. THE YORKSHIRE BASIN

North of the River Derwent the outcrop turns nearly due west, to sweep in a great arc of a circle round the oolitic rocks of the Yorkshire Basin, past Thirsk, Northallerton and Stokesley, to the sea south of Tees Mouth (fig. 27). Throughout the first half of this arc, as far as the neighbourhood of Northallerton (about 30 miles) the Lower Lias is virtually unknown by reason of the dearth of exposures and a thick mantle of Drift. A considerable northerly thickening sets in, and the limestone facies of the lower zones is less evident when the beds are next exposed about Foxton.

At the foot of the escarpment of Cringley and Carlton Moor shales with ironstone nodules, containing ammonites of the *davœi* zone, are seen to a thickness of 150 ft., while near Carlton evidence has been obtained of the *raricostatum, oxynotum* and *bucklandi* zones, all represented by shales and marls.

On the north side of the Guisborough valley, south of Eston Hill, a boring proved 426 ft. of Lower Lias without reaching the bottom. North of this the whole outcrop is again badly hidden by Boulder Clay, and there are no good sections until the coast is reached.[1]

On the coast, the lowest Lias seen is the upper part of the *angulatum* zone, which, together with the *bucklandi* zone, forms the island-like rocks at Redcar, known as The Scars. From Redcar to Saltburn the cliffs show only Drift, but the upper part of the Lower Lias reappears to form the lower part of the cliff and foreshore from Saltburn to Colburn Nab, near Staithes.

The best section of all, however, is to be seen below the Boulder Clay in the lower part of the cliffs of Robin Hood's Bay, between Whitby and Scarborough (fig. 27), and a description of this serves as a good summary of the Lower Lias of North Yorkshire. All the zones down to that of *semicostatum* here travel in a series of arch-like curves round the bay, brought up above sea-level by an anticline, which is cut across almost at right-angles by the shore.

SUMMARY OF THE LOWER LIAS AT ROBIN HOOD'S BAY AND REDCAR[2]

Davœi Zone (= GREEN AMMONITE BEDS OF DORSET), 155 ft.

The *davœi* zone consists of a nearly uniform series of soft shale, for the most part sandy, with numerous rows of clay-ironstone doggers and pyritous

---

[1] C. Fox-Strangways, 1892, *J.R.B.*, pp. 63–4.

[2] Based on C. Fox-Strangways, G. Barrow, and S. S. Buckman, 1915, 'Geol. Country between Whitby and Scarborough', 2nd ed., pp. 7–10, 66–71; *Mem. Geol. Surv.*

nodules, often containing ammonites. The highest 30 ft. is rather sandier than the rest and was formerly grouped on lithological grounds with the Sandy Series of the Middle Lias. At the top is a $4\frac{1}{2}$ ft. band of hard ferruginous sandstone, overlain by a band of ferruginous doggers. The sandstone contains layers of fossils, notably ammonites of the genus *Oistoceras*, with *Protocardia truncata*, *Gryphæa cymbium*, and abundant *Dentalium giganteum* Phil.

The only other feature is an oyster bed, about 30 ft. from the top. This consists of thin sandy laminæ ($1\frac{1}{2}$ ft.) passing into a harder calcareous band composed largely of *Gryphæa cymbium* and *Oxytoma inæquivalvis*, accompanied by *Oistoceras* spp. and other fossils.

Prodactylioceras davœi has not been found in Yorkshire, but the species of *Oistoceras* and *Lytoceras* serve for correlation of these sandy shales with the upper part of the Green Ammonite Beds of Dorset.

The lowest $35\frac{1}{2}$ ft. of sandy shale with bands of doggers yield *Androgynoceras* spp., which characterize the lower stone band and surrounding clays in the Dorset Green Ammonite Beds. Buckman regarded them as of *striatum* date.

Ibex and Jamesoni Zones (= BELEMNITE MARLS OF DORSET), 100 ft.

These zones also consist of soft shales with bands of ironstone and pyritous doggers, with no lithological line of separation from the *davœi* zone above. Belemnites are abundant at certain horizons, but no recent work comparable with that by Dr. Lang in Dorset has been done on them.

The upper half of the series yields ammonites of the genera *Platypleuroceras* (*P. aureum* and *P. rotundum*) and *Polymorphites* (*P. trivialis*), which Dr. Lang has found in Dorset in the *jamesoni* zone, below *Uptonia jamesoni*. It would thus seem that the *ibex* and the upper part of the *jamesoni* zones are unrepresented or unfossiliferous.[1]

The lowest 10 ft. of the series yields *Phricodoceras taylori*, which marks the base of the *jamesoni* zone in Dorset.

Raricostatum, Oxynotum and Obtusum Zones (= BLACK VEN MARLS OF DORSET), 175 ft.

These consist, like the higher zones, principally of soft shales with lines of nodules, but in the lower part indurated calcareous bands are intercalated, and some of the shale becomes harder.

The highest 65 ft. contain abundant *Apoderoceras*, together with occasional *Echioceras*, and comprise the old 'armatum zone'—a series of horizons higher than those formerly called by the same name in Dorset, since these belong above and those in Dorset below the horizon of *Echioceras raricostatum*. The *raricostatum* horizon of Robin Hood's Bay is about 100 ft. from the top, 2 ft. above a conspicuous datum-line called the Double Band. Immediately above the two hard beds which comprise the Double Band is a horizon of Oxynoticerates, indicating the *oxynotum* zone. The *raricostatum* zone is therefore over 100 ft. (about 106 ft.?) thick.

[1] Buckman wrongly supposed that *Platypleuroceras* and *Polymorphites* indicated the *valdani* (= *ibex*) zone, and the rare fragments occurring below, which he supposed to be species of *Uptonia*, were apparently misidentified. See S. S. Buckman, 1915, 'Geol. Whitby and Scarborough', p. 70.

The detailed succession of ammonites in the 65 ft. of shales below the Double Band has not been worked out, but about the middle *Promicroceras planicosta* has been recorded, and *Asteroceras obtusum* at more than one level below. It is therefore assumed that most, if not all, belongs to the *obtusum* zone, but there are many lacunæ in the lower part.

Semicostatum Zone (= SHALES-WITH-BEEF OF DORSET), about 45 ft.

This is the lowest zone exposed in Robin Hood's Bay, where 36 ft. can be seen at low tide. The zone consists of blue calcareous shales with thin bands of limestone, 1–8 in. thick, made of comminuted shells. *Arnioceras semicostatum* amd *Arietites turneri* are rather abundant, together with numerous lamellibranchs, such as *Gryphæa arcuata, Oxytoma inæquivalvis, Cardinia listeri, Hippopodium ponderosum, Nuculana, Lima, Chlamys,* and *Dentalium*.

Bucklandi, Angulatum and **Planorbis Zones** (= BLUE LIAS OF DORSET), about 230 ft. (+ ?).

For the downward continuation of the section it is necessary to pass north along the coast to the Scars, near Redcar. Here the *bucklandi* zone is exposed for its whole thickness of about 160 ft., together with 30 ft. of the top of the *angulatum* zone. The *angulatum* zone is the lowest part of the Lias exposed on the Yorkshire coast, and can only be seen for a distance of 400 yards at low water opposite the battery, where it is brought up in the crest of an anticline.

Lithologically the two zones are similar, consisting of shales in which are numerous oyster bands formed of *Gryphæa arcuata*, and occasional thin beds of limestone. About 42 ft. from the top of the *bucklandi* zone is a 3-in. band made up of a mass of *Cardinia listeri*, together with some *Unicardia* and *Gryphæa arcuata*. Two feet below this is a particularly rich fossil bed, 8 in. thick, containing nearly all the fossils found throughout the rest of the zone, and including occasional corals referred to *Montlivaltia haimei* and *M. guettardi*. *M. haimei* occurs lower down, 20 ft. below the top of the *angulatum* zone, in a band of fossiliferous pebble-like concretions (2½ in.) in association with *Ornithella sarthacensis, Astarte oppeli, Cardita heberti, Plicatula liasina,* and *Serpulæ*.

The existence of the *planorbis* zone beneath the sea is proved by the fact that blocks full of the zone-fossil are often washed up in Robin Hood's Bay. It has also been detected inland in some borings and small exposures, but there are no satisfactory sections.

PRE-PLANORBIS BEDS, 20–40 ft.

In the Northallerton district there are about 40 ft. of Pre-*planorbis* Beds, though not all of them are exposed in any one section. The clays partake more of the nature of paper-shales. *Ostrea liassica* as usual abounds in the upper part, and *Pteromya crowcombeia* is common throughout, but especially in a thin limestone band 25 ft. from the top.[1]

Wherever the Pre-*planorbis* Beds are exposed in other parts of the Yorkshire Basin the same features are displayed, and they are usually known as the Oyster Beds and *Pleuromya* Limestone.

[1] C. Fox-Strangways, 1886, 'Geol. Northallerton and Thirsk', *Mem. Geol. Surv.*, pp. 13–14.

VII. THE CUMBERLAND OUTLIER[1]

The basement beds of the Lower Lias form a plateau-outlier west of Carlisle, extending from near Bellevue for about 7 or 8 miles westward. The ground is covered with a thick mantle of Drift and the Lias is known almost exclusively from the cores of borings. It is at least 210 ft. thick at Great Orten, and consists of the usual shales with bands of limestone, some of them sandy and micaceous. Records of fossils are meagre, *Caloceras johnstoni* (upper part of the *planorbis* zone) being the only ammonite found.[2]

VIII. NORTHERN IRELAND: THE ANTRIM PLATEAU

About 120 miles west of the Cumberland outlier the Lower Lias is again met with, sandwiched between Rhætic and Cretaceous sediments beneath the Tertiary basalt plateau of Antrim. It suffered extensive denudation prior to the deposition of the Cretaceous rocks and again before the outpouring of the basalt. In places the Liassic and Rhætic rocks are overstepped by the Cretaceous, while often the basalt oversteps all three and passes on to the Trias. Nevertheless, the Lower Lias probably has a considerable extension beneath the plateau, for it outcrops at numerous points along the edge. Tate traced it 'on the south of the County Antrim from Collin Glen (Belfast) to Whitehead, also on the Carrickfergus Commons and on the shores of Lough Morne; on the east, around the shores of Larne Lough and on the east coast of Island Magee, Larne, Glenarm, and Garron Point; on the north, at Ballintoy and Portrush; and in the County Londonderry, at Magilligan on the NE., at Aghanloo, and Lisnagrib'.[3]

In places strata at least as high as the upper part of the *raricostatum* zone are preserved, for *Leptechioceras macdonnellii*, a species described and figured by Portlock in his *Report* of 1843, is said by him to be 'common to the Ballintoy marls and the indurated Lias of Portrush'. There is considerable attenuation, for at Collin Glen, Belfast, all the beds intervening between the *obtusum* zone with *Arietites turneri* and the Rhætics occupy only 38 ft.; most of the series consists of marls and shales, in the highest 17 ft. of which occur *A. turneri*, *S. angulatum* and numerous lamellibranchs. Below, *P. planorbis* has been found, and most of the thickness is thought to belong to this zone.[4] At the base are 4 ft. of shelly limestones with abundant *Ostrea liassica* and *Modiola minima*, evidently the continuation of the Pre-*planorbis* Beds of England.

IX. THE HEBRIDEAN AREA

Lower Lias occurs in the islands of Skye, Raasay, Pabay (Pabba), Scalpay (Scalpa) and Mull, and on the mainland at Applecross, in Ardnamurchan, and in Morven. It is conformable with the Trias, which rests with great unconformity on Cambrian and older rocks. Above it the higher members of the Jurassic follow to great thicknesses, the highest, of Corallian, and in two places Kimeridgian, date, being found in Skye and Mull (see map, fig. 19, p. 114).

[1] H. B. Woodward, 1893, *J.R.B.*, p. 183, with references.
[2] No palaeontological evidence for the presence of Rhætic Beds has been obtained, but there is no reason for supposing them absent.
[3] R. Tate, 1864, *Q.J.G.S.*, vol. xx, p. 109.
[4] G. W. Lamplugh, 1904, 'Geol. Belfast', p. 26, *Mem. Geol. Surv.*

As remarked in an earlier chapter, the Jurassic strata form plateaux, usually with gentle dips, and they are mere remnants faulted down among the ancient rocks which they once covered. Folding has affected them but little, but alteration, due to the injection of numerous sills and dykes, has often been profound. In Eocene times vast sheets of basalt and granophyne were extruded through them, covering them in places to depths of hundreds of feet. It is generally owing only to the protective action of these sheets of Tertiary igneous rock that the Jurassic and Cretaceous sediments have been preserved at all.

The Hebridean Trias consists of alternations of grits, pebbly grits and conglomerates, which pass up into red and grey marls. The Jurassic rocks, however, with few exceptions, show little indication of close proximity to any shore-line. It cannot be too strongly emphasized that their present distribution is dependent upon the subsequent interaction of extraneous influences, and bears no relation to their original extent.

(a) Skye, Raasay, Pabba and Applecross

The Lower Lias has a far larger surface-area in the Hebrides than any other part of the Jurassic System. The principal outcrop is nearly 10 miles long and from 1 to 2 miles broad, crossing the central district of Strath, Skye, in a SW.–NE. direction, from Lochs Slapin and Eishort to Broadford Bay (fig. 19). It is the eastern limb of an ancient anticline, initiated before Mesozoic times, the western limb of which has been broken down by the granite intrusion of the Red Hills. In the centre Cambrian limestones, dolomites and quartzite are arched over the granite intrusion of Beinn an Dubhaich, and on to them lap successively the lower zones of the Lower Lias, until on the shores of Loch Slapin, the Durness (Cambrian) Limestone is overstepped by the *semicostatum* zone. At the south end of the anticline the Lower Lias plunges steeply beneath the higher Jurassic rocks of Strathaird, while to the north its prolongation is proved by the remnants forming the flat island of Pabba and the south-eastern extremity of Scalpa.

Into the centre of the northern end of this anticline the sea has eaten Broadford Bay, round which Lower Lias forms the coast for 5 miles. The greater part of the beds exposed belong to the lower part of the formation, consisting of hard limestones and sandstones spanning the Pre-*planorbis* to *semicostatum* zones. This lower division was therefore named by H. B. Woodward in 1896 the BROADFORD BEDS. The upper part of the formation, consisting of shales, builds the whole of the island of Pabba, lying off Broadford Bay, and takes from it the name of PABBA SHALES.

In spite of the hardness of the Lower Lias, it has been planed off near the coast almost to sea-level, an upward limit of altitude being imposed by the 100 ft. beach. In Broadford Bay the exposures are poor and disappointing, consisting of little more than reefs and ledges, for the most part covered by the water at high tide and overgrown with seaweed. Better (and highly fossiliferous) exposures of the Pabba Shales are afforded by the low cliffs of Pabba Island, and again on Raasay. On the coasts of Loch Slapin and Loch Eishort the cliffs are considerable and the exposures extensive, but here the Lias is largely altered by heat and fossils are often destroyed.

SUMMARY OF THE LOWER LIAS IN SKYE, RAASAY, PABBA AND APPLECROSS [1]

Davœi, Ibex, Jamesoni and **Raricostatum Zones:** (PABBA SHALES), 700 ft. in Skye, 600 ft. in Raasay.

The shales as a rule are remarkably uniform throughout their great thickness, being always dark and micaceous, with mudstone nodules and thin lenticular bands of calcareous sandstone and argillaceous limestone at intervals.

Towards the top, to a varying extent in different places, they become arenaceous, forming a lithological passage upward into the Middle Lias. In Mull there is a perfect passage from sandstones containing ammonites of the *jamesoni* and even *raricostatum* zones up into the Scalpa Sandstone of Middle Lias age. In Raasay only the *davœi* zone (represented by *Lytoceras salebrosum* (?) Pomp. sp. and *Androgynoceras* cf. *maculatum* Y. and B. sp.) is developed entirely in a sandstone facies, more conveniently classed with the Scalpa Sandstone. Even in Raasay, however, the Pabba Shales become gradually harder and more sandy in the *ibex* and *jamesoni* zones, and lenticular bands of calcareous sandstone are intercalated. This early incoming of the arenaceous sediment more properly characteristic of the Middle Lias recalls the Yorkshire Basin, where we saw in the last section that the lower half of the sandy series, at least in places, contains capricorn ammonites of the *davœi* zone. Here too, as in Yorkshire, many of the Middle Lias lamellibranchs, such as *Pseudopecten æquivalvis* and *Gryphæa cymbium*, first enter with the arenaceous facies. But the occurrence of *Acanthopleuroceras valdani* indicates that some of the strata belong to the *ibex* zone, while a little lower *Uptonia jamesoni* has been found. Two species of *Spiriferina* and giant Rhynchonellids of the *rimosa* group (*Rimirhynchia* of S. S. Buckman) also occur, recalling the *jamesoni* zone of Radstock.

In the Isle of Pabba the *ibex* and *jamesoni* zones are together 100 ft. thick and largely form the island. They have recently been examined in detail by Dr. L. F. Spath, with a view to establishing the ammonite succession.[2] The lowest beds of the island, seen on the north-eastern coasts, belong to the top of the *raricostatum* zone, but only a fraction of this zone comes to the surface on Pabba Island.

The great bulk of the Pabba Shales elsewhere, as best exposed in Raasay, consists of a greatly expanded *raricostatum* zone, 300 ft. thick. This is much thicker than its development in any English locality, and the genus *Echioceras* is proportionately abundant, in both species and individuals, and provides valuable material for evolutionary study. The late Dr. G. W. Lee's collecting brought to light the interesting fact that there are repeated alternations of Echiocerates and Derocerates. The facts are capable of two very different interpretations, the most rational being that which postulates temporary migration and return of the same species with change of conditions.[3]

Of other fossils in the Pabba Shales, lamellibranchs are fairly abundant, especially in the upper part of the *raricostatum* zone, where nearly all the

---

[1] Based on H. B. Woodward and C. B. Wedd, 1910, 'Geol. Glenelg, Lochalsh, and SE. Skye', pp. 98–113; and G. W. Lee and S. S. Buckman, 1920, 'Mes. Rocks Applecross, Raasay and NE. Skye', *Mems. Geol. Surv.*

[2] L. F. Spath, 1922, *Geol. Mag.*, vol. lix, pp. 548–51.

[3] For further discussion of this, with references, see Chapter I, p. 30.

characteristic species have been obtained. *Hippopodium ponderosum* attains a large size. Gastropods, on the other hand, are unusually scarce, only one species having been recorded.

Echioceras abounds to within a few feet of the base of the Pabba Shales, which rest abruptly and non-sequentially upon the calcareo-arenaceous Broadford Beds. The change from the one type of rock to the other is equally sudden in Raasay, Applecross, Skye and Mull, and points to a widely distributed stratal break. This break seems to represent the *oxynotum* zone and, in most places, the *obtusum* zone.

Obtusum, Semicostatum, Bucklandi, Angulatum and Planorbis Zones (BROADFORD BEDS), 340 ft. in Raasay, 240 ft. in Skye.

At the top the series consists of fissile, micaceous sandstones, which pass down into calcareous sandstones, and finally become, for the greater part of the succession, alternations of black or dark blue limestone (often sandy) with shale partings. In Skye there are some bands of fine quartz conglomerate in the basal 10–15 ft.

The upper half of the Broadford Beds is characterized by abundant *Gryphæa arcuata*, while the basal 10–20 ft. at Applecross are crowded with *Ostrea liassica*, just as are the English Pre-*planorbis* Beds.

In the type-area the higher parts of the Broadford Beds, which crop out in the foreshore north-west of Broadford pier, are obscured by coverings of Drift and peat and are altered by contact with numerous dykes and sills. The abrupt and non-sequential junction with the Pabba Shales is, however, clear in the south of the island, and in Raasay and at Applecross. All round Broadford Bay the exposures are poor and discontinuous, the reefs being largely obscured by recent beach material and the succession being complicated by the numerous dykes. Only the lower zones are well exposed farther east, particularly those of *bucklandi, angulatum* and *planorbis*, which form the reef-girt Ardnish headland enclosing the inlet of Ob Breakish, and continue eastward to the small promontory and islet of Lusa.

Combining the evidence collected by the Survey in the several localities, the following generalized succession can be made out:

OBTUSUM ZONE. On Loch Eishort, Skye, the *Asteroceras* fauna has been found in a thin shale overlying the *semicostatum* zone, but it has not been recognized in Raasay. In Raasay, however, in the cliff north of the waterfall at Hallaig, 40 ft. of flaggy, carbonaceous, rusty-weathering sandstones, with a layer of *Gryphæa arcuata* 6 ft. from the top, have yielded *Microderoceras birchi, Arietites turneri* and other species. In Morven the zone attains a considerable thickness (see below, p. 148).

SEMICOSTATUM ZONE. This zone is the most conspicuous part of the Lower Lias in the cliffs of Loch Slapin, where it consists of baked shales and limestones, which overlap the lower zones on to the Cambrian limestone. Specimens of *Arnioceras* abound at several levels. Near the base are *Gryphæa* beds, so intensely baked by igneous intrusions that the shells have been reduced to the semblance of small clinkers lying within perfect moulds left in the baked shale or limestone. Both here and round the Suisnish promontory on Loch Eishort the zone forms cliffs of considerable height.

BUCKLANDI, ANGULATUM ZONES. East of Broadford, the *bucklandi* zone consists chiefly of sandstones, which crop out on the shore as reefs along the low promontory of Ardnish. Here *Coroniceras* occurs, and other conspicuous fossils are *Gryphæa* cf. *arcuata* and *Lima gigantea*.

The tidal inlet of Ob Breakish separates these sandstones from a limestone series representing the *angulatum* and underlying zones. These last form reefs and broad pavements often weathered into remarkable shapes. The most interesting feature is the Ob Breakish Coral Bed, a prominent bed with *Thecosmilia martini* at the base, associated with several bands of limestone containing corals. This bed has been traced for a considerable distance inland. At Applecross the same coral abounds in a 4 ft. band of limestone, and it is in places so abundant as to resemble a reef. At Applecross there are only 30–40 ft. of Lias below the coral bed, but at Ob Breakish there are 60–70 ft., probably owing to the intercalation of beds belonging to the *planorbis* zone. The same coral bed is also found at Loch Sligachan.

In Raasay and at Applecross ammonites have been collected more carefully than in Skye, and the result has been to show that the exact line of demarcation between the *bucklandi* and *angulatum* zones in the Hebrides still remains to be drawn. As in Skye, the *bucklandi* zone is predominantly arenaceous while the *angulatum* zone consists of limestones, but the fossil evidence is scanty. In Raasay the *semicostatum* zone is underlain by 90 ft. of calcareous sandstone and sandy shale which definitely belongs to the *bucklandi* zone; it has yielded *Coroniceras bucklandi* 25 ft. from the top and *Metophioceras* cf. *conybeari* 40 ft. from the top, while in fallen blocks below, *Paracoroniceras gmuendense* has been found. Below this, 50 ft. of sandy shale and brown sandy limestone, yielding only *Gryphæa* cf. *arcuata*,[1] pass down insensibly into 140 ft. of almost unfossiliferous limestones, sometimes sandy, with shaly partings. At Applecross a specimen of *Scamnoceras* cf. *montanum* (Wähner) has been found a few feet above the presumed equivalent of the Ob Breakish Coral Bed.

PLANORBIS ZONE. From Ob Breakish to Lusa, at the end of the Lias outcrop east of Broadford, the shore reefs show a series of hard, compact limestones and calcareous sandstones, probably 30–50 ft. thick. They are still palaeontologically unclassified, but they may ultimately prove to represent the *planorbis* zone. About 6–8 ft. from the base is a second coral bed, the Lusa Coral Bed, composed of *Isastræa*. It too can be traced for some distance inland, where it is well displayed in a quarry west of the cross-roads on the Kylerhea road; unlike the Thecosmilian coral bed of Ob Breakish, it is not represented at Applecross. Its position may be either in the *planorbis* zone or in the Pre-*planorbis* Beds.

PRE-PLANORBIS BEDS. In the basement portions of the Lower Lias east of Broadford and elsewhere in Skye definite indications of deposits on the horizon of the Pre-*planorbis* Beds are lacking. This may be due to the beginning of the overlap of the higher zones towards the south, which culminates in the absence of all the zones below that of *Arnioceras semicostatum* on the shore of

---

[1] The Gryphæas of this lower part of the Broadford Beds are seldom more than 1½ in. in height, and differ from *G. arcuata* also in their greater breadth and ill-defined sulcus. For comparisons see G. W. Lee, 1925, loc. cit., pp. 12–16.

Loch Slapin. Farther north, on the island of Raasay, and still more on the mainland at Applecross, Pre-*planorbis* Beds with their usual characters are well developed.

At Applecross the basement beds of the Lias are best exposed in a stream-bank and limekiln, described in detail by Lee, ¾ mile south-east of Applecross House. Here the sandy marls of the Trias are succeeded by calcareous sand-stone and limestone, in which *Ostrea liassica* abounds at certain levels up to 30 ft. from the base. From 20 to 30 ft. from the base are bands of oolitic limestone crowded with the oysters, which can be collected in a perfect state of preservation, equalling anything to be found in England. The top of this oolite is only 10 ft. below the presumed equivalent of the Ob Breakish Coral Bed, and close beneath the horizon at which *Scamnoceras montanum* was found, leaving but little room for the *planorbis* zone.

(b) Mull, Morven and Ardnamurchan.

Isolated areas of Lower Lias occur at a number of points around the coasts of Mull, north, east, south, and west; at Loch Aline in Morven; and on both the north and the south coasts of the peninsula of Ardnamurchan (map, p. 114). Some of these occurrences have been known from the earliest times and were described in detail by Judd, but a flood of new information concerning them has recently become available through the work of the Geological Survey of Scotland.

The general succession is the same as in the more northerly islands, although the total thickness is only about half that in the north. Except in Morven there is the same broad division into Pabba Shales (about 400 ft.) and Broadford Beds (70–100 ft.) There are several features of special interest, chief among which are local and richly fossiliferous occurrences of the *ibex* zone at Tobermory, and of the *planorbis* zone and Pre-*planorbis* Beds at Gribun in Western Mull, and a thick development of the *obtusum* zone at Loch Aline in Morven.

The fullest development is probably displayed by the sections in Ardna-murchan, where the Broadford Beds reach their maximum thickness of 100 ft. near Swordle, on the north coast, and are also well seen on the south coast near Mingary Castle. But the Pabba Shales, although well exposed on the south coast, in the west side of Kilchoan Bay, are almost completely unfossiliferous; the Survey found only one ammonite of doubtful identity. This is difficult to account for, since the amount of baking to which the shales have been sub-jected would not have been sufficient to destroy fossils had any been present. The Broadford Beds consist mainly, as in Mull and Morven, of limestones crowded with *Gryphœa arcuata*, and they are no less fossiliferous than else-where.

The fullest particulars of the zonal constitution of the Lower Lias in this district have been obtained from the south-east coast of Mull. Grand displays of the formation may be seen in the cliffs of the peninsulas between Loch Buie, Loch Spelve and Loch Don, and again in Duart Bay, and farther west in the Carsaig district. More remote fragments exposed at Gribun in the west and at Tobermory in the north of the island each have a special feature of interest not met with in the larger areas.

SUMMARY OF THE LOWER LIAS IN MULL, MORVEN AND ARDNAMURCHAN[1]

Davœi ?, Ibex, Jamesoni, and Raricostatum Zones (PABBA SHALES), 400 ft.

As in the northern islands, the Pabba Shales denote remarkable uniformity of sedimentation—excepting only that they become increasingly sandy towards the top, and that this gradual increase in the proportion of sandy sediment sets in earlier in some places than in others. In the coast-sections north-east of Torosay Castle, Duart Bay, the sandstone facies extends to the bottom of the division. There not only is the major part of the Pabba Shales sandy but they pass down into 50 ft. of greenish fine-grained flaggy sandstone, resting directly on the Broadford Beds. In short distances, as on the other side of Duart Bay, they become more argillaceous.

The upper part of the Pabba Shales is well exposed in the long stretch of cliffs between Loch Don and Loch Buie, and again, with the downward continuation, in broad tidal flats on both sides of Carsaig Bay. Sandy shale and ferruginous shaly sandstone alternate down through the *davœi* and *jamesoni* zones into the top of the *raricostatum* zone (*olim armatus* zone). Ammonites are exceedingly rare, but enough have been found in fragmentary condition for approximate dating. In the uppermost 100 ft. *Gryphœa cymbium* and *Pseudopecten æquivalvis* abound, thus ushering in the Middle Lias conditions before their time, as in Raasay and Skye and in Yorkshire. On the tidal flats of Carsaig Bay ammonites are less rare than farther east. This is the type-locality of *Uptonia jamesoni* (Sow.), and large specimens of *U. bronni* (Roem.) can still be collected in fair abundance on the west side of the bay. On the east side the sandy micaceous shales, although much the same as those on the west, are on a lower horizon, and in them many small black nodules harbour over a dozen species of *Echioceras*. This is the only locality in the southern part of the Hebridean area where a glimpse is to be obtained of the rich *Echioceras* fauna of Raasay.

A single ammonite found at Port Donain gave the only indication of the *davœi* zone in the Southern Hebridean area, although there are many exposures of the top of the Pabba Shales. None of these localities has yielded any evidence of the *ibex* zone, except a small and isolated patch of Lower Lias at the north end of the island of Mull, near Tobermory. The Tobermory River has cut a small section of the Pabba Shales, and in the bed of the stream and beside a path along the left bank have been found large numbers of crushed and badly preserved fossils, chief among which are some five species of the genus *Tragophylloceras*.[2]

Obtusum Zone (LOCH ALINE SANDSTONE), 0–160 ft. (Morven only.)

The highest Jurassic beds preserved in Morven consist of a remarkable mass of shaly sandstone passing down into sandy shale, the LOCH ALINE SANDSTONE. On both sides of Loch Aline it is a conspicuous feature, reaching as much as 160 ft. in maximum thickness, but towards the head of the loch it has been

---

[1] Based on G. W. Lee, 1925, 'Pre-Tert. Geol. Mull, Loch Aline and Oban', pp. 75–91; G. W. Lee, E. B. Bailey and J. E. Richey, 1930, in 'Geol. Ardnamurchan', pp. 36–42, *Mems. Geol. Surv.*

[2] Judd mistook them for *Oxynoticeras*, thus wrongly recording the *oxynotum* zone—G. W. Lee, 1925, loc. cit., p. 89.

cut off by erosion preceding the Cenomanian transgression and disappears altogether. Fossils are rare, but the Survey obtained an assemblage indicating that the whole of the Loch Aline Sandstone belongs to the *obtusum* zone: namely *Asteroceras suevicum*? (Quenst.), *Xipheroceras aureum* (Y. and B.), *Xiph.* cf. *planicosta* (Sow.) and *Coroniceras* sp. (*bucklandi* Reynès non Sow.). Lee remarked that, in spite of the apparent localization of this fauna to Morven, 'it would be difficult to believe that the *obtusus* zone is really unrepresented in Mull'[1] and this view receives support from Dr. Spath's discovery of the *obtusum* fauna at Mingary Castle in Ardnamurchan.[2] If it is represented in Mull, it is probably to be sought among the lower arenaceous shaly beds classed for convenience with the Pabba Shales (compare especially the 50 ft. of shaly sandstone immediately overlying the *Gryphæa* limestones of the *semi-costatum* zone north-east of Torosay Castle, above, p. 148). It was probably largely in view of this possibility that the Survey in their memoir of 1925 departed from the original classification founded in Skye and included the *obtusum* zone in the Pabba Shales. In the more recent memoir dealing with Ardnamurchan (1930) the Survey have reverted to the original classification and returned the *obtusum* zone to the Broadford Beds. But in the area covered by the 1930 memoir they did not have to deal with the difficulties with which Lee was confronted in south-east Mull and Morven. Since the Loch Aline Sandstone does not accommodate itself naturally within either the Pabba Shales or the Broadford Beds, and is unknown in the type-area of both, it seems advisable to keep it separate.

Semicostatum, Bucklandi, Angulatum, Planorbis and Pre-planorbis Zones (BROADFORD BEDS), 70–100 ft.

The Broadford Beds, which do not reach a third of their thickness in Raasay, consist chiefly of limestones crowded at many levels with *Gryphæa arcuata*, and alternating with beds and partings of shale. The Gryphæas show a steady enlargement as they are traced upward through the series, and at the same time the shells grow more elongate and more sulcate. The average size in the lower levels falls short of 1 in. in length, and few exceed $1\frac{1}{2}$ in.; at the top the length (height) commonly reaches 3 in. Ammonites are always rare, but such as have been found show that the *Gryphæa* Limestones correspond with the *semicostatum* and *bucklandi* zones.

The lower zones of the Broadford Beds are exceptionally barren in the principal sections. In all the exposures in South and South-East Mull and Ardnamurchan (they are not exposed in Morven) the lowest 20–30 ft. of the Lias consists of almost unfossiliferous limestones, which cannot be zonally classified. On the south coast of Ardnamurchan the Survey infer the presence of the *angulatum* and *planorbis* zones from certain accessory fossils, but no ammonites have been found.

Only in Western Mull, on the coast of the Wilderness, south of Gribun, do the basal beds contain the representative faunas met with in England. Here a small isolated patch of the basement beds of the Lower Lias yields the only representative of the *planorbis* fauna known in Scotland, as well as typical Pre-*planorbis* Beds crowded with *Ostrea liassica* as at Applecross. Since these

[1] G. W. Lee, 1925, loc. cit., p. 86.
[2] L. F. Spath, 1922, *Geol. Mag.*, vol. lix, p. 172.

rest upon the best example of the Rhætic Beds known in Scotland, the interest of the locality is exceptional.

The *planorbis* zone comprises 20 ft. of dark shale with bands of limestone, and from it the Survey collected *Psiloceras planorbis* (Sow.), *P. erugatum* (Bean-Phil.), *P. sampsoni* (Portlock) and *P. hagenowii* (Quenst.), with a rich fauna of lamellibranchs of the typical genera and species. Below are 8 ft. of sandy limestone and calcareous shale, crowded with *O. liassica*, and at the base of all 19 ft. of micaceous and calcareous sandstone devoid of all fossils but plant-remains.

X. EASTERN SCOTLAND[1]

There is a small area of Lower Lias in East Sutherland, near Dunrobin, Golspie (map, fig. 28), but it is deeply covered by Drift and the only exposure is on a short stretch of foreshore between tidemarks, immediately east of the grounds of Dunrobin Castle. The total thickness visible is some 60 ft., within which are many gaps, and although long lists of fossils have been published, the ammonites are not wholly satisfactory for diagnosis. Definite indications of the *davœi*, *ibex* and *raricostatum* zones have, however, been found, in the form of ammonites identified by Buckman as *Prodactylioceras* aff. *davœi*, *Cœloceras* aff. *pettos* and *Apoderoceras leckenbyi*. All these forms occurred in one bed 6 ft. 9 in. from the top of the exposed sequence, whereas in Pabba Island the first and last are separated by at least 100 ft. of shales. The last-named characterizes a level now usually considered as the highest subzone of the *raricostatum* zone. There is therefore a considerable condensation at Dunrobin.

Some 20 ft. lower down another fossiliferous bed has yielded ill-preserved ammonites suggestive of the *raricostatum* zone, but they are not fit for identification with absolute certainty.

Lithologically the strata are very varied, comprising clays and shales with bands of limestone and micaceous sandstone.

IX. KENT[2]

The numerous borings in Kent have proved that the Lias is overlapped towards the north and north-east by the succeeding members of the Jurassic System. Its northern boundary underground is approximately coincident with the main line of railway from Dover to Canterbury and probably onward to Faversham and Chatham. South of this line all the borings have proved the presence of the Lias, but north of it the Lower Oolites overlap on to the rising Palaeozoic floor before being in turn overstepped by the Cretaceous.

The Lower Lias is represented only by the highest portions, no evidence having been obtained in any boring for the presence of beds lower down in the sequence than the *ibex* or *jamesoni* zones, while in the most northerly and north-easterly borings these also are overlapped and the *davœi* zone alone is present. The thickness diminishes steadily northward and north-eastward, from 80 ft. at Brabourne and 40 ft. at Folkestone to such small amounts as 4 ft. at Chilton and Fredville, $3\frac{1}{2}$ ft. at Harmansole, and 2 ft. at Bishopsbourne.

---

[1] G. W. Lee, 1925, in 'Geol. Golspie', *Mem. Geol. Surv.*, pp. 69–74.
[2] Based on G. W. Lamplugh and L. F. Kitchin, 1911, 'Mes. Rocks, Coal Expl. Kent'; and Lamplugh, Kitchin, and Pringle, 1923, 'Concealed Mes. Rocks Kent', *Mems. Geol. Surv.*

The typical lithological constituent of the Lower Lias is a grey shaly clay, often shelly along the bedding-planes, with some thin bands of limestone. At

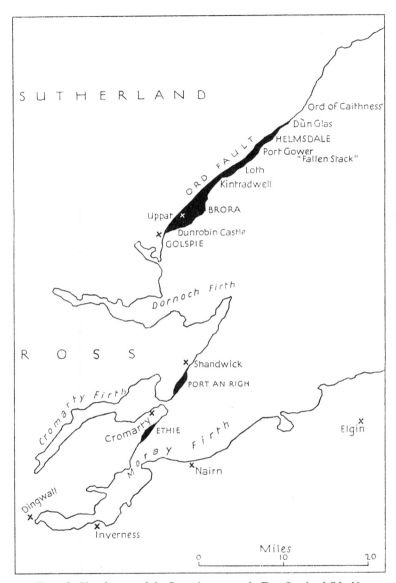

FIG. 28. Sketch-map of the Jurassic outcrop in East Scotland (black).

the base there are frequently small well-rounded pebbles derived from the underlying Palaeozoic sandstones, and sometimes the 'junction suggests that the Liassic basement-bed filled up minor irregularities in the worn surface'.[1] At Chilton the basement-bed was a thin limestone band containing corals,

[1] 1923, loc. cit., p. 82.

apparently *Montlivaltia*, some of which showed definite signs of rolling and had been drilled by boring mollusca.

The greater part of the Kentish Lower Lias belongs to the *davœi* zone, which is about 60 ft. thick at Brabourne, in the south-west, and has the widest extension of all the zones to the north and north-east. Its fossiliferous grey clays have yielded *Androgynoceras maculatum* (Young and Bird), *A.* cf. *latœcosta* (Sow.) and *Oistoceras arcigerens* (Phil.), while among bivalves the genera *Nuculana*, *Astarte* and *Parallelodon* are numerous.

In the south-westerly borings, where earlier beds are preserved, they are relatively thin. At Brabourne and Elham the *davœi* grey clays were found to be separated from the Palaeozoic platform by a few feet of ferruginous marls, some of the layers ironshot. These lowest ironshot beds yielded no ammonites, but at Brabourne and again at Dover *Rimirhynchia* and *Lima antiquata* Sow. were found, while immediately above them an ammonite assigned by Buckman tentatively to the *valdani* hemera would seem to indicate the base of the *ibex* zone. The ironshot marls are therefore believed to represent the upper part of the *jamesoni* zone, if not earlier beds.

PLATE V

Photo.   *Purch.*

Golden Cap (619 ft.) from Seatown.

The three white lines mark the position of the Three Tiers. The capping
(giving the cliff its name) is Upper Greensand.

Photo.   *Purch.*

Down Cliff and Thorncombe Beacon from Eypesmouth.

The nearest cliff on the right is of Middle Lias. The Down Cliff Clay caps
the two farthest cliffs, and above it follow some Bridport Sands and Upper
Greensand on Thorncombe Beacon (in the centre).

CHAPTER VI

MIDDLE LIAS

Stage.	Zones (Faunizones).	Thickness.	Dorset Coast Strata.
DOMERIAN *or* UPPER PLIENSBACHIAN	*Paltopleuroceras spinatum*	8″	Marlstone
		35′–75′	Thorncombe Sands[1]
	Amaltheus margaritatus	1′	*Margaritatus* Bed
		75′	Down Cliff Sands[1] *or* Laminated Beds
		4½′	Starfish Bed
		155′	Micaceous Beds
		25′–35′	The Three Tiers

I. DORSET COAST TO THE MENDIPS

THE Middle Lias of the Dorset coast begins with a prominent basal division called the Three Tiers. These consist of three thick bands of brown, fissile, micaceous and calcareous sandstone, separated by layers of micaceous sandy clay, the whole forming a prominent tripartite ledge 35 ft. thick. The Three Tiers first appear beneath the Greensand capping of Stonebarrow and Westhay Cliffs, east of Charmouth, and fall gently eastward, until they run down to within 80 ft. of the beach under Golden Cap at Seatown, where they are faulted out of sight (fig. 20, p. 119, and Plate V).

The greater part of Golden Cap, excepting the lowest 80 ft., is formed of precipices of Middle Lias marls and sands rising above the Three Tiers, and contrasting sharply with the dark Lower Lias below. The cliff, which is the highest on the South Coast and provides the finest Lias section in England, rising to 619 ft. above the sea, derives its name from a capping of yellow Albian sands.

East of Seatown the Middle Lias forms the greater part of Down Cliffs and the lower precipice of Thorncombe Beacon, thinning eastwards, and finally ending at another fault on the east side of Eypesmouth (fig. 29, p. 154).

The maximum thickness of the Middle Lias in these cliffs is 345 ft., or only just over 100 ft. less than that of the Lower Lias. But palaeontologically it is incomparably poorer, for in contrast to the uniform clays of the Lower Lias, with their scores of ammonite horizons, the more varied succession of the Middle Lias lends itself readily to division into only two zones.

SUMMARY OF THE MIDDLE LIAS OF THE DORSET COAST[2]

Spinatum Zone (MARLSTONE AND THORNCOMBE SANDS), 35–75 ft.

At the top is a highly fossiliferous band of limestone called the Junction Bed, 5 ft. thick, crowded with ammonites. Four-fifths of the Junction Bed belong

---

[1] Named by Buckman, 1922, *Q.J.G.S.*, vol. lxxviii, p. 382. The other names are those used by the Survey many years before.

[2] H. B. Woodward, 1911, 'Geol. Sidmouth and Lyme Regis', 2nd ed., *Mem. Geol. Surv.*, pp. 40–1; and 1893, *J.R.B.*, pp. 195–201; S. S. Buckman, 1922, *Q.J.G.S.*, vol. lxxviii, pp. 395–400.

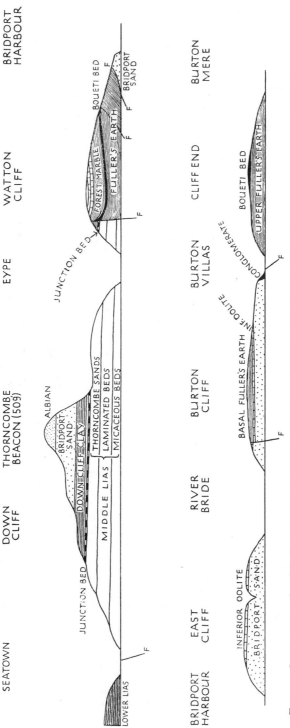

FIG. 29. Section along the cliffs from Seatown to Bridport Harbour and Burton Bradstock. Total distance 5½ miles. Vertical scale exaggerated. The upper section based on H. B. Woodward, 1893, *J.R.B.*, p. 52, fig. 41, with amendments.

to the Upper Lias and will be dealt with in the next chapter, but the lowest 8 in. form a representative of what is known inland as the MARLSTONE. This is itself divisible into two seams, the upper a fine-grained and richly fossiliferous limestone, the lower brown or greenish-grey, iron-shot and conglomeratic, poor in fossils on the coast, but yielding *Tetrarhynchia tetrahedra* a short distance inland. Beneath come 75 ft. (thinning to 60 ft. under Thorncombe Beacon and 35 ft. in Watton Cliff, east of Eype) of light brown and yellow sands, argillaceous at top and bottom, with huge doggers about the central portion. Buckman named them the THORNCOMBE SANDS. They yield *Tetrarhynchia thorncombiensis* and ammonites formerly identified as '*A. spinatus*', but Buckman suggested these might be species of *Amaltheus*.[1]

Margaritatus Zone, 270 ft.

The top of this zone is marked by the MARGARITATUS BED, 1 ft. thick, a hard blue and brown, ferruginous, sandy limestone, yielding *Amaltheus* spp. abundantly, together with *T.? tetrahedra, Pseudopecten æquivalvis, Gryphæa cymbium, Cerithium liassicum* Moore, *Cryptænia* cf. *solarioides* (Sow.), *Ataphrus cinctus* Moore, belemnites, and other less certainly identified fossils. It forms an easily recognized capping to the

LAMINATED BEDS or DOWN CLIFF SANDS: about 75 ft. of micaceous and ferruginous sandy clays and marls with nodules and bands of hard fissile sandstone, poor in fossils, but containing some nests of small cuboidal Rhynchonellids, small Amaltheids, *Gryphæa cymbium, Pseudopecten æquivalvis* and a few other lamellibranchs. A marked layer below the Laminated Beds is formed by the Starfish Bed, 4 ft. 6 in. of hard greenish-grey micaceous and calcareous sandstone with an irregular and hummocky upper surface, noteworthy for containing the starfish *Ophioderma egertoni* and *O. tenuibrachiata*. The 155 ft. between the Starfish Bed and the Three Tiers are occupied by monotonous, bluish-grey, micaceous marl and clay, known as the

MICACEOUS BEDS. They contain nodules of ironstone, iron pyrites, and grey earthy limestone; *Amaltheus* aff. *clevelandicus, A.* cf. *boscensis*, and *Nuculana graphica* are recorded from them. The base is formed by the Three Tiers, already described—three bands of sandstone 2–4½ ft. thick, separated by two 10 ft. bands of clay. They have yielded ammonites recorded as '*A. margaritatus*', '*A. fimbriatus*' and '*A. loscombei*' (the last derived?), and bones of Saurians.

Inland through Dorset and West Somerset[2] the Middle Lias is little known owing to the lack of exposures, but in two or three places the Marlstone is seen, still welded to the under-surface of the Junction Bed. The Three Tiers disappear a short distance from the coast, so that the mapping of the boundary between Lower and Middle Lias in Dorset and Somerset is uncertain, but there are always signs of a considerable thickness of evident Middle Lias. In the Ilminster district are 50 ft. of sands with ironstone nodules, passing down into yellow micaceous marls with sandstones, below which are 100 ft. of blue and grey micaceous marls.

[1] S. S. Buckman, 1922, loc. cit., p. 396; stratal names on p. 382.
[2] H. B. Woodward, 1893, *J. R. B.*, pp. 202–8.

Although the main mass of the beds here attains only half of the thickness on the coast, the Marlstone has thickened to 12 ft. It consists of irregular beds of rusty ironshot limestone, sometimes sandy and micaceous. Long lists of fossils were recorded by Charles Moore, but they are in need of revision. They include *P. spinatum, Pseudopecten æquivalvis, Gryphæa cymbium, T. tetrahedra,* and other brachiopods, with *Ichthyosaurus, Hybodus, Lepidotus,* gastropods, echinoderms, crustacea, sponges, foraminifera and plant-remains.[1] Northward and eastward along the outcrop the Marlstone thins again to 7 ft. at Barrington and 3½ ft. at S. Petherton, until at Yeovil it has a thickness of only 1 ft. 3 in. once more. From there northward the whole Middle Lias thins out steadily towards the Mendip Axis, so that in about 15 miles, before Shepton Mallet is reached, it has disappeared completely.

Along this diminishing outcrop in North Somerset little is seen of the beds, but such exposures as exist show nothing abnormal. They have been worked for bricks on the outliers of Glastonbury and Pennard, where the Marlstone is 1 ft. thick. The farthest outlier forms Brent Knoll, close to the shore of the Bristol Channel, 23 miles from the main outcrop.

II. THE MENDIP AXIS AND DUNDRY HILL

Over the centre of the Mendip Axis, about Nunney and Vallis Vale, south of the Wellow Valley, the Middle Lias is absent, but fossils derived from it are found mixed with others from the Lower and Upper Lias in pipes in the Carboniferous Limestone. Close to the axis on either side, however, the Marlstone is present and unusually thick.

On the south side, at Batcombe and Maes Down, Evercreech, on the main outcrop between Bruton and Shepton Mallet, where the Middle Lias is too thin to appear on the Geological Survey maps, all the strata below the Marlstone are missing, but the Marlstone itself is from 5 ft. to 10 ft. thick. It is a hard, dark, ironshot limestone, and has yielded *P. spinatum, Lobothyris punctata, Tetrarhynchia tetrahedra, Ornithella indentata, Pseudopecten æquivalvis,* and a number of other fossils.[2]

Similarly, on the north side of the axis, it was proved in a boring at Mells to be 9 ft. thick and to rest on Coal Measures.[3]

At the base of the outlier of Dundry Hill, 10 miles west of Bath, the Marlstone takes the form of an impersistent, coarse, yellowish-brown ironshot limestone, usually oolitic, more especially in the upper part, and yielding *Pseudopecten æquivalvis,* &c. The maximum thickness is 3 ft. It rests upon clays, the age of which has not been determined[4] (see fig. 34, p. 196).

In the Radstock district the Middle Lias is again absent altogether, and the Upper Inferior Oolite, sometimes with Upper Lias of very variable thickness intervening, comes to rest on an eroded surface of different horizons in the Lower Lias. These conditions seem to continue as far as Bath, and also down the dip-slope to the east, for Middle Lias was proved to be absent in the Westbury boring.

---

[1] C. Moore, 1853, *Proc. Somerset Arch. Soc.,* vol. iii, pp. 61–76; and 1867, ibid., vol. xiii, pp. 119–244.
[2] L. Richardson, 1906, *Geol. Mag.* [5], vol. iii, pp. 368–9; and 1909, ibid., vol. vi, pp. 540–2.
[3] C. Moore, 1867, loc. cit., p. 150.
[4] S. S. Buckman and E. Wilson, 1896, *Q.J.G.S.,* vol. lii, pp. 686, 705–6.

III. THE COTSWOLD HILLS

North of Bath, on the main outcrop, the Middle Lias gradually returns, thickening northward until it reaches 100 ft. at Wotton under Edge, 150 ft. at Frocester Hill near Stroud, and about 250 ft. in the Mid-Cotswolds. The Marlstone is usually from 10 ft. to 15 ft. thick and forms a fairly conspicuous platform running along the foot of the Cotswold escarpment.[1] It also extends on to a number of outliers, of which the most important are Church Down, Alderton, Dumbleton, and Bredon Hills.[2] It is generally an arenaceous limestone, locally ferruginous, but never sufficiently so to merit working as an iron ore. Here and there it is used as a building stone.

The Marlstone of the Cotswolds contains *Paltopleuroceras spinatum* in its upper part and *Amaltheus margaritatus* in its lower part.[3] It therefore spans a considerable thickness of strata on the Dorset coast.

The lower part of the *margaritatus* zone, consisting of sands and micaceous shales, is poorly fossiliferous. They are best exposed in a brickyard at Wotton under Edge, where, although seen to a depth of 30 ft., they have yielded no fossils.

IV. THE VALE OF MORETON AXIS AND OXFORDSHIRE

East of the Mid-Cotswolds the Middle Lias thins rapidly towards the Vale of Moreton. In the deep boring at Burford it was proved to be only 100 ft. thick, but the greatest reduction takes place farther east, in the Evenlode Valley, where at Fawler it probably does not exceed 30 ft.[4] The Marlstone has a thickness of about 10 ft. and overlies 11 ft. of sands with a blue clay below, the upper part of which is said to yield *Amaltheus margaritatus*[5] and the lower part *Androgynoceras capricornum* at no great depth.

V. THE MIDLAND IRONSTONE DISTRICTS

In the Midlands the Marlstone has been extensively worked as an iron ore in two districts, the first about Banbury, in North Oxfordshire, the second between Wartnaby and Grantham, in Leicestershire and South Lincolnshire. It forms an almost continuous band, giving rise from Edge Hill north-eastward to a broken escarpment overlooking the Midland Plain and the Vale of Belvoir.

On the Edge Hill plateau, where it rises to 700 ft. above sea-level, the Marlstone is a rock band 25 ft. thick, consisting of rusty sandstones, all of which are quarried for road-mending and building. Eastward it thins to 10–12 ft. in the Banbury ironstone district about the Cherwell Valley; to 6 ft. and less in Northamptonshire; and eventually to less than 1 ft. between Keythorpe and Hallaton, 7 miles east-north-east of Market Harborough, approximately on the line of the Charnwood Axis. Here the greater part seems to have been

---

[1] L. Richardson, 1904, *Handbook Geol. Cheltenham*, pp. 47–52.

[2] In old exposures on Church Down the Middle Lias was minutely studied by F. Smithe and at Alderton by F. Smithe and W. C. Lucy; see bibliography.

[3] For revised list of fossils and many details of the Cotswold district see L. Richardson, 1929, 'Geol. Moreton in Marsh', pp. 23–7, *Mem. Geol. Surv.*

[4] F. A. Bather, 1886, *Q.J.G.S.*, vol. xlii, p. 144.

[5] Prof. Trueman informs me that it is doubtful if this identification is correct, and that the species is probably *A. maculatum*.

removed by erosion prior to the deposition of the Upper Lias, and the surviving relic is locally decalcified and impersistent.[1] It may be only coincidence, but it is certainly noteworthy that the Marlstone ceases to be a workable ore just before it approaches the Nuneaton Axis (about 4 miles south of Daventry), and is not again exploited until Tilton in Leicestershire, which is a few miles beyond the Charnwood Axis. The working at Tilton is an isolated one; the main iron-field lies north of Melton Mowbray, occupying the long spur of Middle and Upper Lias and oolites that protrudes westward to Wartnaby, and embracing the main outcrop thence for 25 miles northward to Leadenham in Lincolnshire.

In this Leicester–South Lincs. iron-field the Marlstone is 35 ft. and in places even 40 ft. thick. Only from 9 ft. to 16 ft. at the top, however, yield the ore, the lower part consisting of about 25 ft. of calcareous sandstones as at Edge Hill. The ore gives out altogether at Leadenham, 14 miles south of Lincoln, leaving only the sandstone; at first about 7 ft. thick, this diminishes to 5 ft. between Lincoln and the Humber.

The best Marlstone ore, when unweathered, is a green, densely oolitic limestone, with dark green ooliths set in a darker matrix. It yields from 23 to about 28 per cent. of iron. The poorer varieties are lighter in colour, less densely oolitic, and contain calcite and clusters of the brachiopods *Tetrarhynchia tetrahedra* and *Lobothyris punctata* and other shells. The iron occurs mainly in the form of carbonate, with some silicate. Surface solution and oxidation have partly converted the ore into a soft brown, decalcified oolite, in which both ooliths and matrix consist of ferric oxide. It is in this form that it is usually quarried, at or near the surface, for the natural agencies of surface weathering have enriched the ore by decalcification. Under a thick protective covering of Upper Lias clays it is not worth working.[2]

Palaeontologically the Marlstone is characterized, as in the counties farther south, by the two brachiopods just mentioned, and by the molluscs *Passaloteuthis elongatus* (Miller), *Pseudopecten æquivalvis* (Sow.), *Chlamys liasianus* (Nyst.), *Oxytoma inæquivalvis* (Sow.), *Modiola scalprum* (Sow.) and *Protocardia truncata* (Sow.). The zonal ammonite, *Paltopleuroceras spinatum*, is rare.

The *margaritatus* zone thickens rapidly east of the Evenlode Valley, reaching its maximum, 120 ft., in Leicestershire, and then diminishing towards the Humber. At Grantham it is 55 ft. thick and at Lincoln it has dwindled to 30 ft. In the southern part of the area, in North Oxfordshire and Northants, it consists of sandy, micaceous, blue and brown clays with bands of sandy, fissile limestone at intervals, and occasional layers of cementstone nodules. In the northern part of the outcrop, in Leicestershire and Lincolnshire, the series is, on the whole, rather more argillaceous and in the absence of Amaltheids it is often difficult to distinguish from the Lower Lias.

Locally, in Northamptonshire, the lower boundary is defined by a layer of water-rolled fossils and pebbles, bored, and coated with *Serpulæ* or Polyzoa. In this layer are found *Pseudopecten æquivalvis* and *Plicatula lævigata* and in places there is a green sand largely composed of foraminifera.

---

[1] J. W. Judd, 1875, 'Geol. Rutland', *Mem. Geol. Surv.*, p. 64.
[2] C. B. Wedd, and J. Pringle, 1920, 'Spec. Repts. Min. Resources', vol. xii, pp. 106–140; *Mem. Geol. Surv.*

A mottled rock-band, similarly containing pebbles, rolled fossils and comminuted shells, occurs also in Northamptonshire in the upper half of the *margaritatus* zone. Most of the other rock-bands contain *Protocardia truncata* more or less abundantly, while one immediately underlying the Marlstone near Daventry Railway Station is almost composed of the shells of *Cardinia listeri, C. crassissima, C. crassiuscula, C. lævis, C. concinna*, and gastropods.[1]

V a. The Shropshire Outlier [2]

Interesting evidence of the north-westward extension of the Middle Lias sea is afforded by a Marlstone capping to the Lower Lias outlier at Prees, in Shropshire. The lithological and palaeontological facies on this distant outlier are the same as in the main outcrop, from which it is separated at the nearest point (due west) in Leicestershire by 68 miles.

The Marlstone is known only from graves in Prees churchyard, but it has a thickness of 8–10 ft. and has yielded *Paltopleuroceras spinatum* and all the usual fossils. The underlying micaceous and sandy beds and clays, with *Amaltheus margaritatus*, were formerly worked in an old brickyard and have been seen in various small exposures to 20 ft., but their total thickness is unknown.

VI. THE MARKET WEIGHTON AXIS [3]

The gradual diminution in thickness of the Middle Lias through Lincolnshire is continued beneath the Humber. On its first appearance in Yorkshire, where it was pierced in a boring at Brantingham Grange, near South Cave, the total thickness is not more than $16\frac{1}{2}$ ft. Three miles farther north, in Everthorpe railway-cutting, it has shrunk to 9 ft., and for another 6 miles, as far as Market Weighton, it is just capable of being mapped as a separate formation.

There is here no definite division into Marlstone and *margaritatus* beds, the whole consisting of shales with several thin ferruginous limestone bands at various levels, some of them grey and crystalline. *Paltopleuroceras spinatum* and most of the usual Marlstone fauna have been recorded from several localities, immediately underlain in Everthorpe cutting by Lower Lias with *Androgynoceras capricornum*; from which it appears that the *margaritatus* zone is missing.

From Sancton (2 miles south of Market Weighton) onward for 7 miles the Middle Lias is directly overlain by the Cretaceous. It is everywhere too thin to be mapped, and, after becoming traceable only by occasional patches of ferruginous rubble, it eventually disappears altogether, leaving Cretaceous rocks for about 5 miles resting directly on Lower Lias.

Between the last trace of Middle Lias on the south of the axis and the first point at which it has been detected on the north is a gap of 8 miles. On the north side definite evidence of Upper Lias is obtainable 3 miles earlier than the first signs of Middle Lias. This proves that the Middle Lias thins out altogether against the axis, and that its absence is not due to removal prior to the deposition of the Chalk.

---

[1] B. Thompson, 1889, *The Middle Lias of Northamptonshire*; and 1910, *Geol. in the Field*, pp. 458–61.

[2] H. B. Woodward, 1893, *J.R.B.*, pp. 243–4.

[3] C. Fox-Strangways, 1892, *J.R.B.*, pp. 120–2.

VII. THE YORKSHIRE BASIN (CLEVELAND IRONSTONE DISTRICT)[1]

Across the Derwent thickening sets in rapidly, and along the southern side of the Howardian Hills and in the faulted tracts about Easingwold a Middle Lias terrace-feature is traceable once more. The thickness increases from 30 ft. near the Derwent to 70 ft. about the Coxwold faults, but there are no good exposures. The formation is mainly made up of sandy shales and flaggy sandstones, with calcareous doggers full of the usual fossils—*T. tetrahedra, P. æquivalvis, G. cymbium, P. truncata,* &c., but apparently no ammonites; in places fragments of oolitic ironstone can be detected at the top. West of the Hambleton Hills, except for a few exposures at Feliskirk and elsewhere, Drift covers the outcrop for several miles to Osmotherley.

The neighbourhood of Osmotherley, at the south-western corner of the Cleveland Hills, is one of great interest stratigraphically. Here an important change of facies takes place in the upper part of the Middle Lias, many of the bands of flaggy sandstone passing northward into shale, while at the same time thin bands of ironstone make their appearance. At Feliskirk there are only two thin nodular ironstone bands, 6 and 7 in. thick, but near Osmotherley they have multiplied to four, the thickest measuring 2 ft. At first thin beds of sandstone and sandy shale alternate, but when the escarpment of the Cleveland Hills turns north-eastward the sandstones of the upper part die out, giving place entirely to shales with doggery bands and ironstone. The same change takes place to the east in the numerous inliers exposed in the sides and bottoms of the dales.

Commensurate with this change of facies in the upper part towards the north and east, steady thickening continues. Along the main Cleveland escarpment and in the dales the total thickness is first 120 ft. and later 150 ft.; the formation thus slightly exceeds its thickness in Leicestershire, but barely attains to within 100 ft. of the maximum in the Cotswolds. It is possible, moreover, that the thickest measurements include up to 30 ft. of the *davœi* zone of the Lower Lias, which has been proved, at least on the coast, to take on a sandy facies in its upper part.

The lower half remains almost entirely sandstone all over Yorkshire and is known as the SANDY SERIES or STAITHES BEDS;[2] the upper half, which becomes chiefly a clay-shale formation with ferruginous bands, is known as the IRON-STONE SERIES or KETTLENESS BEDS.[2]

The Ironstone Series of the Middle Lias in the Cleveland Hills is by far the richest source of bedded iron ore in England, yielding the greater part of the total British output. In the dales and in the more southerly sections on the coast, where it crops out round Robin Hood's Bay and Hawsker Bottoms, and at Kettleness and Staithes, the Ironstone Series consists of shales with numerous thin ironstone bands, not worth mining, though some have in the past been worked at their surface outcrop. The most argillaceous development of all is found in the most south-easterly exposures, in Robin Hood's Bay. The general thickness on the coast is about 100 ft.

---

[1] First six paragraphs based on C. Fox-Strangways, 1892, *J.R.B.*, Chapter III; for many other particulars see J. J. Burton, 1913, 'The Cleveland Ironstone', *The Naturalist*, pp. 161–8, 185–94.

[2] Names dating from Young and Bird, 1822.

Northward and westward, towards the edge of the Cleveland Hills overlooking Middlesbrough, the Ironstone Series as a whole becomes thinner but the individual ironstone seams grow thicker and improve in quality. Many of the thin bands expand, closing up and pinching out the shales between them, while new ones come in, until in the outlying mass of Eston Hill, north of the Guisborough Valley, the highest 28 ft. of the series contain 20½ ft. of ironstone. The total thickness is here rather less than 80 ft., with perhaps 60–70 ft. of Sandy Series beneath.

These facts explain the location of the principal ironstone workings along the northern escarpment of Cleveland and Eston Moor, and they account directly for the growth of Middlesbrough, which in 1831, before the coming of the iron industry, was a town with a population of only 154 persons. Ninety years later the census showed a population of 131,103. The rise of this town is a fair measure of the importance of the industry.

The whole of the six million tons of ore that are raised annually along the north slopes of Cleveland are yielded by one seam, known as the Main Seam, which varies in thickness from about 6 ft. to 11 ft. In composition it is chiefly a carbonate of iron, of a greenish-grey earthy appearance, containing a multitude of small ooliths unevenly diffused throughout the mass, and here and there small cavities, sometimes filled with carbonate of lime. The iron content is as high as 30 per cent., but the advantages of this are to some small extent counterbalanced by the lime content being so low (5 per cent.) that large quantities of limestone have to be employed as a flux. It is to this circumstance that the enormous limestone quarries in the Corallian at Pickering and elsewhere are due.[1]

Beneath the Main Seam, which is near the top of the series, there are in the iron-producing districts usually three lesser seams, the general section being as follows:[2]

	ft.
Shale (variable)	0–6
MAIN SEAM IRONSTONE	6–11
Shale	0–5
PECTEN SEAM IRONSTONE AND SHALES . . .	1¾–6
Shale	3–6
TWO-FOOT SEAM IRONSTONE	1¾–21
Shale	20–30
AVICULA SEAM IRONSTONE	0–3

On the north side of Eston Hill the Main Seam and the *Pecten* Seam are in contact, but a shale separates them even before their emergence on the south side of the hill. Gradually towards the south-east, down the dip-slope, the Main Seam in turn comes to be split up by a shale parting of ever-increasing thickness. The NE.–SW. line along which this split in the Main Seam first occurs has been determined with great accuracy and is known as the Shale Line.

The ore of the three minor seams is of inferior quality, though all three have in certain localities been worked in the past. The *Pecten* Seam owes its

---

[1] G. W. Lamplugh, 1920, 'Spec. Repts. Min. Resources', vol. xii, pp. 1–64, *Mem. Geol. Surv.*
[2] C. Fox-Strangways, 1892, loc. cit.

name to the abundance of *Pseudopecten æquivalvis*, the *Avicula* Seam to *Oxytoma* [*Avicula*] *inæquivalvis*.

Palaeontologically the Yorkshire Middle Lias has received the closest investigation outside the ironstone district, at Hawsker Bottoms, south of Whitby (Plate VII), at the hands of S. S. Buckman.[1] The Main Seam and the shales above and below it are the only part of the series that would appear to belong to the *spinatum* zone. Several species of *Paltopleuroceras* have been recorded under the name *spinatum*, the commonest form being a thin-whorled species, *P. hawskerense* (Y. and B.). The lower seams of ironstone, therefore, occur on horizons below the Marlstone over the rest of England except in the Cotswolds, and their inferior quality is thus not to be wondered at.

Buckman placed the upper limit of the *margaritatus* zone at the top of the *Pecten* Seam, giving the zone a thickness of 60 ft., to the base of the Ironstone Series. The highest 23 ft. of the Sandy Series he called the zone of *Seguenziceras algovianum*, assigning the remainder (about 30 ft.) with ammonites of the genus *Oistoceras*, to the top of the Lower Lias (*davœi* or *capricornum* zone). The rest of the fauna, however, is typically a Middle Lias one, consisting of bands of *Gryphæa cymbium*, with *Protocardia truncata*, *Oxytoma inæquivalvis* and belemnites, so that Middle Lias conditions may be said to have begun here at an earlier date than in the rest of England. In North-West Scotland they began even earlier, during *jamesoni* and even locally in the later part of *raricostatum* times.

VIII. THE HEBRIDEAN AREA

(a) Skye, Raasay, and Scalpa[2]

The surface outcrop of the Middle Lias in the Hebrides is small compared with that of the Lower Lias. In South and Central Skye, where the Lower Lias is so conspicuous, the Middle is known only from a small patch of unfossiliferous sandstone north-west of Broadford, and from some bands of sandstone and limestone with *Tetrarhynchia tetrahedra*, *Pseudopecten æquivalvis*, &c., on the beach and in the base of the cliffs on the west side of Loch Slapin. Another small patch occurs on the south-eastern side of the Island of Scalpa, faulted down nearly to sea-level against the Torridonian. Here the only exposures are reefs between tide-marks on the shore near Scalpa House. Both zonal ammonites have been recorded by Judd, as well as the characteristic brachiopods.[3]

The best development of the Middle Lias occurs in Raasay, where it makes a large part of the high eastern cliffs and strikes SW. across the island. It also builds the lower part of the cliffs for 3 miles south of Holm, on the adjoining coast of Trotternish, NE. Skye. Lithologically the whole series consists of a single mass of sandstone, up to 240 ft. in thickness, which has been called, somewhat inappropriately, the SCALPA SANDSTONE.

Owing to the occurrence of fossils, particularly ammonites, being very sporadic, life-zones are difficult to determine. In the course of time, however,

---

[1] S. S. Buckman, 1915, in 'Geol. Whitby and Scarborough', 2nd ed., pp. 68–74, *Mem. Geol. Surv.*

[2] Based on H. B. Woodward and C. B. Wedd, 1910, 'Geol. Glenelg, Lochalsh and SE. Skye', pp. 113–14; and G. W. Lee, 1920, 'Mes. Rocks of Applecross, Raasay and NE. Skye', pp. 25–9, *Mems. Geol. Surv.*

[3] J. W. Judd, 1878, *Q.J.G.S.*, vol. xxxiv, p. 714.

sufficient specimens have been obtained, including *Paltopleuroceras spinatum*, *P. hawskerense* and numerous species of *Amaltheus*, to warrant the supposition that the highest 100 ft. or more of the sandstone belong to the *spinatum* zone, while the rest belongs to the *margaritatus* zone, with doubtfully a local representative of an *algovianum* horizon at the base (said to be indicated by the occurrence of *Amaltheus depressus* (Simpson) in Mull and a *Seguenziceras* sp. in Raasay).

The *margaritatus* sandstones are usually flaggy and micaceous or calcareous, and a striking feature of the *spinatum* portion, especially in Raasay, is a number of huge ellipsoidal doggers, up to 6 ft. in diameter. Otherwise there is little difference between the two zones.

(b) Mull and Ardnamurchan[1]

In the southern part of the Hebridean area the Middle Lias occurs in a number of places in Mull and on the adjoining mainland of Ardnamurchan. The most pronounced feature in this part of the area is a tendency to rapid attenuation in short distances in no constant direction. In Carsaig Bay, on the south coast of Mull, the maximum thickness exceeds 200 ft. Between Port nam Marbh and Port Donain, north of Loch Spelve, the thickness is reduced by a half, while between Loch Spelve and Loch Buie there are only some 40 ft., which is the thickness also in Ardnamurchan (see map, p. 114).

The finest sections are in southern Mull, where the 200 ft. of sandstones build the lower part of the cliffs on either side of Carsaig Bay. There is also a good section between Port nam Marbh and Port Donain. The main mass of the sandstone is presumed to belong to the *spinatum* zone, though no fossils have been found in it. It is white, generally massive, slightly calcareous, and capable of being used as a good freestone, except where large doggers are developed—a feature especially noticeable in Carsaig Bay.

Below the unfossiliferous white sandstone are shaly sandstones, which pass downward into sandy shales. These last have yielded *Amaltheus depressus* Simpson, indicating the *margaritatus* zone; and, just as in the more northerly islands and in Yorkshire, there is a perfect downward passage into the Lower Lias. The physical conditions that gave rise to the type of deposit so characteristic of the Middle Lias here began as early as *jamesoni* times, and with the change of conditions came such characteristic members of the Middle Lias fauna as *Pseudopecten æquivalvis* and *Gryphæa cymbium*.

In Ardnamurchan no fossils have been found. There is only one complete section, on the north coast ⅔ mile east of Rudha Groulin, and to this some disconnected sections on the south coast add nothing. Fine-grained white sandstone makes up the bulk of the formation, and here again it passes downwards gradually through shaly sandstone and sandy shale into the Lower Lias. The passage upwards into the Upper Lias is equally gradual.[2]

IX. EASTERN SCOTLAND

No Middle Lias is exposed in Eastern Scotland; with the Upper Lias and the Inferior Oolite, it is faulted down out of sight near Dunrobin in East Sutherland.

---

[1] Based on G. W. Lee, 1925, 'Pre-Tertiary Geol. Mull, Loch Aline and Oban', pp. 92–4, *Mem. Geol. Surv.*

[2] J. E. Richey, 1930, 'Geol. Ardnamurchan', p. 42, *Mem. Geol. Surv.*

X. KENT[1]

Like the Lower Lias in Kent, the Middle Lias is restricted to the area south and west of the main line of railway from Dover to Canterbury, towards which it thins out and with the whole of the Lias is overstepped by the Lower Oolites. The greatest thicknesses recorded are in the more south-westerly borings: 45 ft. at Brabourne, 20–21 ft. at Elham and Dover, 17 ft. at Folkestone, and 14 ft. at Chilton and Harmansole.

'Broadly speaking', according to Messrs. Lamplugh, Kitchin and Pringle, 'the Middle Lias is developed in a shaly facies in its lower part, with a more pronounced limestone-facies above. It has yielded no ammonities, and the evidence is as yet insufficient for precise zonal subdivision. But it may be permissible to allocate the shales below to the *margaritatus* zone, the limestone series above to the *spinatum* zone, the boundary between them being as yet uncertain.'[2]

Other palaeontological evidence is fairly abundant, especially in the limestone bands, where *Tetrarhynchia tetrahedra* (Sow.) sometimes occurred in large numbers, together with *Rhynch. northamptonensis* Dav. and *Rh. fodinalis* Tate, *Rh. capitulata* Tate, and *Pseudopecten æquivalvis* (Sow.), all fossils highly characteristic of the Marlstone at the outcrop in other parts of England. Only the absence of *Lobothyris* is remarkable.

The shales below are considerably thinner than the limestones and at Harmansole they are wanting. At Brabourne they are micaceous and greenish-grey in colour, showing close lithic resemblance to the ubiquitous micaceous shales of the *margaritatus* zone in other parts of the country. They are characterized by a lamellibranch assemblage, of which the most important members are such species as *Modiola scalprum* Sow., *Protocardia truncata* (Sow.), and some of the large Pseudopectens which abound in the overlying limestones.

At Chilton the base of the Middle Lias was marked by a 1 ft. pebble-bed described as a 'peculiar pebbly and nodular clayey band with streaks of pebbly quartz-grit, mixed with shelly detritus and tubed by boring organisms'.[3] This indicates that the rapid change of facies which came in with or slightly before *margaritatus* times was here preceded by earth-movements, probably along the margin of the London landmass to the north.

---

[1] Based on G. W. Lamplugh and F. L. Kitchin, 1911, loc. cit.; and Lamplugh, Kitchin, and Pringle, 1923, loc. cit.

[2] 1923, loc. cit., p. 197.   [3] 1923, loc. cit., p. 75.

CHAPTER VII

UPPER LIAS

Substages.	Faunizones (Zones). (Plate XXXII).	Teilzones or Epiboles (Subzones).	Dorset Strata.
LOWER AALENIAN [1]	Lioceras opalinum [1]	Cypholioceras opaliniforme	Bridport Sand 40 ft.
		Pleydellia aalensis	
		Dumortieria moorei	
YEOVILIAN [2]	Lytoceras jurensis	Dumortieria spp.	Down Cliff Clay 70 ft.
		Phlyseogrammoceras dispansum	(absent)
		Pseudogrammoceras struckmanni	
		Grammoceras striatulum	Junction Bed 5 ft.
		Haugia variabilis	
		Lillia lilli	
WHITBIAN [2]	Hildoceras bifrons and Dactylioceras commune	Peronoceras braunianum	
		Peronoceras fibulatum	
		Frechiella subcarinata	
	Harpoceras falcifer	Harpoceras falcifer	
		Harpoceras exaratum	(absent)
	Dactylioceras tenuicostatum [3]	Dactylioceras tenuicostatum	
		Tiltonoceras acutum	

THE Upper Lias takes on two principal facies: in the lower part, Upper Lias Clay; in the upper part, usually yellow sands and sandstones, variously known as the Bridport, Yeovil, Midford, Cotswold, and Blea Wyke Sands. Locally, at the top and at the bottom, are slowly-accumulated thin beds packed with ammonites and called the Cephalopod Beds.

S. S. Buckman showed by pioneer researches in the South-West of England that the sands and clays and Cephalopod Beds of various counties and districts were of different dates; that the ammonites crowded into the Cephalopod Bed in South Dorset and North Somerset were spread through a great thickness of Upper Lias Clay in the Cotswolds, and that the ammonites thinly distributed through the Bridport, Yeovil, and Midford Sands of the South-West were to be found collected together in the Cephalopod Bed of Gloucestershire.

[1] The *opalinum* zone (Lower Aalenian) is usually classed with the Inferior Oolite, but, purely for convenience in description, it is here grouped with the Upper Lias, of which it is the natural upward continuation.

[2] The terms Yeovilian and Whitbian, having been coined especially for the upper and lower parts of the English Upper Lias by Buckman, unlike most of the stage-names founded on Continental sequences, apply to the English succession exactly. They are very convenient, since one cannot speak of Upper Upper Lias and Lower Upper Lias, and they are therefore used here.

[3] *Ammonites tenuicostatus* Young and Bird, erroneously called *annulatus* by Tate and Blake. See S. S. Buckman, 1910, *Q.J.G.S.*, vol. lxvi, p. 87.

The basis of these conclusions was refinement in the identification of ammonite species and horizons. The true sequence was established by careful collecting at all available points along the outcrop, and in a series of classic papers published between 1875 and 1922 Buckman showed in more and more detail how these ammonite horizons transgressed the lithic planes previously employed in mapping and stratigraphy. At the same time he led the band of palaeontologists who insisted that the faunas must be trusted as chronological guides and the lithic planes ignored if chaos was to be avoided—a principle which is now hardly questioned. Of all the formations of the Jurassic, the Upper Lias is perhaps the best illustration of this principle, for it provides a unique combination of multiple and widespread ammonite epiboles and differing conditions of sedimentation in different districts (see fig. 30, facing p. 165).

I. THE DORSET-SOMERSET AREA

(a) The Dorset Coast[1]

The Upper Lias first appears at Seatown, on the downthrown (east) side of the fault that lets down the JUNCTION BED between the Middle and Upper Lias to the top of Down Cliff. From Seatown it runs eastward almost horizontally, but rising slightly, along the face of Thorncombe Beacon to Eypesmouth. Beyond this, after appearing for only a short distance at the beginning of Watton Cliff, it is cut off by another fault downthrowing to the east (fig. 29, p. 154).

Above the Junction Bed is 70 ft. of the local Upper Lias Clay, better known as the DOWN CLIFF CLAY from its forming the top of Down Cliff. Finally, capping the clay beneath the Greensand of Thorncombe Beacon, follow the bright yellow Bridport Sands (Plates V, VI). These are better seen on either side of Bridport Harbour and Burton Bradstock, where they build cliffs of curious layered appearance, formed of hard concretionary sandstone bands alternating with soft chrome yellow sand. Their total thickness is 140 ft.

In order to bring out more vividly the changing lithic facies of the various zones of the Upper Lias as they are traced across the country and to aid the memory, they will be described under their lithological headings, the fauna noted, and comparisons drawn with other localities and the standard sequence of Dorset—a departure from the usual practice.

LOWER AALENIAN AND YEOVILIAN, 210 ft.

BRIDPORT SANDS, 140 ft. The highest bed of the Bridport Sands is the *Scissum* Bed, $1\frac{1}{2}$ ft. thick, a hard grey sandstone or sandy limestone here regarded as the base of the Inferior Oolite. This bed is exposed in the cliffs at Burton Bradstock (but is accessible in only a few places) and also near the top of Chideock Quarry Hill farther west.

For about 25 ft. below the *Scissum* Bed are brown sands and sand-burrs with fragmentary and badly-preserved ammonites of the *Pleydellia aalensis* type, and in the highest 6 ft. the brachiopods *Rhynchonella pentaptycta* Buck., *R. stephensi* and *Zeilleria oppeli*.

[1] Based on S. S. Buckman, 1890, *Q.J.G.S.*, vol. xlvi, pp. 518–21; 1910, ibid., vol. lxvi, pp. 80–9; 1922, ibid., vol. lxxviii, pp. 378–436; and J. F. Jackson, 1922, ibid., pp. 436–48; 1926, ibid., vol. lxxxii, pp. 490–525.

PLATE VI

Photo.   *Purch.*

Cliff of Bridport Sands capped with thin Inferior Oolite, West Bay,
Bridport Harbour.

Photo.   *Purch.*

Golden Cap and Thorncombe Beacon (in distance) from Burton Bradstock.
Cliff of Bridport Sands capped with Inferior Oolite in foreground.

PLATE VII

Photo.

Godfrey Bingley.

Cliffs of Whitbian shales overlying Middle Lias, Hawsker Bottom, Whitby.

Photo.

W. J. A.

Cliffs of Yeovilian Sands (Bridport Sands) near Burton Bradstock.

The sands and irregular hard bands are bright chrome yellow.

The next fossiliferous horizon is about 40 ft. below the *Scissum* Bed, at which level are abundant fine-ribbed *Dumortieriæ* indicative of the *moorei* subzone.

Below come 100 ft. of sands and layers of sand-burrs in which no fossils have been found. Buckman assigned them tentatively to the *moorei* subzone.[1] The transition to the clay below is abrupt.

UPPER LIAS CLAY (DOWN CLIFF CLAY), 70 ft. This is a uniform blue-grey clay, becoming sandy towards the base. Buckman records the following ammonites:

At 12 ft. down: *Dumortieria* cf. *striatulocostata*, *D.* cf. *radians*, *D.* cf. *pseudoradiosa*.

At 40 ft. down: *Dumortieriæ* fragments.

At 50 ft. down: *D.* cf. *costula*.

At the base: *D.* cf. *striatulocostata*.

The clay everywhere rests abruptly and non-sequentially, with the absence of the *struckmanni* and *dispansum* subzones found in other parts of England, upon the surface of the Junction Bed, the greater part of which belongs to the Whitbian.

WHITBIAN

JUNCTION BED, $2\frac{1}{2}$–$4\frac{1}{2}$ ft. In the Junction Bed four separate ammonite subzones can be recognized, embracing three of the five zones of the Upper Lias and part of the topmost zone of the Middle Lias: namely the subzones of *striatulum*, *bifrons*, *falcifer* and *spinatum*.

The Upper Lias part contains three layers. Wherever the upper surface of the topmost or *striatulum* layer can be examined it shows evidence of protracted but quiet erosion before the deposition of the Down Cliff Clay. It is planed off perfectly level, so that the ammonites are displayed in section, sometimes 'almost as accurately cut through as if sliced on a lapidary's wheel'.[2] The matrix varies from a hard rubbly buff limestone to a soft earthy marl, with harder lumps and limonitic nodules. The fossils are sometimes rolled and perished.

The *bifrons* layer is a tough limestone with a smoothed top, on which the *striatulum* layer reposes with perfect conformity. It is sometimes wholly fine-grained, but generally contains pebbles of a similar rock, together with a mixture of fresh and derived specimens of *Hildoceras bifrons* and *Dactylioceras commune*. The *bifrons* layer is subdivided by two or three minor planes of erosion, and occasionally the ammonites on these planed surfaces are in the same condition as those at the top of the *striatulum* layer. The removal of the upper parts of ammonites, often standing highly inclined or even vertically, while the lower parts remained held in the matrix, is proof that the rock must have become completely consolidated before the erosion took place.

The division between the *bifrons* and *falcifer* layers is usually rather obscure. The *falcifer* layer is a tough yellowish-pink limestone, often mottled with red, occasionally greenish. It varies from a fine-grained compact rock to a

---

[1] For the sake of simplicity the word 'subzone' is used throughout this chapter, where it occurs with especial frequency. For a discussion of its true meaning (epibole or teilzone) see pp. 17–35. As in the Lower Lias chapter, 'zone' implies faunizone.

[2] J. F. Jackson, 1922, *Q.J.G.S.*, vol. lxxviii, p. 437.

conglomerate of pebbles of similar limestone with broken and rolled Harpo-cerates. As in the *bifrons* layer, there are several minor planes of erosion within the layer. The *falcifer* layer rests upon a planed-off surface of the Middle Lias Marlstone, either welded to it, or separated by a thin ferrugi-nous seam with *Pleurotomaria mirabilis* Desl. and *P. subnodosa* Goldf. (? = *P. precatoria* Day).

Three-quarters of a mile to the east of the point where the Junction Bed runs out on the side of Thorncombe Beacon, it reappears for a short distance on the other side of Eypesmouth, where it abuts against the great fault of Watton Cliff. In this exposure it is thicker (4 ft. 7 in.) and all the various layers have changed their lithic characters to such an extent that they are unrecog-nizable. Buckman attributed the changes in so short a distance to proximity to the Weymouth anticline.[1]

(b) Somerset

Between Bridport on the coast and Ilminster in Somerset, where the outcrop turns eastwards, little is known of the lower part of the Upper Lias owing to the poverty of outcrop and lack of exposures. But from Ilminster it is per-sistent through the rest of Somersetshire to the Mendips.

In general the main divisions of the formation in Somerset resemble those seen on the coast, but there are certain interesting differences.

LOWER AALENIAN AND YEOVILIAN, 185 ft.

YEOVIL (and BRIDPORT) SANDS, 185 ft.[2] The Bridport Sands of the coast can be followed inland past Stoke Knap and Broad Windsor into Somerset, where in the Crewkerne and Ilminster district they thicken to rather over 180 ft. The thickening is due to the extension of the sands downwards into the *Dumortieriæ* subzone, and the replacement of the whole of the Down Cliff Clay by an arenaceous facies continuous with the Bridport Sands above. The Yeovil Sands, therefore, are equivalent to the Bridport Sands plus the 'Upper Lias Clay' of the coast.

In the Ilminster and Crewkerne district[3] the top beds of the sands are usually indurated. At the top is a *Scissum* Bed of hard sandy limestone or calcareous sandstone. Beneath this the next few feet of the sands are also sometimes indurated, or at least form a coherent sandstone, and yield *Cypholioceras opaliniforme*. At Whaddon Hill, near Beaminster, the *opaliniforme* subzone is represented by beds full of brachiopods. The subzone of *Pleydellia aalensis* has been detected at Furzy Knaps, near Seavington, where it yields *P. leura* Buck., *Canavarina venustula* Buck., and *Cotteswoldia subcandida* Buck. At North Perrott *Pleydellia aalensis* occurs in a remanié bed, and somewhere between North Perrott and Yeovil Junction the *aalensis*, *opaliniforme* and *scissum* zones die out, for in the Sherborne district (except for a trace at Marston Road Quarry[4]) they have disappeared, leaving Inferior Oolite lime-stones of the *murchisonæ* zone resting directly on sands of *moorei* date.

---

[1] It should be noted that the section of the Junction Bed at Watton Cliff was first described from a slipped block upside-down, an error which caused embarrassing difficulties and much theoretical discussion. The mistake was discovered and corrected by Mr. Jackson in a later paper (1926, *Q.J.G.S.*, vol. lxxxii, p. 511).

[2] S. S. Buckman, 1889, *Q.J.G.S.*, vol. xlv, pp. 440–73.

[3] L. Richardson, 1918, *Q.J.G.S.*, vol. lxxiv, pp. 148–63.

[4] L. Richardson, 1930, *Proc. Cots. N.F.C.*, vol. xxiv, p. 70.

The main bulk of the Yeovil Sands in Somerset consists of chrome yellow sand with lines of sand-burrs, and belongs to the *moorei* and *Dumortieriæ* subzones. In the Crewkerne and Ilminster district there is an interesting local development of a thick limestone in the *moorei* subzone. It first appears at North Perrott in the form of a hard brown limestone, 18 ft. thick, largely made up of shell-fragments, and called the PERROTT STONE.[1] It thickens rapidly towards the great earthwork-encircled bluff of Hamdon Hill, 5 miles west of Yeovil, where it has been quarried since mediaeval days as a renowned building stone under the name of HAM HILL STONE. The best freestone at Ham Hill occurs in immense blocks consisting of a mass of comminuted shells, and lends itself to sawing and working into elaborate designs. It is 50 ft. thick and the sand for 30 ft. above it contains occasional layers of similar but rougher stone. *Homœorhynchia cynocephala* and very rarely an ammonite of the genus *Dumortieria* occur in the freestone, and in the sands, 55 ft. below, Buckman found a fragment of *Dumortieria rhodanica* in a hard band.[2]

The Ham Hill Stone is purely a local development or facies of the Yeovil Sands, and in a distance of 6 miles towards the east it thins down to 2 ft. at Stoford. Here it is similar in appearance, and is still used for building, but the shells are less comminuted and are identifiable. They include *Dumortieria moorei* (Lyc.), *D. subundulata* (Branco), *Grammoceras mactra* Dum., *Trigonia literata*, *Astarte elegans* and *Ceratomya bajociana*. Still farther east, at Bradford Abbas near Sherborne, the same subzone and probably even the same horizon is identifiable in a foot or two of hard, shelly, sandy limestone called the Dew Bed or Dhu Bed, which yields the same fauna.[3]

Near Yeovil a large proportion of the lower part of the Yeovil Sands passes into hard micaceous sandy shale. The base of the sands is rarely visible, but it was exposed in a recent excavation at Barrington, near Ilminster, where it yielded *Phlyseogrammoceras dispansum* and *Dumortieria*? sp., an indication of the presence of a subzone undetected in Dorset. Below this, resting on a rock-band equivalent to the Junction Bed, were 6 in. of black clay with *Hammatoceras* cf. *insigne*, *Grammoceras* spp., *Pseudogrammoceras* cf. *grunowi* (Dum. non Hauer), *Haugia* sp., and *Hildoceras* sp.[4]

WHITBIAN

UPPER LIAS CLAY, 5–10 ft. As on the coast, the zones of the Whitbian are very much attenuated. They consist of from 5 to 10 ft. of blue, grey and brown marly clay with thin, irregular, nodular and impersistent bands of pale, earthy limestone, rich in ammonites indicative of the *bifrons* and *falcifer* zones. Occasionally the calcareous bands coalesce to form a white limestone. In the recent exposures at Barrington, referred to above, Dactyliocerate ammonites were more abundant and better preserved than at any other locality in England. So great was the wealth of forms collected *in situ*, inch by inch, that one of Buckman's last actions, in the last part of *Type Ammonites* published in his lifetime, was to found upon them twenty new 'genera' of *Dactylioceras*.[5]

[1] L. Richardson, 1918, loc. cit., p. 163.
[2] S. S. Buckman, 1889, *Q.J.G.S.*, vol. xlv, p. 449; and L. Richardson, 1911, *P.G.A.*, vol. xxii, pp. 258–60.
[3] S. S. Buckman, 1893, *Q.J.G.S.*, vol. xlix, p. 485.
[4] J. Pringle and A. Templeman, 1922, *Q.J.G.S.*, vol. lxxviii, p. 450.
[5] *T.A.*, vol. vi, 1926, pp. 41–6.

A note of warning (and of comfort) about these is sounded by Dr. Spath.[1] 'Such Upper Liassic successions as that of Barrington', he writes, 'are now recognized to be of only very limited value for wider correlations on account of local accidents of collecting, horizontal distribution of species, &c.' 'The Dactyliocerates, with their apparently infinite diversity, are a group to mislead those who are insufficiently acquainted with ammonite development as a whole. There are already too many of these ammonite "species". The collection of the late Mr. James Francis, recently presented to the British Museum, contains a large number of similar Whitby forms of doubtful horizons in the Upper Lias that are probably all referable to known species and their varieties. Now the whole outlook on ammonites has changed, it is not considered advisable to name more of these "types". Some authors may disagree, and I am prepared for the usual threadbare arguments to justify the naming of each ammonite individual with a (supposititious) hemera of its own. When the necessary zonal collecting in the Upper Lias has been done it will be found that these numerous names are a hindrance rather than a help . . . in other words, this "modern" method urgently requires bringing up to date'.

The lowest of the nodular limestone bands is of the greatest interest and is known as the SAURIAN AND FISH BED. The nodules, which are up to 6 in. thick and sometimes septarian, often enclose bones and other fossils, of which a magnificent collection was made by Charles Moore and may now be seen in the museum at Bath. The ammonites were considered by Buckman to indicate the *exaratum* subzone, a horizon lower than any detected on the Dorset coast. Among the products of this bed are *Ichthyosaurus acutirostris*, *Pelagosaurus typus*, *P. moorei*, *Lepidotus*, *Leptolepis*, *Hybodus*, *Pachycormus*, *Coleia moorei*, *Penæus* sp., and *Palinurina* sp.; also the belemnite *Acrocœlites ilminsterensis* (Phil.) with ink-sacs preserved, insects and plant remains.[2]

Below this bed in the Ilminster district is 1 ft. 1 in. of green, yellow, and brown clay called the LEPTÆNA BED, with a remarkable fauna of brachiopods: *Leptæna* (*Koninckella*) *bouchardi*, *L. moorei*, *Thecidium rusticum*, *Spiriferina ilminsterensis*, *Zellania liassica*, *Terebratula globulina*, *Rhynchonella pygmæa*; also ostracods and foraminifera.

Finally, the topmost 10 in. of the Marlstone yield species of *Dactylioceras*, which are sufficient to indicate the basal zone of the Whitbian (zone of *Dactylioceras tenuicostatum*) and with this zone, on the strength of evidence obtained farther north, the *Leptæna* Beds should probably be classed. *Harpoceras falcifer* and *H. exaratum* occur together in a clay band between limestones within a foot of the top of the Marlstone at Batcombe, near the Mendip Axis.[3]

II. THE MENDIP AXIS: FROME AND BATH DISTRICT

To within a short distance of the south side of the Mendip Hills the Upper Lias, like the Middle and Lower, is essentially normal and well-developed. On the outliers of Glastonbury Tor and Brent Knoll the Whitbian consists of from 10–15 ft. of compact, earthy limestone bands and clay, full of ammonites of the *bifrons* and *falcifer* zones, overlain by about 30–35 ft. of 'Upper

[1] *Naturalist*, 1926, p. 321.
[2] Charles Moore, 1867, *Proc. Somerset Arch. N. H. Soc.*, vol. xiii, pp. 130–3, gives a figure of the old section at Strawberry Bank, Ilminster, from which the collections were obtained. The macrurous crustacea have been revised by H. Woods, 1925, 'Mon. Foss. Mac. Crustacea', *Pal. Soc.*, pp. 3, 26, &c.
[3] L. Richardson, 1909, *Geol. Mag.* [5], vol. vi, pp. 540–2.

Lias Clay', the zonal position of which is not known with certainty.[1] From ammonites found in an analogous clay on the Dundry outlier on the North side of the Mendips, it may be presumed to be a return of the Down Cliff Clay belonging to the *Dumortieriæ* subzone. The sandy Yeovilian caps the outliers with thick sands and sandstone bands, 175 ft. thick on Glastonbury Tor and about 200 ft. thick on Brent Knoll.

Northward along the main outcrop, however, clays and sands and all traces of the Upper and Middle Lias die out completely. As previously mentioned, all the exposures south of the Wellow Valley show Upper Inferior Oolite resting directly on the *davœi* zone at the top of the Lower Lias, which is itself in some places eroded down until only 8 ft. is left.

At Wellow, 4 miles north-east of Radstock, pockets of earthy ironshot limestone up to $1\frac{1}{2}$ ft. thick are left between the Lower Lias and the Inferior Oolite as the sole relics of the Upper Lias. Four layers can be recognized in these pockets: the topmost is much bored by *Lithophaga*; the next contains *Pseudogrammoceras dœrntense* Denckmann, mixed with *Hildoceras bifrons*; the next contains *Hildoceras* and *Dactylioceras*; the lowest *Harpoceras falcifer* and ammonites perhaps of the *bifrons* zone.

Barely a mile farther north, at Timsbury Sleight (map, fig. 13, p. 70), the mere pockets of Upper Lias have thickened to a continuous stratum containing miniature representatives of most of the zones of the formation. Once again the normal division into sands above and cephalopod limestones below can be applied—5 ft. of yellow sands and 4 ft. of ironshot brown limestones. The topmost bed of the sands (3 in.) is hardened and in places conglomeratic, and yields *Dumortieria subsolaris* Buck. The rest of the sand is unfossiliferous, but from neighbouring exposures it is known to belong to the *dispansum* subzone, and to be the first appearance of the MIDFORD SANDS. The 4 ft. of ironshot oolitic limestone is almost exactly equivalent to the Junction Bed of the Dorset coast, except that its topmost layer contains *Pseudogrammoceras dœrntense* Denckmann, and *Ps. quadratum* (Haug), indicating the *struckmanni* subzone, which is unrepresented in Dorset. The central layers contain *Ps. pedicum* Buck. and other species, and *Esericeras inæquum* Buck, which are sometimes considered to denote separate subzones under the names *pedicum* and *eseri*; the basal layers contain *Grammoceras penestriatulum* Buck. and *Gr. toarcense* (d'Orb.) (*striatulum* subzone), *Hildoceras hildense* (Y. and B.), and allied species, *Dactylioceras commune* (Sow.) and *D. annuliferum* (Simpson) (*bifrons* zone).[2]

At the well-known locality of Midford, north-east of Wellow Railway Station, the Midford Sands have expanded in 2 miles from nothing to 100 ft. They are best seen in a road-cutting south of the viaduct, and at Greenway Lane and Lyncombe railway-cutting to the north, $1\frac{1}{2}$ miles away, at both of which places they are overlain by the Upper Inferior Oolite. The upper 65 ft. are unknown palaeontologically, but lines of sand-burrs in the lower 35 ft. have yielded *Grammoceras fallaciosum* and *Ps. struckmanni* (*dispansum* and *struckmanni* subzones). They rest on a Junction Bed or Cephalopod Bed from $1\frac{1}{2}$ to $2\frac{1}{2}$ ft. thick, consisting of sandy limestone shot with oolite grains. As usual it

[1] H. B. Woodward, 1893, *J.R.B.*, p. 262.
[2] J. W. Tutcher, 1925, *Q.J.G.S.*, vol. lxxxi, p. 621; and L. Richardson, 1907, ibid., vol. lxiii, pp. 413–16.

is divisible into layers, firmly cemented together: the upper layer contains *G. striatulum*; the middle layer *H. bifrons*, *D. holandrei*, and *D. crassum*; and the lowest layer *Harpoceras falcifer*.[1]

A change, then, is found on crossing to the north side of the Mendip Axis, namely the migration of the lower boundary of the Yeovil Sands downward into the *dispansum* and *struckmanni* subzones, which are unrepresented in Dorset but would be expected below the base of the Down Cliff Clay. Nevertheless Prof. Boswell has shown that the sands of different ages on the two sides of the axis cannot be distinguished petrologically; the source of their materials was apparently the same.

III. THE COTSWOLD AREA

(a) Dundry Hill[2] and the South and West Cotswolds[3]

On passing northward into the Cotswolds the downward migration of the sand is found to continue into progressively lower subzones, while the *opalinum* zone and the upper subzones of the *jurensis* zone, the equivalents of the sands in Dorset, thin down to a few feet of limestone, the COTSWOLD CEPHALOPOD BED. This resembles the Junction Bed of the coast, but it contains a completely different suite of ammonites and its position is above instead of below the sands and clay.

The first signs of the Cotswold Cephalopod Bed are met with on Dundry Hill outlier, about 10 miles to the west of the main outcrop near Timsbury Sleight. Here the *opaliniforme* and *aalensis* subzones are compressed into 1 ft. 9 in. of ironshot limestone, continuous with the *murchisonæ* limestones of the Inferior Oolite above. But there is also a remarkable feature, already foreshadowed in the Glastonbury and Brent Knoll outliers south of the Mendips, in the form of a return of the Down Cliff Clay, 60 ft. thick, in its original position in the *Dumortieriæ* subzone. Beneath the clay a second band of ironshot limestone (the Blue Ironshot Bed, 1 ft.–1 ft. 6 in.), as in the Midford district, contains layers representing the *dispansum*, *striatulum*, *bifrons*, and *falcifer* subzones, the last in a pinkish seam at the base (the Pink Bed, 2–3 ft.). The Upper Lias of Dundry, therefore, consists of 60 ft. of *Dumortieriæ* clay with a thin band of ironshot limestone at top and bottom, the upper representing part of the Cotswold Cephalopod Bed, the lower the Dorset Junction Bed.

Along the main outcrop of the South Cotswolds, the whole of the clay and the upper limestone bed of Dundry become absorbed into the Cephalopod Bed, while the zones below expand to form first the Midford Sands, and later the Cotswolds Sands.

THE CEPHALOPOD BED maintains a partly ironshot character along most of the outcrop, consisting typically of layers of fossiliferous ironshot marl and limestone, often crowded with ammonites.

In the southern part of the escarpment, about Bath and as far north as Dodington, the unconformable overstep of the Upper Inferior Oolite cuts out all the highest subzones of the Upper Lias. The top of the sands is of

---

[1] L. Richardson, 1907, *Q.J.G.S.*, vol. lxiii, pp. 406–8, and Table 2.
[2] S. S. Buckman and E. Wilson, 1896, *Q.J.G.S.*, vol. lii, pp. 669–720.
[3] Based on S. S. Buckman, 1889, *Q.J.G.S.*, vol. lxv, pp. 440–6, and 'Mon. Ammonites Inf. Oolite', *Pal. Soc.*; and L. Richardson, 1910, *Proc. Cots. N.F.C.*, vol. xvii, pp. 63–136.

dispansum date, and there is thus a non-sequence corresponding with the whole of the Aalenian and Bajocian Stages.

In a short distance towards the north the gap is rapidly filled, for at Little Sodbury there are 14 ft. of Lower Inferior Oolite and beneath it 11 ft. of Cephalopod Bed—ironshot marls and limestones, yielding ammonites of *opaliniforme*, *aalensis*, *moorei*, *Dumortieriæ* and even *dispansum* dates.[1] Here, therefore, the upper part of the Midford Sands is beginning to turn into Cephalopod Bed.

Six to ten miles farther north, at Wotton under Edge and Nibley Knoll, the *dispansum*, *struckmanni* and *striatulum* subzones have lost their sandy facies entirely and have all become a part of the Cephalopod Bed, which measures some 14 ft. in thickness. Here the layers include ammonites of *striatulum* to *opaliniforme* dates. The deposits of *dispansum*, *struckmanni* and *striatulum* dates, which form the thick Midford Sands farther south, are already very thin, and before reaching Haresfield Beacon they disappear altogether.[2]

The Cephalopod Bed attains its greatest thickness in the neighbourhood of Dursley and Stroud, the measurement at Stinchcombe, near Dursley, being 20 ft. Beyond this it diminishes again, and near Painswick, seven miles to the north-east, only 1 ft. 10 in. is left.

MIDFORD AND COTSWOLD SANDS. At the extreme south of the South Cotswolds the Midford Sands continue, lying within the *dispansum*, *struckmanni* and possibly in part within the *striatulum* subzones, and directly overlain by the Upper Inferior Oolite. Gradually towards the north the sands migrate downward into successively lower subzones while the upper parts pass into new accretions to the Cephalopod Bed. By Little Sodbury the Sands are already COTSWOLDS SANDS and 185 ft. thick, occupying the *striatulum—lilli* subzones. Here the greater part is of *striatulum* date, but 6 miles farther along the escarpment, at Wotton Hill Quarry, above Wotton under Edge, the *striatulum* subzone has joined the Cephalopod Bed, and consists of only 1 ft. 2 in. of ironshot limestone containing *Grammoceras striatulum* and *G. toarciense*, welded to the base of the other layers of the Cephalopod Bed. Here the Cotswold Sands are 123 ft. thick and seem to fall mainly in the *variabilis* and *lilli* subzones, but partly also in the *bifrons* zone, for clay with *Harpoceras falcifer* lies immediately below.

The Cotswold Sands reach thicknesses of 230 ft. at Stinchcombe, 130–40 ft. in Coaley Wood near Nympsfield,[3] and 190 ft. at Haresfield Beacon, west of Painswick. The lowest extension proved is about Frocester Hill, near Nympsfield, south-west of Stroud, where there is a band full of *Hildoceras bifrons* and still 40 ft. of sands below.[4]

As a result of petrological examination of the sands of various ages in the Cotswolds and the West of England, Prof. Boswell has been able to show that the source of the sands, which began to be deposited first about Frocester Hill and spread south over Somerset and Dorset through succeeding hemerae, was always the same, the composition and grade remaining constant throughout. He found that 'the sands differ markedly in petrology from the various

---

[1] L. Richardson, 1910, loc. cit., pp. 93–4.
[2] Ibid., pp. 105–7.
[3] Ibid., 113, 116.
[4] S. S. Buckman, 1889, loc. cit., p. 444.

Palaeozoic rocks of Wales and the West Country, from the Trias, and from the Cretaceous and Eocene of Devon and Dorset',[1] and he concluded that they were probably derived from the south-west, in the direction of Brittany. This result is rather difficult to reconcile with their first appearance in the Cotswolds, while clay was still being deposited along the rest of the outcrop farther south. We may, perhaps, suppose them to have been brought by a river flowing round the Bristol Channel and debouching north of the Mendip Axis into the south of the Cotswold Basin.

UPPER LIAS CLAY AND JUNCTION BED. Beneath the sands in the South Cotswolds are thin argillaceous representatives of the lowest zones of the Whitbian, but owing to lack of sections they are little known palaeontologically.

In the outskirts of Bath, at Primrose Hill, east of Weston, the Midford Sands rest upon a basal Junction Bed of *bifrons* date, as at Timsbury Sleight.[2]

At Little Sodbury the base of the sands is marked by a pyritous layer full of *Hildoceras bifrons*, beneath which is a compact marly limestone with *Dactylioceras commune*, overlying cream-coloured compact marl with *Harpoceras falcifer*. These basal beds, which rest upon the Marlstone, together measure 10 ft. in thickness. They were penetrated by the Sodbury Tunnel.

At Wotton Hill Quarry, Wotton under Edge, the Cotswold Sands are separated from the Marlstone by 10 ft. of grey, sandy clay with nodules, apparently of *falcifer* date. A few miles farther north these basal beds become more predominantly argillaceous and much thicker, forming the 'Upper Lias Clay'. At Cam Long Down, north of Dursley and Uley, the clay is estimated to have a thickness of 70 ft.[3]

(b) The Central and North Cotswolds

North of Frocester the Upper Lias Clay thickens greatly at the expense of the sands, which dwindle from 230 ft. to almost nothing between Stinchcombe and Leckhampton Hill, Cheltenham. The clay is 130 ft. thick at Haresfield Beacon and attains a maximum of about 270 ft. in the Mid-Cotswolds (at Cleeve Hill and on the outlier of Bredon Hill). At first it continues to be capped here and there by traces of sand, probably always of *variabilis* date, directly overlain by the *Scissum* Beds (for instance at Leckhampton Hill).[4] At Bredon Hill, however, where the whole 270 ft. consists of clay with occasional bands of limestone-nodules, the missing zones return and nearly the complete sequence of the Upper Lias is represented in clay facies up to the *moorei* subzone. There are no exposures, but Buckman obtained evidence for the existence of the *moorei*, *Dumortieriæ*, *struckmanni* and *variabilis* subzones in the ammonites obtained from claystone nodules in a gravel pit and a gateway on the hill.[5]

From these facts it may be inferred that a minor anticlinal uplift is crossed in passing from the Stroud district into the Mid-Cotswolds east of Cheltenham, and that its axis corresponds with the Birdlip Anticline, so conspicuous

[1] P. G. H. Boswell, 1924, *Geol. Mag.*, vol. lxi, p. 262.
[2] L. Richardson, 1910, loc. cit., p. 88.
[3] L. Richardson, 1910, loc. cit., pp. 106, 114.
[4] The Geological Survey maps depict a continuous layer of Cotswold Sand as far east as Stow on the Wold, but this is in most places sandy *Scissum* Beds. See L. Richardson, 1929, 'Geol. Moreton in Marsh', *Mem. Geol. Surv.*, p. 28.
[5] S. S. Buckman, 1903, *Q.J.G.S.*, vol. lix, pp. 445–64.

in the Inferior Oolite (see p. 75). Bredon Hill and the Mid-Cotswolds lie in a synclinal trough between the anticlines of Birdlip and the Vale of Moreton.

In this synclinal area, where the Upper Lias and Inferior Oolite attain their fullest development, the *falcifer* and *tenuicostatum* zones are represented by an interesting SAURIAN AND FISH BED and paper-shales with a micromorphic brachiopod fauna—the LEPTÆNA BED—as in the Ilminster district of Somerset. The presence of the same beds was recognized by Charles Upton also in the Stroud district, in the corresponding synclinal area on the south side of the Birdlip Anticline.[1]

These fossiliferous beds were exposed many years ago at Gretton, near Winchcombe, and on the adjoining Dumbleton outlier, whence they were known as Dumbleton Beds. The Saurian and Fish Bed consisted of 6 to 12 in. of laminated nodules of limestone with *Dapedius, Euthynotus, Leptolepis, Pachycormus, Pholidophorus* and *Tetragonolepis* among the fish, and numerous insect remains which are still under investigation.[2] Beneath were 20 ft. of paper-shales with some lines of nodules and the small *Terebratula globulina* and *Rhynchonella pygmæa*; and at the base was a layer with ammonites of the *tenuicostatum* zone, forming the top of the Marlstone. Discoveries by Prof. Trueman of *D. tenuicostatum* in similar paper-shales over the Marlstone in Lincolnshire provide a comparison for the dating of this lower part of the Dumbleton Beds.[3]

IV. THE VALE OF MORETON AXIS AND OXFORDSHIRE

Eastward, towards the Vale of Moreton Axis, the Upper Lias (and the Inferior Oolite) thin out rapidly, and the sandy facies does not return. The Upper Lias Clay is only about 80 ft. thick along the western side of the Vale of Moreton, at the Bourtons, Stow on the Wold and Burford, and the upper zones are shown to be probably missing again, as around Birdlip, by the discovery of ammonites of the *bifrons* zone a few feet below the *Scissum* Beds near Stow on the Wold.[4]

Towards the Evenlode Valley further diminution in thickness reduces the Upper Lias to about 30–40 ft. at Chipping Norton, 20 ft. at Milton under Wychwood, 5–12 ft. at Fawler, near Charlbury, and 14 ft. under Oxford (proved in a boring at Wytham). The most reduced representation of the Upper Lias is to be seen in the old Marlstone workings at Fawler, where little but an unfossiliferous clay-band, 5–12 ft. thick, separates the ironstone from an attenuated representative of the Inferior Oolite. Nearly all the Whitbian zone-fossils are present at the base of the clay, in a thin layer of earthy limestone crowded with ammonites.[5] In a railway-cutting near Bloxham Railway Station, on the west side of the Cherwell Valley, the *acutum, tenuicostatum* and *exaratum* subzones appear to be absent. A thin bed of limestone of *falcifer* date was found welded to the top of the Marlstone, above which followed clays of the *bifrons* zone, yielding *Frechiella subcarinata*. This locality lies some distance to the north of Fawler, and the total thickness of Upper Lias clay is 70 ft.[6] (see discussion on pp. 66–7 above).

[1] C. Upton, 1906, *Proc. Cots. N.F.C.*, vol. xv, pp. 201–7.
[2] For revised list of fossils see Richardson, 1929, loc. cit., pp. 32–3.
[3] See below, p. 179.   [4] L. Richardson, 1929, loc. cit., p. 30.
[5] W. J. Sollas, 1926, in *Nat. Hist. Oxford District*, p. 35; and *J.R.B.*, 1893, p. 268.
[6] L. Richardson, 1921, *Geol. Mag.*, vol. lviii, pp. 426–7.

V. THE MIDLANDS: OXFORDSHIRE TO THE HUMBER

Beyond the Vale of Moreton Axis and the Oxfordshire shallows another great change comes over the Upper Lias. On the South Coast and through the West Country the Yeovilian has played the most important part, the Whitbian being generally represented by only a few feet of strata. From now onwards to the coast of Yorkshire the Whitbian assumes the leading role, so that Upper Lias becomes virtually synonymous with Whitbian. In short, the Upper Lias north and east of the Vale of Moreton Axis is older than the bulk of that to the south and west. Moreover, towards the north successively lower zones thicken at the expense of the higher. In the Midlands, especially in Northamptonshire, the *bifrons* zone increased in thickness to about 150 ft., and it has been found necessary to divide it into three subzones, *subcarinatum*, *fibulatum*, and *braunianum* (from below upwards). Of the total thickness, the lowest subzone, that of *Frechiella subcarinata*, occupies only 5 ft. in Northamptonshire, while the *falcifer* and *tenuicostatum* zones below are together only 6–8 ft. thick. In South Lincolnshire the *subcarinatum* subzone thickens to 51 ft. in the Grantham district, while the *falcifer* and *tenuicostatum* zones together expand to 40 ft. Still farther north, at Lincoln, the *tenuicostatum* zone alone is more than 18 ft. thick, while in Yorkshire it reaches 30 ft.[1]

The Midland outcrop, therefore, well illustrates a principle pointed out by Buckman in 1910, that during the deposition of the Upper Lias the belt of maximum sedimentation migrated southward from Yorkshire towards Dorset. 'Owing to this migration of the area of maximum deposit', he wrote, 'it happens that the strata of the Toarcian in any one English locality do not much exceed 250 ft. in thickness, and are often far less; yet the amount of work done in deposition during that time is equal to 850 ft. or more'.[2] Eight years later Prof. Trueman published the results of a detailed examination of the Lias of South Lincolnshire, and now, by comparison with Mr. Beeby Thompson's accounts of the Northamptonshire Lias, the zones can be followed across the Midlands in considerable detail.

Whatever may have been the nature of the earth-movements causing this migration of the area of sedimentation, the results were considerably affected by the uplifts along certain established axes, of which the principal were those of Market Weighton and of the Vale of Moreton with its adjoining 'Oxfordshire shallows'. This is apparent from the over-all thicknesses: the 12 ft. of clay and the thin fossil bed at Fawler thicken rapidly to 50–70 ft. of clays in the Banbury district (70 ft. at Bloxham), and to as much as 200 ft. in Northamptonshire, Leicestershire and Rutland (average 180 ft. in Northants, 176 ft. in a boring at Oakham, Rutland).[3] In Lincolnshire they thin down to about 120 ft. in the south of the county, around Grantham, but were proved to be 199 ft. thick again 8 miles north of Grantham, in a boring at Caythorpe.[4] In the next 14 miles from Caythorpe to Lincoln there is a rapid diminution to 100 ft. at Lincoln, and the tendency is continued until

[1] A. E. Trueman, 1918, *Geol. Mag.* [6], vol. v, pp. 103–11.
[2] S. S. Buckman, 1910, *Q.J.G.S.*, vol. lxvi, p. 88.
[3] H. B. Woodward, 1893, *J.R.B.*, pp. 272, 284, &c.
[4] H. Preston, 1903, *Q.J.G.S.*, vol. lix, pp. 29–32.

only 50 ft. remains at Appleby and from 50 ft. to 25 ft. between Appleby and the Humber.[1]

Eastward, under Norfolk (Southery), and south-eastward, under Buckinghamshire (Calvert), borings have proved that the Upper Lias is completely overstepped by younger rocks. Early stages of this overstep are to be seen all along the Midland outcrop, where numerous remanié fossils and water-worn pebbles of Upper Lias claystone are embedded in the base of the Inferior Oolite. Prof. Trueman has pictured diagrammatically (fig. 1) the unconformable relations between the Upper Lias and Inferior Oolite from Northampton to Lincoln, and it appears that the unconformity becomes accentuated towards the Market Weighton Axis.

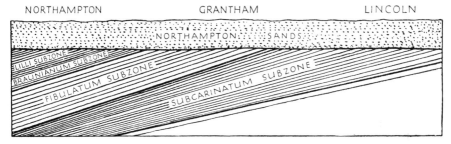

FIG. 31. Diagram showing the overstep of the Upper Lias by the basal Inferior Oolite (Northampton Sands) in Northants and Lincolnshire. (After Trueman, *Geol. Mag.*, 1918, p. 110; redrawn). (Not to scale.)

In consequence of this, the highest beds of the Whitbian are preserved in Northamptonshire, where the subzone of *Lillia lilli* has been detected by Beeby Thompson; he believes also that there may be some deposit of *variabilis* date, completing the Whitbian. The *lilli* beds are overstepped somewhere between Northampton and Grantham, and northwards the Inferior Oolite rests on successively lower horizons of the greatly expanded *bifrons* zone, until at Roxby, in North Lincolnshire, it comes into direct contact with the *falcifer* zone.

The higher clays preserved in Northamptonshire, belonging to the highest or *braunianum* subzone of the *bifrons* zone and to the lowest part of the *jurensis* zone, are especially characterized by the abundance of a small lamellibranch, *Nuculana* [= *Leda*] *ovum*.

Between the Upper Lias and the Middle Lias Marlstone is a Transition Bed with a special ammonite, *Tiltonoceras acutum*, associated with the earliest fine-ribbed Dactyliocerates of the *tenuicostatum* zone.

SUMMARY OF THE UPPER LIAS, NORTHANTS–LINCOLNSHIRE[2]

Jurensis Zone, 13–24 ft. (Northants only).

?*Variabilis*, *lilli* subzones: UPPER *Nuculana ovum* CLAYS (24 ft. west of Northampton, thinning towards the north-east). These comprise micaceous sandy clays, with clay-balls or nodules containing a peculiar assemblage of

---

[1] W. A. E. Ussher, 1890, 'Geol. North Lincs.', *Mem. Geol. Surv.*, p. 55.
[2] A. E. Trueman, 1918, *Geol. Mag.* [6], vol. v, pp. 103–11; and B. Thompson, 1910, *Geol. in the Field*, pp. 461–8.

fossils. A rolled specimen of *Lillia lilli* occurred at Moulton, from which Beeby Thompson infers that the age of the clays may be as late as *variabilis* date. At the base is a layer of waterworn, scratched, bleached nodules, overgrown with oysters and *Serpulæ*.

Bifrons Zone, 150 ft. thick in Northants.

Braunianum subzone: MIDDLE AND LOWER *Nuculana ovum* CLAYS, 70 ft.

The Middle Clays are the chief repository of *N. ovum*, and belemnites are also abundant. In the Lower Clays small gastropods abound, particularly *Cerithium armatum*, together with *Dactylioceras commune, D. crassum, Phylloceras heterophyllum*, &c. The subzone is overstepped probably not far from the northern boundary of Northamptonshire.

Fibulatum subzone: This consists in Northants of a poorly fossiliferous clay, up to 76 ft. in thickness, containing towards the top pyritized ammonites, especially *P. fibulatum, D. commune* and *H. bifrons*. Although Beeby Thompson declares that in Northants the subzone lies below the range of *Nuculana ovum*, Trueman assigns 23 ft. of beds to it at Grantham containing *N. ovum* abundantly in pockets, and he considers that the subzone should include the Lower *Nuculana ovum* Beds of Northamptonshire.[1] At Grantham Trueman records *Peronoceras* cf. *attenuatum, Phylloceras heterophyllum, Porpoceras vortex*, &c. The subzone is probably overstepped completely by the Inferior Oolite not far north of Grantham.

Subcarinatum subzone: In Northants this comprises only 5 ft. of strata, consisting of calcareous clay with numerous small, white concretions and some larger nodules containing immense numbers of Dactyliocerate ammonites, and at the top an exceptionally constant hard band (9–18 in.) full of ammonites:[2] the most conspicuous are *D. commune* and *H. bifrons* (abundant), and *Frechiella subcarinata* (rare). In the Grantham district, S. Lincs., the subzone has expanded according to Trueman to 51 ft. of grey shale with scattered nodules and with 1 ft. of dark earthy limestone about the middle. He records the same ammonites as those found in Northants, with additions, especially *Harpoceras* aff. *mulgravium* in the lower part.

Falcifer Zone, 6–8 ft. in Northants, 25 ft. in S. Lincs.

Falcifer subzone: In Northants all that can be said to belong to this subzone is some 4–5 ft. of marly clay and at the top a hard band, often oolitic, with large Harpocerates of the *falcifer* group. About Grantham Trueman assigns to it 9 ft. of grey shale with nodules, and at the top a 6-in. band of rubbly, ferruginous limestone, also with scattered ooliths, and containing many ammonites: *Harpoceras* aff. *falcifer, H. mulgravium*, Dactyliocerates, and *Harpoceratoides ovatum* (Young and Bird), which occurs at the same level also in Yorkshire.

Exaratum subzone: In Northants Beeby Thompson has recorded *Harpoceras exaratum* at Bugbrooke in an 8-in. band of hard limestone, which he

---

[1] Compare B. Thompson, 1910, *Geology in the Field*, p. 462, and A. E. Trueman, 1918, *Geol. Mag.*, p. 107.

[2] B. Thompson called this the Upper Cephalopod Bed and that in the *falcifer* subzone the Lower Cephalopod Bed, names which ought to be dropped as they may lead to confusion with the Upper Lias Cephalopod Bed of the Cotswolds.

calls the Inconstant Cephalopod Bed. Trueman assigns 15 ft. of grey shales with blue limestone nodules to the zone at Grantham, S. Lincs., where he records the index fossil, with *Dactylioceras vermis* and *Elegantuliceras elegantulum*.

Tenuicostatum Zone, up to 15 ft.

Tenuicostatum subzone: Throughout the area this subzone consists of paper-shales with fish-remains, lithologically identical with the paper-shales and *Leptæna* Bed of Dumbleton and the North Cotswolds, which lie below the Saurian and Fish Bed (? of *exaratum* date). In Northants (e.g. at Bugbrooke) the thickness is from 9 in. to a little over 1 ft., but at Grantham and Lincoln it is as much as 15 ft. At Bracebridge Brick Pit, 3 miles south of Lincoln Cathedral, where the 15 ft. of paper-shales with layers of nodules are the highest beds exposed, Trueman has found them to contain *Dactylioceras* cf. *tenuicostatum* and *D. semicelatum*, both characteristic fossils of the Grey Shales (*tenuicostatum* zone) of Yorkshire. A date is thus arrived at for the paper-shales of Dumbleton and the *Leptæna* Bed of Dumbleton and Ilminster.

Acutum subzone: The *acutum* subzone is present in three areas, forming a Transition Bed between the Marlstone and the Upper Lias: namely, on the Oxon.–Northants border; about Tilton, in Leicestershire; and in the Lincoln district. In between, the succeeding zones cut down into the Middle Lias and in certain places (as about 4 miles south-east of Tilton) the Marlstone has been completely removed.

In the two southerly areas the Transition Bed is merely a thin seam, a few inches thick, more or less joined to the top of the Marlstone, and it contains an admixture of ammonites, especially Dactyliocerates and *Tiltonoceras acutum*.[1] At Lincoln it consists of $2\frac{1}{2}$ to 3 ft. of greenish and grey shale, the lower part passing in one pit into a sandstone, and, although the fine-ribbed Dactyliocerates occur throughout, *T. acutum* appears only in the highest layer. 'Thus we must either conclude', writes Prof. Trueman, 'that only the upper portion of the Transition Bed of Lincoln is homotaxial with the Transition Bed of the Midlands, or else that *T. acutum* did not arrive in the Lincoln area until later.'[2]

VI. THE MARKET WEIGHTON AXIS[3]

The northerly attenuation of the Upper Lias observed throughout Lincolnshire is continued north of the Humber, where, after narrowing greatly for about 5 miles, the outcrop disappears beneath the Chalk at Sancton near Market Weighton, not to reappear for 13 miles to Kirby Underdale. Where last seen, the Upper Lias is from 20–35 ft. thick. At Roxby, 4 miles south of the Humber, the Inferior Oolite has been proved to rest directly on the *falcifer* zone. It is not known whether the overstep becomes any greater over the axis, but although the clays are not adequately exposed north of the Humber, evidence for the *falcifer* zone has been obtained in field rubble and ditches.

---

[1] B. Thompson, 1892, *Rept. Brit. Assoc. for 1891*, pp. 334–51; and E. Wilson and W. D. Crick, 1889, *Geol. Mag.* [3], vol. vi, pp. 296–305.
[2] A. E. Trueman, 1918, *Geol. Mag.* [6], vol. v, p. 106.
[3] C. Fox-Strangways, 1892, *J.R.B.*, pp. 141–4.

VII. THE YORKSHIRE BASIN

The Upper Lias reappears from beneath the Chalk 3 miles before the first signs of the Middle Lias, and thickens rapidly north of the Derwent to 80 ft. at Crayke, 100 ft. at Coxwold, 160 ft. at Swainby Mines and 200 ft. round the northern rim of the Yorkshire Basin. All the way it consists of dark shales and clays, overlain non-sequentially by the Dogger, or basal bed of the Inferior Oolite (*scissum-murchisonæ* zones), which has in places cut deep channels into it. Nearly everywhere the shales on which the Dogger rests are of Whitbian age, but recent discoveries by Mr. W. E. F. Macmillan of ammonites of the *striatulum* and *dispansum* subzones on Danby High Moor, north of the Esk Valley, over an area $3\frac{1}{2}$ miles long by 2 miles wide, show that at least locally the Yeovilian is represented.[1]

Only in the most south-easterly exposure in the Yorkshire Basin, on the coast from Blea Wyke Point to Ravenscar (Peak), are the uppermost Whitbian and Yeovilian beds as a rule fully developed. The Yeovilian there takes on a sandy facies, comparable in lithic character and in age with the Midford Sands.

At Peak, below the Ravenscar Hotel, a fault with a throw of some 400 ft. brings down Lower Estuarine Series (Inferior Oolite) against Middle Lias. South and east of the fault for about a mile along the cliffs, which here rise to a great height above the sea, the Yeovilian sandstones, sands and sandy shales, dipping southward, succeed the Whitbian to a thickness of 150 ft., and are overlain by the Dogger, or basal bed of the Inferior Oolite. North and west of the fault the whole of these Yeovilian strata are absent, together with the highest part (perhaps 60 ft.) of the Whitbian; so that the Dogger, after crossing the fault and being thrown some 200 ft. by it, comes to rest directly on the channelled surface of the Whitbian. Northward, in the direction of Whitby, the general dip is slightly to the north, and higher horizons of the Whitbian gradually make their appearance again below the Dogger.[2]

These facts point to an anticline having arisen north-west of the Peak Fault and to the initiation of a fracture before the end of the Whitbian period. Most of those who have studied the localization of the highest Whitbian and the whole of the Yeovilian strata on the downthrown side have agreed that when these deposits were being formed the Peak Fault must have been already in existence; in fact, that the beds were banked against the gradually rising fault scarp, while denudation proceeded upon the upthrown side. It can definitely be said that there was faulting before the deposition of the Dogger, for while the Dogger can be seen to be thrown only some 200 ft., measurements show that the base of the Upper Lias has been thrown at least 400 ft. In respect of the evidence which it provides for faulting as well as folding in England during Jurassic times, the Peak Fault is thus unique.

Another point of interest is that the ends of the downthrown strata are bent *down* towards the slide plane, while the ends of the upthrown strata curve *upward*. This is the opposite of the usual arrangement. It has been explained by Dr. Rastall as due to a slight final movement in the reverse direction,[3] but

[1] W. E. F. Macmillan, 1925–31, *The Naturalist*, 1925, pp. 236, 316; 1926, pp. 51–3; 1931, p. 345. Since going to press *Dumortieriæ* and *opalinum* horizons have been recorded: see *Proc. Yorks. Geol. Soc.*, 1932, N.S., vol. xxii, p. 122.
[2] See R. H. Rastall, 1905, *Q.J.G.S.*, vol. lxi, pp. 441–60.   [3] Ibid., p. 448.

others, particularly Prof. Kendall, regard it as a survival of the original anticlinal or monoclinal structure ('the preliminary kink') imposed on the strata before (and as a consequence of which) the fracture took place.[1]

Palaeontologically the Blea Wyke Yeovilian strata (BLEA WYKE BEDS) are on the horizon of the Down Cliff Clay of Dorset and the gap beneath it, namely the *Dumortieriæ* and *dispansum* subzones. The Bridport Sands may, however, be represented in some parts of Yorkshire (see note 1 on p. 180).

The Yorkshire Upper Lias Clay contains representatives of all the zones and most of the subzones of the Whitbian, of which it furnishes the type sections around Whitby. Many of the fine exposures are due to exploitation of what were once two commercially valuable products, alum and jet. The alum industry flourished in the eighteenth century, and the series of enormous workings along the outcrop testify to its magnitude. It was killed by the discovery of a cheaper means of production from the shales of the Coal Measures.[2] The jet industry[3] reached its zenith in the nineteenth century, the production for the peak year 1873 realizing £90,000. Its decline was rapid, however, owing to the change of fashion and to the importation of inferior but cheaper jet from abroad, chiefly from Germany and Spain.

SUMMARY OF THE UPPER LIAS OF THE YORKSHIRE COAST (RAVENSCAR, WHITBY AND HAWSKER DISTRICT)[4] (Plate VII)

YEOVILIAN

Jurensis Zone: BLEA WYKE BEDS, 100 ft.: these are divided into Yellow Beds above and Grey Beds below. The YELLOW BEDS consist of even-bedded yellow sandstones, 53 ft. thick. The highest 25 ft. were formerly classed with the Inferior Oolite, but were transferred to the Lias by Dr. Rastall (who has been followed by subsequent writers): they are unfossiliferous. Below comes a fossil-bed full of *Terebratula trilineata* auctt., and in the lowest 25 ft. of the Yellow Beds several species of *Dumortieriæ* have been found (*D. munieri* Haug at the top, *D. externicostata* Branco sp., *D.* aff. *multicostata* Buck.), together with *Pseudolioceras* cf. *gradatum* Buck. and numerous *Homœorhynchia cynocephala*. The ammonites assign at least the lower half of the Yellow Beds to the *Dumortieriæ* subzone, although in the South of England the *Homœorhynchia* is later. The GREY BEDS consist of 35 ft. of grey sandstones overlying 7 ft. of soft grey shale. *Lingula beani* Phil. occurs throughout, and there are many *Serpulæ* in the highest 10 ft., from which the names *Serpula* Beds and *Lingula* Beds have been given them. The ammonites indicate that the Grey Beds belong to the *dispansum* subzone: *P. dispansum*, *Hudlestonia affinis*, *H. wykiensis* and species of *Pseudolioceras* have been recorded. The Blea Wyke Beds contain two special Crustaceans not yet found elsewhere, *Glyphæa prestwichi* Woods and *Eryma birdi* (Bean).[5]

STRIATULUM SHALES. Below the Blea Wyke Beds are 60 ft. of darker grey

---

[1] P. F. Kendall and H. E. Wroot, 1924, *Geol. Yorkshire*, pp. 249–50.
[2] See account in Kendall and Wroot, 1924, ibid., pp. 358–63.
[3] Account in Fox-Strangways, 1892, *J.R.B.*, pp. 455–9.
[4] Based on R. H. Rastall, 1905, loc. cit.; C. Fox-Strangways and S. S. Buckman, 1915, 'Geol. Whitby and Scarborough', 2nd ed., pp. 15–22, 75–83; L. Richardson, 1911, *Proc. Yorks. Geol. Soc.*, N.S., vol. xvii, pp. 188–9; and S. S. Buckman 1911, ibid., pp. 209–12.
[5] H. Woods, 'Mon. Mac. Crust.', *Pal. Soc.*, pp. 51, 74.

argillaceous shale, full of spangles of white mica; they yield *Grammoceras striatulum, G. toarcense* and *Pseudogrammoceras latescens.*

That these Yeovilian zones are not entirely unrepresented in the area on the upthrown side of the Peak Fault is shown by the discoveries of Mr. Macmillan, already mentioned. At Danby Dale and Fryup he found *Phlyseogrammoceras, Hudlestonia* and *Grammoceras* high up on the shale slopes. At one place the ammonites were described as occurring in a 'nest' 10 ft. below the Dogger, and evidently not derived. These occurrences lie about 20 miles west of the Peak Fault.

WHITBIAN

PEAK SHALES. Below the *Striatulum* Shales is an unknown thickness of shales which have yielded species of *Haugia.* Hudleston described them as 60 ft. in thickness, but the exposures are badly concealed by scree and slips and they are involved in a good deal of uncertainty. Buckman named them the Peak Shales. They are the lowest of those beds which at the coast occur only on the east side of the Peak Fault.

Bifrons Zone: ALUM SHALE SERIES, 90 ft. The Alum Shales are grey and crumbling and highly charged with disseminated pyrites. The highest 20 ft. are often known as the Cement Shales on account of their containing lines of nodules formerly made into cement. From this division, excavated in the alum works, most of the rich vertebrate fauna of the Upper Lias has been obtained. Whole skeletons of many of the large saurians were extracted from the workings, and the study of their remains has greatly advanced our knowledge of the Upper Liassic vertebrate fauna. The Reptilia found on the Yorkshire coast comprise three species of *Plesiosaurus,* five of *Ichthyosaurus,* four of *Steneosaurus,* three of *Thaumatosaurus,* two of *Pelagosaurus,* and one each of *Sthenarosaurus* and *Scaphognathus.* There are also some sixteen species of fish of all the usual Liassic genera.

The main mass of the Alum Shales is characterized by ammonites of the *bifrons* zone: *Hildoceras* aff. *hildense, Dactylioceras commune,* D. aff. *holandrei,* &c., together with large quantities of *Nuculana ovum,* as in the Midlands. The basal band of the *bifrons* zone is a double line of pyritous doggers with masses of belemnites and an ammonite peculiar to it: *Pseudolioceras pseudovatum* (with a subzone of its own in the classification of S. S. Buckman).

Falcifer Zone: JET ROCK SERIES, 90–100 ft. This consists of a mass of dense, dark, bituminous shales, characterized by abundant ammonites of the *Harpoceras mulgravium* type, and a certain number of other mollusca, especially *Inoceramus dubius.* Many of the saurians and fish recorded from the Alum Shales occur here also.

The lowest 25–30 ft. constitute the JET ROCK proper, and this, from the occurrence of *Harpoceras exaratum* near the base, Buckman assigned to the *exaratum* subzone. The jet occurs as lumps and lenticles in the hard shale, especially towards the top. It is believed to be a product of water-logged wood so thoroughly altered that all traces of structure have been obliterated. With it occur numerous lines of nodules, usually enclosing fossils filled with bituminous cementstone, or even, when the fossils are ammonites, liquid bitumen in the gas-chambers. The nodules emit a strong odour of mineral oil when broken open.

At the top of the true Jet Rock is a line of large-sized nodules of calcareous shale, up to 15 ft. in diameter. The band forms a reef at sea on both sides of Saltwick, extending eastward, often marked by a line of breakers, until due north of Whitby Abbey.

Tenuicostatum Zone: GREY SHALE SERIES (30 ft.). These, the lowest beds of the Upper Lias, are seldom well exposed, owing to the accumulation of debris fallen from higher up the cliffs. They consist of somewhat featureless grey sandy shales, with nodular earthy limestone bands, in which *Dactylioceras tenuicostatum* occurs. The only other fossils at all common are belemnites.

VIII. THE HEBRIDEAN AREA

(a) Skye and Raasay [1]

The Upper Lias has a considerable outcrop across Raasay and on the adjoining coast of Trotternish in Skye, and it is also found along the east coast of Strathaird, close to the waters of Loch Slapin. It is chiefly remarkable for containing an iron ore, which becomes commercially workable in the island of Raasay. The rest of the formation consists predominantly of shales, in the middle of which the ironstone-band is sandwiched. The ammonites show that most of the Yeovilian and the upper part of the Whitbian Stages are wanting, as in Yorkshire north-west of the Peak Fault, while there is a thick development of the basal Aalenian as in the Dorset–Somerset area.

DUN CAAN SHALES, 70 ft. The upper shaly division, above the ironstone, belongs to the *opalinum* zone. It reaches its greatest development of 70 ft. of micaceous shales at the foot of Dun Caan in Eastern Raasay (Plates X and XIV). In Skye the shales are only some 25 ft. thick, and less micaceous than in Raasay. The best exposure in Skye is south of Holm, but the series has been studied in detail at Dun Caan, where it is best developed, and the sequence there may be taken as typical (fig. 32, p. 184).

The highest 5 ft., with calcareous nodules and hard bands, which form a lithological passage up into the Inferior Oolite, have been identified as the *opaliniforme* subzone. Besides *Cypholioceras opaliniforme*, they have yielded *Pleurolytoceras hircinum* (Quenst.), *Pseudolioceras beyrichi* (Schloenb.), *Walkericeras subglabrum* (S. Buck.), *Pleydellia*, and belemnites.

The remaining 65 ft. of the shales represent a much-expanded *aalensis* subzone, and differences detected by Buckman in the ammonites of the highest 16 ft. induced him to suggest that that portion should be separated as a new subzone of *Canavarina venustula*. By far the commonest fossils throughout are ammonites and belemnites. *Pleydellia aalensis* occurs sparingly, accompanied by *P. leura* (S. Buck.), *Canavarina* spp., *Walkericeras* spp., *Pleurolytoceras leckenbyi* (Lyc.); while most abundant of all are five species of *Cotteswoldia*. The identity of the cephalopod fauna with that of the upper part of the Yeovil Sands of the Crewkerne district of Somerset is noteworthy.

In Strathaird, on the western shore of Loch Slapin, the Yeovil Sands are simulated even in lithological features, for the Dun Caan Shales pass into current-bedded, coarse, calcareous sands and sandstones, inseparable from

---

[1] Based on C. B. Wedd, 1910, 'Geol. Glenelg, Lochalsh and SE. Skye', p. 115; and G. W. Lee and S. S. Buckman, 1920, 'Mes. Rocks Applecross, Raasay and NE. Skye', pp. 30–43, 65–7, *Mems. Geol. Surv.*

the Inferior Oolite above. In this facies of the beds fossils are much scarcer than in the shales of NE. Skye and Raasay.

Below the Dun Caan Shales is a considerable break, in which the *moorei*,

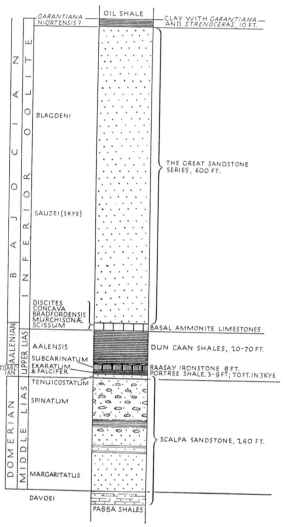

FIG. 32. Section of the Middle and Upper Lias and Inferior Oolite in Raasay. After G. W. Lee, 1920, 'Mesozoic Rocks of Applecross, Raasay and NE. Skye', *Mem. Geol. Surv.*

Dumortieriæ, dispansum, and *struckmanni* subzones are so far unaccounted for. The palaeontological hiatus proved in Raasay is reflected by an abrupt lithic change in Strathaird, Skye, where Wedd noted that 'a knife-blade could separate the two lithological divisions'.[1]

[1] C. B. Wedd, 1910, loc. cit., p. 115. The arenaceous facies of the Dun Caan Shales (presumably) was there treated as Inferior Oolite, in accordance with the practice of the Survey.

RAASAY IRONSTONE, 2–8 ft. Immediately beneath the stratigraphical break just referred to, there is in the Strathaird and the Trotternish districts of Skye a 1 ft. 9 in. to 2 ft. band of dark-blue to greenish oolitic ironstone, in Strathaird little more than a ferruginous oolitic limestone. In the southern part of the Island of Raasay it thickens to 8 ft., at the same time improving in quality and the richness of its fauna. Since its discovery by H. B. Woodward in 1893 it has been worked as an iron ore, and for its shipment a pier has been built at the southern end of the island, connected with the mines by tramway.

The Raasay Ironstone is very variable in composition. In its original state, where the band is thick, it is a dark-green, glossy rock, consisting of green ooliths in a clear matrix, the ooliths containing most of the iron in the form of a green silicate, probably chamosite. But where the ore is thin, as in Skye and the north-east of the Raasay outcrop, the green ooliths have been transformed into black oxides and carbonate (predominantly siderite), leaving only traces of the silicate. The process of formation of the ore has thus been very different from that at one time thought to have accounted for the genesis of oolitic ironstones, in which ordinary oolitic limestone is supposed to have been replaced by carbonate of iron. There is here a strong indication that the iron silicate was an original constituent of the sediment, and modern thought is coming to regard this more and more as the normal process of oolitic ore-formation.

Palaeontologically the ironstone is not rich. The upper part of the ore bed in the main mine, however, has yielded *Hildoceras bifrons* (d'Orb. non Brug.) which indicates the *subcarinatum* subzone of the *bifrons* zone, and mingled with it were found certain Dactyliocerates indicating the *falcifer* zone. Evidence recently obtained by the Survey in Ardnamurchan proves that, as was suspected from their rolled condition, all these fossils are derived from earlier deposits, the date of deposition of the ironstone being *striatulum* hemera or later.

PORTREE SHALES, 3–78 ft.: The rest of the Upper Lias, the portion lying beneath the Raasay Ironstone, displays an extraordinary variability in thickness. Near Portree, the capital of Skye, from which the beds take their name, they consist of dark, micaceous, soft shales with occasional hard bands and limestone nodules. The thickness at the type locality is 48 ft; farther north along the coast of Trotternish a boring at Bearreraig Bay proved 78 ft.; and in S. Skye, on the peninsula of Strathaird, the thickness is only 20 ft. A still greater diminution takes place in Raasay, where the shales are reduced to 9 ft. or less—in places even 3 ft. The presence of the rolled fossils in the overlying ironstone shows that the diminution is due at least in part to intraformational erosion.

In Skye the Portree Shales, although so thick, are less satisfactorily exposed than in Raasay. It has been ascertained, however, that the *falcifer* and *exaratum* subzones are normally represented. The bulk of the shales, down to within 16 ft. of the base, yield numerous Dactyliocerates and Harpocerates of the *falcifer* subzone. Below this *Harpoceras* aff. *exaratum* comes in for a few feet. Of the lowest level of all, nothing is yet known.

On the horizon of *H*. aff. *exaratum*, 14 ft. from the base of the shales near Holm, north of Portree, is a thin band of jet. The associated ammonites prove it to be on the same horizon as the Jet Rock of Yorkshire, a correspondence of a highly abnormal feature that is truly remarkable in two localities so distant as Whitby and Skye.

In Raasay the thin representatives of the Portree Shales are very fossiliferous, abounding in Dactyliocerate ammonites of at least ten species, of which the most abundant is *D. delicatum*, as well as in belemnites (*Megateuthis* and *Dactyloteuthis* spp.). Although both the *falcifer* and *exaratum* subzones are evidently represented, they have not been satisfactorily separated.

The *tenuicostatum* zone has been detected in Raasay in the uppermost 6 ft. of the underlying Scalpa Sandstone, developed in a facies identical with the beds that a few feet lower yield *Paltopleuroceras spinatum*. The lithological continuity of the *tenuicostatum* zone with the Middle Lias thus provides a parallel between these distant islands and the Cotswolds.

(b) Mull and Ardnamurchan [1]

Small patches of Upper Lias occur in Mull in three localities: about Loch Don; between Loch Spelve and Loch Buie; and near Carsaig. On the neighbouring mainland there is a fourth patch near Kilchoan, on the south side of the promontory of Ardnamurchan (see map, p. 114). All these sections have been described in recent memoirs by the Geological Survey of Scotland.

DUN CAAN SHALES: The most complete sections are in the Loch Don district, at Port nam Marbh, where the passage to the Inferior Oolite is seen, but unfortunately the Dun Caan Shales here prove scantily fossiliferous and no ammonites have been obtained. *Cypholioceras* was, however, found at the base of the Inferior Oolite section in Ardnadrochet Glen, thus proving the presence of the *opaliniforme* subzone. The beds, like their counterparts in the northern islands, consist of sandy shales and sandy limestones. In Ardnamurchan the *opaliniforme* subzone was not detected, but a gap was filled by the recognition of both the *aalensis* and the *moorei* faunas, neither of which has been detected in Mull. The *aalensis* subzone is reduced to 5 ft. in thickness. The ammonites were obtained from the upper part of a bed of hard sandy flags on the beach near Kilchoan, and they were identified by Buckman as cf. *Pleydellia aalensis* (Ziet.), cf. *P. fluens* S.Buck., ?*Cotteswoldia* and *Canavarina* sp. The *moorei* fauna is not known at any other point in Scotland. It was discovered by the Survey in the lower part of the same bed of flags as that yielding the *aalensis* fauna, and separated from the Raasay Ironstone by an intrusive dyke. The ammonites identified by Buckman were *Dumortieria* cf. *brancoi* S.Buck., *D.*? *exacta* S.Buck. and *D.* cf. ?*subexcentrica* S.Buck.

RAASAY IRONSTONE: On the shore in the west side of Kilchoan Bay, Ardnamurchan, the Survey have recognized the Raasay Ironstone as a 4-ft. band of bluish ferruginous stone, largely made up of compacted ooliths; the same band also crops out a mile farther west, where it is much baked by the Tertiary gabbro. From the first locality they collected ammonites referred to *Grammoceras* aff. *penestriatulum* S.B., *G.* aff. *toarciense* (d'Orb.), *Alocolytoceras* cf. *perlæve* (Denckm.) and *Dactylioceras* cf. *holandrei* (d'Orb.), together with belemnites. The deposit is therefore at least as late in date as the *striatulum* subzone, while the Dactyliocerates are derived, as in Raasay, from the much-eroded Portree Shales beneath. Before the discovery of these *striatulum*

---

[1] Based on G. W. Lee, 1925, 'Pre-Tert. Geol. Mull, Loch Aline and Oban', pp. 95–112; and G. W. Lee and J. E. Richey, 1930, 'Geol. Ardnamurchan', pp. 43–8; *Mems. Geol. Surv.*

ammonites it was thought, on the evidence obtained in Raasay, that the chief stratigraphical gap lay above the ironstone, but now it seems more probable that it lies mainly below. More intensive collecting may eventually reveal faunas of *struckmanni* or *dispansum* dates in the ironstone; but even the evidence now available shows that the ironstone is Yeovilian.

PORTREE SHALES: The thickest development of the Portree Shales occurs at Port nam Marbh, where the blue shales are 30 ft. thick and very poorly fossiliferous. The Survey obtained two Dactyliocerates and a Harpocerate, indicative of the *falcifer* zone. The other exposures in Mull are both small and fragmentary. In a small patch at Carsaig, however, fossils are more numerous. The Survey collected some seventeen species of ammonites from two horizons; those from the upper horizon were said by Buckman to indicate the *bifrons* zone and those from the lower the *falcifer* zone. Associated with the lower assemblage of ammonites were fish-remains, *Inoceramus dubius* and a small unidentified gastropod said to occur in the Fish Bed and paper-shales of the English Midlands.

In Ardnamurchan the thickness of the Portree Shales is about 20 ft. 'The black shales occur in lenticular masses, a few yards long, amongst a complex of basic cone-sheets and sills'[1] on the shore in the west side of Kilchoan Bay, where the ironstone crops out, and also in the second exposure a mile farther west. They yielded to the Survey a considerable assemblage of Dactyliocerates and Harpocerates of the *falcifer* zone.

IX. EASTERN SCOTLAND

The Upper and Middle Lias, with the Inferior Oolite, are faulted down out of sight in East Sutherland. The only evidence for the occurrence of Upper Lias on the east coast of Scotland is a specimen of *Frechiella subcarinata* in the Hugh Miller collection, said to have been found at Shandwick, Ross-shire (in Drift?).[2]

X. KENT[3]

The Upper Lias, like the Middle and Lower Divisions in Kent, is thickest in the south-westerly borings and thins to the north and north-east, being overstepped by the Lower Oolites approximately along the line of the Dover–Canterbury railway. That the discordance between the Upper Lias and the overlying beds is more in the nature of an overstep than an overlap is shown by the absence of the Yeovilian strata, as in the Midlands and Lincolnshire. The Kentish Upper Lias is exclusively Whitbian, excepting possibly at the single locality of Brabourne, where, in one of the two most south-westerly of all the borings, a single *Grammoceras* was found, which was considered by Buckman to indicate the presence of the *striatulum* subzone. Evidence was insufficient to enable it to be gauged how complete the succession below that level might be, the next highest ammonites detected anywhere in Kent being of *bifrons* date.

The greater part of the Kentish Whitbian consists of clay, sometimes green, sometimes brown, either shaly, silty, or lumpy, and belonging to the *bifrons*

---

[1] J. E. Richey, 1930, loc. cit., p. 43.
[2] G. W. Lee, 1925, 'Geol. Golspie', p. 74, *Mem. Geol. Surv.*
[3] Based on G. W. Lamplugh and F. L. Kitchin, 1911, loc. cit.; and Lamplugh, Kitchin and Pringle, 1923, loc. cit.

and earlier zones. It yields numerous Dactyliocerates of the *bifrons* zone in the highest 8 ft. at Elham and on a corresponding horizon at Brabourne; also *Hildoceras walcotti* (Sow.), *Peronoceras* and *Porpoceras*: these indicate the *subcarinatum* and *fibulatum* subzones. At Elham, Bishopsbourne and Folkestone the *falcifer* subzone was the highest encountered; this yielded *Harpoceras mulgravium* (Young and Bird), *Hildoceras levisoni* (Simps.) and other species. Beneath it, and forming the top of the Upper Lias at Dover and Ropersole, was recognized the *exaratum* subzone, with Harpocerates of the *exaratum* type, *Dactylioceras delicatum* (Phil.) and a number of other fossils, the most distinctive being a carinate *Corbula*. Finally, the *tenuicostatum* zone was easily differentiated by means of its numerous delicately-ribbed Dactyliocerates from the underlying *spinatum* beds, with which it was generally found to be in lithological continuity. Unlike the rest of the Upper Lias it is typically developed as a grey sandy limestone, up to 8 ft. in thickness.

The northerly and north-easterly attenuation is rapid, the thickness of the Upper Lias falling from 41 ft. at Elham and 55 ft. at Folkestone to $8\frac{3}{4}$ ft. at Dover and 4 ft. at Chilton, where it is represented only by the *tenuicostatum* zone.

CHAPTER VIII

LOWER AND MIDDLE INFERIOR OOLITE

Stages.	Zones (Plates XXXIII–IV).		Strata in the Cotswolds.
		VESULIAN TRANSGRESSION	
BAJOCIAN	MIDDLE INFERIOR OOLITE	*Teloceras blagdeni* [1]	Absent from the Cotswolds
		Otoites sauzei	*Phillipsiana* Beds ⎫ of Cleeve Hill *Bourguetia* Beds ⎭
		Witchellia spp.	*Witchellia* Grit Notgrove Freestone
		Shirbuirnia spp.[2]	Gryphite Grit
		Hyperlioceras discites [3]	*Buckmani* Grit Lower *Trigonia* Grit
		BAJOCIAN TRANSGRESSION	
UPPER AALENIAN	LOWER INFERIOR OOLITE	*Ludwigella concava*	Tilestone Snowshill Clay Harford Sands
		Brasilia bradfordensis	Upper Freestone Oolite Marl
		Ludwigia murchisonæ	Lower Freestone Pea Grit
		Ancolioceras spp.[4]	Lower Limestones
		Tmetoceras scissum	*Scissum* Beds

I. THE DORSET–SOMERSET AREA

IN Dorset and Somerset the Lower and Middle Inferior Oolite are much condensed. The zones that are present are but thin representatives of considerable thicknesses of strata elsewhere, while some are to be detected only by their fossils, water-worn and bored, redeposited in conglomerates. In Somerset, over wide areas, the rocks are absent altogether, Upper Inferior Oolite resting directly upon Upper Lias.

In consequence of this attenuation, when the Dorset–Somerset area is viewed without reference to other districts, the impression is gained that palaeontologists have greatly overloaded the formation with unnecessarily

---

[1] Formerly *humphriesiani*, in Buckman's earlier papers. The zone has been subdivided in Germany.

[2] Formerly *Sonniniæ*, but renamed by Buckman, 1910, *Q.J.G.S.*, vol. lxvi, p. 78, on account of generic rearrangement of the *Sonniniæ*. True *Sonniniæ* are said to occur only in the *sauzei* zone.

[3] Certain deposits above those containing *H. discites* were dated by Buckman as post-*discites*. Such deposits, where they occur, will here be treated as part of the *discites* zone.

[4] Founded by Buckman, 1910, loc. cit., p. 79, for certain thin Dorset strata, and now propagated in the literature by Richardson and subsequent workers. No ammonites have been found in the Lower Limestones of the Cotswolds, but *Ancolioceras* has been obtained from the Variable Beds of the Northampton Sands, which are on about the same stratigraphical horizon.

minute zonal subdivisions. But when the Cotswolds are considered, where the same rocks thicken to nearly 300 ft., their value becomes more apparent.

The nature of the zones in use in the Inferior Oolite is of considerable theoretical interest. Most of them are true faunizones, because they are characterized by a special fauna—a whole assemblage of molluscs and brachiopods, the species of which may or may not occur above or below, but which are never assembled in the same proportions outside their particular zone. Over wide tracts of country, therefore, the zones have been traced by means of their general faunal assemblage, often without the discovery of the index-species of ammonite. On the other hand, they were in the first place founded, and ultimately depend, upon the local acmes of certain ammonites selected as index-species. They are therefore epiboles. It may be said, in fact, that in Dorset and Somerset, where ammonites abound, the zones are treated as epiboles, while in the Cotswolds and the Lincolnshire Limestone areas, where ammonites are not autochthonous and are rarely met with, the same divisions have to be treated as pure faunizones (see Chapter I, pp. 34–5). Meanwhile the divisions will be here spoken of simply as 'zones', and the corresponding time-units, in accordance with the universal practice in the literature of this formation, as 'hemeræ', this term taking priority over 'secule' where the two units happen to be identical.

(a) Burton Bradstock to Crewkerne[1]

On the coast the whole of the Inferior Oolite is only 11 ft. thick, forming a capping to the cliffs of Bridport Sands on either side of the mouth of the River Bredy, between Bridport Harbour and Burton Bradstock (Plate VI). Large blocks of this capping have fallen on the beach, where they have formed a collectors' paradise for many years. Six and a half feet of the total thickness belong to the Upper Inferior Oolite, leaving only $4\frac{1}{2}$ ft. to be considered here (figs. 33 and 29, p. 154).

The bulk of the $4\frac{1}{2}$ ft. is taken up by the Middle Division, from the *discites* zone upwards, which is represented by the RED BED, 2 ft. 10 in. thick. The top of the Red Bed is water-worn and pitted, and on it here and there rest patches of conglomerate, largely made up of fragments derived from its destruction elsewhere. The Red Bed contains three distinct layers, all of which may be recognized in the conglomerate. Among the pebbles of the conglomerate have also been found ammonites (for the most part rolled) identified by Dr. Spath as denoting probably *niortensis*, *blagdeni* and *sauzei* zones. Thus somewhere in the district the whole of the Red Bed or local Middle Inferior Oolite seems to have been broken up and eroded immediately before the deposition of the *garantiana* zone—probably during *niortensis* times. A similar erosion occurred at a slightly earlier date (*blagdeni* hemera) on the other side of the Channel, in Normandy, where it is marked by a conglomerate at the base of the famous Ironshot Oolite of Bayeux. These earth-movements seem to have heralded the great Bajocian Denudation, which preceded the transgression of Upper Inferior Oolite (*garantiana*) times.[2]

[1] Based on S. S. Buckman, 1910, *Q.J.G.S.*, vol. lxvi, pp. 52–89; L. Richardson, 1928–30, *Proc. Cots. N.F.C.*, vol. xxiii, pp. 35–68, 149–86, 253–64 for the Burton Bradstock–Broad-windsor district; and idem, 1918, *Q.J.G.S.*, vol. lxxiv, pp. 145–73 for the Crewkerne District. See also idem, Excursion, *P.G.A.*, vol. xxvi, 1915, pp. 47–78.

[2] A curious white lithographic limestone observed by Buckman at Burton Bradstock in

The layers of the Red Bed give evidence of most of the remaining zones of the Middle Inferior Oolite (fig. 33). The upper layer is a hard, crystalline limestone, grey in colour and made up largely of crinoid fragments. Its age is not precisely known, but Richardson considers it of late *Witchelliæ* or *sauzei* date. The middle layer is also a hard crystalline limestone, but it is very distinct on account of its containing numerous ironshot ooliths, and the general colour is brown or pinkish. Fragments of this layer enter largely into the composition of the *niortensis* conglomerate. The date is *Witchelliæ* or

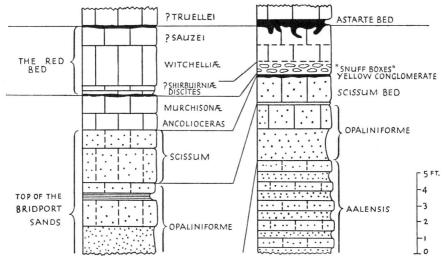

FIG. 33. Comparative sections of the Lower and Middle Inferior Oolite at Chideock Quarry Hill (left) and Burton Bradstock, Dorset. After L. Richardson, 1927, *Proc. Cots. N.F.C.*, vol. xxiii, pp. 51, 60.

perhaps early *sauzei*. The lower layer is also highly oolitic and ferruginous, but it is readily recognized by its large limonitic concretions, locally called 'snuff-boxes'. The nucleus of the snuff-boxes is often a rolled and bored piece of a thick-shelled *Myoconcha* or *Ctenostreon*, fragments of which abound. The date of this lowest layer is probably late *discites* hemera. Its snuff-boxes may be identified in the conglómerate above.

Beneath the Red Bed is another important stratigraphical gap, corresponding with the Aalenian Denudation, the effects of which are widespread in Southern England and the Continent. It is marked by the Yellow Conglomerate, a pebbly layer with an average thickness of only 3 in., but containing derived fossils of the *scissum*, *murchisonæ*, *bradfordensis* and *concava* zones. Its latest fossils are of early *discites* date, from which it can be said that the

association with a *niortensis* fauna, and believed by him to occur as remanié pebbles in a conglomerate of that age, has been explained by Richardson as a secondary deposit resulting from water percolating down fissures, and he has shown that it is present both near the top of the Red Bed and in the Bridport Sands. The white rock is beyond any doubt of secondary origin, due to infiltration. It is associated with the breccia of a large fault, which brings down Forest Marble against Bridport Sands (Upper Lias) and veins of it can be seen running down into the sands. I myself reached this conclusion independently, and it was subsequently confirmed on visiting the spot with Mr. Richardson. (See S. S. Buckman, 1910, *Q.J.G.S.*, vol. lxvi, pp. 69–71; L. Richardson 1915, *P.G.A.*, vol. xxvi, p. 56; S. S. Buckman, 1922, *Q.J.G.S.*, vol. lxxviii, pp. 420–31).

conglomerate was formed at that date, deriving its materials from the destruction of the whole of the Lower Inferior Oolite. Among the many fossils of the Yellow Conglomerate are broken specimens of *Brasilia bradfordensis*, *Burtonia*, *Ludwigella*, *Cirrus nodosus*, and *Astarte (Coelastarte) excavata*.

The only undisturbed remnant of the Lower Inferior Oolite (as here understood) is a portion of the *scissum* zone, consisting of 18 in. of grey sandrock with the zonal ammonite, but lithologically more like the underlying Bridport Sands. To the top of the *Scissum* Bed the Yellow Conglomerate is firmly cemented.

Four to six miles north-east of Bridport, around Powerstock, North Poorton and Mapperton, the whole of the Middle and Lower Inferior Oolite above the *Scissum* Bed is represented by only 2 ft. of strata, and in one quarry at Powerstock it is reduced to 15 in. From this it may be inferred that still farther east, below the Chalk Downs, the formation thins out altogether, as it probably did out to sea, over the Weymouth Anticline.

In the opposite direction, up the dip slope to the west and north-west, the beds expand. The eroded surface below the *discites* zone, marking the 'Aalenian Denudation' (better, Bajocian Transgression), rises and older beds beneath begin to assume separate identity. The change takes place roughly west of a line drawn from Bridport northward to Beaminster.

At Chideock Quarry Hill, west of Bridport, the *blagdeni* and *niortensis* zones are missing, but the remainder have expanded to 9 ft. At the base of the Red Bed, which consists of dark brown, sandy, ironshot oolite with numerous *Witchelliæ* and crinoid fragments, is a trace of the *discites* conglomerate with rolled ammonites of *bradfordensis* date, as at Burton Bradstock (the Yellow Conglomerate). Beneath, however, with an eroded upper surface, instead of the *Scissum* Bed is 2 ft. of yellow limestone crowded with *Ludwigia murchisonæ*. This is locally called the Wild Bed, and its lower half has been separated by Buckman as a separate zone, the *Ancolioceras* zone, on account of its containing a mixture of *Tmetoceras* and Opalinoid ammonites. Beneath, the *Scissum* Bed has expanded to 3 ft.

Similar relations obtain along the western edge of the outcrop for some 6 miles northward from the coast, where a fine section is displayed at Stoke Knap or Waddon Hill, 2 miles west of Beaminster. Here ?*Witchelliæ*, *Shirbuirniæ*, *discites*, *concava*, *bradfordensis* and *murchisonæ* deposits have been recognized, totalling 5 ft. in thickness. The top is planed off at the base of the Upper Inferior Oolite and the surface is overgrown with oysters. The *scissum* zone, though thick, is ill defined, having become a sandy grit continuous with the Brachiopod Beds of the underlying Bridport Sands.

This section is more complete than most in the neighbourhood. In passing north into the Crewkerne District and east, down the dip-slope, a great hiatus becomes general, in which Middle Inferior Oolite disappears. Except possibly at Dinnington, whence labels in the Moore collection at Bath allege that ammonites of the *concava* and *discites* zones were obtained, there is no rock assignable to any zone above that of *bradfordensis*.

The Inferior Oolite hereabouts caps a great thickness of Yeovil Sands, left as isolated patches and blocks between faults. At first the *bradfordensis* zone, removed by the *discites* erosion on the coast, appears in what is for Dorset considerable thickness—3 ft. 7 in. at Beaminster—and is full of ammonites.

At Conegar Hill, Broad Windsor, however, close to Waddon Hill, it has been again removed completely, Upper Inferior Oolite coming to rest on the *murchisonæ* zone. At Misterton Limeworks and South Perrott it reappears, an impersistent bluish-grey ironshot bed, 4 in. thick. Beneath is the *murchisonæ* bed, a hard bluish-grey limestone from 1 to 3 ft. thick, and yielding *Pseudoglossothyris simplex*, *Zeilleria anglica*, *Variamussium pumilum*, and other fossils. The *scissum* zone remains merely the upward continuation of the underlying sands, of which it may include from 3 ft. to 6 ft.

(b) The Sherborne District[1]

In view of the sporadic erosion and partial removal of the Lower and Middle Inferior Oolite in the Beaminster–Crewkerne district, it is not surprising that farther on, where the outcrop turns east towards Sherborne, they have been removed altogether by the Bajocian Denudation. For several miles Upper Inferior Oolite rests directly on Bridport Sands.

The first reappearance, Mr. Richardson informs me, is at North Coker, about 2 miles south-west of Yeovil, where probably—for the sections are not very satisfactory—the sequence is similar to that in the better known section at Stoford, about a mile and three-quarters farther east (fig. 45, p. 233). Here the Fuller's Earth lies only about 3 ft. above the Yeovil Sands. The Lower, Middle and Upper Inferior Oolite then thicken steadily towards Sherborne. At Bradford Abbas, Sandford Lane and other places, the lower zones form the famous fossil beds which have yielded perhaps the richest store of well-well-preserved mollusca of all kinds obtained from any part of the Jurassic System in England.

It was the area embracing Sherborne, Milborne Port and Bradford Abbas that formed the subject of the late S. S. Buckman's earliest researches, carried on from his father's farm at Bradford Abbas. The publication in 1893 of his masterly paper, 'The Bajocian of the Sherborne District', made this classic ground. In the numerous quarries the detailed zonal succession of the Inferior Oolite was first worked out, the crowded ammonites being carefully collected, inch by inch, from their proper beds. The unparalleled richness of the ammonite fauna is best illustrated by the profusion of genera and species figured by Buckman in his later years in *Type Ammonites*, from the world-famous quarries of Bradford Abbas, Sandford Lane, Oborne, Compton and the neighbourhood.[2] Here too, Buckman and his father collected many of the finest of the gastropods, in a superb state of preservation, that were figured in Hudleston's *Monograph of the Inferior Oolite Gastropoda*.

As the result of his refinements in stratigraphy, Buckman was able to show that at the beginning of the *murchisonæ* hemera there was an overstep westward, followed by a gradual recession eastward throughout the rest of Lower and Middle Inferior Oolite times. This may be taken to be the regression which culminated in the Bajocian Denudation, and was followed in Upper Inferior Oolite times by renewed westerly transgression.

Those who wish to acquaint themselves with the district cannot dispense

---

[1] Based on S. S. Buckman, 1893, 'The Bajocian of the Sherborne District', *Q.J.G.S.*, vol. lxix, pp. 479–522; and L. Richardson, 1932, *Proc. Cots. N.F.C.*, vol. xxiv, pp. 35–85.
[2] Space forbids giving lists of Inferior Oolite ammonites, very large numbers of which have been figured by Buckman also in his monograph.

with the detailed descriptions supplied by Buckman's paper and Richardson's invaluable revisionary work; the present object is only to show the general development of the strata. In compiling the following summary I have drawn largely upon a correlation-table very kindly lent me for the purpose by Mr. Richardson in advance of publication.

SUMMARY OF THE LOWER AND MIDDLE INFERIOR OOLITE OF THE SHERBORNE DISTRICT

Blagdeni Zone. Absent in the western part of the district. Appears at Halfway House as a thin seam 0–2 in. thick; 0–6 in. at Louse Hill; absent again at Sandford Lane; thickening in the east to a hard grey-brown or pinkish, irony stone (6 in.–2 ft. 3 in.) at Clatcombe and Oborne. At Milborne Wick the zone consists of 10 in. of white crumbly rock with green grains of glauconite, and is crowded with fossils, of which the commonest is *Astarte spissa* S. Buckman.

Sauzei Zone. Traces at Halfway House and locally to the west; 6–12 in. at Stoford; top part of the Fossil Bed at Sandford Lane; 4 in. at Oborne; white and grey limestone 1 ft. 8 in. at Milborne Wick.

Witchelliæ and Shirbuirniæ Zones. Absent in the west and at Halfway House; first appears as the lower part of the Fossil Bed at Sandford Lane (the total thickness of which is 1 ft. 9 in.); thickening eastward to 3 ft. of hard blue limestone at Oborne and 13 ft. of grey sandy limestone at Milborne Wick.

Discites and Concava Zones. Fossil bed in the west, thickening eastward from 10 in. at Clayton's Quarry to 3 ft. 8 in. at Halfway House. Locally Richardson has been able to separate the two zones, but at Oborne and Halfway House the bed is less fossiliferous and separation is more difficult. Buckman stated that the 3 ft. of blue-centred, yellow, ironshot limestone forming most of this bed at Halfway House yielded *Ludwigella concava* abundantly. Farther east the zones thicken still more. At Sandford Lane they are readily separable—the *discites* zone consists of 1 ft. 3 in. of sandy limestone and sand immediately below the Fossil Bed, and the *concava* zone of 3 ft. of sandy stone with *L. concava* and other fossils, as at Halfway House. Still farther east and north, at Oborne and Corton Downs, the *concava*, *discites* and probably *Shirbuirniæ* and *Witchelliæ* zones become blended in a uniform development of grey, sandy, glauconitic limestone with sandy and marly partings. Here, owing to the rarity of ammonites and other fossils, the different zones cannot be separated.

Bradfordensis Zone. Absent or locally represented only by the thinnest traces west of Halfway House, but there consisting of 1 ft. 2 in. of yellow ironshot limestone, and thickening eastward. Best seen at Marston Road Quarry, where it is separable into two layers: an upper, the *Rhynchonella ringens* Beds (1 ft. 8 in.) and a lower the '*Perisphinctes*' *brebissoni* Beds (1 ft.). The upper layer overlaps the lower westward to Halfway House.

Murchisonæ Zone. Yellow and blue limestone, 10 in.–2 ft., in the west, the upper portion locally called the Paving Bed. Probably represented in thicker development to the east, together with the *bradfordensis* zone, but the exposures are too shallow to reach them. Apparently absent at Stowell, near Milborne Port.

Ancolioceras and Scissum Zones. Generally absent, but recognized by Richardson in a 1 ft. 6 in. band at Marston Road Quarry.

In a boring made in 1907–8 at Stowell, north of Milborne Port, within the Fuller's Earth outcrop, the total thickness of Lower and Middle Inferior Oolite seems to have been about 40 ft. The *Rhynchonella ringens* Beds (*bradfordensis* zone) were proved, within 2 ft. of the base of the limestones, and ammonites of the *sauzei* zone 32 ft. higher up.[1]

Immediately west of this boring, higher up the dip-slope, the Lower and Middle Inferior Oolite disappear, having been removed by the Bajocian Denudation. The beds are last seen at Corton Downs, 4 miles north of Sherborne, beyond which Upper Inferior Oolite oversteps on to Yeovilian sands for the next 6 miles over the North Devon Axis.

(c) The Cole Syncline[2]

Centred at Cole, midway between Bruton and Castle Cary, and between the North Devon and Mendip Axes, is a small syncline, in which Lower and Middle Inferior Oolite is preserved to a thickness of about 10 ft. This outlying patch was first described in detail by Richardson, of whose account the following is a summary. The usual non-sequence was detected at the base of the *discites* zone, the *bradfordensis* and *concava* zones having been removed before the Bajocian Transgression.

SUMMARY OF THE LOWER AND MIDDLE INFERIOR OOLITE OF THE COLE SYNCLINE

Blagdeni Zone. Hard ironshot oolite, 1 ft., with *Teloceras blagdeni*, the top levelled, with oysters attached. Seen only in Lusty Quarry, Bruton.

Sauzei Zone. *Pecten*-bed. Grey limestone with *Entolium*, 1–3 ft., seen in Lusty and Sunny Hill Quarries,[3] Bruton.

Witchelliæ and Shirbuirniæ Zones. Grey limestone, a downward continuation of the *Pecten*-bed, and sometimes combined with it, 1–2 ft.

Discites Zones. Ammonite bed, 2 ft. at Lusty Hill; 3 ft. 3 in. at Pitcombe Rock, ¼ mile south of Sunny Hill Quarry; full of ammonites which are difficult to extract:—*Graphoceras inclusum* S. Buck., *Terebratula eudesii* Oppel.
    (Non-sequence: *concava* and *bradfordensis* zones absent.)

Murchisonæ Zone. Hard grey limestone, 1–3½ ft., passing into a conglomeratic bed at Lusty Hill, 1 ft. 10 in., and containing *Montlivaltia* cf. *lens* Ed. and H., the first sign of corals in the Inferior Oolite.
    (Non-sequence: *Ancolioceras* and *scissum* zones absent.)

II. THE MENDIP AXIS AND DUNDRY HILL

For many miles north of the Cole Syncline there is no trace of Lower or Middle Inferior Oolite. Over the eastern end of the Mendip Axis the Upper Inferior Oolite transgresses over progressively lower beds until it comes to rest on the Carboniferous Limestone, as described in the next chapter.

Nevertheless, on the north side of the Mendips, 22 miles from Cole, there is preserved on the western part of Dundry Hill an outlier of Lower and Middle Inferior Oolite of the Dorset–Somerset type (fig. 34). Consisting

---

[1] J. Pringle, 1909–10, *Sum. Prog. Geol. Surv.* for 1908, pp. 83–6, and for 1909, pp. 68–70.

[2] L. Richardson, 1916, *Q.J.G.S.*, vol. lxxi, pp. 494–503.

[3] Unfortunately the fine exposure at Sunny Hill is fast becoming obscured, houses and gardens having been built in the old quarry.

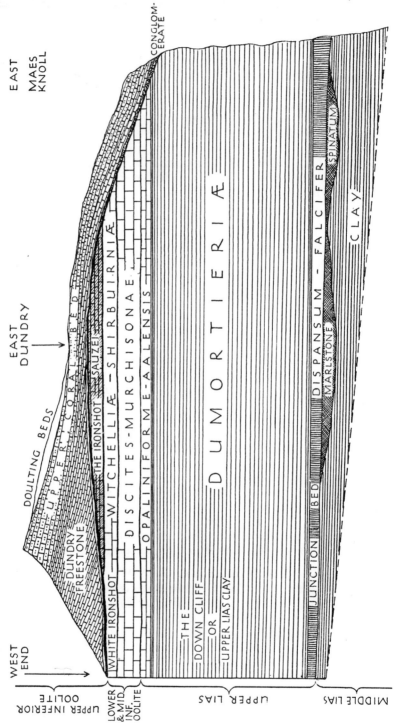

FIG. 34. Diagrammatic longitudinal section through Dundry Hill, Somerset. (After Buckman and Wilson, 1896, *Q.J.G.S.*, vol. lii, p. 695.) (Re-drawn and the terminology brought up-to-date.)

principally of ironshot oolites, it stands in the sharpest possible contrast with the Cotswold type of deposit, and furnishes convincing evidence that, as in Upper Lias times, the shallow belt separating the Cotswold basin of deposition from the Somerset–Dorset basin lay some distance north of the Mendips.

The thickness of the Dundry Beds has never been accurately determined, but it is not very great. At the base are a few feet of hard, massive, irony oolites, representing the *murchisonæ* and *bradfordensis* zones, overlain by grey limestone and marls representing the *concava* and *discites* zones. Above this comes the fossiliferous White Ironshot, chiefly of *Shirbuirniæ* date, but including a *Witchellia* bed, capped by the Ironshot Oolite proper, forming the *sauzei* zone. From both of these ironshot limestones large collections of well-preserved mollusca have been made and some may be seen in most of the museums in England.

The *blagdeni* zone has been everywhere removed from the hill by the Bajocian Denudation. At the eastern end the denudation has also destroyed the whole of the underlying zones, bringing Upper Inferior Oolite down on to Yeovilian sands as on the main outcrop.[1]

III. THE COTSWOLD HILLS[2]

North of Bath, along the Cotswold escarpment, many new types of deposit representing the Lower and Middle Inferior Oolite make their appearance above the Cephalopod Bed, first becoming visible at Horton, 4 miles north of Old Sodbury, and thickening northward. The first to appear is about 6 ft. of sandy, ferruginous limestone yielding *Tmetoceras scissum*—the equivalent of the *Scissum* Bed of Dorset. It thickens to 9 ft. about Stroud and forms a basement-bed to the Inferior Oolite all through the Cotswolds, in places reaching a thickness of over 30 ft.

Above the *Scissum* Beds in the district between Sodbury and Dursley a series of oolitic limestones gradually appears, consisting of layers of freestone alternating with layers of detritus of Crinoidea, Echinoidea, and bivalves, with small oysters attached to some of the beds. These limestones, which reach 25 ft. in thickness, were named by Witchell the LOWER LIMESTONES. Owing to the absence of diagnostic fossils, little progress has been made since Witchell's days in the determination of their exact date. They were usually regarded as forming the base of the *murchisonæ* zone, but Buckman suggested that they may represent the *Ancolioceras* zone.

More oolites succeed the Lower Limestones, reaching a thickness of 45 ft. about Dursley and continuing to thicken northward. These are known as the LOWER FREESTONE. They first come to be separated from the Lower Limestones at Uley Bury, near Durlsey, by a seam of large pisoliths, which thickens later into a richly fossiliferous accumulation known as the PEA GRIT. The two together represent the *murchisonæ* zone. The *bradfordensis* zone follows with a mass of incoherent oolite called the OOLITE MARL, the upper portion of which is consolidated to form an UPPER FREESTONE.

[1] S. S. Buckman and E. Wilson, 1896, *Q.J.G.S.*, vol. lii, pp. 669–720; and 1901, *P.G.A.*, vol. xvii, pp. 152–8.

[2] For the South Cotswolds:—L. Richardson, 1910, *Proc. Cots. N.F.C.*, vol. xvii, pp. 63–136, and 1908, *P.G.A.*, vol. xx, pp. 514–29; for the Mid-Cotswolds:—S. S. Buckman, 1895, *Q.J.G.S.*, vol. li, pp. 388–462, and L. Richardson, 1904, *Handbook to Geol. Cheltenham*; for North Cotswolds:—S. S. Buckman, 1897, *Q.J.G.S.*, vol. liii, pp. 607–29, and idem, 1901, vol. lvii, pp. 126–55, and L. Richardson, 1929, 'Geol. Moreton in Marsh', *Mem. Geol. Surv.*

The top of the Upper Freestone is often bored and overlain by a pebbly and conglomeratic bed, and the *concava* zone is missing. The discordant Bajocian beds then succeed, consisting of variable, fossiliferous, rubbly limestones and marls known as the RAGSTONES, which represent the *discites*, *Shirbuirniæ* and *Witchelliæ* zones.

The Aalenian Denudation preceding the Bajocian Transgression, already familiar in Dorset and Somerset, caused the removal of the *concava* deposits in the Western Cotswolds. Its effects hereabouts are the same as farther south, in spite of the great difference in the facies of the rocks, but farther north and east, in the Central and North Cotswolds, they are not perceptible, and deposits of *concava* date are present.

The whole of the strata in the Cotswold Hills, from the Middle Lias to the Ragstones, were laid down in a gradually sinking syncline between the anticlinal axes of the Malverns and the Bath–Mendip district on the one hand and the Vale of Moreton on the other. Over these axes deposition was probably almost at a standstill, but in between sediments accumulated to a thickness of about 800 ft. After the formation of the Ragstones, however, not only were uplift and denudation renewed along the main axes, but there was also accentuated activity along the Birdlip Anticline (referred to on p. 75), which strikes roughly NW.–SE. through Birdlip Hill and runs along the line of the Roman Ermine Street towards Cirencester. The effect of this axis was now to divide the Cotswold region into two minor synclines, in which the Ragstones were preserved, while they were completely removed in the centre. At the outcrop about Birdlip, upon the crest of the anticline, Upper Inferior Oolite was deposited directly upon Upper Freestone (*bradfordensis* zone) (figs. 35 and 36).

The centre of the southern syncline was located near Painswick, but comparatively little has been left of it owing to recession of the Cotswold escarpment.

The northern syncline is better preserved and in it the succession of rocks is more complete. The axis of depression runs through the plateau promontory of Cleeve Hill, the highest point in the Cotswolds (1,070 ft. O.D.), towards Chedworth, and is then lost down the dip-slope below the Great Oolite. On the Cleeve Hill plateau the highest beds of the Ragstones are preserved, belonging to the *sauzei* zone, not found at any other locality in the Cotswolds. Still higher beds may have existed over the Vale of Gloucester before the retreat of the escarpment, but if so they have entirely disappeared.

The strata that accumulated within the synclinal of the Cotswolds in Lower and Middle Inferior Oolite times reached a thickness of about 300 ft., without making allowance for the *blagdeni* or *niortensis* zones, which may have been deposited and subsequently removed by the Bajocian Denudation. They consist of oolites, pisolites, broken-shell limestones, masses of conglomerated Echinoderm remains, shells, sponges, and coral fragments. They are essentially neritic accumulations, and along the coast to which they bear witness coral reefs and shell banks were forming apparently in a semi-tropical climate.

The coast was certainly that of the old Welsh landmass, but exactly where it lay will probably never be known. The Cotswold Hills are largely built of the debris torn from coral reefs by breakers. Interstratified with the debris are some layers of coral which actually grew upon it, probably from 5 to 25 miles from the shore. The main fringing reefs and the white strand of oolite sand that lay behind them, perhaps along the Malvern and Abberley Hills, perhaps farther west, have been entirely swept away.

The Cotswold deposits, by reason of their origin, are essentially different from the strata of equivalent age in Dorset and Somerset. There all the zones are represented by thin condensed deposits, crowded with well-preserved ammonites, gastropods and lamellibranchs, indicating that deposition of

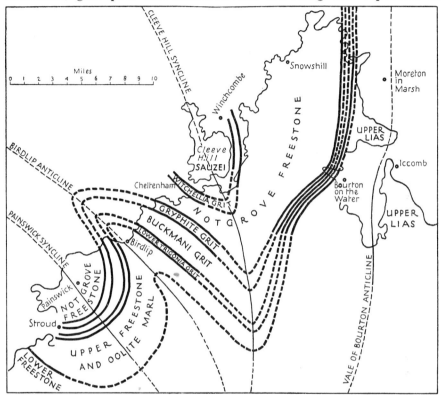

FIG. 36. Map showing the different beds in the Cotswolds upon which the Upper Infer or Oolite was deposited after the Vesulian Transgression. (After S. S. Buckman, *Q.J.G.S.*, 1901, vol. lvii, pl. VI, p. 154, and embodying corrections by L. Richardson in *Q.J.G.S.*, 1903, vol. lix, p. 382.)

sediment was always slow, sometimes virtually at a standstill. In the Cotswolds, on the contrary, sedimentation was rapid and ammonites are rare and confined to certain horizons, where they usually seem to have drifted in from a distance. Correlation by means of ammonites is of so little value, owing to their rarity, that recourse must be had to other forms of life. The brachiopods, with their colonial and sessile habit of growth, seem to have found the banks of debris a congenial habitat, and their numerous well-preserved shells have been largely used in stratigraphical work.

It is only possible to account for the differences between the Dorset–Somerset and the Cotswold strata by supposing that they were laid down in two basins of deposition more or less (but not entirely) separated by a land-barrier or shallows. In considering where such a barrier might have been situated it is necessary to remember that the Bajocian strata of the Dundry outlier, on the north side of the Mendip Hills, are essentially a continuation of the Dorset–Somerset type of deposit. The principal barrier therefore cannot have been the main Mendip Axis. It seems, rather, to have lain somewhere in the neighbourhood of Bath and Keynsham, where we saw that minor folding was operative during the formation of the Lower Lias. S. S. Buckman visualized the Dundry area as connected with the southern sea by way of a circuitous channel passing round the western end of the Mendips and over Sedgmoor (see fig. 15).[1] The evidence of the Westbury boring, however, would seem to suggest a connexion round the eastern end of the Mendips. There, beneath the Corallian outcrop, the Inferior Oolite had the great thickness of about 128 ft., and near the base, at about 1,122 ft. below the surface, was found a thin band of dark grey limestone containing fragmentary ammonites, which Buckman suggested seemed to belong to the *murchisonæ* zone.[2] This seems to indicate that eastward, down the dip-slope beneath the younger formations, are preserved the relics of the marine connexion of Lower Inferior Oolite times between the Dorset–Somerset area and the Cotswold Basin, and it may have been a tongue of this deeper sea that projected westward to Dundry Hill. A similar connexion is demanded also during the Upper Lias period, to account for the resemblance between the Upper Lias of Dundry Hill and that of the outliers of Brent Knoll and Glastonbury Tor, at all of which places the *Dumortieriæ* subzone is represented by a thick clay as in Dorset, while in the Cotswolds it forms part of the Cephalopod Bed.

The elucidation of the sequence, structure and palaeontology of the Inferior Oolite of the Cotswolds was primarily the work of S. S. Buckman, and to it he brought the knowledge and experience gained in his first geological studies of the same formation in the Sherborne district. Methods were invented to meet contingencies as the work progressed, and much was added to the science of palaeontology as a result, particularly in the province of brachiopod morphology and phylogeny. The stratigraphical conclusions were read to the Geological Society in three long papers, published in 1895, 1897, and 1901. These laid the foundations of the minute study of the Inferior Oolite, which has been amplified and extended to the Dorset coast on the one hand and into Northamptonshire on the other by Mr. L. Richardson in the last thirty years. The high value of the results, both theoretical and practical, can hardly be over-estimated, and we now know more of the Inferior Oolite than of any other part of the Jurassic System.

SUMMARY OF THE MIDDLE AND LOWER INFERIOR OOLITE OF THE COTSWOLDS[3]

(*Niortensis* and *blagdeni* zones absent.)

THE RAGSTONES (max. thickness 80 ft.)

Sauzei Zone (PHILLIPSIANA and BOURGUETIA BEDS), max. 24 ft.: These beds are confined to the Cleeve Hill plateau. They consist of some dozen or more beds of hard limestone, some of them shelly, yielding *Heimia phillipsiana* (Walk.), *Rhynchonella quadriplicata* Dav. non Zeit. and other brachiopods,

---

[1] S. S. Buckman, 1890, 'The Relations of Dundry with the Dorset–Somerset and Cotswolds areas during part of the Jurassic Period', *Proc. Cots. N.F.C.*, vol. ix, pp. 374–87; and 1923, *T.A.*, vol. iv, map, p. 52.
[2] J. Pringle, 1922, *Sum. Prog. Geol. Surv.* for 1921, p. 151
[3] Based on the works of Buckman and Richardson; see Bibliography.

PLATE VIII

Photo. *W. J. A.*

Photo. *W. J. A.*

Leckhampton Hill, near Cheltenham.

Showing the inland scarp of the Lower Inferior Oolite (Lower Freestone)
bared by ancient quarrying. In the upper view is the Devil's Chimney. The
Liassic plain with the town of Cheltenham is seen below.

PLATE IX

Photo. W. J. A.

Leckhampton Hill Quarry, Cheltenham.

Showing a section of the Inferior Oolite from the Upper Lias to the Ragstones. In the foreground are ruins of lime-kilns. O. M. = Oolite Marl,
UP. FR. = Upper Freestone.

Photo. W. J. A.

Crickley Hill, near Cheltenham.

Showing quarried cliff of Pea Grit Series, as viewed from the Birdlip road
above Tuffley's Quarry, west of the 'Air Balloon'.

together with large molluscs such as *Bourguetia striata* (Sow.), *Ostrea* (*Lopha*) '*marshii*' (auctt.), *Ctenostreon pectiniforme* (Schloth). Among the ammonites occasionally found is *Emileia vagabunda*.[1]

Witchelliæ Zone (WITCHELLIA GRIT[2] and NOTGROVE FREESTONE), max. 25–30 ft. The WITCHELLIA GRIT consists of 3–4 ft. of grey or brown ironshot limestone, some layers flaggy and shelly, yielding ammonites of the genus *Witchellia*, with *Tubithyris wrighti* (Dav.), an *Acanthothyris* and lamellibranchs. The outcrop forms a ring round the Cleeve Hill plateau.

The NOTGROVE FREESTONE, 15–25 ft. thick, has a wide surface outcrop over the greater part of the main hill-mass of the North and Central Cotswolds, and it is also the highest stratum remaining in the Painswick Syncline. It consists typically of hard fine-textured, white, oolitic limestone, usually unfossiliferous, but locally abounding in the little Pectinid, *Variamussium pumilum* (Lamk.). Ammonites resembling *Sonniniæ* sometimes occur at the base. The freestone[3] takes its name from railway-cuttings at Notgrove, between Bourton on the Water and Cheltenham, where its stratigraphical position was first discovered by Buckman.

Shirbuirniæ Zone (GRYPHITE GRIT), max. about 8 ft. This so-called 'grit' is a massively bedded, somewhat sandy limestone, characterized by an abundance of oysters of the genus *Gryphæa*, with *Entolium* sp. and *Pachyteuthis gingensis* (Oppel). Other fossils are rare. On the Cleeve Hill plateau the thickness is 5 ft., and in the main hill-mass up to 8 ft.

Discites Zone (BUCKMANI and LOWER TRIGONIA GRITS): The BUCKMANI GRIT (max. 17 ft.) is a yellowish, sandy limestone, generally merging gradually up into the Gryphite Grit, but distinguishable by the greater abundance of fossils. In the Mid-Cotswolds it always contains a readily-distinguished sandy layer, which yields the characteristic Terebratulid, *Lobothyris buckmani*.

The LOWER TRIGONIA GRIT (max. 7 ft.) is a highly fossiliferous deposit, consisting of rubbly, often ironshot, limestone and marl. The characteristic *Trigoniæ* are *T. costata* Sow. and *T. formosa* Lyc., with which occur *Pholadomya fidicula*, *Opis cordiformis* and various other lamellibranchs and brachiopods.

Corals, principally *Chorisastræa gregaria* (M'Coy), abound where the beds are well developed. In many places the basal portions are conglomeratic, containing iron-coated and bored pebbles of limestone and calcareous clay or mudstone, the latter perhaps derived from erosion of the Snowshill Clay.

UNCONFORMITY—BAJOCIAN TRANSGRESSION

Concava Zone (TILESTONE, 18 ft.; SNOWSHILL CLAY, 15 ft.; and HARFORD SANDS, 9 ft.). These beds are confined to the eastern limb of the Cleeve Hill Syncline, and best developed between Winchcomb and Chipping Campden on the western side of the main promontory of the North Cotswolds.

The TILESTONE (best exposed in the Holt Quarry, near Blockley) consists of sandy oolitic limestone with a variable proportion of sand, which predominates towards the north and gives rise to a sandy soil. The upper portion locally contains well-rolled pebbles of oolite, denoting a minor penecontemporaneous erosion. Fossils are usually rare, but lamellibranchs sometimes

---

[1] Figured Buckman, *T.A.*, pl. DCCXXIII.
[2] The term 'grit' is a misnomer for this and the other so-called 'grits' of the Ragstones.
[3] Also a misnomer as it is nowhere soft enough for cutting with a saw.

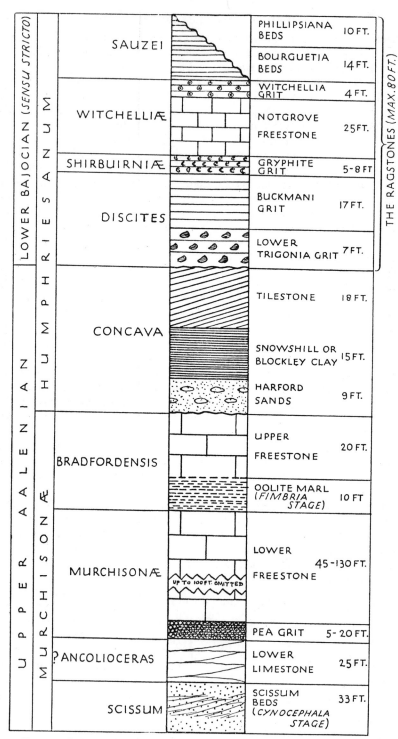

FIG. 37. Column showing the subdivisions recognized in the Lower and Middle Inferior Oolite in the Cotswold Hills.

occur, and a fish tooth and gastropod have been recorded. 'Tilestone' is another misnomer, due to an erroneous early correlation by Buckman with fissile roofing-slates since shown by Richardson to be Chipping Norton Limestone.

The SNOWSHILL CLAY (max. 15 ft.) is a stiff green, brown, chocolate or black clay, thickest about Blockley and feathering out on Charlton Common, near Cheltenham, and also to the E. and S.[1]

The HARFORD SANDS (max. 9 ft.) are of even more local distribution than the Snowshill Clay, thinning out westward before they reach Charlton Common. They are a white and pale brown quartz sand, often cemented by calcium carbonate to form large doggers or sand-burrs. Analyses have been made by both Boswell and Skerl, who state that the composition is markedly different from that of the Cotswold Sands (Upper Lias); the Harford Sands are characterized by abundant sphene and rare kyanite.[2]

Bradfordensis Zone (UPPER FREESTONE, max. about 25 ft.; and OOLITE MARL, 10 ft.).

The UPPER FREESTONE is only the indurated and usually more oolitic upper part of the Oolite Marl, and it varies in thickness accordingly. It is best developed in the Mid-Cotswolds and is the highest bed left in the centre of the Birdlip Anticline. To the east, where the Oolite Marl is well developed (e.g. at Notgrove), the Upper Freestone is thin and ill-differentiated from the marl. The same difficulty in separating them recurs in the extreme west, about Stroud.

The OOLITE MARL is, after the Pea Grit, the most fossiliferous deposit in the Cotswolds. It consists of a soft mass of ooliths in a marl matrix, and is the home of abundant beautifully-preserved brachiopods, the most characteristic of which is *Plectothyris fimbria*, a species confined entirely to the Cotswold Hills. Species of '*Zeilleria*', *Pseudoglossothyris* and *Rhynchonella* also abound, together with lamellibranchs, gastropods and a wealth of microscopic organisms—fragments of Crinoids, Ostracods and seven species of sponges. Locally corals are abundant. The Upper Freestone is usually less fossiliferous. Its characteristic brachiopod is *Rhynchonella tatei*, but since the acme of *Plectothyris fimbria* occurs immediately below, the whole was named by Lycett the '*Fimbria* Stage'. The designation has become obsolete because it was originally used to include also the Lower Freestone, which is best regarded as part of the *murchisonæ* zone.

Murchisonæ Zone (LOWER FREESTONE, 45–130 ft.; PEA GRIT SERIES, max. 30 ft.; and LOWER LIMESTONES of the West Cotswolds, 25–35 ft.).

The LOWER FREESTONE is the thickest deposit of the whole Inferior Oolite and builds the main cliff-like escarpment all along the part of the Cotswolds facing towards Gloucester and Cheltenham. Its maximum thickness of about 130 ft. is attained in the scarp of Leckhampton Hill, above Cheltenham, where it is laid bare in immense quarries, together with the other strata from the Upper Lias to the Lower Ragstones. It consists of a fine white oolite, much sought after as a building stone on account of its even texture, freedom from

[1] Richardson formerly thought that at Blockley its position was below the Harford Sand, but what he then took for Harford Sand has proved to be Tilestone, and so his Blockley Clay is the same as the Snowshill Clay of Buckman (see L. Richardson, 1929, 'Geol. Morton in Marsh', *Mem. Geol. Surv.*, p. 39).

[2] J. G. A. Skerl, 1926, *Proc. Cots. N.F.C.*, vol. xxii, pp. 153–60.

fossils and ease of working. It is not entirely devoid of fossils, however, many bands being largely made up of finely-comminuted shells, echinoderms and corals, while the very ooliths often contain at the centre the microscopic calcareous alga *Girvanella pisolitica*, first detected by E. B. Wethered. Macroscopic organisms, however, are almost entirely confined to two shelly layers, one of which is exposed 18 ft. below the Oolite Marl near the Devil's Chimney, where the Freestone is thickest, and displays a mass of shells and corals, a foot and a half thick. None of the fossils is really characteristic (Plate VIII).

The PEA GRIT SERIES (including the LOWER LIMESTONES of the W.). In the west, about Stroud and Painswick, the strata between the Lower Freestone and the *Scissum* Beds can be clearly differentiated into a coarse pisolitic deposit above, known as the Pea Grit, and white crystalline limestones below—Witchell's Lower Limestones, already mentioned. Farther east, however, on Cleeve Hill, beds of pisolite occur low down in the Lower Limestones, while the higher beds, on the level of the Pea Grit of the western district, become non-pisolitic, massive and sandy. Still farther east, in the main hill-mass of the North Cotswolds, pisolitic structure disappears altogether, and the rock is a beautiful yellow freestone (Guiting Stone). On account of this it is best to consider the whole of the beds in the eastern half of the Cotswolds between the Lower Freestone and the *Scissum* Beds as the Pea Grit Series.

The true Pea Grit, where typically developed, is a remarkable deposit of pisoliths the size of a pea, but somewhat flattened, containing tubules of *Girvanella pisolitica* and sometimes encrusted with Bryozoa. The best exposures are along the escarpment of Crickley Hill and in the Leckhampton Hill Quarry, where thousands of beautifully-preserved shells, brachiopods, echinoids and fragments of coral have been obtained (Pl. IX). The principal brachiopod is the largest Jurassic species in Britain, *Pseudoglossothyris simplex* (J. Buckman), and *Curtirhynchia oolitica* is also characteristic. The sea-urchin tests include *Pygaster umbrella* (*semisulcata* auctt.), *Galeropygus agariciformis*, *Stomechinus germinans*, *Diplopodia depressa*, *Hemipedina* (half a dozen species) and *Cidaris* spp. (*sensu lato*), besides the usual spines of the Regularia.

The sea-urchins are essentially reef-dwelling animals, and the Pea Grit, besides being a repository of their hard parts, provides also the best insight we can obtain into the nature of the corals of which the reefs were composed. As usual in the Jurassic, the chief reef-builders were *Isastræa* and *Thamnastræa* (*I. tenuistriata*, *T. terquemi*, *T. mettensis*, &c.) supplemented by several species of *Latimæandra* and *Montlivaltia*. Ten species of sponges have also been obtained from the Pea Grit, and some are numerous.[1]

In the Lower Limestones, where pisolitic structure is not in evidence, the shells are nearly all comminuted, as in the Freestones above. But as usual there are a few shelly horizons, and one of these at Crickley Hill has yielded fine examples of *Pseudoglossothyris simplex*, *Terebratula withingtonensis*, *Rhynchonelloidea subangulata* and others. Here, too, was obtained the attenuated gastropod *Ptygmatis* (*Bactroptyxis*) *xenos* Hudleston, regarded as the earliest representative of the *Nerineidæ* known in Britain. Eight species of sponges are known, some restricted to this horizon.

A fossil of widespread distribution in the main hill-mass of the North

[1] L. Richardson and A. G. Thacker, 1920, *P.G.A.*, vol. xxxi, pp. 185–6.

Cotswolds is *Variamussium pumilum* (Lamk.), which marks the horizon of the upper part of the Lower Limestones. Buckman suggested that the Lower Limestones, in which no ammonites have been found, might be on the horizon of his *Ancolioceras* zone in Dorset.

Scissum Zone. The *Scissum* Beds expand from 6 to 12 ft. in average thickness towards the Mid-Cotswolds, and in places they reach 30 ft. They consist usually of brown sandy limestones, but in the main hill-mass of the North Cotswolds the lower portions are largely represented by loose brown sand. The most characteristic and abundant fossils are *Homœorhynchia cynocephala* (Richard.), *R. subdecorata* auctt., *Aulacothyris blakei* Walker, *Astarte elegans* Sow., *Modiola sowerbyana* d'Orb., and *Pholadomya fidicula*, but occasionally such ammonites as *Tmetoceras scissum, Lioceras thompsoni* Buck., and *Hammatoceras* are found.

IV. THE VALE OF MORETON AXIS AND OXFORDSHIRE

The broad tongue of Inferior Oolite forming the main North Cotswold hill-mass is bounded abruptly on the east by a N.–S. valley of Lower Lias. Beginning as the wide Vale of Moreton, this forks south of Stow on the Wold into the Evenlode Valley, which runs south-eastward, and the Vale of Bourton, which continues almost due south. In between the two forks of the valley a small triangular wedge of hills projects some eight miles north-north-west of Burford, past the Rissingtons to Iccomb—the Iccomb hill-mass.

On the east side of the Vale of Moreton the Cotswolds are continued, geologically speaking, by a broken plateau (on which Chipping Norton stands) bounded by the Moreton–Evenlode Vale on one side and the Cherwell Valley on the other.

This plateau, though now so much dissected that it is little more than a group of outliers, still extends north-west of Banbury to the borders of Edge Hill, and its total length and breadth are roughly equal to those of the main North Cotswold hill-mass. Its greater dissection is due to the thinness of the Inferior Oolite beds, of which many of the components fail to cross the Vale of Moreton Axis, while others in crossing it become attenuated almost beyond recognition.

The thinning out of the Inferior Oolite against the Vale of Moreton Axis was outlined by Buckman in his study of the North Cotswolds. More recently Richardson has devoted a number of years to the thorough investigation of both sides of the axis, the Iccomb hill-mass and the plateaux, and he has now made it possible to obtain a complete picture of the strata in the disturbed area.

The attenuation of the various components of the formation as they pass eastward through the North Cotswolds is steady and continuous (fig. 35, p. 197). East of a line running NNW.–SSE. from Snowshill to between Bourton on the Water and Aston Blank there is no more Lower *Trigonia* Grit, while the rest of the *discites* zone, the much thicker *Buckmani* Grit, also disappears before reaching the edge of the hills at Bourton on the Hill, Stow on the Wold and Upper Slaughter.[1]

Along this edge there is also no Oolite Marl, and the remaining zones have become so thin that those who have not traced them continuously from the area of their fuller development find it difficult to recognize them. In Aston

[1] L. Richardson, 1929, 'Geol. Moreton in Marsh', *Mem. Geol. Surv.*, Ch. III.

Farm railway-cutting, on the edge of the hills west of Bourton on the Water, the whole of the Ragstones are reduced to 11 ft. of strata. Richardson has recognized a 7-in. band of Notgrove Freestone, with a bored upper surface, overlain directly by the transgressive Upper Inferior Oolite, and below it the following representatives of the underlying Ragstones—which together with the Notgrove Freestone represent a thickness of 100 ft. in the Cotswolds: Gryphite Grit 8 in.; *Buckmani* Grit 2 ft.; Lower *Trigonia* Grit, with pebbles, 1 ft.; *concava* zone or Harford Sands and Tilestone 7 ft. (the Snowshill Clay missing). The surface of the Upper Freestone is bored, and although its full thickness is not seen, it cannot be very thick, for a few hundred yards farther east a quarry shows that it and the whole of the Middle Inferior Oolite have disappeared, leaving Upper Inferior Oolite (*Clypeus* Grit) resting on the lower part of the *murchisonæ* zone (comprising thin representatives of the Pea Grit and Lower Limestones).[1]

Across the Vale of Bourton, on the Iccomb hill-mass, these last traces of Lower and Middle Inferior Oolite are no more seen, and *Clypeus* Grit rests directly on Upper Lias.[2]

The same conditions obtain eastward along both sides of the Evenlode Valley and in the southern part of the Chipping Norton plateau, south of a line running approximately north-east from Chipping Norton towards the Cherwell Valley, between Banbury and Adderbury, and so on towards Towcester. South of this line representatives of the Upper Inferior Oolite or later rocks, whatever they may be (*Clypeus* Grit at first, and later, towards the Cherwell Valley, Chipping Norton Limestone), overstep on to Upper Lias (see fig. 38). North of the line the *Scissum* Beds intervene and, in at least one or two places on the east of the Vale of Moreton, a representative of the *Ancolioceras* zone is also present.

The locality where Richardson has detected the supposed *Ancolioceras* zone east of the Vale of Moreton is Cornwell, $2\frac{1}{2}$ miles west of Chipping Norton, and he believes it to be present also at Oatley Hill, between Hook Norton and Wichford. The highest beds seen, overlying the *Scissum* Beds, are sandy limestones full of *Variamussium pumilum* (the characteristic fossil of the Lower Limestones in the Cotswolds) and numerous other fossils.

Over the rest of the area north of this SW.–NE. line, *Scissum* Beds crop out at the surface or are directly overlain by Upper Inferior Oolite or higher beds. They are in all lithological and palaeontological features a direct continuation of the *Scissum* Beds of the Cotswolds, consisting chiefly of brown sandy limestones, with *Homœorhynchia cynocephala, Pholadomya fidicula, Astarte elegans* (at Compton Wyniates and Burton Dassett), the same ammonites, *Lioceras thompsoni, Hammatoceras* sp., and the same peculiar variety of the coral *Montlivaltia lens*. The usual thickness is about 7–10 ft.[3]

V. THE EASTERN MIDLANDS

(a) Southern Northamptonshire: the Northampton Iron-field

Eastward, towards Banbury, the sandy limestones of the *Scissum* Beds pass into calcareous sandstones and ferruginous sand. On the east side of the

[1] L. Richardson, 1929, loc. cit., pp. 80–1.
[2] L. Richardson, 1907, *Q.J.G.S.*, vol. lxiii, pp. 437–43.
[3] L. Richardson, 1922, *Proc. Cots. N.F.C.*, vol. xxi, pp. 112–13, 118–20; and 1925, ibid., vol. xxii, p. 139.

Cherwell Valley they extend in a broad tract towards Towcester and Northampton, gradually thickening threefold and becoming more ferruginous in the lower part, until in the Northampton iron-field the base yields a rich ore. Throughout this tract, as far as Wellingborough and Kettering and up the valley of the Nene to beyond Oundle, there is no trace of any zone of the Lower or Middle Inferior Oolite higher than that of *Ancolioceras*. This and the *Scissum* Beds are overlain directly by sands of the age of the Chipping

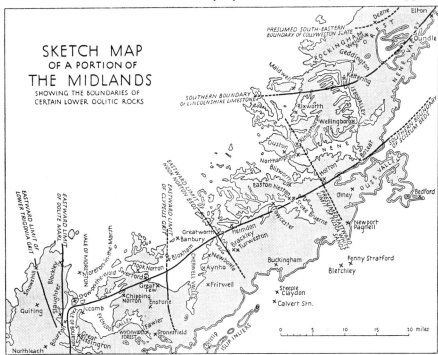

FIG. 38. Sketch-map showing the 'Scissum Line' and certain other boundaries. Outcrop of the Lower Oolites stippled, and the top of the Cornbrash represented by double line.

Norton Limestone and in the east by the Upper Estuarine Series (Great Oolite).

From the Oxfordshire border to the Humber these lowest representatives of the Inferior Oolite, grouped together as the NORTHAMPTON SANDS, are a constant and important feature. Besides being overstepped by various parts of the Great Oolite Series, they in turn rest non-sequentially upon Whitbian clays (the lower part of the Upper Lias) without the intervention of any Yeovilian. At the base they often contain pebbles and fossils derived from the Upper Lias, which in some places form a veritable *bifrons* bed. It was pointed out by Judd that the gas-chambers of these ammonites are filled with Lias, although they now lie in an ironstone matrix.

The Northampton Sands are usually divided into the VARIABLE BEDS or LOWER ESTUARINE SERIES[1] above and the IRONSTONE SERIES below, the total

---

[1] Here restricted to exclude the White Sands of Southern Northamptonshire, which Richardson has shown to be of *fusca* date.

thickness reaching from 20 to 35 ft. and in some places much more (perhaps 60–70 ft.).[1]

The VARIABLE BEDS fully justify their name. As a rule they consist principally of flaggy limestones, which pass by many gradations into calcareous sandstones and brown sands, while at some places, such as Brixworth and Harpole, they seem to become white sands and sandstone. In this last facies they are difficult to distinguish from the much younger White Sands of the *fusca* zone. At New Duston and elsewhere near Northampton, the rich, rusty, sandy limestone of the Variable Beds has been extensively quarried for building, while nearby, in Sandy Lane, west of Old Duston, the same beds pass into deep white sands.

The building stones at New Duston and Northampton are highly fossiliferous in certain bands, particularly in a coarsely granular limestone, varying from crystalline to oolitic, called the Pendle. The fauna conclusively settles the early age of the beds, in spite of their similarity to some of the *zigzag-fusca* strata of the district. It includes *Variamussium pumilum* (Lamk.), *Velata abjecta* (Phil.), *Lima rodburgensis* Whidborne and other early forms, and very occasionally ammonites of *Ancolioceras* date. Buckman has figured from the Pendle of Bass's Pit, Northampton, a specimen of *Ancolioceras mæandrus* (Reinecke),[2] while Richardson has found at New Duston a less well preserved specimen indicating approximately the same date.[3]

At Harlestone, north-west of Northampton, a novel use has been found for a soft reddish-brown sandstone at the base of the Variable Beds, on the horizon of some of the building stone of New Duston: it is ground down, mixed with cement, and made into scouring bricks or 'cotters'.[4]

Mr. Beeby Thompson considers that some of the more calcareous parts of the Variable Beds, restricted to a small area around Northampton, indicate coral growth in the vicinity, and Richardson has pointed out their resemblance to some of the Lower Limestones of the Cotswolds, which, although they have yielded no ammonites, occupy the stratigraphical position of the *Ancolioceras* zone.

The IRONSTONE SERIES is 27 ft. thick at New Duston near Northampton, but it does not usually exceed 12–20 ft., the average thickness worked being from 8 to 12 ft. The main iron-field lies between Wellingborough, Northampton and Market Harborough, with a south-westerly extension to Towcester. There are also workings at intervals farther north, by way of Cottesmore, Market Overton and Leadenham to Lincoln.

The ore is an oolitic green carbonate, resembling that in the Marlstone, but less calcareous than the Banbury ironstone and richer in iron, yielding from 31 to 38 per cent. iron. On the whole it is more sandy, especially towards the base, where it passes down into green ferruginous sandstone. A peculiar result of weathering, which serves to distinguish the ore from that of the Marlstone, is the so-called 'box-structure'. This results from the vertical joint-planes and horizontal bedding-planes becoming filled with hard, dark brown, structureless limonite, dissolved out of higher layers by infiltrated water.[5]

---

[1] B. Thompson, 1928, *The Northampton Sand of Northamptonshire*; and L. Richardson, 1925, 'Certain Jurassic (Aalenian-Vesulian) Strata of the Duston Area, Northamptonshire' *Proc. Cots. N.F.C.*, vol. xxii, pp. 137–52.

[2] 1928, *T.A.*, pl. DCCLXXXVII.   [3] 1925, *Proc. Cots. N.F.C.*, vol. xxii, pp. 147–8.

[4] B. Thompson, 1925, *Journ. Northants N.H.S. and F.C.*, vol. xxiii, p. 48.

[5] C. B. Wedd, 1920, Spec. Repts. Min. Resources, *Mem. Geol. Surv.*, vol. xii, pp. 141–207.

The heavy minerals, studied by Mr. Skerl, have yielded interesting information. An assemblage of minerals of thermal metamorphism (notably garnet, kyanite, sphene, chloritoid, spinel and staurolite) has been traced and they have been found to increase in size and quantity from all parts of the outcrop towards the Kettering district. From this fact, particularly when considered in relation to the known occurrences of the mineral chloritoid, Skerl deduces that the materials composing the Northampton Iron Ore were derived from the London landmass, to the south-east and east of the present outcrop, and directly from some metamorphic area undergoing denudation rather than from any previously-formed sediments.[1]

Palaeontologically the ironstone is at first disappointing, being often entirely barren. At certain localities, however, fossils are not uncommon, though generally preserved only as casts. The most significant are *Nerinea cingenda* Phil., which is found on the same horizon through Northants. and South Lincolnshire and in abundance in the Dogger of Yorkshire, *Astarte elegans*, *Variamussium pumilum*, *Velata abjecta*, *Ceratomya bajociana*, *Trigonia bella*, *T. compta*, *T. striata*, *T. sharpiana*, and some brachiopods not yet worked out.[2] The ammonites are somewhat unsatisfactory and have given rise to considerable controversy, owing to the swamping of contemporaneous species by derived forms from the Upper Lias. Mr. Beeby Thompson has always maintained that they indicate an Upper Liassic age, but Richardson has brought forward incontestable proof that the latest species are identical with those in the *Scissum* Beds of the Cotswolds. The forms on which he bases this conclusion are *Lioceras thompsoni* Buck. and several other species, and *Hammatoceras* spp. The occurrence of *Tmetoceras scissum* itself seems doubtful. One specimen is said to have been found near the top of the Ironstone Series.[3]

The significant SW.–NE. line, south of which the *Scissum* Beds thin out in Oxfordshire and disappear, is continued north-eastward along the southern boundary of the Northampton iron-field (fig. 38). In all the exposures around Brackley and beside the Cherwell Valley at Newbottle and Steeple Aston, and in the Fritwell railway-cuttings, beds younger than the Upper Inferior Oolite rest directly on Whitbian Lias, as in the Evenlode Valley (except for the absence of *Clypeus* Grit). Near Towcester the *Scissum* Beds are seen to be absent at the south end of the Blisworth canal tunnel, where only 5 ft. of sands (probably of *fusca* date) separate the Upper Lias from the Upper Estuarine Clay of the Great Oolite Series. Similarly farther north-east, about Horton, Bozeat and Wollaston, any sands that there may be separating these formations are so thin that they cannot be traced and the Great Oolite Series is mapped as resting directly on the Whitbian Lias.

Comparable indications have been obtained in borings south-eastward, down the dip-slope, at Olney and Stony Stratford. In the deep boring at Calvert, south of Buckingham, on the outcrop of the Oxford Clay, some part of the Great Oolite Series was proved to have overstepped still farther, on to the *algovianum* subzone of the Lower Middle Lias.[4]

If we join all the places at which the last trace of Lower Inferior Oolite disappears, leaving Upper Lias and Upper Inferior Oolite or higher beds in

---

[1] J. G. A. Skerl, 1927, *P.G.A.*, vol. xxxviii, pp. 375–94.
[2] L. Richardson, 1925, *Proc. Cots. N.F.C.*, vol. xxii, p. 149.
[3] B. Thompson, 1928, *Northampton Sand*, p. 257.
[4] A. Morley Davies and J. Pringle, 1913, *Q.J.G.S.*, vol. lxix, p. 311; and see below, p. 325.

contact, we obtain a nearly straight line over 50 miles in length. It can be plotted from near Upper Slaughter, on the west side of the Moreton Axis, passing rather north of Chipping Norton and somewhat south of Towcester (Stoke Bruerne) to Bozeat, near the junction of Buckinghamshire, Bedfordshire and Northamptonshire.

That this line is continued much farther to the north-east, beneath the Chalk, was indicated by a deep boring in the Fens east of Southery, Norfolk, described by Dr. J. Pringle in 1923. Here not only the whole Inferior Oolite was found to be missing, but the Upper Lias also, leaving a thin representative of the Great Oolite Series resting directly on the Marlstone.[1]

(b) The Lincolnshire Limestone Area[2]

At the eastern end of the line just indicated as marking the southern boundary of the *Scissum* Beds, the strike of the rocks turns in a more northerly direction and intersects the line obliquely. Therefore similar conditions are no more to be expected in passing northward along the outcrop. Henceforth, in fact, instead of only *Scissum* Beds, other zones of the Inferior Oolite are met with in regular succession.

Northward through Lincolnshire the outcrop crosses a basin of deposition in which thick deposits of Lower, Middle and probably Upper Inferior Oolite were laid down. They first appear at Kettering and, thickening rapidly northward, reach a maximum of over 100 ft. at Sleaford, 16 miles south of Lincoln. Beyond they become thinner once more, and continue to attenuate steadily towards the Market Weighton Axis.

This great lens of Inferior Oolite deposits is known as the LINCOLNSHIRE LIMESTONE. Its age was not even approximately ascertained until after 1870, when the classic investigations of Samuel Sharp and J. W. Judd began to be published. Those pioneers showed independently that the Lincolnshire Limestone belonged to the Inferior Oolite, and the great bulk of it to the Lower and Middle divisions, but at the point where their investigations left off the subject still stands to-day. No more interesting line of research could be advocated than a detailed study of the Lincolnshire Limestone, but before results of any value could be obtained prolonged field work involving careful collecting would have to be undertaken, and a thorough knowledge of Jurassic palaeontology would be essential.

The difficulties arise from bewilderingly rapid and frequent changes of facies, combined with what may be termed a general stratigraphical homogeneity. Thus the whole thickness, especially about Lincoln, may be a single, indivisible mass of limestone, or any part of it may become a 'rag' composed of corals, or pass into false-bedded banks of rolled shells.

In places, Judd remarked: 'The patches of limestone rock . . . afford ample evidence of having once been coral-reefs; near Castle Bytham a pit is opened in a rock seen to be almost wholly made up of corals.' In other places the limestone

'consists almost wholly of small shells or fragments of shells, sometimes waterworn and at other times encrusted with carbonate of lime. The shells belong to the genera

---

[1] J. Pringle, 1923, *Sum. Prog. Geol. Surv.* for 1922, pp. 126–39.
[2] Based on S. Sharp, 1870–3, 'The Oolites of Northamptonshire', *Q.J.G.S.*, vol. xxvi, pp. 354–93, and vol. xxix, pp. 225–302; J. W. Judd, 1875, 'Geol. Rutland', *Mem. Geol. Surv.*; and H. B. Woodward, 1894, *J.R.B.*, pp. 179–227.

Cerithium, Trochus, Monodonta, Turbo, Nerinea, Astarte, Lima, Ostrea, Pecten, Trigonia, Rhynchonella, &c.; and spines and plates of Echinoderms, joints of Crinoids and teeth of fishes also occur abundantly in these strata, which exhibit much false-bedding. The Gastropods are usually waterworn and the specimens of Lamellibranchs and Brachiopods usually consist of single valves often broken and eroded. These beds it is clear were originally dead-shell banks, accumulated under the influence of constantly varying currents.'[1]

The conditions, therefore, were essentially similar to those so characteristic of the Corallian period, and to a lesser extent of the Great Oolite period also. The water was apparently shallower than in the Cotswold area, for coral reefs were able to grow more luxuriantly. South Lincolnshire is perhaps to be compared rather with the strip that lay between the existing Cotswold deposits and the coast-line to the west, where all traces of Inferior Oolite have been removed by subsequent denudation. The shells in Lincolnshire are nearly all broken and rolled by the pounding of waves on the reefs, sand-banks, or beaches, while those in the Cotswolds are more often perfect, having been quietly entombed not far from where many of them lived and died.

One of the principal obstacles in the way of detailed correlation of the Lincolnshire Limestone with other deposits is the extreme scarcity of ammonites. In this respect there is a likeness to the Cotswolds, but the dearth is far greater. In the Cotswolds the scarcity and frequently worn condition of ammonites give grounds for supposing that they are not autochthonous, but drifted in from outside. The sea in which the Lincolnshire Limestone was formed seems to have been still farther removed from the source of ammonites.

The only contribution towards the dating of parts of the Lincolnshire Limestone based on modern palaeontological methods was made by W. H. Hudleston, in his monograph on the Inferior Oolite gastropods. Exclusively from a study of the gastropods he concluded that the lower portion of the limestone, which abounds in *Ptygmatis (Bactroptyxis) cotteswoldiæ* and sometimes in casts of the massive *Natica cincta*, is on about the horizon of the Oolite Marl and in part slightly earlier (*murchisonæ–bradfordensis* zones).

The highest beds, with masses of small rolled gastropods and other shells, of which the most famous sources are Ponton and Weldon, he assigned to a much later date, remarking that many of the species are not far removed from Great Oolite forms. Certainty in identifications is rarely possible, however, owing to the small size and worn condition of the specimens.[2] Provisionally it may be assumed that the highest part of the Lincolnshire Limestone is probably of Upper Inferior Oolite date.

Recognition of intervening horizons has not proceeded very far. The Survey recorded an ammonite, '*A. polyacanthus*', from a limestone known as the Silver Bed at Lincoln, and this Buckman stated to be a definite indication of the *discites* zone.[3] The same bed yielded *Natica cincta*, which Hudleston took to be an indication of the Oolite Marl (*bradfordensis* zone), but this may possibly have been derived. The level of the Silver Bed is only 8–10 ft. from the base of the Lincolnshire Limestone, and there is a 6-in. pebble-bed at the junction with the Northampton Sands.[4] It therefore seems likely that on this

[1] J. W. Judd, 1875, 'Geol. Rutland', *Mem. Geol. Surv.*, pp. 139–40.
[2] W. H. Hudleston, 1888, 'Mon. Inf. Ool. Gastropoda', *Pal. Soc.*, pp. 71–3, and 196.
[3] S. S. Buckman, 1911, *Proc. Yorks. Geol. Soc.*, N.S., vol. xvii, p. 205, footnote.
[4] H. B. Woodward, 1894, *J.R.B.*, pp. 216, 217.

northern side of the basin of deposition the *discites* zone overlaps the lower zones, evidence once again of the Bajocian Transgression.

Concerning the detailed stratigraphy little can be said, owing to the doubtful relations of the rocks seen in the various exposures.

The Lincolnshire Limestone first makes its appearance as a thin seam above the Northampton Sands near the old mill on the River Ise at Kettering, and it is also first met with westward in about the same latitude in outliers about Maidwell. The places of its first appearance farther east are considerably more to the north, Upper Estuarine Clays (Great Oolite Series) remaining in contact with Northampton Sands along both sides of the Nene Valley as far as Oundle. At Elton, 4 miles north-north-east of Oundle, the limestone first appears on the west bank of the valley only, and it is again found in rudimentary development at Castor. These points, when joined, are seen to lie on a curve, running at first nearly W–E. then SW.–NE., approximately parallel to the feather-edge of the *Scissum* Beds as delineated in the last section (see fig. 38, p. 207).

To the west and north of its first appearance, under the hills of Rockingham Forest, the Lincolnshire Limestone thickens rapidly. At Geddington, 3 miles north-east of Kettering, it is 15 ft. thick, and near Cottingham, 4 miles farther north, it is 25 ft. thick; this, however, is near the edge of the outcrop, where the full thickness is not represented, and at Weldon, the same distance from Kettering but farther down the direction of dip, it measures 30 ft. Only 6 miles east of Weldon, at Oundle, there is no trace of the rock. Farther north-east the Lincolnshire Limestone passes in thick development under the Upper Jurassic formations, showing that the axis of the trough in which it was deposited lay obliquely across the present outcrop.

The earliest horizons to appear seem to be the lowest, and at first, in all the sections in the district of Rockingham Forest, the principal palaeontological feature is a bed full of *Ptygmatis* (*Bactroptyxis*) *cotteswoldiæ* and *Nerinea subcingenda* Hudl. At or below this level occurs the *Natica cincta*, and numerous lamellibranchs and other fossils not sufficiently identified to satisfy modern requirements.

In a small area less than 10 miles in diameter, from Deene (5½ miles west-north-west of Oundle) to Stamford, occur the COLLYWESTON SLATES. These can be regarded either as the base of the Lincolnshire Limestone or as the top of the Northampton Sands. They are worked in underground galleries reached by ladders down vertical shafts. When first raised the rock is a hard, solid stone, frequently blue-hearted, the faculty for splitting along bedding laminæ being induced by exposure to frost. The resulting slates consist of a fine-grained calcareous sandstone, highly fissile, the bedding planes covered with shells, principally *Gervillia acuta* Sow. and *Trigonia compta* Lyc. They bear a strong resemblance to the slates of Stonesfield, and like them contain plant remains and small fish teeth and scales, but none of the interesting mammalian and saurian bones so characteristic at Stonesfield.

The bed from which the Collyweston Slates are obtained is only some 3 ft. in thickness. Between it and the Lincolnshire Limestone are one or more beds of sand with concretions, the detailed succession being extremely variable.

The NORTHAMPTON SANDS beneath are usually from 20 ft. to 30 ft. thick, but in one locality south-west of Stamford, about Barrowden and Wakerley, they thin out completely, leaving Lincolnshire Limestone for a short distance resting on Upper Lias clay.[1] This locality is about 10 miles east-north-east of that where the Marlstone of the Middle Lias is wanting, between Hallaton and Keythorpe, near the line of the Charnwood Axis (see p. 81). No axis has so far been suggested to account for this non-sequence.

The Lincolnshire Limestone of the neighbourhood of Stamford, according to Samuel Sharp, reaches a thickness of 65–75 ft. The upper beds have been extensively quarried round an argillaceous tract situated on Upper Estuarine Clays, called Stamford Lings, and also at Ketton, where they yielded large quantities of excellent freestone. These exposures show horizons at and near the top of the series, and their age has not been determined.

At Barnack, east-south-east of Stamford, a celebrated stone called Barnack Rag was already being obtained 500 years ago and has been used in several abbeys and cathedrals; but there appear to be no noteworthy exposures now left open. The rock consists of masses of corals and shells, and according to Judd it lies at the base of the series, close above the Collyweston Slate. This is borne out by the occurrence of *Natica cincta*.

Farther north, at Great and Little Ponton, south of Grantham, fine sections of the Lincolnshire Limestone were exposed in the middle of last century in the making of the Great Northern Railway. Morris, who described the cuttings,[2] believed the beds to be Great Oolite and consequently, against the better judgement of Lycett, a number of the fossils were figured in their 'Monograph on the Mollusca from the Great Oolite' (1850–4). The principal fossil-beds are near the top of the Lincolnshire Limestone, and Hudleston, from a study of the gastropods, suggested a tentative correlation with the similar beds at Weldon, but pointed out that the faunas showed certain marked differences.[3]

The most extensive freestone quarries anywhere in the Lincolnshire Limestone are still actively worked at Ancaster, north of Grantham. The freestone, like that at Ketton and elsewhere, is at the top of the series and its zonal position is unknown. The Upper Estuarine Clays of the Great Oolite Series as usual lie unconformably upon it. Ancaster and Sleaford lie in the centre of the trough of deposition, where the Lincolnshire Limestone reaches 100 ft. in thickness. A maximum measurement of 104 ft. was obtained in a boring.

Between Grantham and the Humber the Lincolnshire Limestone builds the peculiarly straight and abrupt escarpment of Lincoln Edge, known locally as The Cliff. Running due north and south along its summit is the Roman Ermine Street. About Lincoln the thickness is 60–70 ft., but although there are several deep exposures, the homogeneous succession of limestones has yielded little of interest or value for correlation purposes except the ammonite of *discites* date in the so-called Silver Bed building-stone.

In the strip of outcrop between Lincoln and the Humber[4] the beds begin to be differentiated into buff or cream-coloured oolites above, sometimes false-bedded and fissile, and grey, fine-grained cementstones, in appearance

---

[1] J. W. Judd, 1875, 'Geol. Rutland', *Mem. Geol. Surv.*, pp. 91, 95.
[2] J. Morris, 1853, *Q.J.G.S.*, vol. ix, pp. 324, &c.
[3] W. H. Hudleston, 1887, 'Mon. Inf. Oolite Gastropods', *Pal. Soc.*, pp. 72–3.
[4] Hence based on W. A. E. Ussher and C. Fox-Strangways, 1890, 'Geol. N. Lincs.', *Mem. Geol. Surv.*, pp. 59–79, and A. J. Jukes-Browne, 1910, *Geol. in the Field*, pp. 496–500, &c.

like Blue Lias, below. These two series are separated by from 3 to 8 ft. of marl or clay. During the survey it was found convenient to name the upper series HIBALDSTOW BEDS, after the village of that name down the dip slope, and the lower series KIRTON BEDS, after Kirton Lindsey, situated on the edge of the escarpment. The division is a purely local one, however, and certain parts of the upper series resemble the lower and vice versa. Little has been found to distinguish them palaeontologically, though the occurrence of abundant *Trigonia hemisphærica* Lyc. in the intervening marl is interesting, for this species occurs in the Cotswolds only in the Lower *Trigonia* Grit (*discites* zone) and in Yorkshire in the Millepore Bed, which Richardson has assigned tentatively to the *discites* zone also.[1] The Kirton Beds are worked at several localities for the manufacture of hydraulic cement.

About the type localities north of Lincoln the Hibaldstow and Kirton Beds, constituting the whole of the Lincolnshire Limestone, are together not more than 40–50 ft. thick. Various authors have correlated them tentatively with the upper parts of the limestone south of Lincoln, leaving some 30 ft. of beds below, including the Lincoln Silver Bed, unaccounted for (? overlapped) in the north of the county. No such correlations can yet be considered sufficiently established, for in the absence of palaeontological criteria they are based solely on lithological resemblances, which are notoriously untrustworthy in the Lincolnshire Limestone.

The Northampton Sands with their ironstone beds continue throughout Lincolnshire under the various designations Lower Estuarine Series, Basement Beds, or Dogger. For about 11 miles, between Navenby and Burton, on either side of Lincoln, they contain at the base a rich development of ore, with as much as 40 per cent. of iron. Both to north and south the ore dies away, giving place to ferruginous sands and clays with ironstone nodules.

Near Lincoln the ironstone, 12–16 ft. thick, is worked directly under the Lincolnshire Limestone, without the intervention of any sands. The limestone rests on the ore with a basal pebble bed, and the base of the ore is in turn crowded for a thickness of 2 ft. with phosphatic nodules and pebbles derived from the Lias.

Between Lincoln and the Humber the ironstone soon gives place to clays, shales, and sands, but the series thickens in places to 20–30 ft. A new feature is a 5-ft. band of hard brown sandstone at the base, which has been called the Dogger and supposed to represent the Dogger of Yorkshire. It is probably on approximately the same horizon, but no palaeontological evidence of any value seems to have been obtained.

The extremely variable lithology of the beds on the horizon of and above the ironstone warrants the continued use of the Northamptonshire terms Variable Beds and Ironstone Series. These names are more applicable than Lower Estuarine Series and Dogger and are less committal in the matter of correlation. The typical Lower Estuarine Series of Yorkshire is 200–280 ft. thick and is cut off from the more southerly deposits by the Market Weighton Axis. The Lincolnshire and Northamptonshire strata, on the other hand, belong to the same basin of deposition, their thickness and lithic characters are similar, and their outcrop is continuous.

The Variable Beds of the Northants.–Lincs. Basin are doubtless represented

[1] J. Lycett, 1877, 'Mon. Brit. Foss. Trigoniæ', *Pal. Soc.*, pp. 175–7.

in some part of the Lower Estuarine Series of Yorkshire, but the sequence south of the Market Weighton Axis is condensed and incomplete. The correlation of the Ironstone Series with the Dogger (*sensu stricto*) is much more certain, and provides a useful datum.

VI. THE MARKET WEIGHTON AXIS[1]

The continuity of the Lincolnshire Limestone under the Humber is attested by an outcrop in the river, forming a reef known as Brough Scalp. When it reappears on the opposite side, the limestone is still 20–30 ft. thick and is the most conspicuous and best-known member of the Jurassic rocks between the Humber and Market Weighton, forming a noticeable ridge past South Cave, Newbald and Sancton. It finally disappears beneath the Chalk east of Market Weighton, undiminished in thickness.

Along this ridge the limestone has been extensively quarried under the name Cave Oolite. It is a hard, blue-centred, oolitic limestone when freshly exposed, but it soon weathers to a white, friable, oolitic sand. The upper division frequently shows false bedding, like the Hibaldstow Beds, while the lower division was proved in a boring at Brantingham Grange to consist of highly calcareous mudstone or cementstones like those in the Kirton Beds.

Palaeontologically the limestone yields no more information than in Lincolnshire. The principal fossil is a polyzoan, *Spiropora* (= *Millepora*) *straminea* (Phil.), which in the Cotswolds occurs most abundantly in the Lower *Trigonia* and *Buckmani* Grits (*discites* zone), although it is also found in the Pea Grit.[2]

The lower beds, between the limestone and the Upper Lias, are also a direct continuation of those in North Lincolnshire, and are equally liable to rapid variation. The hard, brown ferruginous basal sandstone was pierced in the Brantingham boring, where it was 2 ft. 8 in. thick, while near by, at Ellerker, it gives place to a thin seam of fossils (none of stratigraphical value). The rest of the Northampton Sands consists chiefly of clays and sandy shale or shaly sandstone. In the northern part of the outcrop south of the axis, about Newbald and Sancton, a conspicuous band of hydraulic clay-limestone appears, similar to a band on the north side of the axis, and probably identifiable with one or more of the bands at the base of the Kirton Beds.

The Inferior Oolite is totally obscured by the Chalk for rather more than 10 miles north of Market Weighton. On first reappearing near Kirkby Underdale its most conspicuous features are again the limestones, which, although thinner than where last seen south of the axis, are similar in appearance. Beneath is a thin series of shales and sands, the beginning of the Lower Estuarine Series, with at the base traces of a ferruginous rock that expands northward into the Dogger. The extreme thinness of the component parts of the Inferior Oolite proves that the axis was an active factor in controlling sedimentation. (For further details see p. 64.)

VII. THE YORKSHIRE BASIN[3]

When traced north-westward away from the Market Weighton Axis, the Inferior Oolite as a whole thickens rapidly, although the limestone divisions become thinner and die out.

[1] General stratigraphy, thicknesses, &c., based on C. Fox-Strangways, 1892, *J.R.B.*
[2] L. Richardson, 1911, *Proc. Yorks. Geol. Soc.*, N.S., vol. xvii, p. 201.
[3] General stratigraphy based on C. Fox-Strangways, 1892, *J.R.B.*

The Lincolnshire Limestone or Cave Oolite reappears at Kirkby Underdale and at first, in the neighbourhood of the Derwent Gorge, it is the most important member of the series. Owing to its having been extensively quarried about Whitwell it is here known as the Whitwell Oolite. The quarries show 30 ft. of thick-bedded, blue-centred oolite, used for road metal and lime, above which are 20 ft. of siliceous limestones and sands, weathering into large doggers or tabular slabs.

Westward along the Howardian Hills the limestone thins down to 20 ft. at Terrington and is only 10 ft. thick at the west end, where all that is left is a hard siliceous limestone-band with marine fossils, weathering into huge tabular masses overhanging the beds below. North of the Coxwold faults it becomes nothing more than a false-bedded white grit with a calcareous cement, containing occasional wedges of comminuted shells and crinoid remains. Beyond Kirkby Knowle, 6 miles north of the Coxwold faults, it is no longer traceable.

The basal bed of the Inferior Oolite, called throughout Yorkshire THE DOGGER, and probably corresponding fairly closely with the Northampton Ironstone Series, does not increase appreciably in thickness. At the Derwent it is 12 ft. thick and consists of ferruginous sandstone, the lower part calcareous and full of pebbles. Attempts have been made to work it for iron, but it has proved too poor an ore to repay commercial exploitation. Along the Howardian Hills its occurrence is very irregular, and for 10 miles it is absent altogether. At the west end it is a massive ferruginous limestone, and beyond the Coxwold faults it is sometimes unrecognizable and sometimes conspicuous as a 5 ft. band of calcareous ironstone. It so continues, with many changes, around the Yorkshire Basin.

The great increase in the thickness of the Inferior Oolite is therefore not due to the calcareous elements, but to the sandstones and shales between and above them—the so-called Estuarine Series. As the limestones die away, so the great 'estuarine' (really deltaic) beds thicken, until on the north side of the basin they dominate completely and become the most characteristic member of the Yorkshire Oolites, forming the greater part of the Cleveland Hills and the North York Moors, with a total thickness of 400–500 ft.

The lithological appearance of these rocks and the scenery to which they give rise both recall strongly the Millstone Grit or Coal Measures. To one accustomed to the subdued and wooded scenery of the oolitic areas of Southern England and particularly to the fertile red fields and stone walls of the Cotswolds, it is difficult to realize that these barren, heather-clad moors are built of rocks of the same age as the familiar Oolites of the South.

It is interesting to trace the expansion of the LOWER ESTUARINE SERIES, which separates the Dogger from the Whitwell Oolite. Almost as soon as it appears on the north side of the axis it is thicker than on the south, being already 50 ft. thick at Acklam, while in the Howardian Hills it soon reaches 120 ft. A band of hydraulic limestone, correlated with that to the south of the axis, forms a conspicuous feature in the series from Burythorpe to the west end of the Howardian Hills. It consists of hard, close-grained argillaceous limestone, light blue-grey in colour, with a conchoidal fracture, and it has a thin band of ironstone below it.

North of the Coxwold faults the hydraulic limestone gradually passes into a sandstone and the ironstone below contains a few marine fossils. Northward among the moors the ironstone band multiplies, and in places has even been worked as an ore, while such marine fossils as *Astarte*, *Trigonia*, *Pleuromya*, *Pholadomya*, and *Ostrea*, unidentifiable specifically, are always associated with it. This marine horizon is called the ELLER BECK BED, after a locality near Goathland, where it is well exposed and especially fossiliferous.

The most important difference marking off the Inferior Oolite north of the Market Weighton Axis from that to the south is the addition of two more series of beds above the Whitwell Oolite.

Between Market Weighton and the Humber some 20 ft. of sands and clays separate the Cave Oolite from the Kellaways Rock. Almost nothing is known of them, but they are considered to represent the Upper Estuarine Beds (Great Oolite Series) which everywhere south of the Humber rest non-sequentially upon the Lincolnshire Limestone.

North of the Market Weighton Axis the Whitwell Oolite is succeeded by similar estuarine beds full of plant remains and coal seams (the MIDDLE ESTUARINE SERIES), and those in turn by a marine horizon (the SCARBOROUGH BEDS or GREY LIMESTONE SERIES) which has yielded on the coast ammonites of *blagdeni* date. Above this follows another thick mass of unfossiliferous sandstones (the Upper Estuarine Series), more than enough to represent the Great Oolite. The full Inferior Oolite sequence is therefore as follows:

LINCOLNSHIRE	YORKSHIRE BASIN	DATES[1]
	Scarborough Beds (max. 70 ft.)	*blagdeni* (in part)
	Middle Estuarine Series (100 ft.)	
Lincolnshire Limestone . .	Whitwell Oolite and Millepore	
(Cave Oolite of S. Yorks)	Series (0–50 ft.)	*discites*
Northants { Variable Beds . .	Lower Estuarine Series (max. 286 ft.) (with Eller Beck Bed)	
Sands { Ironstone Series .	The Dogger (usually 5–20 ft.)	*murchisonæ*

The difficulties of correlation are obvious. The dating of the highest marine series at Scarborough (at least in part) to the *blagdeni* zone provides a useful upward limit. But owing to the lack of significant fossils in the Lincolnshire Limestone we cannot be sure whether the whole of that limestone is represented in the Cave and Whitwell Oolites, or whether the Middle Estuarine Series is an intercalation in its upper portion, accompanied by great expansion. The similarity of the upper part of the Cave and Whitwell Oolites to the Hibaldstow Beds and of the lower part and the hydraulic limestone to the Kirton Beds, however, makes it extremely probable that the Middle Estuarine and Scarborough Series are additions to the Yorkshire sequence which have been removed south of the Market Weighton Axis, or were never deposited.

These higher strata first appear near Westow, between Burythorpe and the Derwent. The Middle Estuarine Series reaches 20 ft. at the Derwent, 30–40 ft. at Terrington, and 60 ft. at the west end of the Howardians. Farther north it thickens less rapidly to its maximum of 100 ft. in North Yorkshire. It is at first a soft sandy rock with lenticles of dark shale and carbonaceous clay, the whole much like Coal Measures. North of Kirkby Knowle more massive beds of sandstone are a feature.[2]

[1] L. Richardson, 1911, *Proc. Yorks. Geol. Soc.*, N.S., vol. xvii, pp. 185–6.
[2] C. Fox-Strangways, 1892, *J.R.B.*, pp. 217–27.

Workable coal-seams are found in many places on the moors both north and south of Eskdale, and in the Howardian Hills, especially about Coxwold. They seldom exceed 6 to 12 in. in thickness, but seams have been known measuring 2 ft., and in both the Lower and Middle Estuarine Series seat-earths frequently occur below them. Plants are the chief interest of the Middle Estuarine Series: several localities where the beds are exposed, such as Marske and Roseberry

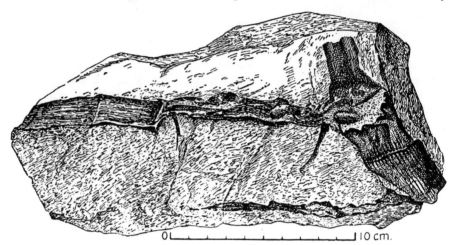

FIG. 39. Stems of *Equisetites columnaris* (Brong.) with rhizomes attached; Lower Estuarine Series, Peak Alum Works, Yorks. (From Halle and Kendall, 1913, *Geol. Mag.* [5], vol. x, fig. 1.)

Topping in Cleveland, and Gristhorpe Bay on the coast, have become famous throughout the palaeobotanical world.

The flora is characterized principally by an abundance of ferns, cycads and certain types of conifers. The Ginkgoales too play a prominent part, as also the Equisetums, closely allied to the modern horse-tails.

The majority of the plants have been drifted (even if only a short distance) to the positions where they are now found, but Dr. Halle and Prof. Kendall have obtained proof that at least some of the well-known upright *Equisetites* stems were entombed by sediment as they stood, rooted in the delta marsh. In the Lower Estuarine Series at the Peak Alum Works they found examples standing upright with long rhizomes growing out from the base in a horizontal direction (fig. 39).[1] Other plants found growing *in situ* at the base of the Inferior Oolite near the Yellow Sands at Whitby have been described as the remains of 'a forest of *Dadoxylon*', and the roots have been traced ramifying in the shales of the Upper Lias.[2] These discoveries throw extremely interesting light on the conditions under which the Estuarine sandstones were laid down; especially since Dr. Bather has explained certain associated tubular markings as undoubtedly the U-shaped burrows of Polychaet worms (fig. 40), and Mr. Stather has even found their heaps of castings. These worms, which have been named *Arenicolites statheri*, lived in colonies on intertidal flats, moving constantly towards the surface as their burrows became covered with

[1] T. G. Halle and P. F. Kendall, 1913, *Geol. Mag.* [5], vol. x, pp. 3–7.
[2] *The Naturalist*, 1930, pp. 186–7; and P. F. Kendall, *Q.J.G.S.*, vol. lxxxv, p. 438.

sand.[1] In the Lower Estuarine Series there are also beds of the earliest British *Unio, U. kendalli* Jackson.[2]

The most important work on the Jurassic plants in recent years has been the investigation by Mr. Hamshaw Thomas into the nature of certain minute carpels, fruits, seeds, anthers and pollen found in the Middle Estuarine plant bed in Gristhorpe Bay, and now shown to belong to a new class, the earliest

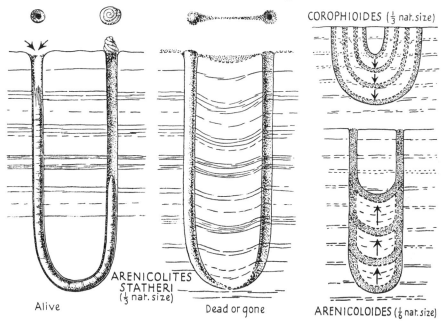

COROPHIOIDES ($\frac{1}{3}$ nat. size)

ARENICOLITES STATHERI ($\frac{1}{3}$ nat. size)

Alive   Dead or gone   ARENICOLOIDES ($\frac{1}{6}$ nat. size)

FIG. 40. U-shaped worm-burrows in the Lower Estuarine Series near Blea Wyke, Yorkshire.
    The left-hand drawing shows the *Arenicolites* worm within the burrow, with its head on the left and tail on the right. A current of water with sand and organic matter is drawn in for food, as shown by the arrows; above the other vent a casting has just been left. In the abandoned burrow the left-hand tube has become filled with sand and the bottom of the U has sunk in, causing the laminæ above to sag and produce a depression at the surface.
    In the compound burrows on the right, the animal has moved to successive levels as shown by the arrows. (From F. A. Bather, 1925, *Proc. Yorks. Geol. Soc.*, vol. xx, p. 188.)

Angiosperms yet known. Under the name Caytoniales the class has been known and its Angiospermous affinities have been suspected for a number of years. Two species of two genera have now been described as *Caytonia sewardi* and *Gristhorpia nathorsti*.

Mr. Hamshaw Thomas has summed up his views on these plants as follows:

'The Caytoniales possess two of the features most characteristic of the modern flowering plants, viz., the closed carpel with a stigma, and the anther with four longitudinal lobes. On this account it seems permissible to group them with the modern Angiospermæ, though they do not seem to resemble any modern family. It is possible that they belong to a line of evolution which was quite distinct from that which gave rise to the modern Dicotyledons and Monocotyledons, and represent a parallel series of forms now completely extinct. . . . The Caytoniales seem to occupy a position between the Palaeozoic Pteridosperms and the recent Angiosperms, and

[1] F. A. Bather and J. W. Stather, 1925, *Proc. Yorks. Geol. Soc.*, vol. xx, pp. 182–99.
[2] J. W. Jackson, 1911, *Naturalist*, pp. 211–14.

thus they suggest a possible solution for one of the great outstanding problems of evolution.'[1]

The constant association of the remains of the Caytoniales with the peculiar leaves known as *Sagenopteris*, which show venation similar to that of some dicotyledonous leaves, is believed to indicate that these were parts of the same plants. The two have also been found together in the Rhætic plant-beds of Greenland.

Mr. Hepworth has described infilled river or stream channels ('washouts') in the Lower and Middle Estuarine Series and notes that the drifted plants are most abundant in the base of these channels.[2] Similar phenomena will be more fully discussed in connexion with the Upper Estuarine Series (p. 315).

The highest member of the Yorkshire Inferior Oolite, the SCARBOROUGH BEDS or GREY LIMESTONE SERIES, named after the best-known occurrence on the coast, begins as a very thin band, scarcely traceable before the Derwent. The first good sections, at Stonecliffe Wood[3] and in Cram Beck, near the Derwent gorge, show soft sands with doggers and several bands of siliceous limestone crowded with fossils. The principal forms are *Pseudomonotis lycetti* Rollier (= *braamburiensis* pars, auct.), *Trigonia signata* Lyc., *Gervillia scarburgensis* Paris, *Pleuromya securiformis* (Phil.), with Pectens, oysters, &c. The first two species are abundant in and characteristic of the Hook Norton Beds of Oxfordshire (see p. 303), and it is significant that *Trigonia signata* has not been recorded from below the Upper *Trigonia* Grit (*garantiana* zone) elsewhere in England.[4] Records do not suffice to show the range of the *Pseudomonotis*, but the combination of these two species may indicate that the beds in part represent the Hook Norton Beds.

In the Howardian Hills the Scarborough Beds consist of two parts. Above are 20 ft. of brown porous grit, full of casts of *Pseudomonotis lycetti*, or a soft sandrock with doggers, abounding in the same fossil; below are 20 ft. of hard siliceous limestone, splitting into slabs, and sometimes so fissile as to have been used as roofing slates.

Beyond the Coxwold faults thickening continues and thenceforth along the escarpment of the Hambleton and Cleveland Hills and across the moors to the sea there are three distinct divisions: a predominantly shaly division above, a predominantly arenaceous division in the middle, and a predominantly calcareous division below. The maximum thickness of the three portions together is probably 90 ft. Each is subject to considerable local variation. Thus at Spindle Thorn, on Spaunton Moor, a noticeable 5 ft. bed of hard blue limestone is intercalated in the lower part of the upper or shaly division. Richardson has called this the Spindle Thorn Limestone. Again, he has distinguished as Lambfold Hill Grit a thick fossiliferous grit at a higher horizon in the shaly division. This forms an important capping to many of the ridges and watersheds between the interior dales.[5]

By far the finest sections of the Yorkshire Inferior Oolite are, of course, on the coast, where the same SE.–NW. thickening can be traced as along the

[1] H. Hamshaw Thomas, 1925, *Phil. Trans. Roy. Soc.*, vol. ccxiii. B, p. 356.
[2] E. Hepworth, 1923, *Trans. Leeds Geol. Soc.*, part xix, pp. 24–8.
[3] Described as one of the principal Cornbrash exposures in Blake's 'Monograph of the Fauna of the Cornbrash', *Pal. Soc.*, 1905, p. 16. But see C. Fox-Strangways, 1892, *J.R.B.*, p. 249.
[4] J. Lycett, 'Mon. Brit. Foss. Trigoniæ', *Pal. Soc.*, p. 29; and Appendix, pp. 1–10.
[5] L. Richardson, 1911, *Proc. Yorks. Geol. Soc.*, vol. xvii, p. 197.

inland escarpment. It should be remembered that the most southerly cliff-sections, in Gristhorpe Bay, between Scarborough and Filey, are little farther from the Market Weighton Axis than are the exposures in the neighbourhood of the Derwent gorge.

To the foregoing account of the formation inland a summary of the sequence displayed along the coast may be usefully appended.

SUMMARY OF THE INFERIOR OOLITE OF THE YORKSHIRE COAST[1]

Blagdeni Zone (Zigzag? to sauzei?) (SCARBOROUGH BEDS). The Scarborough Beds first rise from beneath the Upper Estuarine Series in Gristhorpe Bay. At the east end they measure only 3 ft., but before reaching the west end of the bay they have already more than doubled in thickness. The rock is a grey crumbly shale, with two bands of ironstone nodules, and the characteristic *Pseudomonotis lycetti* abounds.

At White Nab, Cayton Bay, the thickness is 30 ft., and when Cloughton Wyke is reached, north of Scarborough, it exceeds 70 ft. The same *Pseudomonotis* is abundant, together with *Trigonia signata*, while one band near the base is crowded with *Gervillia scarburgensis* Paris.[2] Progressively with the thickening, a tripartite division becomes perceptible, as in the western escarpment. North of Scarborough all the sections show predominantly argillaceous, arenaceous and calcareous divisions, from above downwards. The argillaceous division, consisting of shale with some ferruginous bands, is here the thickest and to it is confined the *Pseudomonotis*. Thirty-seven feet from the top a bed has yielded *Isocrinus*, a fossil which in the moors to the north-west forms a thick grit-band about this horizon, known as the Crinoid Grit.

Although the exact horizons of the ammonites recorded from the Scarborough or Grey Limestone are not known, it may be presumed that most if not all came from the lowest or limestone division. The available specimens in Scarborough Museum and from other sources were investigated in 1909–11 by Buckman, who identified the following species: *?Skirroceras triptolemus* (Bean MS.) (= *Am. braikenridgei* Morris and Lyc. non Sow. et auctt.), *Teloceras coronatum* (Quenst.), *Teloceras* sp. (= *Am. blagdeni* Mor. and Lyc. non Sow.), *Stemmatoceras subcoronatum* (Oppel), *Stepheoceras* cf. *zieteni* (Quenst.), *S.* cf. *pyritosum* (Quenst.), and *?Dorsetensia* spp.[3]

On the whole this evidence suggests approximately a *blagdeni* date, but a belemnite occurs rather commonly (*B. ellipticus* Miller) which is common in the *sauzei* zone in the Cotswolds.

Witchelliæ and ? Shirbuirniæ Zones (MIDDLE ESTUARINE SERIES). The sands, sandstones, and shales comprising this division are already 40–50 ft. thick on first appearing in Gristhorpe Bay, while northward at Cloughton Wyke and Blea Wyke they thicken to 90–100 ft. The passage both upward and downward into the marine beds is gradual. No internal evidence for dating has been obtained, the only fossils being the plants (see p. 218). The best plant locality is in Gristhorpe Bay.

Discites Zone (MILLEPORE SERIES OR WHITWELL OOLITE). The Whitwell Oolite is recognizable palaeontologically in the most southerly sections on the coast, in Cayton Bay, and again at Cloughton Wyke, north of Scarborough.

---

[1] Based on Fox-Strangways and on Richardson, loc. cit.
[2] See E. T. Paris, 1911, *Proc. Cots. N.F.C.*, vol. xvii, p. 255.
[3] S. S. Buckman, 1911, *Proc. Yorks. Geol. Soc.*, N.S., vol. xvii, pp. 205–8.

As in the western escarpment, it dies away towards the north, and cannot be traced beyond Robin Hood's Bay—somewhat north of its disappearance in the west.

On the coast the horizon is not represented by an oolite, but by a variable series of sandstones, with, however, the same fauna as at Whitwell and Cave. The most conspicuous animal of all is the small polyzoan, *Entalophora* (= *Spiropora* or *Cricopora* or *Millepora*) *straminea* (Phil.), from which the series derives its name.

In Cayton Bay the true Millepore Bed is composed of 15 ft. of hard calcareous sandstone, but the marine fossils continue upward through 25 ft. of additional ferruginous sandstone and sandy shale not found elsewhere. North of Cloughton the horizon rapidly becomes so sandy and unfossiliferous as to be rarely recognizable, while north of a line joining Robin Hood's Bay with Kirkby Knowle on the western escarpment the Middle and Lower Estuarine Series are in contact and inseparable.

None of the fauna of the Millepore Series and Whitwell and Cave Oolite is very satisfactory for purposes of correlation. The lamellibranchs are nearly all species which range throughout the Inferior Oolite and have even been recorded under names supposed to show their identity with Bathonian species. Such species as *Gervillia whidbornei* Paris and *G. prælonga* Lyc., which occur in the *discites* zone in the Cotswolds, however, may have some significance.[1] Richardson has also drawn attention to certain spinose Rhynchonellids of the genus *Acanthothyris* which seem to be common to the Whitwell Oolite, the Lincolnshire Limestone and the *Buckmani* Grit.[2]

Concava–Bradfordensis Zones (LOWER ESTUARINE SERIES). This series is more arenaceous than the Middle Estuarines, but its general aspect is similar, and the two floras are closely allied, though not entirely identical.

The Eller Beck Bed, already described, first rises above sea-level at Iron Scar, a mile north of Cloughton Wyke, and thence northward is seen intermittently as far as the Peak Fault, again from Hawsker to Whitby, and at Kettleness and Hinderwell, where it turns inland. The continuity of this marine band as compared with the Millepore Series is remarkable. Its position is usually rather above the middle of the Lower Estuarine Series, which expands to a maximum thickness of 286 ft. at Blea Wyke.

Murchisonæ–? Scissum Zones (THE DOGGER). The basal marine band of the Yorkshire Inferior Oolite is best known on the coast, where its varied lithology has been described in the following words by Fox-Strangways:[3]

'It changes from a sandstone to a limestone or a valuable ironstone, and from a fine-grained shaly bed to a nodular calcareous oolitic rock with little bedding. In some places it seems to form a passage between the Lias and the Lower Oolite, in others it rests on a distinctly eroded surface of the shales, while here and there it is itself cut out entirely by the Estuarine Sandstones, which rest immediately on the Alum Shale.'

Sometimes, as at Blea Wyke, the base of the Dogger contains a bed full of *Nerinea cingenda* Phil., the gastropod found in the Northampton Ironstone as far away as the Oxfordshire border. At other places, as in the Peak Alum

[1] E. T. Paris, 1911, *Proc. Cots. N.F.C.*, vol. xvii, pp. 237–54.
[2] L. Richardson, 1911, loc. cit., p. 202.   [3] C. Fox-Strangways, 1892, *J.R.B.*, p. 154.

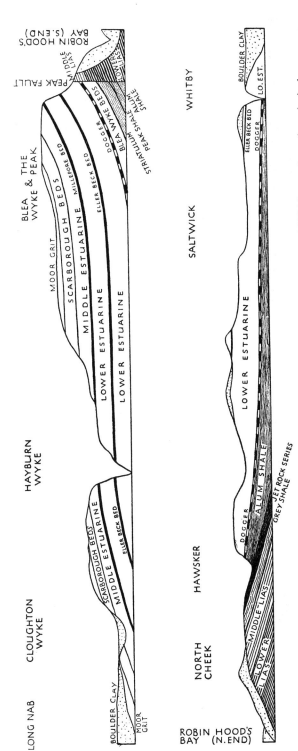

Fig. 41. Diagrammatic section along the Yorkshire coast between Long Nab, north of Scarborough, and Whitby, to show the principal exposures of the Inferior Oolite Series. Total distance about 13 miles. Between the two sections, in Robin Hood's Bay, 1¾ miles of low cliffs of Lower Lias covered by Boulder Clay are omitted. (Based on J. F. Blake, *P.G.A.*, vol. xii, p. 116.)

Works near Blea Wyke, 5–15 ft. of sandstone is entirely unfossiliferous, though there are usually rolled pebbles, and sometimes ammonites, derived from the Upper Lias—a feature familiar in the Eastern Midlands. At least one fragment of an ammonite, of *striatulum* type, has been found partly embedded in a pebble of Upper Lias shale.[1]

A number of channels have been described, cut through the Dogger and deep into the Upper Lias prior to the deposition of the Lower Estuarine Series,

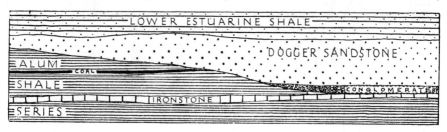

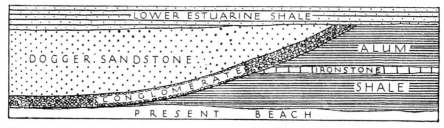

FIGS. 42, 43. Diagrammatic representations of the sections at the eastern quarry, Boulby Alum Works (above), and in East Cliff, Whitby (below), illustrating channelling of the Upper Lias shales by the basement beds of the Inferior Oolite in Yorkshire. (After R. H. Rastall, 1905, *Q.J.G.S.*, pp. 450, 452, re-drawn.)

thus proving renewed or continued earth movements after the formation of the Dogger, and even larger channels were cut earlier and filled by the Dogger (see figs. 42–3). One of the largest was described by Fox-Strangways inland at Vittoria Plantation, Bilsdale. The Dogger, which fills it to a thickness of 50 ft., is there a 'ferruginous echinital limestone', largely composed of the spines of an *Acrosalenia*. Richardson has called attention to the fact that the Lower Limestone of the Edge, Painswick, Gloucestershire (which may be presumed to be of approximately the same age) is also full of an *Acrosalenia* —*A. lycetti*—and the remains of crinoids.

In other places the Dogger contains a workable ironstone from 1 to 4 ft. thick, but in Rosedale as much as 14 ft. thick. It is of inferior quality, however, and has not been exploited to any great extent. An extraordinary development is found in Rosedale, where there are two channels (? or two sections of the same channel), 70 ft. deep, filled with a confused mass of irregular lumps of dark blue oolitic ironstone coated with dark brown or purplish iron oxide, the whole highly magnetic. If echinoid or other organic remains were ever present here, they have been obliterated.[2]

[1] L. H. Tonks, 1923, *Trans. Leeds Geol. Soc.*, part xix, p. 32.
[2] C. Fox-Strangways, 1892, *J.R.B.*, pp. 170–1.

VIII. THE HEBRIDEAN AREA

The Lower, Middle, and Upper Inferior Oolite of the Inner Hebrides are entirely of marine origin, although the predominantly sandy sediments and the presence of a considerable amount of drifted wood indicate that they were formed in proximity to land.

Palaeontologically the strata afford a remarkable contrast with those laid down at the same time in Yorkshire, or indeed anywhere north of the Mendip (or Keynsham) Axis; for some of the beds are crowded with autochthonous ammonites. All the zones of the Lower Inferior Oolite are condensed into a few feet of highly ammonitiferous beds, comparable only, among British deposits, with the ammonite beds of the Dorset–Somerset area and of Dundry Hill. The zones of the Middle Inferior Oolite, on the other hand, are greatly expanded, and fossils are much less plentiful, owing to their being dispersed through unusually great thicknesses of rapidly-deposited rock (600 ft. of sandstone in Raasay, over 300 ft. in Skye); but nevertheless ammonites of nearly every zone are present. The Upper Inferior Oolite (still, at first, with autochthonous ammonites) follows conformably and sequentially, but begins in most places with a rapid lithic change to shale and clay (see fig. 32, p. 184).

The succession has been carefully re-examined in recent years by Lee, whose ammonite collections were submitted to Buckman and are now on view in the Royal Scottish Museum, Edinburgh. The exact lines of demarcation between most of the zones could not be determined, but the general succession of ammonite horizons and the approximate relative thicknesses of the zones were made out. The rest of the fauna has still to be studied.

(a) Raasay and Skye [1]

The outcrops in both Raasay and Skye are extensive. In Raasay, where the formation is thicker than anywhere else in the British Isles, it builds the main precipices of the great eastern cliffs, and the scenery to which it gives rise is also probably the grandest Jurassic landscape in the country. In Skye the sandstones, only half as thick, but still measuring 300 ft., build the long line of cliffs, often capped with basalt, which form the eastern bulwark of Trotternish, both north and south of Portree Bay. In the peninsula of Strathaird the outcrop is chiefly heather-covered, forming a slope between the basalt plateau and the shore of Loch Slapin (map, p. 114).

The bulk of the formation in both islands is made up of the Middle Inferior Oolite, which consists of rather soft, white-weathering yellow sandstone, locally developing large doggers—a type of deposit which never proves highly fossiliferous and undoubtedly testifies to relatively rapid accumulation. The thin ammonitiferous beds of the Lower Inferior Oolite, although so interesting palaeontologically, feature little at the surface and are not often well exposed.

SUMMARY OF THE LOWER AND MIDDLE INFERIOR OOLITE OF RAASAY AND SKYE

Blagdeni Zone. The topmost part of the sandstone in both islands is coarse and gritty and full of broken lamellibranchs. The whole series, although well exposed in the cliffs, is inaccessible, and most of the collecting has

[1] Based on C. B. Wedd, 1910, 'Geol. Glenelg, Lochalsh and SE. Skye', pp. 115–20; and G. W. Lee and S. S. Buckman, 1920, 'Mes. Rocks Applecross, Raasay and NE. Skye', pp. 44–50, 65, 72–9; *Mems. Geol. Surv.*

perforce to be done from fallen blocks. The greater part of the thickness seems to be characterized by Stephanocerates and allied ammonites of the *blagdeni* zone, usually flattened: *Teloceras* cf. *coronatum* (Quenst.), *Stepheoceras* sp., *Stephanoceras* sp., and *Stemmatoceras* cf. *subcoronatum* (Oppel) were recorded.

Sauzei, Witchelliæ and Shirbuirniæ Zones. Below the sandstone cliff on the north side of Bearreraig Bay, Skye (Pl. X), the fallen doggers have yielded large numbers of *Sonniniæ*, denoting the *sauzei* zone, and single specimens of *Witchellia* and *Shirbuirnia* have also been obtained. The *Shirbuirnia* is a gigantic form measuring more than 18 in. in diameter, and was identified as *S. trigonalis* S. Buck. The *Sonniniæ* are comparable with swarms in the ironshot oolite of Dundry and the top part of the Sandford Lane fossil-bed near Sherborne, where they are associated with *Otoites sauzei*. The absence of *Otoites* in Scotland is noteworthy.

Discites Zone. The *discites* zone has not been detected with any degree of certainty in Skye, but in Raasay a large number of small ammonites of *discites* date have been obtained at one locality only, on the Dun Caan path. Their restricted distribution suggests a lenticular deposit.

Concava and Bradfordensis Zones. A few ammonites of the *concava* and *bradfordensis* zones were obtained in Raasay in the same blocks as numerous species of *murchisonæ* date. Four species of *Ludwigella* are recorded, and three of *Brasilia*, including *B. bradfordensis*, the index-fossil. There are similar indications in Skye, below inaccessible cliffs between Holm and Bearreraig. The rocks whence specimens of *Brasilia* have dropped consist here of bluish shaly sandstone with doggers, situated 30–50 ft. above sea-level. It is possible that the zones are less condensed than in Raasay.

Murchisonæ Zone. Holm, in Skye, is the type locality of *Ludwigia murchisonæ*, but it is rare there, although other ammonites of the same date are plentiful. The zone consists of a small thickness of flaggy sandstone, which forms the base of the sandstone series, cropping out between tide-marks on the south side of the mouth of the River Bearreraig. The ammonites, which belong to the genera *Ludwigia, Ludwigina, Hyattina, Welschia, Strophogyria, Crickia,* and *Brasilina,* are extremely well-preserved and abundant, both here and at several points in Raasay (the best locality being beside the Dun Caan path). These outcrops prove that the *murchisonæ* zone is more richly ammonitiferous in the Hebrides than at any other locality in the British Isles, excepting possibly Chideock, Dorset (the Wild Bed). Nevertheless, the absence of certain southern families is remarkable, especially the Hammatoceratidæ— *Hammatoceras, Erycites,* &c.

Ancolioceras Zone. The only indication of deposit of this date is the presence of cf. *Ancolioceras substriatum* Buck. in the *scissum* zone of Raasay.

Scissum Zone. The *scissum* zone of Raasay consists of 8 ft. 6 in. of thin-bedded limestones and shaly limestones, separating the shales below from the great sandstone series above. At the top of the zone on the path one mile south of Dun Caan is a band of limestone with lenticles crowded with small ammonites of *scissum* date. Thirteen species of *Lioceras* and one each of *Canavarella, Ancolioceras,* and *Rhæboceras* are recorded, but the characteristic genus *Tmetoceras* is wanting.

A similar development of the zone occurs in Ardnadrochet Glen, Mull.

PLATE X

Photo. Geol. Survey.

Bearreraig Bay, $6\frac{1}{2}$ m. N. of Portree, Isle of Skye.

Inferior Oolite sandstones overlain by sill of Tertiary columnar dolerite.
Ammonitiferous Lower Inferior Oolite in foreground.

Photo. Geol. Survey.

Dun Caan and the east coast of Raasay from Rudha na'Leac.

The high cliff consists of Inferior Oolite sandstone, with a slope of Great
Estuarine shales above, capped by the Tertiary plateau basalt of Dun Caan.
Lias and Rhætics in foreground.

(b) **Mull and Ardnamurchan** [1]

The Lower and Middle Inferior Oolite cover a small area in the south-east corner of Mull, about Loch Spelve and Loch Don, and they also occur on the south coast of the promontory of Ardnamurchan, as a faulted outlier a mile and a half south-west of Kilchoan. A former extension over the north of Ardnamurchan is proved by the presence of blocks in the agglomerate of a volcanic vent (map, p. 114).

The thickness is much less than in the northern islands, not much exceeding 90 ft. in Ardnamurchan and 80 ft. in Mull, but the succession, although contracted, is in essential features the same. The upper and by far the bulkier part of the formation consists of poorly-fossiliferous sandstones, the greater portion presumably of *blagdeni* age, while the thin lower beds are predominantly calcareous and yield abundant ammonites of many different zones.

The most complete sections are on the shore at Port nam Marbh and Port Donain, Mull, supplemented a short distance inland by a useful section of the basal beds in a stream running through Ardnadrochet Glen, south of Loch Don. The Ardnamurchan exposure is an irregular one along the shore, extending about a mile on either side of Sròn Bheag Point, and it is greatly complicated by intrusions and faulting. Nevertheless some of the zones are there more fully developed than in Mull, and Buckman claimed from them some support for the subdivision of the *discites* deposits which he advocated in his last years.

SUMMARY OF THE LOWER AND MIDDLE INFERIOR OOLITE OF MULL AND ARDNAMURCHAN

Blagdeni Zone. Contrary to the course of events in Skye and Raasay, sandstones continued to be deposited, at least in Mull, until after the immigration of the *garantiana* fauna of the Upper Inferior Oolite. Consequently there is no definite upper boundary to the *blagdeni* zone, which, in the absence of any sign of the *niortensis* zone, may be considered to extend upwards to the first occurrence of the genus *Garantiana*, 10 ft. below the shales of the Great Estuarine Series. Thus, if the same amount be subtracted from the top of the sandstone in Ardnamurchan, where *Garantiana* has not been found, as at Port nam Marbh in Mull, some 50 ft. of almost entirely unfossiliferous sandstone are left in both localities to represent the thick *blagdeni* zone of Skye and Raasay.

Blagdeni, Sauzei, and Discites Zones. In Ardnamurchan the highest fossils found are ammonites of *discites* date, so that an unknown thickness of the unfossiliferous sandstone may possibly represent the intervening zones. That these zones should be developed to any considerable extent is unlikely, however, for at Port nam Marbh they are all compressed into a single bed 1 ft. 6 in. thick. This highest fossiliferous bed consists of grey sandy limestone, in which specimens of *Inoceramus* are conspicuous. The Survey obtained from it three species of *Dorsetensia*, indicative of the *blagdeni* zone,

[1] Based on G. W. Lee, 1925, 'Pre-Tert. Geol. Mull, Loch Aline and Oban', pp. 98–112; and G. W. Lee and E. B. Bailey, 1930, 'Geol. Ardnamurchan', pp. 44–9; *Mems. Geol. Surv.*

a *Sonninia* cf. *buckmani* Haug, doubtfully suggesting the presence of the *sauzei* zone, and ?*Braunsina angulifera* S. Buck., said to be indicative of the *discites* zone.[1]

Discites and Concava Zones. The *discites* zone at Port nam Marbh has a thickness of 3 to 4 ft., since its fauna, represented by a rich assemblage of ammonites, occurs also 18 in. lower, in the highest foot of a 6 ft. block of siliceous limestone, where it is mingled with a fauna of *concava* date. Buckman identified 33 species of ammonites from this level, including most of the principal characteristic forms, too numerous to cite.[2] In the Ardnamurchan section the two zones are much thicker and quite distinct. The *discites* zone comprises 15 ft. of baked blue shale or flags, weathering to a soft sandstone, overlying $4\frac{1}{2}$ ft. of limestones with hard shaly layers. Buckman subdivided this zone into three horizons of *Docidoceras*, *Reynesella*, and *Platygraphoceras*, in descending sequence. The *concava* zone beneath is not continuously exposed, but it is probably at least 7 or 8 ft. thick. It yielded what Buckman considered to be ten species of *Ludwigella*, and four other genera.

Bradfordensis Zone. In Ardnamurchan the *bradfordensis* fauna was recognized, partly segregated in a limestone band 2 ft. thick, and partly mingled in another block, 1 ft. 9 in. thick, with the *Ludwigella* assemblage. Besides *Brasilia bradfordensis*, several species of *Brasilina* were collected. In Mull the zone has not been separated, though some of its ammonites have been detected.

Murchisonæ Zone. The *murchisonæ* zone is very rich in ammonites both in Mull and in Ardnamurchan, and in both places it is confined within a very small compass. In Ardnamurchan it is restricted to $1\frac{1}{2}$ ft. of limestone and at Port nam Marbh to the lowest 1 ft. of the same block of siliceous limestone as that which yields in its highest foot the *concava* and *discites* assemblage. The ammonites include *Ludwigia murchisonæ*, *L. gradata* S. Buck., *Ludwigina*, *Crickia*, *Hyattia*, *Apedogyria*, &c.

Ancolioceras Zone. The only definite evidence for this zone hitherto found is a single specimen of *Ancolioceras* aff. *cariniferum* S. Buck. in Ardnadrochet Glen, and an immature and doubtful specimen of *Ancolioceras* in 8 ft. of limestones in the Ardnamurchan section. But in the Ardnadrochet Glen there is an assemblage of fossils transitionary between *scissum* and *murchisonæ* faunas, which are considered to show that the zone is separately represented.

Scissum Zone. The *scissum* zone is well represented both by ammonites (*Tmetoceras* and *Lioceras*) and by brachiopods in Ardnadrochet Glen, but in the coast section at Port nam Marbh it has only been recognized by brachiopods—*Rhynchonella subdecorata* Dav. The rock (sandy limestone) in which the fossils occur is only some 3 ft. thick, but all or part of another 8 ft. of limestone above, in which no fossils have been found, may also belong to the zone. In the Ardnamurchan section a few feet of sandy beds at the base

---

[1] The Survey were not quite confident that the remarkable assemblage believed to have been obtained from this bed might not have been in some way brought together by faulting. 'Geol. Mull, Loch Aline and Oban', 1925, p. 105, *Mem. Geol. Surv.*

[2] They are listed in the Survey Memoir, 1925, loc. cit., pp. 104–5. The 'species' are to be understood as bearing separate names according to Buckman's conception of genus and species.

of the Inferior Oolite have yielded nine species of *Lioceras* belonging to the *scissum* zone, but no *Tmetoceras*.

IX. EASTERN SCOTLAND

No Inferior Oolite is exposed in Eastern Scotland.

X. KENT

In the Kent borings Inferior Oolite was met with in only a few places and was often apparently absent and never completely developed. Nowhere was evidence obtained for the presence of either Lower or Middle Divisions.

At several places sands with doggers were met with above the Lias, but they yielded lamellibranchs indicative of the Upper Inferior Oolite. The sands pass unconformably across the Lias on to the Palaeozoic platform in the north of Kent, suggesting that the Lower and Middle Inferior Oolite were removed during the Bajocian Denudation.

CHAPTER IX

UPPER INFERIOR OOLITE

Zones (Plate XXXIV).	Dorset.	Sherborne District.	Dundry and South Cotswolds.	North Cotswolds.
Parkinsonia schlœnbachi[1]	Microzoa and Sponge Beds. Massive Beds.		Rubbly Beds & *Anabacia* Limestones.	*Clypeus* Grit.
Strigoceras truellei	*Truellei* Bed.		Doulting Stone. / Upper Coral Bed.	—
Garantiana garantiana	*Astarte obliqua* Bed.	Sherborne Building Stone.	Dundry Freestone. / Upper Conglom.	— / Upper *Trigonia* Grit.
Strenoceras niortensis[2]	Conglomerate.	*Niortensis* Bed.	—	

W E have already seen that the Middle Inferior Oolite period closed with the most important episode of orogenic movements that had occurred in England since the beginning of Jurassic times. After long ages of comparative quiescence, the E.–W. or Armorican anticlines revived their activity, and along the Mendip, North Devon, Weymouth, and Birdlip Axes the Lias and Lower and Middle Inferior Oolite were raised up in gentle saddles, probably in places high above sea-level (especially over the Mendips), everywhere at least within the range of wave or current action. When the epeirogenic subsidence that followed lowered both anticlines and synclines beneath the sea so that normal sedimentation was resumed, the crests of the anticlines had been planed off and the new Upper Inferior Oolite strata were deposited evenly across their basset edges. The period of erosion that occurred during the interval was called by Buckman the Bajocian Denudation, and the epeirogenic movement by which the denudation was brought to an end is known as the Vesulian Transgression.

It is unfortunate that the term Inferior Oolite was universally used for the beds both below and above this important plane of discordance before the discordance was detected. Where Townsend (borrowing it from William Smith) first applied the term, in the country around Bath, the only part of the formation present is that *above* (namely the Upper Inferior Oolite), and so if any restriction were attempted, the remainder of the formation, described in

---

[1] Buckman assigned this to a new genus, *Haselburgites* Buckman; see *T.A.*, 1924, vol. iv, pl. CDXCIII, and for definition of genus *T.A.*, 1920, vol. iii, p. 30; genotype *H. admirandus* S. Buck., figd. 1921, pl. CCIII. It would appear, however, that Dr. Spath rejects the genus; see 'Revis. Jurass. Ceph. Fauna of Kachh', Part IV, *Pal. Indica*, 1931, p. 280.

[2] Formerly considered the topmost zone of the Middle Inferior Oolite, but transferred by Prof. Morley Davies for reasons explained in this chapter. See *P.G.A.*, vol. xli, 1930, pp. 231–3.

the previous chapter, would be the part to require a new name.[1] Certainly no better plane for separating two formations occurs in the Jurassic System, and the failure of our classification to take account of it is one of its worst shortcomings.

D'Orbigny also ignored this important stratigraphical break. In spite of his protests against other geologists for being guided by lithology in grouping the rocks, he was swayed by the very same considerations, for he included in his Bajocian Stage all our Inferior Oolite up to the *zigzag* zone, or basal Fuller's Earth. Indeed, if we subtract the transgressive portion from the Bajocian of the type-locality of Bayeux in Normandy, very little is left. For this reason Norman geologists are emphatic that the Bajocian Stage must be kept intact, as d'Orbigny created it,[2] and the same conclusion is very forcibly impressed upon the observer on the coast of Dorset. When a broader view is taken, however, there seems every justification for dismembering the Bajocian and reviving for the overstepping portion Marcou's term Vesulian. Introduced in 1848, this term is no upstart, but can claim an antiquity equal to that of d'Orbigny's names.

In England the only possible course is to describe the Upper Inferior Oolite separately, in a chapter of its own. Wherever we trace it in the southern counties it behaves independently of the rest, spreading over scores of square miles where the Lower and Middle divisions do not exist, overstepping the edges of the Lias and even the Carboniferous Limestone. In comparison with the thick and complex Great Oolite Series above, it is but a thin and homogeneous formation. It cannot even be traced with certainty beyond Oxfordshire. But it is not on that account by any means insignificant. If we wish to swell its bulk we had rather describe with it the Fuller's Earth and Chipping Norton Limestone, which are perfectly conformable with it, than the Middle and Lower Inferior Oolite; for they are a part of the same formation only in name.

I. THE DORSET–SOMERSET AREA

The full sequence of the Upper Inferior Oolite is exposed in the cliffs between Bridport Harbour and Burton Bradstock, and the section is amplified by quarries inland at Chideock Quarry Hill, Burton Bradstock allotments, Shipton Gorge, Walditch, Loders and other places. As already remarked, the beds here are thick in comparison with the attenuated representatives of the Lower and Middle Inferior Oolite, although they probably took only a fraction of the time to form.

The thickest zone is that of *Parkinsonia schlœnbachi*, which comprises $4\frac{1}{2}$–14 ft. of white limestones, in places full of sponges and resembling the White Sponge Limestone on the same horizon in Normandy. Beyond this, in the cliff-sections, there is little else, and the total thickness is only just over 7 ft.

As in Normandy, the White Sponge Limestone is separated from the Fuller's Earth Clay by two thin rubbly beds of limestone, which here would be more logically grouped, according to lithology, with the Inferior Oolite. They

[1] Cheltenham Beds was used for the whole Inferior Oolite up to the base of the Fuller's Earth, less the Yorkshire Dogger and basal Northampton Sands (*scissum-murchisonæ-Ancolioceras* zones) by Sir A. Geikie in 1885, *Text Book of Geology*, 2nd ed., p. 788.

[2] After a visit to Normandy and a conversation there with Professor Bigot I was so far convinced as to write that 'the abolition of the stage-name Vesulian ... has now been generally decided upon' (*P.G.A.*, vol. xli, 1930, p. 407).

contain *Zigzagiceras* spp. and *Oppelia fusca*, however, and so belong rightly with the Fuller's Earth.

SUMMARY OF THE UPPER INFERIOR OOLITE OF SOUTH DORSET [1]

Schlœnbachi Zone.

The *schlœnbachi* limestones with sponges, ostracoda and foraminifera thicken from 4½ ft. at Burton Bradstock to nearly 14 ft. at Chideock Quarry

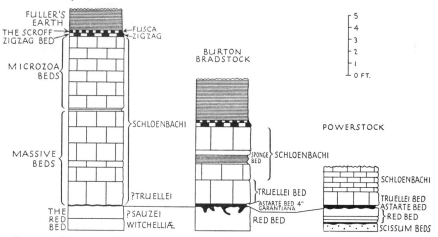

FIG. 44. Comparative sections of the Upper Inferior Oolite in South Dorset. (After L. Richardson, 1927–28, *Proc. Cots. N.F.C.*, vol. xxiii, pp. 51, 60, 171.)

Hill. The lower beds are massive, even crystalline, and yield few fossils, while in the upper half of the zone some are marly and crowded with sponges, microzoa and polyzoa. The richest locality for polyzoa is the village of Shipton Gorge. No less than twenty-eight species of sponges have been recorded from the neighbourhood, many of which were figured for the first time by Hinde in his 'Monograph of the British Fossil Sponges'. Some of the principal genera are *Holcospongia, Leucospongia, Peronidella, Platychonia, Tremadictyon,* and *Corynella.* A number of echinoderms also occur, and the common *Belemnopsis bessina* and *Sphæroidothyris sphæroidalis.*

Truellei Zone.

The botton bed of the massive limestones—1 to 2 ft. thick—is the zone of *Strigoceras truellei.* It is not always distinguishable (e.g. at Shipton Gorge) but it can generally be recognized by an eroded upper surface, to which *Serpulæ* are attached, and by a more or less continuous line of *S. sphæroidalis* in the lower part. It has yielded, besides many other fossils, two ammonites of the genus *Dimorphinites*, which occur in the top of the Ironshot Oolite of Bayeux.

Garantiana Zone.

This zone is usually represented by a mere seam of fossils in a crumbly brown ironshot matrix, the *Astarte obliqua* Bed. It is crowded with lamelli-

[1] Based on L. Richardson, 1928–30, *Proc. Cots. N.F.C.*, vol. xxiii, pp. 35–68, 149–85, 253–64; and 1915, *P.G.A.*, vol. xxvi, pp. 47–78.

branchs, gastropods, and *Belemnopsis bessina*, and generally appears to be joined on to the top of the Red Bed of the Middle Inferior Oolite. A few miles inland, at Vetney Cross, the zone is 1 ft. thick, but on the coast at Burton Bradstock it averages only 4 in.[1]

Niortensis Zone.

Below the *Astarte obliqua* Bed, on the coast, are local patches of a thin conglomerate containing remanié fossils of the underlying Middle Inferior Oolite. It rests upon an eroded and pitted surface of the Red Bed, of which its constituent pebbles are made up, together with 'snuff-boxes' derived from that bed, and rolled ammonites of the *blagdeni* and *sauzei* zones. Besides these, however, there are some ammonites which both Buckman and Spath have identified as belonging to the *niortensis* zone. The conglomerate therefore belongs to the base of the Upper Inferior Oolite or to the time of the Bajocian Denudation which preceded the Vesulian Transgression, and it provides interesting evidence that somewhere in this neighbourhood during the *niortensis* hemera Middle Inferior Oolite was undergoing destruction and redeposition.

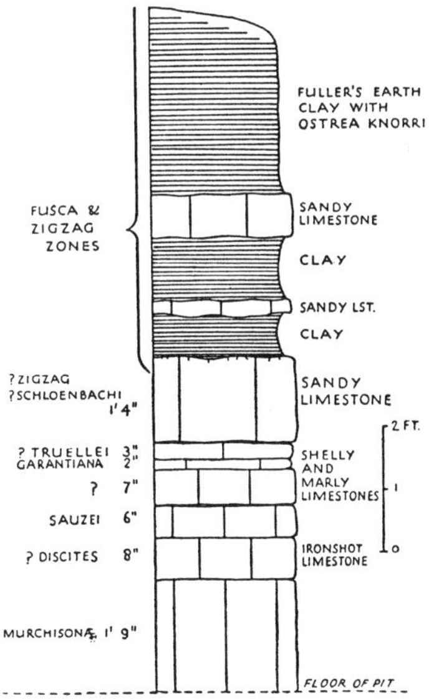

The zones maintain essentially the same characters (except that the *niortensis* zone is absent, apparently overlapped) through North Dorset and over the Somerset border past Crewkerne,[2] until the outcrop bends eastward to Yeovil and Sherborne. The thickest zone throughout the district is always that of *schlœnbachi*, which contains rich ostracod and foraminifera beds at Misterton and elsewhere,

FIG. 45. Section at Stoford, near Yeovil, showing the abnormal reduction of the Inferior Oolite Series. (Based on S. S. Buckman, 1893, *Q.J.G.S.*, vol. xlix, p. 484, with amendments by L. Richardson, 1932, *Proc. Cots. N.F.C.*, vol. xxiv, p. 52.)

and a sponge bed at Haselbury Plucknett. Its thickness varies from about 9 ft. in the west to 3 ft. in the east. At Misterton fourteen species of foraminifera have been recorded. The *truellei* zone can seldom be identified with certainty in the Crewkerne district, but, as on the coast, it is probably represented in the base of the Massive Beds.

In the north-eastern part of the Dorset–Somerset area, from the neighbourhood of Yeovil to the Mendips, the development of the beds is at first sight very different from that on the coast. Although at Stoford the whole Inferior Oolite is almost incredibly compressed, the Upper Division occupying only a couple of feet in the quarry near the station (fig. 45),[3] a sudden thickening of limestones takes place east of Yeovil to 30 ft. at the Halfway House, between

[1] Buckman has figured a number of ammonites from the *Astarte obliqua* Bed of Burton Bradstock: *Garantiana garantiana* (*T.A.*, pl. CCCLVIII), *Parkinsonia rarecostata* (CCCLII), *P. interrupta* (CCCXXXVII), &c., and also uncoiled ammonoids of the genera *Plagiamites* and *Spiroceras*.

[2] L. Richardson, 1918, *Q.J.G.S.*, vol. lxxiv, pp. 145–73; 1915, *P.G.A.*, vol. xxvi, pp. 69–75.

[3] L. Richardson, 1911, *P.G.A.*, vol. xxii, p. 262 and pl. XXXIX.

Yeovil and Sherborne, and 45 ft. at Sherborne. Northward the limestones attenuate with equal abruptness and then maintain a steady 10–20 ft., usually 10–15 ft., to the Mendips.

The chief centre of interest lies naturally in the area of greatest thickening about Sherborne, and this coincides with the region in which some of Buckman's earliest researches were carried out, as mentioned in the preceding chapter.[1] It has more recently been carefully revised by Richardson, with interesting results.[2]

In the early days the whole of the 'Upper Inferior Oolite' of the Sherborne district was spoken of as the 'Top Beds' or *Parkinsoni* Zone, with the implication that it was of one date throughout. One of the first important results of Buckman's earliest stratigraphical work and the application of his new methods was to show that the Top Beds are of different dates in different parts of the district and that in general they are older in the east than in the west. He demonstrated that at the Halfway House, Compton, in the west, the 30 ft. of poorly fossiliferous limestones constituting the Top Beds *overlie* the *truellei* zone, which is a 1 ft. fossil bed and rests in turn upon a thin, ochreous, marly representative of the *garantiana* zone, like the *Astarte obliqua* Bed of Burton Bradstock and full of the same fossils. He further showed that these thin *garantiana* and *truellei* beds expand eastward in the direction of Sherborne and pass laterally into thick yellow, blue-centred building stone (the Sherborne Building Stone) and rubbly limestone, best seen at Combe Limekiln, Clatcombe, and Redhole Lane, Sherborne.

It follows that when the maximum thicknesses of all these zones are added together a very considerable amount of deposit is found to have been formed during a time often represented in any one section by quite insignificant beds.

Buckman believed that the shifting of the belt of maximum sedimentation was not fortuitous, but was the direct result of the westward transgression of the sea. As the Vesulian Transgression advanced from the east or south-east the successive sedimentary belts, with their concomitant faunas, migrated westward. In this way the thin littoral facies of each zone, crowded with fossils, came to be overlain by the thick and poorly-fossiliferous facies of the next; and now the littoral facies proper to every zone is to be found to the west of the area of its maximum development.

It is certain that a variable portion of the 'Top Beds' in the Sherborne district does not belong to the Upper Inferior Oolite at all but to the *zigzag* and *fusca* zones of the Fuller's Earth. This has become increasingly evident as a result of Richardson's revision of the district. In the white and grey argillaceous limestones with marl partings at King's Pit, Bradford Abbas, for instance, Richardson has obtained ammonite evidence that nearly the whole of the 'Top Beds' at that place (about 8 ft. in thickness) belong to the *zigzag* zone, the true Upper Inferior Oolite being represented below by seams only a few inches thick.[3] Of the same age he believes (but on this point there is less certainty) are the 30 ft. of thick limestones above the *Truellei* Bed at the Halfway House and in the area to the west, and he correlates them with 20 ft. of similar limestones above the Sherborne Building Stone and Rubbly Beds

[1] S. S. Buckman, 1893, *Q.J.G.S.*, vol. xlix, pp. 479–522.
[2] L. Richardson, 1932, *Proc. Cots. N.F.C.*, vol. xxiv, pp. 35–85.
[3] Idem, 1911, *P.G.A.*, vol. xxii, p. 262.

of Crackment or Crackmore, south-west of Milborne Port.[1] The *schlœnbachi* zone would seem on this hypothesis to be almost entirely absent.

The most interesting deposit of all is that containing *Strenoceras niortensis.* It is known only at a few localities in the district, and excepting the basal conglomerate at Burton Bradstock and traces in the Hebrides, *Strenoceras* has been found nowhere else in the British Isles. In the Sherborne district the fauna has been recognized in 18 in. of ironshot oolite at Clatcombe, in a thin

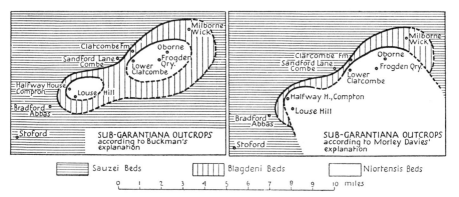

FIG. 46. Maps showing diagrammatically the surface upon which the *garantiana* zone is supposed to have been deposited in the Sherborne district, on the supposition that the principal unconformity lies below the *garantiana* zone (left), and on the alternative supposition that it lies below the *niortensis* zone (right). Continuous lines represent actual boundaries, broken lines hypothetical boundaries. (From A. M. Davies, 1930, *P.G.A.*, vol. xli, p. 232, fig. 22.)

brown ironshot at Frogden Quarry, Oborne, in the lower half of the *Astarte obliqua* Bed at the Halfway House, Compton, and in several other exposures in the immediate neighbourhood.

Such a restricted distribution can be explained in two ways. Either the beds can be regarded as relict patches of a once continuous stratum which has been all but completely destroyed by denudation; or they can be visualized as the first deposits laid down by the invading sea at the beginning of or immediately prior to a transgression, soon to be overlapped by the more widespread deposits of the succeeding zones. On the interpretation chosen depends the position to which the *niortensis* zone should be assigned—whether to the top of the Middle Inferior Oolite or to the base of the Upper Inferior Oolite. The position here favoured, at the base of the Upper Inferior Oolite, has been recently advocated by Prof. Morley Davies (fig. 46).[2]

Prof. Davies points out that the extensive denudation required on the first hypothesis to reduce the *niortensis* zone of the whole of England to the two small patches near Sherborne must have taken a prolonged period of time, during which sediments would have continued to be laid down elsewhere on the deepest parts of the sea-bed. Such sediments would be intermediate in age between the *niortensis* zone and the overstepping or transgressive stratum, the *garantiana* zone. But no such sediments are anywhere found, although in France the *niortensis* and *garantiana* zones are seen over wide areas, always in

[1] L. Richardson, 1923, in 'Geol. Shaftesbury', *Mem. Geol. Surv.*, p. 17.
[2] A. M. Davies, 1930, *P.G.A.*, vol. xli, pp. 230–3.

contact. Further, the *niortensis* fauna as a whole is more closely allied to that of the *garantiana* zone than to that of the *blagdeni* zone below, and Prof. Davies considers it is too closely comparable to admit of any prolonged time interval having elapsed between the two hemeræ.

Consequently it would seem to be best to consider the *niortensis* zone as deposited in a greatly restricted sea during a period of regression, while the Bajocian uplifts and denudation were actually in progress; or, as Prof. Davies puts it, to 'regard the rare occurrence of the *niortensis* zone in England as marking the localities where sedimentation restarted prior to the great transgression in the *garantiana* hemera'. Such views are supported by the occurrence of the conglomerate of *niortensis* date at Burton Bradstock, and also by the position of the patches of beds near Sherborne, far down the dip slope in a re-entrant angle in the outcrop, where they would be likely to appear if they were overlapped rather than overstepped.

In the Sherborne district and northward along the escarpment as far as Blackford[1] the thick limestones of *garantiana* and *truellei* dates, which transgress all the zones of the Middle and Lower Inferior Oolite over the North Devon Axis, form a homogeneous mass. Farther north, from Blackford to the Mendips, they can be more easily separated.

In the southern part of the escarpment the most important feature is the *garantiana* zone, for by Blackford, Woolston and Hadspen the Sherborne Building Stone is continued in the form of the HADSPEN STONE. This is a massively-bedded, ferruginous, brown, building stone, with a maximum thickness of 10 ft., and is full of fossils, especially nests of *Acanthothyris spinosa* and *S. sphæroidalis*, together with lamellibranchs and echinoderms. The characteristic ammonite is *Parkinsonia rarecostata* S. Buck. Occasionally, as at Woolston, the base is seen to be conglomeratic, and everywhere over the axis it rests unconformably upon Upper Lias sands of *Dumortieriæ* or *moorei* date.

Towards the northern part of the outcrop, near where the Lower and Middle Inferior Oolite appear in the Cole Syncline, and beyond the syncline towards the Mendips, the *garantiana* zone becomes recessive. No more is seen of the Hadspen Stone, and the zone consists of bands of ragstone or limestone separated by marl partings and often having bored upper surfaces. The thickness varies from a maximum of about 8 ft. to just over 3 ft. At Batcombe and Lamyatt Hill, north-west of Bruton, the base is again pebbly, and at Strutters Hill, near Cole, there is an *Astarte obliqua* Bed like that at the Halfway House, Compton, and in South Dorset.

The *truellei* zone, where it comes to be separable from the *garantiana* zone north of Blackford, forms a second limestone, white, massive, sparry or flaggy, called the DOULTING STONE, from its great development around Doulting, close to the Mendips. At Blackford 9 ft. of the stone are seen, but the usual thickness in the exposures is 4 or 5 ft. It is generally massive below and flaggy towards the top, and is separated from the Hadspen Stone by a few layers of limestone and marl of uncertain age.

In the northern part of the escarpment the Doulting Stone becomes the dominant surface rock, covering the whole outcrop from Bruton to beyond the Mendips. At Lamyatt Hill a rubbly limestone below the Doulting Stone,

---

[1] Hence based on L. Richardson, 1915, *Q.J.G.S.*, vol. xxxvii, pp. 473–520.

yielding *Parkinsonia densicosta* (Quenst.), indicative of the *truellei* zone, abounds in corals (*Isastræa*), and Richardson considers that it can be identified as the most southerly indication of the Upper Coral Bed of Dundry Hill, Midford and the South Cotswolds.

Close to the Mendips higher beds appear, representative of the *schlœnbachi* zone, but these will be better considered later on.

To summarize the salient features of the Dorset–Somerset area: the Upper Inferior Oolite is thickest in the centre, about Sherborne and Milborne Port, where the expansion occurs in the oldest zones, those of *garantiana* and *truellei*, and where also the *niortensis* zone is present. Here the thickest deposits are the least fossiliferous. The area of greatest thickness lies at the apex of a re-entrant angle in the outcrop, which exposes to view beds more south-easterly (farther down the dip slope) than can be seen at any other point, and therefore beds which may be presumed to have been formed under the deepest water. The thinning of progressively higher zones in a westerly direction, away from Sherborne, points to a transgression of the sea in that direction.

It is interesting to note that an exposure on the most north-westerly projecting spur of the outcrop, at Hinton Park, north-west of Crewkerne,[1] shows a development of the Hadspen Stone, connecting the *garantiana* zone at that point and at Hadspen, on the outcrop north of Milborne Port.

II. THE MENDIP AXIS

Within some 5 miles of the Mendips, at Lamyatt Hill and Batcombe, the base of the *garantiana* zone, as has been remarked, becomes pebbly, and the Upper Lias sands on which it rests are older than the subjacent sands farther south, being here probably of *dispansum* date. On approaching close to the Mendips the *garantiana* zone passes entirely into a conglomerate and oversteps first on to attenuated representatives of various Liassic clays and later on to the Carboniferous Limestone. Eventually it disappears altogether, being overlapped by the *truellei* zone or Doulting Beds.[2]

As in previous periods, the Mendip Anticline did not function as a simple axis of elevation, but was flanked on the north by an area of minor disturbances extending beyond Radstock (map, p. 70). The conglomeratic facies of the *garantiana* zone, which is first met with on the south side of the hills at Doulting, laps round the eastern end, then strikes away to the north-west, where it is seen at Timsbury Sleight and on the outlier of Dundry Hill, in the form of the MAES KNOLL CONGLOMERATE (fig. 34, p. 196).

In general the constituent pebbles and blocks in the conglomerate consist of yellow limestone and pale grey, fine-grained sandstone, much bored and rolled, with *Serpulæ* attached, but near Mells-Road Railway Station there are pebbles of quartz and chert.[3] On Dundry Hill the conglomerate consists of lumps of blue-grey sandstone and irregularly-shaped masses of ironshot oolite (measuring up to 2 ft. 10 in. in diameter), together with rolled fragments of *Pleydellia aalensis* and *Dumortieria*.[4]

In a north-easterly direction, in a very short distance away from the

---

[1] L. Richardson, 1918, *Q.J.G.S.*, vol. lxxiv, p. 161.
[2] L. Richardson, 1907, *Q.J.G.S.*, vol. lxiii, pp. 383–426 for Bath–Doulting district.
[3] Ibid., p. 405.
[4] S. S. Buckman and E. Wilson, 1896, *Q.J.G.S.*, vol. lii, p. 685.

conglomerate belt, but evidently down a rapidly shelving shore, the *garantiana* zone passes into brown, finely ironshot, shelly ragstones and slightly sandy limestones, identifiable as the typical Upper *Trigonia* Grit of the Cotswolds. This facies is seen in a lane section at Wellow (where it is 1 ft. 8 in. thick), and in a road-cutting near Midford (where it is 5 ft. thick), and at both places the base is still somewhat pebbly.[1]

At Midford, Whatley Combe, and wherever the upper surface of the *garantiana* zone is seen, it is in turn bored and oyster-covered, and the *truellei* zone, represented by the Upper Coral Bed and Doulting Stone, rests non-sequentially upon it. The significance of this non-sequence will be seen in the next section.

The Doulting Stone increases greatly in thickness as it approaches the south side of the Mendips, reaching a maximum of 44 ft. in the railway-cutting and adjoining quarries at Doulting. It consists of a brownish, oolitic, usually massively bedded limestone, and 16 ft. of the upper part are an excellent false-bedded freestone. The Doulting Freestone is said to be more durable than the Bath Stone, a view borne out by Wells Cathedral and Glastonbury Abbey, which were mainly built of it.

On approaching the Palaeozoic core of the eastern end of the Mendip Hills, the Doulting Stone, after overlapping the thin *garantiana* conglomerate, passes on with little change over the planed and eroded surface of the Carboniferous Limestone (Pl. XI). Still from 20 to 30 ft. in thickness, it forms a covering which would hide the Palaeozoics beneath but for the fact that the plateau is deeply dissected by valleys, along the sides of which the junction is exposed to view in numerous quarries (map, p. 70).

The extraordinary geological interest of the area about Vallis Vale, near Frome, has perhaps never been better expressed than by W. D. Conybeare in 1822, in the following passage quoted by Richardson in his well-known paper on the Inferior Oolite of the Bath–Doulting District, the source of the facts on which the present account is based.[2] In the words of Conybeare,

'an uniform and elevated plain of the Inferior Oolite spreads over the whole surface, furrowed by valleys about 150 or 200 ft. deep, which expose the Mountain Limestone. The character of many of these valleys (particularly of that between Mells and Frome and its lateral branches) is highly romantic; the streamlets that flow through them being skirted by bold and rocky banks overgrown by feathering woods; while the geologist observes, as a feature of peculiar interest in their precipitous escarpment, the actual contact of the horizonal bed of Inferior Oolite resting on the truncated edges of strata of Mountain Limestone, thrown up in an angle of from 50 to 60 degrees. This line of contact is sometimes perfectly level for a considerable distance (as if the edges if the Mountain Limestone strata had been rendered smooth by some mechanical force abrading them previously to the deposition of the Inferior Oolite), but in other instances it is rugged and irregular; sometimes the contact is marked by a breccia of fragments of the older, cemented by the newer rock, but this is by no means constant.'

In places, in addition, a thin remanié bed representing various zones of the Liassic and Rhætic rocks is interposed, but in most of the area about Vallis

---

[1] L. Richardson, 1907, loc. cit., pp. 408–9.
[2] W. D. Conybeare and W. Phillips, 1822, *Outlines of the Geology of England and Wales*, pt. i, pp. 254–5.

PLATE XI

Photo.   *W. J. A.*

Upper Inferior Oolite (Doulting Beds) overlying Carboniferous
Limestone, Nunney, Eastern Mendips.

The overhanging ledges in the Carboniferous Limestone are pre-Jurassic
thrust-planes.

Photo.   *W. J. A.*

Upper Inferior Oolite (Doulting Beds) overlying Carboniferous
Limestone, Vallis Vale, near Great Elm, Eastern Mendips.

Vale the *truellei* zone rests directly upon the Carboniferous Limestone. As described by Richardson:[1]

'The surface of the Carboniferous Limestone is remarkably even. If one stands upon the platform formed by this rock after the Inferior Oolite has been stripped off by the quarrymen, and looks across the flat-bottomed valley, with its mural sides, at the other extensive quarries, one cannot repress a feeling of astonishment at the excessive evenness of the plane of erosion. This surface is riddled with annelid— and, less commonly, *Lithophaga*—borings, and in places is strewn with oysters.'

The *truellei* zone is the only zone in the whole series of the Jurassic Rocks that can be seen passing in more or less unaltered form directly across the Palaeozoic core of one of the periclines of the Mendip Axis. This fact testifies to the great extent of the subsidence and transgression which preceded its formation. The same, too, is witnessed by the thickening of the Doulting Stone in the immediate vicinity of, and over, the axis. Whether the *truellei* zone in turn thinned out and passed into a conglomerate higher up against the Mendip Hills before the core was stripped bare by Tertiary and Quaternary erosion, or whether the Doulting Beds formerly passed right over the summit, showing that it was entirely submerged, is a difficult question to answer. At Cranmore, however, the surface indications are strongly favourable to the view that the Doulting Beds ended abruptly against a cliff, or at least a sharply rising slope, of the Carboniferous Limestone; moreover, about Whatley the Upper Inferior Oolite is entirely overstepped by the middle part of the Fuller's Earth.

The Upper Inferior Oolite is completed in the neighbourhood of Doulting, Frome and northward by a development of the *schlœnbachi* zone, in places up to 10 ft. thick. But the facies of this zone belongs to the South Cotswold province, of which it is essentially a continuation, and with which it is better treated. At the top a thin white earthy limestone (1 ft. 2 in.), exposed in Doulting Bridge Quarry and the railway cutting, abounds in ammonites of the *zigzag* zone, and this is succeeded by Fuller's Earth just as in South Dorset.

III. THE COTSWOLD HILLS[2]

There are plentiful signs of the great upheavals and erosion connected with the Bajocian Denudation in the Cotswold basin of deposition. Although the signs are less spectacular than those just described over the Mendip Axis, the detailed work of S. S. Buckman showed conclusively that they are present, and of no small magnitude. Their measure may be judged by an examination of the strata upon which the *garantiana* zone reposes between the Mendips and the Vale of Moreton (fig. 35, facing p. 197).

The *blagdeni* and *niortensis* zones, which should normally precede the *garantiana* zone, have been removed from the whole region, or were never deposited. The *sauzei* zone, next below, is confined to a small area in the centre of the deepest syncline, at the summit of the Cleeve Hill plateau. The rest of the eight zones of the Middle and Lower Inferior Oolite are in turn overstepped away from Cleeve Hill, until along the western margin of the Vale of Moreton in the north and between Old Sodbury and the Mendips in the south, *garantiana* beds come to rest directly upon Upper Lias.

[1] L. Richardson, 1907, loc. cit., p. 400.
[2] The sources are the same as in the corresponding section of Chapter VIII, p. 197.

The overstep is not so simple as this plain statement would indicate, however, for, as has been explained in previous chapters, the Lower and Middle Inferior Oolite deposits were thrown into two smaller synclines and an anticline within the main Cotswold Basin. The Birdlip Axis passed NW.–SE. under Birdlip, running approximately along Ermine Street towards Cirencester. Over the crest of this axis in the region of the outcrop the overriding *garantiana* zone was laid down upon an eroded, bored and oyster-covered surface of the Upper Freestone, and pebbles of Upper Freestone are included in its basal layers. In the Cleeve Hill syncline to the north, as we have seen, beds as high as the *sauzei* zone were preserved. To the south of the Birdlip Anticline a smaller syncline was formed about Painswick, but as the greater part of this has been destroyed by the retreat of the escarpment, the highest beds remaining beneath the *garantiana* zone are not so high in the sequence as those preserved upon the projecting plateau of Cleeve Hill (fig. 36, p. 199).[1]

In spite of the enormous amount of disturbance that attended the Bajocian Denudation, the conditions prevailing before it and governing deposition in the Cotswold province continued with surprisingly little alteration after it. The beds comprising the Upper Inferior Oolite are essentially similar in general facies, both lithologically and palaeontologically, to those building the Middle and Lower divisions. This may be taken as a further indication that the bulk of the disturbances were orogenic rather than epeirogenic.

As before, corals grew along the coast of the Welsh land, where the waves broke fragments off the reefs and spread the debris over the sea-bed to the east. The same types of detrital accumulations grew up, composed of the hard parts of echinoderms, corals, brachiopods, sponges, lamellibranchs and gastropods, with occasional seams of marl and banks of false-bedded oolite.

As in the earlier divisions, the largest numbers of recognizable corals are found in the west, along the escarpment of the South Cotswolds from Stroud southward. Here in the *murchisonæ* hemera two beds of corals accumulated, the nearest approach to the remains of a coral reef preserved anywhere in the Cotswold Inferior Oolite; the lowest bed, between the Lower Freestone and the Pea Grit, is from 15 to 25 ft. thick on the sides of the Slad Valley near Stroud, and consists of large masses of coral and coralline limestone embedded in a creamy mudstone, together with a considerable number of sponges. In the same area a similar coral bed, known as the Upper Coral Bed, forms part of the *truellei* zone in the Upper Inferior Oolite. All the genera of the corals are the same, and the general facies is identical. Comparing the corals of the later reefs with the earlier, Martin Duncan wrote:[2]

'The remarkable varieties of *Thecosmilia gregaria*, which resemble the genus *Symphyllia* and *Heterogyra*, are found principally in the lower reef, but they exist in the upper also. Some species appear to be peculiar to the different reefs, but . . . there is evidently a considerable affinity between the [coral] faunas of the reefs, and there is nothing to indicate anything more than a temporary absence from and a return of the species to an area.'

Important changes in the configuration of the coast-line and the relative

---

[1] The uprise of the lower beds in the south-west limb of the Painswick Syncline, far down the dip-slope beneath the Great Oolite, was proved in a boring at Shipton Moyne, where the *garantiana* zone was found to rest upon the Pea Grit. See L. Richardson, 1930, 'Wells and Springs of Gloucestershire', *Mem. Geol. Surv.*, p. 150.

[2] M. Duncan, 1872, 'Supplementary Monograph of the Brit. Foss. Corals', *Pal. Soc.*, p. 11.

depths of the various minor areas of deposition must, nevertheless, have taken place as the result of the Bajocian Denudation and the earth-movements which gave rise to it.

The importance of the ancient Mendip Axis was greatly diminished, the provinces to north and south of it coming to be much less sharply differentiated. In Upper Inferior Oolite times the Cotswold type of deposit was accumulated, not only on Dundry Hill, but also across the eastern end of the axis, and through eastern Somerset far down towards the Dorset border. In *truellei* times, as already remarked, rolled corals reached as far south as Lamyatt Hill, near Bruton.

Thus the two areas came to be blended, and by *zigzag* and *fusca* times (basal Fuller's Earth) autochthonous ammonites were to be found from Normandy at least as far as the South Cotswolds.

SUMMARY OF THE UPPER INFERIOR OOLITE OF THE COTSWOLDS (INCLUDING BATH AND DUNDRY)

The fullest and most varied series of deposits is found in the South Cotswolds, in the Bath district, and on the Dundry outlier. To the north-east, away from the source of the sediment, some of the subdivisions mingle and lose their identity, while others thin entirely away (see fig. 35, p. 197).

Schlœnbachi and Upper Truellei Zones.

RUBBLY BEDS and ANABACIA LIMESTONES. The *schlœnbachi* zone is represented in the South Cotswolds, the Bath district and on Dundry Hill by from 6 to 12 ft. of white oolitic limestone. The upper or Rubbly Beds are usually little more than a rubbly development of the top, but locally, as on Stinchcombe Hill, they merit consideration as a distinct lithic division. Here there is a remarkable bed crowded with Microzoa, closely comparable with that on the Dorset Coast, especially at Shipton Gorge. The remainder, or *Anabacia* Limestones, consists of evenly-grained white oolite, often flaggy, and characterized by an abundance of the little button-coral, *Anabacia complanata* (Defr.).

The *Anabacia* Limestones present constant features in the district from Doulting to Dundry Hill and onward to Old Sodbury, ten miles north of Bath. Beyond this, at Horton, they become temporarily softer and more oolitic, but at Scar Hill, Nailsworth, their original form is resumed for a short distance.

The DOULTING STONE becomes thin north of the Mendips. At Midford it is only some 12 ft. thick, and about the same thickness is maintained into the South Cotswolds, the essential characteristics of the stone being nevertheless preserved almost as far as Stroud.

CLYPEUS GRIT. In the Stroud district the Doulting Stone, *Anabacia* Limestones and Rubbly Beds lose their distinctive features and fuse laterally to form a different deposit, the *Clypeus* Grit.[1]

The *Clypeus* Grit has the widest extension in the Cotswolds of any bed in the Inferior Oolite. The upper third is uniformly rubbly and is crowded with fossils, of which the most conspicuous are the large *Clypeus plotii* and Stiphrothyrids of the various species known collectively as *Terebratula globata*. In the Cheltenham district *Anabacia complanata* abounds, recalling

---

[1] The upper portion of this includes the so-called White Oolite of Edwin Witchell (*Geology of Stroud*, p. 62), which is not of *zigzag-fusca* date as supposed by Buckman (*teste* L. Richardson, in lit.).

the *Anabacia* Limestones on the same horizon farther south. The lower two-thirds are a more massive limestone from which fossils do not weather so easily. Lithologically the distinguishing feature of both divisions is the invariable presence of large ooliths or pisoliths.

The full thickness of the *Clypeus* Grit is rarely seen, owing to its forming the capping to most of the exposures. It was, however, completely sectioned in the region of its maximum thickness in the North Cotswolds by the railway west of Notgrove Railway Station, where the thickness is 40 ft. 6 in. One of the finest complete sections is in a quarry at Rodborough Hill, Stroud, where part of the rock resembles a pudding-stone from the abundance of brachiopods (mainly Stiphrothyrids).

Lower Truellei and Upper Garantiana Zones.

UPPER CORAL BED and DUNDRY FREESTONE. The non-sequence below the Doulting Stone in the Mendip region, denoted by an eroded and oyster-covered surface of the *garantiana* beds, disappears north of the axis, where two strata, the Upper Coral Bed and the Dundry Freestone, take its place. These beds are believed to complete the lower part of the *truellei* zone and the upper part of the *garantiana* zone.

The UPPER CORAL BED is a useful stratum owing to its persistent characters and widespread distribution. It is crowded with Isastræan corals, and in addition contains a rich fauna of micromorphic brachiopods[1] (at least 17 species, of which the commonest and most characteristic is *Spiriferina? oolitica*), ostracods (11 species), foraminifera (16 species) and a few remains of sponges, polyzoa, echinoids, and holothurians. It is therefore a typical debris derived from the destruction of growing coral reefs. The bed is traceable from Lamyatt Hill near Bruton by way of Dundry Hill and the Bath district into the South Cotswolds as far north as Rodborough Hill and Stroud. Beyond this it dies away, but it may be represented by sporadic occurrences of corals as far as Hook Norton, in Oxfordshire.

The age of the Upper Coral Bed was determined by the finding of a specimen of *Lissoceras psilodiscum* (Schlœnbach) at Coombe Hill Quarry, near Wotton under Edge.[2]

The interest of the bed lies in the proof that it affords of the survival or revival of coral growth along an extended length of coast-line after the disturbances connected with the Bajocian Denudation.

The DUNDRY FREESTONE[3] is much less widely distributed than the Upper Coral Bed, being restricted in anything like its maximum thickness to the west end of Dundry Hill (near the church), where it is 27 ft. thick. At the eastern end it has dwindled to 4 ft. (in the quarry near the Butcher's Arms), and to 3 ft. 4 in. in another quarry east of the road to the southern edge of the hill. It is continued in the outlier of Timsbury Sleight, where it is 4 ft. 3 in. thick, and on the main outcrop it extends from near the village of English Combe in ever-diminishing thickness into the South Cotswolds as far as Stroud, where Richardson believes it to be identifiable as a thin limestone band 4 in. thick in Stanley Wood Quarry, and 2 in. thick at Wotton under Edge.[4]

---

[1] For a list of microzoa from Timsbury Sleight and neighbourhood see C. Upton, in Richardson, 1907, *Q.J.G.S.*, vol. lxiii, p. 413.
[2] L. Richardson, 1910, *Proc. Cots. N.F.C.*, vol. xvii, p. 86.
[3] S. S. Buckman and E. Wilson, 1896, loc. cit.   [4] L. Richardson, 1910, loc. cit., pp. 121, 104.

As its name implies, the Dundry Freestone is a good building stone, and it has been much used locally. Like other rapidly-accumulated deposits, it has yielded little or no palaeontological information.

North of the Stroud district, over the rest of the Cotswolds, the Dundry Freestone and Upper Coral Bed are overlapped, and the *Clypeus* Grit usually rests non-sequentially upon the lower part of the *garantiana* zone—the Upper *Trigonia* Grit.

Lower Garantiana Zone.

UPPER TRIGONIA GRIT. The least variable and most persistent member of the Upper Inferior Oolite, the only zone to extend without appreciable change from the Mendips near Midford and Wellow across the entire Cotswold Basin to the flank of the Vale of Moreton Axis, is the Upper *Trigonia* Grit (see fig. 35, p. 197).

It attains no great thickness, the maximum measurement even in the North Cotswolds being 9 ft., while 6–8 ft. is the average. The 'grit' (a misnomer) is a hard, grey, rough-textured limestone, often somewhat flaggy, and full of fossils. The difficulty of extracting the fossils contrasts markedly with the ease with which perfect specimens may be obtained from the overlying *Clypeus* Grit. The typical species are *Trigonia costata* Sow., which often occurs as casts in great profusion, and the Scaphoid species, *Trigonia duplicata* Sow. Equally characteristic are *Acanthothyris spinosa* and various forms of *Stiphrothyris*.

The unconformable relations of the Upper *Trigonia* Grit to the underlying Middle and Lower Inferior Oolite and earlier beds have already been described.

IV. THE VALE OF MORETON AXIS AND OXFORDSHIRE

The Vesulian transgresses on to Upper Lias across the Vale of Moreton Axis, just as over the other axes that took part in the revived orogenic activity at the end of Middle Inferior Oolite times. Over the Mendip Axis it is the Doulting Beds that overstep the older formations without themselves undergoing any great change; here it is the same two zones, those of *truellei* and *schlœnbachi*, in the form of the *Clypeus* Grit. Over both axes the *garantiana* zone is overlapped.

The UPPER TRIGONIA GRIT thins rapidly towards the eastern margin of the North Cotswold hill-mass and is absent over approximately the same marginal strip of country, overlooking the Vale of Moreton, as that in which the Oolite Marl is wanting (fig. 38, p. 207). It is last seen in the railway-cuttings west of Bourton on the Water, where its thickness is a little over 3 ft. In the last cutting eastward (Aston Farm Cutting) it has disappeared and *Clypeus* Grit has overlapped on to Notgrove Freestone.[1] No positive indication of the *garantiana* zone is known between this point and the Hebrides.

The CLYPEUS GRIT passes on, changing little in its lithology or in its organic contents, but gradually becoming thinner. On the eastern plateau about Chipping Norton and in the Iccomb hill-mass it is still about 20 ft. thick, and it rests upon the Whitbian portion of the Upper Lias clay everywhere south of the boundary of the *Scissum* Beds, as shown in fig. 38. The thickness is never made up again east of the Vale of Moreton Axis, and in the Evenlode

---

[1] L. Richardson, 1929, 'Geol. Moreton in Marsh', *Mem. Geol. Surv.*, p. 80.

Valley and about Chipping Norton there is a further slight diminution as compared with the Vale of Bourton. The rock is still readily recognized by its large ooliths and pisoliths and its perfect fossils, especially of *Clypeus plotii* and Stiphrothyrids, while the quarries show that the upper 8 ft. are still rubbly and the lower 12 ft. more massive, as in the Cotswolds.[1]

The last exposures where the *Clypeus* Grit is well displayed are the old Marlstone workings at Fawler, near Charlbury, in the Evenlode Valley. Here, to the last, it is quite typical, with its upper portion rubbly and its lower portion more massive, and it has yielded many fine *Clypei* and other fossils to several generations of collectors. Resting directly on it is the Chipping Norton Limestone, of *zigzag-fusca* date, while welded to its under side is a hard band, a few inches thick, containing corals—possibly a representative of the Upper Coral Bed of the Cotswolds. Below the coral bed has been seen a still thinner seam of conglomeratic sandstone with rolled pebbles of limestone, resting on a bored and eroded surface of a thin argillaceous limestone band, which is presumably the top of the Upper Lias.[2]

Towards the south the *Clypeus* Grit can be traced along the Evenlode Valley as far as Stonesfield, where it is lost beneath the younger rocks.

Towards the north and east it dies out with surprising rapidity. The Hook Norton railway-cutting displays a section on almost exactly the same line of longitude as Fawler but 9 miles farther north, and here the *Clypeus* Grit as such has entirely disappeared. All that is left between the Hook Norton Beds (lower part of the Chipping Norton Limestone) and the *Scissum* Beds is about 20 in. of hard, ferruginous limestone, in which are embedded an irregular, loose, conglomeratic band and locally an impersistent band of corals. Contrary to what is seen at Fawler, however, the pebbles lie above the band of corals. The pebbles are of limestone and broken shells, rolled, bored, encrusted with *Serpulæ* and oysters, and enclosed in a brown marl.[3] From the coral bed, which also suggests itself as a possible representative of the Upper Coral Bed of the South Cotswolds, Tomes recorded *Isastræa limitata*, *Thamnastræa defrancei* (Mich.), *Clausastræa conybeari* (Ed. and H.), *Latimeandra lotharinga* (Mich.) and other species.[4]

A similar conglomeratic band of the same thickness, but without corals, has been described at Fern Hill, near Bloxham.[5] This is the most easterly known occurrence in this part of the country of any rock that can be ascribed to the Upper Inferior Oolite. The formation dies out completely before reaching the Cherwell Valley, being overlapped by the equivalents of the Chipping Norton Limestone, as it so nearly is at King's Pit, Bradford Abbas.

[1] L. Richardson, 1907, *Q.J.G.S.*, vol. lxiii, p. 441.

[2] My principal authority for the Fawler section is L. Richardson, 1910, *Proc. Cots. N.F.C.*, vol. xvii, p. 31, and for the description (but not the interpretation) of the beds at the junction of the *Clypeus* Grit and Upper Lias C. J. Bayzand, in Professor Sollas's 'Geology of the Country Around Oxford' in *The Natural History of the Oxford District*, 1926, p. 34, fig. 1. Mr. Bayzand appears to have been misled by the more massive nature of the lower part of the *Clypeus* Grit into calling it Upper *Trigonia* Grit; it is quite typical *Clypeus* Grit, however, and there seems no reason for departing from Mr. Richardson's interpretation of sixteen years earlier. Similarly there seems no evidence for identifying the pebbly seam as 'Gryphite Grit', or the bored limestone as '*Scissum* Bed'; the presence of either would be contrary to all expectations. Richardson compares the conglomeratic bed with the *garantiana* conglomerate of the Radstock district and Maes Knoll, Dundry.

[3] L. Richardson, 1911, *Proc. Cots. N.F.C.*, vol. xvii, p. 214.

[4] R. F. Tomes, 1879, *P.G.A.*, vol. vi, pp. 152–65.

[5] L. Richardson, 1922, *Proc. Cots. N.F.C.*, vol. xxi, p. 131.

The overlap of the Upper Inferior Oolite clearly has no connexion with the Vale of Moreton Axis, but is much more widespread, of the class of phenomena resulting from epeirogenic movements. It is, in fact, part of a general overlap against a rising coast-line to eastward (fig. 47). The overlap that is begun by the Upper Inferior Oolite is continued by the equivalents of the Chipping Norton Limestone (as will be seen in the next chapter) and then in turn by several other members of the Great Oolite Series.

V. LINCOLNSHIRE AND YORKSHIRE[1]

The uppermost shelly beds of the Lincolnshire Limestone at Weldon and Ponton yield a fauna of rolled and principally micromorphic gastropods, which, as W. H. Hudleston pointed out, show marked differences from those in the lower beds and to some extent even display Bathonian affinities. It is possible that these highest beds of the Lincolnshire Limestone represent part of the Upper Inferior Oolite, but their age must remain open until the stratigraphy of the Lincolnshire Limestone has been investigated in detail.

In Yorkshire the uppermost or predominantly argillaceous division of the Scarborough Beds (above the Scarborough Grey Limestone with ammonites of *blagdeni* date) abounds in *Pseudomonotis lycetti* Rollier, with which is associated *Trigonia signata* Lyc. These are the two most characteristic fossils of the Hook Norton Beds in North Oxfordshire, and so a provisional correlation of these beds with the Hook Norton Beds or *zigzag* zone may be tentatively suggested. Such a correlation implies the absence or thin and disguised representation of any Upper Inferior Oolite.

Both of these correlations are attended with such uncertainty that the rocks have been described in Chapter VIII with the Lower and Middle Inferior Oolite, of which they are usually considered to form a part.

VI. THE HEBRIDEAN AREA

In the islands of Raasay and Skye the thick Inferior Oolite sandstones, of which the

[1] For references to the Lincolnshire and Yorkshire areas see preceding chapter.

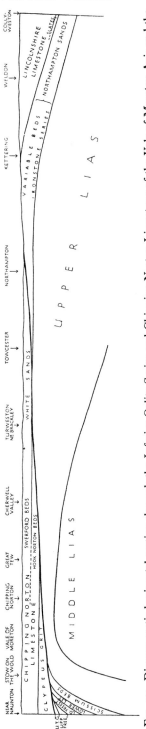

Fig. 47. Diagrammatic horizontal section through the Inferior Oolite Series and Chipping Norton Limestone of the Vale of Moreton Axis and the southern part of the Northants–Lincolnshire Basin. The top of the *Clypeus* Grit is reduced to horizontality and the effects of modern erosion are eliminated. Horizontal Scale, 1 in. = about 8 miles; vertical scale, 1 in. = about 140 ft.

highest several hundred feet belong to the *blagdeni* zone, are abruptly overlain by up to 10 ft. of plastic clay and shale, yielding ammonites of the *garantiana* and possibly also of the *niortensis* zones. The outcrop of the clay, following so suddenly upon the thick sandstones, is easily traceable in both islands.

In Skye the shales are indifferently exposed in a number of places along the heathery slopes of Strathaird, where the black ammonite-bearing shales pass up into dark-grey shaly, micaceous flags, the whole about 30 ft. thick. There is a gradual change upward through the flags into the grey and white, pebbly sandstone of the Great Estuarine Series (see Chapter X). Specimens of *Garantiana* have been found both in Strathaird and in Trotternish, north of Bearreraig, where the plastic clay passes laterally into soft black sandstone.[1]

The most fossiliferous locality yet discovered is the bank of a stream north-east of Storav's Grave, in Raasay, the only satisfactory exposure in that island.[2] The Survey collected *Garantiana* cf. *garantiana* and four other forms, and in addition *Strenoceras bifurcatum* (Ziet.), *S. minimum* Wetzel, and *S. subfurcatum* (Schloth.). Especial interest attaches to this discovery in view of Prof. Davies's suggestion, adopted above, to consider the *niortensis* zone as the lowest zone of the transgressive Vesulian. The idea is strongly supported by this occurrence, in so remote a locality, of *Strenoceras* and *Garantiana* in close association in a bed differing completely from the underlying *blagdeni* and lower zones. Evidently the Vesulian Transgression is marked here by an abrupt change of lithology, and the important point is that *Strenoceras* entered *after* the change of conditions (which was apparently one to deeper water).

It is strange that in the south of the Hebridean area no sudden alteration in the type of sedimentation appears to have taken place with the Vesulian Transgression, for at Port nam Marbh, in SE. Mull,[3] *Garantianæ* have been found 10 ft. below the top of the Inferior Oolite sandstone. Both in Mull and in Ardnamurchan the sandstone continues right up to the base of the Great Estuarine Series (which is considered to be of Great Oolite age). No evidence for the presence of any higher ammonite zones than that of *garantiana* has been found anywhere in Scotland, where estuarine or deltaic conditions prevailed until the time of the Cornbrash.

VII. KENT[4]

In some of the borings in Kent sands and sandy limestones were found to rest upon the Lias and to overstep it northward on to the Palaeozoic platform. These sandy strata are not all of the same age; at Tilmanstone and Snowdown Collieries they yielded fossils typical of the lower part of the Great Oolite, but at Brabourne and Dover they proved to belong to the Upper Inferior Oolite.

Most information was obtained at Dover, where a shelly sand, with hard bands of calcareous grit, and containing numerous small pebbles and phosphatic nodules, was proved resting on the *falcifer* zone of the Upper Lias. Some twenty-five species of lamellibranchs were obtained and these agree as

[1] C. B. Wedd, 1910, 'Geol. Glenelg, Lochalsh and SE. Skye', *Mem. Geol. Surv.*, p. 116.
[2] G. W. Lee, 1920, 'Mes. Rocks Applecross, Raasay and NE. Skye', *Mem. Geol. Surv.*, pp. 47, 50–1.
[3] G. W. Lee, 1925, 'Pre-Tert. Geol. Mull, Loch Aline and Oban', *Mem. Geol. Surv.*, p. 99.
[4] Based on Lamplugh, Kitchin and Pringle, 1923, loc. cit.

a whole with the fauna of the Upper Inferior Oolite. The most abundant were *Trigoniæ* of the group of *T. duplicata* Sow., a species common in the Upper *Trigonia* Grit (*garantiana* zone) of the Cotswolds, and also such typical shells as *Pholadomya fidicula* Sow. and *Gresslya abducta* (Phil.).

Messrs. Lamplugh, Kitchin, and Pringle wrote:[1]

'The condition of the fossils in the sand-bed at Dover suggests that the sea-floor on which they lay was scoured by sand-laden currents, as many of the shells show signs of considerable abrasion by which most of their ornament has been obliterated . . . in an example of a costate *Trigonia* all the costæ have been worn away on the anterior part of the shell, while they are unworn posteriorly.'

The unconformable relations of these sandy beds to the Lias, without the intervention of any Lower or Middle Inferior Oolite strata, is in harmony with the palaeontological evidence in pointing to their having been laid down during or after the Vesulian Transgression. Their being overlapped northward by the lower part of the Great Oolite Series, of which the higher portion in turn overlaps the lower and passes on to the Palaeozoic platform, lends strong support to this correlation, for it is an additional point of agreement with Oxfordshire and Northamptonshire.

[1] Ibid., p. 22.

CHAPTER X

GREAT OOLITE SERIES

Ammonite Zones.[1]	Brachiopod Zones.	Cotswold Strata.[2]	Dorset Strata.
?	Epithyris marmorea	Wychwood Beds	Forest Marble
	Epithyris bathonica and Ornithella digona	Bradford Beds	
Tulites subcontractus and Morrisiceras morrisi[3]	Epithyris oxonica	Kemble Beds	(Gap of Uncertain Extent)
		White Limestone	
		Hampen Marly Beds	
		Taynton Stone	Upper Fuller's Earth
		Stonesfield Slate	
		Upper Fuller's Earth	
	Ornithella bathonica, O. pupa and Stiphrothyris globata	Fuller's Earth Rock	
Oppelia fusca	?	Lower Fuller's Earth	Lower Fuller's Earth
Zigzagiceras zigzag		Chipping Norton Limestone	Crackment Limestone

THE Great Oolite Series, less the Cornbrash, is here described as a single formation in the same sense as in all the memoirs of the Geological Survey. It includes the Fuller's Earth and *Zigzag* Beds (which latter probably pass laterally into the lower part of the Chipping Norton Limestone), the Fuller's Earth Rock, and the Great Oolite Limestones with their numerous successive variants of facies—the White Limestone, Forest Marble, Bradford Clay, Hinton Sands, Bath and Taynton Freestones, Hampen Marly Beds, Upper Estuarine Series, Stonesfield Slate Beds, &c. These constitute a bewildering but palaeontologically interesting complex of strata, of a thickness and importance out of all proportion to the thin and relatively straightforward Upper Inferior Oolite, upon which they follow with perfect conformity. The lower half (up to the Stonesfield Slate Beds) of the Great Oolite Series, as here

[1] Previous writers have sometimes proposed more ammonite zones: the reasons for the selection here made are discussed in the text. One or two ammonites have been found in the Bradford and Kemble Beds, but it would be absurd to use them as zonal indices (Pl. XXXV).

[2] Each of these stratal divisions (with 2 exceptions) represents a well-characterized fauni-zone, recognizable by an assemblage of mollusca and brachiopoda. (The exceptions are the Fuller's Earth clays, which are largely unfossiliferous). The significance of this has been discussed in Chapter I, pp. 34–5.

[3] *Morrisites* (Buckman, 1921) is considered by Dr. Spath to be a synonym of *Morrisiceras* (Buckman, 1920), and the genus is assigned to the Macrocephalitidae, not the Tulitidae as was done by Buckman. See 1932, *Med. om Grønland*, vol. LXXXVII, no. 7, pp. 10, 15.

understood, forms the upper part of the Vesulian Stage of Marcou, overlapping with d'Orbigny's Bathonian.

The different facies of the formation have been known since the earliest times, as is shown by their names, most of which date from William Smith and Buckland; but it is only in the last few years that any real progress has been made in elucidating the true correlation. There are still many points that remain undecided and some are likely to remain so for many years to come, for the difficulties appear insurmountable. They arise from the peculiar changeability of both the fauna and the lithology and the absence or extreme rarity of ammonites throughout the greater part of the succession. In these respects the Great Oolite Series is analogous with the Corallian Beds, in which also ammonites are abundant in certain thin beds but entirely absent or extremely rare through great thicknesses of rock, while facies-faunas introduce complications at all horizons. Neither formation can be described on a strictly palaeontological basis in the same way as the Inferior Oolite or Lias. We cannot draw up a sequence of zones and, having described their development on the South Coast, proceed to trace them inland across the country, following out their changes of facies. Assemblages that remain constant over such wide areas are too scarce and there are too few trustworthy index-species. We cannot even define a succession of partly palaeontological and partly lithological divisions that we can be sure of recognizing along more than a small fraction of the outcrop. This is possible in the Corallian Beds and the task is proportionately easier. Instead we are reduced, in dealing with the Great Oolite Series, to describing the succession in each district on its own merits, endeavouring as we go along to draw comparisons and establish correlations, but usually without sufficient certainty to justify any simplification of the stratigraphical nomenclature.

At the base of the formation the zones of *Zigzagiceras zigzag* and *Oppelia fusca* provide a convenient datum, for in them, at least in the south, ammonites usually abound; but the upward limit of the *fusca* zone is vague. Another horizon yielding ammonites in fair abundance is the Fuller's Earth Rock of Dorset, but farther north specimens are too rare to be of much service. They belong to lævigate and simplified types, some of which seem to have a very long range: as mentioned in Chapter I, their biozones seem to exceed several faunizones in thickness. Brachiopods often abound and they are usually the best guides in correlation. Nevertheless, when a wide area is considered their colonial habits become a serious drawback, and it is often necessary to fall back upon a comparison of the general assemblage of lamellibranchs and gastropods. In local correlations in Oxfordshire certain gastropods of the family Nerineidæ have been found extremely useful, the epiboles of four species maintaining a constant relation over a considerable tract of country where other fossils fail. Oysters also are often valuable.

The master problem in the stratigraphy of the Great Oolite Series is the complete disappearance of the Great Oolite limestones, excepting the Forest Marble, south of the region between Bath and the Mendip Axis. The extent to which any of the Upper Fuller's Earth south of this is equivalent to any of the limestones to the north is still an open question, but such evidence as there is will be reviewed in this chapter.

I. THE DORSET–SOMERSET AREA

(a) The Dorset Coast

The Great Oolite Series on the coast is predominantly a clay formation. It comprises the following divisions:

	ft.
[Cornbrash above.]	
Forest Marble	80–90
Goniorhynchia boueti Bed	1
Upper Fuller's Earth Clay	?125
Fuller's Earth Rock	25
Lower Fuller's Earth Clay	?140
The Scroff with *Oppelia fusca*	$\frac{1}{4}$
Zigzag Bed	$\frac{1}{2}$
[*schlœnbachi* limestone below.]	
Total about . .	380

The best section is the very fine one in Watton Cliff (Plate XII) between West Bay (Bridport Harbour) and Eypesmouth, where the formation is let down in a great trough fault between Middle Lias on the west and Bridport Sands on the east. The Forest Marble forms the capping to the cliff [1] and the projecting crags of the highest 80–90 ft. Next comes the *Boueti* Bed, which, with a parallel band of white cementstone at the top of the Fuller's Earth, are easily recognized datum lines, about 100 ft. above the beach. Below, an almost perfectly vertical face of nearly homogeneous conchoidal grey clay extends to the foot of the cliff.

Watton Cliff is unscaleable from above or below, but at the west end there is a deep recess, known in geological literature as the Fault Corner, where a path leads up to the top. The path is only made possible by the melting down of the clays in a tumbled mass, completely hiding the Fuller's Earth, but it gives welcome access to the *Boueti* Bed and the Forest Marble above. These are bent up steeply against the faulted Middle Lias in the corner and can be studied with ease (fig. 29, p. 154).

East of the centre of Watton Cliff a third fault, downthrowing north-west, enters the cliff obliquely and cuts out the Forest Marble and Upper Fuller's Earth, bringing up Lower Fuller's Earth in their place. Thence eastward to the other boundary fault, which brings up Bridport Sands close to the esplanade, the cliff consists of Lower Fuller's Earth, dipping inland at a steep angle and much slickensided. If any fossils existed in this clay they have been almost entirely destroyed. There is much secondary gypsum, obviously connected with the disturbance.

The fault in the centre of the cliff brings up a number of hard bands of argillaceous limestone, which are the only representative of the Fuller's Earth Rock known on the coast. Unfortunately, besides being nearly vertical, they are much disturbed, and the succession is very difficult to interpret. Buckman considered that the rock-bands were merely bent over at a high angle towards the fault-plane, so that the beds below them could be measured off in regular sequence. In a number of visits to the spot, however, both separately and together, Mr. L. Richardson and I have come to the conclusion that the

---

[1] There was formerly a tiny patch of Cornbrash on the summit, since removed by coast erosion; the whole thickness of the Forest Marble is therefore present.

PLATE XII

Photo. *W. J. A.*

Watton Cliff, Bridport, Dorset.

The vertical part is Upper Fuller's Earth Clay; B.B. = *Boueti* Bed, with
Forest Marble above.

Photo. *W. J. A.*

Fuller's Earth Rock, Shepton Montague railway-cutting, Somerset.

whole of the rock-bands visible in the cliff are out of place; are in fact parts of a fault-breccia of very large lumps, caught up in the fault. On this view the succession is by no means perfect and it is impossible to measure the total thickness. Nevertheless, it is probably safe to say that samples of all the types of Fuller's Earth present are represented in the jumbled rock-masses of the fault-breccia.

In certain states of the beach it is possible to see further exposures of the rock-bands, which are full of brachiopods, appearing from beneath the shingle, both here and along the line of the great fault, opposite Fault Corner.

The basal part of the Fuller's Earth is not exposed at Watton Cliff. A few feet of clay immediately above the Inferior Oolite, with The Scroff and *Zigzag* Bed, may be seen in the brow of the vertical cliff of Bridport Sands and Inferior Oolite between the mouth of the River Bredy and Burton Bradstock (fig. 29), and again in a more accessible position in a quarry at Burton Bradstock allotments.

The Upper Fuller's Earth is best seen at Cliff End, east of Burton villas, with the *Boueti* Bed in the brow of the small cliff and the Forest Marble outcropping in the partly grassgrown bank above. It is again exposed indifferently at several points in the low banks of the Fleet Backwater about Langton Herring, together with the *Boueti* Bed and the greater part (in discontinuous sections) of the Forest Marble. (See map, p. 342.)

The sections along the shores of the Fleet, although at first unpromising, provide the key to the interpretation of the disturbed succession at Watton Cliff.

The classic exposure of the *Boueti* Bed, the type-locality of the name-fossil, is the little promontory of Herbury (or Herbeyleigh), south of Langton Herring. On the north-west extremity of the promontory the bed descends to water-level and runs gradually out under the shore. The gentle lapping of the waves of the backwater has washed out the fossils from the soft matrix until the beach has come to be largely composed of brachiopods. Perfect specimens of *G. boueti*, together with numerous other large Rhynchonellids and abundant *Ornithella digona* and Avonothyrids may be picked up by the hundred. Eastwards the lower and middle portions of the Forest Marble are well exposed and in places highly fossiliferous. Owing to the discontinuity of the sections, however, it is difficult to determine the position of the several fossil-bands in the sequence.

Below the *Boueti* Bed, in the small cliff on the north side of the promontory of Herbury, about 12–15 ft. of the highest beds of the Fuller's Earth can be seen. The unfossiliferous white cementstone band so conspicuous in Watton Cliff has here multiplied to several bands. The beds rise northward in the southern limb of the Weymouth Anticline, the axis of which runs out to sea about ¾ mile to the NW., beyond Langton Herring coastguard station. The greater part of the Upper Fuller's Earth is concealed in the low grassy shore around the little bay of Herbury, but on the north side the banks begin to show small sections intermittently for a mile, and these afford some indication of the lowest beds that come to the surface in the anticline. Nearly the lowest horizon seen is a remarkable bank of oysters. At one place, west of the coastguard station, the shells form a solid mass 10 ft. thick, marked on the Survey map as 'masses of *Ostrea acuminata* in the Fuller's Earth', and recorded as that species by H. B. Woodward, S. S. Buckman and others. This misidentification has been the source of much of the trouble in the correlation of the

Fuller's Earth of Dorset; for the species bears no resemblance to the true *Ostrea acuminata* Sowerby, which marks a constant horizon *below* the Fuller's Earth Rock in North Dorset and Somerset. It is a large and peculiar variety or mutant of *Ostrea sowerbyi* Mor. and Lyc., to which the name var. *elongata* has been given by Dutertre, and its position is above and not below the local representative of the Fuller's Earth Rock. If the rock were present above, it would certainly be exposed somewhere along the shore. Instead there is a positive indication that it lies below ground, for one mile west of Langton Herring a small section shows the *Elongata* Bed (3 ft. thick) overlying 2 ft. of clay with nodules of argillaceous limestone and numbers of *Rhynchonelloidea smithii*, which is a fossil indicative of the Fuller's Earth Rock (and of the upper part in particular).

Loose blocks of the *Elongata* Bed are also caught up in two of the faults at Watton Cliff. In Fault Corner an isolated patch was first discovered by Buckman, close to the wall of Middle Lias, and dragged up almost level with the *Boueti* Bed. Larger blocks are associated with the Fuller's Earth Rock in the middle fault, and in 1931 there was some evidence that the oyster bed lay almost immediately on the top of a large mass of the rock-bands, as is to be expected from the indications at Langton Herring. Buckman came to the same conclusion prior to 1922, although he misidentified the oysters as *O. acuminata*.[1]

SUMMARY OF THE GREAT OOLITE SERIES AS EXPOSED ON THE DORSET COAST

'Forest Marble': Wychwood Beds and Bradford Beds, 80–90 ft.

Detailed stratigraphical research on the Forest Marble south of the Mendips still remains to be done, and a start needs to be made with the splendid sections on the Dorset coast. Meanwhile it is necessary to continue using the old facies-term Forest Marble, although it is highly probable that a large proportion of the rock so called will in time be found to yield the distinctive fossils of the Bradford Clay (or better Bradford Beds, since, as we shall see later, the clay at the type-locality passes laterally into limestones of Forest Marble facies).

The Forest Marble of the coast falls into three divisions of approximately equal thickness: a central block of hard, flaggy, massive, false-bedded, broken-shell limestone, blue-centred and oolitic, sandwiched between two predominantly argillaceous divisions. The clays, however, are very different from those of the Fuller's Earth, for they are usually greenish or brown, sandy, shaly, or micaceous, with numerous thin and impersistent laminæ of hard, fissile shale or broken-shell limestone. The limestones are largely made up of broken valves of *Ostrea sowerby*, Pectinidæ, &c. It is remarkable that this peculiar shallow-water facies persists almost unaltered near the top of the Great Oolite Series as far as Buckingham and recurs even farther north. Lately, however, evidence has been obtained showing that the conditions which determined the facies set in earlier in some districts than in others.

The central mass of limestone is used locally for building and road-mending, so that the coast-sections are amplified by numerous quarries about Burton,

---

[1] S. S. Buckman, 1922, *Q.J.G.S.*, vol. lxviii, pp. 381–2. *O. sowerby*, including var. *elongata*, often assumes a Gryphæate form and shows nascent ribbing. The presence of such specimens at Watton Cliff no doubt accounts for Buckman's recording the *Ostrea knorri* Beds as well; repeated search by Mr. Richardson and the writer has failed to produce a single specimen of *O. knorri* from the coast-sections.

Swyre, and Bothenhampton. The immense quarries at the last village, however, cited by H. B. Woodward as some of the finest sections of Forest Marble
in the country, have long ago been abandoned. The stone is full of rolled and
bored pebbles of argillaceous limestone, up to several inches in diameter, as
well as ochreous clay galls. The pebbles have evidently been derived from the
erosion of some previous deposit, but it is not clear what this deposit was.
The most cursory examination reveals the presence of myriads of shells of
Ostrea sowerbyi, Chlamys obscura, Camptonectes laminatus, Lima cardiiformis,
and a *Gervillia* or *Perna* which has been dissolved away, leaving numerous
cavities. There are also species of *Navicula* and other fossils, and more rarely
Rhynchonellids (the valves separated and very difficult to detach) and an
Epithyris (?*marmorea* Oppel).

On the west end of the Herbury Promontory the lower middle part of the
Forest Marble contains some highly fossiliferous beds, reminiscent of some
in the Oolithe Blanche de Langrune on the opposite side of the Channel.
They contain abundant valves of *Oxytoma costata*, with nests of a rather small
Rhynchonellid, sponges, and ossicles of *Apiocrinus*.

The Boueti Bed: Bradford Beds, pars.

The richness of the *Boueti* Bed has already been remarked on. Although
only 1 ft. thick, it has supplied almost enough specimens of *Goniorhynchia,
Kutchirhynchia, Ornithella*, and *Avonothyris* at Herbury to pave the beach.
It also contains numerous well-preserved valves of *Chlamys vagans, Placunop-
sis socialis* and a small oyster (rarely more than 2 or 3 mm. in diameter). At
Cliff End, Burton Bradstock, the brachiopods are less abundant, and at
Watton Cliff, 12 miles from Herbury, they are considerably scarcer, though
G. boueti can always be found in moderate numbers. At the latter place the
Boueti Bed is overlain by a separate bed crowded with *Placunopsis* and the
micromorphic oysters, and yielding an occasional *Dictyothyris*.

It is noticeable that the fossils of the *Boueti* Bed are commonly encrusted
with Polyzoa, minute oysters, or *Serpulæ*, sure signs of slow deposition.

Upper Fuller's Earth Clay.

In the mural face formed by the Upper Fuller's Earth of Watton Cliff about
100 ft. are exposed, but the cliff is extremely dangerous and yields little information, for the fallen blocks cannot be properly located in the sequence. The
clay falls away in large masses with a conspicuously conchoidal fracture, whence
Buckman called it the Large Conchoidal Bed. Well preserved but fragile shells
of *Grammatodon, Nucula, Astarte*, &c., may be collected from the debris.

At Cliff End, Burton Bradstock (fig. 29, p. 154), the highest 50–60 ft. can
be studied at leisure and the following section is obtainable:

UPPER FULLER'S EARTH AT BURTON BRADSTOCK
[*Boueti* Bed above.]   *ft.*

4. Prominent band of hard white laminated cementstone .   .   .   2–3
3. Unfossiliferous grey clay   20
2. Unfossiliferous grey clay with two thin bands of red claystone 4 ft. apart .   4
1. Fossiliferous grey clay with numerous pale crumbling concretions and
    nests of fossils: abundant small gastropods—'*Eulima*' sp., *Nucula waltoni*
    Mor. and Lyc., *Nucula* sp., *Grammatodon* sp., *Trigonia* sp., *Oxytoma*
    sp., *Lucina* sp., *Goniomeris* sp.: seen to .   .   .   .   .   .   *c.* 30

At the base of the Upper Fuller's Earth is the oyster bank of *Ostrea sowerbyi* var. *elongata*, already referred to. It thins rapidly from about 10 ft. near Langton Herring coastguard station to 3 ft. a mile farther north-west, but it is still several feet thick 12 miles away, at Watton Cliff. The peculiarity that characterizes the oysters is a ventral elongation (more correctly 'heightening') so that full-grown specimens become the shape and size of a human finger. There is at the same time every gradation to the normal form so common in the Forest Marble. M. Dutertre informs me that he has found the same association of this form with the typical shapes in the Boulonnais, and has been accustomed to regard it as only a variety of *O. sowerbyi*. There is no real resemblance to the *O. acuminata* of a lower level.

Fuller's Earth Rock.

The Fuller's Earth Rock has previously been alleged to be absent on the Dorset coast, but there can be little doubt that it is represented by the 25 ft. (?) of argillaceous limestone and clay bands cautiously called by Buckman the 'Brachiopod Beds'. There seem to be (so far as can be seen from the jumbled exposures) about 6 or 8 bands of stone, each from 6 in. to 1 ft. thick, separated by as many clay bands. A collection of the brachiopods has been examined by Miss Muir-Wood, who finds that the most abundant species, *Stiphrothyris wattonensis*, is peculiar to this locality, although associated with it are specimens of *S. nunneyensis*, which is the common species along the inland outcrop. In addition *Rhynchonelloidea smithii* and *Acanthothyris powerstockensis* abound, with occasionally an *Ornithella* sp. Buckman also recorded a large ammonite (*?Parkinsonites*).[1]

Lower Fuller's Earth Clay and Zigzag Bed.

As already remarked, the Lower Fuller's Earth Clay of Watton Cliff is much squeezed and disturbed and yields few if any fossils. Buckman distinguished three lithological divisions as follows:[2]

	ft.
Umber Bed; umber-coloured clay with a nodular band	20
Small Conchoidal Bed; grey clays with small conchoidal fracture, and a band of ochreous nodules (1 ft.) at base	40
Laminated Clays, with occasional nodules of pyrites	50
	110

These thicknesses are very hypothetical, and might be double or half the true thicknesses, since it is impossible to make accurate measurements. The *Ostrea knorri* Clay, said by Buckman to be at the top and 5 ft. thick, is probably below beach-level, and there is no evidence that any specimen of *O. knorri* has ever been found at Watton Cliff. Certainly it would be unlikely to occur so high in the series, for elsewhere it marks a constant horizon near the base of the Fuller's Earth, and Richardson has found it only a few feet above the *Zigzag* Bed as near as Beaminster.[3] It is probable that Buckman saw a specimen of *Ostrea sowerbyi* with incipient ribbing, a not uncommon variety.

Also below the beach-level at Watton Cliff are some few feet of clays seen

[1] S. S. Buckman, 1922, *Q.J.G.S.*, vol. lxxviii, p. 382.
[2] S. S. Buckman, 1922, loc. cit.
[3] L. Richardson, 1915, *P.G.A.*, vol. xxvi, p. 70.

at Burton Bradstock, yielding *Belemnopsis bessina* (d'Orb.), a belemnite especially common in the equivalent Marnes de Port-en-Bessin. Below this, and best seen in the Allotments Quarry, is the 2 or 3 in. of indurated marl known as The Scroff, characterized (but rarely) by *Oppelia fusca* and *Aulacothyris cucullata*.[1] At the base of all is the *Zigzag* Bed.

The *Zigzag* Bed consists of about 6 in. of white limestone containing abundant ammonites. Buckman has figured the following from Burton Bradstock (the names are those under which Buckman figured the specimens and the numbers refer to plates in *Type Ammonites*):

Planisphinctes planilobus	CCCXXVII
Polysphinctites polysphinctus . . .	CCCXXII
Polysphinctites replictus	CCCLIX
Procerites tmetolobus	CDXVI
Ebrayiceras rursum	DCCLVIII
Ebrayiceras jactatum	DCCLXIX
Harpoxyites fallax	CDXCIX

(b) The Inland Outcrop to the Mendips

Forest Marble: Wychwood Beds, Hinton Sands and **Bradford Beds**, 100–130 ft.

The Forest Marble is exploited in numerous quarries all along the outcrop, but they are shallow in relation to the whole formation and show only a small proportion of the total thickness. Although the details vary in every quarry, so far as can be seen the tripartite arrangement—a central limestone block sandwiched between two predominantly argillaceous divisions—is maintained throughout Dorset and Somerset.

The greatest thickness of Forest Marble in England, about 130 ft., is developed in the neighbourhood of Sherborne. In its effect upon the scenery hereabouts and for many miles to the north the formation is the most important member of the Oolites. Its escarpment builds Lillington Hill and the rest of the high ridge that divides the waters flowing into the Bristol Channel by the River Yeo from those destined for Blackmoor Vale and the Stour, which takes them into Bournemouth Bay. Most of the extra thickening seems to take place in the upper argillaceous division, which expands to over 60 ft. Below this, from 61–70 ft. from the top, a sandy element appears, in the form of foxy sands enclosing a hard calcareous grit.[2] The sands can also be seen to a thickness of 10 ft. at Charlton Hill, north-east of Charlton Horethorne, with 25 ft. of the main limestone division below, consisting of typical dark, blue-grey shelly Forest Marble, with partings of loam, sand and clay. These are the first appearances of the Hinton Sands, described by William Smith at Hinton Charterhouse, near Bath, and traceable as far as Cirencester.

The lower clay division has recently been well exposed in widening the main road up the hill a mile north-east of Bruton, where its thickness was estimated by Richardson and myself to be about 25 ft. At the base was shown a thin seam of brown, shaly, platy Forest Marble limestone containing some of the fauna of the Bradford Clay: *Rhynchonella* cf. *obsoleta*, ossicles of

---

[1] See also *Gonolkites vermicularis* S. Buck., *T.A.*, pl. DXLVII.

[2] H. B. Woodward, 1894, *J.R.B.*, p. 347. He describes a complete section in a road-cutting at West Hill, south of Sherborne, showing the details.

Apiocrinus, and abundant *Oxytoma costata*, *Chlamys vagans* and *Nuculæ*. Near by, at Wincanton, numerous specimens of *Goniorhynchia boueti* were found in this bed in a boring.[1] This is the last inland record of the species. The distance north of Herbury is 30 miles. The Wincanton Boring also showed the total thickness of the Forest Marble to be 103 ft., and this thickness seems to be maintained fairly steadily across the Mendip region into the Cotswolds.

Upper Fuller's Earth, 130–240 ft.

According to the interpretation of the Wincanton Boring the Upper Fuller's Earth clay has there the prodigious thickness of 240 ft., while at Scale Hill, south-east of Batcombe, De la Beche estimated its thickness to be still 133 ft. Considerably more than 120 ft. was known to be present in the neighbourhood of Milborne Port, for the Stowell Boring passed through that amount although it was started some distance from the top of the formation.

In spite of this great thickness, scarcely anything is known of the Upper Fuller's Earth of the district. Woodward remarked that it was 'well exposed in a brickyard north-east of Maperton, and south-east of the Cock Inn, Holton. About 25 ft. of grey marly clay was exposed, the beds, where dry, being pale and hard, and much resembling those exposed in [Watton] cliff near Eype'.[2] From this description it is not to be expected that the Upper Fuller's Earth would yield much of interest; but it would be useful to obtain a complete section in order to know whether the *Elongata* Bed is present.

Fuller's Earth Rock, 18–35 ft.

From the neighbourhood of Thornford, 3 miles south-east of Yeovil, the Fuller's Earth Rock becomes an important surface-feature, its escarpment, although overlooked by that of the Forest Marble, being a conspicuous element in the landscape. Old limekilns and quarries, some of them still worked, become numerous, and there are in addition three nearly or quite complete sections in road- and railway-cuttings, at Shepton Montague (Plate XII), at Laycock, north of Milborne Port, and near Blackford Lake, between Maperton and Charlton Horethorne. The thickness of the Rock is 35 ft. in the three neighbouring localities of Laycock Cutting, Stowell and Wincanton, but thins northward to 18 ft. at Shepton Montague and 25 ft. at Scale Hill near Batcombe.

The rock is a grey or buff, cream-weathering, non-oolitic, argillaceous limestone, the lower part surprisingly hard and massive, and strongly suggesting certain of the limestones in the Great Oolite proper in Oxfordshire and Gloucestershire. The upper part is usually more rubbly, like Lower Cornbrash, and is crowded with *Ornithella bathonica*, *O. pupa*, and *Stiphrothyris nunneyensis* and *S. globata*.[3] Towards the top, at Charlton Horethorne and Maperton, Richardson has also described a bed of marl with *Diastopora*, which he calls the Polyzoa Marl.[4]

---

[1] W. Edwards and J. Pringle, 1926, *Sum. Prog. Geol. Surv.* for 1925, pp. 183–8.

[2] H. B. Woodward, 1894, *J.R.B.*, p. 238.

[3] For identifications of all Fuller's Earth brachiopods I am indebted to Miss H. M. Muir-Wood, who has very kindly given me much information in advance of publication.

[4] See L. Richardson, 1909, *P.G.A.*, vol. xxi, pp. 212–13, for descriptions of sections and lists of fossils; also in 'Geol. Shaftesbury', *Mem. Geol. Surv.*, 1923, p. 22.

The Ornithellids in the upper part of the Rock or *Ornithella* Beds show great diversity of form and belong to a number of different lineages. They may be collected with the Stiphrothyrids in a perfect state of preservation at numerous places—Shepton Montague, Charlton Horethorne, Maperton, Alham Lane, Haydon, &c.[1]

The fauna of the Fuller's Earth Rock, and of this upper portion especially, is rich and interesting. Besides the wealth of beautifully-preserved brachio-pods it comprises some forty-five species of lamellibranchs, of which at least fifteen are familiar Lower Cornbrash species. In any collection from one of the Fuller's Earth Rock quarries of Somerset may be noticed such typical Lower Cornbrash shells as *Lima duplicata* (Sow.), *L. rigidula* Phil., *Entolium rhypheum* (d'Orb.), *Ostrea* (*Lopha*) *costata* Sow., &c., while even a species usually considered so characteristic of the Lower Cornbrash as *Pseudomonotis echinata* (Sow.) seems to be absolutely indistinguishable when it occurs at this lower horizon. This striking community of fauna may well be connected with the strong lithic resemblance between the two rocks.

A number of peculiar and highly characteristic Cadicone ammonites are also tolerably abundant. They have been studied comprehensively by Buck-man, who founded for their reception the family Tulitidæ, with genera *Tulites*, *Tulophorites*, *Madarites*, *Pleurophorites*, *Rugiferites*, *Sphæromorphites*, *Bulla-timorphites*, *Morrisites*, and *Morrisiceras*.[2] Unfortunately he had no field-knowledge of the circumstances or order of their occurrence and no practical familiarity with the Fuller's Earth Rock, but he arranged them in a hypo-thetical order according to their matrices.[3] This risky procedure, which he had attempted in 1913 with the museum specimens from the Yorkshire 'Kellaways Rock' (see below, p. 363), led to results incongruous with the field-evidence obtainable in the exposures throughout Somerset and North Dorset, and it deprives the stratigraphical results of any value. In spite of the impressive table of hemeræ and their corresponding matrices, by which the epiboles were supposed to be recognizable, it must be understood that there is at present no field evidence whatever for the numerous ammonite-horizons within the Fuller's Earth Rock.

Without describing a single section in support of his conclusion, Buckman divided up the Fuller's Earth Rock into two portions, Milborne Beds above and Thornford Beds below.[4] The Milborne Beds were said to be 'brown (ironshot) limestones' and the Thornford Beds 'whitish, chalky limestones'. It must not be thought, however, that the upper or *Ornithella* Beds are to be called Milborne Beds and the more massive lower portion Thornford Beds. If we look through Buckman's records of the localities of the ammonites studied, we find that he used the names in quite a different sense.[5] He assigned to 'Milborne Beds' all the ammonites, from whatever horizon, labelled as from any of the principal Fuller's Earth Rock localities of Somerset, includ-ing all the specimens from Milborne Wick (Laycock railway-cutting near Milborne Port Railway Station) and Shepton Montague railway-cutting, at

---

[1] See note 4 on previous page.
[2] S. S. Buckman, 1921, *T.A.*, vol. iii, p. 43. For figures of *Tulites* and *Morrisiceras* from Haydon, Milborne Wick and Shepton Montague see *T.A.*, plates CLXVII, CCLXIX, CCLXXIV, CCLXXXV, CCCLXXVII.   [3] 1921, loc. cit., pp. 50–1.
[4] S. S. Buckman, 1918, *Pal. Indica*, iii (2), p. 237; and 1921, *T.A.*, vol. iii, p. 51.
[5] 1921, loc. cit., pp. 43–9.

both of which places the whole of the Fuller's Earth Rock is exposed. At these places the ammonites occur as commonly in the lower part of the Rock as in the higher; therefore the whole of the Fuller's Earth Rock throughout the main escarpment of Somerset must be classed as 'Milborne Beds'. The only locality where Buckman recognized 'Thornford Beds' was Troll Quarry, near Thornford, Dorset, $3\frac{1}{2}$ miles south-west of Sherborne.

The Fuller's Earth Rock of Troll Quarry is certainly peculiar. Buckman published a description of the quarry five years later,[1] and it was the only exposure of Fuller's Earth Rock he ever described. From his experience of the museum material from other Fuller's Earth Rock localities he concluded that 'the ammonoid fauna yielded by Troll quarry is almost, if not quite, unique in England'. It would seem, therefore, that this one exposure affords a glimpse of beds that are not developed in fossiliferous facies anywhere along the Somerset outcrop; and from such field evidence as there is it would appear also that they may be, as Buckman believed, on a somewhat lower horizon than the ordinary Fuller's Earth Rock, or, as he called it, Milborne Beds.

Troll Quarry is only some 8 ft. deep. It displays five bands of nodular, somewhat rubbly to very hard, whitish, chalky limestone, separated by grey, white-weathering marl. As a whole the beds are monotonous and poorly fossiliferous (any attempt to recognize the horizon of specimens by the matrix being of very doubtful value) and the only common fossils are casts of *Pholadomya lyrata*. One band contains small *Rhynchonelloidea* sp. and a *Pseudomonotis*. Mr. Richardson and I have found two specimens of the almost spherical *Sphæromorphites* (the lowest hemeral index of Buckman's hypothetical sequence) *in situ* only 2 ft. from the top of the section. It is therefore very doubtful if the existence of any distinct epiboles will ever be established in this quarry; it seems more probable that the peculiar ammonites lived contemporaneously and together form a single faunizone.[2]

Unfortunately the quarry is an isolated one. The nearest other exposure is about 2 miles away on the opposite side of Thornford, where a cutting on the Sherborne road shows 5 ft. of the base of the Fuller's Earth Rock and about 10 ft. of the top of the Lower Fuller's Earth Clay. Both rock and clay appear to be unfossiliferous except for some minute Rhynchonellids, but the rock is of a similar type to that at Troll. Another exposure of similar rock, even more unfossiliferous, resting on Fuller's Earth Clay, has also been revealed an equal distance to the WNW., in a cutting on the main road south of Yeovil. Perhaps it is significant that no trace of any *Ostrea acuminata* Bed appears below the rock in either of these sections, although a thick bed crowded with this oyster everywhere underlies the normal Fuller's Earth Rock or 'Milborne Beds' throughout Somerset, for at least 30 miles from the neighbourhood of Milborne Port to the district around Bath. It is possible, therefore, that the Thornford Beds are below the main horizon of *Ostrea acuminata*.

Lower Fuller's Earth Clay and **Zigzag Beds**, 120–?200 ft.

The Lower Fuller's Earth Clay was completely fathomed by the borings at Wincanton and at Stowell. At Wincanton it proved to have a thickness of

---

[1] S. S. Buckman, 1927, *T.A.*, vol. vi, p. 50.

[2] For figures of the Troll ammonite fauna see *T.A.*, pl. CCCLXVII–CCCLXXI and CCCXXXVIII. Dr. Spath accepts far fewer genera than Buckman; see *Med. om Grønland*, 1932, vol. LXXXVII, no. 7, pp. 9–15.

120 ft. At Stowell it was at least 57 ft. thicker, but there has been some confusion regarding the interpretation of the relevant part of the record, for the lower part of the clay yielded ammonites said by Buckman to be suggestive not only of the *zigzag* but also of the *truellei* zone,[1] while nothing definite was found in the top of the thick Inferior Oolite limestones below. Since, however, the *truellei* and *zigzag* (and presumably *schlœnbachi*) zones are definitely represented by limestones all along the neighbouring outcrop, which passes only a mile from the Stowell Boring,[2] it seems highly probable that the ammonite supposed to be indicative of the *truellei* zone was misidentified. The suggestion might be made that in a core-sample any Oppelid might be mistaken for a *Strigoceras*; in fact a specimen figures in the same list as '*Oekotraustes*? (or *Strigoceras*)', from the Fuller's Earth Rock.

In these circumstances we may probably conclude with safety that there was nothing unusual about the Lower Fuller's Earth in the Stowell Boring, and that, as elsewhere, the upper part of the *zigzag* zone was represented by clays and the lower part by limestones. As the limestones which appear to belong to the *zigzag* zone in the near-by outcrop are probably about 25 ft. thick, we ought to add about that amount to the Lower Fuller's Earth, making the total thickness some 200 ft.

About 5 miles north of Stowell is the Wincanton Boring, where the thickness of the Lower Fuller's Earth Clay was proved to be 120 ft. (+an unknown amount for limestone of *zigzag* date?), and 6 miles farther at Scale Hill, near Batcombe, De la Beche estimated it to be only 21 ft. (fig. 48, p. 264). With this low estimate H. B. Woodward agreed, although he was not able actually to check it.[3] Over the end of the Mendips, as we shall see, the thickness certainly diminishes still further. Therefore it appears that, unlike the Upper Fuller's Earth Clay, the Lower Clay diminishes rapidly northward along the Somerset outcrop.

At the top of the Lower Fuller's Earth, immediately beneath the Rock, is the most characteristic palaeontological horizon in the whole formation, the *Ostrea acuminata* Beds. As much as 6 or 7 ft. of the top of the clay may be crowded with shells, massed together as a typical oyster-bank. At this horizon they are the true *O. acuminata* of Sowerby, very sickle-shaped and uniformly small, altogether different from the oysters of the *Elongata* Bed in Dorset. The true *Acuminata* Bed does not seem to have been recognized south of Milborne Port; but thence northward it is continuous to beyond the Mendips, having been seen, always in the same position, at Laycock[4] and Stowell,[5] Holton,[6] Wincanton,[7] Shepton Montague,[8] Alham Lane near Batcombe,[9] and close to the flanks of the Mendips, in the escarpment of the Fuller's Earth Rock in Chesterblade Lane, south of Doulting Bridge Quarry.[9]

The bulk of the Lower Fuller's Earth Clay, like the Upper, is hardly ever

[1] S. S. Buckman, in J. Pringle, 1910, *Sum. Prog. Geol. Surv.* for 1909, p. 70.
[2] For a detailed account of neighbouring exposures see L. Richardson in H. J. Osborne White, 1923, 'Geol. Shaftesbury', *Mem. Geol. Surv.*, pp. 11–18.
[3] H. B. Woodward, 1894, *J.R.B.*, p. 238.
[4] L. Richardson, 1923, 'Geol. Shaftesbury', *Mem. Geol. Surv.*, p. 22.
[5] J. Pringle, 1910, loc. cit.
[6] H. B. Woodward, *J.R.B.*, p. 236.
[7] Edwards and Pringle, loc. cit.
[8] A fine exposure, where *O. acuminata* may still be collected in hundreds (Pl. XII).
[9] L. Richardson, 1909, *P.G.A.*, vol. xxi, p. 212.

exposed and is almost unknown. It was completely penetrated in the Stowell and Wincanton Borings, but yielded nothing of interest.

The basal layers, like the topmost, are better known, for they are sometimes present in the brow of quarries opened for the underyling Inferior Oolite. Some of these exposures show another interesting oyster-bank, the *Ostrea knorri* Bed. This little oyster was first described from the borders of Germany and Switzerland, and it forms conspicuous oyster beds in many places in South Germany, Lorraine, &c. It is almost gryphæate in form, the left valve very convex and the right valve small and flat, while an altogether distinctive appearance is given it by the presence of numerous delicate radial ribs on the left valve only. It has been considered so distinct from all other oysters that one palaeontologist has made it the type of a new genus, *Catinula*.[1]

The *Knorri* Bed has been seen at the base of the Fuller's Earth resting on the *Zigzag* Bed, and this in turn on the limestones of the Microzoa Beds (*schlœnbachi* zone) in the side of a road at North Poorton, only 6 miles inland from the coast at Burton Bradstock, and again 5 miles farther NW., in the roadside midway between Beaminster and Broad Windsor.[2] Day recorded it as long ago as 1864 at Powerstock, only 5 miles from the coast.[3] It occurs again at Stoford, south of Yeovil,[4] still close down upon the *Zigzag* Bed. Thence it does not seem to have been met with again until the Bridge Quarry, Doulting, where Richardson has called attention to the enormous abundance of it, once more at the very base of the Fuller's Earth Clay.[5]

At the extremities of the district, where the *Knorri* Bed is found, the base of the Lower Fuller's Earth is formed by the 3–6 in. band of white limestone with ammonites of *zigzag* date, as at Burton Bradstock; it is rich in ammonites of the genus *Zigzagiceras* (*Z. zigzag*, *Z. subprocerum*, *Z. pollubrum*, *Z. rhabdouchus*, *Z. phaulomorphus*, *Z. crassizigzag*, &c.) with *Oppelia fusca* (Quenst.) and its allies.[6] The most richly fossiliferous localities are Grange Quarry, Broad Windsor, Crewkerne Railway Station, Haselbury Mill Quarry,[7] and Doulting Bridge Quarry. Immediately succeeding this but below the *Knorri* Bed can usually be recognized 2 or 3 in. of hardened marl known as The Scroff, which occasionally yields *Oppelia fusca*.

In the central area around Sherborne and Milborne Port, the *zigzag* zone seems to be partly represented by up to 25 or 30 ft. of whitish limestones, which have been called the CRACKMENT LIMESTONES.[8] They first appear, in argillaceous development and interbedded with clays, at King's Pit, Bradford Abbas, where, as already mentioned, Richardson has obtained ammonite evidence that they belong to the zone of *Zigzagiceras zigzag*.[9] Limestones which he believes to be mainly of this age overlie the *truellei* zone to a thick-

---

[1] L. Rollier, 1911, *Faciès du Dogger*, p. 272. This should serve to indicate its distinctness from *Ostrea (Lopha) costata* Sow. which some have doubted. *Lopha* is entirely different, both valves being convex and strongly plicated.

[2] L. Richardson, 1928, *Proc. Cots. N.F.C.*, vol. xxiii, pp. 173, 181.

[3] E. C. H. Day, in K. von Seebach, 1864, *Der Hannoversche Jura*, p. 93.

[4] Shown me by Mr. L. Richardson (see fig. 45, p. 233).

[5] L. Richardson, 1907, *Q.J.G.S.*, vol. lxiii, p. 393.

[6] For figures of ammonites from the *Zigzag* Bed of the first two localities see S. S. Buckman, *T.A.*, pls. CLIII, CLXXIII–IV, CCLIX, CCC, CCCXXXV, CCCLI, CCCLXXVI, DXLVI, DXCV, DCXXIII–IV, DCXLIII, and below, Pl. XXXV.

[7] L. Richardson, 1918, *Q.J.G.S.*, vol. lxxiv, p. 166.

[8] L. Richardson, 1923, in 'Geol. Shaftesbury', *Mem. Geol. Surv.*, pp. 16–18.

[9] L. Richardson, 1911, *P.G.A.*, vol. xxii, p. 262.

ness of 25 ft. at the Halfway House, between Yeovil and Sherborne (p. 234), and succeed the Rubbly Beds above the Sherborne Building Stone north and east of Sherborne. Here, in the type-sections at Crackment (or Crackmore) they are 20 ft. thick, and yield *Pholadomya* sp. in some abundance, *Belemnopsis bessina*, *Sphæroidothyris* and *Pleuromya*; and near by, at Goathill, he obtained a specimen of *Zigzagiceras procerum* S. Buck., indicating the *zigzag* zone.[1] As usual, however, where the limestones are thickly developed they are much poorer in fossils than where they are attenuated.

The restriction of the thick development of the Crackment Limestones to the centre of the area around Sherborne and Milborne Port, while towards the Dorset coast they thin down to a 6-in. fossil-band or *Zigzag* Bed, is remarkable. It suggests that the Mendip and Weymouth Axes were exercising some control on the sedimentation during the *zigzag* hemera. At present evidence seems to be lacking to show whether the North Devon Anticline and the Cole Syncline had any similar effect.

II. THE MENDIP AXIS

The Mendip Axis certainly continued to cause paucity of sedimentation, accompanied by overlaps, until the end of Fuller's Earth times. The northerly overlap of the Lower Fuller's Earth by the Fuller's Earth Rock, indicated by the reduction of the former from 120 ft. at Wincanton to 21 ft. at Scale Hill near Batcombe, is continued over the axis until at Bonneyleigh Hill, 2 miles north-east of Frome, the Lower Fuller's Earth is only 8 ft. thick.[2] At the same time the Fuller's Earth Rock is reduced in thickness to 10 ft. and the Upper Fuller's Earth to about 20 ft. (as compared with 25 ft. and 133 ft. at Scale Hill), and so about Frome the whole formation is not much more than 35 ft. thick.[3] For a considerable distance near Whatley it is mapped as overstepping on to Carboniferous Limestone (fig. 13, p. 70).

The Rock continues to be highly fossiliferous, yielding at Egford Bridges, Whatley and Bonneyleigh Hill (map, fig. 13) the usual abundance of *Ornithella bathonica*, *O. pupa*, *Stiphrothyris globata*, *Rhynchonelloidea smithii* and lamellibranchs. The *Acuminata* Bed seems to disappear over the crest of the axis, but the oyster occurs sparingly in the Lower Fuller's Earth.

About three miles north of the Mendip Axis (3 miles north-west of Bonneyleigh Hill) a boring at Hemington was started in Upper Fuller's Earth Clay and, after penetrating about 30 ft., proved a thickening of the Fuller's Earth Rock to 15 ft. and of the Lower Fuller's Earth Clay to 40 ft.[4] These thicknesses, so quickly regained, are maintained with but little increase northward to beyond Bath. Immediately below the Fuller's Earth Rock in the Hemington Boring the *Acuminata* Bed, 1½ ft. thick, was met with again, as on the south of the axis.

It is instructive to compare with these figures the thicknesses proved in the Westbury Boring. This is situated only a mile or two south of the presumed underground continuation of the Mendip Axis, but some miles down the dip slope, 4½ miles east of Bonneyleigh Hill. It proved the Lower Fuller's

[1] L. Richardson, 1923, in 'Geol. Shaftesbury', *Mem. Geol. Surv.*, pp. 17–18.
[2] L. Richardson, 1909, *P.G.A.*, vol. xxi, p. 219.
[3] H. B. Woodward, 1894, *J.R.B.*, p. 239.
[4] C. Cantrill and J. Pringle, 1914, *Sum. Prog. Geol. Surv.* for 1913, pp. 98–101.

Earth Clay to have the same thickness (40 ft.) as at Hemington, but the Fuller's Earth Rock to be thicker (21½ ft.). The Upper Fuller's Earth Clay, however, showed the surprising thickness of 146 ft., or rather more than at Scale Hill.[1] Thus it would appear that the attenuation of the Fuller's Earth from east to west, up the dip-slope, is of greater magnitude than the local thinning at right angles to this direction, over the Mendip Axis.

A tabulation of the figures discussed in this and the foregoing sections may make the matter clearer (the thicknesses are in feet):

	South.		Mendip Axis.		North.	East.
	Stowell.	Wincan-ton.	Scale Hill.	Bonney-leigh Hill.	Heming-ton.	Westbury.
Upper Fuller's Earth	(120+)	240	133	c. 20	(32+)	146
Fuller's Earth Rock	35	35	25	c. 8	15	21½
Lower Fuller's Earth	?200	120	21	8	40	40

The Forest Marble seems to undergo no noteworthy change in its course over the axis. It maintains the same thickness of 90–100 ft. and forms a prominent escarpment from near Cranmore, past Marston Bigot to Frome. The Hinton Sands have been exposed at Marston Bigot, and the Bradford Fossil Bed with *Ornithella digona* was met with at the base in a boring at Buckland Denham, immediately north of the axis.[2] The thickness of the Forest Marble in the Westbury Boring was 91 ft.

III. THE COTSWOLD HILLS

About 2 miles north-east of Hemington, and some 4 miles north of the nearest point on the Mendip Axis, the Great Oolite limestones begin to make their appearance between the Fuller's Earth and the Forest Marble.

At first they comprise only a few feet of rubbly oolite seen in old quarries between Wellow and Norton St. Philip, north-east of Hassage and east of Lower Baggeridge (map, fig. 13, p. 70). They rest upon the clays and earthy limestone of the Fuller's Earth and are overlain by Forest Marble.[3]

Within 3 or 4 miles to the north, around the south and west of Bradford-on-Avon, in the ancient quarries of Murrel or Murhill near Winsley, Avon-cliff near Upper Westwood, and at Monkton Farleigh, from 45 to 80 ft. of oolites, freestones and ragstones are to be seen. Hereabouts and for some miles northward the Great Oolite yields the famous Bath Stone, and hence the formation, with the Forest Marble above, ranges continuously through the Cotswolds and Oxfordshire into the Eastern Counties. Indeed, in the Cotswolds it has by far the widest outcrop of all the Oolites, covering the entire dip-slope from the Cornbrash fringing the Oxford Clay Vale to the

[1] J. Pringle, 1922, *Sum. Prog. Geol. Surv.* for 1921, pp. 147–53.
[2] H. B. Woodward, 1894, *J.R.B.*, p. 350.
[3] Ibid., p. 260.

highest plateaux in the centre of the North and Mid-Cotswolds and lapping up to the edge of the escarpment above Bath, Wotton under Edge and Stroud. The average width from Bath to Burford is 8–10 miles and tongues extend northward to Condicote, 15 miles from the Cornbrash.

The question, to what the Great Oolite is equivalent over and south of the Mendip Axis, has given rise to numerous speculations. If exposures of the Fuller's Earth in the Cotswolds were as numerous and complete as those displaying the Great Oolite, the problem might be easily solved. But unfortunately wells and borings and chance outcrops where cattle have worn paths down steep banks provide the only sections. The commercial exploitation of the Fuller's Earth is confined to a small area round Bath and the workings are restricted to the Upper Fuller's Earth Clay.

In recent years it has been fashionable to correlate the Great Oolite limestones with the Fuller's Earth Rock. The basis of this correlation is the announcement by Buckman that certain ammonites are common to the two formations. In 1921 he wrote: 'The family [Tulitidæ] is of particular interest as showing the contemporaneity of the Fuller's Earth Rock of South England with the Great Oolite of Gloucestershire–Oxfordshire, and thus that the Fuller's Earth Rock is later, not earlier, than the Stonesfield Slate.'[1] Buckman's reputation at that time was so high that the statement, unsupported by any field-evidence, soon came to be repeated as if it were an expression of proven fact. In reality the question is an extremely difficult one—by far the most difficult remaining in British Jurassic stratigraphy—and any dogmatic pronouncements are the last thing needful.

William Smith regarded the Great Oolite as distinct from both the Forest Marble and the Fuller's Earth, as shown by a table made in 1819, illustrating the strata beneath the Cornbrash in Dorset:[2]

	Clay
[Forest Marble]	Forest Marble
	Clay
	Place of the Upper [Great] Oolite
	Clay
[Fuller's Earth]	Fuller's Earth Rock
	Clay

Since the critical area for any inquiry is that about Bath and the South Cotswolds, the country which William Smith mapped in detail and knew so well, his conclusion is not to be lightly dismissed.

A thorough examination was made by H. B. Woodward and published in *The Jurassic Rocks of Britain*. Having satisfied himself that the Great Oolite could not pass into the Forest Marble, as some had supposed, he made a careful investigation of its relations to the Fuller's Earth, and in particular to the Fuller's Earth Rock.

'The notion that the two might be portions of one formation possessed me for some time,' he wrote, 'but it was dispelled when I came to examine the ground at Bath. In several places where the Fuller's Earth Rock had become too attenuated

[1] S. S. Buckman, 1921, *T.A.*, vol. iii, p. 43.
[2] *Geol. View and Section through Dorset and Somerset*, 1819; quoted by H. B. Woodward, 894, *J.R.B.*, p. 260.

to be shown on the map it is nevertheless present; as I found . . . between Wellow and Norton St. Philip; as the Rev. H. H. Winwood pointed out to me on the slopes of Lansdown, and as Prof. Hull has shown to be the case at Slaughterford, N.E. of Bath. It is therefore clear that the Great Oolite overlies the Fuller's Earth Rock in the neighbourhood of Bath and Bradford-on-Avon. At the same time this rock maintains a fairly uniform character of white marly limestone, *and contains a similar assemblage of fossils*,[1] in its range from Dorsetshire to Somersetshire, while it merges upwards and downwards into the marly clays of the Fullonian formation, and is of varying thickness and importance.

'Hence', he concluded, 'it is quite possible that south of Bradford-on-Avon the lower portion of the Great Oolite may be replaced to some extent by the Upper Fuller's Earth Clay. More than this I am not prepared to say. . . . The evidence is in favour of the mass of the Great Oolite (over part at any rate of the Dorsetshire region) having been eroded, and there is consequently a local break between the Forest Marble and Fullonian formation, marked by the rich fossil-bed [the *Boueti* Bed] which has been identified with the Bradford Clay.'[2]

Evidence subsequently collected confirms these opinions. The most telling contribution, ignored by Buckman, was Richardson's announcement that the Fuller's Earth Rock not only underlies the Great Oolite and a considerable thickness of Upper Fuller's Earth Clay as far along the Cotswold scarp as Dyrham, 2 miles north-west of Marshfield, but that it abounds in the same characteristic assemblage of Ornithellids and Stiphrothyrids and *Rhynchonelloidea smithii* as throughout Somerset.[3] More recently traces of a similar rock, still with Stiphrothyrids but less the Ornithellids and Rhynchonellid, and still closely associated with and apparently overlying a bed densely packed with *Ostrea acuminata*, have been recognized 13 miles farther north, at Symond's Hall Hill above Wotton under Edge.[4] At both of these places the rock is succeeded by a considerable thickness of Upper Fuller's Earth Clay, above which follows the steeper slope of the Great Oolite limestones, with their numerous quarries, some of which show the Stonesfield Slate Series at the base. Sandy and fissile beds on the same horizon as the Stonesfield Slates have long been known to extend as far south as Bath and Midford, always overlain by the Great Oolite and underlain by the Upper Fuller's Earth[5] (see table, fig. 48). Thus it can be said to have been established ten years before Buckman made his new correlation, that the Fuller's Earth Rock, recognizable by a rich assemblage of brachiopods as well as by its lithology, is a continuous stratum from the neighbourhood of Sherborne to north of Marshfield in the South Cotswolds. This stratum appears to be as fully entitled to be considered synchronous throughout its area of occurrence as almost any bed containing mollusca and brachiopods; and for at least 10–12 miles it is overlain by Upper Fuller's Earth Clay, Stonesfield Slate Series and Great Oolite (Bath Freestone), one above another in constant succession (see fig. 48).

The reappearance of a few species of ammonites in the Great Oolite limestones, apparently identical with some of those in the Fuller's Earth Rock, is

---

1 The italics are mine.
2 H. B. Woodward, 1894, *J.R.B.*, pp. 258, 259, 260.
3 L. Richardson, 1910, *Proc. Cots. N.F.C.*, vol. xvii, p. 78.
4 L. Richardson and W. J. Arkell, field-notes.
5 H. B. Woodward, 1894, *J.R.B.*, p. 266.

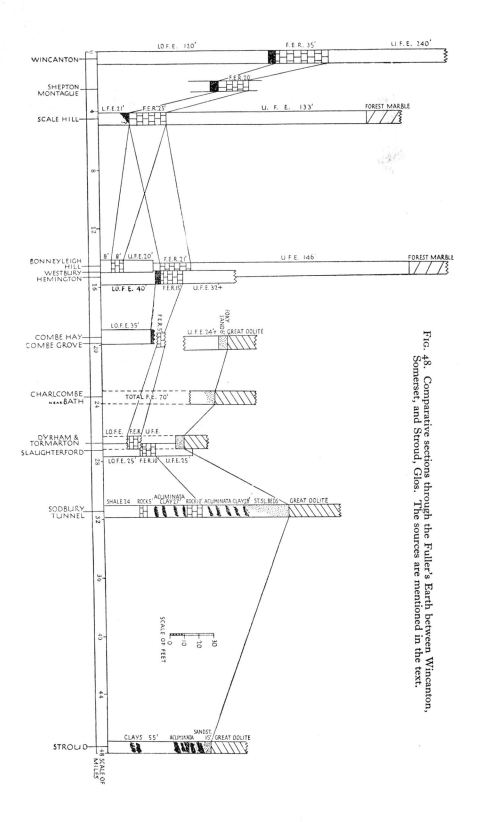

FIG. 48. Comparative sections through the Fuller's Earth between Wincanton, Somerset, and Stroud, Glos. The sources are mentioned in the text.

remarkable, but it is doubtful whether it has the particular significance attached to it by Buckman. The ammonite fauna of the Fuller's Earth Rock of Dorset and Somerset is a rich one, of which the majority of the species are *not* found in the Great Oolite. The possibility has not yet been explored that the forms in the Great Oolite may be unrelated homœomorphs of those in the Fuller's Earth Rock; but more probably they may be long-ranged 'zone-breakers'.[1] Either supposition involves less difficulties than the theory that the Fuller's Earth Rock, with its rich brachiopod and lamellibranch assemblages distributed from end to end of its outcrop, is a diachronic stratum. The apparent absence of ammonites from the Fuller's Earth Rock north of the Mendips is at least partly explicable by the poverty of exposures. If quarries and railway-cuttings were as numerous as in Somerset and Dorset it is highly probable that ammonites would have been found as well as the far more abundant lamellibranchs and brachiopods. In the Great Oolite north of the Mendips, in which exposures are abundant and actively worked, on the other hand, the Fuller's Earth Rock brachiopods have never been found and the characteristic lamellibranchs are replaced by a different assemblage.

A description of the Great Oolite Series of the Cotswold province becomes largely a description of the Great Oolite limestones. A great deal is known of their detailed development, thanks to numerous quarries and a series of long railway-cuttings. In the south abundant information is supplied by the great quarries and mines in the Bath Freestone south and east of Bath, about Corsham, Box and Bradford-on-Avon, as well as by the approach cuttings to Box Tunnel, on either side of Corsham Railway Station. This area is classic ground since the work of William Smith and J. Sowerby. A few miles farther north are the long and instructive cuttings on the South Wales line at Acton Turville and Alderton, described so lucidly by Prof. Reynolds and the late A. Vaughan, and by H. B. Woodward. Then comes the classic region of Minchinhampton, the scene of the lifelong labours of J. Lycett, where a freestone like the Bath Stone has been worked in extensive quarries and whence Lycett and others obtained an unrivalled collection of perfectly-preserved mollusca, figured in Morris and Lycett's *Monograph of the Mollusca from the Great Oolite* (1850–3). Only a few miles beyond this are the extensive railway-cuttings on the Swindon and Stroud line north-west of Kemble, with the Sapperton Tunnel. Another 8 miles farther the Swindon and Cheltenham line provides yet another complete section between Chedworth and Cirencester, which has been minutely described by Richardson. Finally, the development of the formation in the North Cotswolds is clear from the Hampen cutting, near Notgrove, on the Banbury and Cheltenham railway, also described by Richardson. In these sections and in numerous lesser quarries, the following subdivisions are to be recognized:

---

[1] Apparent increase in the range may be the effect of accelerated deposition during the biochron of the species, as explained in Chapter I, p. 35. It is not necessary to suppose that the biochron was any longer than the average although the biozone is much larger and embraces several faunizones.

THE GREAT OOLITE SUCCESSION IN THE COTSWOLDS

[Cornbrash above]

Stratal Divisions.		Thickness.	Principal Invertebrate Faunas.
FOREST MARBLE	WYCHWOOD BEDS with HINTON SANDS	Ft. 50–100	*Epithyris marmorea* (Oppel) and Forest Marble facies-fauna.
GREAT OOLITE LIMESTONES	BRADFORD BEDS and ACTON TURVILLE BEDS	0–50	*Ornithella digona* (Sow.), *Dictyothyris coarctata* (Park.), *Rhynchonella obsoleta* Dav., *Epithyris bathonica* Buck., *Apiocrinus parkinsoni*, &c., &c.
	KEMBLE BEDS	10–30	*Epithyris oxonica* Arkell, corals, &c. Also F. M. facies-fauna.
	THE WHITE LIMESTONE	40–?60	*Epithyris oxonica, Nerinea eudesii* Mor. and Lyc., corals, &c., *Aphanoptyxis bladonensis* Ark. near top. *Burmirhynchia hopkinsi* (Dav.) and *Ornithella digonoides* Buck.
	HAMPEN MARLY BEDS (in the north only)	0–30	*Rhynchonella concinna* (Sow.) and *Ostrea sowerbyi* Mor. and Lyc.
	TAYNTON and MINCHINHAMPTON (? BATH) FREESTONES	15–30	Rich fauna of lamellibranchs and gastropods; the main source of the fossils figured by Morris and Lycett.
SANDY 'SLATE' BEDS	STONESFIELD SLATE BEDS	5–30	*Ostrea acuminata, O. sowerbyi, Rhynchonella* spp., and *Stiphrothyris* spp. in 2 bands at the top. In the slate, *Trigonia impressa* Sow., *Gervillia ovata* Sow., *Gracilisphinctes gracilis*, &c.
UPPER FULLER'S EARTH CLAY (with the Fuller's Earth of commerce)		?10–?50	
FULLER'S EARTH ROCK (in the south only)		0–15	*Ornithella bathonica, Stiphrothyris globata, S. nunneyensis, Rhynchonelloidea smithii*, &c.
LOWER FULLER'S EARTH CLAY, WITH UPPER ESTUARINE CLAY IN THE NORTH-EAST		5–?70	*Ostrea acuminata* (without *sowerbyi*) in a band at the top.
CHIPPING NORTON LIMESTONE (in the north only)		0–20	*Oppelia fusca, Ostrea knorri* (both very rare). *Zigzagiceras zigzag* at the base of the clay in the south.

[*Clypeus* Grit below]

SUMMARY OF THE GREAT OOLITE SERIES OF THE COTSWOLDS

Wychwood Beds with Hinton Sands = Forest Marble Proper.

The Forest Marble enters the district and passes on through the Cotswolds with but little change. Despite innumerable minor variations its peculiarities render it generally recognizable at a glance. The thickness is 80–90 ft. in the south, in the Bradford-on-Avon district, 105–120 ft. in the Alderton cuttings, and probably at least 70 ft. about Cirencester and Fairford, but thence eastward it becomes thinner rather rapidly, until at Witney, over the Oxfordshire border, only 8 ft. are left. This seems to be due to overstep by the Cornbrash. As a rule the three major divisions, a central limestone mass sandwiched between two predominantly clayey series, are recognizable over the southern part of the area. In addition, as far north as Cirencester, a fourth division is highly characteristic, the Hinton Sands of William Smith.

The Hinton Sands, as we have seen, first make their appearance at Sherborne, but are discontinuous. They can be traced from Wanstrow and Marston Bigot on the south side of the Mendip Axis, round the eastern end of the Carboniferous tract by Frome, until they thicken on the north side of the axis to 30 ft. about Hinton Charterhouse, Buckland Denham and Rudge. Pits show up to 25 ft. of white and buff sands with large doggers of fissile calcareous sandstone, devoid of fossils except some microzoa, and looking more like the Lower Calcareous Grit than part of the Forest Marble. Farther north they are exposed east of Castle Combe, at several places about Malmesbury, and in the first railway-cutting on the main line north of Kemble. Formerly they were also to be seen at Sandy Lane, south of Cirencester.

The stratigraphical position of the sands is in the upper part of and above the central limestone division of the Forest Marble. Above come clays with flaggy and sandy or broken-shell limestones, generally full of *Ostrea sowerbyi*, *Camptonectes laminatus*, *C. rigidus*, &c. About Malmesbury, however, much of the limestone becomes sandy, for exposures near the station and at Brokenborough show thick beds of white sands and flaggy, concretionary sandstone, mixed with bands of broken-shell limestone, resting on 20 ft. of ripplemarked, flaggy calcareous gritstones and shales. In some places, as at Foxley Road Quarry, Malmesbury, Pickwick, near Corsham, and in the Kemble railway-cutting, the upper argillaceous division is either wanting or is almost entirely replaced by sands, which extend up to the Cornbrash. The most likely explanation is that the Cornbrash has begun its overstep and has cut out the superior beds.

Down the dip-slope, at Bradford-on-Avon and Corsham, in the direction in which deeper water presumably lay, there is little or no sign of the Hinton Sands, the only representative being a relatively small thickness of sandy shales with thin bands of sandstone, as seen at Pound Pill, near Corsham Railway Station.

A highly interesting feature, found only at Pickwick, near Corsham, is a thick band of grey, shelly and lignitiferous Forest Marble limestone full of the largest Bathonian brachiopod, *Epithyris marmorea* (Oppel). This is the type-locality of the species,[1] and it is not found abundantly or well-preserved anywhere else. The bed yielding it is at the very top of the Forest Marble, within 6 ft. of the Cornbrash, from which it is separated by clay and sandy shale. Buckman in consequence gave to the Forest Marble at this locality a special stage-name 'Corshamian', the meaning of which is not clear, since the bed with the *Epithyris* seems to occur in the upper part of the equivalents of the Hinton Sands, which in the same work he called 'Hintonian'.[2] Moreover, large implicate Epithyrids, perhaps identical with *E. marmorea*, occur rarely in the central limestone block of the Forest Marble in Dorset. It is not improbable, however, that these beds seen about Corsham and Hinton Charter-

---

[1] The type is the specimen figured by T. Davidson, *Mon. Brit. Foss. Brachiopoda*, pt. 3, pl. IX, fig. 4.					[2] S. S. Buckman, 1927, *Q.J.G.S.*, vol. lxxxiii, p. 7.

house are overstepped by the Cornbrash before they reach Oxfordshire, where *E. marmorea* does not occur, and that therefore the Wychwood Beds of Oxfordshire are equivalent to the lower part of the Forest Marble of this district, as Buckman suggested.[1] However, for the present there seem insufficient grounds for complicating the nomenclature by retaining more than one designation for the Forest Marble, and Wychwood Beds is open to the fewest objections.[2]

That there is considerable overstep of the Great Oolite by the Forest Marble, at least over the Mendip Axis, seems probable from the occurrence of many derived ooliths and pebbles of oolitic limestone in the Forest Marble limestones. Moreover, in the Cotswolds the Forest Marble is often seen to rest, without the intervention of any Bradford Clay, upon an irregular and channelled surface of the Great Oolite. Such a junction was well displayed in the Stow-Road railway-cutting, between Chedworth and Cirencester, where it was remarked on and figured by Richardson.[3] Here, locally, the erosion has entirely removed the Kemble Beds, though they appear in the next cutting on the same line. Another particularly fine section is to be seen at Eastleach, where some Kemble Beds remain, with a deeply channelled upper surface.

A complete section of the Forest Marble opened in the railway-cuttings between Alderton and Hullavington, on the South Wales line, was described by Woodward as follows:[4]

FOREST MARBLE AT ALDERTON

[Cornbrash unconformable above.]

	ft.
Clays with thin layers of calcareous grit and a bed of *Ostrea sowerbyi* at top	20
Shales and limestones	5
Sands and hard calcareous sandstones [Hinton Sands]	20
Oolitic and shelly and gritty limestones with lignite, and shales, false-bedded, replacing one another [the main limestone block] . .	35–50
Blue clay with occasional bands of limestone	25
Total . . .	105–120

[Pale oolite below, seen to 18 ft.]

Bradford Beds, or Bradford Clay and Acton Turville Beds.

The fame of the Bradford Clay is due, not to any considerable thickness, but to its extraordinarily rich and constant fauna. Although the fauna is found only intermittently at the base of the Forest Marble, often being absent over considerable areas, it recurs repeatedly at the same horizon from Buckland Denham, close to the Mendips, to the eastern border of Oxfordshire, at Islip, and it was met with in the deep boring at Swindon. In most of the places where the fauna occurs, the bottom of the Forest Marble consists of a clay, which, in the typical area around Bradford-on-Avon, is about 10 ft. thick; but it does not follow that wherever a clay is present the fauna will be found also.

[1] See previous page, footnote 2.
[2] W. J. Arkell, 1931, *Q.J.G.S.*, vol. lxxxvii, pp. 563–629, for full discussion.
[3] L. Richardson, 1911, *P.G.A.*, vol. xxii, pp. 96, 103–4, pl. xv, xvi.
[4] H. B. Woodward, 1902, *Sum. Prog. Geol. Surv.* for 1901, pp. 59–60.

The Bradford Clay assemblage comprises colonies of the brachiopods *Ornithella digona* (Sow.), *Dictyothyris coarctata* (Park.), *Eudesia cardium* (Lamk.), *Epithyris bathonica* Buck., *Avonothyris bradfordensis* (Dav.), *A. langtonensis* (Dav.), *Rhynchonella obsoleta* Dav., *Kutchirhynchia morieri* (Dav.), and numerous small Rhynchonellids not yet systematically studied. With them are associated *Cidaris bradfordensis* Wr., abundant *Oxytoma costata* (Sow.), and sometimes complete specimens of the crinoid *Apiocrinus parkinsoni*. The assemblage grew at Bradford-on-Avon in clear water, rooted upon an eroded surface of the underlying Great Oolite, and it appears to have been choked and killed by an influx of mud, which laid the crinoids undisturbed full-length on the sea-floor amongst the brachiopods. Subsequently there was a pause long enough for all the shells and crinoids to become encrusted with Polyzoa and *Serpulæ*.

In one of the principal quarries at Bradford-on-Avon, between the Upper Westwood road and the tithe barn, the Bradford Clay can be seen to pass laterally into flaggy, false-bedded, typical Forest Marble, with the shells still abounding in the basal layers, but for the most part broken.

The fossiliferous Bradford Clay has been detected in a boring at Buckland Denham, north-west of Biddestone, near Yatton Keynell, at West Keynton, Tiltups End near Nailsworth (in the form of a marl full of *Rhynchonella obsoleta* encrusted with *Serpulæ* and Polyzoa), at Tetbury Road Railway Station near Kemble, at Ewen, and at Cirencester College. The close similarity of the faunas at these places renders the bed one of the most useful datum-lines in making correlations.[1] At Cirencester it yielded a unique ammonite, the only specimen of *Clydoniceras* found in England between the Cornbrash and the Stonesfield Slate—the type specimen of *Clydoniceras hollandi* (J. Buckman).[2]

It is not always that the fauna of the Bradford Clay is confined to so thin a seam and is divided by so sharp a boundary from the Great Oolite. In Normandy the fauna is spread through a very considerable thickness of rock—clays, marls, and limestones like our Great Oolite (the Pierre de Taille de Ranville and the second Caillasse).[3] Similar conditions recur in the South Cotswolds, between Marshfield and Corsham in the south and Malmesbury and the Wotton under Edge district in the north. Passing north from Bradford-on-Avon, the first signs of the change have been noted by H. B. Woodward at Yatton Keynell; he stated that 'The upper beds of the Great Oolite here contain many of the characteristic Bradfordian fossils, and show the intimate connexion between the Great Oolite and Forest Marble'. Again of a section at West Keynton he remarked: 'It is noticeable that pale false-bedded and fissile oolites, resembling beds of Great Oolite, occur above the fossiliferous Bradford Clay at this locality. Stratigraphically there is no real break in the series.'[4]

A complete section of these interesting beds was opened up in the approach cutting at the east end of the Sodbury Tunnel, near Acton Turville,

---

[1] S. S. Buckman stated that the Bradford Clay of Tetbury Road Railway Station was earlier, at least in part, than that of Bradford-on-Avon, but he published no reasons. *T.A.*, vol. v, 1924, p. 28.

[2] S. S. Buckman assigned it to a different genus, *Harpoceratidarum*; see full discussion, *T.A.*, vol. v, 1924, pp. 25–9; but if it is to be considered generically distinct, there will have to be at least half a dozen 'genera' made for the Cornbrash forms.

[3] W. J. Arkell, 1930, *P.G.A.*, vol. xli, pp. 403–5.

[4] H. B. Woodward, 1894, *J.R.B.*, pp. 268, 270.

during the construction of the South Wales line. Fortunately it was described in detail by Reynolds and Vaughan, who collected the fossils bed by bed, and thus a full record of this unique section is preserved. Resting on fine-grained, white, poorly-fossiliferous limestones (?Kemble Beds), were no less than 45 ft. of massive, wedge-bedded limestones, with lenticular seams of sandy clay, the whole abounding in the fossils of the Bradford Clay—*Ornithella digona, Rhynchonella obsoleta, Eudesia cardium, Avonothyris* aff. *bradfordensis, Epithyris bathonica, Cidaris bradfordensis,* with *Oxytoma costata, Lima cardiiformis, Terebellaria ramosissima* and the other characteristic Polyzoa of Bradford-on-Avon. Above came thick shales or clays with bands of limestone, indistinguishable from Forest Marble, but still with the Bradfordian fauna abundant towards the base.

In collecting fossils from these remarkable beds, Messrs. Reynolds and Vaughan noticed an increasing resemblance to the assemblage at Bradford-on-Avon from below upwards. Although they grouped the strata on lithological grounds with the Great Oolite, they remarked that, except for their having failed to find either *Dictyothyris coarctata* or *Apiocrinus* 'it would be utterly impossible to separate the beds . . . from the Bradford Clay on palaeontological grounds . . . the uppermost beds are homotaxial with the clay at Bradford-on-Avon'.[1] To these strata Buckman later gave the name Acton Turville Beds.[2] It is possible that the lower portions of the 45–50 ft. of strata containing the Bradfordian fauna may be somewhat older than the fossil-bed at Bradford-on-Avon. On the other hand, the Bradford fossil-bed was obviously accumulated and covered with sediment very slowly, and it is probable that the more rapidly accumulated strata in the South Cotswolds were laid down during the same secule. If that is so, the name Acton Turville Beds is synonymous with Bradford Beds; but it may perhaps be usefully retained to distinguish the different facies.[3]

The Great Oolite Limestones.

The detailed succession of the Great Oolite limestones changes so completely from place to place, and the stratigraphy and palaeontology have as yet received so little close study, that it is not practicable to give an ideal sequence, which might be recognizable throughout the Cotswolds. Certain broad correlations seem highly probable, but they are by no means certainly established. It is therefore necessary to give a brief résumé of the succession first in the Bath district (the South Cotswolds) and then in the Minchinhampton, Cirencester and Chedworth district (Mid- and North Cotswolds).

(a) THE BATH DISTRICT

?Kemble Beds, with **Coral Bed,** 12–20 ft.

At Bradford-on-Avon the Bradford Clay rests upon 12–20 ft. of massive, obliquely-bedded to false-bedded and flaggy white or cream oolite, the lowest

---

[1] S. H. Reynolds and A. Vaughan, 1902, *Q.J.G.S.*, vol. lviii, pp. 742–6.

[2] S. S. Buckman, 1924, *T.A.*, vol. v, p. 28.

[3] The observation that the Acton Turville Beds became palaeontologically more like Great Oolite downwards and more like Bradford Clay upwards seems to have been mainly based on the brachiopod recorded as '*Terebratula maxillata*'. Owing to many previous misleading records, this was thought to be a Great Oolite species—what is now called *E. oxonica*. But in reality the species is *E. bathonica,* which is common in the Bradford Clay. (*Teste* Mr. L. Richardson, who saw some of the specimens.)

5–10 ft. of which are known as the Coral Bed. The corals are drifted, and when weathered out they leave ramifying caverns. They include *Calamophyllia* (*Eunomia*) *radiata, Anabacia complanata, Convexastræa waltoni, Isastræa limitata, Microsolena excelsa, Montlivaltia caryophyllata, Oroseris slatteri, Stylina plotii,* and *Thamnastræa.*[1] With them are sponges and, somewhat rarely, *Epithyris oxonica.*

? White Limestone: Ancliff Fossil Bed (15 ft.), and Upper Rag Beds (0–8 ft.).

Below the Coral Bed is the highly fossiliferous oolite of Ancliff (Avoncliff, near Upper Westwood, west of Bradford-on-Avon), from which Sowerby received from local collectors many of the minutest fossils figured in the *Mineral Conchology.* It is a thin-bedded, coarse oolite, which is readily recognized by the abundance of shell-fragments and minute specimens of lamellibranchs and gastropods. The adults of the same species are found elsewhere, and the accumulation of juvenile individuals may be ascribed to the sorting action of currents. Among the types in the Sowerby collection obtained from this bed at Ancliff are those of *Nucula variabilis, Nuculana lachryma, N. mucronata* (a rubbed shell of the same), *Parallelodon rudis, Navicula* (*Eonavicula*) *minuta, Barbatia pulchra, Limopsis minima, Trigonia pullus,* &c. The gastropods include *Cerithium costigerum, Exelissa formosa, Rissoina acuta, R. duplicata, Turbo burtonensis* and *Solarium turbiniformis.*

At Monkton Farleigh, farther south, towards the Mendip Axis, the Ancliff Fossil Bed is only 2 ft. thick and is more or less blended with the Coral Bed.

The Fossil Bed at Monkton Farleigh rests directly on the Bath Freestone, but in the Bradford-on-Avon district, in the large quarries at Avoncliff, Murhill (Murrel), &c., it is separated from the freestone by 3–8 ft. of poor quality oolite known as the Rag. Part of this usually forms the roof-bed of the stone mines.

Bath Freestone, 8–30 ft.

The Bath Freestone is an even-grained, poorly fossiliferous, fine oolite, which is still extensively mined in the district south-west of Corsham. Its excellent qualities are said to have been first generally realized when it had to be pierced for the Box Tunnel. Previously the demand had been only local, but after the coming of the railway Bath Stone soon began to be sent to all parts of the country, wherever a smooth, white freestone was in request.

The freestone is reached by inclined shafts as much as 70–100 ft. deep, driven through the Forest Marble and Kemble Beds, and the galleries ramify for miles underground. Here, on the windy plateau, amid the stacks of huge blocks and the clink of chisels, it is easy to understand why William Smith used the term Great or Bath Oolite. Indeed, the name is peculiarly appropriate from here all through the South Cotswolds, to Minchinhampton and Stroud. As in all good freestones, fossils are scarce and small, and there seem to be none of any value for correlation.

Lower Rag Beds, 10–40 ft.

Under the freestone is a variable thickness of shelly and marly limestones, passing down into fissile oolite, with occasional bands of coral, known as the

[1] R. F. Tomes, 1885, *Q.J.G.S.*, vol. xli, pp. 174, 189.

Lower Rag Beds. Exposures are nowadays extremely rare, and little or nothing is known of them palaeontologically. H. B. Woodward records '*Ammonites subcontractus*' in the basal bed near Charlcombe, Bath, but unfortunately the section in which he mentions it is a composite one.[1] Farther north, along the escarpment, quarries at Tolldown near Dyrham and near Tormarton show that the base of the Great Oolite, directly above the Stonesfield Slate Beds, consists of false-bedded, shelly white oolite suggestive of the Taynton Stone or the Ragstone Beds of the Stonesfield Slate about Northleach.[2]

(b) THE MID- AND NORTH COTSWOLDS

Kemble Beds, 10–30 ft.

Underneath the Bradford Clay with its rich fossil-bed at Tetbury Road Railway Station, north of Kemble, are exposed to the base of the quarry and railway-cutting 10 ft. of massive, false-bedded, buff-coloured, oolitic, detrital limestones. The downward relations of the same beds are displayed north-west of Kemble, in a long cutting on the Tetbury branch line. These are the Kemble Beds of H. B. Woodward and subsequent writers, and they rest upon an evenly planed surface of White Limestone. In the type-locality they are about 30 ft. in thickness and are well exposed in the railway-cuttings north, north-west and south of the town. In the southern cuttings the junction with the Forest Marble clay is exposed for upwards of half a mile, the clay there being about 20 ft. thick but not yielding the Bradfordian fauna.

In these cuttings there is displayed, rather below the middle of the Kemble Beds, an irregular fossil-bed crowded with *Lima cardiiformis*, *Epithyris oxonica*, *Ostrea sowerbyi*, and many other lamellibranchs, with corals of the genera *Thamnastræa*, *Isastræa*, *Cladophyllia*, and *Cyathophora*. The bed varies in thickness from 2 ft. to 10 ft.[3]

These beds (with which those in a corresponding position under the Bradford Clay around Bradford-on-Avon agree well enough) have a wide extension all over the Cotswolds. They are exposed at numerous places, showing below the Bradford fossil-bed at Tiltups End near Nailsworth, and at Ewen, and in force around Tetbury and Cirencester. Eastward they become more flaggy and assume a facies indistinguishable from Forest Marble. In this facies they spread all over the Cotswolds between Cirencester and the Oxfordshire border and have been mapped as Forest Marble. The change of facies even deceived H. B. Woodward, who wrote that 'The Kemble Beds evidently become thinner when traced from Kemble to Cirencester; while onwards in a north-easterly direction they become overlapped, near Baunton Downs, by the Forest Marble, which then rests directly on the White Limestone'.[4] However, the conclusion that this was no overlap but a lateral change of facies was arrived at by Richardson in Gloucestershire and by the writer in Oxfordshire simultaneously and independently, and it is confirmed by the Oxfordshire section shortly to be described, where the Forest Marble facies of the Kemble Beds is succeeded by the Bradford fossil-bed.

---

[1] H. B. Woodward, 1894, *J.R.B.*, p. 266.
[2] L. Richardson and W. J. Arkell, field-notes.
[3] H. B. Woodward, 1894, *J.R.B.*, pp. 273–5; and W. J. Arkell, 1931, *Q.J.G.S.*, vol. lxxxvii, pl. LI and field-notes.
[4] H. B. Woodward, 1894, *J.R.B.*, p. 285.

White Limestone, 40–?60 ft.

The White Limestone is the most variable of all the members of the Great Oolite. The division comprises two principal types of rock: white, buff, or pinkish limestones, in which ooliths are often rare or absent, alternating with pale marls; and intensely hard, splintery, sublithographic limestone, often riddled with ramifying cavities. The sublithographic rock is known as Dagham Stone (after Dagham or Daglingworth Downs, near Cirencester, where it occurred at the surface and was quarried for rustic work and rockeries). Bands occur at several horizons in the series, but the principal development is generally at or near the top. In the railway-cuttings between Cirencester and Chedworth there are three bands, the highest and lowest 40 ft. apart. The majority of the cavities are probably due to the solution of branching corals

Besides this there are often bands of corals which have not been dissolved away, the best known being that at Fairford, which had yielded some of the best-preserved fossil corals in the country—*Isastræa*, *Thamnastræa*, *Thecosmilia*, *Stylina*, *Montlivaltia*, *Microsolena*, *Cryptocœnia*, and *Bathycœnia*. Most of the specimens were obtained from a ploughed field and special excavations some miles north-west of the town, at Honeycomb Leaze, in the neighbourhood of which some quarries still show traces of the bed. The Fairford Coral Bed is traceable as a definite horizon near the top of the White Limestone into Oxfordshire, having been exposed at Milton under Wychwood and Burford as well as around Kemble. *Nerinea eudesii* Mor. and Lyc. is generally abundant in association with the corals.

Another type of bed characteristic of the White Limestone is a lenticular band of marl crowded with brachiopods. The commonest species is *Epithyris oxonica*, which is confined to the upper part of the division. There are, however, a number of other smaller Terebratulids (*Stiphrothyris* and *Cererithyris?*) which still remain to be studied. The most interesting of these brachiopod marls are exposed in the fine railway-cuttings at Stony Furlong and Aldgrove, near Chedworth. Here Richardson detected and described an *Ornithella* Marl, 10 ft. thick, with abundant *Ornithella digonoides* S. Buck. and *O.* cf. *minor* (Martin), superficially indistinguishable from *O. obovata* and its many varieties from the Cornbrash. At a higher level in another cutting on the same line he found a marl with *Burmirhynchia hopkinsi* (Dav.), and lower down a limestone full of the calcareous alga, *Solenopora jurassica*, which retains its original red coloration and is known in consequence as 'beetroot stone'.[1] The marl with Ornithellids has also been found above the Fairford Coral Bed by Richardson.[2]

Beds crowded with gastropods are another characteristic feature of the White Limestone: *Nerinea eudesii* frequently forms *Nerinea* beds, usually in association with corals; *Ptygmatis* (*Bactroptyxis*) *bacillus* (d'Orb.) abounds below the *Ornithella* Marl at Aldgrove cutting; a '*Nerinea* bed' near Condicote is formed of myriads of *Aphanoptyxis ardleyensis* Ark., while another at Leach Bridge near Aldsworth is composed of *A. bladonensis* Ark.[3]

It was apparently from the lower part of the White Limestone (the equivalent of the Ancliff Fossil Bed) that Morris and Lycett obtained at Minchin-

---

[1] L. Richardson, 1911, *P.G.A.*, vol. xxii, pp. 103, 110.
[2] L. Richardson, 1933, 'Geol. Cirencester', *Mem. Geol. Surv.*
[3] W. J. Arkell, 1931, *Q.J.G.S.*, vol. lxxxvii, pp. 615–22 and pls. LXIX, L.

hampton a large proportion of the beautifully preserved mollusca illustrated in their monograph. One of the principal sources of the specimens was 8–15 ft. of wedge-bedded, shelly white oolite, called the 'Planking', underlying a coral bed. At the base of the Planking the gastropod genus *Purpuroidea* makes its first appearance.

Two fossils which cannot be passed over in describing the White Limestone are Reptilian eggs (of *Teleosaurus*?) found at the Hare Bushes Quarry, north-east of Cirencester, by James Buckman, and now in the British Museum;[1] and the ponderous lamellibranch, *Pachyrisma grande* Mor. and Lyc., which has been found in abundance, but in only two localities, Bussage and Chalford, south-east of Stroud.[2]

The base of the White Limestone, finally, has yielded some of the few ammonites of which the exact horizons are known. From a level at or near the base ½ mile west-north-west of Salperton Church, in the Cotswolds, came a specimen identified by Buckman as *Tulophorites tulotus* Buck. (a species found in the Fuller's Earth Rock at Troll).[3] Similarly from the White Limestone (if not higher) must have come a unique specimen of *Bullatimorphites bullatimorphus* Buck. found at Tiltups Inn, south of Nailsworth,[4] since the quarry there now exposes Forest Marble, Bradford Clay and Kemble Beds, and is unlikely to have been formerly worked very much deeper. In view of these and two Oxfordshire records, it seems probable that it was the Planking at Minchinhampton that in the past yielded most of the fairly numerous ammonites collected there, and not the Shelly Beds or Weatherstones lower down, as sometimes assumed. From Minchinhampton came the types of *Tulites subcontractus* (Mor. and Lyc.), *Morrisiceras morrisi* (Oppel), *Madarites madarus* Buck., *Oxycerites waterhousei* (Mor. and Lyc.) and *Suspensites suspensus* Buck.,[5] but unfortunately Morris and Lycett paid no attention to the exact horizons from which they came and the large quarries have long since been abandoned.

Hampen Marly Beds, 0–30 ft.

In the North Cotswolds the White Limestone rests upon 20–30 ft. of grey marls with soft bands of earthy, marly limestone, characterized by beds crowded with large specimens of *Ostrea sowerbyi* and the typical form of *Rhynchonella concinna* (Sow.). The type-section is the Hampen cutting, near Salperton, on the Cheltenham–Banbury railway, described by Woodward and in greater detail by Richardson.[6] Here 28 ft. of the marls are exposed, overlain by 36 ft. of White Limestone and underlain by the Taynton Stone (30 ft.) (see fig. 49). The boundaries, both above and below, are quite distinct, and the division can be recognized at many places to the north and east as far as eastern Oxfordshire. Towards the south-west, however, the beds seem to die out or pass laterally into limestones, losing their distinctive fossils. Only

[1] J. Buckman, 1860, *Q.J.G.S.*, vol. xvi, pp. 107–10.

[2] H. B. Woodward, 1894, *J.R.B.*, p. 280.

[3] L. Richardson, 1929, 'Geol. Moreton in Marsh', *Mem. Geol. Surv.*, p. 119.

[4] J. Lycett, 1863, *Mon. Moll. Great Oolite*, Supplt., pp. 3–4 and pl. XXXI, fig. 1; also Buckman, *T.A.*, pl. CCLXXII.

[5] See S. S. Buckman, 1921, *T.A.*, vol. iii, pp. 43–52 and pls. DV, CDLXXVI, CCCXLVI, CCLXX, CCLXVIII A, CCLXXI–CCLXXIII, for discussion and figures of Minchinhampton ammonites.

[6] H. B. Woodward, 1894, *J.R.B.*, p. 292; L. Richardson, 1929, 'Geol. Moreton in Marsh', p. 104, *Mem. Geol. Surv*.

6 miles away in this direction, in the complete section afforded by the Stony Furlong cutting south of Chedworth, the White Limestone would appear to have expanded to nearly 80 ft. and there is scarcely anything that can be correlated with the Hampen Marly Beds. There are a few feet of marls and

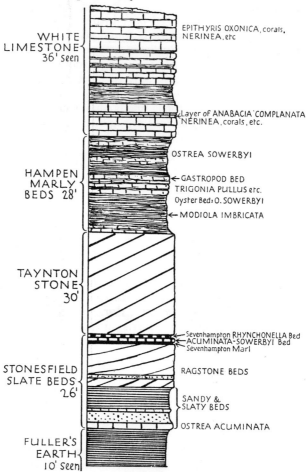

WHITE LIMESTONE 36' seen — EPITHYRIS OXONICA, corals, NERINEA, etc

← Layer of ANABACIA COMPLANATA NERINEA, corals, etc.

OSTREA SOWERBYI

HAMPEN MARLY BEDS 28'
← GASTROPOD BED
TRIGONIA PULLUS etc.
Oyster Bed: O. SOWERBYI
← MODIOLA IMBRICATA

TAYNTON STONE 30'

← Sevenhampton RHYNCHONELLA Bed
← ACUMINATA-SOWERBYI Bed
Sevenhampton Marl

STONESFIELD SLATE BEDS 26' — RAGSTONE BEDS

SANDY & SLATY BEDS

OSTREA ACUMINATA

FULLER'S EARTH 10' seen

FIG. 49. Section of the Great Oolite Series at Hampen railway-cutting, near Notgrove, Glos. (Based on L. Richardson, 1929, 'Geol. Moreton in Marsh', pp. 103–6, *Mem. Geol. Surv.*)

marly limestones towards the base,[1] one yielding *Ostrea sowerbyi*, but no certain correlation is possible.

Taynton Stone, 20–30 ft.

The downward continuation to the Fuller's Earth is obscure in the cuttings on the Chedworth line and the lowest beds of the Great Oolite can no longer be seen, but the Hampen cutting affords a clear section. Beneath the Hampen Marly Beds is 30 ft. of white, false-bedded, shelly oolite, locally fissile and in places coarsely oolitic, and having a very distinctive appearance from the large

[1] Harker's 'Organic Bed'; see Richardson, 1911, *P.G.A.*, vol. xxii, p. 111.

numbers of small and broken shells, very difficult to identify. This stone was extensively worked as a freestone about Windrush, Barrington, Taynton, Burford and Swinbrook at least as early as the middle of the seventeenth century, and of it St. Paul's Cathedral was largely built. Woodward seems to have introduced the name Taynton Stone, but it was known to Plot and Hooke, in the time of Charles II, as the Burford Stone. At Hampen it rests on the Stonesfield Slate Series, but in most parts of Oxfordshire it overlaps that series and comes to rest on the local equivalents of part of the Fuller's Earth.

The principal objective at the great quarries at Minchinhampton was likewise a freestone, which rests in some places upon a thin representative of the Stonesfield Slate Beds and in others directly upon Fuller's Earth. It was called by Morris and Lycett the Weatherstones and Shelly Beds, and its thickness is 16–20 ft.[1] The stone is current-bedded and hard with shelly layers. The shells often constitute a considerable proportion of the whole mass and, being converted into crystalline calcium carbonate, enhance the good weathering qualities of the stone.

That at least some of the ammonites found at Minchinhampton came out of these beds seems to be shown by a label on a specimen of *Morrisiceras morrisi* (Oppel) in the Lycett collection, reading 'Minchinhampton, base of Great Oolite'.[2] They certainly supplied a large proportion of the other mollusca, which seem therefore to be less comminuted here than farther north-east.

Owing to the disappearance of the Hampen Marly Beds to the south-west, the Weatherstones and Shelly Beds at Minchinhampton are directly overlain by the White Limestone ('Planking'). It is tempting to correlate the freestones of Taynton and Minchinhampton with the Bath Freestone. There are still the 10–40 ft. of Lower Rag Beds under the Bath Freestone to account for, and a possible equivalent for these occurs in the North Cotswolds, around Northleach, where freestones and 'ragstones' are developed to a considerable thickness between the Taynton Stone and the true slate-bed of the Stonesfield Slate.

The Stonesfield Slate Beds.

With the Stonesfield Slate Beds we again reach a stratum that can confidently be followed as a constant datum throughout the Cotswold Hills, from south of Bath into Oxfordshire. Its lithological peculiarities and special fauna make it the most valuable guide for elucidating the stratigraphy. It everywhere divides the Great Oolite above from the Fuller's Earth below, and there is always an intimate blending of the base of the sandy slate series with the top of the Fuller's Earth Clay.

The Stonesfield Slate Beds consist typically of thin sands and sandy limestones, often in the form of spheroidal doggers called 'pot-lids' or 'burs', some of which split under the weather into fissile roofing-tiles. The principal fossils are abundant *Ostrea acuminata* mixed with *O. sowerbyi* (small), the ammonite *Gracilisphinctes gracilis* (J. Buckman), and *Trigonia impressa* Sow. and crushed Rhynchonellids.

[1] J. Morris and J. Lycett, 1850, 'Mon. Mollusca Great Oolite', *Pal. Soc.*, pp. 2–3, and J. Lycett, 1857, *The Cotteswold Hills*, pp. 93–4; and H. B. Woodward, 1894, *J.R.B.*, pp. 278–9 (beds 2 and 3).

[2] S. S. Buckman, 1921, *T.A.*, vol. iii, p. 49.

The most southerly known occurrence of sandy strata assignable to the Stonesfield Slate Beds seems to be indicated by a record of William Smith's at Combe Grove Fuller's Earth Pit (now closed), presumably somewhere on the slopes of the Combe Hay valley, south of Bath.[1] Here he described 6 ft. of 'sand and burs' separating the shelly Lower Rag Beds of the Great Oolite from the Upper Fuller's Earth (seen to 23 ft. 9 in.). No fossils were recorded, nor were any seen in sections near Tormarton, opened in 1931 during the laying of a water main. These showed grey, flaggy, sandy limestones, 5 ft. thick, between the white oolites like Taynton Stone and the Fuller's Earth Clay (seen to 25 ft.).[2]

Two miles north of this is the Sodbury Tunnel, where Reynolds and Vaughan record 36 ft. of compact sandy limestones, with thin clays, passing eastward into clays with subordinate hard, sandy limestones, and yielding the zonal index-fossil of the Stonesfield Slate, *Gracilisphinctes gracilis*.[3] Above were the usual coarse-grained oolitic limestones of the Great Oolite; below, the Fuller's Earth (90 ft. thick). The lateral passage from sandy limestones to clays corroborates the earlier writers, all of whom insisted upon the 'intimate blending' of the Stonesfield Slate Series with the Upper Fuller's Earth. The same alternation of bands of sandstones and clay, with abundant *Ostrea acuminata*, was recorded by Witchell in a trial shaft at Stroud.[4] Here the thickness was 15 ft., and of the Fuller's Earth below 55 ft. About Minchinhampton the Stonesfield Slates are sometimes absent and sometimes present, but they thicken in the country north-east of Stroud and become typical in both lithology and palaeontology. They were formerly worked for roofing-slates about Througham near Bisley, where the series is 15 ft. thick. The main slate bed at the base, consisting of 2 ft. of fissile micaceous sandstone, contains *Trigonia impressa*,[5] the most characteristic fossil at Stonesfield and nowhere found at any other horizon. Slates were also worked in this district at Miserden, Nettlecomb near Birdlip, and at Rendcomb, 3 miles south-west of Chedworth. Thus they occur in workable development on each side of and over the Birdlip Axis of Inferior Oolite times.

A few miles farther north-east, in the North Cotswolds, the Stonesfield Slate Beds attain their maximum development and are still worked in several places under the name of Cotswold Slates. The principal pits—open workings —are at Eyford Hill, Summerhill and Kineton Thorns, between Naunton and Condicote. The activity of the industry in the past may be judged by the fact that at Kineton Thorns alone no less than 120,000 slates were made in one season. Formerly workings also extended much farther west, to Salperton Downs and Sevenhampton Common, on the Cleeve Hill outlier above Cheltenham.

The whole of this area has been carefully revised by Richardson, who has established the relations of the various types of rock to one another, to the Taynton Stone above and the Fuller's Earth below. As a type-section the Hampen railway-cutting is again useful (fig. 49). The total thickness of the Stonesfield Slate Beds is here 26 ft., constituted as follows:

[1] Printed in Phillips's *Memoirs of W. Smith*, 1844, p. 60.
[2] L. Richardson and W. J. Arkell, field-notes.
[3] S. H. Reynolds and A. Vaughan, 1902, *Q.J.G.S.*, vol. lviii, pp. 739–41.
[4] E. Witchell, 1882, *Geology of Stroud*, p. 69.
[5] H. B. Woodward, 1894, *J.R.B.*, p. 281.

STONESFIELD SLATE BEDS OF HAMPEN CUTTING AND DISTRICT [1]

[Taynton Stone above, 30 ft.]

	ft.	in.
Sevenhampton *Rhynchonella* Bed: yellowish marl with *Kallirhynchia* spp. (as in the marl below) and *Stiphrothyris* sp. At Sevenhampton this bed also contains corals, *Isastræa* and *Anabacia*, *O. acuminata*, *O. sowerbyi*, &c.	1	0
Acuminata-Sowerbyi Bed: limestone crowded with oysters	1	0
Sevenhampton Marl. At Sevenhampton this comprises 4½–6 ft. of sandy indurated marl, with *O. acuminata* and many Rhynchonellids—*R. obtusa*, *R. decora*, *R. communalis*, *R. deliciosa*, &c. At Hampen it is only .		6
Ragstone Beds: Limestones, mostly fine-grained and compact, sparsely oolitic, locally fissile, locally shelly and coarsely oolitic, with 9 in. band of soft-weathering sandstone towards the bottom	12	3
Slate Beds: bluish paper-shales with fissile sandy layers	6	3
Yellow sand, with 6 in. band of fissile sandy limestone at top .	3	9
Hard, blue-hearted, slightly sandy limestone with claystone inclusions, *O. acuminata* and a few other fossils	1	3
	26	0

[Fuller's Earth below, consisting of bluish paper-shales with layers of hard grey marl and occasional thin seams of fissile sandy stone; seen to 10 ft.]

The Sevenhampton *Rhynchonella* Bed and the *Acuminata-Sowerbyi* Bed[2] are very persistent over the North Cotswolds and form a useful index for distinguishing the Taynton Stone above from the Ragstones of the Stonesfield Slate below; and in the country around Northleach Richardson has shown that these two stones become extremely similar.[3] At Fosse Quarry, Farmington, for instance, a 20 ft.-face of limestones and freestone closely resembling Taynton Stone is exposed, but he shows that the *Acuminata-Sowerbyi* Bed at the top indicates that it is the Ragstone Beds in a freestone facies. These beds may well correspond with the Lower Rag Beds beneath the Bath Freestone.

The fauna of the actual slate beds is very rich and characteristic, leaving no room for doubt that the slates are on the same horizon as those at Stonesfield. On the surfaces of the split slabs of fissile, sandy oolite may be seen abundant *Trigonia impressa* and *Placunopsis socialis*, while from the slatters have been obtained *Gracilisphinctes gracilis*,[4] *Micromphalites oxus* S. Buck.,[4] and remains of pterodactyles (*Rhamphocephalus prestwichi* Seeley), deinosaurs (*Megalosaurus* sp.), crocodiles (*Steneosaurus* and *Teleosaurus*), fish (*Lepidotus*, *Mesodon*, *Hybodus*, *Ischyodus*, *Strophodus*), belemnites (*Belemnopsis bessina* d'Orb. sp. and *B. aripistillum* Lhwyd sp.) and other mollusca, a cirripede, a starfish, an annelid, insects, and plants.[5]

Eastward, before reaching the edge of the Vale of Bourton, the Stonesfield Slate Beds die out, being overlapped by the Taynton Stone (as in Roundhill

[1] Condensed from L. Richardson, 1929, 'Geol. Moreton in Marsh', *Mem. Geol. Surv.*, pp. 102–16 with references to all previous literature.

[2] This has usually been called the *Acuminata* Limestone, but it is important to keep it distinct from the pure *Acuminata* Bed at a lower level farther south.

[3] L. Richardson, 1933, 'Geol. Cirencester', *Mem. Geol. Surv.*

[4] Figured S. S. Buckman, *T.A.*, pl. CXCIII, DCXLIV.

[5] For more complete list, and references, see Richardson, 1929, 'Geol. Moreton in Marsh', p. 114.

railway-cutting) so that Taynton Stone rests directly on Fuller's Earth; or remaining very thin and yielding no characteristic fossils or workable slates (as at Oddington near Stow on the Wold).

The Fuller's Earth (with Upper Estuarine Clay and Chipping Norton Limestone).

The Fuller's Earth of commerce is still mined above the village of Combe Hay, south of Bath, the method being to drive adits into the steep hill-side. Formerly there were pits at Wellow, Midford, South Stoke, Combe Monkton, Lyncombe, Widcombe, and elsewhere. The commercial product is a soft, grey, slightly greasy clay, with a bluish or greenish tinge when fresh, but weathering buff or brown. It contains slight quantities of lime, iron, magnesium, sodium and potassium, but its fulling qualities are due to its peculiar physical properties, one of which is disintegration in water. The commercial Fuller's Earth occurs only in the Upper Fuller's Earth Clay.[1]

The complete succession of the formation between the Mendip Axis and Bath may be pieced together from a number of records. The Lower Rag Beds below the Bath Freestone (resembling Taynton Stone or the Ragstone Beds of the Stonesfield Slate Beds) may be seen well exposed in the hill-side about 12–15 ft. above the entrance of the present working mine. We know from William Smith's record at Combe Grove Pit (already quoted) that, at least locally, up to 6 ft. of sands with 'potlids' assignable to the Stonesfield Slate intervene between the Great Oolite and the Upper Fuller's Earth, and that the workable seam of Earth occurs there 11 ft. below the sands. It is itself 5 ft. thick and rests upon a 2-ft. band of stone, with more marls below. At Midford Woodward recorded $17\frac{1}{2}$ ft. of clays above the workable Earth, which there also is 5 ft. thick.[2]

The Hemington boring tells us that the Upper Fuller's Earth is more than 32 ft., the Fuller's Earth Rock about 15 ft., and the Lower Fuller's Earth 40 ft. thick, while the wells about Bath show a total thickness of 70 ft., only slightly less than at Hemington. At Hemington the band packed with *Ostrea acuminata* was encountered immediately below the Fuller's Earth Rock, just as south of the Mendips; and although no fossils were recorded from the Rock itself in the boring, the exposures a few miles away, at Bonneyleigh Hill and Egford Bridges, show that it is highly fossiliferous, and the fossils are specifically identical with those in the Fuller's Earth Rock farther south, and form the same assemblage. Finally, the Lower Fuller's Earth is known in detail from the railway-cutting at Combe Hay, put on record by Richardson.[3] The cutting showed 35 ft. of Lower Fuller's Earth and the junction with the Inferior Oolite. By comparison with the Hemington boring, the top of the cutting must have reached within a few feet of the base of the Fuller's Earth Rock, a conclusion borne out by the fact that the highest level exposed was crowded with *Ostrea acuminata*. The 35 ft. of clay contained a few thin bands of limestone, but no real development of 'rock'. At the base *Ostrea knorri* was abundant, as at Doulting and in Dorset, and at the junction with the Inferior Oolite an indication of the *Zigzag* Bed was given by a large ammonite of the *zigzag* zone.[4]

[1] H. B. Woodward, 1894, *J.R.B.*, p. 242.   [2] Ibid.
[3] L. Richardson, 1909, *P.G.A.*, vol. xxi, p. 426.
[4] Figured, S. S. Buckman, 1922, *T.A.*, pl. cccii*.

There are some indications that westward the Fuller's Earth diminishes in thickness, just as it does west of the Westbury boring. Lycett estimated the total thickness penetrated in the Box Tunnel at 148 ft., an estimate which Woodward rejected on the ground that the wells at Bath do not prove more than 70 ft.;[1] but in the light of the Westbury boring, where $207\frac{1}{2}$ ft. was proved, it appears quite possible. The discovery of typical Fuller's Earth Rock fossils (*Stiphrothyris, Ornithella cadomensis, Rhynchonelloidea smithii,* and *Ostrea acuminata*) in pockets in the Upper Inferior Oolite on Dundry Hill, and indeed not on the top of the Inferior Oolite but upon the Upper Coral Bed, suggests that the westward attenuation may involve an overlap of the Lower Fuller's Earth by the Fuller's Earth Rock.[2] A similar conclusion is suggested by a statement of Woodward's that he saw Fuller's Earth with '*Waldheimia ornithocephala* [*bathonica*] resting on a bored surface of the Inferior Oolite' near Severcomb Farm.[3] Again at Dyrham, on the edge of the escarpment, where Richardson discovered the Fuller's Earth Rock crowded with brachiopods (*Ornithella bathonica* and *O. pupa*) as in Somerset, it can be seen to be not more than about 25 ft. above the Inferior Oolite.

About 7 miles due east of Dyrham, Hull described the following section in a lane east of Slaughterford:[4]

	ft.
White marls with occasional stony bands	25
White and grey limestone and marlstone (Fuller's Earth Rock) . . .	10
White and blue clays with *Terebratula perovalis* and *T. maxillata* . .	30
	65

Five miles north-east of Dyrham is the Sodbury Tunnel, where Reynolds and Vaughan estimated the thickness of the Fuller's Earth to be 90 ft. From their study of the materials brought up from the shafts and the records of strata, they concluded that 'when traced laterally, the lithological character is very inconstant, for the clays pass, on the one hand into shales, and on the other into beds of hard shelly marl. In the middle of the series, however, there are one or more beds of argillaceous limestone, which mark a fairly constant horizon'. Their generalized section is as follows:[5]

SUMMARY OF THE FULLER'S EARTH AT SODBURY TUNNEL

	ft.
[Stonesfield Slate with *Gracilisphinctes gracilis* above.]	
Clay and shale, full of *Ostrea acuminata*	28
Argillaceous limestone . . . about	10
Clay with *Ostrea acuminata*	27
Compact blue limestone with [*Stiphrothyris*] and *O. acuminata* . . .	5
Shale	24
	94

[Oolitic limestone of Inferior Oolite below.]

---

[1] 1894, *J.R.B.*, p. 243.
[2] L. Richardson, 1907, *Q.J.G.S.*, vol. lxiii, pp. 421–2; and Reynolds and Vaughan, 1902, *Q.J.G.S.*, vol. lviii, p. 740.
[3] H. B. Woodward, 1894, *J.R.B.*, p. 241.
[4] E. Hull, 1858, 'Geol. Parts of Wilts. and Gloucester', *Mem. Geol. Surv.*, p. 12.
[5] S. H. Reynolds and A. Vaughan, 1902, *Q.J.G.S.*, vol. lviii, pp. 739–41.

They also recorded *Rhynchonell[oidea smithii]* as common, with *Ornithella* [*bathonica*] and *Pseudomonotis echinata*, but without stating at what horizon [presumably in the 'hard shelly marl'?].

In the base of the Fuller's Earth at Lansdown near Bath and at Kingscote, south-west of Nailsworth, *Zigzagiceras* cf. *procerum* has been recorded,[1] and Richardson has noted occasional specimens of *Ostrea knorri* as far north as Slad Valley, Stroud, and Cooper's Hill, Gloucester,[1] indicating that in the South Cotswolds the basal part of the formation is probably the same as at Combe Hay.

I have dwelt on the available evidence in this Bath and South Cotswold region with a disproportionate amount of detail, because it is a critical area. The Great Oolite Limestones have already reached complete development while the Fuller's Earth with its subdivisions is still the same as south of the Mendips. The two parts of the formation therefore show themselves to be sequential and not contemporaneous, even though both may lie within the biozone of certain Tulitid ammonites.

Farther north information concerning the Fuller's Earth becomes more scanty and uncertain. The total thickness was proved to be 70 ft. at Stroud and at Sapperton Tunnel, and 50 ft. in the railway-cuttings through Chedworth Woods; but near Bisley, at Througham and Lypiatt, it is said to be as little as 10 ft., and only 30 ft. at Miserden.[2] Palaeontologically, all along the outcrop as far as Chedworth, the most conspicuous feature is a doggery band of hard limestone almost entirely composed of *Ostrea acuminata*. It can be traced along the escarpment about Wotton under Edge, Kingscote, Nailsworth and Stroud, and was thrown up in large quantities on the tip-heaps round the Sapperton Tunnel shafts. Distinct from the *Acuminata* Bed, but not far removed from it in distance, both occupying approximately the middle of the outcrop, can sometimes be recognized a band or bands of argillaceous limestone like Fuller's Earth Rock, with specimens of *Stiphrothyris*, &c. This rock has been identified above Wotton under Edge and in the Sapperton Tunnel tip-heaps, but its stratigraphical relations with the *Acuminata* Bed remain to be proved.[3]

In the North Cotswolds considerable changes set in and much of the Fuller's Earth is missing. The stratigraphy is complicated in the extreme, but thanks to Richardson's investigations in the past twenty years, the problems at last show signs of nearing solution.

The Fuller's Earth alone continues to separate the Stonesfield Slate Beds from the *Clypeus* Grit as far as the Guiting Valley (the valley of the Upper Windrush). Two sections in the western half of the North Cotswolds show it resting directly upon *Clypeus* Grit: one at Lime Hill Wood Quarry, 2 miles north-north-west of Hawling, the other the first railway-cutting west of Notgrove Station.[4] At the base, reposing on the surface of the *Clypeus* Grit, is a tough green, brown, chocolate and black clay of highly distinctive appearance. Richardson has traced this clay at the base of the local Fuller's Earth throughout the North Cotswolds and across the Vale of Moreton into North

[1] L. Richardson, 1910, *Proc. Cots. N.F.C.*, vol. xvii, p. 79.
[2] H. B. Woodward, 1894, *J.R.B.*, pp. 244–5, with full references.
[3] Richardson and Arkell, field-notes.
[4] L. Richardson, 1929, 'Geol. Moreton in Marsh', *Mem. Geol. Surv.*, pp. 54, 75.

Oxfordshire, and he calls it the Upper Estuarine Clay, for it closely resembles the clay at the base of the Upper Estuarine Series in Southern Northampton-shire, with which he has connected it by a chain of quarries. It is a highly important and easily recognized datum.

Beyond the Guiting Valley, east of a line drawn N.–S. from Snowshill, through the Guitings to Notgrove, a new feature makes its appearance between the Upper Estuarine Clay and the *Clypeus* Grit—the Chipping Norton Lime-stone.[1] It spreads over the hills in the North Cotswolds as far north as Snowshill and Bourton on the Hill, attaining a thickness of 20–25 ft., and con-tinues as a surface-feature of considerable importance across Iccomb Hill into Oxfordshire. It was originally confused with and mapped as Great Oolite, but its true position has now been established by Richardson.

The best section illustrating the succession in this eastern part of the North Cotswolds was formerly to be seen in Roundhill railway-cutting, about a mile south of Naunton. The following is a condensation of the section (now much obscured), amplified by a neighbouring quarry at Roundhill Farm:[2]

SECTION AT ROUNDHILL RAILWAY-CUTTING AND FARM

[Taynton Stone above, seen to about 8 ft.]
[Non-sequence: Stonesfield Slate overlapped.]

	ft.	in.
FULLER'S EARTH: clays and marls	20	0
UPPER ESTUARINE CLAY (seen at Roundhill Farm Quarry): heavy clay, chocolate above, green in middle, black at base	3	7
PEBBLE-BED: brown marly layer with waterworn pebbles of hard brown oolite, encrusted by oysters, *Serpulæ* and polyzoa	4	
CHIPPING NORTON LIMESTONE: obliquely laminated oolite, about . .	12	0
with at the base a seam of clay (Roundhill Clay)	1	6

[*Clypeus* Grit below, seen to 5 ft. 9 in.]

Here the Fuller's Earth, including the thin Upper Estuarine Clay at the base, is still 23½ ft. thick; but towards the north-east, before reaching the Vale of Moreton, it is overlapped by the Taynton Stone, so that on the hills around Condicote the Great Oolite limestone (Taynton Stone) rests in places directly on the Chipping Norton Limestone (as at Stonehill Quarry, near Stow on the Wold).[3] Generally, however, a few feet of Fuller's Earth remain, comprising little else but the Upper Estuarine Clay, which in turn rests upon a water-worn surface of the Chipping Norton Limestone, with the pebble-bed at the junction. Many of the quarries in the Chipping Norton Limestone show this pebble-bed, with a few feet of Upper Estuarine Clay above, and in one or two places (especially Bolton's Ground Quarry, ½ mile south of Condicote[4]) there is at the base of the Upper Estuarine Clay, mixed with the pebble-bed, up to 1 ft. 8 in. of pale greenish-grey fossiliferous marl. Richardson correlates this with a freshwater bed containing *Viviparus* in North Oxfordshire, to be described in the next section.

---

[1] So named by Hudleston in 1878.
[2] L. Richardson, 1929, loc. cit., pp. 112, 98.
[3] L. Richardson, 1929, loc. cit., p. 113.
[4] L. Richardson, 1929, loc. cit., p. 99.

Richardson's detailed field-work has therefore shown that in the North Cotswolds there are two discordances within the lower part of the Great Oolite Series. Towards the east the Taynton Stone overlaps or oversteps the whole or the greater part of the Fuller's Earth; towards the west the Upper Estuarine Clay oversteps the Chipping Norton Limestone on to *Clypeus* Grit. The significance of the latter overstep depends on the discovery of *Oppelia fusca* in the upper part of the Chipping Norton Limestone at several places in the North Cotswolds and at Oakham, near Chipping Norton; the limestone therefore belongs to the *fusca* and probably the *zigzag* zones and so is equivalent to the basal portion of the Fuller's Earth in the South Cotswolds and to the Crackment Limestone in the Sherborne District. Richardson has thus proved an overstep of the basal Fuller's Earth by the Upper Estuarine Clay— itself an horizon within the Lower Fuller's Earth. How far west and south this overstep continues is still unknown, for the Upper Estuarine Clay is still unidentified within the main mass of the Fuller's Earth in that direction. Towards the north-east, however, the Upper Estuarine Clay, as the base of the Upper Estuarine Series, remains a highly transgressive horizon throughout the eastern counties.

This section must be concluded with a brief description of the Chipping Norton Limestone.[1]

The typical aspect of the rock is a brownish or white oolite, often black-specked with tiny flecks of ground-up lignite, but otherwise difficult to distinguish from Great Oolite. In this facies it is best seen near Stow on the Wold and at Newpark Quarry, Longborough, where it has yielded vertebræ of *Megalosaurus*, *Teleosaurus subulidens* Phil. and *Steneosaurus*.

Oblique lamination is frequent in the upper part, and local accentuation of this feature produces flaggy stone, which may become sandy and sufficiently fissile to be used as roofing slates or tilestones. The fissile, sandy facies is especially well developed in the north-west, towards Snowshill, where the slates have been extensively worked at Hyatt's Pits and used for roofing many of the buildings in the village. It is also developed south-west and south-east of Condicote, where old mines can be seen at Flagstone, midway between Condicote and Upper Swell. The slates are usually thick and heavy, but locally they are of better quality and may be virtually indistinguishable from the Stonesfield Slates quarried in the same area.

The sandy element is especially marked in the west, near Temple Guiting, where, probably as a product of decalcification, layers of true sand are intercalated in the upper part of the limestone. Over a considerable area of the outcrop the soil resulting from the decomposition of the beds is so sandy as to be quite unlike the soil to which a limestone usually gives rise. Mr. J. G. A. Skerl has analysed samples of the sand and found it to contain upwards of 70 per cent. of calcium carbonate, with smaller proportions of kyanite and sphene than are found in the sands at lower horizons.

Some peculiar features in the Chipping Norton Limestone, especially marked at Guitinghill Quarries near Condicote, are lines of intensely hard calcareous lumps or 'Node beds' near the top, and a pebble bed about the centre of the series. The pebble-bed, 1 ft. thick, is a marl, crowded with *Ostrea sowerbyi* and some other fossils, especially the corals *Isastræa limitata*

[1] See L. Richardson, 1929, loc. cit., pp. 86–98.

and *Cyathophora pratti*, *Trigonia pullus*, *Lima cardiiformis*, and gastropods. Few species besides these occur in the Chipping Norton Limestone of the Cotswolds, but, taken in conjunction with the vertebrate remains, they indicate much closer affinity with the Great Oolite proper than with the Inferior Oolite.

At the base of the series, separating it from the *Clypeus* Grit, there is locally a thin band of clay named by Richardson the Roundhill Clay. Its greatest thickness is 6–20 in.

IV. OXFORDSHIRE

The behaviour of the Upper Inferior Oolite, with which the Great Oolite Series is usually conformable, has already prepared us for what is to be expected in the country east of the Cotswolds. So far as known, there is no thinning of the Great Oolite compensated by further thickening eastward over the Oxon. border and therefore directly attributable to the Vale of Moreton Axis. Instead there is a general, though irregular, easterly attenuation throughout Oxfordshire and into Northamptonshire. In order to trace the several stages of this attenuation, which is a complicated one, affecting nearly all the members of the series, it is necessary to describe as a separate province the county of Oxfordshire. It is only by following out every stratigraphical subdivision from the Cotswolds through Oxfordshire, noting the lateral changes and the disappearances of certain strata, that we can hope to arrive at a true correlation of the greatly modified and attenuated Great Oolite Series of the Northamptonshire to Lincolnshire area.

A more convenient province for our purpose than that enclosed by the county boundary could scarcely be defined. It is divided into three parts by the valleys of the Evenlode and the Cherwell (see map, fig. 38, p. 207). The most westerly part is a triangular area south of the Evenlode, the base formed by the county boundary from Eastleach to the Iccomb hill-mass, the apex at Handborough. In the centre is the main bulk of the outcrop, the dissected plateau stretching from Woodstock past Chipping Norton and Hook Norton to Banbury and Epwell, in the extreme north of the county; the original resemblance of this plateau to that of the North Cotswolds has already been noticed in Chapter VIII. Finally, east of the Cherwell Valley lies another triangular area like the first, the apex at Enslow Bridge near Kirtlington, the base defined by the valley of the Ouse, from Buckingham to Brackley.

Sections are plentiful in all three areas. In the south the quarries expose only the upper beds, the 'Forest Marble' and the upper part of the White Limestone. These strata may be studied in great detail, in vast excavations at the Oxford Portland Cement Works at Shipton-on-Cherwell and Kirtlington, and in numerous quarries at Bladon, Handborough, Witney, Asthall and over all the area south of Burford. Lower beds, especially the Taynton Stone and Chipping Norton Limestone, formerly much in demand for building, are exposed in the north of the county. The Stonesfield Slate cannot be seen anywhere *in situ* and all the old workings are abandoned.

Within the area thus defined, the Great Oolite Series comprises the following subdivisions:

[Cornbrash above.]

	Stratal Divisions.		ft. thickness.
FOREST MARBLE	WYCHWOOD BEDS		0–12
	BRADFORD BEDS		0–7
	KEMBLE BEDS		10–20
GREAT OOLITE LIMESTONES	FIMBRIATA-WALTONI BEDS (Green clays and fossiliferous greenish marls)		3–14
	THE WHITE LIMESTONE		20–30
	HAMPEN MARLY BEDS		17–25
sandy slate beds	TAYNTON STONE		10–25
	STONESFIELD SLATE BEDS		0–6
UPPER ESTUARINE CLAY			0–3
SHARP'S HILL [NEÆRAN] BEDS			0–10
CHIPPING NORTON LIMESTONE	SWERFORD BEDS and WHITE SANDS		10–25
	HOOK NORTON BEDS		

[*Clypeus* Grit overlapped below.]

SUMMARY OF THE GREAT OOLITE SERIES OF OXFORDSHIRE[1]

Wychwood Beds, 0–12 ft.   ⎫   ⎧FOREST MARBLE AND BRADFORD CLAY
Bradford Beds, 0–7 ft.   ⎬ = ⎨FARTHER S. AND W.

After an interval of about 16 miles from the Cirencester district, the fossil-bed of the Bradford Clay reappears in the south-west corner of Oxfordshire, west of Carterton cross-roads, near Shilton. It consists of 1 ft. of argillaceous limestone, hard in the centre, but marly above and below, from which weather numerous examples of *Ornithella digona* and *Rhynchonella obsoleta* Dav., together with other Rhynchonellids and *Dictyothyris coarctata* (Park.). The fossil-bed lies in the centre of 6 ft. of buff to grey clay, with limestones of Forest Marble facies above and below. Thence north-eastward it has been described at three other places, Witney, Bladon and Islip, extending over a total distance of 17 miles.

In the area of Wychwood Forest, where the Forest Marble acquired its name, there are now only a few shallow sections not overgrown, and the stratigraphical position of the rock quarried in William Smith's time as 'marble', for fashioning into polished mantelpieces and door-jambs, is in some doubt. But by comparison with the quarries at Witney, on the southern out-skirts of the forest, and with other quarries in the county, there seems little doubt that the actual 'marble' was obtained, not from the thin-bedded shaly limestones above the Bradford Clay, universally known as Forest Marble farther south, but from the Forest Marble facies of the Kemble Beds, below the horizon of the Bradford Clay. According to Plot the marble industry had

---

[1] Forest Marble to Taynton Stone based on W. J. Arkell, 1931, 'The Upper Great Oolite, &c. of South Oxfordshire', *Q.J.G.S.*, vol. lxxxvii, pp. 563–629.

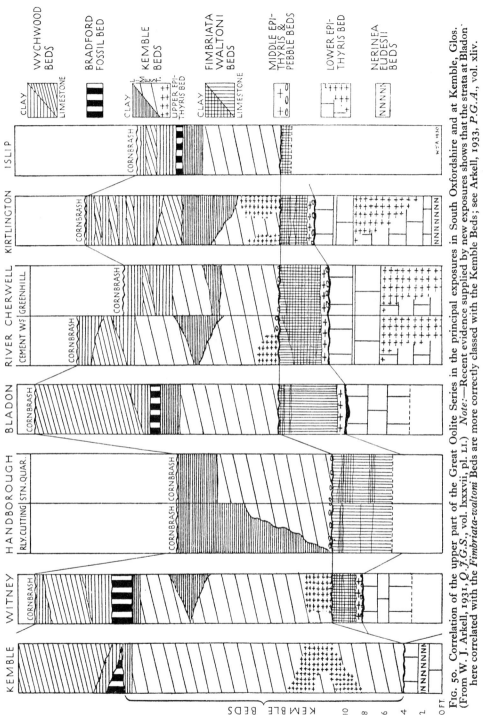

FIG. 50. Correlation of the upper part of the Great Oolite Series in the principal exposures in South Oxfordshire and at Kemble, Glos. (From W. J. Arkell, 1931, *Q.J.G.S.*, vol. lxxxvii, pl. LI.) *Note*:—Recent evidence supplied by new exposures shows that the strata at Bladon here correlated with the *Fimbriata-waltoni* Beds are more correctly classed with the Kemble Beds; see Arkell, 1933, *P.G.A.*, vol. xliv.

already begun nearly a century and a half earlier in the parish of Bletchington (presumably at the ancient Greenhill Quarries, near Enslow Bridge), where 'a sort of Grey Marble' was made into chimney-pieces and pavements for Bletchington Park and Cornbury Park; it was also used, he tells us, in the pillars of the portico of St. John's College, as well as in the making of tombstones, tables, and millstones.[1] Here the only available rock is certainly a part of the Kemble Beds, as may be seen in the neighbouring quarry at Islip, where identical stone is succeeded by the Bradford Fossil Bed, which in turn lies only 4 ft. below the Cornbrash.

Thus in and about Wychwood Forest the *stratigraphical* term Forest Marble, as used by William Smith and his successors for over 130 years, embraces both the strata between the Cornbrash and the Bradford Clay and the Kemble Beds; moreover, the rock to which the *petrological* term was formerly applied is part of the Kemble Beds. For this reason the name should be abolished and the strata between the Bradford Fossil Bed and the Cornbrash in future be called the Wychwood Beds (from Buckman's Wychwoodian, which was defined as post-Bradfordian). Detailed stratigraphical research, however, has not yet been carried far enough in Wilts., Somerset and Dorset to enable the name to be applied along the whole outcrop, and there the old term 'Forest Marble' has still to be retained.

The Wychwood Beds in Oxfordshire are simply a continuation of those in Gloucestershire, comprising the same thin-bedded, flaggy, broken-shell limestones, with lignite, clay-galls, ripple-marks and worm-tracks, interspersed with greenish-grey shaly clay. The fauna consists almost entirely of lamellibranchs, especially *Camptonectes laminatus* and *C. rigidus*, *Ostrea sowerbyi*, &c., and locally *Corbula islipensis* Lyc., *Gervillia islipensis* Lyc., *Cuspidaria ibbetsoni* (Mor. & Lyc.), tests of *Acrosalenia hemicidaroides*, &c. The great reduction in the thickness of the beds as compared with Gloucestershire points to a north-eastward overstep by the Lower Cornbrash. Still farther north-east, in Buckinghamshire, Northants. and beyond, this overstep is continued, the Wychwood Beds and Bradford Beds being cut out altogether. Even in Oxfordshire there were several local uplifts and the Cornbrash sometimes rests directly on Kemble Beds. These places were mapped by E. Hull and described in a Survey Memoir in 1859.[2] One is at Handborough, another at Greenhill Quarries near Enslow Bridge, in the Cherwell Valley, while between them, at Bladon and Campsfield, the Wychwood Beds are 12 ft. thick.

Kemble Beds, 10–20 ft.

The Kemble Beds enter the county from Gloucestershire in their Forest Marble facies and cover the greater part of the triangle south of the Windrush to a thickness of about 20 ft., under Holwell and Alvescot Downs. There are some deep exposures in hard oolitic flags, and near Holwell a thick intercalated band of clay was formerly worked for brick-making. The best sections are in Crawley Road and Ducklington Lane Quarries, Witney, where the complete sequence may be seen. The Kemble Beds are 19 ft. thick and are overlain by the Bradford Fossil Bed, with *Ornithella digona* and the usual fossils, succeeded by 8 ft. of Wychwood Beds and then the Cornbrash (fig. 50). The bulk of the

[1] R. Plot, 1676, *Natural History of Oxfordshire*, p. 79.

[2] E. Hull, 1859, 'Geol. Woodstock', *Mem. Geol. Surv.*, p. 24 and accompanying map; and Arkell, 1931, loc. cit., pp. 578, 586.

PLATE XIII

Photo.   *W. J. A.*

Great Oolite Series at Kirtlington Cement Works, near Oxford.

CB = Lower Cornbrash; WB = Wychwood Beds; KB = Kemble Beds;
FW = *Fimbriata-waltoni* Beds (green clay); WL = The White Limestone.

Photo.   *W. J. A.*

Chipping Norton Limestone-equivalents, Fritwell railway-cutting, Oxon.

North side of cutting at east end of Fritwell Tunnel, showing black and white
sands of the Swerford Beds, resting on a deeply channelled surface of the
Hook Norton Beds. A channel is shown by the white lines.

Kemble Beds consists of massive, false-bedded, coarsely-oolitic, blue-hearted limestones, but near the top there is a 4 ft. lenticle of dark blue clay and at the base a 2 ft. 4 in. block of highly fossiliferous 'Cream Cheese' or sublithographic limestone, with a splintery fracture. The Cream Cheese Bed is the highest bed in Oxfordshire containing *Epithyris oxonica* in abundance, and it has therefore been called the Upper *Epithyris* Bed. With the brachiopods are masses of the branching corals *Convexastræa waltoni* and *Thamnastræa lyelli*, together with *Stiphrothyris capillata*, *Lima cardiiformis* and other fossils. An identical rock is found also at the base of the Kemble Beds farther east, at Shipton-on-Cherwell Cement Works near Enslow Bridge, at Kirtlington Cement Works, and at Blackthorn Hill railway-cutting near Bicester.

At both Shipton-on-Cherwell and Kirtlington the Cream Cheese Bed (there 4 ft. thick), crowded with fossils, can be seen to pass laterally along the quarry face into the unfossiliferous, false-bedded, oolitic facies, and then at Shipton, by the intercalation of seams of clay and sandy oolite, into typical 'Forest Marble'. At Greenhill Quarries, opposite the Shipton-on-Cherwell Cement Works, the Forest Marble facies is indistinguishable from Wychwood Beds, having all the clay-galls, ripple marks, sandy texture and the common facies-lamellibranchs.

At Kirtlington (Plate XIII and fig. 51) the Cream Cheese Bed is overlain by 6½ ft. of grey-blue clay with three pale mudstone layers, into which the upper part of the unfossiliferous Kemble Beds is transformed by lateral passage. The only common fossil in the clay is *Placunopsis socialis*, which abounds. The same clay, with abundance of the little *Placunopsis*, overlies the limestones of the Kemble Beds at numerous places, such as Blackthorn Hill, Handborough, &c., while at others, such as Witney and Shipton-on-Cherwell, it occurs as a lenticle enclosed in the limestones. At Handborough, in the railway-cutting, the whole of the limestones are replaced by clay to a thickness of 14 ft., and the passage may be traced out in a continuous section.[1]

Pebbles and other signs of erosion have been seen at the base of the Kemble Beds at Handborough and Shipton-on-Cherwell.

Fimbriata-Waltoni Beds, 3–9 ft. and in N. Oxon. over 13 ft.

With great regularity all through the county, along at least 30 miles of outcrop, the Kemble Beds are underlain by green clays with argillaceous fossil-beds and lignite, which have at the base a thin bed of drifted, loose or crushed valves of *Epithyris oxonica* and *Ostrea sowerbyi* (the Middle *Epithyris* Bed) resting on an eroded surface of the underlying limestones, often with rolled and bored pebbles. These strata have been called after the two commonest fossils, *Astarte fimbriata* Lyc. and *Gervillia waltoni* Lyc., which often occur in myriads, together with abundant *Protocardia subtrigona* (Mor. and Lyc.), *Corbula hulliana* Mor., *Aphanoptyxis bladonensis* Ark. and other species. The fossils readily weather out of their argillaceous matrix and may be collected in a more perfect state of preservation than those from any other part of the Great Oolite Series. The seam of drifted Epithyrids and pebbles is also the place of sepulture of the huge bones of the land deinosaur, *Ceteosaurus oxoniensis* Phillips, which were found at Enslow Bridge, Bletchington Railway Station and Kirtlington Cement Works. Unfortunately the modern method of

[1] W. J. Arkell, 1931, loc. cit., p. 587.

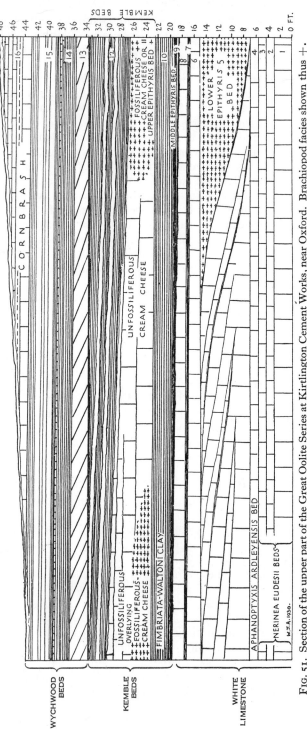

FIG. 51. Section of the upper part of the Great Oolite Series at Kirtlington Cement Works, near Oxford. Brachiopod facies shown thus +. (Horizontal scale much compressed.) (From W. J. Arkell, *1931, Q.J.G.S.*, vol. lxxxvii, p. 571.)

working, by blasting and steam excavators, precludes the recovery of any more bones, and doubtless many skeletons have already been turned into cement.

As we have seen, the fauna of the *Fimbriata-waltoni* Beds is found in Gloucestershire in the top of the White Limestone, and it seems that there is a lateral change of facies eastward into the green clay. Transitionary stages may be seen at various places. At Asthall and at Bladon the characteristic gastropods (*A. bladonensis*) pass up into the Kemble Beds; and at Asthall they occupy nearly 5 ft. of hard sublithographic rock.

In North Oxfordshire again, the beds containing the characteristic assemblage and immediately underlying the Kemble Beds are developed predominantly as limestones. At Sibford Mill, near Sibford Ferris, there are two beds crowded with *A. bladonensis* and the lower can be seen to pass laterally into green clay; while in the eastern limekiln quarry at Rollright Railway Station there are three beds containing the characteristic gastropod, spread through a thickness of over 13 ft., with abundant *Gervillia waltoni* and *Astarte fimbriata* in some of the beds.[1] In the neighbouring Pest House railway-cutting east of Rollright Railway Station the main *Fimbriata-waltoni* Bed can still be seen, crammed with these two fossils, and also *Aphanoptyxis bladonensis*, *Amberleya nodosa*, *Corbula hulliana* and *Protocardia subtrigona*, though nearly everything else is now overgrown.[1] When the cutting was first made, Hudleston described this bed as 'a complete museum', but owing to a misidentification of the *Astarte* he called it the *angulata* bed.[2]

White Limestone, 20–30 ft.

Still the most important and the most persistent of all the subdivisions of the Great Oolite Series, and from Gloucestershire northwards the Great Oolite *par excellence*, is the White Limestone. The upper beds are exposed in innumerable quarries all over the county, but the lower layers are rarely seen.

As in Gloucestershire, the White Limestone often contains an abundance of corals, brachiopods, lamellibranchs, and gastropods. In Oxfordshire there is frequently near the top a band largely composed of *Epithyris oxonica* (the Lower *Epithyris* Bed), which at Kirtlington and about the Cherwell Valley fills the rock to a thickness of up to 8 ft. The brachiopods clearly had colonial habits, however, and the rock in which they lie is wedge-bedded (fig. 51, p. 290), so that in some exposures there may be no sign of a brachiopod, while in others specimens can be collected by the sackful. The gastropods occur more persistently, and three species have local epiboles always in constant sequence and of considerable help in making correlations. Below the *Aphanoptyxis bladonensis* epibole, which corresponds with the *Fimbriata-waltoni* beds, is nearly always a conspicuous band of a more slender and elongate gastropod, *A. ardleyensis*, the position of which is immediately below the Lower *Epithyris* Bed where that is present. Lower still is an equally conspicuous epibole of *Nerinea eudesii*, but this last has such a long local range (its teilzone is so large) that the species cannot be relied upon to have achieved its acme at the same time even in the confines of Oxfordshire. A short distance below this a fourth species, *Ptygmatis bacillus* d'Orb., abounds in certain places, recalling the *Ptygmatis* Bed of Stony Furlong and Aldgrove cuttings. Near Burford, over 5 ft. of

---

[1] Field-notes.   [2] W. H. Hudleston, 1878, *P.G.A.*, vol. v, p. 386.

rock are crowded with *A. ardleyensis*, and it seems highly probable that the bed is on the same horizon as that containing the species so abundantly at Condicote, in the Cotswolds, 10 miles to the north-west. Hence it has been traced in numerous quarries for 18 miles north-eastward to Ardley cutting, where it forms the so-called *Nerinea* Rock.

The lower part of the White Limestone is exposed only in four places: the top of a quarry in the Taynton Stone at Milton under Wychwood, a quarry in the side of the Evenlode Valley at Whitehill Wood, a railway-cutting near Hook Norton, and the Ardley–Fritwell railway-cutting. The only complete exposure is the last, where the total thickness of the White Limestone is 32 ft.

The lower half of the division is especially characterized by *Clypeus mülleri*, which rarely occurs at higher levels, and it has also yielded, very occasionally, ammonites identifiable as *Tulites* cf. *subcontractus*. Specimens have been found at Asthall, at a level of about 16–18 ft. from the top[1] (i.e. slightly below the middle) and in the Fritwell cutting, in the basal bed.[2]

The lower portion of the White Limestone may be very poorly fossiliferous, as at Whitehill Wood, where a total thickness of 25 ft. yields hardly any fossils of note (neither gastropods nor brachiopods) or it may contain extremely rich bands of brachiopods and echinoderms. At Ardley Railway Station Quarry there is near the base a thick marly bed crowded with *Stiphrothyris* sp. and some large Rhynchonellids, *Clypeus mülleri*, &c., but the bed dies out before reaching the Fritwell cutting. At Milton under Wychwood the lowest levels contain abundance of a small *Epithyris*. It is remarkable, however, that neither the *Ornithella* nor the particular Stiphrothyrid so abundant in the *Ornithella* Marls of Gloucestershire has been found in Oxfordshire.

Hampen Marly Beds, 17–25 ft.

The Hampen Marly Beds at Whitehill Wood railway-cutting, in the Evenlode Valley, attain a thickness of rather more than 20 ft., consisting, as in Gloucestershire, chiefly of grey and buff marls and marly clays, with some bands of marly or false-bedded limestone. There is a gradual passage downward into the Taynton Stone. No less than 6 ft. of the beds, both at the top and near the bottom, are crowded with large Rhynchonellids of the *R. concinna* group and large specimens of *Ostrea sowerbyi*, which thickly strew the slopes of the cutting.[3]

The whole of the Hampen Marly Beds, the thickness amounting to 17–18 ft., is well exposed at Milton under Wychwood, beneath 22 ft. of White Limestone and above 10 ft. of Taynton Stone. This fine section was described in detail by Richardson.[4] As at Whitehill Wood cutting, there are two beds in the series crowded with the large form of *Ostrea sowerbyi*, like those in the Kemble Beds, and *Rhynchonella concinna*. About 4 miles south of Milton the full thickness of the beds is again seen at Swinbrook, overlying the Taynton Stone, with a skimming of White Limestone in the soil above. The thickness is again about 17 ft., but *R. concinna* and oysters, though present, are rarer.

[1] W. J. Arkell, 1931, loc. cit., pp. 607–8.
[2] M. Odling, 1913, *Q.J.G.S.*, vol. lxix, p. 490. See *Madarites glabretus* S. Buck., *T.A.*, vol. iii, p. 52.
[3] W. J. Arkell, 1931, loc. cit., p. 612.
[4] L. Richardson, 1910, *Geol. Mag.* [4], vol. vii, pp. 537–42. The Hampen Marly Beds are his Beds 13 to 17 inclusive.

The recognition of the Hampen Marly Beds, with all their characteristic features, as far east as the Evenlode Valley within two miles of Stonesfield, has thrown new light on the record of the section ascertained by the British Association at Stocky Bank, Stonesfield, in 1894–6,[1] to determine the stratigraphical position of the Stonesfield Slates. The scarped bank, part of which is still open to view, showed immediately below the White Limestone 17 ft. of 'marls and limestones with oysters and *Rhynchonella concinna*' (Walford's Beds 2–9) which now fall into line as the Hampen Marly Beds of the neighbouring Whitehill Wood cutting.

There are no further exposures for 11 miles in the direction of strike until the Ardley–Fritwell railway-cuttings. Here the White Limestone is again underlain by a series of marls and predominantly argillaceous beds, with abundant *Rhynchonella concinna* and allied forms, and thousands of large *Ostrea sowerbyi*. Underneath are false-bedded oolites like the Taynton Stone. The marly series, which is here about 27 ft. thick, has been misidentified as 'Neæran Beds' by Odling[2] and by Thompson,[3] while the false-bedded limestones beneath have been mistaken by the same writers for Chipping Norton Limestone. By reason of both the fossils and the lithology, however, the marls are to be correlated with the Hampen Marly Beds and the limestone with the Taynton Stone. A conspicuous feature at the bottom of the Hampen Marly Beds, peculiar to these cuttings, is an argillaceous limestone full of black, shiny, pseudo-oolitic grains, called the Bird's Nest Rock. It yields abundant *R. concinna* and is the source of a specimen of *Zigzagites imitator* S. Buck.[4]

About 4 miles north-west of these cuttings the beds crowded with *R. concinna* and *O. sowerbyi* were formerly worked in the Allotments Quarry, Aynho, and this was the source of the type specimen of *R. concinna* (Sowerby).[5]

Taynton Stone, 10–25 ft.

The great freestone quarries at Taynton, north-west of Burford, worked since at least the sixteenth century for the 'Burford Stone' mentioned by Robert Hooke and Plot, are now sadly overgrown, but fresh sections may still be seen at Milton under Wychwood and Swinbrook. The freestone at these places is the same as at Barrington, over the Gloucestershire border, and the Hampen Marly Beds are well exposed above. The downward sequence, however, is not visible until the Evenlode Valley near Stonesfield.

In the Whitehill Wood railway-cutting and the adjoining Ashford Mill railway-cutting, the Taynton Stone is seen to have a thickness of about 20 ft.[6] It consists as usual of white, false-bedded, more or less fissile, broken-shell oolite, with an impersistent marly zone near the top, by which it merges up into the Hampen Marly Beds. *Ostrea sowerbyi* is abundant, though usually broken, but other fossils are for the most part small, fragmentary and indeterminable.

[1] E. A. Walford, 1894–6, *Repts. Brit. Assoc.*, Oxford, Ipswich and Liverpool; and 1917, *Lower Oolite of North Oxford*, p. 17.

[2] M. Odling, 1913, *Q.J.G.S.*, vol. lxix, pp. 484–511.

[3] B. Thompson, 1931, *Q.J.G.S.*, vol. lxxxvi, pp. 436–8.

[4] S. S. Buckman, *T.A.*, 1922, pl. ccci.

[5] E. A. Walford, 1917, *Lower Oolite of North Oxford*, p. 10. Specimens may still be collected in the soil of the allotments. Buckman wrongly assigned Sowerby's species to the Cornbrash,— see Douglas and Arkell, 1932, *Q.J.G.S.*, vol. lxxxviii, p. 152.

[6] W. J. Arkell, 1931, loc. cit., p. 614.

About 9 miles to the north-north-east there is another complete exposure of the Taynton Stone in the railway-cutting at Langton Bridge, north of Chipping Norton, with the Hampen Marly Beds above. Here the thickness is 17½ ft., and the upper part (about 6 ft.) is very fissile and sandy. The topmost bed is a fissile calcareous and slightly micaceous sandstone, suggestive of the Stonesfield Slate: in fact H. B. Woodward grouped the Taynton Stone here as 'Stonesfield Slate Series'. But he remarked 'it is noteworthy that we have current-bedded shelly oolites below the slaty beds instead of above them as in other localities near Burford and Notgrove. It is not unlikely therefore, as suggested by Mr. Walford, that the slaty beds here are developed at a somewhat higher horizon in the series than elsewhere.'[1] A closer examination confirms the suggestion that this sandstone is not on the same horizon as the true Stonesfield Slate, for it is not only far too irregular to be used for roofing, but the most careful search fails to reveal any of the fossils so characteristic at Stonesfield and in the Cotswolds. By stratigraphical position as well as by the negative fossil-evidence, it is a sandy development of the upper part of the Taynton Stone.

A similar sandy and fissile facies of the topmost beds is seen at Castle Barn Quarry, between Churchill and Chadlington, which otherwise shows a section closely comparable with that at Ashford Mill cutting.[2]

In the extreme north and north-east of the county the Taynton Stone is nowhere exposed, and it is possible that locally it is overlapped by the Hampen Marly Beds. On the east side of the Cherwell Valley, however, it is present again in the Fritwell cutting, where it is about 15 ft. thick and highly false-bedded and fissile.

Stonesfield Slate Beds.

We have seen that in the North Cotswolds the Stonesfield Slate Beds thin out eastward before reaching the Vale of Bourton and are entirely absent from the Roundhill railway-cutting and Stow on the Wold, where Taynton Stone overlaps them and rests on Fuller's Earth.

East of the Vale of Moreton the Stonesfield Slate Beds continue to be absent all over North Oxfordshire and even down the Evenlode Valley to Charlbury and Ashford Mill cutting, within a mile of Stonesfield. The first good section, Castle Barn Quarry between Churchill and Chadlington, on the edge of the Vale of Moreton, shows almost the same conditions as Roundhill cutting, except that the Taynton Stone has overlapped not only the Stonesfield Slate Beds but also the Fuller's Earth. At one end of the quarry it rests on the Upper Estuarine Clay and at the other end upon the Chipping Norton Limestone.[3] An almost duplicate section is to be seen in the Ashford Mill cutting, about 6 miles to the south-east. Here, however, there is at the base of the Taynton Stone a bed crowded with large specimens of *Ostrea sowerbyi*, *Rhynchonella deliciosa* Buck., *R. decora* Buck., and corals. The *Rhynchonellæ* and the stratigraphical position indicate that this is the Sevenhampton *Rhynchonella* Bed of the Cotswolds, and the top of the clays beneath contains the true *Ostrea acuminata*. It therefore appears that the topmost layers of the

[1] H. B. Woodward, 1894, *J.R.B.*, p. 332.
[2] More will be said of this section below (see fig. 52, p. 299).
[3] L. Richardson, 1911, *Proc. Cots. N.F.C.*, vol. xvii, p. 223, and field-notes.

Stonesfield Slate Beds take part in the overlap, for they cut out the ragstone beds and the slate beds before being themselves overlapped by the Taynton Stone.

In all the exposures in North Oxfordshire the Taynton Stone and Upper Estuarine Clay are separated by these thin and highly fossiliferous representatives of the topmost part of the Stonesfield Slate Beds. Both at Langton Bridge and at Sharp's Hill the Upper Estuarine Clay is immediately succeeded by a thin clay with *Ostrea acuminata* and *O. sowerbyi* (the *Acuminata-sowerbyi* Bed), overlain by the Sevenhampton *Rhynchonella* Bed, crowded with fossils. At Langton Bridge especially, the same small *Stiphrothyris* that occurs at Sevenhampton is abundant; and here a poorly-fossiliferous band of limestone intervenes between the two fossil-beds. Eastward this band thickens to $1\frac{1}{2}$–$2\frac{1}{2}$ ft. and overlaps the *Acuminata-sowerbyi* Bed, for at Tadmarton and near Deddington it rests directly upon the White Sands equivalent to the Chipping Norton Limestone.[1]

In the Fritwell cutting, also, there is no sign of the *Acuminata-sowerbyi* Bed or of any part of the Stonesfield Slate Beds. Gritty and marly limestones, containing only the large *Ostrea sowerbyi* of higher levels and fragmentary Rhynchonellids, intervene between the Taynton Stone and the representative of the Upper Estuarine Clay.

Thus the development at Stonesfield of true sandy slate beds like those of the Cotswolds is a peculiar feature for Oxfordshire. They are absolutely restricted to a small area immediately under the village of Stonesfield and extending perhaps for 2 or 3 miles towards the north and north-east, where they seem to have been preserved in a shallow basin-shaped depression and to have been overlapped on all sides. This is less surprising when the extreme thinness of the actual slate bed is realized. The overwhelming palaeontological interest of the stratum sometimes gives rise to a false impression of its stratigraphical importance, but it is not more than 1–$1\frac{1}{2}$ ft. thick, as in the Cotswolds. Sometimes there may be two beds, but they are no thicker and only a few inches apart. The greatest thickness of slate and associated sandy beds is 6 ft., recorded by Fitton, and in the shaft descended by Woodward there was only 4 ft. The slate is identical with that in the Cotswolds, consisting of dark grey, very fissile, micaceous, calcareous sandstone, occurring in concretionary masses known as 'pot-lids'. Ooliths are curiously distributed in streaks and patches, and there is a conspicuous band of well-rounded pebbles of hard oolite, as in the Cotswolds. According to Fitton the position of the pebble seam is in the best slate bed, the upper of two layers of pot-lids, lying in a sand.[2]

The use of the Stonesfield Slate dates back to an antiquity even more venerable than that of the Forest Marble.

Plot tells us that already in the middle of the seventeenth century 'the Houses are covered, for the most part in Oxfordshire, not with Tiles but Flatstone, whereof the lightest, and that which imbibes the water least, is accounted the best. And such is that which they have at Stunsfield, where it is dug first in thick Cakes, about Michaelmas time, or before, to lie all the winter and receive the Frosts, which make it cleave in the Spring following into thinner Plates.'[3]

---

[1] L. Richardson, 1922, ibid., vol. xxi, p. 130, and field notes.
[2] W. F. Fitton, 1828, *Zool. Journ.*, vol. iii, p. 412; and transcribed 1894, *J.R.B.*, p. 312.
[3] R. Plot, 1676, *Nat. Hist. Oxfordshire*, p. 78.

All the slate accessible at the surface was long ago worked out and for probably more than a century there has been no visible surface outcrop or exposure. Originally adit galleries 6 ft. high were driven into the hill-sides,[1] but for many years latterly the material was reached only by vertical shafts, down which the slatters descended with ropes to depths of from 20 to 70 ft. below the surface. H. B. Woodward found that it was almost impossible for a geologist to study the succession as he descended these shafts, for with one hand he had to hold on to the rope and with the other he held a candle. Now the last mine has been abandoned and only the vast piles of debris remain to mark the old workings.

The fauna is as peculiar as the lithic characters, the abundant little *Trigonia impressa* being found at no other horizon in the Jurassic System. Some of the surfaces of the slabs on the tip-heaps at Stonesfield show hundreds of specimens, the valves open but still united by the ligament.

The fossils that have made the Stonesfield Slate famous the world over are the jaws and bones of small ancestral mammalia, occasionally found when the pits were in work. According to the latest reviser, Dr. G. G. Simpson, there are known altogether four species belonging to four genera, classed as follows:[2]

MULTITUBERCULATA	*Stereognathus ooliticus* Charlesworth.
TRICONODONTA	⌠*Amphilestes broderipi* Owen.
	⌡*Phascolotherium bucklandi* (Broderip).
PANTOTHERIA	*Amphitherium prevosti* (V. Meyer).

Dr. Simpson writes that '*Stereognathus* appears to be a last survivor of a group already found in the Rhætic. The others as true mammals appear for the first time. *Amphitherium* is probably the most significant single genus of Mammalia known, for it represents a very early, very generalized stock, which ... provides an ideal structural ancestor for all known post-Paleocene mammals except monotremes.' The known mammalian faunas of the Mesozoic, he remarks, 'stand out like lights in the vast darkness of the Age of Reptiles— and very dim lights most of them are. This mammalian prehistory is two to four times as long as the "historical period" which followed it, and yet the materials for the latter are literally many thousandfold those for the former. This ... makes close scrutiny of the Mesozoic mammalia which are known the more necessary, and the results which are to be obtained from them the more precious.'[3]

The first discovery of these remarkable fossils was made by W. J. Broderip in 1812 or 1814, when he was a young man studying law at Oxford and at the same time attending Buckland's lectures on geology. One day two of the jaws were brought to his rooms in Oriel College by an old stonemason who collected for him, and both he and Buckland were convinced that they were jaws of mammals. But it was an established dictum that no mammals existed before the Tertiary, while these were embedded in unmistakable Stonesfield Slate; and even though in 1818 Cuvier visited Oxford and confirmed their opinion, it was not until 1824 that Buckland ventured on publication. He was followed

---

[1] W. D. Conybeare, 1822, in Conybeare and W. Phillips, *Outlines of Geol. England and Wales*, p. 204.

[2] G. G. Simpson, 1928, *Cat. Mesozoic Mammalia Brit. Mus.*, p. 198.

[3] Ibid., p. 199 and p. 7.

in 1825 by Prévost, who had spent some time in Oxford. Both Prévost and Cuvier referred the bones to an unknown genus allied to the opossums.

Critics immediately arose declaring that the jaws must be reptilian or even piscine, some of the fiercest opposition coming, in spite of Cuvier's verdict, from Paris. De Blainville, the loudest of the critics, mistook the internal groove shown on Buckland's figures for a suture. In 1838 Buckland made a pilgrimage to Paris with two specimens and converted all the French naturalists who saw them—Valenciennes, Dumeril, and Geoffrey St. Hilaire. On his return he handed over the jaws to Owen, who described, discussed and figured all the available material in 1838 and 1842. Owen finally settled the matter by showing the supposed suture to be a groove with an entire bottom, and pointing out numerous other mammalian features.

The land fauna associated with the Stonesfield mammalia comprises two pterosaurs, *Megalosaurus bucklandi* (found also in the White Limestone and the Chipping Norton Limestone), and insects. The plants include ferns, cycads, conifers, and an undoubtedly dicotyledonous leaf, figured by Prof. Seward.[1] Before the Caytoniales were described by Dr. Hamshaw Thomas from the Yorkshire Inferior Oolite (see p. 219) this leaf was the earliest known specimen of an Angiosperm recovered from the stratified rocks.

These remains were all drifted into a shallow sea from neighbouring land, for they are far outnumbered by the marine fauna, both vertebrate and invertebrate. There are remains of *Teleosaurus*, *Steneosaurus*, *Plesiosaurus*, a Chelonian, and about forty species of fish—including sharks, ganoids and a species of *Ceratodus*.

Cephalopods are rare, but include *Gracilisphinctes gracilis* (J. Buck.), *Micromphalites micromphalus* (Phil.),[2] an unidentifiable species of *Clydoniceras*, and *Nautilus baberi* Mor. and Lyc. Long lists of lamellibranchs and gastropods have been published,[3] but the only really abundant and characteristic species are *Trigonia impressa* Sow., *Gervillia ovata* Sow., *Ostrea acuminata* and *O. sowerbyi*, with crushed Rhynchonellids and numerous specimens of *Chlamys* and *Camptonectes* of the common Great Oolite species.

Upper Estuarine Clay, 0–3 ft., and **Sharp's Hill Beds**, 0–10 ft.—'NEÆRAN BEDS'.

As already remarked, Richardson has traced the Upper Estuarine Clay from quarry to quarry through the North Cotswolds and North Oxfordshire into Northamptonshire, where it forms the base of the so-called 'Upper Estuarine Series'. Throughout this district, although it is unfossiliferous, it is unmistakable on account of its peculiar lithological appearance, its green and chocolate colours and heavy, greasy, quality. At the base there is frequently a black carbonaceous layer, and below that numerous pebbles and sometimes phosphatic nodules, which rest upon an uneven surface of the different beds below.

Locally in North Oxfordshire the black carbonaceous layer that usually forms the base of the Upper Estuarine Clay is underlain by a pale seam of marl (usually 0–8 in.), mixed with pebbles but overlying the main pebble-bed. This

---

[1] A. C. Seward, 1904, *Cat. Mes. Plants in Brit. Mus.*, pt. 2, p. 152 and pl. XI, figs. 5, 6
[2] S. S. Buckman, 1923, *T.A.*, pl. CDLII.
[3] H. B. Woodward, 1894, *J.R.B.*, pp. 314–17.

pale band is remarkable for containing in abundance the freshwater gastropod *Viviparus langtonensis* (Hudleston)[1] and more rarely *Valvata comes* Hudl.,[1] after which it is called the *Viviparus* Marl. The *Viviparus* superficially resembles the species in the Purbeck Marble, which, in fact, it approaches more nearly than that found in approximately contemporaneous beds in the Department of Indre, France, and in Scotland.[2]

This freshwater bed, with the pebble bed below, or where the freshwater fossils are absent, the Upper Estuarine Clay, is markedly discordant in its relations to the strata beneath. At Sharp's Hill, near Hook Norton, Richardson describes it as resting non-sequentially on an eroded surface of the lower beds. 'It indicates a change', he adds, 'and in places the non-sequence between it and the underlying deposits is far greater than at Sharp's Hill. At Castle Barn, for example, it rests directly upon the Chipping Norton Limestone.'[3] Subsequent extension of the quarry at Castle Barn has shown that the pebble bed there is very marked, comprising some 6 in. of rolled and bored lumps of hard Chipping Norton Limestone, intimately blended with the *Viviparus* Marl (fig. 52).

Before the widespread erosion which gave rise to the pebble bed and preceded the formation of these freshwater and estuarine deposits (and the extent of the erosion is denoted by the effects already referred to in the North Cotswolds) a series of highly fossiliferous marine deposits was accumulated over the central plateau of Oxfordshire, These were grouped by H. B. Woodward and other writers as Fuller's Earth and little attention was paid to them except by a local geologist, E. A. Walford of Banbury. Walford discovered the great palaeontological interest of the beds, and his valuable researches have been carried on and extended by Richardson.

Walford grouped together all these beds, including the Upper Estuarine Clay and the *Viviparus* Marl, under the name Neæran Beds, and Richardson at first followed this plan. Later, however, as the significance of the higher clays came to be realized, he restricted the name to the beds below the Upper Estuarine Clay, and if it is to be used at all there is no doubt that it should only apply to the strata below the plane of discordance.[4]

Apart from this the name Neæran Beds is an unfortunate one for several reasons. In the first place specimens of *Neæra* are only very occasionally to be found and are much more common in the Upper Estuarine Series of the Eastern Counties; the same species, *N. ibbetsoni* Mor. and Lyc., also occurs in the Hampen Marly Beds and I have even found it in the 'Forest Marble' of Kirtlington.[5] In the second place the correct name of the genus, by the rules of priority, should be *Cuspidaria*. For these reasons it is proposed to drop the name Neæran Beds altogether, in favour of Upper Estuarine Clay and Sharp's Hill Beds.

The succession is best illustrated by the type-section at Sharp's Hill,

[1] Figd. W. H. Hudleston, 'Mon. Inf. Oolite Gastropoda', *Pal. Soc.*, pp. 488–9, pl. XLIV, figs. 1 a, b, 2 a, b; and pl. XLIII, fig. 27.

[2] See M. Cossmann, 1899, *Bull. Soc. géol. France* [3], vol. xxvii, pp. 136–43, 543–85, and pls. XIV–XVII.

[3] L. Richardson, 1911, loc. cit., p. 210.

[4] Richardson included in the Neæran Beds the *Viviparus* Marl, but since it contains freshwater fossils while the Neæran Beds are purely marine, and since it moreover succeeds the pebble bed, it is here grouped with the Upper Estuarine Clay.

[5] M. A. P. Dutertre informs me that it has an equally long range in the Boulonnais.

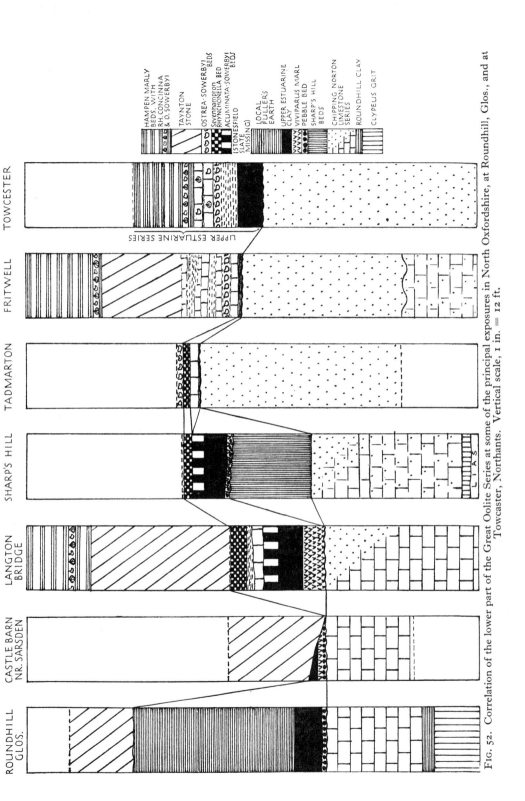

FIG. 52. Correlation of the lower part of the Great Oolite Series at some of the principal exposures in North Oxfordshire, at Roundhill, Glos., and at Towcester, Northants. Vertical scale, 1 in. = 12 ft.

2 miles north-west of Hook Norton, minutely described by Walford and by Richardson. The sequence of fossil beds and grey and greenish clays is as follows (fig. 52):

SECTION AT SHARP'S HILL, NEAR HOOK NORTON[1]

ft. in.

SEVENHAMPTON RHYNCHONELLA BED: crowded with *Rhynchonella communalis, R. decora, R. ? deliciosa* and *Ostrea sowerbyi* 1 0

Clay 1 0

ACUMINATA-SOWERBYI BED: clay, full of whitened *O. acuminata* and some *O. sowerbyi* 6

(non-sequence)

UPPER ESTUARINE CLAY: tough, dark green, brown, and bluish clay, the basal 6 in. black and bituminous, almost resembling a coal seam[2] . . 3 0

VIVIPARUS MARL AND PEBBLE BED: pale whitish marl with *V. langtonensis, Ataphrus labadyei* (d'Arch.) and *Nerinea* spp., and numerous pebbles and concretions, some phosphatic 8

(plane of erosion)

SHARP'S HILL BEDS: *total about 10 ft.*

Upper *Nerinea* Bed; limestone passing into marl; *N. eudesii* Mor. and Lyc., *Aphanoptyxis tomesi* Ark., &c. 1 9

Coral Bed with *Cyathophora bourgueti* (?*Cryptocœnia delauneyi*) . . 2 1

Lower *Nerinea* Bed; limestone as above, 0–1 ft. 8

Astarte oxoniensis Limestone, 0–8 in. 5

Marl parting 3

Exelissa Limestone; grey argillaceous limestone crowded with tiny black gastropods (*Exelissa*) and small lamellibranchs, 0–10 in. . . . 6

Limestone (impersistent) 5

Perna Bed: pale yellow and greenish marl and stone, full of *Isognomon* [*Perna*] *oxoniensis* (Paris), &c., &c., 1 ft. to 1 6

Reddish-brown sandy layer with nodules of shaly limestone . . . 8

Black clay 2 0

[Chipping Norton Limestone below, 19 ft.]

It is exceptional for the Sharp's Hill Beds to be so fully represented. Usually the thickness is not more than 5–8 ft., owing to overstep by the *Viviparus* Marl and Upper Estuarine Clay, as well as to minor non-sequences within the series. Nevertheless the beds have been traced over nearly all the central plateau of Oxfordshire, between the valleys of the Evenlode and the Cherwell, from Sharp's Hill and Temple Mill north of Hook Norton, by Chipping Norton, Chastleton, Enstone and Chadlington, to Charlbury and the Ashford Mill railway-cutting near Stonesfield. Only towards the east are they absent, where at Tadmarton, Duns Tew and Fritwell, as already mentioned, either the Sevenhampton *Rhynchonella* Bed or a band of limestone immediately below it rests directly upon White Sands equivalent to the Chipping Norton Limestone. At Fritwell the Upper Estuarine Clay may be represented by a thin black peaty seam, only a few inches thick.

Where the Sharp's Hill Beds are well developed the fauna is extremely interesting. The pale argillaceous limestone crowded with tiny black gastro-

---

[1] Condensed from L. Richardson, 1911, loc. cit., pp. 206–9.

[2] Below this in Walford's record come 3 ft. of limestones and marls, but Richardson and Paris were unable to find them in 1911 and they have not appeared since, although the quarry is still in work.

pods of the genus *Exelissa* is unique in the British Jurassic. At Enstone, Sharp's Hill and a few other places, the *Perna* Bed has also yielded a unique assemblage of *Isognomon* and *Gervillia*, collected by Richardson and described by Paris—*I. oxoniensis* (Paris), *G. enstonensis* Paris, *G. richardsoni* Paris.[1] The *Nerinea* Beds contain also a unique gastropod, *Aphanoptyxis tomesi* Ark., found at Sharp's Hill, near Chipping Norton, and in Ashford Mill cutting, associated with the common *Nerinea eudesii* Mor. and Lyc., all superbly preserved.[2] Most of the sections also show clays crowded with *Placunopsis socialis*.

To any one familiar with these beds in their type-localities, it is difficult to see how they can have been confused with the Hampen Marly Beds at Fritwell, and so with the Upper Estuarine Series of Northamptonshire. The whole fauna presents a marked contrast with those beds, with their characteristic masses of *Ostrea sowerbyi* and *Rhynchonella concinna*.

Chipping Norton Limestone, 10–25 ft.

The Chipping Norton Limestone continues into Oxfordshire from the North Cotswolds by way of Oddington, near Stow on the Wold, and its distribution and stratigraphy have been carefully worked out by Richardson. About the type-locality it forms the chief surface-feature. It is quarried extensively at numerous places on the plateau to the south and west of the town of Chipping Norton about Chastleton, Sarsden, Chadlington, Enstone and Charlbury. It can also be well seen in Langton Bridge railway-cutting, west of Rollright Railway Station, while southward it is co-extensive with the Sharp's Hill Beds, down the Evenlode Valley to Cornbury Park, Fawler and Stonesfield.

Over all this area it consists of a fine-grained, cream-coloured or white, oolitic limestone, some 15–20 ft. thick, often with black carbonaceous specks or flecks of pulverized lignite, and very poorly fossiliferous. Only occasionally a band of *Pseudomelaniæ* occurs near the top. From time to time, however, collectors watching the extensive quarries have made important finds. Bones of *Cetiosaurus oxoniensis* were long ago found at Sarsden and Padbury's Quarry, Chipping Norton.[3] Of ammonites, a single specimen of *Oppelia* cf. *fusca* was found by Richardson at Oakham Quarry, between Little Rollright and Little Compton, in the upper part of the series.[4] This guiding ammonite is of the highest value in correlation and enables it to be said that the Chipping Norton Limestone of the type-locality is contemporaneous with the lower part of the Lower Fuller's Earth Clay farther south—as might be expected on stratigraphical grounds, from its occurrence above the *Clypeus* Grit. Since the upper part of the Chipping Norton Limestone can be dated to the *fusca* hemera, or equivalent to the Scroff and some of the superincumbent clay, the lower part may be inferred to be of *zigzag* date, or equivalent to the *Zigzag* Bed and the Crackment Limestone of Dorset and Somerset. An instructive parallel may therefore be drawn between the Chipping Norton Limestone and the Calcaire de Caen.

---

[1] E. T. Paris, 1911, *Proc. Cots. N.F.C.*, vol. xvii, pp. 233–5 and pl. xxvii.
[2] W. J. Arkell, 1931, loc. cit., p. 619, and pl. L, figs. 15–17.
[3] J. Phillips, 1871, *Geol. Oxford*, pp. 164, 245.
[4] L. Richardson, 1911, *Proc. Cots. N.F.C.*, vol. xvii, p. 228, and 1922, ibid., vol. xxi, p. 116.

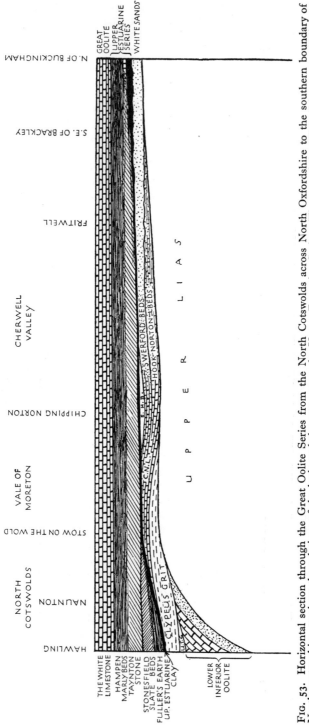

FIG. 53. Horizontal section through the Great Oolite Series from the North Cotswolds across North Oxfordshire to the southern boundary of Northamptonshire, to show the relations of the beds and the passage to the Upper Estuarine Series. The top of the White Limestone is adjusted to the horizontal. C.N.L. = Chipping Norton Limestone; S.H.B. = Sharp's Hill Beds; U.T.G. = Upper *Trigonia* Grit; the thick black line = the Upper Estuarine Clay, with basal pebble bed and locally the *Viviparus* Marl.

About Chipping Norton and over the plateau to the south and west the Limestone is divided into two roughly equal parts by a very hard band graphically named the Knotty Bed, but little difference is to be detected in the nature of the stone above and below it. To the north and east of Chipping Norton, however, the Knotty Bed begins to separate two very distinct series of strata. At the same time it becomes itself highly fossiliferous and was named by Walford after its principal fossil, the *Trigonia signata* Bed.

The upper portion of the Limestone is the first to change, passing towards the Cherwell Valley into sands, sandstones and sandy limestones, which alternate about Hook Norton, Swerford, Sharp's Hill and Tew, in a manner reminiscent of the same beds in the North Cotswolds. This arenaceous facies has been named by Richardson the Swerford Beds. They are 10–15 ft. thick and are especially well seen at Swerford, at Langton Bridge road-cutting near Rollright, and at Great Tew, where perfect gradations from siliceous limestone to white sand may be studied.

The lower portion, below the Knotty Bed or *T. signata* Bed, likewise takes on a more arenaceous facies east and north of Chipping Norton, but it is more massive than the upper. Instead of passing into white sands like the Swerford Beds, it develops into massive brown sandstones and flaggy, sandy limestones ('planking') with brown sand and occasionally marly bands. Often the lowest layers are highly fossiliferous, yielding in particular multitudes of *Pseudomonotis lycetti* Rollier and plant-remains and lignite. To this series, including the *Trigonia signata* Bed above, the name Hook Norton Beds has been applied by Walford and subsequent writers. The average thickness is 10–15 ft.

Just as the *Clypeus* Grit overlaps the Upper *Trigonia* Grit at the Vale of Moreton, so it is itself overlapped between the Valleys of the Evenlode and Cherwell by the Hook Norton Beds, and they in turn are overlapped by the Swerford Beds east of the Cherwell. At the same time both the Hook Norton and the Swerford Beds become progressively more sandy. The stages of this process have been elucidated by Richardson by means of prolonged and careful field-work, and the subject here merits attention in a little more detail than this passing statement (fig. 53).

At Sharp's Hill and in the Hook Norton railway-cutting the Hook Norton Beds consist of hard, blue-hearted, brown sandstone and siliceous limestones, which overlie the attenuated representative of the *Clypeus* Grit. On the east side of the Cherwell Valley, however, they have become more sandy and rest directly upon Upper Lias clay. This is well seen in the Fritwell railway-cutting, near the tunnel mouth, and at Newbottle Spinney, near Aynho.[1] At these places the lowest beds are highly ferruginous and marly and contain, besides abundant fossils, numerous small pebbles of dark Upper Lias claystone.

Still farther north and east the Hook Norton Beds attenuate rapidly, being last seen at Turweston, near Brackley, as a 4-in. conglomeratic band with pebbles of Upper Lias claystone. A few miles to the north again, in the railway-cutting between Greatworth and Helmdon, they have disappeared entirely.[2]

The beginning of the overlap of the Hook Norton Beds by the Swerford Beds is seen at Swerford and Great Tew, where the two series are divided by

[1,2] L. Richardson, 1923, 'Certain Jurassic Strata of Southern Northamptonshire', *P.G.A.*, vol. xxxiv, pp. 99, 103.

a conglomeratic band, containing rolled and waterworn shells and pebbles of Hook Norton Beds, bored and encrusted with oysters and *Serpulæ*.[1] Towards the Cherwell Valley the slight discordance between the two series becomes accentuated. The Swerford Beds at the same time pass entirely into sands, which weather white, but may be bluish or black when freshly exposed and often contain much carbonaceous matter towards the top. Such sands are seen at Tadmarton and Banbury on the west side of the Cherwell Valley, and on the east side they reappear in the Fritwell railway-cutting, at Charlton near Aynho, about Brackley, and at numerous other places.

The most instructive section of all is that in the Fritwell cutting, where the black and white sands of the Swerford Beds fill deep channels hollowed out in the underlying Hook Norton Beds (see Plate XIII), sometimes to a depth of at least 10 ft. This prepares us for the section at Turweston near Brackley, 7 miles to the north-east, where only a thin conglomeratic band separates the white sands of the Swerford Beds from the Upper Lias. Beyond this, as just stated, the Hook Norton Beds are nowhere seen (see map, fig. 38, p. 207).

The further continuation of the Swerford Beds, although they encroach into Northamptonshire, is more appropriately dealt with here, since they do not come into any more than the southern fringe of the Northants.–Lincs. area.

Richardson has traced them from across the Cherwell Valley into the Brackley and Towcester districts, at Turweston, Helmdon, Easton Neston, Blisworth and Gayton, and so into the country immediately surrounding Northampton, where they are the well-known White Sands, long known at New Duston and many other places.[2] Here, in the north, they rest upon the Variable Beds of the Lower Inferior Oolite, but, as explained in Chapter VIII, in the south, about Brackley, they repose directly upon Upper Lias (fig. 38, p. 207, and fig. 47, p. 245; and fig. 53, p. 302).

The White Sand of Southern Northamptonshire was formerly regarded as part of the Lower Inferior Oolite and was grouped with the 'Lower Estuarine Series'; in fact similar white sand undoubtedly passes under the Lincolnshire Limestone farther north. On this account Thompson[3] did not accept Richardson's correlation of any of the White Sand of S. Northamptonshire with the *fusca* zone. But that all the white sand formerly called 'Lower Estuarine' or 'Northampton Sand' could not be synchronous was realized by Hudleston so long ago as 1878, when he pointed out that in the Hook Norton and Banbury district similar sand overlies the *Clypeus* Grit.[4] Richardson has now carried the matter farther by showing that on the east side of the Cherwell Valley, on the borders of Oxon. and Northants., White Sand overlies the Hook Norton Beds (as may be verified in the Fritwell cutting, where the Hook Norton Beds are highly fossiliferous, as at Newbottle Spinney). Since he has also traced it laterally into the upper part of the Chipping Norton Limestone via the Swerford Beds, the case seems to be conclusively proved.

In the district around Northampton, however, the two white sands of different ages come into juxtaposition (see fig. 47, p. 245), and much detailed

[1] L. Richardson, 1922, 'Certain Jurassic Strata of the Banbury District', *Proc. Cots. N.F.C.*, vol. xxi, p. 125; and 1911, ibid., vol. xvii, p. 218.

[2] L. Richardson, 1923, loc. cit.; and 1921, 'Excursion to the Banbury and Towcester Districts', *P.G.A.*, vol. xxxii, pp. 109–22.

[3] B. Thompson, 1924, *P.G.A.*, vol. xxxv, pp. 67–76.

[4] W. H. Hudleston, 1878, *P.G.A.*, vol. v, p. 382.

work will be required to separate them in every exposure. At New Duston and elsewhere a well-defined, even surface marks off the softer *fusca* White Sand from the harder brown Variable Beds or 'Planking' of the Lower Inferior Oolite, but the actual passage of brown sandstones of the Variable Beds laterally into white sand was described and figured by Woodward at Spratton Ironstone workings near Brixworth.[1]

V. NORTHANTS—LINCOLNSHIRE

Compared with the complex development in Oxfordshire, the Great Oolite Series of the long outcrop from Northants and Bucks. through Rutland and Lincolnshire is a simple formation. It consists of the following three divisions:

	ft.
Blisworth Clay	5–30
Great Oolite [White] Limestone	15–30
Upper Estuarine Series [= mainly Hampen Marly Beds] . . .	10–35

Some of this apparent simplicity may be due only to the insufficiency of our knowledge of the detailed succession of the faunas. But although modern refinements of stratigraphy might lead to the detection of further possibilities of subdivision, we already have thorough and accurate descriptions of exposures all along the outcrop, from the pens of Judd, Samuel Sharp, and others.

Blisworth Clay, usually 10–30 ft.

The Blisworth Clay, so named by Samuel Sharp in 1870, has been called the Great Oolite Clay; but that name was first applied by Sharp to the Upper Estuarine Series, and since it is also less explicit, it should be dropped.

The clays are variegated bluish, greenish or purplish grey, black or yellow, with impersistent sandy bands, ironstone nodules, or thin oyster-beds composed of *Ostrea sowerbyi*, *O. subrugulosa* Mor. and Lyc. or *Placunopsis socialis* Mor. and Lyc. Other fossils are rarely to be obtained. In their general appearance the strata much resemble the Upper Estuarine Series below, and like them are suggestive of fluvio-marine deposits.

In the south of Northamptonshire the Blisworth Clay was found by the Survey to be too thin to be shown on a map. It thickens in the north of the county and in Lincolnshire, being 12 ft. thick at Thrapston, about 14 ft. at Peterborough and 20–30 ft. about Lincoln. North of Lincoln, at Brigg, it is still $24\frac{1}{2}$ ft. thick, beyond which the whole Great Oolite Series undergoes great changes towards the Market Weighton Axis.

The explanation of the thickening is to be found in the quarries about the borders of Oxfordshire, Buckinghamshire, and Northamptonshire. Here it can be seen that the Blisworth Clay is a lateral development of the 'Forest Marble', and many degrees of transition occur from one facies to the other. The change does not take place once and for all, but there are alternations of limestone and clay. For instance, at Bradwell, 2 miles east of Stony Stratford, a section described by Woodward[2] showed between the Cornbrash and the Great Oolite Limestone $13\frac{1}{2}$ ft. of Blisworth Clay with an oyster-bed at the top, while about 10 miles to the north-east, at Newton Blossomville east of Olney, a large modern quarry shows a like thickness of massive Forest Marble. The two facies are interchangeable, just as in the Kemble Beds of Long Hand-

[1] H. B. Woodward, 1894, *J.R.B.*, p. 186.   [2] Ibid., p. 392.

borough and other places in Oxfordshire. Where the clay is thin it is in part represented by Forest Marble, which has been grouped with the Great Oolite Limestone. The marble facies recurs at intervals for many miles northward, being seen, with some Blisworth Clay above it, about Oundle and Peterborough. In the Peterborough district it was formerly worked for the same purposes as in Oxfordshire, under the name of Alwalton Marble, and was used for the slender shafts on the front of Peterborough Cathedral (erected in the thirteenth century).[1] The workings lay along the steep escarpment of Alwalton Lynch, but Judd records that they were all closed before the time of his visit in the 1870's. The 'marble' appears to have taken a good polish but not to have been very durable.[2]

There can be no doubt that the bulk of the Blisworth Clay in most places represents the Kemble Beds. The sections at Long Handborough and at Islip show that towards the north-east the Wychwood Beds and Bradford Beds either thin out and are overlapped by, or are overstepped by, the Cornbrash, and it follows from this, as well as from the lithology of the beds, that the Forest Marble to the north and east is most likely to be Kemble Beds. Indeed, the massive limestone of Newton Blossomville and the Alwalton Marble are indistinguishable from the Kemble Beds of parts of Oxfordshire and Gloucestershire, while the clays with abundant *Placunopsis socialis* and *Ostrea sowerbyi* agree well with those into which the Kemble Beds pass at Handborough, Kirtlington and Blackthorn Hill, near Oxford. The only difference is that the clays become more fluvio-marine, with their intercalations of sand and their variegated colours, while the ribbed form of *Ostrea sowerbyi*, named *O. subrugulosa* Mor. and Lyc., not met with south of Oxfordshire, becomes predominant. This shell seems to be a local race of *O. sowerbyi*, in which radial ribs, such as are seen in incipient stages on certain specimens of var. *elongata* in Dorset, have become fixed.

The possibility of Bradford Beds and even Wychwood Beds recurring north of Oxfordshire, at least locally in pre-Cornbrash synclines, cannot be excluded, for old records of fossils said to have been collected by the Rev. Griesbach in the Cornbrash of Rushden, and certain brachiopods preserved in the Davidson collection, ex Griesbach collection, if they really came from Rushden, indicate the presence of the Bradford Fossil Bed. They include *Kutchirhynchia morieri*, identical with Bradfordian specimens.[3]

Another subdivision of the Great Oolite Series which probably contributes to form the basal portion of the Blisworth Clay is the *Fimbriata-waltoni* Beds. They remain in the form of marly limestones, full of the typical fossils, alternating with greenish clays, at least as far as the Ouse Valley between Buckingham and Stony Stratford, where they are exposed near Thornton.[4] At Oundle, in a large quarry near the George Inn, there is some indication that they have almost completely passed into the base of the Blisworth Clay, for the only sign of the fossiliferous limestone facies left is a 4-in. band of soft argillaceous marlstone, full of small specimens of *Corbula hulliana* and *Gervillia waltoni*. This lies 8 in. above the local bottom of the Blisworth Clay.[4] Above it is a

[1] S. Sharp, 1873, *Q.J.G.S.*, vol. xxix, p. 278.
[2] J. W. Judd, 1875, 'Geol. Rutland', *Mem. Geol. Surv.*, p. 209.
[3] See J. A. Douglas and W. J. Arkell, 1932, *Q.J.G.S.*, vol. lxxxviii, p. 154.
[4] Field-notes.

widespread feature of the clay in this district, a band of ironstone nodules. They are so abundant in some places that they have been dug for iron, between 100 and 200 tons having once been raised at Bottlebridge, near Overton Longville, and sent to Wellingborough to be smelted.[1]

White Limestone, 15–30 ft.

If the Forest Marble, which we have seen to belong more rightly with the Blisworth Clay, be excluded, the Great Oolite Limestone of this area corresponds with the White Limestone. Both lithologically and palaeontologically there is close agreement. The rock is very rarely oolitic and generally of the creamy or pinkish-grey marly type, which already predominates in Oxfordshire. It is incomprehensible how Lycett, after visiting the Great Oolite quarries at Kingsthorpe, Duston and Blisworth under the guidance of Samuel Sharp, could have said that it seemed to him to correspond with the Forest Marble of East Gloucestershire and North Wiltshire. He supported this statement by quoting a number of fossils all characteristic of the White Limestone and (in part) Kemble Beds.[2]

From the Ardley–Fritwell cuttings the White Limestone may be traced in numerous minor exposures along the sides of the Ouse Valley and through the Towcester and Northampton country so well described by Sharp. The thickness is about 30 ft. in the south and is gradually reduced to about 20 ft. at Peterborough and northwards to Lincoln. There are good sections showing the greater part of the series at Kingsthorpe, Blisworth and in the Thrapston district. The detailed palaeontological succession has still to be worked out, but it is clear that in general it is the same as in Oxfordshire. Towards the top are beds of *Ostrea sowerbyi* and *O. subrugulosa* and in the middle portions fossil-beds full of corals, Epithyrids, lamellibranchs and gastropods. *E. oxonica* is met with intermittently in typical colonies over all this area, at least as far north as Thrapston, while corals occur in abundance as far as Belmesthorpe.[3] At many places in the Peterborough and Stamford districts (e.g. Helpston) typical White Limestone is quarried, full of *Isastræa* or *Epithyris oxonica* or such characteristic mollusca as *Parallelodon rugosum* (Buck.), *Barbatia pratti* (Mor. and Lyc.), *Modiola imbricata* Sow. and the Nerineidæ.

At the base of the White Limestone at Kingsthorpe and at Blisworth three 'large smooth Ammonites, some 15 inches in diameter' and identified by Etheridge as 'old and smooth individuals of *A. gracilis* Buck.', were obtained.[4]

It is interesting to notice the reappearance of the *Ornithella digonoides* of Gloucestershire at Wootton Hall south of Northampton, and at Wollaston, where it was recorded by the earlier geologists as *O. digona* and taken for 'occasional evidence of beds that represent the Bradford Clay'.[5] But, as we now know, it is somewhat different from *O. digona* and occurs near Cirencester and Fairford below the Kemble Beds, in the White Limestone.

---

[1] J. W. Judd, 1875, loc. cit., p. 217.

[2] J. Lycett, in Sharp, 1870, *Q.J.G.S.*, vol. xxvi, p. 379.

[3] J. W. Judd, 1875, loc. cit., p. 209. Corals still occur sporadically as far north as Lincoln; see *J.R.B.*, p. 427.

[4] S. Sharp, 1870, loc. cit., p. 361.

[5] H. B. Woodward, 1894, *J.R.B.*, pp. 396 and 402–3. Note that Woodward was led to group the bed in which it occurs as 'base of the Great Oolite Clay' and this misled Buckman into considering the Great Oolite Clay as '*digonoides* hemera'. But it occurs in a bed with *Lima cardiiformis* and '*T. maxillata*', lithologically and palaeontologically White Limestone.

In places the rock, by reason of the total absence of ooliths and a rubbly quality acquired on weathering, becomes very difficult to distinguish from shelly Lower Cornbrash. This has been emphasized by Judd and other Surveyors,[1] and it led to a number of typical White Limestone fossils, which never occur in the Cornbrash, having been figured as Cornbrash species by Sowerby in the *Mineral Conchology* over a century ago. About Oundle there is a rock crowded with *Astarte* and other shells, which bears an unmistakable resemblance to Lower Cornbrash; but the *Astarte* is *A. rotunda* of the Bath Great Oolite, not *A. hilpertonensis* of the Cornbrash.

Upper Estuarine Series, 10–35 ft.

From the Oxfordshire border all along the outcrop to North Lincolnshire 'the lowest division of the Great Oolite consists of clays, occasionally very sandy, of various colours, light blue being the predominant one, but bright tints of green, purple, &c. being not uncommon'.

'Interstratified with the clays', wrote Judd, in his famous Rutland Memoir,[2] 'are bands of sandy stone, with vertical plant-markings and layers of shells, sometimes marine, as *Pholadomya, Modiola, Ostrea, Neæra,* &c., at other times fresh-water, as "*Cyrena*", *Unio,* &c. Beds full of small calcareous concretions and bands of "beef" or fibrous carbonate of lime also frequently occur, and the sections sometimes closely resemble those of the Purbeck beds. In its lower part this series consists usually, but not always, of white clays passing into sands. At the base of these clays there is always found a thin band of nodular ironstone, seldom much more than one foot in thickness (fig. 54); this "ironstone junction-band" is everywhere conspicuous and marks the limit between the Great and Inferior Oolite Series. . . . There is very decided evidence of a break, accompanied by slight unconformity, between these two series in the Midland area. All the characters presented by the beds of the Upper Estuarine Series point to the conclusion that they were accumulated under an alternation of marine and fresh-water conditions. . . .'

'The fresh-water fossils are for the most part too badly preserved for specific determination,' adds H. B. Woodward; 'the only fresh-water species identified is "*Cyrena*" *cunninghami.*[3] None of the plant-remains, so far as I am aware, have been determined. The "rootlets" occur at different horizons, and do not in themselves mark any important break.'[4]

The thickness is usually between 20 and 30 ft., but in the area south of Lincoln, at the Essendine railway-cuttings and towards Peterborough, it is slightly more (up to 35 and possibly in at least one place 40 ft.).

Since the work of Sharp and Judd our knowledge of the Upper Estuarine Series in Northamptonshire has been greatly increased by the publications of Mr. B. Thompson, and the details he has supplied of this most southerly part of the outcrop have rendered correlation with the subdivisions of the Great Oolite Series in Oxfordshire more feasible.

Judd stated simply 'The Upper Estuarine Series is on the same geological horizon as the Stonesfield Slate', and this view was tacitly accepted for half a century.[5] But there is no foundation for the correlation except a certain

---

[1] J. W. Judd, 1875, 'Geol. Rutland', p. 202, &c.
[2] Ibid., pp. 188–9.
[3] He also records *Paludina* [*Viviparus*]. Thompson recorded *V. langtonensis* (Hudl.), but as he believed it to be a *Natica* the identification is open to doubt (*Q.J.G.S.*, vol. lxxxvi, p. 439).
[4] H. B. Woodward, 1894, *J.R.B.*, p. 382.
[5] J. W. Judd, 1875, loc. cit., pp. 186 and 90.

lithological resemblance in some of the bands, and since the Stonesfield Slate bed (the only part of the series yielding the plants and land vertebrates which qualify it to be considered of 'estuarine' facies) is only 1–1½ ft. thick, it would require strong palaeontological evidence to maintain such a correlation. In reality none of the characteristic fauna of the Stonesfield Slate (as enumerated above) occurs in the Upper Estuarine Series. On the other hand, there are undeniable palaeontological and lithological reasons for correlating the greater part of the series with the Hampen Marly Beds.

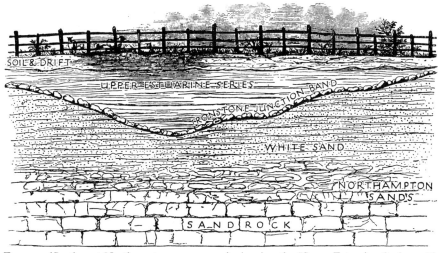

Fig. 54. 'Section at Northampton race-course', showing the Upper Estuarine Series, with the Ironstone Junction Band at the base, resting unconformably upon the White Sands (*fusca* zone?). (From J. W. Judd, 1875, 'Geol. Rutland', *Mem. Geol. Surv.*, p. 34.)

As the result of a comparative study of a very large number of exposures all over Northamptonshire, Beeby Thompson has ascertained that the Upper Estuarine Series comprises three main stratigraphical divisions: a central block of sandy and shelly limestones, generally crowded with large Rhynchonellids and *Ostrea sowerbyi*—the Upper Estuarine Limestones—2½–15 ft. thick, sandwiched between two clay divisions. The details according to him are as follows:[1]

IDEAL SECTION OF THE UPPER ESTUARINE SERIES
IN NORTHAMPTONSHIRE
(Condensed from Beeby Thompson.)

BEDS ABOVE THE ESTUARINE LIMESTONES: 5–16½ ft.

Upper Freshwater Beds: variegated clays, blue, green and red, with abundant vegetable matter and vertical (?plant) markings. Fossils rare, generally absent. 3 ft. at Finedon to 8½ ft. at Moulton Park Farm.

Corbula and *Astarte* Beds: pale green or grey clays with many fossils as casts; *Corbula attenuata*, *C. hulliana*, and *Astarte angulata* predominant, with *Cuspidaria ibbetsoni*, *Placunopsis socialis*, &c. Vertical markings. 2½ ft. at Brigstock to 7 ft. at Roade.

Placunopsis, *Nucula*, and '*Cyprina*' Beds: clays, 0–2½ ft. thick, detected at a few places only (Hopping Hill, Moulton Park Farm).

[1] B. Thompson, 1930, *Q.J.G.S.*, vol. lxxxvi, pp. 431–4.

UPPER ESTUARINE LIMESTONES or RHYNCHONELLA BEDS: 2½–15 ft.

There are two bands of limestone, each containing large *Rhynchonella concinna* and oysters, with an oyster bed, largely composed of *Ostrea sowerbyi*, above the upper course and below the lower. In the middle and west of Northamptonshire the two courses are divided by a soft parting only a few inches in thickness, but in the east of the county, as at Brigstock, the parting grows into 6 ft. of beds —unfossiliferous ferruginous sandstone and dark pyritous clay, with vertical markings.

BEDS BELOW THE ESTUARINE LIMESTONES: 3–18 ft. [UPPER ESTUARINE CLAY of Richardson.]

Lower Freshwater (or Brackish-water) Beds: variegated clays, usually containing abundant vegetable matter and vertical markings. Fossils rare. Thompson believed this division to be the home of the supposed '*Cyrena*' often mentioned farther north by Judd, because nowhere in Northamptonshire did he find any fossil that he could so identify.

Ironstone Junction-bed: ¼–1 ft.

Mr. Thompson's work has confirmed the view that the series becomes attenuated down the dip-slope in SE. Northamptonshire, in the exposures at Rushden, Higham Ferrers, Raunds, Denford, &c., east of the Nene Valley. As Judd perceived, this is doubtless due to increasing proximity to the London landmass, and Thompson has shown that it is brought about by overlap of the Upper Freshwater Beds across all the subjacent portions of the series on to Upper Lias. Thompson believed that the overlap was due to a local anticline with a Caledonian strike, parallel to the Nene Valley, but he based this view upon two doubtful borings farther east, one of which he himself queried, and both of which are capable of other explanations.[1] To say the least, it is extremely doubtful whether the Upper Estuarine Series ever thickens again south-east of the outcrop.

Thompson's views on the probable correlation of the Upper Estuarine Series with the strata to westward, over the Oxfordshire border, may be summed up in his statement that 'The Upper Estuarine Beds above the [Estuarine] Limestones in Northamptonshire correspond to the Neæran Beds of Northern Oxfordshire, *as defined by Odling in his description of the Ardley section* (1913, p. 491) *but not as defined by Walford in his detailed description of Sharp's Hill*.'[2] He supports this statement by means of comparative columns, which show a very fair degree of correspondence between Ardley and Northants, and by a note from Buckman to the effect that 'the similarity (rather identity) of Rhynchonellas in the Estuarine Limestone of Northamptonshire and the Bird's Nest Rock of Ardley and top of Bed (2) shows that they can be correlated as contemporaneous'.

As we have seen, the strata in the Ardley cuttings with the Bird's Nest Rock at the base, misidentified as 'Neæran Beds' and Chipping Norton Limestone by Odling, correspond in reality with the Hampen Marly Beds, and have nothing whatever to do with Walford's Neæran Beds in North Oxfordshire. The abundant large *Rhynchonella* of the Bird's Nest Rock is identical with that so common in the Hampen Marly Beds in the Evenlode Valley, at Milton

---

[1] B. Thompson, 1930, loc. cit., p. 452.
[2] B. Thompson, 1930, loc. cit., p. 436. The italics are mine.

under Wychwood and Swinbrook, and it is the same as the typical *R. concinna* of Aynho.[1]

The invaluable but misinterpreted section in the Ardley–Fritwell cuttings therefore enables a definite correlation between the Upper Estuarine Series and the Great Oolite succession of Oxfordshire to be made for the first time with a proper palaeontological and lithological backing. On Thompson's and Buckman's (*unconscious*) showing, the beds above and including the Estuarine Limestones in Northamptonshire are equivalent to the Hampen Marly Beds. Below, the Taynton Stone is well developed in the Fritwell cutting but has disappeared in Northamptonshire, where we are left with nothing but the 'Lower Freshwater Beds' or Upper Estuarine Clay (3–18 ft.) and the Iron-stone Junction Bed to correspond with it and all the strata below. The plane of unconformity below the Ironstone Junction Bed and Upper Estuarine Clay is the one already familiar, since Richardson's work in North Oxford-shire; but here there is another non-sequence directly above it, for strata equivalent to the Hampen Marly Beds, with the same Rhynchonellas and the same large oysters, seem to have overlapped the Taynton Stone and the Stonesfield Slate Beds entirely. (See fig. 53, p. 302.)

VI. THE MARKET WEIGHTON AXIS[2]

In North Lincolnshire the Great Oolite Limestone gradually becomes thinner northward, the last known occurrence being at Brigg, where $11\frac{1}{2}$ ft. of rock-bands separated by shales were penetrated in a boring, with 24 ft. of Blisworth Clay above and 24 ft. of Upper Estuarine Series below. The Blis-worth Clay so far retains its normal characters, with beds of *Ostrea sower-byi* at the top, but the Upper Estuarine Series becomes predominantly sandy.

North of Brigg the outcrop is almost entirely obscured by alluvium and Boulder Clay to the Humber, except in the neighbourhood of Appleby Railway Station and Thornholme Priory, where the limestone has disappeared and the total thickness is much reduced. Thence northward to the Market Weighton Axis the Lincolnshire Limestone is separated from the Cornbrash by a single series of clays and sands, 20–30 ft. thick, which are continued in a narrow outcrop north of the Humber, past South Cave and Newbald, until they disappear under the Chalk near Sancton.

The final reduction in thickness from about 60 ft. to 30 ft., and the dis-appearance of the limestone, take place quite suddenly in the 4 miles under alluvium between Brigg and Appleby Railway Station. The changes seem to follow a temporary expansion at Brigg, for, according to the accepted interpreta-tion of the Brigg Boring, the Upper Estuarine Series below the limestone is 24 ft. thick, while about Waddington and Redbourn, a few miles to the south, it has already been reduced to little more than 10 ft. The explanation of this may be that Brigg lies on the Caistor–Louth–Willoughby Axis of Kendall and

---

[1] Buckman gave to Thompson's Rhynchonellas from the Estuarine Limestone the names *Burmirhynchia dromio* and *B. patula*, but it must be remembered that he mistook the identity of *R. concinna*, fancying it to be a Lower Cornbrash species and calling it a *Kallirhynchia* (see p. 293, note 5).

[2] W. A. E. Ussher, 'Geol. North Lincolnshire', *Mem. Geol. Surv.*, pp. 81–6; H. B. Wood-ward, 1894, *J.R.B.*, pp. 427–30; C. Fox-Strangways, 1892, *J.R.B.*, pp. 259–60.

Versey, which we have seen was a synclinal axis of deposition in the Jurassic period and an anticline in the Cretaceous (see pp. 84–5).

The correlation of the thin clays north of Brigg with their counterparts farther south is well-nigh impossible, owing to the absence of sections and the dearth of fossils. The question has been discussed by Ussher, who, although remarking that the clay is of about the same thickness as the Blisworth Clay south of Brigg and resembles it in appearance, points out that there is no proof that the Upper Estuarine Series is unrepresented. An overlap of the Blisworth Clay on to the Lincolnshire Limestone is therefore not proved. But there seems no doubt that it overlaps the Great Oolite Limestone; and in view of the correlation already traced between the Blisworth Clay and the Kemble Beds and Bradford Beds, and of the transgressive behaviour of these beds both eastward in the Southery Boring, Norfolk, and south-eastward under London (as will be seen in a later section of this chapter), an overlap here is to be expected.

On the north side of the Axis the strata between the supposed Inferior Oolite and the Kellaways Rock as far as the Derwent comprise nothing but a thin and insignificant strip of sands, devoid of fossils. The sand, indeed, is sometimes difficult to distinguish from the Kellaways Rock, with which it is in contact. Whatever the exact age of this deposit, it provides clear proof of differential uplift along the Market Weighton Axis through almost the entire Great Oolite period.

VII. THE YORKSHIRE BASIN

In the Yorkshire Basin deltaic conditions on a grand scale held sway during the Great Oolite period more completely than ever before.[1] Several times during the Inferior Oolite period the sea encroached northward over the delta-fans and laid down marine sediments and fossils all over Yorkshire; but during the formation of the Great Oolite there were no such interludes. As was pointed out in Chapters VIII and IX, the last marine episode, that of the Scarborough Beds, may have lasted from the hemera *sauzei* to the hemera *zigzag*—the time of formation of the Hook Norton Beds. Afterwards delta-formation continued, uninterrupted by any incursion of the sea, until the Callovian (Upper Cornbrash) Transgression.

The series lends itself to subdivision into two broad lithological groups, a level-bedded group above, consisting predominantly of shales and mudstones, and a current-bedded group below, comprising false-bedded sandstones, mostly impure and silty in the upper part and changing laterally into shale, pure and sometimes very hard and siliceous in the lower part (the MOOR GRIT). The total thickness in the central area of the basin amounts to 200–220 ft. The only fossils besides plant-remains are casts of the freshwater bivalves

---

[1] The deposits formed during the Great Oolite period were known for more than half a century, since the days of the pioneers, Phillips and Williamson, as the Upper Sandstone and Shale, but in 1880 the Geological Survey introduced the term Upper Estuarine Series, which has now unfortunately become current. Since it was already preoccupied by Judd in 1867 and by Sharp in 1870 for the Northamptonshire strata, with which probably only about one third of the Yorkshire division is to be correlated, the term has no legal status in Yorkshire. Further than this, 'estuarine' is a misnomer, for the deposits are deltaic like the Millstone Grit. In a consistent stratigraphical nomenclature the whole of the Yorkshire 'Estuarine Series' would need to be renamed, and there seems no reason why Upper, Middle and Lower Deltaic Series should not be applied.

Unio distortus Bean and *Unio hamatus* Brown, found in some abundance in certain beds towards the base of the series in Gristhorpe Bay and about Scarborough. Although for many years their true affinitites were in doubt, they are now definitely considered to be Unios and are even said to be closely related to the modern *Margaritana*.[1] The plant remains are less common than in the Middle and Lower 'Estuarine' Series, and they are nearly all drifted.

North of the Market Weighton Axis, both in the inland outcrop and on the coast, the series thickens gradually. It is thin and of little importance as a surface feature all through the Howardian Hills. Differentiation has already begun to show, however, for the upper portion is more shaly and the lower more sandy, in places almost a grit.

The Moor Grit begins to make a feature in the south of the Hambleton Hills, sometimes covering large areas of moorland and running out in wide spurs, the ground covered with white siliceous blocks. Here the total thickness of the series increases to 100 ft. and then 200 ft., then dwindles temporarily to 100 ft. at the north end of the Hambletons, before regaining 200–220 ft. throughout the Clevelands and the moors both north and south of the Esk.

On the Yorkshire Moors the Upper Estuarine Series forms barren ground. The more shaly upper part gives rise to a cold, wet surface, on which the heather has a struggle for existence, while the hard Moor Grit supports but little soil. Locally the rock, which is much quarried for roadstone, is called White Flint. It is intensely hard, has a glassy surface, and is translucent in thin flakes. Under the microscope it is seen to be composed of sub-rounded grains of transparent colourless quartz, cemented by a secondary deposit of transparent crystalline quartz. The whole is so compact as to merit the name of a quartzite rather than a grit.[2] Towards the coast, as about Cloughton, it is more tractable and makes a good building-stone.

The coast displays by far the finest sections in the county, but owing to the monotony of the series there is nothing to be gained by describing them here in detail. As in the inland outcrop, the thickness increases northward, from about 125 ft. in Gristhorpe Cliff to 200 ft. at Wheatcroft and farther north. The Moor Grit does not attain its characteristic development until the neighbourhood of Cloughton, and it is barely distinguishable south of Scarborough. The best sections of most of the series are seen in the cliffs from Cloughton to Scalby Ness, but the highest part of the even-bedded group is better shown at White Nab, south of Scarborough.

Very interesting work on the conditions of deposition of the Upper Estuarine Series and the origin of the plant-beds has been done by Mr. Maurice Black.[3] He has demonstrated that in the continuous cliff-sections the Upper 'Estuarine' Series shows the deltaic origin of the beds even more clearly than do the earlier 'Estuarine' Series. He likens the lower or false-bedded group to the foreset beds formed by a modern delta as it advances across a depression and the even-bedded group to the topset beds laid down after the depression has been nearly filled up. The whole sequence seems to indicate one major cycle

---

[1] J. W. Jackson, 1911, *The Naturalist*, pp. 104–7, 119–22.
[2] C. Fox-Strangways, 1892, *J.R.B.*, p. 257, and for the three preceding paragraphs ibid., pp. 254–9, and pl. IV.
[3] M. Black, 1929, *Q.J.G.S.*, vol. lxxxv, pp. 389–437; and 1928, *Geol. Mag.*, vol. lxv, p. 301.

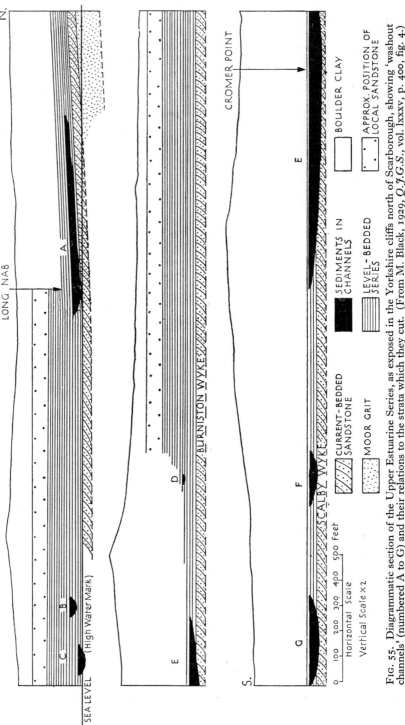

Fig. 55. Diagrammatic section of the Upper Estuarine Series, as exposed in the Yorkshire cliffs north of Scarborough, showing 'washout channels' (numbered A to G) and their relations to the strata which they cut. (From M. Black, 1929, *Q.J.G.S.*, vol. lxxxv, p. 400, fig. 4.)

of depression and elevation. Further he has pointed out a number of 'washouts' or filled-up channels similar to those described in the Lower Series by Hepworth, cut at various levels in the even-bedded group and sometimes down into the false-bedded group below (figs. 55–57). He explains these as the channels of streams which flowed over the surface of the delta.

Mr. Black considers that, with the exception of a few subordinate beds containing only Equisetales and a few ferns, the majority of the plant-remains have been drifted; and since they have been sorted by water in common with other sedimentary materials, they do not represent natural floral assemblages.

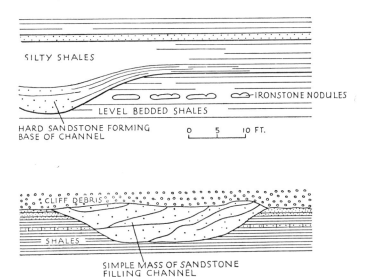

Figs. 56 and 57. Details of channels C and D in fig. 55. (From M. Black, 1929, loc. cit., figs. 5 and 6, p. 403.)

This applies to all the plant-beds containing ferns, conifers and Ginkgoales found in the level-bedded group and also to those with Ginkgoales and Cycads in the washouts. He contrasts the floras of these beds, presenting fragmentary and scattered specimens of different assemblages, with the Middle Estuarine Gristhorpe Plant Bed, in which all the parts of an individual are found together, not far from the position of growth, and even the most delicate structures are preserved. The sorting by mechanical means, by floating and by differential oxidation, all tending to destroy the more delicate tissues, may give an altogether misleading impression of the original assemblage. In this way many of the differences between the assemblages met with at different horizons, and especially those distinguishing the plant-beds of the Upper, Middle and Lower Estuarine Series, may be explained.

Mr. Black notes that in the one locality (between Cloughton and Scalby Ness) where exposures are extensive enough to admit of the courses of the 'washout' channels being mapped, the dip of the current-bedding shows that the water flowed along the channels from north to south. This hint as to the origin of the sediments and the location of the mouth of the river which formed the delta points in the same general direction as the conclusions long ago reached by Sorby in studying the false-bedding in the Lower Estuarines. Both Sorby and Hepworth regarded the origin of the materials as in the north-west, where the deposits become more coarsely sandy.[1] The heavy minerals are still under investigation and have not yielded any positive results, except that they appear to have been re-derived from older sediments rather than first hand from an igneous or metamorphic region.

VIII. THE HEBRIDEAN AREA[2]

On both sides of Scotland, in the Hebridean area and in East Sutherland, the same peculiar freshwater conditions prevailed in the Great Oolite period. At the top of the *garantiana* zone the marine succession is suddenly broken off and is not finally resumed until the Cornbrash. During the interval there were formed in the Hebridean area up to about 450 ft. of shales and mudstones, passing laterally into sandstones, of the same facies as the deposits in Eastern England. In the Hebrides the series is interspersed with beds of estuarine and freshwater shells; and on both sides of Scotland Unios have been found which Jackson believes are probably identical with the Yorkshire *U. distortus* Bean.[3] But towards the top there are marine oyster banks.

Murchison originally termed the series in Western Scotland the Loch Staffin Beds, after a locality in the north-east extremity of Trotternish, Isle of Skye, but since the work of Judd it has generally been known as the Great Estuarine Series. (Map, p. 114.)

Judd pointed out that the Great Estuarine Series of Scotland is developed in two main facies, an argillo-calcareous and an arenaceous; the former he likened to the Purbeck Beds of Dorset (for which previous to 1850 they were mistaken), the latter to the Hastings Sand. Very rapid lateral changes of facies and thickness take place in short distances. The series attains its greatest development (about 460 ft. according to the recent survey),[4] in the south of the area, in the islands of Eigg and Muck. It crops out from below the basalts on the north-west and the east coasts of Eigg, and on the south coast of Muck. Here it consists of three major divisions, a central mass of sandstones, according to Judd more than 500 ft. in thickness but according to the modern survey only 200 ft., sandwiched between two predominantly shaly series, the upper 50–60 ft. and the lower 200 ft. thick. In Mull only the very basal part of the series is known to occur in one small exposure.

The upper shaly division consists of black shale, locally full of *Cypris*, alternating with thin bands of argillaceous limestone. Some of the bands are crowded with '*Cyrena*' and *Cyclas*, others with *Viviparus scoticus* (Tate).[5] The most conspicuous feature of the division, however, consists in thick

---

[1] H. C. Sorby, 1852, 'On the Direction of Drifting of the Sandstone Beds of the Oolitic Rocks of the Yorkshire Coast', *Proc. Yorks. Phil. Soc.*, vol. i, pp. 111–13; and E. Hepworth, 1923, *Trans. Leeds Geol. Soc.*, pt. xix, p. 27. For fuller discussion see below, p. 580.

[2] Based on G. W. Lee, 1920, 'Mes. Rocks Applecross, Raasay and NE. Skye', pp. 52–7; C. B. Wedd, 1910, 'Geol. Glenelg, Lochalsh and SE. Skye', pp. 121–7; G. Barrow and A. Harker, 1908, 'Geol. Small Isles of Inverness', pp. 19–33; *Mems. Geol. Surv.*

[3] J. W. Jackson, 1911, *The Naturalist*, p. 119.

[4] Judd greatly over-estimated the thickness, his records showing a total of 850 ft.

[5] This species is entirely different from the Oxfordshire species, *V. langtonensis* (Hudl.), but associated with it in Skye is a form comparable with *V. aurelianus* Cossm., found in the department of Indre, but not in Oxfordshire.

oyster-beds, completely made up of the shells of *Ostrea sowerbyi*, which is known in all the literature as *O. hebridica* Forbes.[1] These are found wherever the upper part of the Great Estuarine Series occurs, all over the Western Isles, though in Strathaird, Skye, there are some 117 ft. of beds between them and the Callovian or Oxfordian as compared with only 20 ft. in Eigg.

The central sandstone division consists mainly of white and grey, false-bedded, ripple-marked calcareous sandstones, often covered with sun-cracks or worm-tracks, and locally developing enormous doggers. The sandstones locally become coarser and contain stringers of little quartz pebbles, but usually the grains are very uniform in size; sometimes so much so that they give rise to musical sands. Plant-remains are abundant and in places they form thin coal-seams, as in Yorkshire. Other fossils are rare and, except for casts of *Cyclas* and '*Cyrena*', unidentifiable.

The lower shaly division resembles the upper and contains many of the same fossils, but towards the base it develops a series of conglomerates and shelly limestones, abounding in freshwater mollusca. It also yields numerous remains of ostracods, insects, fish, and reptiles, including *Plesiosaurus*, *Acrodus*, *Hybodus*, *Lepidotus*, *Saurichthys* and fragments of bones, vertebræ and teeth of crocodiles, deinosaurs, and pterodactyls.

In the northern islands of Skye and Raasay this tripartite arrangement does not everywhere hold. In the promontory of Strathaird, in Southern Skye, where the thickness is 400 ft. and the distance from Eigg only 13 miles, the sandstone dies out and the two shaly divisions come together. In the north of the same island and the adjoining Isle of Raasay, however, although the total thickness diminishes to 250 ft., the sandstone returns and is as much as 50–100 ft. thick.

The basalt plateaux of Strathaird and Trotternish have been responsible for the preservation of large areas of the Great Estuarine Series in Skye, and an unknown amount may also underlie the plateaux forming the western promontories of Vaternish and Duirinish, for small fragments crop out round the coasts at various points. Some of the chief exposures lie along the north-east coast of Trotternish in the line of cliffs between Bearreraig and Loch Staffin. This is the classic locality where the age of the beds was first proved by E. Forbes in 1850, during a cruise especially planned for the purpose. Here he found the estuarine strata, which Murchison had believed to be Wealden or Purbeck, cropping out from beneath the Oxford Clay.[2]

Perhaps the best sections of the series in the Hebrides are afforded by the south-western cliffs of Strathaird. North of Elgol the Great Estuarine Series succeeds the Inferior Oolite and dips away north-west towards the Cuillins, followed in orderly succession by the Oxfordian sandstones, representing the Oxford Clay discovered by Forbes at Loch Staffin. Since the sequence is typical of that developed all over Strathaird and is of the greatest interest for comparison with the development in other districts, it may be quoted in full.

---

[1] My original impression, formed in Skye, that these names referred to one and the same species, has been confirmed by the kind loan of a quantity of material by Mr. Malcolm MacGregor. In one locality he has found *Rhynchonella* cf. *concinna* and other marine fossils with the oysters. The name *hebridica* (E. Forbes, 1851, *Q.J.G.S.*, vol. vii, p. 110) has priority over *sowerbyi* by two years, but in the interests of convenience it is desirable to deviate from the rules of nomenclature in this instance.   [2] loc. cit.

ft.

VI. Blue shaly marl with blue or white calcareous nodules: up to 30 or .   .   **40**

V. *Viviparus scoticus* Limestones. Blue fine-grained, smooth, argillaceous limestones or cementstones, weathering cream-coloured, containing gastropods, alternating with shales, fibrous carbonate of lime ('beef') and thin beds of calcareous sandstone **37**

IV. Black and blue shales and mudstones with occasional thin limestones: thickness doubtful, perhaps as much as **40**

III. *Ostrea 'hebridica'* Beds. Calcareous shales or limestones crowded with *O. 'hebridica'* Forbes (=*O. sowerbyi* Mor. and Lyc.)   .   .   .   **17**

II. *'Cyrena'* Limestones. Massive blue sandy and often crystalline limestones and calcareous sandstones, full of small lamellibranchs—*'Cyrena'* auct. (=*Neomiodon?*)—generally crushed together, alternating with dark shales and occasional bands of 'beef': sometimes a bed of *Viviparus* cf. *aurelianus* Cossmann occurs near the top. About.   .   .   .   **70**

I. *'Cyrena'* Shales. Black laminated shales with numerous beds of *'Cyrena'* throughout, and occasional thin bands of blue limestone and calcareous sandstone. About **200**

Total   .   .   .   **404**

In Raasay and the adjoining coast of Trotternish the Great Estuarine Series exhibits some peculiar features, chief of which is the development of a true oil-shale at the base. The series is much thinner, not exceeding 250 ft., but there is nevertheless, as just mentioned, a partial return of the sandstone facies immediately above the oil-shale. The outcrop in Raasay is oblong in shape, occupying the highest part of the island, from the basalt caps of Dun Caan (Plate XIV) to the boundary fault which brings the Mesozoic rocks down against the Torridon Sandstone in the north. No sections are available showing the whole succession in unbroken continuity, but the Survey have been able to piece together sufficient information from scattered openings to give a general picture.

The higher parts of the series, with the *Ostrea sowerbyi* Beds and *Viviparus* and *'Cyrena'*, are well developed and together some 200 ft. thick. Below is a mass of white sandstone, 50 ft. thick in Raasay and in the north of Trotternish, but double that thickness farther south, near Holm. The lower part of the sandstone is carbonaceous and contains thin coal-seams as in Eigg and Muck, also *Estheriæ* and fish-remains. It passes downwards through increasingly argillaceous strata into the oil-shale.

The Dun Caan Oil-shale is a true oil-shale, much resembling those of Carboniferous age worked in the Central Lowlands. Its thickness is 8–10 ft., and its yield of crude oil round about 12 gallons per ton. It follows conformably and gradually upon the clay of the *garantiana* zone, and it may therefore be as old as some part of the Upper Inferior Oolite.

IX. EASTERN SCOTLAND[2]

(a) The Brora Coal-Field, Sutherland

A series of similar beds, the counterpart of the Great Estuarine Series of the Western Isles, is found also in the same stratigraphical position on the east

[1] After Wedd, loc. cit.
[2] Based on G. W. Lee, 1925, in 'Geol. Golspie', pp. 74–9, *Mem. Geol. Surv.*

PLATE XIV

Photo. *Geol. Survey.*

Dun Caan and the east side of the Isle of Raasay.

The top slopes are of Great Estuarine Series shales, capped by Tertiary basalt on Dun Caan (top left); the main cliff is Inferior Oolite sandstone, with a slope of Upper Lias below, and close to the sea on the right is the top of the Middle Lias (Scalpa Sandstone). The tumbled ground in front towards the left is a landslip.

Photo. *W. J. A.*

Fascally Coal Mine, Brora, Sutherland.

The coal at the top of the Great Oolite Series is worked 250 ft. below river-level. The brick-pit in the Brora Brick Clay (Lower Oxford Clay) is immediately behind and to the right of the derrick. Fascally Sandstone (Upper Oxford Clay) is seen in the river bank on right.

coast of Scotland, in Sutherland and Ross. It is comparatively little known, however, for only about 60 ft. of the highest part is exposed. It is overlain abruptly by the marine equivalent of the Kellaways Clay, but how far downwards it extends is a matter of conjecture, for it abuts against a large fault, which throws down out of sight, not only the major portion of the Estuarine Series, but also any representatives there may be of the Inferior Oolite and the Upper and Middle Lias.

The outcrop is small—only about ⅓ mile wide and 2 miles long—and is entirely covered with Drift. It strikes nearly W–E. under Dunrobin Mains and curves round to the north along the shore at Strathsteven, following a fault, which, after taking it out to sea for a short distance, brings it across the beach again, greatly narrowed, at a point south-south-west of Brora (map, p. 366). This last is the only locality at which it can be at all favourably seen, and there the exposure consists of little more than reefs covered by the sea at high tide.

Most of the series consists of bands of sandstone and more or less sandy clay and shale, from at least one of which fifteen species of plants have been recorded by Dr. Stopes[1] and Carruthers. The highest part consists of 26 ft. of laminated, black, bituminous shales, at the top of which is the well-known Brora Coal. The bituminous shales contain *Estheria, Unio distortus* Bean and '*Cyrena*' cf. *jamesoni* Forbes (both Hebridean species), and three other species of '*Cyrena*', together with such marine forms as *Isognomon, Pleuromya* and '*Potamomya*'. Judd also recorded scales and teeth of fish, including *Lepidotus, Semiotus, Pholidophorus, Hybodus* and *Acrodus*.

The Brora Coal main seam averages about 3 ft. 6 in. in thickness, and there are several thin and impersistent seams in the underlying shales. According to Dr. Lee,

'The coal is black, lustrous and brittle, breaking into small fragments. It contains a high percentage of disseminated pyrites, besides a lenticular band of solid pyrites towards the middle of the seam. The ash content is also very high, a large bulk being left when burnt. Spontaneous combustion . . . is of frequent occurrence, both in the mine and in the waste heap. The water from the working galleries is so acid that special arrangements had to be made to drain it directly into the sea, as, when drained into the river, it killed the fishes.'[2]

The only mine at present working is at Fascally in the valley of the River Brora (Plate XIV, and map, p. 366). About 30 tons daily are raised for use in the village and the immediate neighbourhood.

According to Sir John Flett the first mine was opened in 1598 by the Countess of Sutherland, and this was followed by the sinking of four or five new pits by the Earls of Sutherland in the early part of the seventeenth century. The workings were never more than barely remunerative and they have been frequently abandoned and reopened only to be again abandoned. This is attributed to several objectionable properties, chief of which are a sulphurous smell, a high percentage of light ash which blows all over the room when the coal is used domestically, and liability to spontaneous combustion owing to the presence of the pyrites. The surviving mine was opened in 1810, and it

---

[1] The title of Dr. Marie Stopes' paper, *The Flora of the Inferior Oolite of Brora (Sutherland)* is misleading, since the beds are likely to be much more nearly equivalent in age to the upper part of the Great Oolite Series.

[2] G. W. Lee, 1925, loc. cit., p. 75.

reaches the coal at a depth of 250 ft. Some of the earlier openings were sur-
face-workings at the only outcrop, on the beach south-west of Brora, but the
coal can no longer be seen there. Another mine was sunk in 1872 at Strathsteven, near the shore, 2 miles south-west of Brora, but this failed owing to
the coal being cut out by a fault.[1]

(b) Ross-shire

Small remnants of the highest part of the Estuarine Series have been
described by Lee at Port-an-Righ, south of Balintore, Ross-shire, at the south
end of the section which is dealt with in the next chapter in connexion with
the Oxford Clay. The Brora Coal is represented by a carbonaceous layer
4 in. thick, beneath which can be seen some 9 ft. of shales and sandstones in
discontinuous section. '*Cyrena*' *jamesoni* and *Isognomon* are recorded.

X. KENT[2]

In view of the marked transgression of the Great Oolite at the outcrop
from the Oxfordshire–Northants border eastward and north-eastward to the
Humber, it is not surprising to find the formation transgressive also to the
south-east, in Kent. The Kent borings have shown, in fact, that the Great
Oolite period witnessed the most important transgression of Jurassic times
against the southern border of the London landmass. The series soon overlaps
the thin sandy representative of the Upper Inferior Oolite (p. 246) upon
which it rests in the most south-westerly borings, at Brabourne and Dover,
overstepping on to Upper Lias at Fredville, on to Middle Lias at Chilham,
Harmansole, &c., on to Coal Measures still farther to the north and east, at
Oxney, Bere Farm, Tilmanstone, &c., and eventually on to Silurian in the
extreme north, at Bobbing. The Great Oolite Series therefore spreads a
considerable way beyond any of the earlier Jurassic rocks before being itself
overstepped by the Cretaceous, and there is a strip of country under which it
alone separates the Cretaceous rocks from the Palaeozoic platform (fig. 8,
p. 52).

At the same time important lithological changes take place and the series
thins out piecemeal from 150 ft. at Brabourne and 144 ft. at Chilham to 82 ft.
at Bobbing in the north and 51 ft. at Oxney in the east.

From the detailed records and excellent palaeontological data published by
the Survey, it is interesting to attempt to distinguish which of the subdivisions
present at the outcrop in the west can be recognized.

All the borings (excepting that at Brabourne, to be mentioned later) showed
three main lithological divisions: at the top 7–18 ft. of clays and limestones of
Forest Marble facies and grouped as 'Forest Marble'; in the middle a thick
mass (40 ft. to over 100 ft.) of predominantly white oolitic limestones; and at
the base a very variable thickness of sands, sandy limestones and sandy clays.

Forest Marble, 7–18 ft.

The presence of beds classified according to their lithological and palaeontological facies as 'Forest Marble' in all the borings and the relative uniformity

---

[1] J. S. Flett, 1922, 'Special Reports, Mineral Resources', xxiv, *Mem. Geol. Surv.*, pp. 32–6.
[2] Based on Lamplugh, Kitchin and Pringle, 1923, 'The Concealed Mesozoic Rocks in
Kent'; and Lamplugh and Kitchin, 1911, 'The Mesoz. Rocks in some Coal Explorations
in Kent', *Mems. Geol. Surv.*

of their thickness is of interest, for it indicates that the north-easterly attenuation of the Great Oolite Series as a whole is not due to overstep by the Cornbrash. Rather it is a piecemeal overlap and thinning against a shore-line.

Another matter of interest is the small thickness of the 'Forest Marble' in all the borings: no higher reading than ?18 ft. was obtained anywhere. This suggests comparison with East Oxfordshire and Bucks., and as there were well over a dozen borings and no trace of Bradfordian fossils was found in any one of them, it seems likely that the 'Forest Marble' is older than the Bradford Clay; that is, is Kemble Beds, as in most of East Oxfordshire and Bucks. Further, it would appear that at least in some of the borings part of the 'Forest Marble' belongs to the *Fimbriata–waltoni* Beds, for at Tilmanstone and at Harmansole the Survey remarked on the similarity of the greenish-black, shelly and lignitiferous clays to those at Blackthorn Hill and Ardley railway-cuttings near Bicester.[1] With this suggestion the occurrence of *Astarte fimbriata* Lyc., Epithyrids, and a number of other fossils not usually found in the Wychwood Beds agrees. There seems, therefore, to be a non-sequence below the Cornbrash all over Kent, just as over the more northerly Eastern Counties.

Great Oolite Limestones, 40–100 ft. + ?

As a whole the Great Oolite Limestones are rather featureless and are not susceptible to subdivision. For instance at Harmansole the record is obliged to treat as a single bed 79 ft. of 'creamy white and bluish-grey, finely oolitic limestones, evenly bedded, and shelly in places'. This sounds all like White Limestone, but below there are only 7 ft. of sandy beds, and then the Middle Lias. It is probable, therefore, that it also contains representatives of the freestones below the White Limestone—the 'Lower Division' of the Great Oolite. At one place, Tilmanstone, there are indications of Hampen Marly Beds below obvious White Limestone. This record is so interesting that it is worth repeating with suggested correlations, as illustrative of the Kent Great Oolite. (*See next page.*)

The record of *Ornithella*, suggestive of the *Ornithella* Beds of Chedworth, is especially interesting when considered in conjunction with records of *Burmirhynchia hopkinsi* (Dav.) at Dover[2] and of *Solenopora jurassica* at Bobbing[3]—all three especially characteristic of the middle and lower part of the White Limestone in the Chedworth cuttings. The clays coming below, crowded with *Ostrea sowerbyi*, strongly suggest the Hampen Marly Beds.

As is to be expected, there are some elements in the Great Oolite fauna which are common to the Boulonnais but foreign to the inland outcrop in England. Of such the most noticeable is *Trigonia clavulosa* Rig. and Sauvage, obtained rather low down in the series in several of the borings.

Fuller's Earth and Chipping Norton Limestone.

The most south-westerly of all the borings touching Great Oolite, that at Brabourne, passed through 13 ft. of 'Forest Marble', then 114 ft. of limestones, the lowest 33 ft. becoming somewhat sandy, and finally through 23 ft. of 'dark grey or bluish calcareous shales', . . .'passing down into 44 ft. of grey-blue, somewhat muddy oolitic limestone of medium texture . . . apparently unfossiliferous'. This last, which was queried as Inferior Oolite, rested upon

---

[1] 1923, loc. cit., pp. 116, 143. [2] 1911, loc. cit., p. 140.
[3] 1923, loc. cit., p. 160.

the Upper Lias clay.[1] The grey shales unfortunately yielded only a few ill-preserved fragments of *Astarte*, *Oxytoma*, &c. and nothing to connect them definitely with the Fuller's Earth. This is the only boring in Kent that has

TILMANSTONE BORING.

Suggested Correlation.		*Record.*[2]	*ft.*	*in.*
		[Cornbrash above.]		
F.M. of the Record 7 ft.	KEMBLE BEDS	Whitish limestones, somewhat clayey in places.	5	0
	FIMBRIATA–WALTONI BEDS	Brown, black and greenish-black clays, as at Blackthorn Hill and Ardley, with several pale fossiliferous bands	2	0
Great Oolite of Record 65½ ft.	WHITE LIMESTONE	Coral and fossil bed: greyish-white earthy limestone; *Isastræa, Thamnastræa, Montlivaltia, Epithyris [oxonica], Parallelodon, Lima cardiiformis*, &c., &c.	1	0
		White and creamy oolitic limestones with marly bands	15	8
	(*Ornithella* Beds of Chedworth represented here?)	Dark grey clayey limestones; *Rhynchonella* spp. (*concinna* group), *Ornithella* sp., *Nerinea eudesii*, *N.* sp., *Natica* spp., *Pholadomya* spp., *Lucina bellona, Lima cardiiformis*, &c. .	20	8
		Limestones, with a thin band of dark clay full of *Ostrea sowerbyi*	5	0
	HAMPEN MARLY BEDS	Soft dark brown shelly clay full of *Ostrea sowerbyi*	1	5
	?	Hard grey limestone. All shells dissolved out of fossiliferous band in middle of bed; *Trigonia clavulosa, T. painei, Parallelodon, Gervillia monotis, G. subcylindrica*, '*Terebratula*' sp., *Nerinea* sp., &c. . . .	2	6
	CHIPPING NORTON LIMESTONE	Sands, with some bands of clay . . .	19	3

given any indication of the Fuller's Earth facies so characteristic of Dorset and Somerset.

Farther north and east the Great Oolite Limestones are underlain by sands and sandy limestones which, in their lithology and palaeontology and their unconformable relations to the underlying strata, correlate with the sandy facies of the Chipping Norton Limestone (the Hook Norton and Swerford Beds). In some places, such as at Harmansole, where they rest on Middle Lias, there is at the base a pebble-bed composed of sub-angular detritus of limestone; at others, where the sands repose on a fresh surface of Coal Measures, as at Tilmanstone, they are composed almost entirely of quartz, the grains ranging up to the size of a marble. At Bobbing, where the Silurian is the bed-rock, the base contains slate pebbles.[3]

An invaluable link between these sandy basement-beds and the Hook Norton Beds of Oxfordshire is the occurrence of *Trigonia signata* Ag. at Elham and Snowdown. On the same horizon, however, there are foreign elements, the most conspicuous being *Rhynchonella lotharingica* Haas, which is rather abundant in some of the borings.

---

[1] 1911, loc. cit., p. 48.
[2] 1923, loc. cit., pp. 140–1; some of the fossils and notes added from pp. 143–5.
[3] 1923, loc. cit., pp. 113, 144, 153.

XI. THE EASTERN MARGIN OF THE LONDON LANDMASS

The very pronounced transgression in Great Oolite times proved by the borings in Kent was long ago known round the eastern margin of the London landmass, for Great Oolite had already been found under Richmond and Streatham Common in Surrey and at Meux's Brewery in the Tottenham Court Road, in the heart of London. Like the most northerly and easterly of the Kentish borings to penetrate any Jurassic rocks, these London borings lie in the belt where nothing but Great Oolite intervenes between the Palaeozoic platform and the Lower Greensand (see Dr. Rastall's map, fig. 6, p. 45).

It is interesting to note that two of these borings, those at Richmond and Tottenham Court Road, disclosed richly fossiliferous Bradford Beds. Indeed, it was owing to the highly distinctive assemblage of brachiopods, polyzoa and other fossils brought up from these beds that the presence of the Great Oolite Series in so unexpected a position was recognized beyond possibility of doubt as early as 1880. J. W. Judd sent collections of brachiopods to Davidson and of polyzoa to Vine—all 'exquisitely preserved'—without telling the specialists whence they had been obtained. Both reported them to be of the age of the Bradford Clay. The brachiopods and other fossils included *Dictyothyris coarctata*, *Epithyris bathonica*, *Cidaris bradfordensis* and *Apiocrinus*, and Vine recognized fourteen species of Bradfordian polyzoa.[1]

The most satisfactory record was obtained at Richmond, where the section was as follows (greatly condensed):

RICHMOND BORING

		ft.	in.
WYCHWOOD BEDS and BRADFORD BEDS 57 ft.	Oolitic limestones, some of the bands with many foraminifera	53	6
	Bradford Clay: blue clay with bands of limestone and many Bradfordian fossils	3	6
WHITE LIMESTONE 17 ft.	Oolitic and shelly limestones, with 6 in. marly bed at base	17	6
?TAYNTON STONE and STONESFIELD SLATE BEDS 13 ft.	Fine-grained oolitic limestone with much pyrites, becoming sandy lower down; and at the base a fissile calcareous and micaceous sandstone, like Stonesfield Slate; *Acrosalenia* and other fossils .	9	0
	Oolitic limestones with shell-fragments, *Ostrea sowerbyi*, &c., and a few grains of quartz and particles of anthracite [derived from Coal Measures] in basal 6 in.	4	0

The difference between this record and those in Kent, with their small thickness of 'Forest Marble', and that apparently all pre-Bradfordian, is most striking.

The Great Oolite in the Streatham boring[2] seems to have been more conformable to the Kentish type. No Bradfordian fossils were encountered, and only $38\frac{1}{2}$ ft. of Great Oolite strata were present, or only 8 ft. more than the amount below the Bradford Clay at Richmond. Within $8\frac{1}{2}$ ft. of the top an oyster queried as *O. acuminata* was recorded, and between $8\frac{1}{2}$ and 11 ft. from the top were 'sandy rock' and 'hard grey calcareous sandstone', by which Woodward was reminded of the Stonesfield Slate of Througham near Bisley

[1] J. W. Judd, 1884, *Q.J.G.S.*, vol. xl, pp. 724–64. As Davidson reported that the *Terebratula maxillata* was the same as those in the Bradford Clay, it must have been *E. bathonica*.
[2] W. Whitaker, 1889, *Rept. Brit. Assoc.* for 1888, p. 656; and H. B. Woodward, 1894, *J.R.B.*, p. 362.

(Glos.). Below this the remainder of the beds ($27\frac{1}{2}$ ft.) were mostly clay, with occasional bands of limestone, and *Ostrea acuminata* towards the top. These facts suggest that the Greensand here rests on Stonesfield Slate Beds and Fuller's Earth. There is more affinity between the borings at Streatham and Brabourne than between those at Streatham and Richmond or Tottenham Court Road. Apparently the Fuller's Earth dies out towards the north just as along the outcrop in the West of England; though the complete disappearance of over 27 ft. of it between Streatham and Richmond is remarkably abrupt.

The nearest connecting links between these London borings and the outcrop are those at Calvert, south of Buckingham, on the Oxford Clay. Although so much nearer the outcrop, they may be most appropriately considered here in connexion with the other borings.

The succession at Calvert agrees with the Kentish succession in the absence of the Bradfordian fauna, but the Forest Marble is at least twice as thick (38 ft. 9 in.). The record agrees especially well with that at Tilmanstone, described on p. 322, for there is a thick development of Hampen Marly Beds— in fact the thickest known. Two marked non-sequences are shown: at the top the Lower Oxford Clay with *Kosmoceras* rests directly on the Forest Marble, without the intervention of any Cornbrash or Kellaways Beds; at the bottom the Great Oolite Series rests directly on the lower part of the Middle Lias— the same transgression as that found in Kent, Surrey, and Middlesex, and more marked than at the outcrop. As there is some doubt as to what part of the series is represented by the transgressive bed at the bottom, a brief condensation of the record may be given, with the interpretation now suggested. (*See next page.*)

In their interpretation Prof. Morley Davies and Dr. Pringle bracket the lowest 7 ft. 6 in. as Chipping Norton Limestone. They admit that this is suggested on purely lithological grounds, and apparently the chief reason is that a part of the stone when examined microscopically resembled the Hook Norton Beds of Sharp's Hill.[1] But as the borings are only 7 miles from the Great Oolite outcrop, they must be compared with the sequence at the neighbouring outcrop; and there are complete sections only 10 miles away, in the Ardley–Fritwell railway-cuttings.

The presence of Hampen Marly Beds in force at Calvert agrees with the Ardley–Fritwell cuttings and also with the nearest other exposures, in the Evenlode Valley; but at Calvert the beds are much thicker—indeed even thicker than in the North Cotswolds—and they seem to have thickened at the expense of the White Limestone, which is unusually thin. In the Ardley–Fritwell cuttings the Hampen Marly Beds rest on well-developed Taynton Stone, with a condensed representative of the Stonesfield Slate Beds below, and as we know the Taynton Stone to be itself transgressive (see p. 302) it is hardly likely to have been overlapped entirely by the Hampen Marly Beds in so short a distance (no such overlap being known anywhere at the outcrop). It is more probable that the Taynton Stone and possibly Stonesfield Slate Beds here overlap the lower parts of the series on to the Lias. The fossils in the basal limestones are in harmony with this view—fragments of broken *Rhynchonella*, *Terebratula*, *Ostrea sowerbyi* and *O.* aff. *acuminata*—not at all suggestive of Chipping Norton Limestone. The chief objection to correlating

[1] A. M. Davies, 1913, *Q.J.G.S.*, vol. lxix, pp. 317, 19.

anything in the Calvert boring with Chipping Norton Limestone is that if any strata of that age were present they should, by analogy with all the exposures on the neighbouring outcrop, be in the facies of White Sands. The locality is well to eastward of Turweston, where the Hook Norton Beds are entirely overlapped (see p. 303 and fig. 38, p. 207), and so nothing but White Sands (if

CALVERT BORING

Interpretation now suggested.	Summary of Record after A. Morley Davies and J. Pringle.[1]		
?Trace of Wychwood Beds (perhaps). KEMBLE BEDS FIMBRIATA–WALTONI BEDS Together 38 ft. 9 in.	Oolitic 'Forest Marble' limestones, with a 3-in. band of bright green clay . . . . Grey marly clays, passing down into brown and greenish clays, with a bed of green sandstone and a hard band of grey limestone . . .	ft. 25 16	in. 6 9
WHITE LIMESTONE 12 ft.	Limestones; top bed compact, bottom bed with *Epithyris* [*oxonica*]	12	0
HAMPEN MARLY BEDS 42 ft. 3 in.	Dark grey marly clay full of *Ostrea sowerbyi* and *Rhynchonella concinna*; 6 in. of grey marl at top . Grey earthy limestone . . . . . Soft dark grey sandy shales and marls; no core seen Grey limestones and marls . . . . Grey marly clay full of *O. sowerbyi* and *R. concinna*	3 5 25 6 2	0 0 9 6 0
TAYNTON STONE and ?STONESFIELD SLATE BEDS 12 ft. 9 in.	Grey limestone with an *Ostrea* band near top . Dark grey marl . . . . . . Grey thinly-bedded limestone with lignite fragments . . . . Yellowish oolitic limestone . . . . Sandy limestone with abundant ooliths and fragments of *Rhynchonella*, *Terebratula*, *Ostrea sowerbyi* and *O.* aff. *acuminata* . . . Yellowish false-bedded oolitic limestone . .	3 1 2 5	6 3 6 0 0 6

[Middle Lias below.]

anything at all) should be left. If the thread of pp. 303–5 be followed on a map, it will readily be seen that the occurrence of oolitic limestones down to the surface of the Lias in the Calvert boring can almost be taken as a certain indication that nothing of the age of the Chipping Norton Limestone is present.[2]

One other place where borings have proved the Great Oolite Series to overstep eastward against the London landmass is much farther north, at Southery, Norfolk, midway between Newmarket and King's Lynn, near the eastern edge of the Fens. Here only 8 ft. of rock of Forest Marble facies was found beneath the attenuated Cornbrash, resting again on Middle Lias.[3] The beds consisted of dark, greenish-black, shelly clays, enclosing 3 ft. 3 in. of hard, bluish-grey, earthy limestone. Insufficient fossils were recognized for the strata to be dated more accurately, but Bradfordian forms were definitely absent. Lithologically the description of the rocks suggests the *Fimbriata–waltoni* Beds.

[1] A. M. Davies and J. Pringle, 1913, *Q.J.G.S.*, vol. lxix, pp. 310–19.
[2] At the time when the Calvert boring was described, of course, sufficient detailed work on the outcrop had not been done to enable satisfactory comparison to be made.
[3] J. Pringle, 1923, *Sum. Prog. Geol. Surv.* for 1922, pp. 129–31.

CHAPTER XI

CORNBRASH

Ammonite Zones.	Brachiopod Zones.		Stratal Divisions.
Macrocephalites macrocephalus	Rhynchonelloidea cerealis Buck. group	Microthyris lagenalis (Schloth.)	Upper Cornbrash
		Microthyris sidding- tonensis (Dav.)	
Clydoniceras discus	Kallirhynchia yaxleyensis (Dav.) group	Ornithella obovata (Sow.)	Lower Cornbrash
		Cererithyris intermedia (Sow.)	

THE Cornbrash, like the Upper Inferior Oolite, is a transgressive deposit which on stratigraphical grounds demands description in a separate chapter. If this be thought insufficient reason for alienating it from its parent formation, the Great Oolite Series, then justification is readily forthcoming from a study of the palaeontology. Eighty years ago Lycett suggested that the Cornbrash should be ranked as equal in importance to the whole of the rest of the Great Oolite Series; he wrote: 'Basing our generalizations upon the zoological characters of the deposits, we are led to the conclusion that the Great Oolite should be arranged into two stages or fauna[s], and that all the subordinate groups of deposits older than the Cornbrash constitute a single and lower stage of the formation.'[1]

This is not the whole story, however. Although the Cornbrash as a whole is transgressive, there is a still greater transgression and a far greater change of fauna in the middle of the formation. William Smith noticed this before 1816, for he stated, 'The Cornbrash, though altogether but a thin rock, has not its organized fossils equally diffused, or promiscuously distributed. The upper beds of stone which compose the rock, contain fossils materially different from those in the under.'[2] This remarkable observation was overlooked for more than a century, while authors preferred to speak of the Cornbrash as if it were a single homogeneous stratum, indivisible, and remarkable only for its wide distribution and lithic uniformity.

In reality the Cornbrash consists of two essentially distinct parts, an Upper and a Lower, which differ widely in lithology as well as in their fauna.

The Lower Cornbrash is characterized by the ammonites Clydoniceras discus (Sow.), C. hochstetteri (Oppel) and other allied forms, and can generally be subdivided into two brachiopod zones, the lower characterized by Cererithyris intermedia, the upper by Ornithella obovata. It is linked by these genera and by most of its lamellibranch fauna to the Great Oolite Series, though there is relatively seldom specific identity (except with the lamellibranchs of the Fuller's Earth Rock, see pp. 256–7). The richest profusion of lamellibranchs occurs at the top of the obovata zone, where there is usually an ASTARTE– TRIGONIA BED, crowded with Astarte hilpertonensis Mor. & Lyc., Trigonia

[1] J. Lycett, 1857, The Cotteswold Hills, p. 84.
[2] William Smith, 1816, Strata Identified by Organized Fossils, p. 25.

angulata Sow., *T. rolandi* Cross, *Entolium rhypheum* (d'Orb.) and the button-coral, *Anabacia complanata*. Besides the zonal brachiopods, Rhynchonellids of the group of *Kallirhynchia yaxleyensis* (Dav.) are especially characteristic.

The lithology of the Lower Cornbrash varies from brown or grey marly rubble to massive beds of blue-hearted limestone, but it nearly always stands in marked contrast with the Forest Marble below and there is seldom any sign of passage between the two. Although it is only rarely that pebble beds or other signs of erosion are perceptible at the base, the change of lithology is nearly everywhere abrupt and complete, and observation of a long stretch of outcrop shows that there is also gradual overstep of the Forest Marble.

The Upper Cornbrash is less variable in lithic characters, consisting usually of sandy limestones or ferruginous sandy marls with doggery limestone bands, and occasionally massive beds of hard, pink or purplish-centred limestone weathering into flags. Palaeontologically it provides a much more marked contrast with the Lower Cornbrash than does the Lower Cornbrash with the rest of the Great Oolite, for its fossils are essentially of Callovian aspect. Instead of the Bathonian genus *Clydoniceras*, which seems to have become totally extinct, we find the Callovian family *Macrocephalitidæ* (*Macrocephalites*, *Kamptocephalites*, &c.). The Bathonian *Ornithella* and *Cererithyris* have disappeared, giving place to the Callovian brachiopod genus *Microthyris*, of which two zones can be recognized, those of *M. siddingtonensis* below and *M. lagenalis* above; while the abundant Kallirhynchias of the Bathonian have been replaced entirely by the little *Rhynchonelloidea ccrealis* Buck., closely allied to the *Rh. socialis* of the Kellaways Beds. It is impossible to distinguish the specimens of *M. lagenalis* by external characters from those of the Kellaways Rock.

With the new brachiopods enters a fresh suite of lamellibranchs, such as *Trigonia cassiope* d'Orb., *T. elongata* Sow., *T. scarburgensis* Lyc., and the oysters *O.* (*Liostrea*) *undosa* Phil. and *O.* (*Lopha*) *marshii* Sow.

At the base of the Upper Cornbrash there is sometimes a pebble-bed marking the junction with the Lower Division, and the line of demarcation is nearly always plainly discernible. Upwards, on the contrary, the passage into the Kellaways Beds is more or less gradual. The logical line of separation between the Bathonian and Callovian stages should therefore, beyond any question, be drawn between the Lower and Upper Cornbrash. In the South of England this line marks one of the most easily detected conformable transgressions in the Jurassic System, while northwards, in Yorkshire, it becomes very conspicuous. There the Upper Cornbrash is spread over the unfossiliferous top of the Upper Estuarine Series as the first stratum of the marine Upper Jurassic succession.

Had d'Orbigny realized that neither in France nor in England do the Macrocephalitan and the Clydoniceratan faunas occur together, he would not have included the *macrocephalus* zone in his Bathonian Stage. Oppel regarded it as Callovian; but he did not realize that *Macrocephalites* occurs in the Cornbrash, for he chose as type-locality Stanton and Chippenham, where only Lower Cornbrash is developed. He admitted failing to find *M. lagenalis* (his zone-fossil for the Cornbrash) in the Chippenham district, but he found it at another place in Wiltshire; and, little suspecting that it came out of *macrocephalus* beds, he made the *lagenalis* zone Bathonian.[1]

[1] A. Oppel, 1856–8, *Die Juraformation*, p. 456; and see also p. 509.

I. THE DORSET—SOMERSET AREA [1]

The Cornbrash attains its greatest thickness of 25–30 ft. in Dorset, and of this one-third belongs to the Lower and two-thirds to the Upper Division.

The complete sequence is exposed in the low cliffs of the Fleet backwater, at Chesters Hill, near Abbotsbury Swannery. The greater part of the thickness consists of alternate bands of cream-coloured argillaceous marl and hard concretionary or doggery limestone, most of it belonging to the *siddingtonensis* zone, and the topmost beds to the *lagenalis* zone. Large gerontic Macrocephalitids may sometimes be found (map, fig. 61, p. 342).

The Lower Cornbrash consists of 4 ft. of rubble full of fossils belonging to the *obovata* zone, with at the base 3 ft. of hard pinkish limestone of the *intermedia* zone. The usual fossils, *Pseudomonotis echinata* (Sow.), *Chlamys vagans* (Sow.), *Chl. hemicostata* (Mor. & Lyc.), *Pholadomya deltoidea*, *Pleuromya decurtata*, *Nucleolites clunicularis*, &c., abound here, as at almost every exposure of Lower Cornbrash.

The section is repeated by a fault about three-quarters of a mile to the south-east along the shore, and there the Lower Cornbrash has expanded to about 11 ft., of which 10 ft. are assigned to the *obovata* zone. This furnishes an example of the great variability in thickness of the zones in short distances, a feature characteristic all over England. The palaeontological vagaries in the Cornbrash, resulting from the local distribution of colonial brachiopods, are well illustrated in two inland sections not far from the coast. In one, at Swyre, the Lower Cornbrash is capped by a band crowded with *Kallirhynchia multicosta* Douglas and Arkell, a species met with only rarely at two or three other Dorset localities, while at East Fleet there is another quarry containing only an occasional *Kallirhynchia*, but instead, hundreds of specimens of *Ornithella classis* Dougl. and Ark., a species hitherto found in no other district, and rare even in the coast sections. Similarly, at Yetminster in North Dorset, on the other side of the Chalk Downs, the Lower Cornbrash contains a band abounding in *Ornithella rugosa* Dougl. and Ark., which is very rare elsewhere, while at Corscombe, near by, the largest Cornbrash Rhynchonellid, *Kutchirhynchia idonea* Buckman, makes its only known appearance in England, in a thin seam of marl.

All these occurrences are in the *obovata* zone of the Lower Cornbrash, and they are the exception rather than the rule. Usually the fauna of this zone is monotonously typical and may be recognized at a glance.

The dominance of the Upper Cornbrash with its thick sandy and marly beds and bands of concretionary (doggery) limestone is continued throughout Dorset and into the southern part of Somerset, being well seen at Corscombe, Closeworth, Holwell, Stalbridge and Wincanton. In the same district, in quarries at Yetminster and Bishop Caundle, the *intermedia* zone is also seen at the maximum development attained in England, namely 5–6 ft. of hard, massive, blue- or inky-hearted limestone, having much of the appearance of Kemble Beds.

At Cards Farm near South Brewham, on the other hand, and at South Cheriton, the *intermedia* zone has shrunk to its usual thickness, not exceeding

---

[1] Based on J. A. Douglas and W. J. Arkell, 1928, 'The Stratigraphical Distribution of the Cornbrash', Part I, *Q.J.G.S.*, vol. lxxxiv, pp. 143–55.

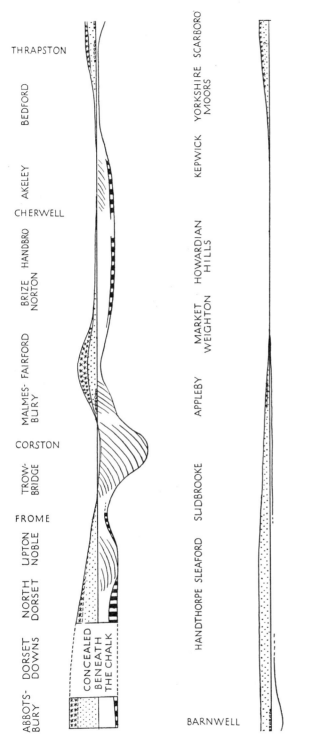

FIG. 58. Diagrammatic longitudinal section of the Cornbrash along the outcrop from Abbotsbury on the Dorset coast to Scarborough on the Yorkshire coast, showing the changes in thickness and the distribution of the brachiopod zones. The boundary between Lower and Upper Cornbrash is reduced to horizontality. Horizontal scale, 1 in. = 22½ miles; vertical scale, 1 cm. = 18 ft.

2 ft., and the Upper Cornbrash has diminished to a thin bed of marly rock with *M. lagenalis*. Here, and henceforth for many miles, until beyond Malmesbury in North Wiltshire, the principal role is taken by the *obovata* zone. At South Cheriton and South Brewham it is from 7 to 8 ft. thick, consisting of the *Astarte–Trigonia* Bed at the top and flaggy to massive, white, poorly fossiliferous limestones below. These limestones reach their maximum thickness at Corston, Wilts., from which they take the name CORSTON BEDS. The *Astarte–Trigonia* Bed at South Brewham is a remarkable repository of perfectly preserved lamellibranchs, from which Sowerby figured the type specimen of *Trigonia angulata*.

II. THE MENDIP AXIS (FROME DISTRICT)

Two of the finest sections of the Cornbrash in England were opened up in 1931–2 in the cuttings for the Great Western Railway's new loop-line to avoid Frome Railway Station. The cuttings have exposed the whole of the beds over the continuation of the Mendip Axis, in a region where exposures were particularly to be desired but had previously been few and meagre.[1]

In the southern cutting, immediately beneath a sandy clay forming the base of the Kellaways Beds, the brachiopods of the *siddingtonensis* zone, *M. siddingtonensis* and *Rhynchonelloidea cerealis*, abounded in about 2–3 in. of ferruginous rottenstone. In the northern cutting the bed had expanded to 1 ft. of hard, purplish, doggery limestone. This was all that represented the Upper Cornbrash. The *lagenalis* zone was entirely absent.

The bed beneath formed the principal feature of the Cornbrash, consisting of a solid block of very hard, grey to purplish, sandy limestone, 3 to 4 ft. thick, the upper portion deeply piped by solution cavities. Palaeontologically it was of great interest, for the lower part contained in abundance the fauna of the *Astarte–Trigonia* Bed, and the upper yielded numerous Perisphinctids of the '*Homœoplanulites*' (*Choffatia*) *subbakeriæ* type, associated with abundant *Pseudomonotis echinata* and an occasional specimen of *Ornithella obovata*. The lithology suggested Upper Cornbrash and similar Perisphinctids are common at Stalbridge in association with *M. siddingtonensis* and *R. cerealis*, but the brachiopod and the *Pseudomonotis* prove beyond doubt that at Frome this bed is part of the *obovata* zone.

The upper part of this bed had been long exposed near by at Berkley, north-east of Frome, and at Wanstrow and Cards Farm, South Brewham, to the south-west. At Cards Farm it is seen in the road-cutting, occupying a position between the *Astarte–Trigonia* Bed and the *lagenalis* zone. On stratigraphical grounds it had, therefore, been assigned to the *siddingtonensis* zone, though the rare Ornithellids (or Microthyrids) found in it only at Berkley could not be identified with certainty.

Beneath the *Astarte–Trigonia* Bed the railway-cutting showed normal marls with rubble and abundant *Ornithella obovata* (3–4 ft.), and at the base a hard band of limestone (1–2 ft.) with *Cererithyris intermedia*. The Lower Cornbrash is similarly developed at Hilperton, on the Trowbridge inlier (except for the presence of Corston Beds and the absence of the solid block of limestone of Frome), and again at Wincanton to the south.

---

[1] The following account is from my field-notes; a description is to be published by Dr. F. B. A. Welch, in the *Sum. Prog. Geol. Surv.*

At Chatley (a Sowerbyan type locality), Tellisford and Rode, a short distance north of Frome, shallow pits expose Lower Cornbrash, in which the faunas of the *intermedia* and *obovata* zones are mingled in a single bed of rubble. These exposures are somewhat to the north of the line of the Mendip Axis, and in view of the more regular development of the beds in the Frome cuttings, the admixture of faunas which they show can scarcely be attributed to the influence of the axis.

III. THE DIP-SLOPE OF THE COTSWOLD HILLS [1]

In the Cotswold district, from the neighbourhood of Bath to the neighbourhood of Oxford, the Cornbrash marks the natural boundary between the hills and the Oxford Clay vale. The distribution of the several zones in this district, so well demarcated in Inferior Oolite times by the axes of uplift along the Mendips on the one hand and the Vale of Moreton on the other, admits of certain generalizations, but there are local anomalies which show that by this time the movements controlling deposition had grown more feeble and perhaps more complex.

We have seen how in the Dorset–South Somerset area the dominant role is played by the Upper Cornbrash, which becomes thinner at South Brewham and Frome and disappears near Trowbridge. Conversely it may be said that in general in the Cotswold province, and from Frome as far north as Bedford, the Lower Cornbrash plays by far the most important part, while the Upper Division is recessive or absent. An exception must be made, however, of the central region from Malmesbury to Fairford; for here the Upper Cornbrash returns in force and the Lower is reduced to a rubble with mixed *intermedia* and *obovata* faunas, or to a condensed fossil-bed.

(a) Trowbridge to Malmesbury

From south of Trowbridge to Malmesbury the Upper Cornbrash is represented at most by a few feet and usually by only a few inches of brown marl, containing crushed specimens of *Microthyris lagenalis*, with occasional *Rh. cerealis*. The bulk of the Cornbrash is made up of the CORSTON BEDS (*obovata* zone), which consist of bands of massive limestone, originally dark-centred but weathering white, and forming an important surface feature about Chippenham, Clanville, Stanton St. Quintin and Corston. At the type locality of Corston the limestones have an apparent thickness, measured at right angles to the bedding, of 50 ft. or more, but they were evidently deposited with a deposition dip and their true thickness may be no more than 25 ft. Fossils are usually limited to clusters of *Pseudomonotis echinata*, but at Corston and one or two other places there are considerable numbers of sea-urchin tests (*Pygurus michelini* and *Acrosalenia hemicidaroides*) and a peculiar brachiopod, found in no other part of England, *Ornithella foxleyensis* Dougl. and Ark.

Both at Chippenham and at Bancombe Wood, south of Rodbourn, the Corston Beds have an eroded upper surface to which oysters (*Lopha marshii*) are attached, and upon the surface rests the thin representative of the Upper Cornbrash, without the intervention of any *Astarte–Trigonia* Bed.

On approaching Malmesbury, however, as may be seen in the Foxley Road

[1] Based on Douglas and Arkell, 1928, loc. cit., pp. 127–42; and 1932, *idem*, part 2, *Q.J.G.S.*, vol. lxxxviii, pp. 123–7.

Quarry, the Corston Beds lose their characteristic identity, the *Astarte–Trigonia* Bed returns, and *Ornithella foxleyensis* is found below in ordinary *obovata*-zone rubble. The *intermedia* zone is nowhere seen in the vicinity of Malmesbury.

To the north-east, at Garsdon and Charlton, the Corston Beds are still well developed, but a few miles farther in the same direction, at Shorncote, the whole of the Lower Cornbrash is reduced to a single condensed bed 2 ft. thick, crowded with fossils of both *obovata* and *intermedia* zones. In between these places, at Poole Keynes and Murcott, a glimpse is seen of the *Astarte–Trigonia* Bed resting upon Corston Beds.

(b) Malmesbury to Fairford and Witney

The chief interest from north of Malmesbury as far as Fairford centres in the Upper Cornbrash, which is about 6–8 ft. thick and contains *M. siddingtonensis* in the lower part and *M. lagenalis* in the upper. Certain bands of purplish sandy limestone, reminiscent of those in Dorset, also yield abundant *Macrocephalites macrocephalus*, *Kamptocephalites herveyi* and occasionally specimens of *Choffatia*. The brachiopods occur in dense clusters in perfect preservation in a soft marl, and a small quarry at Milton End, Fairford, was long renowned for its unique yield of *M. lagenalis*, *M. sublagenalis*, &c. in flawless condition. The excavation is now unfortunately filled up and grass has been sown over it.

The base of the Upper Cornbrash at Charlton is a boulder-bed of rolled and bored lumps of limestone around which are loose shells of *M. siddingtonensis* and *Rh. cerealis*. Some of the boulders when broken open are found to contain Lower Cornbrash fossils, but others are composed of hard purplish limestone containing *M. siddingtonensis*. In a neighbouring quarry at Garsdon the hard *siddingtonensis* bed, from which some of the boulders seem to have been derived, is still present beneath the boulder bed and rests in turn upon another eroded surface at the top of the Corston Beds (fig. 59). There is thus clear evidence of erosion and removal of the *Astarte–Trigonia* Bed prior to or at the beginning of the deposition of the *siddingtonensis* zone and also the continuation or repetition of the erosion during the *siddingtonensis* hemera, with the breaking up and rolling of previously-solidified *siddingtonensis* beds.

About Cirencester [1] and Fairford the Upper Cornbrash forms a considerable surface feature, consisting chiefly of purplish limestone of an almost

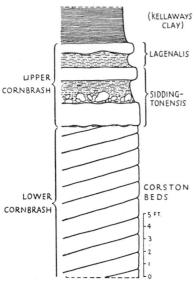

FIG. 59. Section of the Cornbrash at Garsdon, near Malmesbury, Wilts. (From J. A. Douglas and W. J. Arkell, 1928, *Q.J.G.S.*, vol. lxxxiv, p. 138.)

Labels in figure: (KELLAWAYS CLAY); LAGENALIS; UPPER CORNBRASH; SIDDING-TONENSIS; LOWER CORNBRASH; CORSTON BEDS; 5 FT. 4 3 2 1 0

[1] Near here is the village of Siddington, which gave its name to the zonal brachiopod.

flinty hardness. Fossils are rare, but at Poulton the limestone yields *Micro-thyris sublagenalis* and *M. siddingtonensis*, while farther east, at Broughton Poggs, it is overlain by a marl containing dense clusters of the typical *Micro-thyris siddingtonensis* as at Fairford, together with *Kamptocephalites herveyi*. The latter, indicative of the Upper Cornbrash, has been found as far east as Witney.

(c) Oxfordshire: Witney to Buckingham

The outcrop from Witney eastward through Oxfordshire to Buckingham-shire resembles that between Trowbridge and Malmesbury in the absence or

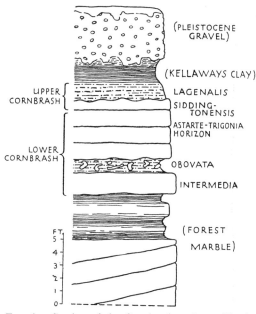

meagre representation of the Upper Cornbrash and the thick development of the Lower.

The Upper Cornbrash is ex-posed at Long Handborough, in the Swan Inn Quarry, where it is only 1 ft. thick (fig. 60). Under the Kellaways Clay are marl and crushed *M. lagenalis* and ill-preserved small Macro-cephalitids, resting upon a 6-in. band of hard purplish limestone with *M. siddingtonensis, Rh. cerealis* and other typical fos-sils, which reposes in turn upon the *Astarte–Trigonia* Bed. The Upper Division seems to be somewhat thicker towards the Cherwell Valley, for in the Shipton-on-Cherwell railway-cutting (now obscured) Wood-ward recorded 3 ft. 8 in. of beds containing *M. lagenalis*.[1]

FIG. 60. Section of the Cornbrash at Long Hand-borough, Oxfordshire. (From J. A. Douglas and W. J. Arkell, 1928, loc. cit., p. 128.)

On the opposite side of the valley, however, at Greenhill, Enslow Bridge, Upper Cornbrash is absent altogether, and it is again wanting in an exposure of the junction with the Kellaways Clay at Akeley Brickyard, north of Buckingham.

The Lower Cornbrash, on the other hand, is well developed, usually from 6 ft. to 10 ft. thick. It forms broad slopes of typical red stone-brash land, especially characteristic of the country between Oxford, Woodstock and Bicester, where it caps the plateau between the valleys of the Rivers Cherwell and Evenlode. Along these valleys are numerous sections, most of them incidental to the exploitation of the underlying Great Oolite, from which the Cornbrash has to be removed as overburden. Supplementary information is afforded by the line of inliers along the Islip Anticline, rising through the Oxford Clay lowlands of Otmoor, at Islip, Charlton-on-Otmoor, Ambrosden and Blackthorn Hill.

[1] H. B. Woodward, 1894, *J.R.B.*, p. 447.

All these exposures show a rather monotonous alternation of rubble with rubbly or platy limestones, most of which belong to the *obovata* zone. *O. obovata* is abundant, together with *Ps. echinata, Nucleolites clunicularis* and casts of *Pleuromya, Homomya, Pholadomya, Gresslya,* &c. At certain levels *Clydoniceras discus* and allied forms are not uncommon, while the *Astarte–Trigonia* fauna is also frequently in evidence. Here and there, especially at Stratton Audley and other places near Bicester, some of the zone is a hard, poorly-fossiliferous flaggy limestone recalling the Corston Beds.

At the base is normally a band of cream-coloured limestone crowded with *Cererithyris intermedia.* It is conspicuous as far west as Southrop, and at Brize Norton, Witney, Handborough, Kidlington, Charlton-on-Otmoor, and Akeley near Buckingham. It, too, occasionally yields *Clydoniceras.* Along the sides of the Cherwell Valley and in the Islip inlier, however, the bed is absent, and *C. intermedia* is found disseminated through the *obovata* zone at levels several feet higher, while the *Astarte–Trigonia* assemblage, usually character-istic of the top of the Lower Cornbrash, appears near the bottom. This ad-mixture of faunas may result from penecontemporaneous erosion and re-deposition, but it is more likely to have an ecological explanation. In view of the long range of allied Terebratulids in the Great Oolite, it is unreasonable to assume that *C. intermedia* became extinct after attaining its acme at the beginning of the Lower Cornbrash; it is more probable that the unfavourable conditions which terminated its acme and caused it to be succeeded by *Ornithella obovata* drove off straggling survivors to other parts of the sea-bed, not previously colonized. Once again we must be careful to distinguish the epibole from the biozone.

IV. BEDFORDSHIRE AND DISTRICT [1]

Near Buckingham the outcrop of the Cornbrash enters a Drift-covered area, in which exposures are few and poor. Besides the mantle of Drift, an additional reason for the failure of exposures is the thinning of the Cornbrash and the thickening of the argillaceous strata beneath it, so that it becomes a thin rock-band between two clays. In the past the Kellaways Clay was worked in several small brickyards, which penetrated to the underlying rock, but with modern centralization nearly all have been abandoned. One such brickyard was situated north of the Ouse at Felmersham, and from it and from Great Oolite quarries south of the river, a former rector, the Rev. T. O. Marsh, sent a number of fossils to Sowerby to be figured in the *Mineral Conchology*. One was named *Ostrea marshii* after its donor, who is thus commemorated in one of the most conspicuous and abundant fossils of the Upper Cornbrash.

The old brickyards along the Ouse Valley at Bedford, at Bourne End, Bletsoe, and West End, Stevington, were visited and described by H. B. Woodward before they fell into ruin. From his descriptions it is evident that, although the total thickness is no more than 2–3 ft., fossils of both Upper and Lower Cornbrash are present. The 'band of tough shelly limestone' was described as yielding *O. obovata* and *Clydoniceras* as well as *M. lagenalis, Ostrea (Lopha) marshii* and *Macrocephalites.* [2]

---

[1] Based on Douglas and Arkell, 1932, loc. cit., pp. 127–9.
[2] H. B. Woodward, 1894, *J.R.B.*, p. 451.

Judging by the narrowness of the outcrop along the banks of the Ouse Valley and in the outer escarpment between Bozeat and Rushden, the Cornbrash seems to be thinly developed over the whole of the area where Dr. Rastall has shown the continuation of the Charnwood Axis to lie.

On the prolongation of the Nuneaton Axis at Stowe Nine Churches, 8 miles west of Northampton and some 13 miles from the main outcrop, is a small outlier of Cornbrash owing its preservation to faulting. Here the thickness is again only 3 ft. All of it is Lower Cornbrash, for *Clydoniceras* occurs at the top.

In connexion with the changes in thickness of the different parts of the formation in the Midland Counties it should be noted that the boring at Calvert, 6 miles south of Buckingham, seemed to show that the Cornbrash and Kellaways Beds were absent altogether, although the site of the boring is only 3 miles from the nearest inlier at Marsh Gibbon and barely a mile farther from the main outcrop.

V. NORTHAMPTONSHIRE AND LINCOLNSHIRE TO THE HUMBER [1]

Along the 90 miles of outcrop from Higham Ferrers in the Nene Valley to the Humber, the Upper Cornbrash is paramount once more and is continuously developed, while the Lower Division is seldom seen and makes little of a surface feature.

West of Peterborough, although the Cornbrash is much dissected, the total area covered by it (largely as inliers and outliers) is considerable. Farther north the outcrop soon narrows, and for the last thirty miles, along the eastern foot of the slope leading up to Lincoln Edge, it is a mere strip.

The best sections are situated in the vicinity of Thrapston, Northants, where the Great Oolite beneath is exploited as a flux in the iron furnaces, but about Peterborough and in South Lincolnshire the Upper Cornbrash has been quarried extensively as a road stone and wall stone. A quarry at Thrapston was visited by William Smith, who noticed that the rock shown was Upper Cornbrash, because it contained *O. marshii*.[2]

At the top is generally a thin representative of the *lagenalis* zone, best seen at Thrapston L.M.S. Railway Station Quarry, where it is 1 ft. 8 in. thick and richly fossiliferous. It consists of ferruginous marls and impersistent rusty limestone, crowded with very perfect specimens of the zonal brachiopod, together with large oysters, *Ostrea undosa* and *Lopha marshii*, the former often bearing the imprint of other bivalves upon which they grew, especially *Trigonia scarburgensis* Lyc., *Gervillia* sp. and *Chlamys* (*Radulopecten*) *anisopleurus* (Buv.). The bed also yields vertebrae of *Murænosaurus*, and has furnished a wealth of Polyzoa, which have formed the subject of a special memoir by G. R. Vine.[3]

The principal feature of the Cornbrash in the Peterborough district and northward through Lincolnshire lies immediately below the *lagenalis* zone. It is a solid band of tough, grey limestone, weathering flaggy and creamcoloured. At first sight it recalls the solid block of limestone of Berkley and Frome, but it yields a different fauna, which ranges it with the hard purplish

---

[1] Based on Douglas and Arkell, 1932, loc. cit., pp. 129–37.
[2] In Phillips's *Memoirs of William Smith*, 1844, p. 75.
[3] G. R. Vine, 1892, *Proc. Yorks. Geol. Soc.*, N.S., vol. xii, pp. 247–58.

flaggy limestone of Poulton. At Thrapston and in numerous other exposures farther north search reveals *Microthyris sublagenalis*, *M. siddingtonensis* var., *Lopha marshii* and other definitely Upper Cornbrash fossils. Although the typical *M. siddingtonensis*, such as occurs in clusters in Gloucestershire and elsewhere, has never been found north of Oxford, this bed is taken to represent the *siddingtonensis* zone.

At the base of the zone at Thrapston is a pebble-bed like that at Charlton and Garsdon near Malmesbury. The pebbles consist of hard purple-centred Cornbrash, well bored by *Lithophaga* and encrusted with *Serpulæ*, but unlike those at Charlton they have yielded no contemporary fauna. Associated with them are numerous small Macrocephalitid ammonites, especially *Doliceph-alites typicus* and *Kamptocephalites hudlestoni*. This bed rests directly on the Lower Cornbrash, a single band of grey and white chalky and rubbly limestone, only 5 in. thick, packed with fossils, and closely resembling, both lithologically and palaeontologically, the development at Shorncote near Cirencester. It yields both *Cererithyris intermedia* and *O. obovata*, with *Clydoniceras*.

The pebble-bed is only a local feature, for at Islip Ironworks, near Twywell, it has disappeared, and the Lower Cornbrash has thickened to about 2 ft. To the north, at Barnwell, near Oundle, the Lower Division is exceptionally developed, reaching nearly 5 ft.; but such a thickness is purely local.

The faulted inlier between the villages of Stilton and Yaxley, south of Peterborough, formerly yielded numerous fossils characteristic of both Upper and Lower Cornbrash, but the only exposure has been completely obscured for at least thirty years. Here the types of *Kallirhynchia yaxleyensis* (Dav.) and *Ornithella stiltonensis*. (Dav.) were obtained. The exact horizon of the latter has not yet been determined. Numerous Macrocephalitid ammonites have been collected about Peterborough, especially at Castor, Chesterton and Stanwick, and may be seen in the Peterborough and other museums.

About Bourn and Sleaford, all the exposures show the *siddingtonensis* zone, which is generally divisible into three distinct elements. At the top is the hard, purplish-grey limestone with *M. sublagenalis* and *M. siddingtonensis* var., as about Peterborough, which forms the principal surface feature; in the middle is a yellow sandy marl, crowded with perfect specimens of *Lopha marshii* and *Ostrea undosa* and other fossils; while at the base is a solid block of tough, grey-centred, poorly fossiliferous limestone. Such tripartite sections are to be seen at Hacconby, Handthorpe and other places in the vicinity of Bourn, and at Quarrington and Roxholm near Sleaford. The yellow sandy marl is the home of a rare brachiopod, *Tegulithyris bentleyi* (Dav.), which is restricted to this part of England.[1]

In North Lincolnshire exposures are few and far between. The best is at Sudbrooke Park, north-east of Lincoln, where the whole Cornbrash can be seen, the Upper Division 3 ft. 6 in. thick, the Lower Division 1 ft. thick. *Microthyris lagenalis* has not been found here, but it was recorded from old workings no longer open, at Appleby. The chief element at Sudbrooke is still a purplish flaggy limestone, yielding the typical Upper Cornbrash fauna— *Kamptocephalites herveyi*, *Trigonia scarburgensis*, *T. cassiope*, *Lopha marshii*,

---

[1] The type-specimen came from Handthorpe.

&c. The thin representative of the Lower Cornbrash is quite distinct, a hard, blue-hearted shelly band with *Ornithella obovata, C. intermedia, Kallirhynchia yaxleyensis, Trigonia rolandi, Clydoniceras legayi* and many other fossils.

VI. THE MARKET WEIGHTON AXIS

Where last seen south of the Humber, at Appleby, the Cornbrash is only 3 ft. thick, and under the estuary it thins out altogether. In the oolite tract south of Market Weighton there is no sign of the rock or of the fossils associated with it, and it has been sought in vain all along the Howardian Hills and the southern part of the Hambletons. The first traces have been recognized at Kepwick, about 50 miles from the last vestiges south of the Humber.[1]

VII. THE YORKSHIRE BASIN

From Kepwick round the north end of the Hambleton Hills a thin sandy limestone has been followed inferentially at the base of the Kellaways Rock, but no exposures have indicated its nature or its thickness. At Shaken Bridge on the Rye some ironstone is believed to have been formerly obtained from the Cornbrash. The first definite indications, however, begin to be visible east of Bilsdale. From Bilsdale the rock strikes due east across the moors, thickening steadily towards Newton Dale, where it attains its maximum of about 5 ft., and it so continues to the coast at Scarborough.[2]

The steep sides of the glacial gorge of Newton Dale and its tributaries provide the best sections in Yorkshire, and are the finest natural sections of the Cornbrash in England. The rock consists of black, shelly limestone, sometimes oolitic, overlying an intensely hard, shelly, mottled purple and red band. Both parts contain in abundance the typical lamellibranchs of the Upper Cornbrash found south of the Humber, such as *Lopha marshii, Ostrea undosa, Trigonia scarburgensis, T. cassiope, Chlamys (Radulopecten) anisopleurus*, &c., but the brachiopods are unfamiliar and ammonites extremely rare; the principal brachiopod is *Burmirhynchia fusca* Doug. and Ark.

The 5 ft. of Cornbrash immediately overlies about 7 ft. of blackish sandstone, the two together forming a conspicuous hard band, which stands out of the shale as a prominent feature along the cliffs. The sandstone was included in the Cornbrash by Fox-Strangways, but although it may represent an estuarine development of the Lower Cornbrash, it does not appear to yield any fossils other than traces of possibly *Serpulæ*. Until diagnostic species have been found, it should be grouped with the Upper Estuarine Series.

Towards the coast the thickness is maintained. The best section in East Yorkshire is the large quarry, formerly known as Peacock's Quarry, in the side of the hill near Scarborough Railway Station. The Cornbrash is still a solid block with the same fauna as inland, but the upper part is red and the lower part dark blue-grey. The same arrangement holds in the coast sections between Scarborough and Filey, and everywhere a thin marly layer at the top abounds in crushed specimens of *M. lagenalis*. This passes down imperceptibly into the limestone, of which it may be the decalcified top layer. Above are the grey shales belonging to the Kellaways Beds, but formerly called the 'Shales of the Cornbrash'.[3]

[1] H. C. Versey, 1928, *Naturalist*, p. 117.   [2] C. Fox-Strangways, 1892, *J.R.B.*, pp. 270–2.
[3] J. A. Douglas and W. J. Arkell, 1932, loc. cit., pp. 137–41.

The best natural section on the coast at present is in Cayton Bay, where the rock outcrops through the sand for a distance of several hundred yards along the beach below Red Cliff. Most of the protruding fossils have long been removed, but a century ago local collectors, chief among whom was Bean, obtained rich harvests here. The other source from which the large collections now in Scarborough, York, the Sedgwick and other museums were made was the beach below Scarborough Castle. Blocks of fossiliferous limestone strewed the foreshore before the building of the undercliff road, but as they were derived not only from the Cornbrash, but also from various parts of the Kellaways and Hackness Rocks, and the Corallian, a number of mistakes in the location of the various species were inevitably made.

Between Scarborough and Cayton Bay the Cornbrash appears for a short distance at the top of the cliff at Wheatcroft, and to the south it dips down below sea-level in Gristhorpe Bay, where it is, however, much obscured by landslips. From Scarborough southward there is a steady diminution in thickness, from 5 ft. to 18 in., the conclusion being that a few miles farther south, towards the Market Weighton Axis, the Cornbrash thins out altogether as in the western escarpment.

The zonal position of the Yorkshire Cornbrash is difficult to determine for the same reasons as are various parts of the Kellaways Beds and Oxford Clay, namely, that many of its fossils are not found south of the Market Weighton Axis. The lamellibranch fauna is, with few exceptions, the same as that in the Upper Cornbrash of Lincolnshire, and most of the principal species occur in the Upper Cornbrash all over England. The brachiopods, however, are more restricted in their distribution. The presence of *Microthyris lagenalis* indicates the highest zone, and it seems to occur only at the top of the rock as in other parts of England. On the other hand, there is no *M. siddingtonensis* or *M. sublagenalis*, or any of the familiar Rhynchonellids, while the Macrocephalitid ammonites for the greater part belong to different species from those in the Upper Cornbrash in other counties. Neither *Clydoniceras* nor *Cererithyris* have been obtained, but there are in the York collection a few specimens of *Ornithella obovata*. If the label proclaiming these to have come from Yorkshire is to be relied upon, the whole of the deposit may not belong everywhere to the Upper Division, as would seem to be the conclusion from the field evidence.

VIII. THE HEBRIDEAN AREA

Only one small patch of Cornbrash is known in Scotland, in the central upland of Raasay.

'This was discovered during the recent survey 9/10 mile NE. of Storav's grave, on the right bank of the middle branch of the Glam Burn, 300 ft. west of the basalt boundary. The ground is exceedingly difficult of interpretation owing to faulting, drift and peaty covering, which makes it impossible to estimate the extent of the Cornbrash area, though there is no doubt that the beds referred to it follow upon the Great Estuarine Series.

'What is seen of the sequence consists of 15 feet of gritty white limestone, strikingly granular and unlike any other rock in Raasay, with comminuted fossils, resting on some 6 or 8 feet of gritty, flaggy limestone, darker and with red ironstone nodules. This lower limestone is full of comminuted fossils, but has also yielded a few that are entire and sufficiently determinable to give a clue to its age. Most of the fossils obtained are brachiopods, which were submitted to Mr. Buckman, who determined them as follows: *Ornithella*, two species, *Rhynchonella*, three or four species, *Terebratula intermedia* Sow.' [1]

Lithologically the fossiliferous limestone bears a striking resemblance to some of the flaggy forms of Lower Cornbrash of the *obovata* zone in Southern England. Its presence provides an interesting additional link between the Hebridean area of deposition and that of South and Central England, and a further contrast with the Yorkshire Basin in Jurassic times.

In Skye and Eigg either Kellaways Beds (*kœnigi* zone) or some part of the Oxford Clay everywhere rest directly upon the Great Estuarine Series, without the intervention of any Cornbrash.

IX. EASTERN SCOTLAND

If there is any deposit of the age of the Cornbrash in Eastern Scotland it is of 'estuarine' origin and indistinguishable from the Estuarine Series below. The first horizon above the Estuarine Series with marine fossils is the Brora Roof Bed, of the age of the Kellaways Clay (*kœnigi* zone).

X. KENT [2]

Most of the typical fossils of both Upper and Lower Cornbrash were obtained in the borings in Kent, but there are insufficient data on which to base any conclusions as to the development of the various zones. The cores showed varying alternations of marls and limestone bands, the basal limestone being usually the most fossiliferous and having often a pisolitic structure. *C. intermedia*, *O. obovata*, *M.* cf. *siddingtonensis* and *M. lagenalis* were recorded, as also Macrocephalitid ammonites and a large number of lamellibranchs, but no succession of faunas has been made out. At Guildford the total thickness is about 10 ft., but at Tilmanstone it is much greater, totalling 24 ft., or about the same as in the area of maximum thickness at the outcrop, on the Dorset coast. Here 7 ft. of nodular and pisolitic limestones are overlain by 17 ft. of hard grey limestone with *Macrocephalites*, suggesting some of the Upper Cornbrash limestones of Dorset and Gloucestershire. According to the records, *M. lagenalis* occurred at Tilmanstone to within 1 ft. of the base, mixed with *O. obovata* and *C. intermedia*, and rearrangement is suggested by 'numerous flat nodules of irregular shape, some of which exceed two inches in length'.[3]

The Cornbrash was found to be overlain in nearly all the Kent borings by a sandy clay with a *Pseudomonotis* (less inflated and less strongly ribbed than *P. echinata*), which was correlated with the shales overlying the rock in Yorkshire and other parts of England. In the official records this shale was included in the Cornbrash, but as has already been stated, it is better grouped with the Kellaways Clay.

[1] G. W. Lee, 1920, 'Mesozoic Rocks of Applecross, Raasay, &c.', *Mem. Geol. Surv.*, p. 58.
[2] Based on Lamplugh and Kitchin, 1922, loc. cit., and Lamplugh, Kitchin and Pringle, 1923, loc. cit.
[3] Lamplugh, Kitchin and Pringle, 1923, loc. cit., p. 142. Could the brachiopods at the base recorded as *M. lagenalis* have been *Ornithella classis*? This species might be expected in Kent, where the total thickness and lithology of the Cornbrash are much the same as on the Dorset coast.

TABLE SHOWING THE ZONES INTO WHICH THE OXFORD CLAY AND KELLAWAYS BEDS HAVE BEEN DIVIDED BY DIFFERENT AUTHORS

Ages.	Buckman 1913 and 1915.	Morley Davies 1916.	Neaverson 1925.	Spath 1926.	Morley Davies 1929.	Brinkmann 1929.	Pringle 1930.	Adopted Here.
base of CARDIOCERATAN (cordatum zone s.l.)	scarburgense	præcordatum		præcordatus (with vernoni and renggeri)	præcordatum	tenuicostatum	scarburgense	præcordatum (= scarburgense auct.)
	vernoni				vertumnus			
	gregarium							
QUENSTEDTOCERATAN (= Vertumniceratan) (lamberti athleta zone s.l.)	vertumnus			mariæ (with vertumnus and renggeri)			mariæ	
	renggeri / lamberti	renggeri	lamberti	lamberti (with lalandei and bicostata)	renggeri	lamberti	renggeri / lamberti	renggeri / lamberti
	athleta	athleta	athleta		athleta	spinosum (= ornatum auct.)	proniæ / athleta	athleta (with proniæ and spinosum)
	ornatum	duncani	duncani	duncani (with athleta and proniæ)	duncani		duncani	duncani
KOSMOCERATAN (ornatum or athleta zone s.l.)		reginaldi (with castor and pollux)	castor	fraasi (with castor and pollux and coronatum)	reginaldi	castor and pollux (with coronatum)	castor	reginaldi (with coronatum, castor and pollux)
	coronatum	coronatum	coronatum		stutchburii		coronatum	
		elizabethæ or jason	elizabethæ				elizabethæ	
	anceps		conlaxatum (with jason)			jason (= conlaxatum)	conlaxatum	jason (= conlaxatum, with elizabethæ, &c.)
REINECKEIAN (anceps zone s.l.)	calloviense			anceps (with calloviense and gulielmi)	calloviense		calloviense	calloviense
	koenigi			rehmanni (with königi)	koenigi		koenigi	koenigi

CHAPTER XII

OXFORD CLAY AND KELLAWAYS BEDS

Zones	Strata.		
(Plates XXXVI–VII).	Dorset–Humber.	Yorks.	
Cardioceras præcordatum [1]		Basal Lower Calc. Grit.	
Creniceras renggeri [2]	Clays with ammonites chiefly as pyritic casts.	OXFORD CLAY	
Quenstedtoceras lamberti [3]			
Peltoceras athleta			
Kosmoceras duncani			
Erymnoceras reginaldi [4]	Shaly clay with compressed Kosmocerates		
Kosmoceras jason [5]			
Sigaloceras calloviense	KELLAWAYS ROCK	?	
Proplanulites kœnigi	KELLAWAYS CLAY	KELLAWAYS ROCK & SHALES 50 ft. +	

(In the Dorset–Humber column: OXFORD CLAY runs vertically. In the Yorks. columns: HACKNESS ROCK and 'Kelloway Rock' auctt. run vertically.)

I. DORSET TO THE MARKET WEIGHTON AXIS
(a) The Dorset Coast

THE Oxford Clay, like the other Jurassic rocks outcropping at Weymouth, covers a small area completely separated from the main outcrop to the north by the overstepping Chalk of the Downs. The low ground floored by the soft Oxford Clay has been hollowed out from the east to form Weymouth Bay, behind which lie the Lodmoor Marshes and the Radipole Backwater.

[1] This species was described by R. Douvillé, first as a *Cardioceras*, then as a *Quenstedtoceras*; but, although its characters are said to be to some extent intermediate, Douvillé's figures show it to be a typical cordate *Cardioceras*; indeed, he said that it passed by insensible gradation into *Cardioceras cordatum* (1912, *Mém. Soc. géol. France*, vol. xix, p. 62, pl. IV, figs. 10–20); see also Morley Davies, *Geol. Mag.*, 1916, p. 397, and Buckman, *T.A.*, vol. v, 1924, p. 32. As a zonal index it replaces 'scarburgense' auctt., the true *scarburgense* having proved to be something different (Buckman, *T.A.*, vol. v, 1924, p. 32). The zone was originally called the Pre-*cordatus* zone by Buckman (1913, *Q.J.G.S.*, vol. lxix, p. 159).

[2] Between the *renggeri* and *præcordatum* zones Buckman placed three zones, *vertumnus*, *gregarium* and *vernoni*, found only in Yorkshire. It is not fully known to what extent any or all of these are contemporary with the faunas south of the Humber, but the *vernoni* zone may safely be regarded as synonymous with the *præcordatum* zone and probably the other two with the *renggeri*.

[3] Made genotype of *Bourkelamberticeras* Buckman, 1920, *T.A.*, vol. iii, p. 17. (Synonym.)

[4] Introduced by Morley Davies, *Geol. Mag.*, 1916, and retained by him in the *Handbook of the Geology of Great Britain*, 1929. It is equivalent to Buckman's zone of *Erymnoceras coronatum*, which corresponds according to Brinkmann with the maximum of *Kosmoceras castor* and *K. pollux*.

[5] According to Brinkmann (1929), *Kosmoceras conlaxatum* Buckman, which was used as index of the basal zone of the Oxford Clay by Neaverson, is a synonym of *K. jason* Reinecke, which therefore takes priority. The zone as used here includes the *elizabethæ* zone of Morley Davies (1916) and Neaverson (1925), *K. elizabethæ* (Pratt) being according to Brinkmann (1929) synonymous with *K. ornatum* (Schloth.) *non* auctt. Morley Davies (1929) chose *Kosmoceras stutchburii* (Pratt) as index, but according to Brinkmann that species is a synonym of *K. gulielmii* (Sowerby). Note: the *lamberti* to *præcordatum* zones comprise Oppel's zone of *Am. biarmatus*.

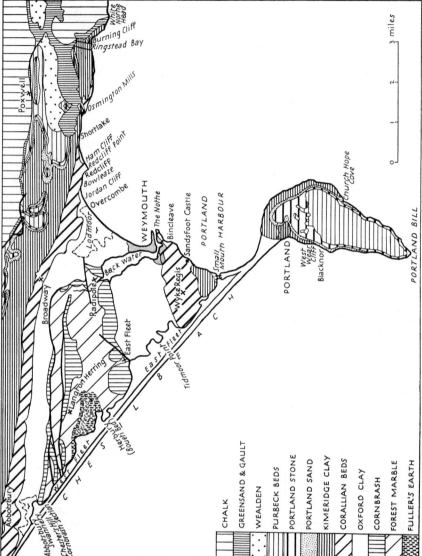

FIG. 61. Sketch-map of Weymouth and Portland.

The clay underlies the bay from Redcliff and Jordan Hill in the north to the entrance to the harbour in the south, and in both directions it dips below sea-level under the Corallian Beds. Thus the lowest zones rise to the surface in the centre of the bay, where there are no exposures, only an esplanade separating the sea from the town and the marshes.

Only at the north end of the bay, under Jordan Hill (between Overcombe and Bowleaze) and on the east side of Redcliff Point (Ham Cliff) are there any natural cliff-sections; here the upper half of the clay is exposed, from the junction with the Corallian down to the *lamberti* zone. These zones occupy about half the total thickness of the Oxford Clay and Kellaways Beds, which has been estimated to reach about 500 ft. The old exposure at the south end of the bay has been obliterated by the fortification of the Nothe.

The base of the Lower Calcareous Grit, and so of the Corallian formation, falls in the midst of the Cardioceratan deposits. The grits become bluish-grey and argillaceous downwards, but there is nevertheless a sharp lithic change to the grey, soapy Oxford Clay below. Immediately beneath the grit at Ham Cliff and Jordan Cliff, and faulted up to sea-level at the end of Redcliff Point, are the RED NODULE BEDS.[1] These clays contain, as their name implies, lines of red claystone nodules, which occasionally enclose fossils: the types of *Modiola bipartita* Sowerby and *Thracia depressa* (Sow.) were obtained from them over a century ago. The beds are most conspicuous, however, by reason of the large quantities of *Gryphæa dilatata* which they contain, often of great size.

Below the Red Nodule Beds, and forming the greater part of Jordan Cliff, are thick clays which Buckman named the JORDAN CLIFF CLAYS. These are also full of large specimens of *Gryphæa dilatata*, of which they are, indeed, the chief repository. They yield in addition a fauna of small Cardiocerates, denoting the lower part of the *præcordatum* zone, and the large *Peltoceratoides hoplophorus* (S. Buck.).[2] The Red Nodule Beds are presumed to be the upper part of the same zone.

The lowest clays of all exposed on the coast north of Weymouth are seen for a short distance beneath the fault east of Redcliff Point. They belong to the *lamberti* zone (see fig. 65, p. 382).

Lower zones, down to the Kellaways and Cornbrash, are to be seen sporadically in small banks around the north shore of Radipole Backwater, but it is difficult to make out the sequence.

Westward the Weymouth Anticline brings Cornbrash, Forest Marble and Fuller's Earth to the surface, dividing the Oxford Clay into two strips, dipping away north and south. The northern strip is narrow and strikes the West Fleet backwater at Abbotsbury Swannery, where the shore is shelving and affords no exposure. The southern strip is shorter and twice as broad, but the coast-sections along the East Fleet are poor and discontinuous owing to the protective action of the Chesil Beach; while the harder rocks often form low cliffs, the clays tend to melt down in grassy banks. There are a number of small exposures, however, the best of which is at Tidmoor Point, behind

---

[1] Called by Buckman the Red Beds, an inappropriate name, preoccupied by Damon for Corallian deposits higher in the same cliffs.

[2] *T.A.*, 1925, pl. DLXIV, and 1927, pl. DCCIII (named 'Peltomorphites', a synonym of *Peltoceratoides* according to Spath, 1931, *Pal. Indica*, N.S., vol. ix, p. 558).

the rifle-range. Here the clays of the *lamberti* zone crop out in a low, slipped cliff, and yield pyritic Quenstedtocerates.[1] There is also a band of small *Oxytoma*, belemnites, &c. Buckman distinguished the clays here as the TID(E)MOOR POINT BEDS, and correlated them with the lowest clays seen east of Redcliff Point.[2]

The lower zones are ill-exposed on the shore and the Kellaways Beds are probably carried below sea-level by a fault at East Fleet village. Two large brickyards less than a mile inland, south of Chickerell, however, display magnificent sections of several zones. The more westerly brickyard, half a mile due south of Chickerell Church, shows a 60 ft. face of shaly clays with crushed Kosmocerates and small bivalves (*Nucula*, &c.), principally belonging to the *duncani* zone. At many levels there occur sporadically large flattened spheroidal septaria, some of which contain badly preserved ammonites suggestive of *Erymnoceras reginaldi*. One dogger of intensely hard stone found somewhere low down in the pit, of much harder, more splintery matrix than the rest, was crowded with *Erymnoceras coronatum* in good preservation (*reginaldi* zone).

The more easterly brickyard (that of the Dorset Brick, Stone and Ball Clay Co.), at a lower level, shows one of the best sections of the Kellaways Beds in the South of England. At the top of the pit is the Kellaways Rock—sand and yellow sandy clay with doggers and lenticles of hard sandstone, crowded with *Gryphæa bilobata* Sow. Below are dug some 10 ft. of blue Kellaways Clay full of *Proplanulites* spp., *Cadoceras sublæve* (Sow.) and *Cadoceras tolype* Buck. There are also numerous small round nodules, some of which are crowded with beautifully preserved shells of an *Oxytoma*.

Small faulted tracts of Oxford Clay occur a short distance inland almost as far west as Bridport and from a faulted patch beneath the Fleet many fossils weather out and are cast up on the beach west of Langton Herring.[3] At Compton Valence, in the midst of the downs, a diminutive inlier has been brought to light by the erosion of a valley in the Chalk.

(b) North Dorset and Wiltshire: the Vale of Blackmoor to the Vale of the White Horse

From its emergence on the north side of the Dorset Downs the Oxford Clay forms all along its outcrop a continuous low-lying vale, most of which was until a few centuries ago thickly forested. The Blackmoor Vale in the south merges northward into the timbered vales of Penselwood and later of Braydon Forest, which leads on to the Vale of the White Horse, forming the valley of the Upper Thames as far as Oxford. Braydon Forest, lying between Malmesbury, Cricklade and Chippenham, was not cleared until the reign of Charles I, and both there and in Penselwood extensive areas of woodland still remain.

Throughout this tract the Oxford Clay and Kellaways Beds vary little in thickness, totalling from 550 to 600 ft., or a little more than on the Dorset coast. The estimated thickness in the south is 560 ft., while the Westbury boring proved 600 ft., and the Swindon boring 573 ft.

---

[1] Specimens are figured in *T.A.*, 1920, pl. CLIV; 1922, pl. CCCXXXIX; 1925, pl. CLIVA.
[2] S. S. Buckman, *T.A.*, 1923, vol. iv, p. 41; and 1925, vol. v, p. 66.
[3] W. J. Arkell, 1932, *Geol. Mag.*, vol. lxix, pp. 44–5.

Exposures are nowadays neither numerous nor very instructive, being limited to a few scattered brickyards. In the past, however, more or less complete sections have been furnished by railway-cuttings, long since grassed over. Three of these were fully described, and two became famous for the abundance of perfectly preserved ammonites and belemnites which they yielded. The main G.W.R. line from Swindon to Bath, built in 1840-1, passed over Oxford Clay for 10 miles between Wootton Bassett and Chippenham. For most of this distance the line was raised on an embankment and to obtain materials long pits were dug beside the railway. The pits are now lakes filled with rushes and water-lilies, but during their excavation they yielded many hundreds of fossils to local collectors.

The richest harvest of all was reaped in a cutting near Christian Malford, 4 miles from Chippenham, where the lower shales of the Oxford Clay (*jason–duncani* zones) were laid bare. S. P. Pratt [1] described some of the ammonites from here, and scores of specimens, flattened but otherwise perfectly preserved, have found their way into museums all over the country. Most of Pratt's types have recently been figured by Buckman in the last parts of *Type Ammonites*. They include the familiar species *Kosmoceras elizabethæ*, *K. stutchburii*, and *K. sedgwicki*, all of which Brinkmann has recently declared to be nothing more than varieties of other forms previously named (*K. ornatum* Schloth., *K. gulielmii* Sow. and *K. jason* Reinecke). The belemnites, *Cylindroteuthis puzosi* (d'Orb.) [= *B. owenii* auct.] and *Belemnopsis hastatus* (Blainv.) were sometimes found with their ink-sacs preserved.

The second cuttings were made on the Weymouth branch of the G.W.R. at Trowbridge, and were described by R. N. Mantell and the fossils by G. A. Mantell and John Morris in 1848-50. [2] From here Morris figured the type of *Erymnoceras reginaldi* (named after Reginald Mantell). Although the deepest cuttings average only 14 ft. in depth, they exposed about 45 ft. of the Lower Oxford Clay, up to at least the *reginaldi* and probably the *duncani* zone, as well as the Kellaways Beds. From the latter large quantities of *Sigaloceras calloviense* and *Proplanulites kœnigi* were obtained, while in the overlying shales the crushed ammonites and belemnites were as abundant as at Christian Malford, and much lignite was found.

The most recent cuttings were made in 1898, during the construction of the G.W.R.'s South-Wales-Direct line, which crossed Oxford Clay for 9 miles west of Wootton Bassett. For most of this distance the line was raised on an embankment, but this time no pits were dug as abundant material was available from the cuttings and tunnel which were being driven through the Forest Marble and Great Oolite farther west. Several cuttings were, however, made in the Oxford Clay and were described by H. B. Woodward, and again four years later by Messrs. Reynolds and Vaughan. [3]

Immediately west of Wootton Bassett Railway Station, under the bridge, the highest 9½ ft. of the Oxford Clay was seen below the Corallian, but the only fossil found was a *Thracia*. This clay may be a lateral replacement of what is

[1] S. P. Pratt, 1841, *Ann. Mag. Nat. Hist.*, N.S., vol. viii, pp. 161–5. The best account of the cuttings and their fossils is by Oppel, in *Die Juraformation*, pp. 535–6.
[2] R. N. Mantell and J. Morris, 1850, *Q.J.G.S.*, vol. vi, pp. 310–19; G. A. Mantell, 1848, *Phil. Trans. Roy. Soc.*, vol. cxxxviii B, pp. 171–83.
[3] H. B. Woodward, 1899, *Sum. Prog. Geol. Surv.* for 1898, pp. 188–92; S. H. Reynolds and A. Vaughan, 1902, loc. cit.

elsewhere Lower Calcareous Grit sands. Ammonites indicating the Cardioceratan zones, and also *Quenstedtoceras lamberti*, were found in a cutting near
Brinkworth, while farther west, 1½ miles east of Somerford Railway Station,
another cutting showed the shaly Lower Oxford Clay, with *Kosmoceras jason*
and other species of the zone. A brick pit at Rodbourne near by still displays
the *reginaldi* zone, with crackers from which large specimens of the index
fossil, measuring over 1 ft. in diameter, may be obtained. Another brickpit
at Purton is also still worked, but there the clay is singularly barren. Prof.
Morley Davies tells me that prolonged search resulted in his finding one
ammonite, identified by Buckman as *Cardioceras dieneri*,[1] and denoting the
præcordatum zone.

The best cutting on this line was that at the west end of the outcrop, passing
through Bancombe (or Bincombe) Wood, between Rodbourne and Kingway
Barn, near Corston. This laid bare the whole of the Kellaways Beds. The
Kellaways Rock was represented by 10–20 ft. of poorly fossiliferous sandy and
loamy beds, overlying 20 ft. of Kellaways Clay with septaria. From the
latter abundant *Proplanulites kœnigi* were obtained, with *Gowericeras gowerianum* (also denoting the *kœnigi* zone).

The thickest development of the Kellaways Rock exposed in the West of
England was seen in a short cutting on the Midland and South-Western
Junction Railway at South Cerney, near Cirencester, described by Harker.
Here 22 ft. of yellow sands and ferruginous brown sands with nodules and
large doggers were passed through. Fossils were again almost entirely absent.[2]

In other places where temporary openings have been made in the Kellaways
Rock it has sometimes proved highly fossiliferous, although it is frequently
much thinner—a variability remarked on by William Smith. An altogether
peculiar development was described by Richardson[3] at Calcutt, near Cricklade, where a well proved above the Cornbrash 18 ft. of Kellaways Clay,
followed by 18 ft. of Kellaways Rock (very hard calcareous sandstone), overlain in turn by 13 ft. of clay and then 25 ft. of fine-grained grey sand. If all
these are rightly included in the Kellaways Beds, their total thickness there
is 74 ft. The only comparable record is that of the deep boring at Swindon,
where the Kellaways Beds were said to be 60 ft. thick, and where both of the
zone fossils were found.[4]

The best locality for studying the fossiliferous facies of the rock is still the
old type-locality at Kellaways, 2 miles north-east of Chippenham. The
exposures are disappointing on a casual visit, consisting of mossy banks and
projecting rocks in the side of the River Avon, and to study them it is necessary to stand in the river with waders. The best sections are in the banks of
an artificial watercourse north-east of Peckingell Farm, about half a mile
below Kellaways Bridge, and also in the main river near by. The rock is from
3 ft. to 6 ft. thick, the upper part being a mass of fossils embedded in brownish
micaceous sandstone. Both the palaeontological and the lithological facies
are very suggestive of some of the Upper Cornbrash. In order of abundance
the principal fossils are: *Microthyris ornithocephala* (Sow.), *Gryphæa bilobata*

---

[1] 1925, *T.A.*, vol. v, p. 66. (The lower record attributed to him on the same page is a
mistake and should be deleted; see the *errata*.)
[2] A. Harker, 1884, *Proc. Cots. N.F.C.*, vol. viii, pp. 176–87.
[3] L. Richardson, 1922, *Geol. Mag.*, vol. lix, pp. 354–5.
[4] H. B. Woodward, 1895, *J.R.B.*, p. 37.

Sow. (some very small, not much larger than the Microthyrid), *Pleuromya recurva* (Phil.), *Rhynchonelloidea socialis* (Dav.), *Modiola bipartita* (Sow.), *Camptonectes lens* (Sow.), *Pseudomonotis* sp. and *Oxytoma expansa* (Phil.), with others less common. In William Smith's time the rock was quarried for road-metal in several small pits about Kellaways, but these have long since been filled in or obliterated. Two brickyards south of Chippenham still show sections of the upper part, more sandy and less fossiliferous. *Gryphæa bilobata* is nearly everywhere the commonest fossil.[1]

(c) Oxford to Bedford and Huntingdon

The Oxford Clay continues from the Vale of the White Horse with little appreciable change past Oxford and under the low-lying tracts of Bedfordshire to the Fens. Everywhere it forms flat and heavy land, except where along the northern or western fringe the Kellaways Rock is developed, rising into a low escarpment above the Cornbrash. Through Wiltshire and Berkshire the highest zones build the base of the ridge of Corallian sands and limestones, and here their outcrop is very narrow and exposures are lacking. These conditions are changed somewhat to the east of Oxford owing to the disappearance of the Corallian, or rather its passage into clay, and the clays immediately underneath it have a wider outcrop and are occasionally seen in brickyards.

There are no natural lines of demarcation between adjacent parts of the clay belt from the Dorset coast to the Market Weighton Axis, but since this tract is too large to describe as a whole, arbitrary divisions have to be chosen. A convenient district for separate treatment is the type locality of Oxford and the adjoining counties to the east, where the zonal sequence has been elucidated in some detail. It is a district too, which is remarkable for an easterly overstep of the upper zones of the Oxford Clay by the equivalents of the Corallian (Argovian). This unconformity was detected by Prof. A. Morley Davies, who showed in 1916 that about Ampthill, Sandy and Woburn the *martelli* zone of the Argovian rests directly on the *renggeri* zone of the Oxford Clay, the whole of the Cardioceratan and part of the Quenstedtoceratan strata having been cut out (fig. 62, p. 348).

At Oxford the total thickness has been estimated at about 450 ft. It is probable that the easterly overstep of the Corallian in Buckinghamshire and Bedfordshire reduces the Oxford Clay considerably, but still farther east it probably thickens again, for 300 ft. was proved at Bluntisham, and H. B. Woodward considered 500 ft. a reasonable estimate for the Huntingdon district.[2]

Immediately north-east of Oxford the Oxford Clay forms the low-lying tract known as Otmoor, which was an undrained swamp until 1825–30, and is still waterlogged for a large proportion of the year. The highest zones crop out round the southern fringe of Otmoor and about the projecting hills at Brill.

---

[1] A representative selection of the ammonite fauna of the Kellaways Rock type-locality has been figured from time to time by Buckman, *T.A.*, pls. CCXXVIII, CCLV, CCLXXV, CCXC, CCXCIII, CCXCIV, CCCLXXIX, DVII, DXXXVII, DXXXVIII, DLXXXVI, DLXXXVII, DCXIV, and a revised list of all the fossils by J. Pringle and S. S. Buckman is to be found in the 'Geol. Marlborough', 1925, *Mem. Geol. Surv.*, pp. 11–12. Another revision is needed already, however.

[2] 1895, *J.R.B.*, p. 56.

SUMMARY OF THE OXFORD CLAY ABOUT OXFORD

Præcordatum Zone. The actual junction with the Lower Calcareous Grit can nowhere be seen, though it was formerly exposed in a brick-pit at New Marston, to the west of the Cherwell (in the outskirts of Oxford), and has several times been pierced by wells. One well was described by Prof. Morley Davies at Studley,[1] where the Arngrove Stone (basal Corallian, *cordatum* zone) rested on clay with finely-ribbed Cardiocerates of the *tenuicostatum* type, which he found in several places to characterize the upper part of the *præcordatum* zone. He also described the same horizon in the G.W.R.

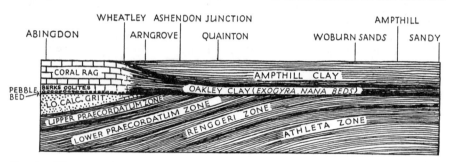

Fig. 62. Diagrammatic section from the borders of Berkshire and Oxfordshire into Bedfordshire showing the unconformity at the base of the *martelli* zone. (Adapted from Morley Davies, *Geol. Mag.*, 1916, p. 399, with alterations to the Corallian part.)

cutting immediately north of the tunnel in Rushbeds Wood, near Brill, where the ammonites[2] were associated with *Gryphæa dilatata* Sow. and its variety *discoidea* Seeley.[3]

The best section of the *præcordatum* zone is opened at a lower level in the zone, at Studley (or Horton, or Horton-cum-Studley) brickyard, by the roadside ½ mile south-east of the village. Ten feet of clay are seen, in which are hundreds of large Gryphæas. When collected in large numbers they can be arranged in lines connecting up several forms usually regarded as distinct species, especially *G. dilatata* Sow., *G. discoidea* (Seeley), *G. controversa* Roem., and *G. exaltata* Rollier. The pit is principally noteworthy, however, for yielding numerous well-preserved Cardiocerate ammonites in the form of attractive brown or ochreous casts, which display the suture lines to advantage. They include *C. præcordatum* R. Douv., *C. cardia* Buck., *C. costellatum* Buck., *C.* cf. *quadratoides* Nikitin, all species of rather stout proportions as compared with the fine-ribbed *tenuicostatum*-like species found in the higher part of the zone. There are also minute Perisphinctids, and the genus of *Cardioceras*-like ammonites which Buckman named after this locality, *Hortoniceras*. Prof. Morley Davies records that the *præcordatum* zone was also exposed in the brickyard near Quainton Road junction, and in a cutting on the Great Central Railway north of Wotton Railway Station (ENE. of Brill).[4]

Renggeri and Lamberti Zones. These zones are the least often

[1] A. M. Davies, 1907, *Q.J.G.S.*, vol. lxiii, p. 40.
[2] Originally recorded as *C. cordatum* and distinguished later.
[3] A. M. Davies, 1907, *P.G.A.*, vol. xx, p. 185.
[4] A. M. Davies, 1916, *Geol. Mag.* [6], vol. iii, p. 397.

seen of all in the district, but they were studied by Prof. Davies in the Ludgershall railway-cutting north of Brill and numerous fossils were collected. There was a well-marked zone of *Creniceras renggeri*, but no evidence for a separate *lamberti* zone was seen, *Quenstedtoceras* and *Creniceras* being mixed throughout.[1] Prof. Davies pointed out that, except for the absence of Phylloceras, the greater abundance of *Quenstedtoceras* and the presence of *Kosmoceras*, there was a remarkable resemblance between the fauna of this zone at Ludgershall and equivalent beds in the Bernese Jura, described by de Loriol.[2] The Bernese fauna was later obtained at the same horizon in the borings for coal in Kent.

The *renggeri* zone is also worked in the upper part of the brickfield at Woburn Sands and either the *renggeri* or the *lamberti* zone again at Sandy, Beds., in the region where the higher strata are missing and the *martelli* clay of the Corallian rests upon it non-sequentially. At Oxford the *renggeri* and/or *lamberti* zones were at one time exposed at Summertown, but the great clay pit is now derelict. There are still indications of the base of the *lamberti* zone in the form of rare occurrences of *Quenstedtoceras* in the very top of the pit at Wolvercote.

Athleta Zone. The *athleta* zone is well exposed in the large brickyard at Wolvercote, on the north side of Oxford.[3] *Peltoceras athleta* occurs near the present base of the pit, in a band of clay 2½ ft. thick, overlying a 6-in. band of cementstone. Dr. J. Pringle restricts the zone to these two beds, giving it a thickness of only 3 ft. The characteristic fauna besides the zone fossil comprises the small, much incoiled, and almost equilateral *Gryphæa lituola* Lamk., together with the brachiopods *Aulacothyris bernardina* (d'Orb.) and *Rhynchonella spathica* (Lamk.). The same *Gryphæa* and *Aulacothyris* were found by Prof. Davies to characterize the zone in the Ludgershall railway-cutting and again in Woburn Green brickyard. At Wolvercote 18 ft. of clays with a band of cementstone succeed the horizon at which this fauna is found, thus comprising the greater part of the section now exposed. The fauna includes three species of *Peltoceras*, but not apparently *P. athleta*; and as it is chiefly characterized by Kosmocerates of the *proniæ* and *subnodatum* types Dr. Pringle has considered it a separate zone of *K. proniæ* Teisseyre. Its separation has not, however, been substantiated by studies in other districts, and it is more convenient to consider it, on the strength of its Peltocerates, as a part of the *athleta* zone. The range of *Kosmoceras proniæ*, according to Brinkmann, extends from the top of the *athleta* zone down to the upper limit of the true *castor* and *pollux* (our *reginaldi*) zone.

Duncani to **Jason Zones.** The Lower Oxford Clay or typical Kosmoceratan, from the *duncani* zone down to the top of the Kellaways Rock, is nowhere well exposed near Oxford, though the top of the *duncani* zone is sometimes reached in the base of the Wolvercote brickyard and was formerly exposed in the Summertown pit. The presence of the *jason* zone was proved in 1924 by a boring at Gosford Hill, Kidlington, where Dr. Pringle found *Kosmoceras stutchburii* (Pratt) and *K. effulgens* Buck.[4] (? = *jason*).

---

[1] A. M. Davies, 1916, loc. cit., p. 396.
[2] 1898–1900, *Mém. Soc. pal. Suisse*, vols. xxv–xxvii.
[3] J. Pringle, 1926, 'Geol. Oxford', 2nd ed., *Mem. Geol. Surv.*, p. 36.
[4] J. Pringle, 1926, loc. cit., p. 35; and S. S. Buckman, 1925, *T.A.*, pl. DXCVII.

The best exposures of the Oxford Clay in the central part of England (south of the Peterborough district) are in the lower zones at Calvert, near Claydon, 5 miles south of Buckingham (Pl. XV) and at Bletchley. Here enormous brick-pits have grown up since the coming of the railway in 1898, when a branch brickworks was started at Calvert by Messrs. Itter of Peterborough. The beds dug are the shaly clays of the *reginaldi* and *jason* zones, with their hundreds of crushed and iridescent ammonites. The deep borings, for which Calvert is renowned, were started, one in the floor of the main pit, about 70 ft. below surface, the other nearly at ground level.[1] They passed through 32 ft. of clays below the floor of the pit, making 97 ft. of Oxford Clay in all. The only ammonites encountered were *Kosmoceras sedgwickii* (Pratt) (? = *jason* Rein.), and *K.? stutchburii* (Pratt); and judging by these fossils and because the clays rested directly on the Bathonian limestones, it was presumed that the Kellaways Beds were missing, as was definitely the Cornbrash. In view of the great thickness of the clays, however, the possibility must not be lost sight of that the Kellaways Beds may be represented but unfossiliferous.

The lower zones of the Oxford Clay are repeatedly well exposed in extensive brickworks at Charndon and Bletchley, Bucks., and Wootton Pillinge (Stewartbury), Beds. At the last two places the pits or 'knot holes' are of vast extent and present a vertical face 70 ft. deep, but in spite of these facilities for study, the elucidation of the successive faunas still remains unattempted. The type specimen of *Kosmoceras duncani*, figured by Sowerby, came from St. Neots.

Calloviense and **Kœnigi Zones** (KELLAWAYS BEDS). The Kellaways Beds seem to be feebly developed from the neighbourhood of Cirencester through the Thames Valley, for they have hardly ever been exposed.

In the Shipton-on-Cherwell railway-cutting north of Oxford, 5 ft. of yellow and grey sands overlay 10 ft. of clay above the Cornbrash, and a similar section is still visible at Akeley brickyard, north of Buckingham. At the latter place A. H. Green many years ago recorded *Proplanulites kœnigi*.[2]

About Bedford the Kellaways Beds regain considerable importance. Numerous sections have shown in the past that the sands are about 10 ft. thick and contain large concretionary doggers of grey-hearted sandstone like those at South Cerney near Cirencester, and yielding *S. calloviense*, *Gryphæa bilobata*, *Oxytoma expansa*, *Pleuromya recurva*, *Camptonectes lens*, *Entolium demissum* and lignite. Beneath is about the same thickness of stiff Kellaways Clay, from which *Gowericeras gowerianum*, mentioned in the old lists, probably came. The beds can still be seen in the waterworks pit, but one of the best exposures was in the Midland Railway cutting north of Oakley Railway Station, now overgrown.

A boring at Bletchley in 1887 encountered pebbles and blocks of granitic rocks in what Jukes-Browne, Cameron and later H. B. Woodward supposed to be the Kellaways Rock, thickened to 54 ft. This view was accepted until 1913, when the Calvert borings, not many miles away, enabled Prof. Morley Davies to revise the interpretation of the Bletchley record and to show that the granitic rocks are much more likely to have belonged to the base of the

[1] The site of the boring from which gas rose under pressure has been covered up by tip-heaps.
[2] A. H. Green, 1864, 'Geol. Banbury, Woodstock, &c.', *Mem. Geol. Surv.*, p. 41.

PLATE XV

Photo.

Geol. Survey.

Lower Oxford Clay, Calvert Brickyard, Bucks.

The photograph shows the face being worked by hand in stages, a system now replaced by steam excavators (of which a small one is seen on the top, removing overburden).

attenuated Lower Lias, on, or not far above, the Palaeozoic floor.[1] This was in accordance with the expectations of Prof. Kendall, who had in 1905 suggested that formations lower than the Kellaways were probably present.[2]

(d) Peterborough and the Fens to the Humber

The Oxford Clay underlies large areas of the Fens, sometimes protruding through the alluvium as islands, as at Whittlesey and Ramsey. Northward through Lincolnshire the outcrop gradually narrows, but there is only a gradual decrease in thickness towards the Humber, for in North Lincolnshire the thickness is still in the neighbourhood of 300 ft. Towards the east attenuation is more rapid. A boring at Southery, in the Norfolk Fens, proved only 200 ft., with complete absence of the Kellaways zones.[3]

A large proportion of the outcrop is masked by deep deposits of alluvium and Drift. This applies especially to the Fenland and its borders, and also to the valley of the Ancholme, draining into the Humber in the north. The few railway-cuttings of the past have all been opened along the western margin of the outcrop, and have shown only the lowest zones of the clay and the Kellaways Beds. On the whole, exposures of the Upper Beds are small and scattered, and with the extinction of local brick-making industries most of the small pits mentioned by the earlier writers have become grassed over.

The Middle and Lower Oxford Clay, at least in the Peterborough district, are now exposed far more completely than ever before. This is due to the growth within the present century of an enormous brick-making trade, centred around that town on account of the good railway transport which it affords. Fletton, Kingsdyke, Yaxley, and indeed the whole landscape in the vicinity of Peterborough, are now dominated by a forest of chimneys, and the very name of Peterborough recalls bricks. The huge clay pits, like those at Calvert but duplicated many times, have to some extent compensated the geologist for the loss of the local exposures whose decline and extinction all this centralization has brought about. But the compensation is not all that could be desired, for whereas the old local industries gave information concerning many zones over the length and breadth of the outcrop, the new excavations, although far larger, all show about the same part of the formation and they are concentrated around one locality. However, the opportunities for studying the succession of ammonite faunas which the Peterborough brick industry has provided are held to be the best in Europe, and a study of them has revolutionized our knowledge of the lower zones of the Oxford Clay.

For records of the highest zones, in the absence of recent research, and in view of the probability that investigations would not now yield much new information, it is necessary to resort to the accounts of Judd and Roberts and other writers of the last century.

J. W. Judd recognized an upper group of clays 'with Ammonites of the group of the *cordati*', but stated that it was usually concealed by Drift. This division was exposed to a depth of 40 ft. north of Bardney Railway Station, east of Lincoln, and judging from the ammonites recorded (allowing for archaic

[1] H. B. Woodward, 1895, *J.R.B.*, p. 49, with refs.; A. M. Davies, 1913, *Q.J.G.S.*, vol. lxix, p. 332.

[2] P. F. Kendall, 1905, *Report of Royal Commission on Coal Supplies*, p. 195.

[3] J. Pringle, 1923, *Sum. Prog. Geol. Surv.* for 1922, pp. 126–39.

nomenclature) it represents the *præcordatum* zone and perhaps some slightly earlier strata.[1] '*Ammonites cordatus*' and '*A. lamberti*' were also recorded from a small brickyard at Langworth, north-east of Lincoln, where the base of the *præcordatum* and the top of the *renggeri* or *lamberti* zones may be presumed to have been seen. The *lamberti* zone also appears to have been worked at Timberland, Lincs. There are records of '*Ammonites*' [*Hecticoceras*] *hecticus* at Langworth [2] and near Cambridge, in which neighbourhood also Cardio-cerates of the *præcordatum* zone abound in the Oxford Clay; while Prof. Davies informs me that he has seen specimens of *C. renggeri* from St. Ives.

The old records are too untrustworthy to base on them any conclusions as to the presence or absence of separate zones of *renggeri* and *lamberti*, but modern evidence is forthcoming at Eye Green, north-east of Peterborough. Here Dr. Neaverson has described the clays at the top of the pit as yielding a fauna of the coarser-ribbed and feebly-rostrated Quenstedtocerates of the type of *Q. lamberti*, without *Creniceras* but with *Hecticoceras*.[3] This he inter-prets as confirmation of Buckman's suggestion that there should be a *lamberti* horizon below and separate from the *renggeri* zone. Thus the evidence at Eye Green contradicts that of the Buckinghamshire railway-cuttings examined by Prof. Morley Davies and illustrates how different zonal schemes are likely to be evolved by the several palaeontologists working in different districts.

The whole of the remaining zones down to the base of the Oxford Clay are exposed several times in the great brick pits around Peterborough, at Eye Green, Kingsdyke, Fletton, Farcet, &c. There is wide difference of opinion concerning the interpretation of the detailed ammonite succession. In 1925 Dr. E. Neaverson published a valuable paper in which he set forth a zonal classification as follows, in descending order: [4]

> *lamberti*
> *athleta*
> *duncani*
> *castor*
> *coronatum*
> *elizabethæ*
> *conlaxatum* (with *jason*)

In the following year Dr. R. Brinkmann, having obtained a grant from the Academy of Sciences at Göttingen, spent some two months at Peterborough collecting the ammonites *in situ* and studying them by the most painstaking statistical methods. In 1929 he published a lengthy report on the results, followed by a monograph on the genus *Kosmoceras*.[5] The report contains 250 pages, 129 tables and 55 diagrams, in which the object, methods and con-clusions are set forth. Since they concern the whole principles of the zonal classification of the Oxford Clay, they must receive attention before any more can be said.

Brinkmann set out with the object of determining accurately the succession

[1] In confirmation of this Prof. Davies informs me that he has purchased *renggeri* forms bearing Bardney labels.

[2] J. W. Judd, 1875, 'Geol. Rutland', *Mem. Geol. Surv.*, pp. 232–9.

[3] E. Neaverson, 1925, *P.G.A.*, vol. xxxvi, pp. 27–37.

[4] E. Neaverson, 1925, loc. cit.

[5] R. Brinkmann, 1929, *Abh. Gesell. Wiss. zu Göttingen*, N.F., vol. xiii, parts 3 and 4, 373 pp., and see review in *Geol. Mag.*, 1931, vol. lxviii, pp. 373–6.

of the innumerable forms of *Kosmoceras* contained in the clays, using minute quantitative methods to define them. He confined his attention to three of the pits, those of the London Brick Co. south of Fletton and between Kingsdyke and Whittlesey, and that of the Itter Brick Co. at Kingsdyke. These overlap so as to provide a continuous section of 13 metres (43 ft.), which he divided into centimetres for purposes of collecting, thus obtaining 1,300 horizons. From numbered horizons in this section he collected and measured 3,000 specimens. At the same time he paid particular attention to the lithology of the various beds with the object of discovering, if possible, the conditions under which they were deposited; and, as already mentioned in Chapters I and III, his conclusions are of considerable interest, especially because he was able to work with lineages.

Methods of determining ammonite evolution upon a quantitative basis naturally depend upon the measurement and documentation of the characters of large numbers of individuals at each horizon. The average variation in any particular character at a given horizon is taken by Brinkmann to indicate a definite phase, and comparison of the figures for successive horizons shows the rate of evolution per unit thickness of strata. The two chief potential sources of error in this operation are: first, that the measurements are largely made upon crushed material; secondly, that the assumption is involved that the assemblage of individuals preserved in any particular bed is a truly representative selection of the living population. Brinkmann argues that neither source of error is so serious as to interfere with the conclusions. In the first place he finds that the errors in measurements due to distortion of the shells are negligible in comparison with the degree of individual variation. In the second place he believes that, although the assemblage preserved at any given level may be selected posthumously by wave or current action, it is safe to assume as a working hypothesis that it truly represents the living population. He emphasizes that the greater concentration of ammonites in the shell-beds is purely an artificial one, due ultimately to tectonic disturbances and not to fluctuations in the population.

The conclusions bearing upon evolution and ammonite systematy to which these considerations lead are of the highest importance. Brinkmann recognizes at Peterborough only thirteen species of Kosmocerates, belonging to four lineages, which he regards as natural subgenera of the single genus *Kosmoceras*. In the synonymy of the genus he places twenty-one of Buckman's genera.

This simplification results from the rational conception of the subgenus as an evolutionary chain or lineage, in which the species are successive links, of arbitrarily-chosen but uniform length. The four subgenera, *Anakosmoceras*, *Zugokosmoceras*, *Spinikosmoceras* (all due to Buckman) and *Kosmoceras sensu stricto*, he finds evolved continuously side by side. The Oxford Clay could therefore be rationally zoned by any of these subgenera, successive species being taken to mark successive zones (true biozones, as defined above, p. 22), but not by all of them together, as is usually done.

The principal systematic rearrangements made by Brinkmann, affecting Dr. Neaverson's zonal table, are the following: he regards *K. conlaxatum* S. Buck., the zonal index to Dr. Neaverson's lowest zone, as synonymous with *K. jason* (Rein.); he regards *K. elizabethæ* (Pratt), the index to the second zone, as synonymous with the true *K. ornatum* Schloth., of which he selects and figures a lectotype; *K. castor*, used as index to Dr. Neaverson's fourth zone, he considers a misidentification, and declares that the true *K. castor* (Rein.) lived contemporaneously with *Erymnoceras coronatum* (index of Dr. Neaverson's third zone). Further, he states that *K. ornatum* has hitherto been misidentified and that *K. ornatum* auctt. is synonymous with *K. spinosum* (Sow.); finally that *K. stutchburii* (Pratt), which has recently been chosen by Prof. Davies as index of his lowest zone, is to be regarded as a synonym of *K. gulielmii* (Sow.). (Buckman placed these last not only in different genera, but even in different families.)

Clearly if these conclusions are to be accepted, profound modifications of the previous zonal partition of the Lower Oxford Clay must be made. Brinkmann retains the following zonal table, using the various overlapping subgenera, but eliminating overlapping species so far as possible:

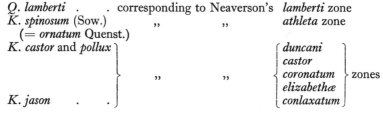

Q. lamberti . . corresponding to Neaverson's	*lamberti* zone	
K. spinosum (Sow.) ,, ,,	*athleta* zone	
(= *ornatum* Quenst.)		
K. castor and *pollux* ⎫	⎧ *duncani* ⎫	
⎪	⎪ *castor* ⎪	
⎬ ,, ,,	⎨ *coronatum* ⎬ zones	
⎪	⎪ *elizabethæ* ⎪	
K. jason . . ⎭	⎩ *conlaxatum* ⎭	

Every geologist cannot go into the subject and decide for himself which system is the right one, and the reader on laying down Brinkmann's exhaustive report and monograph may feel at first inclined to accept his conclusions without further question. Nevertheless, those who by diligent research (in many more parts of England besides that with which Brinkmann is acquainted) have gradually built up the existing zonal table are entitled to more consideration than this first impulse would accord them. The succession of faunas described by Dr. Neaverson at Peterborough is no myth, and it accords in general with that determined by Prof. Davies in Buckinghamshire and Bedfordshire, and by Buckman in Yorkshire and Scotland. When we make the necessary changes in nomenclature it remains substantially the same as before, only under a new guise. The further changes advocated by Brinkmann, such as the elimination of a *duncani* zone, and the substitution of a *spinosum* zone for the old *athleta* zone, serve no purpose and only increase confusion. The choice of *K. spinosum* as a zonal index is especially undesirable in view of the fact that it is a 'rediscovered' species, which takes the place of the *K. ornatum* of authors, the true *ornatum* of Schlotheim being, according to Brinkmann, something altogether different. The only justification that can be put forward for such changes is that they make for an ideal zonal scheme of universal application and eliminate 'teilzones' (see pp. 33–4). But this is generally an impossibility in geology, and so long as we cannot work with biozones in other formations, or even in the Oxford Clay in other places, we might as well be consistent and not overhaul our zonal scheme in order to introduce them at Peterborough. The best we can generally hope to do is to determine the succession in a given area or country, noting the teilzones which there succeed one another, and later to compare the sequence with those in other regions. Moreover, the fossils that form useful zonal indices in one country may often be valueless in another, so that it is impossible to restrict ourselves to a given lineage.

Thus we can admit that neither of Pratt's names, *elizabethæ* and *stutchburii*, should be retained, and that Buckman's *conlaxatum* is a synonym of *jason* Reinecke, and still legitimately adhere to the slightly modified zonal table adopted here, viz.:

Zone of *C. præcordatum*
,,   *C. renggeri*
,,   *Q. lamberti*
,,   *P. athleta* $\Big\}$ (= *spinosum* zone of Brinkmann)
,,   *K. duncani*
,,   *E. reginaldi* (= *coronatum* zone of Buckman and Neaverson)
,,   *K. jason* (= *conlaxatum* + *elizabethæ* zones of Neaverson
                   = *castor* and *pollux* + *jason* zones of Brinkmann).

SUMMARY OF THE SUCCESSION AT PETERBOROUGH [1]

Of the succession of clays and their faunas comprised within these zones as exposed at Peterborough, Dr. Neaverson has given a clear if brief account.

Athleta Zone. The clays with large Peltocerates, some of which can be referred to *P. athleta* (Phil.), contain the same fauna as at Oxford. The most abundant fossils are the *Gryphæa*, best identified as *G. lituola* Lamk., and also the brachiopod *Aulacothyris bernardina* (d'Orb.). It is noteworthy, however, that Kosmocerates of the *proniæ* type do not here occur above this horizon as at Oxford but below, in the clays assigned by Neaverson to the *duncani* zone. This seems to furnish a good reason against the retention of a *proniæ* zone, and indeed Brinkmann declares that the species ranges down to the base of Neaverson's *elizabethæ* zone.

Duncani Zone. The clays of this zone are principally characterized by abundant pyritic casts of ammonites, among the most common of which are species of the subgenus *Kosmoceras sensu stricto* (*K. spinosum* (Sow.), and *K.* cf. *gemmatum* (Phil.), *K. duncani*, &c.), together with *K.* cf. *proniæ* Teiss. There are also some pyritic casts of lamellibranchs, especially *Nuculana* cf. *phillipsi* (Morris), *N.* cf. *longiuscula* (Merian) and *N. cottaldi* de Lor. *Belemnopsis hastata* begins a little below this zone and ranges upwards.

Reginaldi Zone. Dr. Neaverson says: [2]

'The only substantial evidence obtained for the presence of this zone is the occurrence of large specimens of *Erymnoceras reginaldi* (Morris) in Messrs. Itter's brickyard 1¼ miles west of Whittlesey. The specimens occur in shales similar to those of the *elizabethæ* zone but were not *in situ* and the position of the zone in the section could not be located. A poor fragment from similar shale at Fletton is also referred to this species.'

According to Brinkmann *K. duncani* (Sow.) ranges down to below the *reginaldi* horizon; but he misinterpreted *K. duncani* (see note opposite Plate XXXVII, fig. 6). About this level is the maximum of *Cylindroteuthis puzosi*.

Jason Zone. This zone chiefly consists as usual of a considerable thickness of shaly clay full of compressed Kosmocerates. The lithology of the shales,

---

[1] Based on E. Neaverson, loc. cit., 1925.
[2] 1925, loc. cit., p. 32.

and also many of the species of ammonites and their preservation, are the same as in the Christian Malford railway-cutting and the brickworks at Calvert. *Nucula nuda* is also abundant and was used by Judd as index species. Dr. Neaverson called this the *elizabethæ* zone (= *ornatum* Schloth. *non auctt.* according to Brinkmann) and separated as the *conlaxatum* (= *jason*) zone some basal clays which form the floor of the brickfields at Fletton and Kingsdyke. These basal clays are not favourably exposed for study, but they contain large septaria which are thrown out during the construction of shallow drainage trenches, and from them many ammonites can be collected, uncrushed and excellently preserved. The principal forms are varieties of *K.* (*A.*) *gulielmii* (Sow.) and the true *K.* (*Z.*) *jason* (Reinecke). 'Speaking generally,' Dr. Neaverson remarks, 'the series contains forms which are highly ornamented, and others that show loss of sculpture correlated with a close approximation of the septa as growth proceeded.' *Cylindroteuthis puzosi* also occurs in the septaria.

Calloviense and **Kœnigi Zones** (KELLAWAYS BEDS). Dr. Neaverson describes a small brick-pit north-west of the railway-bridge near Werrington, on the main road north-north-west of Peterborough, where about 4 ft. of dark-blue unfossiliferous clay passes up into a thin band of soft concretionary sandstone, which contains part of the fauna of the Kellaways Rock. *Gryphæa bilobata*, *Oxytoma sp.* and *Entolium demissum* are recorded, but no ammonites.

Far better sections of the Kellaways Beds have been seen from time to time along the western margin of the outcrop, where the Kellaways Rock makes a surface feature and has been cut through during the construction of some of the railways. In a cutting south-west of Bourn, on the Bourn and Saxby line, H. B. Woodward in 1892 saw the 'dense grey shaly clay with selenite, lignite and numerous flattened ammonites' of the *jason* zone, with a band of septaria 10 ft. from the base (perhaps corresponding to that at Peterborough), resting on 2 ft. of calcareous, flaggy, blue-hearted sandstone with abundant *Gryphæa bilobata* and belemnites—the Kellaways Rock. This was in turn separated by a clay from the Cornbrash.[1] A nearly similar section was described by Morris in 1853 in the Casewick cutting. Here *Sigaloceras calloviense* was recorded from the Kellaways Rock, together with most of the characteristic lamellibranchs. The rock, which was ferruginous and passed into brown sandy clay, was separated from the Upper Cornbrash by 10 ft. of 'dark, laminated, unctuous' Kellaways Clay, in which abundant Macrocephaloid ammonites were said to occur.[2]

The sandy development of the Kellaways Rock, always with *Gryphæa bilobata* and certain belemnites, persists at about the same horizon throughout Lincolnshire to the Humber. It continues to be separated everywhere from the Cornbrash by a clay, which varies from 2 ft. to 18 ft. At Brigg a boring proved 2 ft. of Kellaways Rock, with 18 ft. of clay below, while a little to the south a considerable portion of the clay becomes sand, locally consolidated into rock, which may reach a thickness of 20 ft. The greatest recorded thickness of 25 ft. was proved at Sudbrooke, near Lincoln.[3]

[1] H. B. Woodward, 1895, *J.R.B.*, pp. 63–4.
[2] J. Morris, 1853, *Q.J.G.S.*, vol. ix, p. 333.
[3] H. B. Woodward, 1895, *J.R.B.*, pp. 65–7, with refs.

VERTEBRATA OF THE OXFORD CLAY OF PETERBOROUGH

For over thirty years two brothers, C. E. and A. N. Leeds, of Peterborough, assiduously collected the bones of vertebrates from the Oxford Clay of the rapidly-growing brickfields. The collections so formed became renowned, providing subject-matter for numerous papers by Seeley, Lydekker, Hulke, C. W. Andrews and Sir Arthur Smith Woodward.[1] They were eventually purchased by the British Museum and have been described in two magnificent volumes by the late C. W. Andrews.

The exact zonal positions of the specimens are not recorded, but the whole rich fauna must have been yielded by some part of the *jason–athleta* zones inclusive and the majority undoubtedly came from the *jason* zone. The skeletons are said to have 'occurred spread over well-defined old floors in the clay, and could thus with care be recovered almost in their entirety'.[2] This suggests that they occurred on the horizons corresponding with breaks in sedimentation, described by Brinkmann.

An excursion of the Geologists Association visited Peterborough and the Leeds Museum in 1897, under the direction of A. N. Leeds and Sir Arthur Smith Woodward. In the report Sir Arthur gave a valuable summary of the features of the vertebrate fauna,[3] which he has very kindly re-written, bringing it up to date (1932), especially for insertion here:

'The land-reptiles (Dinosauria)', he writes, 'are represented only by imperfect skeletons without the head, and by one portion of lower jaw. They include one Sauropod (*Cetiosauriscus leedsi*) about 60 ft. in length, and two species of an armoured Orthopod which have been referred to *Stegosaurus*, a genus best known from the Upper Jurassic of North America. The lower jaw may belong to one of the latter species; and a small femur seems to be referable to an Orthopod related to the Wealden *Iguanodon*. The Crocodiles *Steneosaurus* and *Metriorhynchus* are represented by many nearly complete skeletons of several species.[4] Both are marine, and *Metriorhynchus* has no bony armour. The Pterodactyles, as might be expected from their mode of life, are scarcely known; but it is very curious that no trace of a Chelonian has hitherto been discovered. The Sauropterygia predominate, and include both the small-headed Plesiosaurians and the large-headed Pliosaurians, of several genera. There are skeletons of individuals of all ages, from the very young to the extremely old. The Ichthyopterygia are represented by an almost toothless ally of *Ichthyosaurus* with comparatively broad and flexible paddles (*Ophthalmosaurus*).

'Fish remains are very abundant, and they are important as supplementing our knowledge of the Upper Jurassic fauna in the Lithographic Stone of Germany, France, and Spain. Whereas the Continental fossils display the general contour of the various fishes embedded in very hard limestone, the specimens in the Oxford Clay were originally macerated and buried in soft clay, from which the different bones and teeth can be washed and examined separately. There are fine groups of teeth of *Hybodus* found in association with the jaws of this shark. There are also many still finer groups of the teeth named *Strophodus*, in undoubted association with the fin-spines named *Asteracanthus* and the hooked head-spines named *Sphenonchus*. Chimæroids are known only by dental plates and fragments of spines.

[1] See bibliography.
[2] A. N. Leeds, 1897, *P.G.A.*, vol. xv, p. 189.
[3] A. Smith Woodward, 1897, loc. cit., p. 190, with full bibliography.
[4] Andrews has described four new species (1909, *Ann. Mag. Nat. Hist.* [8], vol. iii, pp. 299–308).

The remains of Ganoid fishes are especially fine and numerous, and those of *Lepidotus*, the Pycnodont *Mesturus, Caturus, Eurycormus,* and *Hypsocormus* have added much to our knowledge of the skull in these primitive types. *Leedsia problematica* is a gigantic Pachycormid fish, with a forked tail 9 feet in span.'

LIST OF THE OXFORD CLAY VERTEBRATA FROM PETERBOROUGH[1]

Reptilia.

Cetiosaur[isc]us leedsi (Hulke)
Omosaurus durobrivensis Hulke
A large Stegosaurian
Camptosaurus leedsi Lyd.
Sarcolestes leedsi Lyd.
Rhamphorhynchus sp.
Ophthalmosaurus icenicus Seeley
Murænosaurus leedsi Seeley
M. durobrivensis Lyd.
M. platyclis Seeley
Picrocleidus beloclis (Seeley)
P. sp.
Tricleidus seeleyi Andrews
Cryptocleidus oxoniensis (Phil.)

Pliosaurus ferox (Sauv.)
Simolestes vorax Andrews
Peloneustes philarcus Seeley
P. evansi (Seeley)
Metriorhynchus superciliosus Desl.
M. brachyrhynchus Desl.
M. sp.
Suchodus durobrivensis Lyd.
Dacosaurus sp.
Steneosaurus edwardsi Desl.
S. leedsi Andr.
S. durobrivensis Andr.
S. obtusidens Andr.
Mycterosuchus nasutus Andr.

Pisces.

Hybodus obtusus Ag.
Asteracanthus ornatissimus Ag.
  var. *flettonensis* A. S. Woodw.
Pachymylus leedsi A. S. Woodw.
Brachymylus altidens A. S. Woodw.
Ischyodus egertoni (Buckland)
I. beaumonti Egerton
Lepidotus leedsi A. S. Woodw.
L. latifrons A. S. Woodw.
L. macrocheirus Eg.

Heterostrophus sp.
Mesturus leedsi A. S. Woodw.
Caturus sp.
C. sp.
Osteorachis leedsi A. S. Woodw.
Eurycormus egertoni (Eg.).
Hypsocormus leedsi A. S. Woodw.
H. tenuirostris A. S. Woodw.
Leedsia problematica A. S. Woodw.
Pholidophorus sp.

II. THE MARKET WEIGHTON AXIS

North of the Humber the Oxford Clay becomes very thin. Its total thickness is not known, but the amount of shaly clays between the top of the Kellaways Rock and the top of the Kimeridge Clay measures rather over 100 ft. near South Cave, and probably not more than half of this is accounted for by Oxford Clay. Only the basal portion has been exposed, and from the recorded Kosmocerates (*K. elizabethæ* and *K. comptoni*) this would seem still to belong to the *jason* zone.

The section which showed the base of the Oxford Clay is a deep cutting on the Hull and Barnsley Railway, made in 1883, at Drewton, north of South Cave. It also laid bare a fine section of the Kellaways Rock, which was described in detail by Keeping and Middlemiss.[2] Immediately below the clay were 10 ft. of hard, brown, ferruginous sandy beds, crowded with fossils, especially *Gryphæa bilobata, Pseudomonotis* sp., *Rhynchonelloidea socialis* and belemnites, below which were 45 ft. or more of sands, the upper part brownish

---

[1] C. W. Andrews, 1910, *Descr. Catal. Marine Reptiles Oxford Clay* (*B.M.*), vol. i, p. viii.
[2] 1883, *Geol. Mag.* [2], vol. x, pp. 215–21.

or yellow and containing two lines of large doggers, the lower parts pure white and highly micaceous. The sands were nearly barren except for a band about the middle which was riddled with hollow casts of belemnites.

The section of Kellaways Rock in the Drewton cutting is typical of all the region between the Humber and Newbald, where the outcrop is lost beneath the Chalk. The only variation of note is in the thickness. The sands are, or have been, dug in several pits about Newbald. The largest has been disused for some years, but at the top can still be seen the base of the reddish-brown soft sandstone, crowded with *Gryphæa bilobata* and belemnites. In a pit near Brough were found a number of bones of a saurian, identified as *Cryptocleidus*.[1]

The zonal position of these sands long remained doubtful. Messrs. Keeping and Middlemiss recorded '*Ammonites modiolaris*, *A. kœnigi* and *A. gowerianus*' from the fossiliferous upper 10 ft., but they also mentioned '*A. duncani*' and even '*A. mariæ*', so that probably not much reliance can be placed upon their records. Buckman in 1922 figured a Macrocephalitid form from the Drewton cutting under the new generic name *Catacephalites*, and he considered it to indicate that the Kellaways Rock of South Cave was older than that in other parts of England.[2] Three years later, however, he announced that the same forms had been collected by the Geological Survey in the base of the Kellaways Rock of Wilts., and therefore did not indicate a Macrocephalitan date.[3] In view of the occurrence of typical shales of the *jason* zone immediately above, it seems certain that the Kellaways Rock is a direct continuation of that in Lincolnshire and other parts of England.

The outcrop is concealed beneath the Chalk for a distance of over 13 miles, from between North Newbald and Sancton to north of Kirby Underdale. The disappearance is proved by the occurrence of derived Kellaways ammonites in the base of the Cretaceous to be due to overstep by the Red Chalk. Blake described a section of Red Chalk at Great Givendale with 'here and there nodules full of [*Sigaloceras*] *calloviense* and [*Proplanulites*] *kœnigi*, some of which have been broken up and rolled about in the Red Chalk sea, till their crevices are filled with the latter matrix'.[4]

The importance of the Market Weighton Axis during Callovian and Oxfordian times is attested by a number of facts. Not only does the Oxford Clay attenuate towards it from the south and the north, but in all the Yorkshire Basin north of the axis the facies of the Oxford Clay differs markedly, both lithologically and palaeontologically, from the development over the rest of England. As far as the axis the shales of the *jason* zone are the same from the Dorset coast, while there is reason to suppose that at least the majority of the succeeding zones also are normally developed. North of the axis only the uppermost part (corresponding roughly with the *præcordatum* zone) is a clay, and even that is usually to be likened more to a sandy shale than to the familiar clay of Southern England. All the rest of the zones, or such of them as are present, have become thin ferruginous sandstones or oolitic sandy limestones of varying lithology, often crowded with fossils. Many of the ammonite species are peculiar to Yorkshire and some show affinities with Russian forms.

[1] T. Sheppard, 1900, *Geol. Mag.* [4], vol. vii, pp. 535–8; and see *The Naturalist*, 1931, p. 87.
[2] S. S. Buckman, 1922, *T.A.*, vol. iv, p. 54, pl. CCLXXXIII.
[3] 1925, *T.A.*, vol. v, p. 73.
[4] J. F. Blake, 1878, *P.G.A.*, vol. v, p. 248.

III. THE YORKSHIRE BASIN

(a) The Inland Escarpments of the Howardian, Hambleton and Tabular Hills[1]

No modern palaeontological work has been carried out on the Oxford Clay and Kellaways Beds of the inland escarpments of the Yorkshire Basin; the most recent account available is that written by Fox-Strangways in his Memoir on the Jurassic Rocks of Yorkshire (1892). Until research on modern lines is undertaken, the equivalence of the inland rocks with those on the coast must remain in doubt, and it is safest to continue to use the old terms 'Oxford Clay' and 'Kellaways Rock', although researches on the coast have shown that these names have quite different meanings there from those attaching to them in the rest of England. It is with this proviso that they are now used, their probable zonal equivalence being discussed in a later subsection.

The formation first appears from beneath the Red Chalk at Garrowby, and both the clay and the rock below are traceable from the first, forming a small feature along the foot of the Wolds. The Oxford Clay is not more than 20 ft. thick, and the 'Kellaways Rock' probably from 15 ft. (or less) to 30 ft.

The Oxford Clay thickens to 70 ft. in the neighbourhood of the Derwent, but seems to diminish again beneath the Howardian Hills. The outcrop along these hills can be traced sporadically by a band of wet ground, but there are no good sections, and the thickness therefore cannot be satisfactorily estimated. The Kellaways Rock in the Howardians is a more important feature: it consists of soft sandstone with harder, more siliceous lenticles, which weather out of the surrounding sands, giving rise to prominent headlands or nabs.

A short distance north of the Coxwold faults, the Oxford Clay seems to disappear entirely. Intraformational erosion may have removed some of it, but it is certain that a considerable portion passes into sandstones indistinguishable from the Lower Calcareous Grit. Immediately north of the faults, about Wass and Ampleforth, the clay is still normally developed, though probably not half its thickness at the Derwent; a few miles away, in the cliff-like scarp of Roulston Scar,[2] sandstones with ammonites at least as early as *athleta* date,[3] and with a basal pebble-bed, rest directly on the Kellaways Rock. The 'Kellaways Rock' is here a peculiar, ferruginous reddish variety, crowded with *Gryphæa bilobata* and belemnites, and may be of any age from Proplanulitan to Upper Kosmoceratan, such as that on the coast undoubtedly is (see below).

Along the foot of the great inland cliff of the Hambleton Hills, at least as far as Kepwick, the outcrop is much obscured by landslips and talus. At the north-west end, below Black Hambleton, and in the small outlier on Osmotherley Moor, natural exposures become somewhat clearer, showing that the Oxford Clay returns again for a time and is about 50 ft. thick. The thickness of the 'Kellaways Rock' here is 60–70 ft., and it has become a thick-bedded massive sandstone, partly siliceous, partly soft, with a ferruginous band towards the top, as on the coast.

Eastward across Yorkshire to the sea the outcrop follows the edge of the

---

[1] Based on C. Fox-Strangways, 1892, *J.R.B.*, pp. 283–99. See map, fig. 27, p. 138, above.
[2] P. F. Kendall, 1915, *Proc. Yorks. Geol. Soc.*, N.S., vol. xix, pp. 284–5.
[3] *Kosmoceras rowlstonense* (Young and Bird), figured by Buckman, 1923, *T.A.*, pl. CDXXXVII

Tabular Hills, where sections are principally to be seen in the sides of becks. Although the general direction of the outcrop along the Tabular Hills is W.–E., it is repeatedly cut back in a zigzag fashion, the characteristic feature of the formation being a series of nabs formed by the Kellaways Rock.

In the west, as in the western escarpment at Roulston Scar, although the total thickness of Oxford Clay and Kellaways Rock thickens to about as much as in the east (120–50 ft.), the proportion of sandstone is much greater— more than 100 ft. This is again due to the lower portion of the Oxford Clay passing into sandstone, which is often separated from the rest of the rock below by a thin clay-band.

In the almost complete absence of palaeontological evidence little can be said as to the zonal significance of these changes. Hudleston visualized the whole of the Middle Oolites as deposited around a shore lying to the north-west, from which the materials were derived (largely from the Millstone Grit? But see p. 581). The nearer the deposits lay to the shore, the higher would be the proportion of arenaceous material, and he believed that over the Vale of York, before the retreat of the escarpment, there might have been seen a continuous succession of sands and sandstones from the Estuarine Series to the Coral Rag.

Towards the east end of the Tabular Hills, on the Hackness outlier and on the coast, the major portion of the 120 to 150 ft. consists of clay, but both its upward and its downward boundaries are indefinite—Fox-Strangways wrote: 'The junction of the upper portion of the Oxford Clay with the base of the Calcareous Grit is so gradual that no exact line can be drawn.' Certainly on the coast the base of the Grit falls within the *præcordatum* zone.[1]

The best sections anywhere inland are afforded by the precipitous gorges of Newton Dale and its tributaries, which cut deep into the heathy plateau formed by the Kellaways Rock. Miles of more or less vertical cliffs expose up to 90 ft. of sandstones comparable with some of the Kellaways Rock of the rest of the county, above which are from 30 ft. to 50 ft. of the softer sandstones representing the lower part of the Yorkshire Oxford Clay (probably about *athleta* date). At the base of all are· 10–15 ft. of grey shales full of small crushed lamellibranchs, especially a species of *Pseudomonotis*.

(b) The North Yorkshire Outliers

North of the valley of the Esk a syncline striking E.–W. through Whitby has given rise to the preservation of a large elongated outlier, or rather series of outliers, of the Kellaways Rock from 8 to 12 miles from the main outcrop, forming Roxby and Danby High Moors, Moorsholme Moor and neighbouring heights. The sandstones on these outliers are close-grained and well bedded, with seams of quartz pebbles. The most interesting part is the upper, which is hard and siliceous, and is riddled with the hollow casts of belemnites, the guards completely dissolved away.

'Sometimes as many as 50 of these casts occur in a cubic foot of the bed, and blocks of from three to six inches thick will have a dozen cylindrical perforations right through them. This clearly proves that the Kellaways Rock . . . at some depth below the surface must be a calcareous sandstone, its extremely porous nature being in part due to the dissolution of lime.'[2]

---

[1] Judging by ammonites in the gritstone matrix in the Leckenby collection.
[2] C. Fox-Strangways, 1892, *J.R.B.*, pp. 284–5.

(c) The Yorkshire Coast

The Oxford Clay and Kellaways Beds are well exposed at intervals in the cliffs from Gristhorpe Bay to Scarborough. The longest section is in Gristhorpe Bay, where the Oxford Clay forms the greater part of Gristhorpe Cliff, but both it and the Kellaways Beds are much obscured by slips, talus and Drift. The whole series is magnificently displayed in Red Cliff, Cayton Bay, but as the cliff is vertical and very high the rocks are scarcely accessible. There is a more accessible but less clear exposure of Oxford Clay on the north side of Cayton Bay, where the Kellaways Rock is faulted down below beach-level. The rock and the shale beneath extend on to Wheatcroft Cliff, and the whole series is again brought down by faults at North Cliff, Scarborough, and in the Castle Hill (fig. 76, p. 430).

The thickness of the Oxford Clay in Gristhorpe and Cayton Bays is about 120 to 150 ft. It is a grey sandy shale, nearly uniform throughout, and it passes upwards gradually into the Lower Calcareous Grit; fossils are everywhere scarce except in the bottom layers at Scarborough. According to Hudleston a thick mass of sandstone like the Lower Calcareous Grit comes in in the lower part of the clay in the north side of Cayton Bay, but the cliff has since become obscured by landslips. It would, however, be only in accordance with expectations, for a similar change certainly occurs inland when the clay is traced towards the north.

The few ammonites that have been collected from the Oxford Clay have been discussed briefly by Buckman,[1] whose conclusions seem to point to the greater part of the clay belonging approximately to the *præcordatum* zone. At least two species, *Klematosphinctes vernoni* (Young and Bird)[2] and *Neumayriceras oculatum* (Phil.)[3], are peculiar to Yorkshire and so afford no help in correlation. Buckman provisionally spoke of the *vernoni* zone, but this had only local value and his subsequent attempt[4] to fit it into sequence with other hemeræ was little more than guesswork. Dr. Spath has since concluded that it is a synonym of the *præcordatum* zone.[5]

The so-called Kellaways Rock, where it first rises from the beach at Newbiggin Wyke, at the south end of Gristhorpe Bay, is only 12 ft. thick. It thickens steadily northward, reaching about 50 ft. in Cayton Bay and 76 ft. in North Cliff, Scarborough. Two essentially different parts can be made out: an upper portion (in the south about half the rock, in the north 25 ft.) crowded with fossils embedded in a variety of matrices, some grey, some blue, some red, often oolitic, but usually more or less ferruginous; and a lower portion consisting of thick beds of almost barren sandstone. The upper portion yielded a wealth of fossils to the Scarborough collector of the middle of last century, John Leckenby, most of whose collections are now in the Sedgwick Museum, Cambridge.[6] They were nearly all obtained from boulders on the beach at the foot of Scarborough Castle Hill, from temporary excavations for buildings on the North Cliff, and from the cliffs and beach in Gristhorpe Bay. Unfortunately the Scarborough locality was long ago covered up by the

[1] S. S. Buckman, 1913, *Q.J.G.S.*, vol. lxix, p. 159.
[2] *T.A.*, 1922, pl. cccxxxiii.　　　　　　　　　[3] *T.A.*, 1921, pl. ccxxiv.
[4] *T.A.*, vol. v, 1925, p. 72.
[5] L. F. Spath, 1926, *The Naturalist*, p. 324.
[6] J. Leckenby, 1859, *Q.J.G.S.*, vol. xv, pp. 4–15.

extension of the marine parade round the foot of the hill, while the outcrop in Gristhorpe Cliff has been for many years badly obscured by landslips.

From the numerous ammonites collected by Leckenby and his contemporaries, it is evident that the term Kellaways Rock is a misnomer, and it was to distinguish the Yorkshire rock from that south of the Humber that the mis-spelling Kelloway Rock was formerly retained, an expedient tending only to add to the confusion. Dr. Spath has recently suggested as alternatives HACKNESS ROCK (a term introduced by William Smith as early as 1829-30, which it is appropriate to his memory to revive)[1] or Castle Hill Beds.

Oppel first pointed out that Phillips's 'Kelloway Rock' in Yorkshire contained fossils indicative of the *athleta* zone, which was called Oxford Clay elsewhere; and he used the fact as an argument for including the Lower Oxford Clay in the Callovian Stage.[2] In 1875, also, Hudleston wrote:[3]

'. . . of the petrological group known as the Kelloway Rock, the Upper, or Cephalopoda division, seems to contain ammonites belonging to two geological divisions, of which *A. modiolaris* (Lhuid) and *A. gowerianus* (Sow.) represent callovian forms, whilst *A. duncani* (Sow.), *A. jason* Reinecke (*A. gulielmi* Sow.), *A. lamberti* (Sow.), and many others, are characteristic of the *Ornatus*-clays, or Lower Oxfordian' . . . 'in order to prevent mistakes arising from names it is necessary to point out that the fauna of the upper part of the Kelloway Rock of Yorkshire, as indicated by its ammonites, embraces much of the Oxfordian of English geologists.'

Hudleston also realized that, like so many ferruginous deposits full of Cephalopods, these beds were slowly-formed accumulations representing thick strata elsewhere, and probably of very different ages in different places.

'Are the fossiliferous (Cephalopoda) beds in the upper part of the Kelloway Rock at Scarborough, at Red Cliff, and at Gristhorpe, wholly contemporaneous deposits? At Scarborough the ornati seem to be very plentiful; at Red Cliff, *A. gowerianus*, &c.; at Gristhorpe the cordati, as *A. vertumnus* Leck. (*A. mariæ* d'Orb.), and *A. flexicostatus* Phil. (*A. lamberti* Sow.), &c. This question can only be settled by a very close attention to the precise position of specimens collected during a considerable lapse of time.' And again: 'The deposition of the upper portions alone extended over a period sufficiently long to intercept and entomb amongst other remains the ammonites of two horizons which seem to have been distinguished in other districts.'[4]

Unfortunately the prolonged and careful field investigation advocated by Hudleston has been rendered almost impossible for the reasons just mentioned. In 1913 Buckman published the results of an examination of many of Leckenby's, Young and Bird's and Phillips's types and attempted to arrange the ammonites in sequence according to their matrices, identifying the matrices so far as possible with the beds in Cayton Bay, as described by Leckenby and by Fox-Strangways.[5] The sources of error in such an arm-chair method are so great as to render the results of little practical value. In the first place, the ammonites examined came chiefly from Scarborough and Gristhorpe, at which places, as pointed out by Hudleston, the faunas and matrices are both very different. An attempt to recognize these in the

---

[1] 'Memoir on the Stratification of the Hackness Hills', 1829-30, printed in *J.R.B.*, vol. i, 1892, pp. 507-14.
[2] A. Oppel, 1856-8, *Die Juraformation*, pp. 538-44.
[3] W. H. Hudleston, 1875, *P.G.A.*, vol. iv, p. 372.
[4] Ibid.
[5] S. S. Buckman, 1913, *Q.J.G.S.*, vol. lxix, pp. 152-68.

inaccessible Red Cliff at Cayton Bay, and that only secondhand from descriptions, was unlikely to prove a profitable task. Nevertheless, the results obtained have a theoretical interest and are suggestive of possibilities open to local geologists who might be able to devote a long time to the task.

According to Fox-Strangways,[1] the uppermost 19 ft. at Red Cliff, Cayton Bay, consists at the west end of one indivisible mass of brownish-red ferruginous rock, which passes eastward into several locally distinct bands. At the top towards the east may be distinguished a very oolitic, hard band, 1 ft. thick, which stands out from the cliff, and overlies 9 ft. of calcareous shale like Oxford Clay. From this clay or its representative at Scarborough were doubtless collected many of the anomalous ammonites recorded from the base of the Oxford Clay of that locality. The hard band or its equivalent were thought by Fox-Strangways and by Buckman to be the same as a fossiliferous band spoken of by Leckenby as the 'calcareous pisolite'. With this Buckman identified the matrix of *A. gregarium* Bean-Leckenby,[2] *A. turgidum* Bean, and the associated fauna, and he named it the *gregarium* zone. The shale below, which at Osgodby Nab is described as oolitic, Buckman identified as the provenance of *Vertumniceras vertumnus* (Bean-Leckenby) and *Quenstedtoceras lamberti* (Sow.) with an associated fauna. Finally, the old records of *Ammonites crenatus* he took to indicate the possible representation of the zone of *Creniceras renggeri*. Thus he divided this topmost 10 ft. of the Hackness Rock into the zones of *gregarium, vertumnus* and *lamberti* in descending order, and suspected the presence of a representative of the *renggeri* zone in unknown relation to the rest. Since *Q. gregarium* and *V. vertumnus* are peculiar to Yorkshire and have never been collected *in situ* it is not known whether they attained their acmes separately, or contemporaneously with *Q. lamberti*. For present purposes it is safe to assume only, as does Dr. Spath, that these 10 ft. of beds were probably deposited in the hemerae *lamberti* and *renggeri* (or *mariae*).

Next below come ferruginous beds to a maximum thickness (at Red Cliff) of 9 ft. Fox-Strangways describes these as comprising 3 ft. of soft sandstone overlying 6 ft. of red, irony, partly oolitic rock full of Gryphæas; Leckenby divides them into three, a bed of soft sandstone sandwiched between two fossiliferous irony bands. In various matrices identified with these beds Buckman records a number of ammonites, including *Peltoceras athleta, Kosmoceras duncani*, and a number of Kosmocerates typical of the basal shales of the Oxford Clay farther south (the *jason* zone).

Perhaps the most peculiar feature of the sequence is the record of *Ammonites kœnigi* by Leckenby in both of his irony bands, and also his records of *A. sublævis, A. gowerianus* and *A. chamusseti* in the lower. Buckman identified in the collections the true *Proplanulites kœnigi*, and also several species of *Cadoceras* and two species of *Sigaloceras*, all in a light-brown calcareous matrix with small ooliths. These he believed to have come from Leckenby's lower irony bed.

Thus at Red Cliff between the fauna of the *kœnigi* zone and the Cornbrash there are about 35 ft. of strata, which become still thicker towards the north. At the base are shales formerly associated with the Cornbrash (6–15 ft.) which pass up gradually through yellow sandy shale into sandstone. All the

---

[1] 1892, *J.R.B.*, p. 280.

[2] Made genotype of *Prorsiceras* Buckman, 1918, *T.A.*, vol. ii, pl. cxvii.

rest of the rock consists of almost barren sandstone with occasional plant-remains. Near the base at Scarborough, however, Hudleston described a shelly bed containing *Trigonia rupellensis*, *Pseudomonotis* sp., *Pholadomya murchisoni* and other bivalves.

Since no ammonites have yet been found in these basal sandstones and shales they must be for the present regarded as part of a greatly-expanded *kœnigi* zone, and so equivalent to the Kellaways Clay and possibly also to the lower part of the true Kellaways Rock of Wiltshire. The presence of a *calloviense* zone is doubtful, for although species of *Sigaloceras* have been found, they have not been proved to occupy any separate stratum of rock, being apparently mixed with the fauna of the *kœnigi* zone.

Thus the greater part of the so-called Kellaways Rock of the Yorkshire coast is still apparently a true representative of at least the lower part of the Kellaways Beds of the rest of England. The name Hackness Rock should therefore be restricted to the highly fossiliferous topmost part, containing the fauna of the Lower and Middle Oxford Clay, and for this hitherto unnamed but important series of beds it is a valuable addition to stratigraphical nomenclature.

IV. EAST SCOTLAND

(a) The Brora District of Sutherland

In East Sutherland rocks of the age of the Oxford Clay and Kellaways Beds cover an area about 3 miles square and attain a thickness of 350–400 ft. (fig. 63, p. 366). They occupy the central and widest part of the narrow coastal strip of Jurassic strata about Brora, bounded inland by the great Ord Fault, which throws them down against the Old Red Sandstone mountains. The softer Jurassic rocks form a low coastal plain, largely obscured by glacial debris and raised beaches.

The coast-sections are at first sight disappointing. Such cliffs as exist are nearly all cut in raised beach material or in blown sand, while the Jurassic substratum protrudes only as reefs and ledges, overgrown with seaweed and covered at high tide.

The Oxford Clay is more favourably exposed for study than the other formations, however, owing to the River Brora having cut a steep-sided valley across the outcrop. Natural and artificial sections along the valley yield considerably more information than those on the shore. In the village of Brora the valley becomes a gorge, whose cliff-like sides are composed of sandstones equivalent to the upper part of the Oxford Clay. Higher upstream, behind Fascally Coal Mine, a large clay-pit has been excavated for the exploitation of the Lower Oxford Clay for a brickworks near by.

The deposits are divisible into three well-marked lithic groups. The highest 200 ft. (approximately) consist of sandstone with sandy shale at the base, united under the term BRORA ARENACEOUS SERIES; they correspond to the Upper Oxford Clay (*præcordatum*, *lamberti* and *athleta* zones) and to an unknown extent also to the Lower Calcareous Grit. Below is a series of argillaceous shales and clays, the BRORA ARGILLACEOUS SERIES, which is from 150 to 225 ft. thick, and spans the Lower Oxford Clay (*duncani* to *jason* zones). At the base of all is a hard band of rock forming the Roof-Bed of the Brora Coal and yielding the fauna of the Kellaways Clay (*kœnigi* zone).

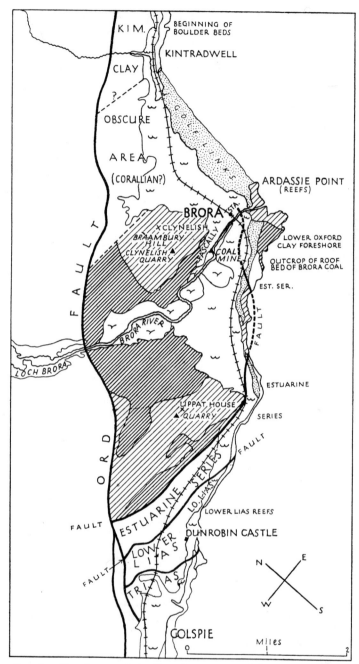

FIG. 63. Sketch-map of the Jurassic area around Brora, Sutherland, showing the position of Fascally Coal Mine, Clynelish Quarry, Uppat Quarry, &c. (Based on the 1-inch map of the Geological Survey of Scotland, Sheet 103.)

Wide hatching = Brora Argillaceous Series; close hatching = Brora Arenaceous Series. Blown sand dotted; alluvium thus: Y; raised beach material thus: W. (See fig. 28, p. 151.)

Præcordatum Zone (? and **Cordatum Zone** pars) (BRORA SANDSTONE), *c.* 120 ft.

The thick sandstones which form the cliffs of the gorge below the bridges in Brora were estimated by Lee to have a thickness of about 100 ft.[2] Unfortunately they have not in recent times yielded any fossils, and they therefore cannot be exactly dated. The upper portions probably correspond largely with the lower part of the English Lower Calcareous Grit, and the lower portions with the upper subzone of the *præcordatum* zone. A definite Lower Calcareous Grit fauna is found above and an ammonite assemblage dated as *præcordatum* a short distance below.

Three-quarters of a mile north-west of Brora Railway Station the Brora Sandstone rises into one of the highest eminences on the coastal plain, the low rounded Braambury Hill, immediately south of the hamlet of Clynelish. Old quarries scar the south side of the hill, the highest showing 20 ft. of unfossiliferous white sandstone, a lower one showing another 15 ft. of sandstone, chiefly white, with pebbles of quartz and nests of Limas and Pectens. At the top of the lower quarry is a 6 ft. band crowded with casts of *Pseudomonotis braamburiensis* (Phil.).[3] The rock is in places a spongy mass of the casts, but they are poorly preserved and good specimens are not easily obtained.

Præcordatum Zone (pars) (CLYNELISH QUARRY SANDSTONE), *c.* 20 ft.

Farther south, down the hill-side towards the River Brora, ½ mile north-west of Fascally Coal Mine, a large excavation called Clynelish Quarry is still worked, and yields a rich fauna. The sandstone is white and fine-grained, and, though originally soft, it is silicified to a depth of 10 ft. from the surface in such a way as to resemble a fine quartzite, and it is much used for building. The softer lower portion contains abundant casts and moulds of fossils, of which by far the commonest are lamellibranchs of typical Lower Corallian species. The most abundant are *Chlamys fibrosa*, *Chl. splendens*, *Lima rigida*, *L. mutabilis*, *Lucina* sp., *Exogyra nana* and *Gryphæa dilatata* var. *exaltata*. Besides these there are a few species not known elsewhere, such as *Trigonia joassi* Lycett, and a brachiopod figured by Davidson as *Terebratula bisuffarcinata* Schloth., but probably a new species. The ammonites are of uncertain date, for scarcely any of them have been found elsewhere; Buckman considered them to be of late Vertumnicertan age, equivalent to what is here grouped in the *præcordatum* zone. The commonest is *Aspidoceras silphouense* (Y. & B.), which may be found in fair abundance as casts. Of rarer occurrence are casts of *Sutherlandiceras sutherlandiæ* (Sow.),[4] *S. albisaxeum* S. Buck.,[5] *Cardioceras* cf. *cardia* S. Buck. and *Quenstedtoceras* cf. *macrum* (Quenst.).

Lamberti and **Athleta Zones** (FASCALLY SANDSTONE, *c.* 20 ft., and FASCALLY SHALE, 50 ft.).

The beds below the Clynelish Quarry Sandstone are exposed in the banks

---

[1] Based on G. W. Lee, 1925, 'Geol. Golspie', *Mem. Geol. Surv.*, and observations.
[2] Judd, owing to certain errors, gave an estimate of 400 ft. But see Lee, 1925, loc. cit., p. 86.
[3] The name, derived from this hill, was first used by Sowerby, but as a *nomen nudum*, a Yorkshire specimen being later figured by Phillips and becoming *ipso facto* the type.
[4] *T.A.*, pl. CCCLXIV.   [5] *T.A.*, pl. CCCXX.

of the River Brora east of Fascally Coal Mine (Pl. XIV). The cliffs which there rise sheer from the water show about 15 ft. of brown, grey and rusty FASCALLY SANDSTONE, in which ochreous impressions of *Lucina lyrata* Phil. and *L. discoidalis* Buv. abound. The Survey obtained from the cliffs on the sides of the river a number of specimens of *Quenstedtoceras*, which Buckman compared to *Q. lamberti*, *Q. prælamberti*, and four other species. With them were found some Kosmocerates. A correlation was thus established with the Tidmoor Point Beds. No sign of the *renggeri* zone has been found.

At the base of the sandstones is a row of doggers, from which have been obtained fragmentary Kosmoceratids. The late Dr. Lee and Buckman did not name them, but Dr. Pringle considers them to indicate the horizon of *K. proniæ*,[1] which is here considered to be part of the *athleta* zone (see table, p. 340).

The row of doggers divides the sandstones from about 50 ft. of efflorescent sandy shales, lithologically intermediate between the Brora Arenaceous and the Brora Argillaceous Series. Only the highest part (5 ft. or more) is visible on the banks of the river, in the side of the road leading down to Fascally Coal Mine, and this highest part Buckman originally included in the Fascally Sandstone. It seems advisable, however, to keep it separate, under the name of FASCALLY SHALE; on the foreshore, where the full thickness can be seen, Dr. Lee obtained a measurement of 50 ft.

From the upper part of the Fascally Shale the Survey collected three species of *Peltoceras*, indicative of the *athleta* zone.

Jason–Duncani Zones (BRORA BRICK CLAYS AND BRORA SHALES).

Below the Fascally Shale are at least 200 ft. (225 ft. measured by Lee on the shore) of clays and shales representing the lower part of the Oxford Clay. The entire succession crops out south of the Brora estuary, where it forms muddy platforms accessible at low tide. This is not, however, a favourable locality for studying the fauna, and the best exposures are again to be found on the north bank of the Brora River, near Fascally.

The highest clays, which may be distinguished as the BRORA BRICK CLAYS,[2] are dug in a large open brick-pit immediately above Fascally Coal Mine. The only common fossils are *Cylindroteuthis puzosi* and *Gryphæa* cf. *lituola*, but the Survey have recorded *Kosmoceras* cf. *jason* (Rein.), *K.* cf. *stutchburii* (Pratt) and *K. sedgwickii*? (Pratt), the first two of which Buckman figured as *K. zugium* S. B. (genotype of *Zugokosmoceras*) and *K. interpositum* S. B. Brinkmann has since declared both to be synonymous with *K. grossouvrei* Douv.[3]

Beneath the brick clay is a thick series of shaly clays, little of which is exposed except on the foreshore; they are distinguished by the Survey as the BRORA SHALES. From the shore platforms the Survey have collected numerous Kosmoceratids, to which nearly all the current names have been applied: *duncani, elizabethæ, castor, jason, stutchburii, sedgwickii*, &c. Until very careful revision of the Kosmocerates collected at Brora has been undertaken, in the light of modern work, the most that can safely be said about these beds is that

---

[1] J. Pringle, 1930, *P.G.A.*, vol. xli, p. 74.

[2] Buckman referred to the Brick Clay as the 'Fascally Shales (Brickyard Beds)' (*T.A.*, 1923, vol. iv, p. 41), a misnomer, since of all parts of the series this deposit most deserves to be classed as a clay. The Survey use the term Brora Clays (G. W. Lee, 1925, loc. cit., p. 81).

[3] *T.A.*, pls. CCCLXXXIX and CDXIX; Brinkmann, 1929, loc. cit., p. 50.

the whole of the Brora Argillaceous Series corresponds in general with the Lower Oxford Clay (*jason–duncani* zones).[1]

Kœnigi Zone (KELLAWAYS BEDS), 3–5 ft. (+ ?).

Judd recorded *Ammonites calloviense* from Brora, but modern collecting has failed to confirm his record, and it seems probable that the *calloviense* zone is wanting.

The *kœnigi* zone, or stratal equivalent of the Kellaways Clay, forms the ROOF BED of the Brora Coal and is one of the best-known fossiliferous beds of the district. It is a very hard, ferruginous, sandy limestone, weathering red, from 3 ft. to 5 ft. thick, and passing down into grey calcareous sandstone with lenticles of coal. In the early days of the coal-mining at Brora large quantities of the Roof Bed were broken up and brought to the surface, so that fossils could easily be collected. Several were figured in Sowerby's *Mineral Conchology* and so became types of familiar species; notably *Gowericeras gowerianum, Pholadomya murchisoni, P. acuticostata, Protocardia striatula* and *Anatina undulata.*

Nowadays it is difficult to study the Roof Bed. The only exposure, except in the working mine, is in the form of a ledge exposed at low tide on the foreshore half a mile south of Brora. The stone is much overgrown with seaweed and barnacles, and is very untractable with the hammer.

The ammonite fauna, comprising *Gowericeras*[2] and two species of *Proplanulites*, leaves no doubt as to the correspondence of the Roof Bed with the Kellaways Clay.

(b) Ross-shire[3]

Twenty miles south of Brora, in the diminutive patch of Jurassic rocks beside the Moray Firth, on the coast of the peninsula which separates the Dornoch from the Cromarty Firth, two shore-sections of Oxfordian beds have been studied by Judd and more lately by Lee. They are situated at Port-an-Righ, $1\frac{1}{2}$ miles south of Balintore (Shandwick), and consist of reefs visible only at low tide. The best exposure stretches along the beach northeast from the Fishermen's hut, while the other, less complete but more comprehensive, lies at Cadh'-an-Righ, a few hundred yards farther south. Both are near the boundary fault, which throws the Oxfordian strata against the Old Red Sandstone at a high angle.

The more northerly section displays 50 ft. of shaly beds containing the fauna of the *præcordatum* zone at top and bottom, and passing up into Corallian Beds. The more southerly exposure also shows part of the *præcordatum* zone, and in addition 14 ft. of shales belonging to the *lamberti* zone, and, after a considerable gap, another 14 ft. of clay and shale of the *jason* zone, resting directly upon the Brora Roof Bed.

The chief interest of the sections lies in the shaly development of the *præcordatum* and *lamberti* zones, which present a marked contrast with the sandstones of the Brora Arenaceous Series only 20 miles to the north. Here

---

[1] Buckman in 1923, without sufficient knowledge of the conditions under which the strata occur, constructed a hypothetical sequence, introducing several unfortunate stratal terms, not adopted by the Survey and better dropped. *T.A.*, 1923, vol. iv, p. 41, and see explanatory note at top of p. 40.

[2] See *T.A.*, pls. CCLXXXVII* and CDIV.

[3] G. W. Lee, 1925, loc. cit., pp. 99–101, 85–6.

the beds are almost entirely shales, for the most part grey and soft, but occasionally harder and calcareous, forming a shaly-weathering limestone. The Survey have collected a rich fauna of Cardiocerates, belonging to species characteristic of both the upper and the lower subzones of the *præcordatum* zone at Horton-cum-Studley, Oxfordshire; among them are *C. cardia*, *C. præcordatum*, and *C.* cf. *tenuicostatum*. Associated with them are a few lamellibranchs, such as *Gryphæa dilatata*, *Pinna lanceolata* and other familiar species.

From the shales of the *lamberti* zone in the southerly section they obtained *Q. lamberti* and several other species; and from the basal layers of the Lower Oxford Clay an assemblage of the usual Kosmocerates. The absence of the *calloviense* fauna here corroborates the negative evidence obtained in Sutherland.

The Brora Roof Bed, 1–2 ft. thick, is represented by tough calcareous shelly sandstone, weathering olive green, and having at the top a band of ironstone nodules and abundant belemnites. No ammonites were found in it.

V. THE HEBRIDEAN AREA[1]

The extension of the Oxfordian sea over the Hebridean area is proved by relics of typical Oxford Clay, and in two places of Kellaways Beds, in the three islands of Skye, Eigg, and Scalpa. There is reason to believe that the formation underlies the basalts of a large part of the northern end of Trotternish in the north of Skye, for it crops out from underneath the volcanic covering on both sides of the peninsula and round the north end, around Loch Staffin in the east and also on the west coast at Duntulm, Mugsted or Monkstadt and Uig. It was first described at Staffin by E. Forbes in 1851, and in 1878 Judd discovered 'the same beds, with precisely similar characters and fossils' cropping out at low water in Laig Bay, on the island of Eigg. In the present century the Survey have discovered a considerable outcrop around the edge of the basalts of the Strathaird peninsula in the south of Skye, and also small faulted remnants at Strollamus, Skye, and on the adjoining coast of Scalpa.

A remarkable feature of these occurrences is the differences in the zonal position of the lowest members of the Oxfordian formation, resting upon the Great Estuarine Series, in the different places. In Strathaird and Eigg the whole of the formation, with the exception of a localized thin development of the Kellaways Beds, consists of Cardioceratan deposits of *præcordatum* date, grading up into *cordatum* deposits referable to the Lower Calcareous Grit. In Trotternish in the north, however, the sequence appears to have been more fully developed, and Kosmoceratan and Proplanulitan faunas intervene between the Cardioceratan deposits and the Great Estuarine Series.

In Trotternish the formation consists of blue clays and shales with subordinate bands of argillaceous limestone and septarian nodules. The only passable section is on the foreshore in Staffin Bay, for a description of which I am indebted to Mr. Malcolm MacGregor, M.Sc., whose important researches in Northern Trotternish it is hoped will soon be published. He

---

[1] Based on C. B. Wedd, 1910, 'Geol. Glenelg, Lochalsh and S.E. Skye', pp. 128–31; G. Barrow, 1908, in 'Geol. Small Isles of Inverness-shire', pp. 26–8; G. W. Lee, 1925, in 'Pre-Tert. Geol. Mull, Loch Aline and Oban', pp. 113–14, *Mems. Geol. Surv.*

informs me that the section (which is the one mentioned by Forbes), although the lower beds can only be consulted at low water during spring tides, shows about 100 ft. of dark shales with some thin bands of concretionary limestone. In the basal 20 ft. of the shale, which rests with a 1 ft. pebble bed on the sands and sandstones of the Great Estuarine Series, he found *Quenstedtoceras lamberti* and *Vertumniceras vertumnus*. About 20 ft. from the base is a double band of concretionary limestone, above which he found abundant Cardiocerates of very various styles, *præcordatum*, *cordatum*, *vertebrale* and *excavatum* types being apparently represented; also *Peltoceratoides*, 'Belemnites oweni', *Gryphæa dilatata*, *Nucula*, &c.[1]

A specimen collected by the Survey and named by Buckman *Kosmoceras degradatum*[2] would seem to indicate the presence of Kosmoceratan deposits also at Staffin.

On the west side of the peninsula, at Uig, Bryce and Tate in 1873 collected *A. lamberti* and *A. jason*, with '*Belemnites oweni*' in dark shales and clays; and near by, at Monkstadt (Mugstok) Murchison as early as 1829 had recorded *A. kœnigi*.[3]

In 1925 the Survey reported the discovery of a tiny faulted patch of Oxfordian strata in Mull, in the right bank of a stream which flows into Duart Bay, at a point 300 yards above the bridge. The rock is a marly, fine-grained sandstone, containing a fossil-bed, from which were collected ammonites identified by Buckman as *Kosmoceras elizabethæ-spoliatum* Quenst., *Reineckeia* spp. and *Phlycticeras* sp. I have seen the material (of which there is a considerable quantity) and, as Dr. Spath points out, the supposed Kosmocerates have keels and are undoubtedly *Amœboceras kitchini* Salf., while the 'Reineckeias' are nothing else than *Raseniæ*. *Reineckeia* therefore still remains undiscovered in Northern England or Scotland (see below, p. 478).

The small faulted fragments overlying the Estuarine Series at Strollamus and Scalpa are much metamorphosed and have yielded no new information. They appear to be Cardioceratan in date.

The occurrences in Strathaird and in Eigg are important and the study of them in recent years by the Survey has added greatly to our knowledge of the Scottish Oxford Clay.

In Strathaird the Oxfordian beds occur over a large part of the promontory, cropping out from below the basalt in the western cliffs, north of Elgol. Here they may reach a thickness of as much as 300 ft., of which an unknown portion near the top should be assigned to the Lower Calcareous Grit; northwards they are overstepped by the basalt. Lithologically the beds are more sandy than in the north of Skye. The upper portion still consists of dark grey or blue micaceous shale, probably altered from a clay, but downwards it becomes

---

[1] Buckman figured from Staffin *Peltoceratoides torosus* (Oppel), *T.A.*, pl. DLXIII, and the genotype and holotype of *Korythoceras korys*, pl. CCCLXI.

[2] *T.A.*, pl. CDXXXVII.

[3] R. I. Murchison, 1829, *Trans. Geol. Soc.* [2], vol. ii, p. 311. Bryce (1873, *Q.J.G.S.*, vol. xxix, p. 332) misquotes Murchison as recording '*A. kœnigi* in masses . . .' The passage actually reads: 'I . . . found several fossils in blue shale through which a deep canal has recently been cut by Lord Macdonald, to drain the lake of Mugsted. Among the shells are the ammonites königi, ostreæ in masses, many belemnites, flattened tellinæ? &c.' Jones showed that the '?flattened tellinæ' were the Ostracod, *Estheria murchisoni* Forbes, of the Great Estuarine Series, and the masses of *Ostreæ* also indicate that the *Ostrea hebridica* Beds of the Great Estuarine Series were cut through.

more sandy until, in about the lowest 80 ft., it has become a succession of shaly, micaceous or calcareous sandstones and grits. The basement beds are coarse-grained and pebbly, in places forming a striking-looking conglomerate of white quartz and quartzite pebbles, often more than an inch in diameter.

From the base upwards the sandstones yield an unmistakable Upper Oxford Clay fauna, with (according to the specific identifications of 1910) distinct Lower Calcareous Grit elements. Wedd records *Cardioceras cordatum, C. ? nikitinianum, C.* cf. *rotundatum, C. rouilleri, C. suessi, C. tenuicostatum, C.* cf. *vertebrale, Chlamys fibrosa* and *Pseudomonotis* cf. *ovalis* as making their appearance near the base of the formation. There is here no indication of any other part of the Oxford Clay but *præcordatum* and *cordatum* beds.

Apparently similar conditions would appear, from Barrow's account, to obtain in Eigg, except that there the sandstones are much thinner. The exposures are limited to shore-reefs between tide-marks in Laig Bay, and to a little gorge east of Laig Farm. The upper part of the sequence, exposed on the shore, consists of dark grey shales, like those at Loch Staffin but more friable than those of Strathaird. They have yielded fossils identified as *Cardioceras cordatum* (2 forms, coarsely and finely ribbed), *C. excavatum, Aspidoceras perarmatum* and a number of Corallian lamellibranchs, and there can be no doubt that they should be referred to the Lower Calcareous Grit (*cordatum* zone) at least in part, as in Strathaird. Underneath, in the gully near Laig Farm, is a bed of limestone about 8 ft. thick, beneath which are about 12 ft. of calcareous sandstone, neither yielding diagnostic fossils, the latter resting on the Great Estuarine Series.

The presence of Kellaways Beds in at least one locality in Strathaird is proved by large fallen blocks of sandstone lying on the beach, north of Elgol. They contain nests of a small Callovian *Rhynchonelloidea,* cf. *socialis,* together with *Ornithella kellawaysensis, Pseudomonotis* sp., and occasional ammonites assigned to *Gowericeras gowerianum,* marking the *kœnigi* zone. The bed from which the blocks fell has not been identified *in situ,* and it is not known whether its position is above or below the basal conglomerate.

The relations of the Oxford Clay to the underlying rocks in the Hebrides are shown by these few remaining relics to be far more non-sequential and irregular than anywhere in England. The isolated patch of Cornbrash in Raasay, the equally isolated occurrences of Kellaways Beds at Elgol and Monkstadt, the restricted area of Lower Oxford Clay in north-east Trotternish, and the overlapping of late *præcordatum* deposits on to the Great Estuarine Series in Eigg and in parts of Strathaird, testify to greater disturbances than were experienced at the period in any other parts of the British Isles.

VI. KENT[1]

The concealed outcrop of the Oxford Clay beneath the Cretaceous rocks under Kent forms a strip from 1 to 3 miles wide, running parallel to the thin Lower Oolites and dipping south with them. Beginning at the coast on the south side of St. Margaret's Bay, it underlies Canterbury, Faversham and Sittingbourne, and passes close to the north of Gillingham and Gravesend.

---

[1] Based on Lamplugh and Kitchin, 1911, loc. cit.; and Lamplugh, Kitchin and Pringle, 1923, loc. cit.

On this belt and to the south of it the Oxford Clay and Kellaways Beds have been pierced in several borings. The bulk of the Oxford Clay consists of uniform blue-grey clay, of fine texture, with subordinate brown bands, while the *præcordatum* zone is altogether peculiar, consisting of ironshot marls such as are developed on the other side of the Channel. The Kellaways Beds, as usual, proved exceedingly variable.

The Oxford Clay thickens rapidly westward from about 130 ft. at Dover to 200–210 ft. at Brabourne and Chilham, near the line of the railway south-west of Canterbury.[1] The Kellaways Beds also thicken from the coast towards Canterbury and thin out again farther west. The maximum thickness recorded is 43 ft. at Fredville, almost midway between Canterbury and Dover, while farther west, at Brabourne and Harmansole, less than 20 ft. was proved.

SUMMARY OF THE OXFORD CLAY IN KENT

Præcordatum Zone.

At the top of the formation, in and just above the ironshot marls, a rich fauna of Cardiocerates was found. The highest of these were still of *præcordatum* date and, as no palaeontological proof of the *cordatum* zone or Lower Calcareous Grit was found, the *præcordatum* zone is probably directly overlain by the *martelli* beds, as at Quainton. The zone is, however, very differently developed from the typical clays familiar in other parts of England, consisting largely of marls and marlstones full of brown ironshot ooliths, such as are found on the same horizon in the cliffs between Villers and Houlgate in Normandy. For this reason the Survey classed them in their Memoirs as Lower Corallian.

The highest bed in which Cardiocerates were found was a band of hard grey marlstone, about 4 ft. thick, crowded with black-coated fossils. The stratum was recognized at Fredville, Snowdown Colliery, Bere Farm, Brabourne, Chilton, Chilham, and probably also at Guildford. Its ammonites were recorded as *Cardioceras* cf. *excavatum* (Sow.), *C.* cf. *pingue* (Rouill.) and undescribed species, two of which were compared with *C. cordatum* Lahusen (non Sow.) and *C. cordatum* de Loriol (non Sow.); *Aspidoceras* sp., resembling *A. faustum* Bayle; *Peltoceras arduennense* (d'Orb.) in abundance, *P.* cf. *eugenii* (d'Orb.); *Perisphinctes bernensis* de Lor. (pars.), *P.* cf. *variocostatus* (Buckland) and *P.* sp. nov. The commonest lamellibranchs were *Isognomon* cf. *cordati* (Uhlig) and *Unicardium sulcatum* Leckenby, with which were associated such more typically Corallian species as *Chlamys fibrosa*, *C. inæquicostata* (Phil.), *Camptonectes lens* (Sow.), and the brachiopods *Terebratula farcinata* Douv. and *Rhynchonelloidea thurmanni* (Voltz.). This mixed fauna and the abundance of the fossils point to the bed being a product of slow deposition, perhaps spanning parts of the *præcordatum* and *cordatum* zones.

A few feet beneath this interesting bed was an almost equally persistent band of marly, crumbly, locally pisolitic clay, which also yielded a number of fossils of Corallian facies—*Nucleolites dimidiatus* (Phil.), *N. scutatus* Lam., *Holectypus* cf. *oblongus* Wright, a large *Gryphæa*, numerous valves of *Chlamys fibrosa*, and fragments of a large *Perisphinctes* cf. *variocostatus*. Were it not

---

[1] These figures include the ironshot marls of the *præcordatum* zone, which the Survey classed with their Lower Corallian.

for the Cardiocerates in the marlstone above it, one would certainly group this bed with the Corallian.

Below are 25–35 ft. of olive green and brown marlstones and marls, more or less filled with ironshot grains and on the whole richly fossiliferous. Cardiocerates were again most conspicuous, including *C. cordatum* Lah. (non Sow.), *C. excavatum* Lah. (non Sow.), *C. rouillieri* Lah., *C. tenuicostatum* (Nik.) and undescribed species. Near the base were found *C. nikitinianum* Lah., *Aspidoceras œgir?* (Oppel) and (?)*Peltomorphites williamsoni* (Phil.). *Gryphææ* were not numerous, but such as were found were of the large, massive varieties of *G. dilatata* so characteristic of the *præcordatum* zone inland. Among the most interesting of the fossils were fragments of the stems of *Millericrinus*, suggestive of the Marnes à *Millericrinus* on the same horizon in the Boulonnais.

It is perhaps noteworthy that, although no ironshot oolite occurs in the *præcordatum* zone in Dorset, yet the presence of a quantity of iron in the zone is attested by the red coatings of the nodules in the Red Nodule Beds.

Renggeri and Lamberti Zones.

Beneath the *præcordatum* zone numerous Quenstedtocerates were found, as usual in the form of pyritized casts, and among them *Q. lamberti* (Sow.), *Q. leachi* (Sow.), *Q. macrum* (Quenst.) and *Q. goliathum* (d'Orb.) were identified. With these were some species of the genus *Creniceras*, indicating the *renggeri* zone, and indeed the name *Renggeri* Beds was originally applied to this middle division. The lamellibranch fauna provided interesting evidence for the connexion with the Bernese Jura, suggested by the Perisphinctids in the overlying zone, and bearing out Prof. Davies's remarks on the fauna from this zone collected at Ludgershall railway-cutting, Bucks. *Parallelodon* (*Grammatodon*) *concinnum* (Phil.), *P.* (*G.*) *montanayensis*, *Lima* cf. *trembiazensis*, *Nuculana rœderi*, *Perisphinctes* cf. *billodensis*, all species described by de Loriol from the Bernese Jura, were identified. As in other parts of the country and on the Continent, two typical Middle Oxford Clay fossils occurred, the Decapod Crustacean, *Mecocheirus socialis* (Meyer), and the subequilateral and well coiled *Gryphæa*, *G. lituola* Lamk.

Duncani–Jason Zones.

The Lower or Kosmoceratan Oxford Clay is condensed and much of it is full of bivalves, especially *Astarte*, *Nucula* and *Cucullæa*, thus showing that the conditions of deposition were not of the anaerobic type normal at the time in most other parts of England. Probably, too, sedimentation was much interrupted by penecontemporaneous erosion. No evidence of the *athleta* zone was found, but the numerous specimens of *Kosmoceras* in the usual pyritized or crushed condition, and some of *Erymnoceras*, gave evidence of *duncani*, *reginaldi* and *jason* horizons, in descending order. At Fredville and Chilham specimens of *Erymnoceras* occurred a few feet from the base.

Calloviense and Kœnigi Zones (KELLAWAYS BEDS).

The Kellaways Rock is sharply demarcated, both lithologically and palaeontologically, from the overlying clays, having always a fauna and certain lithic characters entirely its own. It is more sandy than the Oxford Clay and usually ferruginous, and throughout it *Gryphæa bilobata* and allied species are abundant. The Gryphæas become larger towards the top, those in the lower part being very diminutive, like the small form so common at Kellaways. Other common species, also recalling the fauna at Kellaways and in Yorkshire, are *Oxytoma* sp., *Pseudomonotis* sp., *Pleuromya recurva* and *Entolium demissum*. The ammonites show that the bulk of the rock belongs to the *calloviense* zone. No examples of *Gowericeras* or *Sigaloceras* were found, while only one imperfect fragment of an ammonite resembling a *Proplanulites* was obtained, in the basal bed at Oxney. This is all the palaeontological evidence for the *kœnigi* zone, but there is between the Kellaways Rock and the Cornbrash a bed of clay a few feet thick which may represent the Kellaways Clay. It was classed by Messrs. Lamplugh, Kitchin and Pringle with the Cornbrash, but they likened it to the wrongly-styled 'Clays of the Cornbrash' in Yorkshire, which we have seen are more properly classed with the Kellaways Clay.

CHAPTER XIII

CORALLIAN BEDS

Zones (Plate XXXVIII).	Stratal Divisions.	Dorset Strata.
Ringsteadia anglica [1] and Perisphinctes (Dichotomosphinctes) wartæ [2]	Upper Calcareous Grit	Ringstead Coral Bed
		Ringstead Waxy Clay
		Sandsfoot Grit
	Glos Oolite Series [6]	Sandsfoot Clay
		Trigonia clavellata Beds
Perisphinctes (Dichotomosphinctes) antecedens [3] and Perisphinctes martelli [4]	Osmington Oolite Series	Osmington Oolite Series
	Berkshire Oolite Series	Bencliff Grit
		Nothe Clay
		Trigonia hudlestoni Bed
Cardioceras cordatum [5]	Lower Calcareous Grit	Nothe Grit

[1] Salfeld's zone of *Ringsteadia anglica* Salf. and *R. pseudocordata* (Bl. & Hudl.), including his rather doubtful zone of *Perisphinctes decipiens* (Sow.), a species which has not yet been properly determined. *P. decipiens* (Sow.) has been recorded in the Sandsfoot Grit, but a separate zone for this is unnecessary, as it contains *Ringsteadiæ* from base to summit. As alternative zonal indices to *P. decipiens*, Salfeld used *P. achilles* d'Orb., a foreign species not yet found in this country and [*Prionodoceras*] *serratum* (Sow.) (the former in his two zonal tables, 1914, *N. Jarb. für Min.*, BB., vol. xxxvii, and the latter in the text, loc. cit., p. 129). *P. serratum*, however, seems to range up to the very top of the *Ringsteadia* zone and possibly also into the zone of *Pictonia baylei* at the base of the Kimeridge Clay (as at Swindon, Salfeld, loc. cit., p. 196); Buckman considered the genus Kimeridgian.

[2] The ammonites (and other fauna) of the Sandsfoot Clay are very little known. Salfeld placed it in his zone of *P. wartæ* and [*Amœboceras*] *alternans* von Buch; but the only specimens truly assignable to *P. wartæ* that I have seen from England are one in the Geological Survey collection from the Westbury Iron Ore (No. 25484) and some collected by Prof. Davies from the Ampthill Clay (see p. 410).

[3] Forms identical with *Perisphinctes antecedens* Salf. are found in the Osmington Oolite Series inland, and this series is stratigraphically so distinct from the Berkshire Oolites that it was formerly for the sake of convenience considered to be a separate zone. It is probable, however, that it is only a subdivision of the broad *martelli* zone, for there are two specimens of *P.* cf. *martelli* in the British Museum from some part of the Osmington Oolite Series near Weymouth. *P. antecedens* occurs in the *Trigonia clavellata* Beds.

[4] The Berkshire Oolite Series had not been differentiated at the time when Salfeld evolved his zonal scheme; it is the chief repository of the *martelli* or *transversarius* fauna. The index fossil is very rare, but allied species abound; Schindewolf coined for them a new generic name, *Martelliceras*, which Dr. Spath declares precisely synonymous with *Perisphinctes sensu stricto* (1931, *Pal. Indica*, NS., vol. ix, p. 401).

[5] *Aspidoceras perarmatum* (Sow.) is sometimes made alternative zonal index; but it is only known from Yorkshire, where it is very rare. *A. catena* (Sow.) is better.

[6] From Glos, near Lisieux, Normandy, where the Glos Sands link up the *T. clavellata* Beds (Grès d'Hennequeville) with the Sandsfoot Clay (Argile noire); see Arkell, 1930, *P.G.A.*, vol. xli, p. 399.

I. DORSET

(a) The Northern Limb of the Weymouth Anticline: Ringstead to Abbotsbury

THE Corallian Beds crop out along both limbs of the Weymouth Anticline, dipping away to north and south beneath the Kimeridge Clay.

The northern outcrop, 13 miles in length and seldom more than ½ mile wide, is marked by a ridge of downs running from the coast at Ringstead and Osmington to the Chesil Bank west of Abbotsbury. The ridge rises westward to a maximum height of 316 ft. above the sea at Linton Hill, near Abbotsbury, and near the end it is crowned by the ancient chapel of St. Catherine; southward lie the Oxford Clay lowlands and to the north the deep valley of the Kimeridge Clay (map, p. 342).

At the west end of the outcrop there are no natural sections exposed by the sea, owing to the protective action of the Chesil Bank, but the east end provides the finest exposures of Corallian rocks in Britain. The entire succession may be measured and studied in detail, through a total thickness of rather more than 200 ft., extending, with several repetitions, along some three miles of cliffs (fig. 64).

In Ringstead Bay in the east, the beds first appear gradually from beneath the Kimeridge Clay in low banks, rising into cliffs as the harder middle portions of the formation are brought up above sea-level. For a considerable distance a steep precipice is formed of the Osmington Oolite Series, capped by the *Trigonia* Beds, dipping gently to the east (Pl. XVI). At the foot is the Bencliff Grit. At Osmington Mills some complicated faulting interrupts the succession, and the Nothe Clay and much of the Lower Calcareous Grit are obscured.

From Osmington Mills westward the Corallian Beds are tilted steeply to the north, and the hard bands of the Osmington Oolite Series run in parallel ridges along the beach, standing nearly on edge. At Shortlake Chine they rise in the cliff and a complete section is provided down to the Oxford Clay, but the Lower Calcareous Grit here is somewhat obscured by vegetation. Towards Redcliff Point an anticlinal flexure brings down the Lower Calcareous Grit almost to sea-level once more, and there are magnificent exposures of this division and of the Nothe Clay on both sides of the Point.[1] At Bowleaze Cove at the west end of Redcliff the base of the Lower Calcareous Grit passes finally out of the cliffs, which thence turn southward and expose only Oxford Clay under Jordan Hill (fig. 65, p. 382).

The succession here and south of Weymouth was systematically described during the last century by Blake and Hudleston[2] and again, with amendments, by H. B. Woodward.[3] In 1926 I re-measured and described all the exposures and during recent years have collected fossils carefully, bed by bed.[4]

---

[1] The Point itself consists of a small upthrown block of Oxford Clay, largely obscured by slips.
[2] J. F. Blake and W. H. Hudleston, 1877, *Q.J.G.S.*, vol. xxxiii, pp. 260–75.
[3] H. B. Woodward, 1895, *J.R.B.*, pp. 82–94.
[4] A brief correlation was published in 1927, *Phil. Trans. Roy. Soc.*, vol. ccxvi B, pp. 156–9.

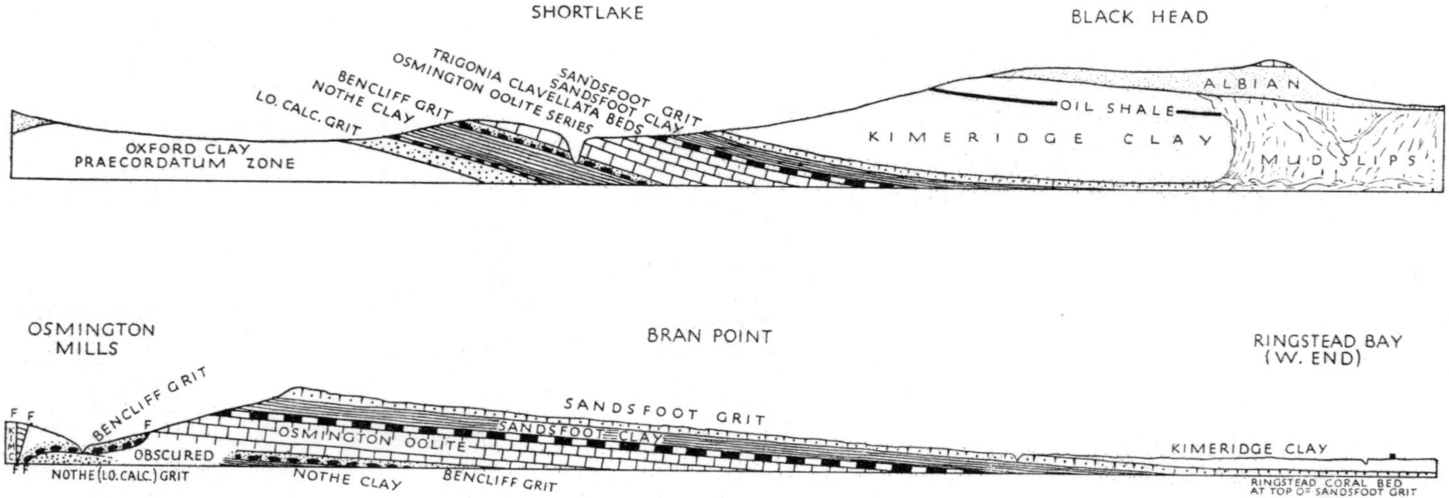

FIG. 64. Sections of the Corallian Beds in the cliffs west of Osmington (above) and east of Osmington (below). Length of each 1·1 miles. The two sections are separated by a ¼ mile of cliffs in which the Corallian Beds dip almost vertically and the strike coincides with the line of the coast. The upper section is continuous with the lower one in fig. 65. (Outline based on Strahan, 1898, 'Geol. Isle of Purbeck and Weymouth', *Mem. Geol. Surv.*, Plate IX.) Vertical scale exaggerated.

SUMMARY OF THE CORALLIAN SERIES, RINGSTEAD TO OSMINGTON[1]

Upper Calcareous Grit, 18–20 ft.

The stratum taken to be the topmost bed of the Corallian is the RINGSTEAD CORAL BED, known as the Kimeridge Grit by Damon and as the Upper Coral Rag by Blake and Hudleston and by Woodward. It rises gradually from the beach in the western side of Ringstead Bay and, though it appears to have been often obscured in the past by slipped Kimeridge Clay and gravel, it has been well exposed in recent years (Pl. XXI). It consists of 8 in. of tough, green, argillaceous limestone, largely made up of shells and broken corals. The commonest fossils are the large *Ctenostreon proboscideum* (Sow.), with which are associated *Chlamys splendens* (Dollf.), *C. nattheimensis* (de Lor.), *Camptonectes lens* (Sow.), *Entolium demissum* (Phil.), *Velata hautcœuri* (Dollf.), *Lima rigida* (Sow.), *Myoconcha texta* (Buv.), *Ostrea delta* Smith, and a number of other lamellibranchs and gastropods, with spines of *Cidaris florigemma* and the corals *Thecosmilia annularis*, *Thamnastræa concinna*, &c.

In a westerly direction, towards Osmington, the fossils become scarcer. Where the bed reappears beyond Osmington, under Black Head, the only species at all abundant is *Ctenostreon proboscideum*. The rock becomes at the same time less tough and is shot with ferruginous ooliths, while farther still, on the Weymouth promontory to the south, it passes into a true ironshot oolite and all fossils are comparatively rare. From this it may be deduced that the coral reef, to which the coral fragments and shells at Ringstead bear witness, lay under the Chalk Downs to the east.

Beneath the Ringstead Coral Bed are the RINGSTEAD WAXY CLAYS, 10 ft. of waxy, ferruginous clay, interrupted by seams of claystone nodules towards the top, and numerous seams of laminated claystone or clay-ironstone throughout, with layers of *Ostrea delta* towards the base. Immediately subjacent to the Coral Bed is a 3-in. band of especially prominent, red clay-ironstone nodules, containing occasional specimens of *Ringsteadia anglica* Salf.

These beds may be correlated approximately with the Westbury Iron Ore of Wiltshire. Recognizable ammonites do not seem to have been obtainable in recent years from the Ringstead Coral Bed itself,[2] but undoubted fragments of *Pictonia* are common in the clay immediately above, while *Ringsteadia* is found immediately below.

The SANDFOOT GRIT, some 25 ft. thick at Sandsfoot Castle, south of Weymouth, measures only about 7 ft. at Ringstead Bay. Where it rises from the beach beneath the Ringstead Waxy Clays, at the west end of the bay, it consists only of a 2-ft. band of bright red sandstone, speckled with white ooliths and quartz grains, and containing fucoid markings, overlying 5 ft. of brown sandy marls. Farther west, at Black Head, Osmington, the single hard band has thickened to 4 ft. and has split up into four bands of red and green ferruginous sandstone, with the same speckling of ooliths and quartz grains. The commonest fossils are *Chlamys midas* (d'Orb.), *Goniomya*, *Pleuromya*, and *Nautilus hexagonus* (Sow.). Blake and Hudleston recorded '*Ammonites decipiens*', and at Sandsfoot *Ringsteadiæ* are not uncommon.

---

[1] Based mainly on unpublished manuscript.
[2] Damon recorded a dozen species, but most of them, if correctly identified, certainly came from beds above and below. (*Geol. Weymouth*, 2nd ed., 1884, pp. 65–6.)

Glos Oolite Series, 50 ft.

The SANDSFOOT CLAY, probably about 20 ft. thick at Ringstead[1] and about 30 ft. thick at Black Head, consists of grey, blue and brown clay with varying amounts of ferruginous staining and greenish sand. Layers of *Ostrea delta* are abundant near the top and are liberally distributed throughout, the shells being indistinguishable from those in the basal portion of the Kimeridge Clay, except that on an average they died before attaining such an advanced stage of growth.

At the west end of the outcrop, near Abbotsbury, the thickness is much diminished. The deposit is therefore somewhat lenticular.

The TRIGONIA CLAVELLATA Beds, 14½ ft. These beds are the most remarkable in the whole Corallian Series, for they are veritable fossilized shell-banks. They consist of strong courses of tough red and purplish-brown shelly limestone (the Red Beds of Damon) separated by marly partings, the whole crowded with bivalves and other fossils. On some of the slabs valves of *Trigonia clavellata* are massed so closely together that they touch and form a continuous pavement; on other slabs Pectens, oysters, Astartes, &c. predominate. As a source of well-preserved fossils these beds were early renowned, and Sowerby figured several of the species in *The Mineral Conchology*: namely *Trigonia clavellata*, *Gervillia aviculoides*, *Mytilus pectinatus* and *Lopha solitaria*. Besides these, type-specimens of species founded by subsequent investigators have come from the same beds—those of the abundant *Plicatula weymouthiana* Damon, *Pteria pteropernoides* (Blake and Hudleston) and *Chlamys superfibrosa* Arkell. Other abundant species are *Cucullæa contracta* Phil., *Pteroperna polyodon* (Buv.) and *Cerithium muricatum* Sow. Small pieces of coral occur and, very rarely, Perisphinctids. All these seem to belong to the narrow-whorled group of which *P. wartæ* Bukowski is a member, but they resemble Salfeld's 'mut' *antecedens* of North-West Germany rather than the type-form from Poland, which has a marked forward sweep of the secondary ribs on the periphery (see Plate XXXVIII).[2]

Osmington Oolite Series, 63–5 ft.

Blake and Hudleston drew the boundary between the *T. clavellata* Beds and the Osmington Oolite Series far below the occurrence of the *Trigoniæ*. There is a gradual lithological passage between the two divisions, but it is essential to be guided by the palaeontology and, following H. B. Woodward, to restrict the name *T. clavellata* Beds to the beds actually containing this fauna.[3] Below comes a highly variable series of strata, for the most part comparatively barren of fossils. They form the cliffs to the east of Osmington, where they are magnificently exposed and can be studied in detail. Of the total thickness of 63–5 ft. between Ringstead and Osmington in the east, only 6 ft. 3 in. consist of the typical white oolite, such as is used for building.

The true Osmington Oolite is a solid, white, even-grained, oolitic freestone, with some clay-galls and vertical tubiform markings. It occurs in two blocks,

---

[1] H. B. Woodward (1895, *J.R.B.*, p. 85) states 10–12 ft., but this seems to be an underestimate. The thickness is difficult to determine owing to slips.

[2] There are a few specimens in the British Museum, collected by Damon, and a fragmentary example in my own collection, none of them the true *wartæ*.

[3] H. B. Woodward, 1895, *J.R.B.*, pp. 85, 86, 91.

PLATE XVI

Photo. *W. J. A.*

Osmington Oolite Series, east of Osmington Mills, Dorset.

SC = Sandsfoot Clay capping cliff; TCB = *Trigonia clavellata* Beds; OF =
Oolite facies and LCF = Littlemore Clay facies of Osmington Oolite Series;
BG = Bencliff Grit.

Photo. *W. J. A.*

Osmington Oolite Series, Bran Point, east of Osmington Mills, Dorset.

The lowest ledge is of gritty oolite and rests directly on the Bencliff Grit; next
above it is (Q) the *Chlamys qualicosta* Bed, and (P) the pisolite seam. The
strong bands near the top of the cliff towards the point are *Trigonia clavellata*
Beds, and the capping is Sandsfoot Clay.

PLATE XVII

Redcliff, near Weymouth.

LCG = Lower Calcareous Grit; THB = *Trigonia hudlestoni* Bed; NC =
Nothe Clay. A crest of Osmington Oolite is seen behind and above the slope
of Nothe Clay and Bencliff Grit (which is mostly concealed).

Redcliff, near Weymouth.

The large blocks on the right have fallen from the *Trigonia hudlestoni* Bed.

2 ft. and 4 ft. 3 in. thick. Most of the rest of the sequence consists of blue and grey oolitic or pisolitic marly clays with bands of nodular, grey, argillaceous limestone. From its occurrence in Littlemore railway-cutting, near Oxford, this facies is known as the Littlemore Clay Beds. At the top are over 20 ft. of conspicuously nodular, concretionary, partly oolitic grey and white limestones; this facies too is typically represented in the Midlands, especially about Oxford, where it is known as the Nodular Coralline Rubble.

Near the base is a $1\frac{1}{2}$ ft. band of hard, shelly, oolitic and pisolitic, gritty limestone, crowded with *Chlamys qualicosta* (Etall.), *C. fibrosa* (Sow.), and *Exogyra nana* (Sow.), and above it, separated by 2 ft. of clay, is a seam of loose pisolite. The pisolite (2 ft.) is a very persistent feature, marking the base of the Osmington Oolite Series throughout Dorset and as far as South Wiltshire.

Westward, along the strike, the solid white oolite-facies increases in importance; west of Osmington the 2-ft. seam divides into two, the 4-ft. seam doubles in thickness, and a third seam (4 ft. thick) appears below the others. At the top of Redcliff the white oolite division takes on a facies closely resembling Forest Marble: false-bedding is intensified, much of the oolite grows sandy and fissile, and it is full of clay-galls. In the face of the cliff, 6 ft. out of the total thickness of 13 ft. pass in a short distance into sandy, shaly, greenish clay, with thin films of fissile limestone. This facies is only local, for in a quarry on the top of Jordan Hill, $\frac{1}{2}$ mile to the west, the division is a solid false-bedded oolite once more.

At the west end of the outcrop, about Abbotsbury, a high proportion of the Osmington Oolite Series has passed into hard white oolite, which has been extensively quarried. A thickness of 22 ft. may be seen in a single quarry-face; but fossils are very scarce.

Palaeontologically the Osmington Oolite Series yields little information. It may be said to be characterized principally by the admixture of *Chlamys fibrosa* and *Chlamys qualicosta*. The former seldom, if ever, ranges above it, and the latter has not been found below it. Ammonites are extremely rare, but in the British Museum (Damon Collection) there are two large specimens of *Perisphinctes martelli* Oppel, labelled as from the Osmington Oolite of Osmington, and in a matrix that seems to bear out the labels. Inland, about Oxford, the commonest of the rare ammonites in these beds belong to the group of *P. antecedens* Salfeld; but associated with them are other forms of *Perisphinctes sensu stricto*, which are certainly very closely allied to *P. martelli* —itself a rare species in this country.

Berkshire Oolite Series, 55–60 ft.

The BENCLIFF GRIT (called after Bencliff or Bincleave at Weymouth) is about 14 ft. thick east of Osmington and only about 10 ft. thick farther west, at Shortlake. It consists of yellow and white sands, locally false-bedded, with a few doggers and seams of interlaminated clay, and at the bottom a band of huge doggers of gritstone; these measure over 6 ft. in diameter, and are composed of intensely hard, blue, false-bedded, indurated calcareous grit, with occasional *Gryphæa dilatata* and *Serpulæ*. They are a conspicuous feature on the beach wherever the grits crop out in the cliff—even when, as at Redcliff, the outcrop is completely hidden by slipped ground.

The NOTHE CLAY (from Nothe Point, Weymouth) is 35–40 ft. thick. East

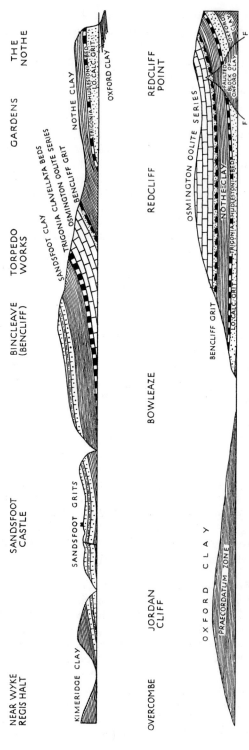

Fig. 65. Sections of the Corallian Beds in the cliffs south of Weymouth (above): distance slightly less than 2 miles (based on H. B. Woodward, 1895, *J.R.B.*, p. 83); and north of Weymouth (below): distance 1·1 miles (outline based on Strahan, 1898, 'Geol. Isle of Purbeck and Weymouth', *Mem. Geol. Surv.*, Plate IX). Vertical scale exaggerated.

and west of Osmington the Bencliff Grit is the lowest bed clearly exposed in ordered sequence, and to study the downward continuation it is necessary to pass westward to Shortlake and Redcliff. At Redcliff the Nothe Clay forms tumbled and slipped ground in the upper part of the cliff on both sides of the Point. About 38–40 ft. may be measured. It consists principally of grey-blue shaly clay, with two thin bands (8–10 in.) of dark-brown ferruginous lime-stone in the lower part, containing fucoid markings and fossils—*Chlamys fibrosa, Lima subantiquata* Roem., *Gervillia aviculoides, Modiola bipartita, Gryphæa dilatata, Pholadomya* and *Pleuromya*. There are also several bands of nodules or nodular limestone.

The PRESTON GRIT or TRIGONIA HUDLESTONI BED (from Preston village, near Redcliff) is 5–6 ft. thick. Blake and Hudleston and other observers in-cluded this bed in the Nothe [= Lower Calcareous] Grit, but as it is one of the most fossiliferous beds in the Dorset Corallian, and one of the most valuable for purposes of correlation, it is important to keep it separate, either under Buckman's name, Preston Grit,[1] or as the *Trigonia hudlestoni* Bed.[2] It is a massive band of hard, brownish-grey, gritty, speckled, doubtfully oolitic limestone, full of fucoid markings, which forms a prominent feature west of the point (Pl. XVII). The huge fallen cubes on the foreshore provide good opportunities for collecting, and they have yielded all the essential fauna of the *Trigonia hudlestoni* Beds of the inland Berkshire Oolite Series: *Peri-sphinctes helenæ* De Riaz, *Cardioceras excavatum* (Sow.), *Cardioceras* cf. *verte-brale* (Sow.), *Aspidoceras* sp., *Goliathiceras* sp., *Trigonia hudlestoni* Lyc., *Camptonectes lens, Chlamys fibrosa, Chlamys splendens, Cucullæa contracta, Lima rigida, L. subantiquata, Ostrea quadrangularis, Lopha gregarea, Gryphæa dilatata, Oxytoma expansa, Pseudomonotis ovalis*, &c.[3]

Lower Calcareous Grit, 27 ft.

Blake and Hudleston called the Lower Calcareous Grit the Nothe Grit, after the Nothe promontory, Weymouth, but this term may be suppressed as a synonym. At Redcliff the whole thickness is conveniently exposed on both sides of the Point. It consists of yellow and grey sands, soft, bluish, argillaceous sand, and somewhat shaly and sandy marl, with several beds of doggers or more or less constant bands of gritstone. Towards the west, as pointed out by Blake and Hudleston, the grits and sands die out and pass laterally into clay 'till at Broadwey, due north of Weymouth, they are scarcely discoverable, unless represented by a single 2-ft. block of ferruginous sandstone, lying in the midst of clays; and only a thin band can be identified in the valley south of Abbotsbury, which is excavated down to the Oxford Clay'.[4]

The only abundant fossils in the Lower Calcareous Grit are large specimens of *Gryphæa dilatata*. Most of the fossils recorded by authors undoubtedly came from the *Trigonia hudlestoni* Bed. Even Buckman, as late as 1925, seems to have fallen into the error of identifying fossils in blocks of grit on the shore, broken from the *T. hudlestoni* Bed, and recording them as from the Lower Calcareous Grit.[5]

[1] S. S. Buckman, 1925, *T.A.*, vol. v, p. 64.
[2] W. J. Arkell, 1927, *Phil. Trans. Roy. Soc.*, vol. ccxvi B, p. 158.
[3] W. J. Arkell, 1927, loc. cit., p. 158.
[4] Blake and Hudleston, 1877, loc. cit., p. 264.
[5] S. S. Buckman, 1925, *T.A.*, vol. v, p. 65, bed 7.

Buckman named three subdivisions of the beds at Redcliff, in descending order, Radcliff Grit, Jordan Grit, and Ham Cliff Grit.[1] But the misidentifications just mentioned vitiate his palaeontological distinctions, while his failure to publish measurements makes it difficult to obtain even an approximate idea of what he intended to include under the three terms. The ephemeral nature of the minor lithological units, as demonstrated by their westerly passage into clay, renders any subdivisional terms of doubtful value.

At a level 4 ft. below the base of the *Trigonia hudlestoni* Bed (as well as in it) I have obtained several specimens of *Cardioceras* cf. *excavatum, in situ* in the cliff west of Redcliff Point, near Bowleaze Cove. A short distance lower begin curious bands of large nodules of intensely hard, dark-grey, argillaceous gritstone, some of which resemble cannon-balls protruding from the larger doggers or grit-bands. Buckman noticed the highest and most conspicuous layer of these nodules (in his Jordan Grit) and said that they were 'obviously derived', perhaps from the Patellate Layer in the Oxford Clay.[2] Closer inspection, however, makes it certain that they are not derived, but are of concretionary origin. They occur at several levels in the Lower Calcareous Grit, the lowest layer being only 5 ft. from the base, and both the nodules and the surrounding grits sometimes contain the same large Gryphæas. Identical concretions are a conspicuous feature of the Lower Calcareous Grit in Yorkshire, where they are especially well seen at Cayton Bay.

(b) The Southern Limb of the Weymouth Anticline: Weymouth

The southern outcrop, though only $2\frac{1}{2}$ miles in greatest length, is classic ground. The area is triangular, with the base of the triangle running E.–W. from the Nothe to the East Fleet Backwater and the apex near Small Mouth, where the road and the railway to Portland leave the mainland, crossing the mouth of the Fleet to Chesil Bank. On it are built the southern part of the town of Weymouth, with the upland parishes of Rodwell and Wyke Regis, and along its eastern coast are the type-sections at the Nothe Point, Bincleave or Bencliff, and Sandsfoot.

The deterioration of these sections has been so great during the present century that it is no longer possible to study the complete succession. The building of the breakwaters enclosing Portland Harbour has arrested coast erosion, and the cliffs of Sandsfoot Clay have now nearly reached stability, overgrown with thickets and trees; the construction of the torpedo works at Bincleave has closed the exposure of the Osmington Oolite Series to the public, and since the making of the road leading to the works the type-section of the Bencliff Grit has become totally obscured. The fortification of the Nothe long ago destroyed the type-section of the Nothe Grits, while that of the Nothe Clay received its *coup de grâce* in 1931 by the laying out of the cliff as gardens. Finally, a supplementary section of the Osmington Oolites and *Trigonia clavellata* Beds in Rodwell railway-cutting, described by Blake and Hudleston, is completely overgrown.

Only the Sandsfoot Grits can still be clearly seen, forming the striking red cliffs below the perched ruin of Sandsfoot Castle. Even of these the uppermost beds are obscured, but a continuous exposure can be seen in the low banks of the Fleet, south of Wyke Regis. The locality affords an interesting

---

[1] 1925, loc. cit., p. 65.   [2] 1925, loc. cit., p. 65.

comparison with the northern limb of the Weymouth Anticline, for, although according to measurements published by H. B. Woodward in 1895 the total thickness of the Corallian Beds is 196 ft., or 7 ft. less than at Osmington, the Upper Calcareous Grit has expanded to 40 ft. Of this the highest 16 ft. comprise the Ringstead Waxy Clays with ironstone-nodules, correlated at Ringstead and Osmington with the Westbury Iron Ore; and here, in fact, a 1-ft. band of typical ironshot oolite like that at Westbury is found at the top, occupying the stratigraphical position of the Ringstead Coral Bed.

The Sandsfoot Grits proper, 24 ft. thick, consist of ferruginous sand and sandstone, the harder portions crowded with fossils; and in the upper part is a 5-ft. band of clay. The lamellibranch fauna of the grits is exceedingly interesting. It comprises only a small number of species, but they have a restricted range and constitute a highly characteristic assemblage. The most fossiliferous bed is near the base: it is packed with the valves of *Chlamys midas*, *Ostrea delta* and *Pinna sandsfootensis* Arkell, with which are associated the large *Camptonectes sandsfootensis* Ark., *Modiola*, and other forms. The ponderous *Ctenostreon proboscideum* is abundant throughout.[1] Some of the bands are a mass of ramifying fucoid markings, like those seen on a smaller scale in the Nothe Grits. Blake and Hudleston considered that these denoted the actual spot where a colony of algæ grew, but although the suggestion is probably correct, no organic structure has been found in them. The ammonites belong principally to the genus *Ringsteadia*, and the locality yielded a number of the specimens described by Salfeld in his monograph of the genus.

Since this type-section of the British Upper Calcareous Grit has been variously interpreted by Waagen, Blake, Blake and Hudleston, Woodward, and Salfeld, so that considerable doubt concerning the succession has crept into the literature (largely owing to the obscure language of Blake, who considered the uppermost a series of 'Passage Beds'—and what beds are not 'Passage Beds'?), I append the following record of the section, measured along the bank of the Fleet south of Wyke Regis and checked so far as possible at Sandsfoot.

UPPER CALCAREOUS GRIT SOUTH OF WYKE REGIS

Kimeridge Clay.

		ft.	in.
13.	Blue-grey shaly clay, with layers of *O. delta*; seen to . . .	6	0
12.	*Exogyra nana* and *E. virgula* Bed: brown sandy band locally hardened like Forest Marble, crowded with small *Exogyræ*; as at Ringstead, &c.		8
11.	*Rhactorhynchia inconstans* Bed; grey clay, with *Rh. inconstans*, *Exogyra prævirgula* Jourdy, and *Modiola durnovariæ* Ark.; *Pictonia* sp. found at Ringstead	1	8

Upper Calcareous Grit.

WESTBURY IRON ORE BEDS (16 ft.)

		ft.	in.
10.	Prominent band of red clay-ironstone nodules		3
9.	IRONSHOT OOLITE: dark ferruginous ooliths in a cream-coloured rubbly limestone matrix; large *Serpulæ*, *Modiola* sp., *Exogyra nana* and *Ctenostreon proboscideum* [=Ringstead Coral Bed] . . .	1	0

[1] It was here that Dean Buckland of Oxford collected the type-specimens figured by Sowerby in *The Mineral Conchology*.

8. RINGSTEAD WAXY CLAY: Brown and red waxy clay, seen to 3 ft., below   ft. in.
   which is an unmeasurable gap of perhaps 3–5 ft., and then 6 ft. of
   blue-grey clay with a few clay-ironstone nodules. [*Ringsteadia
   anglica* Salf. found near the top of this clay *in situ* at Ringstead.]
   Total of clay according to Blake and Hudleston .   .   . 15 0

SANDSFOOT GRITS (24 ft.)

7. Sand, yellow and ferruginous, full of rubble and fucoid markings, with
   Pleuromya and many broken *O. delta* 4 6
6. Sand, more consolidated, with some prominent hard bands, containing *Pinna sandsfootensis* and full of fucoid markings .   .   . 3 0
5. Clay, blue and brown 5 0
   passing down into
4. Sand, impure, ferruginous, with *Pleuromya* and *O. delta* and impersistent hard grit bands full of fucoid markings. *Ringsteadia pseudo-
   yo* Salf. found *in situ* at Sandsfoot 8 6
3. Clay, sandy, passing into sand under Sandsfoot Castle .   .   . 1 6
2. Fossil Bed: hard ferruginous gritstone, a mass of *Chlamys midas*,
   Ctenostreon proboscideum, *Ostrea delta*, &c. This band is very
   noticeable at the base of the grits at Sandsfoot Castle, where a large
   Ringsteadia was seen *in situ* 2 0

Glos Oolite Series.

1. SANDSFOOT CLAY, much concealed by slips and vegetation, with
   abundant *Ostrea delta* at the top. [Total 38–40 ft. at Sandsfoot, according to authors]: seen to about 12 0

It is fairly certain that the ironshot oolite (9) is the equivalent of the Ringstead Coral Bed, although a prominent band of red clay-ironstone nodules
occurs above the one and below the other, for near Osmington the coral bed
is seen to pass into ironshot oolite westward.

The line here adopted as the top of the Corallian formation (the top of the
Ringsteadia zone) agrees with that drawn by Waagen and by Blake and
Hudleston, but Woodward classed all above Bed 7 with the Kimeridge Clay.
The whole of the Sandsfoot Beds, including the Sandsfoot Clay, were stated
by Sedgwick in 1826 to be 'within the limits of the Kimeridge Clay',[1] and this
truly expresses the affinities of their fauna if the beds are to be parcelled into
stages, using the terms Kimeridgian or Oxfordian. Waagen's subsequent
conclusion that the Upper Calcareous Grit belongs with the Oxfordian is not
upheld by the palaeontology.[2]

The remarks at the beginning of this section show that little remains to be
said regarding the downward continuation of the sequence.

The *Trigonia clavellata* Beds form long shore-platforms beneath the cliff
of Sandsfoot Clay south of Bincleave. They are 12 ft. thick and highly fossiliferous, but extremely hard. There is no exposure of this horizon on the shore
of the Fleet.

The Osmington Oolite Series is somewhat thinner than at Osmington,
probably about 45 ft. thick. On the shore of the Fleet it contains interesting
beds crowded with *Isodonta triangularis* (Phil.), not noticed elsewhere.

According to Blake and Hudleston the total thickness of the Bencliff Grit

[1] 1826, *Ann. Phil.*, vol. xi, p. 349.
[2] 1865, *Versuch einer allgemeinen Classif.*, p. 24.

in the type-section was 21 ft.[1] It consisted of sand and sandstone, rather more argillaceous in the lower part, with the usual band of hard sandstone doggers at the base. The Nothe Clay was estimated to be 40 ft. thick, and the Nothe Grits about 30 ft. This last estimate included any representative there may have been of the *Trigonia hudlestoni* Bed. No details are preserved, except that Blake and Hudleston stated that the downward passage to clay was abrupt. The hardness of the gritstone bands is the direct cause of the long, narrow promontory of the Nothe. On the shores of the Fleet the lower beds are but obscurely exposed in small discontinuous sections.

(c) North Dorset

The Corallian Beds reappear from beneath the Chalk Downs at the villages of Mappowder and Wootton Glanville, whence they strike north for 16 miles, the outcrop having a width of from one to three miles, and forming pleasantly undulating and wooded country rising above the clay vales on either side. At Mere, near the north end of this outcrop, the Corallian Beds and Kimeridge Clay are cut off by the fault bounding the north side of the Vale of Wardour, so that they end abruptly against the steep down of Greensand and Chalk. Northward for 12 miles they are hidden beneath it.

In the southern half of the area the rocks are known only from a few unsatisfactory openings at Wootton Glanville, Mappowder and Hazelbury Bryant, most of them showing only a few feet of the *Trigonia clavellata* Beds. In the northern half of the area, however, the Osmington Oolite Series yields a good building stone, which is exploited in large quarries under the name of the Marnhull and Todber Freestone. A complete section near the centre of the area is afforded by the railway-cutting at Sturminster Newton, supplemented by a deep road-cutting by the river-side, south of the town. By combining the information derived from all these exposures, and from other quarries at Stour Provost, East Stour, Langham, Cucklington and Silton, the following sequence can be made out:

SUMMARY OF THE CORALLIAN SERIES IN NORTH DORSET [2]

Upper Calcareous Grit.

The only exposure known is in the Sturminster railway-cutting, at the top of which are 14 ft. of rusty yellow sand, much mixed with clay and soil.[3] Traces of the same beds are evident at Silton and elsewhere.

Glos Oolite Series.

The Sandsfoot Clay is 8 ft. 6 in. thick in the Sturminster railway-cutting, and the base appears above the *Trigonia clavellata* Beds in East Stour Quarry.

The *Trigonia clavellata* Beds are an important feature over the whole area, usually consisting of soft, shelly, rather marly limestones, crowded with well-preserved lamellibranchs, as in the Weymouth district. In the Sturminster railway-cutting their total thickness is 15 ft.

At Stour Provost and East Stour, 5 and 6 miles to the north, the total

---

[1] H. B. Woodward, in enlarging the Bencliff Grit to 35½ ft. (1895, p. 89) included in it 16 ft. of Littlemore Clay Beds clearly belonging to the Osmington Oolite Series and so classed by him at Shortlake and Black Head.
[2] Based mainly on field-notes and Blake and Hudleston.
[3] Blake and Hudleston, 1877, loc. cit., p. 276, fig. 2.

thickness has shrunk to 2½ ft., and the dark, soft, shelly beds are sharply de-marcated from the underlying courses of white oolite, belonging to the Osmington Oolite Series. At these places *Trigonia clavellata* abounds, together with *Chlamys qualicosta*, *Chlamys superfibrosa*, *Ostrea delta*, *Astarte morini*, *Nucula menkei*, *Cucullæa contracta*, *Trigonia reticulata*, *Gervillia avicu-loides*, *Nerinea* sp., and numerous other bivalves as in the Weymouth district.

At the north end of the district, at Silton near Bourton, and Preston, north-west of Gillingham, rolled corals become conspicuous, mixed with the same fossils as at the Stours. The commonest species belong to the genus *Stylina*, together with *Thamnastræa*, *Thecosmilia*, &c. Another interesting feature here is the development of massive, doggery, blue-hearted calcareous grits, full of casts and impressions of *Trigonia clavellata*. This arenaceous facies is similar to the Grès d'Hennequeville and the Sables de Glos at the type-locality of the Glos Oolite Series on the coast of Normandy. At Silton there is also a local mussel-bank composed of the diminutive *Mytilus varians* Roem., elsewhere recorded only from Germany.[1]

Osmington Oolite Series.

The downward passage from the *Trigonia clavellata* Beds, though more abrupt than in the Weymouth district, is not marked by any interruption in the bedding, the uppermost 10 ft. or so of the white freestones being even-bedded. Beneath this, however, the bulk of the Marnhull and Todber Free-stone (15–20 ft. thick) is steeply false-bedded. It is a solid, creamy, fine-grained, hard, oolitic limestone, blue where unweathered. Fossils are scarce, except for micromorphic forms (*Isodonta triangularis*, *Limatula elliptica*, *Chlamys qualicosta*, *Exogyra nana*) and drifted fragments. In the rare shelly bands gastropods, as usual, predominate.

Underneath the Marnhull and Todber Freestone is a variable development of the Littlemore Clay Beds—bands of whitish rubbly claystone alternating with black clay. They are about 8 ft. thick at Sturminster, but thicken south-ward, where they seem to replace the freestone. In these beds, especially in the road-cutting at Sturminster, *Nucleolites scutatus* abounds, together with the rarer *Acrosalenia angulata* and *Hemicidaris intermedia*.

At the base are a few feet of loose, marly, large-grained oolite and a band of coarse pisolite reminiscent of the Pea Grit of the Cotswolds; the pisoliths are flattened, and measure a third of an inch in diameter. The band floors the village square at Stour Provost, giving it a remarkable appearance. At Cucklington, in the west, the pisolitic beds thicken to about 12 ft. and yield a highly characteristic, coarsely pisolitic stone.[2] Fossils in these beds are uncommon (except for the ubiquitous *N. scutatus* and *E. nana*). Indecisive Perisphinctids occur, but rarely and in bad preservation.

Berkshire Oolite Series and Lower Calcareous Grit.

In the Sturminster neighbourhood the pisolite has been seen in several clear exposures (during road widening in 1927 on both sides of the town, and in the railway-cutting) to rest directly on grey clay, indistinguishable from Oxford Clay. It is possible, however, that this is an argillaceous development

---

[1] W. J. Arkell, 1929, 'Mon. Corall. Lamell.', *Pal. Soc.*, pp. 52–3.
[2] A piece is figured, *P.G.A.*, 1916, vol. xxvii, p. 132, pl. XXIII.

of the Lower Calcareous Grit, for a short distance to the north grits appear, and, in the eastern entrance to the tunnel west of Gillingham, Blake and Hudleston saw 18 ft. of 'dark blue sandy marl, containing at various levels immense spheroidal doggers of calcareous grit, and sometimes thin layers of alternating sand and clay'.[1] Above this were noted 33 ft. of beds which also may represent parts of the Lower Calcareous Grit, or the Berkshire Oolite Series, or both, but the description and lists of fossils are insufficient for certain correlation. There do not seem to be any other exposures of beds below the level of the pisolite.

II. THE WILTS.–BERKS.–OXON. RIDGE [2]

(a) The Westbury and Calne Area

Just as in North Dorset an isolated range of Corallian Beds closes the mouth of the Vale of Wardour, so again at the mouth of the Vale of Pewsey the formation reappears for a stretch of 9 miles from Westbury to Seend; north of this is another gap, occupied by the Lower Greensand hills of Spye Park and Bromham. Beyond this the main outcrop begins near Calne and continues to Oxford.

The area about Westbury, Seend and Calne contains such special features that it is best described separately. It includes the famous coral reef of Steeple Ashton, perhaps the most noted locality for corals in Britain, and also works at Westbury where the Corallian iron ore is smelted.

SUMMARY OF THE CORALLIAN SERIES, WESTBURY TO CALNE

Upper Calcareous Grit.

Immediately beneath the Kimeridge Clay is the WESTBURY IRONSTONE, 11–14 ft. thick. It consists of an oolitic, more or less argillaceous ore, varying in colour from dark reddish-brown to green, and yielding from 30 to 35 per cent. of iron. The fossils link it to the Sandsfoot Grit in general, and in particular to the Ringstead Waxy Clay and Coral Bed; and on the other side of the Channel to the ironshot oolite of Hesdin l'Abbé, near Boulogne. Several species of *Ringsteadia—R. anglica* Salf., *R. pseudocordata* (Bl. and H.), &c.—are abundant, and there is in the Geological Survey collection a fragmentary Perisphinctid[3] resembling the true *P. wartæ* Bukowski. *Ostrea delta, Ctenostreon proboscideum, Chlamys midas, Camptonectes sandsfootensis,* &c. also occur. The surface-extent of the ore is rather limited, but it is said to have been worked also in old pits on the high ground round the church at Steeple Ashton. At Calne there is no sign of it.

Underneath the ore are from 4 ft. to 10 ft. of pale greenish-grey, buff and ferruginous sands, containing only casts of *Pleuromya.*

Glos Oolite Series.

At Westbury the Upper Calcareous Grit rests directly on rubbly grey and white oolite and pisolite apparently belonging to the Osmington Oolite Series. Farther north, at Broad Mead, near East Ashton, a reddish shell-bed appears

---

[1] Blake and Hudleston, 1877, loc. cit., p. 278.
[2] Based on W. J. Arkell, 1927, *Phil. Trans. Roy. Soc.*, vol. ccxvi B, pp. 67–181; and 1928, *Journ. Ecology*, vol. xvi, pp. 134–49; and Blake and Hudleston, 1877, *Q.J.G.S.*, vol. xxxiii, pp. 283–313.   [3] No. 25484.

above the white oolite, containing drifted corals, *Trigonia clavellata, Perna mytiloides, Chlamys intertexta, Limatula elliptica, Trigonia ashtonensis, Prorokia problematica* and perisphinctid fragments; this is evidently a continuation of the *Trigonia clavellata* Beds of North Dorset. Near Steeple Ashton the bed passes into a true coral reef—one of the reefs from which the rolled corals of East Ashton and North Dorset were derived. No excavation seems to have been made in the reef, all the specimens in various museums having been picked up on the surface of a field on the north side of a road turning SE. off the high road from Steeple Ashton to Bratton. The best specimens have long since been taken away, but every stone on the field is still a coral. The association is peculiarly rich in species, and the state of preservation, in a red earthy matrix, is better than at any other locality in England. Besides the common species, *Isastræa explanata* (Goldf.), *Thamnastræa* spp., *Thecosmilia annularis* (Flem.), several other corals are abundant, notably *Comoseris irradians* and *Stylina tubulifera*.

Osmington Oolite Series.

The local reef at Steeple Ashton grew upon a foundation of thick white oolites, which can be seen in an old lime-kiln near by, beside the road leading south from the village. Here and at Westbury, as in North Dorset, the upper part of the oolites is even-bedded and the lower and larger part is false-bedded. The fossils, which are chiefly fragmentary, are all worn and coated with calcareous matter, and the beds are looser and more pisolitic than at Marnhull and Todber.

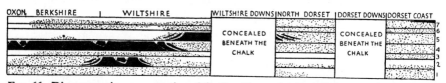

FIG. 66. Diagrammatic representation of the Corallian rocks along the outcrop in the South of England, to show the horizontal and vertical distribution of coral rock. Coral: black; other deposits: dotted. Not to scale. (From W. J. Arkell, 1928, *Journ. Ecology*, vol. xvi, p. 135.)

7 Upper Calcareous Grit		3 Bencliff Grit and Nothe Clay ⎫ Berkshire
6 Sandsfoot Clay ⎫ Glos Oolite	2 *Trigonia hudlestoni* Beds ⎬ Oolite Series
5 *Trigonia clavellata* Beds ⎬ Series	1 Lower Calcareous Grit ⎭
4 Osmington Oolite Series ⎭	

The best section in the Osmington Oolite is a large quarry (now unfortunately closed) behind the workhouse at Calne, which formerly showed 25 ft. of false-bedded oolites, yielding the CALNE FREESTONE, a material at one time much used locally for building. In the base of this quarry was encountered a remarkable colony of fossil sea-urchins. The tests of some species, especially *Hemicidaris intermedia*, which are usually found only singly, were congregated by the hundred, spread thickly over the bedding-planes of the stone. The species represented were *H. intermedia, Cidaris florigemma, C. smithii, Acrosalenia decorata, Glypticus hieroglyphicus, Diplopodia versipora, Stomechinus gyratus, Pseudodiadema pseudodiadema*, and *P. mamillanum*.[1]

The Calne Freestone with its hosts of echinoderms was laid down on the fringe of the great chain of coral reefs that extended through North Wiltshire

[1] A fine collection may be seen in the Wiltshire Museum, Devizes.

and Berkshire to Oxford. Near by can be studied the passage from the Dorset type of oolite to the coral rag so typical for the next 40 miles, and a very illuminating spectacle it provides for the geologist. First Lonsdale and after him Blake and Hudleston described it, beside a brook that runs down from Quemerford and thence towards the town. The oolite and pisolite gradually pass laterally into beds of more obviously detrital materials, with which lenticles of coral rag are intercalated, until eventually the coral rag predominates.

Another facies, that of the Littlemore Clay Beds, is developed 3 miles north of Calne, at Hilmarton, where bands of white argillaceous limestone and clay, containing the Osmington Oolite fauna, recall the typical development at Littlemore. The occurrence of the clay beds coincides with an excessive narrowing of the outcrop, and at this point a brook has cut an opening through the Corallian escarpment—features commonly associated with the local passage of the relatively resistant oolites and coral rag into soft clays.

Berkshire Oolite Series.

A deep boring at Westbury proved 40 ft. of clays beneath the Osmington Oolite and above the Lower Calcareous Grit.[1] Some of this clay (22 ft. out of the 40 ft.) may be accounted for as an argillaceous development of the lower part of the Osmington Oolite Series—a partial passage into Littlemore Clay Beds such as is usual in Dorset. Two quarries almost due west of Hilmarton, upon an outlying ridge of Corallian rocks, however, give a definite indication that the Nothe Clay and at least locally the Bencliff Grit are well developed, as on the Dorset coast. This ridge, known as Spirt Hill, is the key to the sequence in this part of Wiltshire. The high plateau of Bradenstoke and Lyneham to the north is formed of the coral rag resting upon a thin bed of oolite and pisolite. But at Spirt Hill the Osmington Oolite Series, having probably consisted of Littlemore Clay Beds as at the neighbouring village of Hilmarton, has been entirely stripped from its summit; instead, the crest of the hill is formed of yellow sand perhaps 15 ft. thick. Below the sand are about 12 ft. of grey-green mottled clay, the lowest 8 ft. of which are seen in a quarry, resting on a remarkable Pebble Bed, with the coarse, false-bedded sands and doggers of the Lower Calcareous Grit beneath. At the base of the clay is an impersistent band of tough oolitic limestone, yielding only *Chlamys fibrosa*. The general appearance and poverty of fossils recall the Nothe Clay of Redcliff Point, and the presence of the representative of the Bencliff Grit at the top of the hill, exposed in a small sand-pit, completes the analogy.[2]

The PEBBLE BED, which henceforth provides a valuable stratigraphical datum, is an interesting stratum. Consisting of a seam of large water-rolled pebbles, packed close together in a soft clay matrix, it has here a thickness of only 4 in., but in many places it is up to 1 ft. thick, and it is traceable in almost every exposure at which the horizon is to be seen as far as Oxford, a distance of 40 miles. The pebbles are well bored by *Lithophaga inclusa* and often encrusted with *Serpulæ* and *Exogyra nana*. They are of three sorts. The largest and most conspicuous, which vary in diameter from about 1 in. to

[1] J. Pringle, 1922, *Sum. Prog. Geol. Surv.* for 1921, pp. 146–53.
[2] W. J. Arkell, field-notes. After this book had gone to press my attention was called to a cutting on the new loop-line at Westbury, where the Pebble Bed has been exposed at its most southerly point, combined with a *Gervillia* and *Trigonia* bed, separated from the oolite by about 12 ft. of clay and marl.

4 in., averaging perhaps $2\frac{1}{2}$ in., consist chiefly of hard, compact claystone, recalling the hard crackers so common in the Oxford Clay. The derivation of at least some of them from the denudation of Oxford Clay is proved by the discovery of a specimen of *Nucula* in one at Littleworth, Berkshire. Locally some of the largest consist of indurated calcareous gritstone, evidently the product of erosion of already-consolidated Lower Calcareous Grit. The third sort are much smaller and consist of extremely hard black 'lydite' and white vein-quartz, of indefinite Palaeozoic origin. These last are identical with many to be seen in the sands of the Lower Calcareous Grit, whence they have doubtless been rederived.[1]

The Pebble Bed at Spirt Hill is especially interesting on account of its containing a number of oysters. The commonest are *Gryphæa dilatata*, together with *Lopha gregarea* and *Ctenostreon proboscideum*, all of which are common to the Corallian and the Upper Oxford Clay. In addition, however, there are specimens of *Gryphæa lituola* Lamk., the much more incurved and narrower species characteristic of the Middle Oxford Clay, and not known elsewhere above the *athleta* zone. The presence of this species and also the segregation of so many oysters of different species in a thin seam with water-worn pebbles suggests that the fossils have been derived from the erosion of Oxford Clay, but against this must be set the fact that they do not appear to have been rolled.

In North Wiltshire, immediately succeeding the Pebble Bed, coral reefs occur, separated by representatives of the Nothe Clay and Bencliff Grit from the Osmington Oolite coral rag above. Coral growth of this early Berkshire Oolite period may be represented in the Calne district by a reef at Westbrook, which Blake and Hudleston believed to be of earlier date than the rest of the coral rag in the neighbourhood. Their belief was based on the observations that *Cidaris florigemma*, usually the predominant urchin of the rag, was here replaced by *C. smithii*; and that the reef grew directly upon sand, which they took to be the Lower Calcareous Grit. This correlation is probably correct, but in view of the presence of Bencliff Grit at Spirt Hill, careful work will be required before the matter can be finally settled. The old limekilns on Sand-ridge Hill, Westbrook, are now overgrown by a wood, but fortunately they were vividly described by Blake and Hudleston, and whether the rocks they showed should properly be classed with the Berkshire Oolites or with the Osmington Oolites above, they illustrated so well the nature of the Jurassic coral reefs that Blake and Hudleston's description may appropriately be quoted.

'Layer upon layer of large masses of *Thamnastræa concinna* and *Isastræa explanata*,' they wrote, ' bored by the characteristic *Lithophaga inclusa*, and changed not seldom into crystalline limestone in which the organic structure is no longer visible, here spreads over the surface, resting immediately upon a bed of sand. . . . The spaces between the coral growths are filled with rubbly brash, made up of comminuted materials, and sometimes with clay charged with fragments of shells. These intercoralline accumulations obtain the mastery here and there; corals disappear, and we have great rubbly beds of shelly clay and limestone brash forming the whole reef.'[2]

---

[1] For full discussion of the Pebble Bed see Arkell, 1927, loc. cit., pp. 80–4.
[2] Blake and Hudleston, 1877, loc. cit., p. 288.

Lower Calcareous Grit.

The Lower Calcareous Grit covers several square miles of ground in Bowood Park, where it probably attains its greatest thickness. From 25–30 ft. of yellow sands with bands and doggers of hard grey or blue-hearted gritstone are worked at Seend Cleeve, Derry Hill, Conygre Farm and Bremhill. The hard bands contain *Cardioceras cordatum* and *Nautilus hexagonus*, and at Seend an unusually large fauna of lamellibranchs is found: *Gryphæa dilatata*, *Ostrea quadrangularis*, *Chlamys fibrosa*, *Gervillia aviculoides*, *Pseudomonotis ovalis*, *Oxytoma expansa* and a number of other species, principally from a shelly band at the top.

(b) The Purton Axis

A short distance north of Spirt Hill and Hilmarton the Corallian rocks form a marked feature in the landscape, rising about Lyneham and Bradenstoke well above the 400 ft. contour as an elevated plateau, nearly 4 miles wide. This change is directly due to two causes: to the thick development of coral rag and oolite in the Osmington Oolite Series, and to the disappearance of the soft substratum of Nothe Clay. Typical sections may be seen at Tockenham Wick, beside the main road from Lyneham to Wootton Bassett. The coral rag, with some strong bands of white oolite, rests directly on the Lower Calcareous Grit, with the intervention only of the Pebble Bed.

A short distance south-west of Wootton Bassett the outcrop rapidly narrows once more and the Oxford and Kimeridge Clays are mapped as if they were in contact, the Corallian formation having entirely disappeared. The gap in the Corallian ridge is utilized by the Wilts. and Berks. Canal and the Great Western Railway. Fortunately, when the new railway to the Severn Tunnel was constructed, a fresh cutting was made close to the gap, west of Wootton Bassett Railway Station, and was described by Messrs. Reynolds and Vaughan before it became grassed over. From their description it seems that the gap is caused by the lateral passage of the Osmington Oolite Series into Littlemore Clay Beds, as at Hilmarton.[1] At the same time, from north of Spirt Hill the Lower Calcareous Grit has disappeared or become thinner and passed into clay.

North of Wootton Bassett the outcrop widens once more towards Purton, becoming again a high plateau with a steep escarpment overlooking the clay vale occupied by Braydon Forest. Quarries are numerous, and they all show massive limestones built up of comminuted shells and coral fragments, passing here and there into coral reefs. The detrital limestones, from their great development at Wheatley near Oxford, have been called the Wheatley Limestones, and in this part of England they are a striking facies of the Osmington Oolite Series. Quarries south of Purton Church (fig. 67) show how lenticles of growing coral became established from time to time upon the accumulating banks of debris, only to be repeatedly smothered by fresh accretions at every change in the prevailing currents. Near by, other openings show coral islets, built up of masses of well-preserved *Isastræa explanata*, *Comoseris irradians* and *Thamnastræa concinna*, still in the positions in which they grew.

The patches or islets of corals in the Purton district seem to be completely isolated among detrital limestones, the products of the grinding of the coral

[1] S. H. Reynolds and A. Vaughan, 1902, *Q.J.G.S.*, vol. lviii, p. 751.

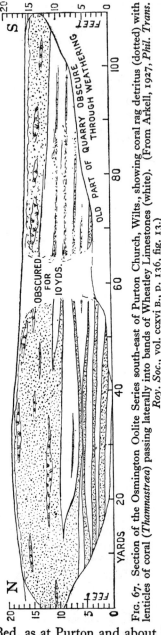

FIG. 67. Section of the Osmington Oolite Series south-east of Purton Church, Wilts., showing coral rag detritus (dotted) with lenticles of coral (*Thamnastraea*) passing laterally into bands of Wheatley Limestones (white). (From Arkell, 1927, *Phil. Trans. Roy. Soc.*, vol. ccxvi B., p. 136, fig. 13.)

reefs by the waves. In some places the corals grew out over a substratum of detritus; in others the debris covered them partially or closed above them. The reefs are of fringing reef type, but are difficult to classify more accurately in terms of modern coral structures. Their thickness probably never exceeds 25 ft. and is usually less, and their distribution seems to have been haphazard. Darwin wrote concerning certain modern fringing reefs: 'It follows . . . that where the sea is very shallow, as in the Persian Gulf and in parts of the East Indian Archipelago, the reefs lose their fringing character and appear as separate and irregularly scattered patches, often of considerable size.'[1] That the sea was shallower than usual in the Purton district is certain on stratigraphical grounds, and this passage of Darwin's seems singularly applicable.

Two miles east of Purton the outcrop for the third time narrows suddenly and there is a gap in the Corallian escarpment, utilized by another branch of the canal and by the railways from Swindon to Cheltenham and Cirencester. Here again the Oxford and Kimeridge Clays have been mapped as if in contact. But one of the railway-cuttings was seen by H. B. Woodward, who described about 20 ft. of clay with bands of 'grey earthy limestone', 'compact coral rock' and 'rubbly irregular grey marly and septarian limestone', the whole full of typical Corallian fossils.[2] Evidently the gap is again caused by a lateral change of facies of the Osmington Oolite Series into Littlemore Clay Beds.

To the east another high plateau of coral rag and Wheatley Limestones juts northward into the vale, about Broad and Little Blunsdon. A useful section in a well sunk into it showed that the Osmington Oolite Series is still separated from the Lower Calcareous Grit by nothing more than a thin Pebble Bed, as at Purton and about Lyneham and Tockenham.

It is concluded, therefore, that over this area deposition was arrested while the Berkshire Oolites were being laid down from Highworth north-eastward on the one hand and from Spirt Hill south-westward on the other. The best

[1] Charles Darwin, *Coral Reefs*, 3rd ed., 1889, pp. 77–8.
[2] H. B. Woodward, 1895, *J.R.B.*, p. 117.

explanation seems to be that the sea was shallower owing to an axis of uplift (the Purton or Wootton Bassett Axis) and it may be significant that the area coincides with the continuation of the Woolhope–May Hill line of uplift in the Palaeozoic rocks far to the north-west (see p. 80).

The Upper Calcareous Grit of the area of the Purton Axis is very little known. Sands, bright red in colour and separated from the coral rag by clay, begin north of Hilmarton and continue to Wootton Bassett. An old brickyard at Wootton Bassett, long since obscured, yielded a number of *Ringsteadiæ*, in addition to *Pictoniæ* (indicating the basal zone of the Kimeridge Clay). A predominantly argillaceous development of the Upper Calcareous Grit was also proved by the deep boring at Swindon. At the base of the Kimeridge Clay were 6 in. of 'grey and brown earthy limestone with patches of ironshot grains', containing *Prionodoceras serratum*, and evidently a representative of the Westbury Iron Ore, separated by 21 ft. 6 in. of clay from the top of the Osmington Oolite Series;[1] and an identical succession was encountered by a boring on Red Down, near Highworth.[2]

The Lower Calcareous Grit, which forms such a conspicuous feature from the Seend and Calne districts, on by Bremhill to Spirt Hill, is absent over most of the area affected by the Purton Axis. It was proved entirely absent by the railway-cutting at Wootton Bassett and again by a well at Purton, and it is not represented on the Survey map except in two patches; one patch, seen under the Pebble Bed in the quarries at Tockenham Wick, also extends on to the outlier of Grittenham Hill, while the other patch forms Paven Hill, Purton, and the bluff on which Ringsbury Camp is cut. This last patch is of great interest, for it consists largely of a non-calcareous deposit of siliceous sponge-spicules ('*Rhaxella* chert') similar to the Arngrove Stone of Oxfordshire, to be described in a later section. It differs from the Oxfordshire rock, however, in the presence of large scattered ooliths, converted to limonite.

(c) Highworth–Faringdon–Marcham (The Faringdon Ridge)

Only 2 miles to the west of the well in the middle of the Blunsdon plateau, where the Berkshire Oolite Series seems to be represented solely by the

---

[1] The Osmington Oolite Series at Swindon is in the facies of Littlemore Clay Beds; see Arkell, 1927, loc. cit., p. 146; the account by Whitaker and Edmunds there referred to ('Water Supply of Wilts.', *Mem. Geol. Surv.*, 1925, pp. 85, 86) is the old misleading version, in which the 21½ ft. of clay and overlying 6 in. of ironshot oolite are grouped as Kimeridge Clay; for the revised grouping see Messrs. Chatwin and Pringle's account, 1922, *Sum. Prog. Geol. Surv.* for 1921, p. 164. The ironstone band was encountered by William Smith in the boring for water in the Vale between Swindon and Wootton Bassett, for which he was adviser in 1816–17 (J. Phillips, 1844, *Memoirs of William Smith*, p. 85, where it is duly entered in the diagram).

Salfeld misquotes Woodward as saying that this ironstone band with *P. serratum* was found in Oxford Clay in a cutting on the M.S.W. Jn. Rly. (now G.W.R.) west of Rodbourne Cheney. He then remarks 'was hier Oxford Clay soll, ist mir völlig unklar'. Woodward says nothing of the sort; he is clearly referring to the deep boring at Swindon and he definitely states 'Towards the base of what we included in the Kimeridge Clay at Swindon, there was a bed of iron-shot earthy limestone, six inches in thickness, which yielded *Ammonites cordatus*, var. *excavatus* [*serratus*], a well-known Lower Kimeridge fossil, though found here [also] in the Oxford Clay' (i.e. in the deep boring; see p. 37, where a Cardiocerate is listed from the Oxford Clay under this name). Woodward makes no mention of the ironstone band having been found in the Rodbourne railway-cutting, which was entirely in earlier beds. (H. B. Woodward, 1895, *J.R.B.*, pp. 37, 118; H. Salfeld, 1914, *Neues Jahrb. für Min.*, B–B, xxxvii, p. 196.)

[2] For the Highworth boring see W. J. Arkell, 1927, *Wilts. Arch. Nat. Hist. Mag.*, vol. xliv, pp. 43–5.

Pebble (or Shell-cum-Pebble) Bed, 1 ft. 6 in. thick, the full sequence is developed about Highworth. For the next 20 miles the Berkshire Oolite Series is the chief centre of interest in the geology. It builds the greater part of the hills about Highworth, Hannington and Coleshill, reaching 30 ft. in thickness, and thence, thinner but maintaining its characteristics, it appears in all the quarries along the Faringdon ridge as far as Cumnor and Marcham, on the borders of Oxfordshire. Its rich shell-beds yield some of the best specimens of the Corallian fauna to be obtained in the country. The Pebble Bed may be seen at the base, resting discordantly upon the false-bedded sands of the Lower Calcareous Grit. The Osmington Oolite Series coral rag continues unchanged above, but with its outcrop receded somewhat down the dip-slope, instead of lapping up to the edge of the escarpment as it does over the Purton Axis.

SUMMARY OF THE CORALLIAN ROCKS OF THE
HIGHWORTH–MARCHAM AREA

Upper Calcareous Grit and Glos Oolite Series.

The Upper Calcareous Grit is most fully developed at Shrivenham, near Highworth, where the Berkshire Oolites also attain their thickest development. There are now no good exposures, but the ferruginous sands form a distinct escarpment rising above the coral rag plateau at Sandhill Farm and round the north and west of the village. Several wells at Shrivenham proved up to 35 ft. of ferruginous sandy beds interdigitating with clays before the coral rag was reached.[1] During the construction of the Wilts. and Berks. Canal, and probably also in the excavation of the pits to obtain materials for the railway-embankment south-east of South Marston, an oolitic ironstone like that at Westbury and Swindon was encountered. *Ringsteadiæ* collected here, some of them by Dean Buckland, are preserved at Oxford and in London, and were mentioned by Salfeld in his monograph of the genus. One species was named *Ringsteadia marstonensis* Salf.[2]

The correspondence of this oolitic ironstone with the Westbury Iron Ore can hardly be questioned, and the 35 ft. of underlying beds together represent the Sandsfoot Grit and Sandsfoot Clay. To the north and west the argillaceous element predominates. An outlier at Red Down, near Highworth, is capped with 20 ft. of clay giving a bright-red soil, overlain by a thin band of Westbury Ironstone, just as at Swindon.[3]

Towards the east the Upper Calcareous Grit is rapidly overstepped by the Kimeridge Clay. At Faringdon all that remains is about 2 ft. of 'reddish-brown and chocolate-coloured ferruginous earth, with black stains from dissolved fossils, and lumps of calcareous clay, ironstone, and fragments of *Ostrea delta* and *Serpula* towards the base', resting upon the surface of the coral rag.[4] A reddish soil is noticeable above the coral rag about Shellingford and Stanford, but nothing is seen east of Shrivenham that could be called undisturbed Upper Calcareous Grit.

Although the *Trigonia clavellata* Beds have nowhere been definitely proved,

---

[1] 'Water Supply of Berks.', *Mem. Geol. Surv.*, pp. 73–4.
[2] H. Salfeld, 1917, *Palaeontographica*, vol. lxiii, p. 83.
[3] Compare Arkell, 1927, *Wilts. Arch. Nat. Hist. Mag.*, vol. xliv, pp. 43–5, and Chatwin and Pringle, 1922, loc. cit., p. 164.
[4] Blake and Hudleston, 1877, loc. cit., p. 301.

records of a shell-bed immediately above the Osmington Oolite Series in the Swindon boring and again in one of the wells at Shrivenham are suggestive.

Osmington Oolite Series.

The Osmington Oolite Series is continuous throughout the area and consists for the most part of coral rag of somewhat monotonous appearance. The corals are often in the position of growth and seem to have formed a thin fringing reef. Some of the mollusca may be obtained in exceptionally perfect condition, especially the large gastropods, which lie where they dropped in the interstices among the corals. At Kingsdown a small colony of *Terebratula kingsdownensis* Ark. has been found.[1]

The majority of the mollusca associated with the corals belong to a few species, and they are met with over and over again, in all the quarries. The principal constituents of the coral-dwelling fauna are *Chlamys nattheimensis* (de Lor.), *Lima zonata* Ark., *Lithophaga inclusa* (Phil.), *Lopha gregarea* (Sow.), *Exogyra nana* (Sow.) and *Littorina muricata* Sow. Of these the first two, although extremely abundant in their natural habitat, have scarcely ever been found unassociated with corals, while the oysters and *Littorina* are much more abundant about the reefs than elsewhere. The corals that enter most conspicuously into the composition of the reefs are *Thecosmilia annularis* (Flem.), *Thamnastræa arachnoides* (Park.), *Th. concinna* (Goldf.), and *Isastræa explanata* (Goldf.).[2]

In two areas, the one about Highworth, the other from Faringdon to Kingston Bagpuize, the coral rag rests discordantly upon thinly false-bedded, fissile, sandy oolites—the PUSEY FLAGS. About Pusey and Buckland these flags reach a thickness of 10–15 ft., and contain a rich micromorphic fauna, comprising *Protocardia dyonisea* (Buv.), *Isodonta triangularis* (Phil.), *Opis phillipsi* d'Orb., *Chlamys fibrosa* (Sow.), *Pseudomonotis ovalis* (Phil.), &c. Rarely Perisphinctids of the type of *P. martelli* Oppel are found.

In 1927 I regarded the Pusey Flags as a link between the Bencliff Grit and the Osmington Oolite Series;[3] but as they pass down gradually into the Highworth (= Bencliff) Grit, and as the junction between them and the coral rag above is abrupt and discordant, I grouped them with the Berkshire Oolite Series. When the relations of the Pusey Flags to the underlying beds are studied over the whole area, however, it is seen that they overstep the other members of the Berkshire Oolite Series unconformably and come to rest at Pusey directly upon the Lower Calcareous Grit (fig. 68). Moreover, locally they lose their sandy character and become almost a pure oolite, suggesting rather the false-bedded lower portion of the Osmington Oolite Series farther south (the Marnhull and Todber Freestone). Throughout North Dorset and South Wiltshire the false-bedded oolitic freestone is overlain by an even-bedded series, the line of junction being very straight and strongly-marked. It may well be that only the upper or even-bedded portion of the North Dorset Osmington Oolite Series is represented by the coral rag of the Highworth–Faringdon–Marcham district, the Pusey Flags representing the lower, false-bedded portion.

[1] W. J. Arkell, 1927, *Phil. Trans. Roy. Soc.*, vol. ccxvi B, pl. I, fig. 6.
[2] For an account of the coral fauna and some aspects of coral reef ecology in the Corallian, see W. J. Arkell, 1928, *Journ. Ecology*, vol. xvi, pp. 134–49.
[3] 1927, loc. cit., p. 159, table.

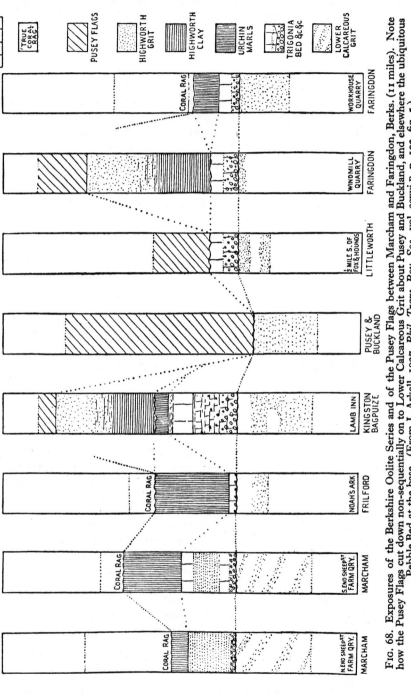

FIG. 68. Exposures of the Berkshire Oolite Series and of the Pusey Flags between Marcham and Faringdon, Berks. (11 miles). Note how the Pusey Flags cut down non-sequentially on to Lower Calcareous Grit about Pusey and Buckland, and elsewhere the ubiquitous Pebble Bed at the base. (From J. Arkell, 1927, *Phil. Trans. Roy. Soc.*, vol. ccxvi B, p. 102, fig. 5.)

It appears from surface-indications that south-west of Sevenhampton the coral rag oversteps all the underlying beds and comes to rest on the Oxford Clay.[1]

Berkshire Oolite Series.

The Bencliff Grit and the Nothe Clay, last seen at Spirt Hill on the far side of the Purton Axis, reappear at Highworth, where they have been called the HIGHWORTH GRIT and HIGHWORTH CLAY. Together they measure 10–12 ft. in thickness, consisting of fine grit-sands, with some doggers, passing down gradually into clay. Most of the water-supply of the old part of the town was derived from the top of the clay, being tapped by means of numerous shallow wells sunk through the coral rag and the subjacent sands. The clay, too, supported two rival brick industries, on the north and south sides of the town. The ruins of the extensive excavations may still be seen beside the road to the station and north of Redlands Court (fig. 69). The upper portion of the sands is false-bedded and can be seen to interdigitate with the Pusey Flags, which at Highworth are only 2 ft. thick. Occasional specimens of *Ostrea quadrangularis* and *Gryphæa dilatata* are almost the only fossils.

For some 3 miles about Faringdon the Highworth Grit and Clay are absent, apparently having been overstepped by the coral rag, but they reappear in the Windmill Pit, 1½ miles east of Faringdon, in the same thickness as at Highworth. In another 2 miles to the east they have been again overstepped by the Pusey Flags, about Pusey and Buckland, but they come in again at the Lamb Inn Quarry, Kingston Bagpuize, before finally disappearing towards Marcham and Oxford.

Beneath the Highworth Clay, in all the quarries where it is seen, are some thin beds known as the URCHIN MARLS. These consist of drab grey marls, variously divided into harder and softer bands, all largely composed of well-formed ooliths and always distinguishable by being crowded with the small urchin, *Nucleolites scutatus*. Their characteristic colour, their densely oolitic texture and the inevitable urchin, render them an easily recognized division from Highworth through Berkshire (including Shellingford where the Highworth Grit and Clay are absent) to Marcham and Garford, a distance of 18 miles. The thickness is 4 ft. at Highworth, 5 ft. at Marcham and the Noah's Ark, Garford (where also *Pygaster semisulcatus* is frequently associated with the more common *Nucleolites*).

Finally come the TRIGONIA HUDLESTONI LIMESTONES,[2] the most interesting part of the Berkshire Oolite Series, with the Pebble Bed at the bottom. They attain their maximum thickness of about 8–10 ft. at Highworth, and may be followed in a score or more of quarries along the Berkshire ridge to Marcham, Garford and Cumnor, everywhere crowded with a rich assemblage of fossils.

The best exposures were until recently the South Quarry, Highworth, and a second large quarry opened since the Great War at Hangman's Elm, beside the road to Shrivenham, 1½ miles south-east of Highworth. From these excavations, and from others actively worked at various points along the outcrop, especially at the Lamb Inn, Kingston Bagpuize, many mollusca have

---

[1] W. J. Arkell, 1927, loc. cit., geological map.

[2] Formerly called the *Trigonia perlata* Limestones, owing to an error in identification by Lycett; see Arkell, 1930, 'Mon. Corall. Lamell.', *Pal. Soc.*, part 2, p. 73.

been collected. The almost endless variety of Perisphinctids still remain to be worked out, but it is evident that they belong to the assemblage of the *transversarius* zone of the continental Argovian. Perhaps the most commonly represented is *Perisphinctes helenæ* De Riaz, with which are associated numerous forms approaching *P. plicatilis*, *P. martelli*, *P. cloroolithicus*, *P. aripripes*, *P. maximus*, *P. orientale*, *P. parandieri*, and the true *P. biplex* Sowerby (a much misinterpreted species). The Cardiocerates include the *excavatum* type (*C. highmoori* and others), the *maltonense* type, and the *vertebrale* type (*Vertebriceras* of Buckman). There are also Aspidocerates of several species.

Besides the ammonites, shell-banks composed of masses of lamellibranchs provide an attractive field for the palaeontologist. *Trigonia hudlestoni* and *Trigonia reticulata* form a *Trigonia* Bed at Marcham and Kingston Bagpuize, and with them are always associated *Gervillia aviculoides*, *Isognomon mytiloides*, the large Limas, *L. rigida* and *L. mutabilis*, *Chlamys splendens*, *C. fibrosa*, *Astarte ovata* and many other shells.

Many of these fossils, including the highly characteristic *Perisphinctes helenæ* De Riaz, have been collected in the *Trigonia hudlestoni* Bed at Redcliff Point, near Weymouth, and the lithological resemblance of that bed to the gritty, sparsely-oolitic, greyish, shelly 'fucoidal' limestones at Highworth is unmistakable.

In the old South Quarry, Highworth, two beds of rolled Thecosmilian corals, together about 2 ft. thick, are intercalated in the shelly limestones (fig. 69). When the new quarry at Hangman's Elm was opened these were found to have coalesced and thickened to a bed nearly 3 ft. thick, chiefly composed of the corals *Thecosmilia annularis* and *Montlivaltia dispar*, unrolled and in fine preservation among clay. It was then found that near by, at Upper Farm, the *Trigonia hudlestoni* Limestones pass laterally in a short distance into a small coral reef or islet. Detailed mapping has revealed that several islets exist in the *Trigonia hudlestoni* Limestones in the neighbourhood of Highworth; another is situated on the north side of Red Down, and a third (perhaps the continuation of the same one) on Staplers Hill, north-west of Hannington Railway Station. They are only a few fields in extent, but they can infallibly be recognized by the abundance of corals on the freshly-ploughed land, while the normal assemblage of lamellibranchs and ammonites is noticeably wanting. Small quarries give good sections of two of the reefs, showing typical coral rag, not only composed of the same species of corals as the coral rag of Osmington Oolite date, but also containing the same molluscan 'coral fauna'—*Lima zonata*, *Chlamys nattheimensis* and the rest, with spines of *Cidaris florigemma* and *C. smithii*. From internal evidence alone, therefore, these earlier reefs are indistinguishable from the more widespread rag of Osmington Oolite date, although the two are separated by the Urchin Marls, the Highworth Grit and Clay, and the Pusey Flags.

In the immediate neighbourhood of Faringdon and Coleshill the *Trigonia hudlestoni* Limestones take on a somewhat deeper-water facies, consisting of uniform white oolite, with the shells more sparsely distributed. This has been termed the Faringdon Facies. The limestones of the Highworth district bear a closer resemblance to those of Kingston Bagpuize than to those at Faringdon, and at Kingston fragments of coral are again found in the *Trigonia* bed.

At the eastern extremity of the district, at Marcham, Cothill and Cumnor, the *Trigonia hudlestoni* Limestones become highly charged with sand, and the *Trigonia* Bed is split into two by 2–4 ft. of interlaminated sand and clay. In

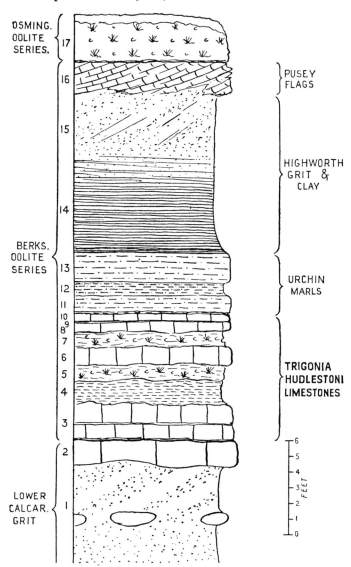

FIG. 69. Section of the Berkshire Oolite Series in the old South Quarries, Highworth, showing two coral beds in the *Trigonia hudlestoni* Limestones. (From Arkell, 1927, *Phil. Trans. Roy. Soc.*, vol. ccxvi B, p. 108, fig. 6.)

this part of the district, too, a new feature is introduced, in the form of an impersistent *Natica* Band. It is best developed at Cumnor and near Cothill, and consists of up to 6 ft. of hard calcareous gritstone enclosing myriads of casts of *Natica (Ampullina) arguta, Ceritella costata, Cloughtonia condensata,* and other gastropods. It was classed with the Lower Calcareous Grit until 1930, when the enlargement of the quarry at Cumnor revealed lenticles of the Pebble Bed beneath it. Previously a second impersistent layer of rolled pebbles above had been mistaken for the true Pebble Bed, but it is now established that the peculiarly fine conglomerate sometimes obtained in parts of the quarry occurs cemented on to the under side of the *Natica* Band.[1]

The Pebble Bed here is typical of that exposed in nearly every quarry from Oxford to Spirt Hill. The commonest pebbles are the usual rounded to rather flattish lumps of grey claystone, together with gritstone, lydite 'and vein-quartz derived from the Lower Calcareous Grit. A unique pebble, about $2\frac{1}{2}$ in. long, was found at the Lamb Inn Quarry, Kingston Bagpuize: a microporphyritic soda-rhyolite, consisting of small crystals of albite in a microlithic, fine-textured, black, silicified matrix.[2] Indications of a special fauna associated with the Pebble Bed are found only at the Lamb Inn Quarry, Kingston Bagpuize, where the pebbles are disseminated through 2 ft. 8 in. of pisolitic rock, containing round-whorled Perisphinctids of Buckman's genus *Cymatosphinctes* (? = *Kranaosphinctes*), not found in the overlying beds, and also a peculiar Aspidocerate, *A.* (*Euaspidoceras*) *crebricostis* Ark.[3]

Lower Calcareous Grit.

The Lower Calcareous Grit consists of the usual yellow sands, with impersistent bands and doggers of hard gritstone. False-bedding is common and is often observed to pass through the doggers, thus proving their secondary origin by some chemical process, which has either concentrated calcareous cement, or partly decalcified and removed cement from the surrounding material, reducing it to sand. The top of the Lower Calcareous Grit is seen in large numbers of sections, but deeper exposures are rare. The best are at Marcham, where for many years large collections of fossils have been obtained, especially the well-known *Aspidoceras* (*Euaspidoceras*) *catena*, *Nautilus hexagonus*, *Pachyteuthis abbreviatus*, and *Cardioceras excavatum*. Drifted wood, often in logs of considerable size, fruits of *Carpolithes* and vertebrae of *Teleosaurus* occur. The false-bedding is especially noticeable at Marcham and in Tubney Wood, as is also the quantity of small pebbles of lydite and vein-quartz. Both the proportion of pebbles and the degree of false-bedding diminish westwards, from which it is deduced that the source of the sands lay in the south-east. The deposit is a sandbank, the materials of which were presumably brought down by some large river, flowing from the landmass occupying the site of the London Basin and the North Sea. It dropped the greater part of its heavier materials on entering the sea, the lighter constituents being carried by a long-shore current south-westward into Wiltshire.

The thickness of the Lower Calcareous Grit is very variable in the east, about Oxford. A boring proved the thickness to be only 11 ft. at Wootton Waterworks, but it is probably nearer 50 ft. at the edge of the escarpment about Cumnor, and a boring at Frilford proved $52\frac{1}{2}$ ft. Thence it increases gradually towards Faringdon, where three borings at Faringdon, Goosey and Shrivenham proved 70 ft., 79 ft., and 76 ft. respectively. This is the area of

---

[1] W. J. Arkell, 1931, *P.G.A.*, vol. xlii, pp. 44–9.
[2] Ibid., p. 51. Identified by Dr. H. H. Thomas.
[3] W. J. Arkell, 1927, loc. cit., pp. 96–7.

maximum thickness, beyond which a more rapid diminution sets in towards the Purton Axis. In 3 miles, at Highworth, the borings prove 20–30 ft.; 4 miles farther, on the slopes of the Blunsdon plateau, 10–20 ft. is the usual thickness; while in another mile the Lower Calcareous Grit dwindles to a parting of sand and finally disappears against the axis.

It is noteworthy that the area in which the Lower Calcareous Grit attains its maximum thickness coincides with the area in which the Berkshire Oolite limestones take on the Faringdon Facies, thought to indicate deposition in deeper water.[1]

(d) The Oxford Axis

The Corallian rocks in the area immediately surrounding Oxford bear a strong resemblance to those about Purton. The thickness of the Lower Calcareous Grit becomes very variable: although in some places it is as much as 50 ft., in others it diminishes to less than 12 ft. The Berkshire Oolite Series, so important a feature in the area to the south-west, thins down to a mere Shell-cum-Pebble Bed, sometimes only 1 ft. thick. The Osmington Oolite Series develops thick masses of detrital Wheatley Limestones, among which the coral rag occurs once more in 'separate and irregularly scattered patches' as in the Purton district, and it advances again up the dip-slope to cover the plateau to the edge of the escarpment. Finally, the Upper Calcareous Grit and Glos Oolites are entirely absent, the *Rasenia* zones of the Kimeridge Clay resting directly upon the Osmington Oolite Series.

All these considerations point to the action of an elevatory force like that which operated in the Purton district, and it is expressed by calling it the Oxford Axis. It may perhaps be significant that the area lies upon the prolongation of the Sedgley–Lickey Axis (see Chapter III, p. 80).

The absence of the Upper Calcareous Grit and Glos Oolite Series, such conspicuous features on the Dorset coast, called forth some of the earliest observations on non-sequential strata, and these invest the Oxford Axis with considerable historical interest.

The junction of the Kimeridge Clay with the underlying coral rag and Wheatley Limestones could for upwards of a century be seen in a large quarry below the brickyard at the foot of Shotover Hill, but the section has now unfortunately become almost entirely overgrown. As early as 1812 W. D. Conybeare noticed that, when the Kimeridge Clay was stripped off in this quarry, the surface of the stone showed signs of being waterworn and was pitted with small cup-shaped depressions.[2] Four years after the publication of Conybeare's observation, Sedgwick made the following memorable correlation:

> May not the coral rag and superincumbent freestone of Headington Hill together represent the central group of the Weymouth and Steeple Ashton sections? The conjecture seems to be confirmed by the appearance of the beds in Headington Quarries. In that place the top freestone supports the Kimeridge Clay; and the separation between the two is as well-defined as a geometric line. Now the instantaneous passage from one formation to another frequently indicates the absence of

---

[1] For authorities as to thicknesses, and discussion of the possible source of the sand and pebbles, see Arkell, 1927, loc. cit., pp. 73–82.

[2] W. D. Conybeare, 1822, in Conybeare and Phillips, *Outlines Geol. Engl. and Wales*, p. 189.

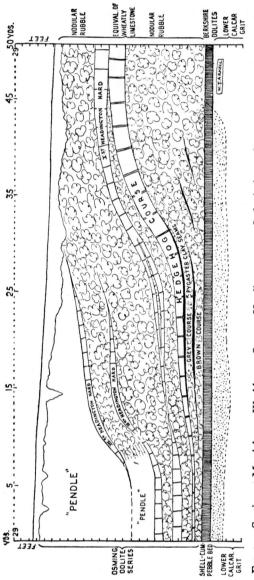

FIG. 70. Section at Magdalen or Workhouse Quarry, Headington, near Oxford, drawn from measurements at 10 yds. intervals, showing the different forms of coral rag debris (with the quarrymen's and other names) false dipping from the reef. At the base is the horizontal Shell-cum-Pebble Bed, yielding abundantly the Berkshire Oolites fauna. (From Arkell, 1927, *Phil. Trans. Roy. Soc.*, vol. ccxvi B, p. 128, fig. 12.)

certain beds or deposits. May not then the upper part of the Weymouth section be wanting near Oxford?[1]

In this passage the Wheatley Limestones and equivalent coral rag were for the first time correlated with the Osmington Oolite Series, and the true position of the Upper Calcareous Grit and Glos Oolite Series ('the upper part of the Weymouth section') was evidently realized. Ten years later Buckland and De la Beche[2] expressed their agreement with Sedgwick's correlation and the ensuing conclusions, and later Phillips also added his confirmation. Finally, one hundred years after the appearance of Sedgwick's memoir, the widening of the London road near Shotover Lodge afforded Dr. Pringle a temporary exposure enabling him to state that the lowest part of the Kimeridge Clay is of *Rasenia* date.[3] The hiatus was thus increased by the addition of the *Pictonia* zone of the Kimeridge Clay to the missing strata.

The Osmington Oolite Series forms an important surface-feature, capping the plateau about Headington and Bullingdon, around the foot of the high ridge of Shotover, and similarly encircling the heights of Boar's Hill and Cumnor Hurst on the other side of the Thames. Northward, as a number of faulted outliers, it caps the hills at Wytham and about Elsfield and Beckley. Thanks to a much greater number of exposures than exist in the Purton district, the lateral passage from true coral rag to the various types of detrital deposits can be followed out in detail. At Headington Quarries, at Wheatley, and on the west side of the Thames at Wootton and Sunningwell, immense quarries have been worked in the WHEATLEY LIMESTONES.[4] They consist of alternate harder and softer bands of the same white detrital material as at Purton, usually non-oolitic and composed entirely of finely-ground fragments of corals and shells. The harder bands were much worked in the eighteenth century as a freestone, to the undoing of some of the colleges in Oxford which employed it in their buildings: it has decayed so rapidly that all have had to be refaced after less than two centuries.

The softer bands have in places a curious nodular structure, hitherto unexplained, and are known as the Nodular Coralline Rubble. At Wheatley, in the weathered face of the old quarries, it is seen to be interbedded with the normal Wheatley Limestone, the two types of rock being composed of identical materials, differing only in the degree of their subsequent cementation. In some of the quarries at Headington tabular masses of coral are intercalated in the detrital material, and they gradually increase and coalesce towards the reefs, the broken fragments at the same time becoming larger. The dip is steeply away from the reefs, and is shown to be only a deposition dip by the fact that the series everywhere rests upon the horizontal Shell-cum-Pebble Bed (fig. 70).

Failure to comprehend the nature of these deposits, owing to insufficient

---

[1] A. Sedgwick, 1826, *Ann. Phil.*, vol. xi, p. 350.
[2] 1836, *Trans. Geol. Soc.* [2], vol. iv, p. 25.
[3] J. Pringle, 1926, 'Geol. Oxford', 2nd ed., *Mem. Geol. Surv.*, p. 63.
[4] First called the Oxford Oolite by Buckland (1818, Table in W. Phillips's *Geol. England & Wales*) where it was said to be worked as a freestone at Headington, and marked as overlying the Coral Rag; but in other quarries identical rock underlies coral rag, and the two facies undoubtedly alternate (see Arkell, 1927, pp. 132–3). Apart from the fact that oolitic structure is very rare, it seems undesirable to retain in its original restricted sense a term that has been so widely used with very different meanings (see above, p. 7). Buckman called the Wheatley Limestones Holton Beds (*T.A.*, vol. v, p. 53).

field-work, led Buckman a few years before his death to subdivide them into several zones. His misunderstanding of the deposition-dip is shown by a statement concerning the principal quarry at Headington (Magdalen Quarry, fig. 70) that 'there is a non-sequence—stratal failure—of about 6 feet in the face of the pit on the E. as compared with the W.'.[1] In reality such an inference is no more justifiable from these beds than from the Forest Marble or the Triassic sandstones, or any other formation in which false-bedding is manifested on a large scale. The true meaning of the false dip was explained by Blake and Hudleston in 1877.[2]

Unbroken fossils are rare, but occasionally they include ammonites, the most frequent being *Perisphinctes antecedens* Salf., though in the Lye Hill Quarry, near Wheatley, some badly-preserved giants of the *P. parandieri* style, like forms in the Berkshire Oolites, are found. Doubtless these floated in from outside the coralline area. Some of the echinoderms, such as *Nucleolites scutatus* and *Pygaster semisulcatus*, lived among the accumulating debris, for their tests are found perfectly preserved in thin argillaceous seams, although all the shells, with the exception of *Exogyra nana*, are finely comminuted.

South of Oxford, in the Thames Valley about Littlemore, Sandford and Kennington, the Osmington Oolite Series takes on the facies of the LITTLEMORE CLAY BEDS, of which this is the type-locality. The clay facies is well displayed in the railway-cutting at Littlemore, consisting of 20 bands of clay separated by as many of nodular, grey, white-weathering argillaceous limestone, the thickness seen being about 17 ft. (fig. 71). A few small fragmentary and crumbly casts of Perisphinctids have been found, but they do not suffice for specific identification and appear to be the young of common *antecedens*-like forms.[3] The other fossils are of common, undiagnostic, Osmington and Berkshire Oolite species, the commonest being *Exogyra nana*, massed together with *Serpulæ* in the form of shoal-like banks which interrupt the otherwise regular bedding. In their lithic and palaeontological characters the beds in every way resemble those at Hilmarton.

Mr. E. S. Cobbold first suggested that the lateral passage of the coral rag into the Littlemore Clay Beds might best be explained by supposing that the muddy waters of a river or stream flowing from the London landmass here found their way through a channel between the reefs to the open sea.[4] This suggestion best fulfils all the conditions, for mud is well known to inhibit coral growth, and the area occupied by the argillaceous beds is always narrow.[5] Moreover, it receives confirmation from a quarry on the side of the Thames Valley near North Hinksey; this seems to show us a deposit formed on the edge of the mud-bearing currents, for it comprises about 10 ft. of impure clay with an admixture of rubbly limestone, enclosing huge disconnected blocks of Isastræan and Thamnastræan corals, all of them encrusted with *Serpulæ* and *Exogyræ*.[6] The section is unique and when freshly worked it suggests very forcibly a reef that has been continually smothered by pollution of the water with muddy sediment.

[1] S. S. Buckman, 1925, *T.A.*, vol. v, p. 50.
[2] Blake and Hudleston, 1877, loc. cit., p. 311, and see Arkell, 1927, loc. cit., pp. 122–8.
[3] For Buckman's identification of these and criticism, see Arkell, 1927, loc. cit., p. 143.
[4] E. S. Cobbold, 1880, *Q.J.G.S.*, vol. xxxvi, p. 319.
[5] Here the width cannot be more than two miles, and at the other occurrences it is probably less.   [6] W. J. Arkell, 1927, loc. cit., p. 144.

The breaking down of the hard limestones into soft clays has enabled the modern drainage to find an easy passage through the Corallian ridge; and so, if Mr. Cobbold's suggestion is correct, as seems probable, the Jurassic river has here determined the course of the Thames.

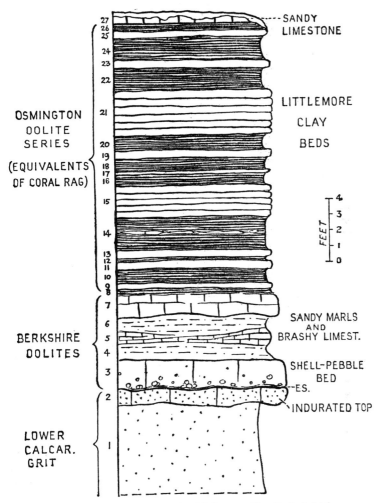

FIG. 71. The type-section of the Littlemore Clay Beds (which pass laterally into coral rag) at Littlemore railway-cutting, near Oxford. (From Arkell, 1927, *Phil. Trans.*, vol. ccxvi B, p. 141, fig. 14.)

The Berkshire Oolite Series, as has been stated, grows much thinner as it approaches the Thames south of Oxford, while the various subdivisions so clearly differentiated for 20 miles to the west are no longer recognizable.

It has already been remarked that the sections at Cumnor, Cothill and Marcham show that before its disappearance under the Kimeridge Clay of the Boar's Hill ridge the Berkshire Oolite Series becomes highly charged with sand. On the beds emerging upon the other side, the same feature is conspicuous in the more southerly of the sections, at Bagley Wood, Little-more, Cowley and Horsepath. At these places the series consists of some 5–7 ft. of shelly sand and sandy, shelly limestone, variously interbedded, and at the base is the ubiquitous Pebble Bed. The sandy layers contain per-fectly-preserved shells of *Pseudomonotis ovalis*, *Chlamys fibrosa*, *Oxytoma expansa*, *Exogyra nana*, &c. The harder bands are often shell-beds, crowded with the common lamellibranchs and ammonites of the *Trigonia hudlestoni* Limestones.

Owing to the high proportion of sand in all the strata below the coral rag in these sections it is often difficult to draw the boundary-line between the Berkshire Oolite Series and the Lower Calcareous Grit. Generally, however, the Pebble Bed provides the required datum.

It has been suggested that the continued influx of sand into the district south of Oxford may be explained as due to a survival of the activity of the river that brought down the Lower Calcareous Grit. Perhaps it was the same river that, when nearing the completion of its cycle, introduced into the district the mud to form the Littlemore Clay Beds.

Towards the north and east, in a short distance, all the quarries about Headington, and also those on the other side of the Thames at North Hinksey, show that the Berkshire Oolite Series has once more been reduced to the Shell-cum-Pebble Bed, only 1–2 ft. thick (see fig. 70). It is a bluish-grey, tough, oolitic limestone, locally dark-centred, largely made up of fossils. Like the shell-beds of the Inferior Oolite, it is the repository of a rich and varied fauna and undoubtedly owes its formation to gradual accumulation on a bottom kept free of sediment by currents. The large and fragile valves (often 4–5 in. in length) of *Lima*, *Chlamys*, *Trigonia*, *Gervillia*, &c. are gener-ally unworn and unbroken, a condition in which they would never have survived had they been subjected to wave-action on a beach or to re-derivation from one deposit to another. The ammonites belong to a score or more of species of the genera *Perisphinctes*, *Aspidoceras* and *Cardioceras*, and include many giants of the *parandieri* and *cloroolithicus* types. Most of them can be matched among the ammonites of the *Trigonia hudlestoni* Limestones of Highworth and Kingston Bagpuize, but there are also new elements. Buck-man formerly identified them with the species of foreign authors, such as Siemiradski and de Riaz,[1] but more recently he had begun to give them all new names and to found for them new genera based on differences of detail in the septal sutures. A revision of this rich ammonite assemblage still remains to be undertaken. A start has been made by Dr. Spath, who places most of the species either in *Perisphinctes sensu stricto* (as represented by *P. martelli* and its allies), or in the closely allied *Dichotomosphinctes* (the *ante-cedens-plicatilis* group), and affirms that most of Buckman's genera may be ignored.[2]

The Lower Calcareous Grit is composed, as usual, of a mass of yellow sands having all the characters of a sandbank, irregularly piled by currents. The thickness varies from a maximum of about 50 ft. to 12 ft. (or possibly less) in short distances. Except for the uppermost layers, which are seen in

[1] A case full, arranged and named by Buckman, is exhibited in the Oxford University Museum.

[2] L. F. Spath, 1931, *Pal. Indica*, N.S., vol. ix, pp. 401–2.

the bottom of nearly all the quarries and are worked at Littlemore for a moulding-sand to a depth of 15 ft., they are palaeontologically a *terra incognita*. As elsewhere, the highest 1–2 ft. is often indurated, and has the Pebble Bed cemented firmly to it. *Pachyteuthis abbreviatus* and casts of *Cardioceras cordatum* are sometimes found, but most of the ammonites recorded from the Lower Calcareous Grit of Cowley and Horsepath by Buckman and others came in reality from the sandy facies of the Berkshire Oolite Series. The immense hemeral table put forward by Buckman for these beds is purely fictitious.[1]

III. THE AMPTHILL CLAY REGION

(a) Oxford to the Fens

Five miles east of Oxford, close to the great limestone quarries of Wheatley and Lye Hill, the Corallian formation seems to come abruptly to an end. Looking eastward one beholds a broad clay vale, and on consulting the Survey maps (even the latest colour-printed edition) it appears that with the exception of a diminishing outcrop of Corallian Beds from Wheatley to Quainton (12 miles) and a small patch at Upware near Cambridge, the Oxford and Kimeridge Clays are in contact from here to the Yorkshire Basin.

The Upware patch represents an isolated reef, which will be described in the next subsection. The intervening area has received considerable attention in the past forty years from the late Thomas Roberts, Mr. C. B. Wedd, and Prof. A. Morley Davies, who have shown that except for a stretch of 30 miles north-east of Leighton Buzzard where the Lower Greensand oversteps on to the Oxford Clay, there is a continuous outcrop of beds of Corallian age; but they are developed in an entirely different facies.

The most complete section in this area was a railway-cutting at Ampthill, Bedfordshire (long since overgrown), where the Corallian Beds had a total thickness of 61 ft. After this exposure, and on account of their predominantly argillaceous facies, Seeley called the beds the AMPTHILL CLAY.[2]

The name Ampthill Clay has been commonly used by Woodward and others for the whole of the equivalents of the Corallian, but more recently it has been found necessary to exclude a basal portion, 6–20 ft. thick, which contains oolitic limestone and marl, often ironshot, with derived fossils and a great many oysters and *Serpulæ* and other shells. This basal division, called the Oakley Clay, or better OAKLEY BEDS, in Oxon. and Bucks., and ELSWORTH ROCK or Elsworth Rock Series about Cambridge, is the equivalent of the 'Upper Corallian' of the Oxford district (the Berkshire and Osmington Oolite Series or *martelli* and *antecedens* zone). From this it follows that the true Ampthill Clay as restricted is younger than any of the Corallian Beds preserved on the Oxford Axis or for a considerable distance farther west.

The Ampthill Clay.

For our knowledge of the Ampthill Clay we are indebted chiefly to Thomas Roberts, who in the latter part of the last century undertook a systematic examination of the numerous small clay-pits on the borders of Huntingdon

[1] S. S. Buckman, 1925, *T.A.*, vol. v, pp. 67–8; see Arkell, 1929, 'Mon. Corall. Lamell.', *Pal. Soc.*, p. 7.

[2] He earlier suggested and abandoned first Bluntisham Clay and then Tetworth Clay. See bibliography.

and Cambridge. Since many of these pits are now abandoned and are unlikely to yield any further information, his careful records are invaluable.

Roberts showed that the Ampthill Clay can readily be distinguished, both lithologically and palaeontologically, from the clays above and below it. It is dark and tenaceous, often black, and contains abundant selenite. Wherever the junction with the Kimeridge Clay is seen it is marked by a band of phosphatic nodules, regarded as the basement-bed of the Kimeridge. The fossils differ markedly from those in the Oxford Clay by their state of preservation, for they are never pyritized, while the ammonites and some other shells in the Oxford Clay are nearly always pyritized; moreover Wedd remarked that while the oysters and other shells in the Oxford Clay are freshly preserved, those in the Ampthill Clay are generally bored and encrusted, both inside and out, with *Serpulæ*. There are several thin bands of limestone, usually hard, nodular and argillaceous, never oolitic.

Roberts described a number of pits from Everton and Gamlingay on the Bedfordshire border eastwards along the outcrop at Great Gransden, Elsworth, Boxworth, St. Ives and Bluntisham to Fenton, where the clay disappears beneath the alluvium of the Fens. Records of Cardiocerates and Perisphinctids recur repeatedly; especially *'Ammonites plicatilis'* or *'A. biplex'*, *'A. excavatus'*, *'A. vertebralis'* and *'A. cawtonensis'*. The *'A. excavatus'* probably signifies a *Prionodoceras* (e.g. *P. serratum*), as also does probably *'A. cawtonensis'*, while the other names may include *Perisphinctes* and *Dichotomoceras*. A revision of old material and any new that can be collected in existing exposures is needed before detailed correlations can be established. Meanwhile, however, it is evident from the presence of *Ostrea delta* that beds as high as the Glos Oolite Series are represented in the Ampthill Clay. One of the most fossiliferous horizons described by Roberts is a double band of hard, compact, grey limestone, 2–3 ft. thick, called the Boxworth Rock.[1] Besides *'A. plicatilis'* and *'A. excavatus'* this yielded to Roberts *Trigonia clavellata, Nucula menkei* and *Alaria bispinosa*—three species highly suggestive of the *Trigonia clavellata* Beds.[2]

The type-section at Ampthill was unfortunately never described with the care that it would have received had it lain within Roberts's area, and little information can now be gained from the record of the cutting. The old lists of fossils[3] are useless, for they include the rich fauna of the underlying Oakley Beds. This has given rise to an idea that the Ampthill Clay contains a mixed Oxford Clay and Kimeridge Clay assemblage.

A more useful section was described by Prof. Morley Davies in a cutting made in 1907 on the Great Western Railway from Princes Risborough to Banbury, between Ashendon Junction and Rushbeds Wood Tunnel.[4] Here 50 ft. of Ampthill Clay was exposed, and also the Oakley Beds below. The highest 30 ft. consisted of the usual selenitiferous clay, while the lowest 20 ft. was more varied, with some thin stone-bands. The highest and thickest stone-band, 20 ft. from the base, yielded narrow-whorled fine-ribbed Perisphinctids with forward sweep on the periphery, indicating the *wartæ* zone

[1] T. Roberts, 1892, *Jurass. Rocks Neighb. Cambridge*, p. 43.
[2] Though the identification of the *Trigonia* needs verifying.
[3] e.g. as given by H. B. Woodward, 1895, *J.R.B.*, p. 137.
[4] A. M. Davies, 1927, in Arkell, loc. cit., p. 152.

and correlating with strata at least no lower than the *Trigonia clavellata* Beds of Dorset. This showed that the bulk of the Ampthill Clay—that portion lying above the stone-band (=the Boxworth Rock?)—was to be assigned to the upper part of the Glos Oolite Series and Upper Calcareous Grit, or even higher beds. The 40 ft. of the Sandsfoot Clay in Dorset, with *Ostrea delta*, affords a suggestive comparison.

Simultaneously with my publication of this correlation, and independently of it, Buckman set down a brief note in *Type Ammonites* to the effect that 'The Ampthill Clay . . . which contains *Dichotomoceras variocostatum* Buckland sp., is not Corallian as usually stated—that is to say, it is later than Perisphinctean Age. It is of Prionodoceratan Age—towards the lower part of the Kimeridgian as generally understood.'[1] But Buckman apparently had not seen the Perisphinctids of *wartæ* style found by Prof. Davies 20 ft. above the base of the Ampthill Clay; moreover the necessarily Kimeridgian age of the genus *Prionodoceras* is highly questionable, since *P. serratum* signalizes the *decipens* zone of Salfeld's classification (= Upper Calcareous Grit, *pars*) and '*Dichotomoceras*' *variocostatum* (Buckland) seems to be a true *Perisphinctes sensu stricto*. But Buckman's suggestion is valuable in so far as it draws attention to the fact that the bulk of the Ampthill Clay occupies a higher stratigraphical position than was usually supposed.

The Oakley Beds and Elsworth Rock.

On the edge of the coralline area, at Stanton St. John, Woodperry and Stow Wood, the Wheatley Limestones are seen, with increasing distance from the reefs, to become diluted with marl and clay. A type of rock develops which at first sight resembles the Littlemore Clay Beds in its alternation of hard and soft bands, but a closer examination reveals that it is largely composed of detrital materials and small oysters (*Exogyra nana*). This interesting change did not escape Blake and Hudleston. They described a quarry behind the school at Stanton St. John as 'a very remarkable quarry . . . composed of alternate layers of hard doggery bands and marl . . . a perfect nest of echinoderms.' 'We see here', they wrote, 'the spot where the corals did *not* grow, as at Headington we saw where they did. Here we have the débris of the ground-up, variably hardened reef, along with the echinoderms that lived in the neighbourhood; and we thus have a natural termination of the Coral Rag in this direction.'[2]

An argillaceous deposit crowded with the shells of *Exogyra nana* was long ago mapped by the Survey across the country east of Stanton St. John and Wheatley, towards Quainton, but no good sections were known before the making of the railway-cutting near Ashendon Junction. Below the true Ampthill Clay in that cutting Prof. Davies found 6–10 ft. of beds full of *Exogyra nana*, varying in lithological composition from bluish clay to white limestone and marls, full of brown ooliths, like the Elsworth Rock (presently to be described) but unconsolidated. They contained also some masses of *Serpulæ* and oysters suggestive of the Elsworth Rock of Gamlingay and other places in Cambridgeshire. Prof. Davies remarked that the abundance of *Exogyra nana* and *Cidaris* spines linked these beds with the 'Upper Corallian'

---

[1] S. S. Buckman, 1927, *T.A.*, vol. vi, p. 49; compare Arkell, 1927, loc. cit., p. 153.
[2] Blake and Hudleston, 1877, *Q.J.G.S.*, vol. xxxiii, p. 310.

F f

around Oxford (the Osmington Oolite Series).[1] The recognition in his collection of *Chlamys nattheimensis*, a mollusc of the coral fauna, seldom if ever found dissociated from corals, indicated that reefs were not, in fact, far distant.[2] But among the fossils collected by Prof. Davies from the *Exogyra nana* Beds, or as Buckman called them, Oakley Clay,[3] there are also some fragmentary Perisphinctids identical with specimens from the Shell-cum-Pebble Bed of the Oxford district. The Berkshire Oolite Series, therefore, is also represented.

Prof. Davies determined that in the Ashendon cutting the Oakley Beds rest directly upon the upper *præcordatum* zone of the Oxford Clay. In the 8 miles from the edge of the escarpment at Stanton St. John, therefore, the Lower Calcareous Grit has been entirely overstepped. He further proved that this overstep continues and increases towards the north-east, for he recognized the Oakley Beds at Quainton apparently resting on the lower *præcordatum* zone, and farther still, at Sandy, upon the *renggeri* or *lamberti* zone[4] (fig. 62, p. 348). A source is thus provided for the great spread of Oxford Clay pebbles, 40 miles in length, comprising the Pebble Bed; especially if we suppose a similar overstep to have taken place towards the south-east or south.

The recognition of the Oakley Beds as far away as Sandy, beyond Ampthill, brings into line the basal beds with many fossils in the Ampthill cutting, described by Woodward as a 'rubbly rock-bed like the basement-bed at Gamlingay'.[5] It also leads to the threshold of another area, where the deposits have been studied independently by Cambridge geologists—the area of the Elsworth and St. Ives Rock.

The basement-beds of the Corallian in Huntingdon and Cambridgeshire consist everywhere of some 10–15 ft. of highly fossiliferous strata known as the Elsworth Rock or Elsworth Rock Series; and, as in Buckinghamshire and Bedfordshire, they rest directly upon the Upper Oxford Clay. The beds consist of two rock-bands, the lower 3–7 ft. and the upper 1–1½ ft. thick, separated by about 5 ft. of clay. The rock composing both bands is, when fresh, a hard, grey, ironshot, shelly limestone, but it weathers readily to a yellowish-brown colour and disintegrates to a yellow, calcareous marl, full of dark-brown ooliths. The upper band is usually more decomposed than the lower, but otherwise they appear to be identical, lithologically and palaeontologically.

Mr. C. B. Wedd mapped the outcrop continuously for about 5 miles from Houghton Hall eastwards, past St. Ives to Holywell. On the south side of the Ouse Valley he mapped it again for over 6 miles past Elsworth to beyond Croxton, thus establishing what Roberts contended on palaeontological grounds, that the Elsworth Rock and St. Ives Rock are one and the same stratum.[6] As Croxton lies only 7 miles from Sandy, where Prof. Davies recognized the Oakley Beds, and as both Prof. Davies at Ashendon and Woodward at Ampthill likened the Oakley Beds to the Elsworth Rock, there cannot be much doubt that the two are in stratigraphical continuity—

[1] A. M. Davies, 1916, *Geol. Mag.*, vol. liii, p. 398.
[2] W. J. Arkell, 1927, loc. cit., p. 153.   [3] 1927, *T.A.*, vol. vi, p. 49.
[4] A. M. Davies, 1916, loc. cit., p. 399.
[5] H. B. Woodward, 1895, *J.R.B.*, p. 136.
[6] C. B. Wedd, 1901, *Q.J.G.S.*, vol. lvii, pp. 73–85.

though it does not follow that they are necessarily of precisely the same age throughout.

The Sedgwick Museum, Cambridge, contains a large collection of fossils from the Elsworth Rock (the greater number from the lower band) of Elsworth, St. Ives, Holywell and other places. The lamellibranchs comprise the same assemblage as the *Trigonia hudlestoni* Beds of the Berkshire Oolite Series, with additional species of earlier date.[1] The commonest are such familiar forms as *Trigonia hudlestoni* (resembling those found in Yorkshire more closely than the typical Berkshire and Oxfordshire forms), *T. reticulata*, *Astarte ovata*, *Cucullæa contracta*, *Chlamys fibrosa*, *C. splendens*, *Camptonectes lens*, *Velata anglica*, *Oxytoma expansa*, *Lima mutabilis*, *L. subantiquata* and many more. Earlier elements are *Nucula oxfordiana* de Lor., *Parallelodon* (*Beushausenia*) *keyserlingii* (d'Orb.), *Astarte cordati* Trautsch. and *Pinna lanceolata* Sow. Since the deposit must be dated by its latest fossils, there seems no alternative but to correlate it with the Berkshire Oolite Series. The ammonites comprise Perisphinctids of the *martelli* zone mixed with Cardiocerates of the *cordatum* zone; and there are in addition a number of forms, including some Oppelids, not known elsewhere in the British Isles.

This correlation was already advocated by Mr. Wedd, who wrote in 1901 that he would place the Elsworth Rock 'on a somewhat higher horizon than the Lower Calcareous Grit'.[2] In the discussion which followed the reading of Wedd's paper Hudleston remarked that, judging by the fossils, 'it was impossible to avoid the conclusion that the rock at Elsworth occupies a very low position in the Corallian Series'. H. B. Woodward made the ambiguous suggestion that 'the occurrence in the Elsworth Rock of the Lower Corallian *Ammonites perarmatus* and the Upper Corallian *A. plicatilis* might be taken to indicate a local blending of the two zones'—an explanation which seems to be certainly the right one, although not in the sense in which Woodward doubtless intended it.

Wedd noted that, all along the outcrop of the Elsworth Rock, even as far away as Gamlingay and Upware, there is a hard blue rock crowded with *Serpulæ* and *Exogyra nana*, which he took to indicate slow deposition. Further, he observed that the Elsworth Rock abounds in casts of ammonites to which, after the complete removal of the shell, *Serpulæ* have attached themselves. Thus penecontemporaneous erosion has undoubtedly taken place, and the 'blending of the two zones' is explained. Sedimentation may have been so slow that no appreciable thickness of rock separated the two faunas, which could then become mingled with very little disturbance or reworking of the strata.

Lower Calcareous Grit: The Arngrove Stone.

In the area between the Corallian escarpment at Stanton St. John and the projecting Portlandian spur of Brill and Muswell Hills, before the overstep of the Osmington and Berkshire Oolite Series on to the Oxford Clay is complete, a thin representative of the Lower Calcareous Grit survives, and has been named by Prof. Davies the ARNGROVE STONE. Lapping round the south of Otmoor, it forms an escarpment which, although subdued, is noticeable in

[1] W. J. Arkell, 1929, 'Mon. Corall. Lamell.', *Pal. Soc.*, p. 11.
[2] C. B. Wedd, 1901, loc. cit., p. 83.

an area otherwise composed of clays. The rock, which is worked as a 'gravel' in some shallow pits near Arngrove, is peculiarly light and absorbent.

Prof. Davies discovered that the Arngrove Stone is composed of spicules of the siliceous sponge, *Rhaxella perforata* Hinde, thus resembling *Rhaxella* cherts in the Lower Calcareous Grit of Yorkshire.[1] It also resembles the chert at Purton except that it lacks the large, scattered, brown ooliths. The rest of the fauna is composed of lamellibranchs of Lower Calcareous Grit species and small Cardiocerates of *cordatum* and *vertebrale* styles. It is a much richer assemblage than any found in the neighbouring sandbank, but it is considered to correspond only to the lowest, unexposed, portion of the Lower Calcareous Grit (see fig. 62, p. 348).

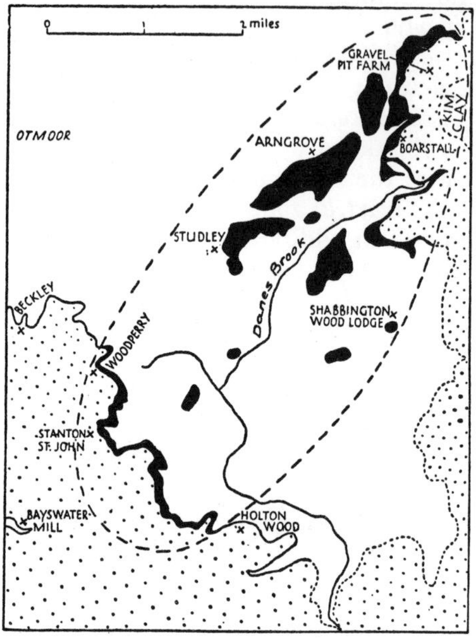

The outcrop of the Arngrove Stone is restricted to an ellipse about 7 miles in greatest length (SW.–NE.) and $2\frac{1}{4}$ miles in width, extending from the slopes of the hills around Stanton St. John and Woodperry to about a mile beyond Boarstall (see fig. 72). Prof. Davies believes that this corresponds with the original area of deposition. Down the dip-slope to the southeast it was believed by Buckman to pass into a white clay, seen in the fields about Oakley and noted by him at Worminghall in a well. In the well the white clay was underlain by a 6-in. band of fossiliferous, yellow, marly sandstone, which Buckman named the WORMINGHALL ROCK, recording from it *Cardioceras* cf. *zenaidæ*, *Miticardioceras mite*, *Perisphinctes* cf. *intercendens*, 'immense *Gryphææ* and numerous lamellibranchs'. Beneath were $26\frac{1}{2}$ ft. of blue Oxford Clay.[2]

FIG. 72. Sketch-map of the outcrop of the Arngrove stone, by Prof. A. Morley Davies, 1931. (Based on the Oxford Special sheet Geol. Survey Map of 1908.) Oxford Clay, white; Arngrove Stone, black; later Corallian Beds, dotted.

The base of the Arngrove Stone was also seen by Prof. Davies in a well at Studley, but there no fossiliferous representative of the Worminghall Rock was found.

The relations of the deposit to the overlying beds have never been clearly exposed and the only place where they could be studied would be on the flanks of the hill for a short distance on either side of Stanton St. John. Prof. Davies noted a considerable thickness of calcareous sandstone above the Arngrove Stone in a road-cutting at Woodperry,[3] but at Stanton Great Wood he found indications of a clay occupying the same position.[4]

[1] A. M. Davies, 1907, *Q.J.G.S.*, vol. lxiii, pp. 37–43.
[2] S. S. Buckman, 1925, *T.A.*, vol. v, p. 54.
[3] A. M. Davies, 1907, *Q.J.G.S.*, vol. lxiii, p. 42.   [4] Idem, 1909, *P.G.A.*, vol. xxi, p. 235.

(b) The Upware Reef

At Upware, 10 miles north-east of Cambridge, is a small reef, entirely isolated among the Fens and seventy miles from the nearest coral-bearing rocks at Wheatley. It rises above the level alluvium as a low hill, 3 miles long and less than a mile wide. As it produces the only limestone in the district, it has been opened up in two large quarries at the north and south ends of the hill, and in addition in several minor excavations. Concerning these openings and their interpretation as bearing upon the structure of the hill and the succession of the strata, an extensive literature has grown up. The locality has naturally always attracted the attention of Cambridge geologists: Sedgwick, Seeley, Bonney, Roberts and Keeping each made contributions to the stratigraphy. After a full description had been published by Blake and Hudleston in 1877 there followed an acrimonious discussion between them

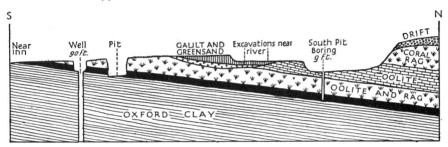

Fig. 73. Diagrammatic section through the Corallian Beds on the south side of the hill at Upware. The coral rag and oolite facies probably interdigitate more than shown. After P. Rigby, in R. H. Rastall, 1909, *Geol. in the Field*, p. 137, fig. 26 (re-drawn).

and Bonney; but Blake and Hudleston's palaeontological conclusions were in the main upheld by Wedd, who collected much new information and published an exhaustive account in 1898. Finally, in 1901 certain remaining difficulties were cleared up by Prof. W. G. Fearnsides, by means of a special excavation made through the floor of the south pit, and the results were published by him in 1904 and summarized in an account by Dr. R. H. Rastall in 1909.[1]

A brief description of the rock-succession suffices for present purposes, without entering into the history of the various controversies. At the same time the opportunity is taken to offer some further suggestions as to the correlation of the component strata.

The highest rock on the hill is a typical coral rag, composed of fine reef-building corals. It is exposed in the upper part of the north end of the south pit to a depth of about 10 ft., dipping gently north into the hill (see fig. 73). The rag is chiefly built up of layers of *Isastræa explanata* and *Thamnastræa arachnoides*, with which are associated *T. concinna*, *Montlivaltia dispar*, *Stylina tubulifera*, and numerous mollusca. It forms a creamy white limestone, the crystalline masses of coral alternating with pockets of whole and broken shells in an earthy matrix. In some of the beds the shells have been

---

[1] For the numerous earlier references see the Bibliography; the two principal accounts are by C. B. Wedd, 1898, *Q.J.G.S.*, vol. liv, pp. 601–9; and R. H. Rastall, 1909, *Geol. in the Field*, pp. 133–8.

dissolved away, leaving hollow moulds and casts. Blake and Hudleston wrote that, 'It is not so rubbly as many of the rags are, but compares well in some portions with that which occurs in Yorkshire, at North Grimston. It has, however, a peculiar creaminess about it that is hard to match else-where. . . . The irregularity of the bedding is an indication of its reef-like character.'[1]

The fauna is remarkable, for, although the typical coral-dwelling assem-blage is present, it is mingled with a large number of mollusca not usually found in any abundance in the coral rag, but more characteristic of the shell beds. In addition there are several peculiar species not found at any other English locality, or only very rarely in Yorkshire, but described from the Continent: these are *Isoarca texata* Quenst., *Isoarca multistriata* Étall., *Opis (Cœlopis) arduennensis* Buv. and *Opis (Trigonopis) virdunensis* Buv.[2] The absence of *Thecosmilia annularis*, the commonest coral in the Wilts.–Berks.–Oxon. ridge, is also striking.

Besides these forms there are two species which are highly significant because they are abundant in the *Trigonia clavellata* Beds of Dorset, but have never been found in the long outcrops of the Wilts.–Berks.–Oxon. ridge, where those beds are absent: *Mytilus (Arcomytilus) pectinatus* Sow. and *Prorokia problematica* (Buv.).

Below the coral rag, and exposed over the southern three-quarters of the south pit, is an oolite, locally loose and pisolitic and very shelly, but becoming less fossiliferous towards the bottom. The thickness seen is probably over 20 ft. The upper portions of this oolite contain most of the shells found in the overlying coral rag. The following lamellibranchs are most prominent in both: *Chlamys nattheimensis, Plicatula weymouthiana, Mytilus ungulatus*; and also *Chlamys intertexta, C. inæquicostata, C. fibrosa* (rare), *Navicula quadri-sulcata, Velata anglica, Lima rigida, L. læviuscula, Trigonia reticulata, Myo-concha texta* and many others, as well as an unusually rich assemblage of echinoderms. The typical *Perisphinctes martelli* also occurs.[3]

A trial hole dug through the floor at the south end of the south pit, made during Prof. Fearnsides's investigations in 1901, passed through 4 ft. more of the oolite (making nearly 30 ft. in all) into a hard band of coral rag. Below this it penetrated another 5 ft. of variable rock, composed of a mixture of coral and marl, until, at a depth of 9 ft. it struck a hard oolitic rock stained with iron, which was considered to be undoubtedly the Elsworth Rock.[4] The boring was not carried any farther, but a continuation of the sequence was seen some years previously during the sinking of a well near the inn at Upware, to the south of the quarry, and nearby in another small opening recorded by Wedd. Here coral rock rested on ferruginous oolite with corals (2 ft. 3 in. thick) taken to be the top of the Elsworth Rock, which was com-pletely penetrated in the well. Beneath the band of ironshot oolite were 9 ft. 9 in. of clay, the lower part full of *Exogyra nana*, below which again was a lower and thicker ironshot rock-band, as at Elsworth, here 4 ft. thick. The lower rock-band, which rested directly on Oxford Clay with *Cardioceras*,

---

[1] Blake and Hudleston, 1877, *Q.J.G.S.*, vol. xxxiii, p. 314.
[2] See Arkell, 1929, 'Mon. Corall. Lamell.', pp. 11, 47.　　　　[3] Sedgwick Museum.
[4] W. G. Fearnsides, 1904, in Marr and Shipley, *Geol. Cambridgeshire*, p. 16; and Rastall, 1909, loc. cit., p. 136.

Peltoceras athleta(?), *P. eugenii*, and '*Ammonites hecticus*', contained the fauna of the Elsworth Rock, the ammonites recorded being *C. cordatum, C. vertebrale, C. mariæ* and *Peltoceras eugenii*.[1] If the last two were correctly identified they must have been derived, and they would prove that the unconformable overstep of the Oxford Clay by the basement-bed of the Corallian, as determined by Prof. Davies in Buckinghamshire and Bedfordshire, is here still more pronounced and has penetrated down to the *athleta* zone.

However this may be, the presence of *Cardioceras cordatum* and *C. vertebrale* and the absence of the plicatiloid Perisphinctids so abundant, in fact preponderant, in the more westerly exposures, suggests that the Elsworth Rock may be not an isochronous deposit—that it may here be earlier in date than at St. Ives and Elsworth, in spite of the close lithic resemblance.

The white oolite is again exposed in the north quarry, where the total thickness was estimated by Wedd to be 40 ft. or more. The coralline facies of the basal portion does not appear here, and the rock is less fossiliferous. Blake and Hudleston long ago placed it, purely on palaeontological grounds, below the main or topmost coral rag, which was at that time the only rock exposed in the south pit.

The upper, more fossiliferous portion, of the oolite, at any rate of the south pit, and the superincumbent coral rag may be regarded as later in date than any of the coral rag of Oxon., Berks. and Wilts. excepting only the reef at Steeple Ashton (above, p. 390). This conclusion is warranted by the peculiarities of its fauna as well as by its general appearance—both features remarked on by Blake and Hudleston. It would thus probably be about equivalent to the Steeple Ashton coral rag and the *Trigonia clavellata* Beds of Dorset (perhaps including the uppermost or rubbly portion of the Osmington Oolite Series and/or a part of the Sandsfoot Clay); and so to the lower part of the Ampthill Clay of Ashendon cutting.

(c) From the Fens to the Market Weighton Axis

The isolation of the Upware reef with its shallow-water coralline facies has long been known from borings at March, Wicken and other places in the Fens, which penetrated from Kimeridge Clay into Oxford Clay without meeting with any Corallian Beds as developed at Upware. The Survey maps show nothing more of the formation south of the Yorkshire Basin, denoting again from their colouring that throughout the long county of Lincolnshire the Oxford Clay and Kimeridge Clay are in contact. This is entirely misleading, however, for Roberts demonstrated in 1889 that the Ampthill Clay is continued through a large part of North Lincolnshire, and there is every reason to suppose that it extends from Upware at least as far as the Humber (see fig. 74, p. 418).

In common with the Oxford and Kimeridge Clays, the Ampthill Clay would seem to diminish in thickness northward along the Lincolnshire outcrop as it approaches the Market Weighton Axis. In two borings at Southery, in the Fens 14 miles north of the Upware reef and just over the Norfolk border, Dr. J. Pringle assigns about 50 ft. of clays to the Ampthill Clay.[2] In the lower portion were encountered several thin bands of argillaceous

---

[1] C. B. Wedd, 1901, loc. cit.
[2] J. Pringle, 1923, *Sum. Prog. Geol. Surv.* for 1922, pp. 126–39.

limestone, but none of them bore any resemblance to the Elsworth Rock. Below the lowest band was a small Cardiocerate, suggestive of the *præcordatum* zone. In the lower parts of the Ampthill Clay no fossils were found. The upper limit was drawn below a bed of clay containing fragments of large smooth ammonites identified as probably *Pictoniæ* and so indicating the basal zone of the Kimeridge Clay. At a level 23 ft. below this in one boring and

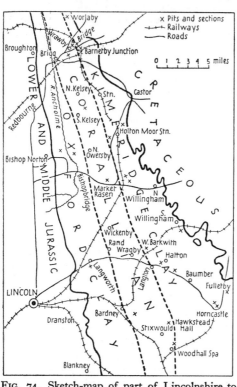

16 ft. below it in the other, specimens of *Prionodoceras serratum* (Sow.) were found.

The work of Roberts was carried out in North Lincolnshire, between Bardney, on the latitude of Lincoln, and Brigg, a distance of about 25 miles.[1] By the same careful methods as those which he had followed in the Cambridge district, examining all the small brickyards now for the most part abandoned, and helped by several railway-cuttings, he established that a continuous belt of Corallian clays could be traced, differing both lithologically and palaeontologically from those above and below (fig. 74). As in the Cambridge district, he found that the ammonites in the Oxford Clay are pyritized, while those in the Ampthill Clay are never pyritized; the Ampthill Clay, moreover, can be readily recognized by its black colour and the abundance of selenite crystals.

FIG. 74. Sketch-map of part of Lincolnshire to show the approximate position of the outcrop of the clays of Corallian date as determined by Roberts. (After T. Roberts, 1889, *Q.J.G.S.*, vol. xlv, p. 546.)

The deepest continuous section of Ampthill Clay recorded was one of 17 ft. seen in a brickyard west of Hawkstead Hall. Other sections were described at Bardney, South and North Kelsey and Wrawby, in the railway-cutting.

Roberts found that the black selenitiferous clay contained Corallian fossils, and out of 23 species he recognized 16 that occur in the Ampthill Clay. The ammonites recorded are such typically Corallian names as '*A. cordatus, A. vertebralis, A. excavatus, A. cawtonensis, A. plicatilis, A. achilles* and *A. decipiens*', but further material and more detailed identifications would be necessary before their significance could be assessed. The records of '*A. excavatus*' and '*A. cawtonensis*' may both refer to *Prionodoceras*. Again there appears to be the same admixture of *Gryphœa dilatata* and *Ostrea delta*, but here Roberts states that *G. dilatata* is commonest in the lower part of the clay, while *O. delta* is more abundant in the upper part. *Alaria bispinosa* also

[1] T. Roberts, 1889, *Q.J.G.S.*, vol. xlv, pp. 545–60.

appears in the lists, but so do such typical Lower Calcareous Grit and Berkshire Oolite fossils as *Chlamys fibrosa* and *Pachyteuthis abbreviata*.

There is thus insufficient evidence for stating which subdivisions of the Corallian formation are represented, and it is unknown how thick the deposits may be. On the whole, however, there seems no reason to doubt that the bulk of the clay is of the same age as the bulk of the Ampthill Clay between Oxford and Cambridge, belonging high in the Corallian Series. No hard basement-beds comparable with those that are such a conspicuous feature in the more southerly area have been detected, but the junction with the Oxford Clay has nowhere been seen.

IV. THE MARKET WEIGHTON AXIS

Between the Humber and the Market Weighton Axis evidence of Corallian Beds has been recognized only in a well at Melton, near North Ferriby, close to the Humber.[1] They are still clays and they must be very thin and imperfect, for the total thickness from the base of the Oxford Clay to the top of the Kimeridge does not exceed 100 ft. It is probable that as they approach the axis they wedge out altogether.

We have to pass at least six miles to the north of the Market Weighton Axis before we find any Upper Jurassic rocks preserved beneath the Chalk, and by that time the normal limestone and sandstone facies of the greater part of the series has already returned. The transition from the one facies to the other, or the thinning out of both against a dividing ridge of older rocks, will never be seen by geologists, for over the critical region the overstep of the Chalk has obliterated all the evidence. Nevertheless, the country immediately to the north holds the key to the region on the axis and enables us to tell what becomes of the Corallian Beds as well as if we could see them.

The first trace of the beds is found above Ley Field, near Garrowby, where a small pocket of normal Lower Calcareous Grit has been quarried beneath the Chalk.[2] North of this, about Kirby Underdale, there is another considerable gap, owing partly to obscuring talus fallen from the Chalk, but principally to minor pre-Cretaceous folds, from the summits of which the Upper Jurassic rocks were eroded before the Chalk was deposited. In the valley of the Gilder Beck, south-east of Acklam, the Chalk rests again on Lower Lias, and on the crest of another sharp anticline between Acklam and Leavening it reposes on the Estuarine Series. These minor axes recall those parallel to and north of the main Mendip Axis, about Keynsham and Radstock.

Farther north, the Corallian Beds emerge fully from the Cretaceous covering and it becomes possible to discern the movements that took place during the Corallian period. The rocks are at first very incomplete. They thicken to the north and north-west into the Yorkshire Basin, not only by increments to every member of the series, but also by the incoming of some new and important subdivisions, of which at least one is not met with anywhere in the South of England.

Until almost as far north as Birdsall Kimeridge Clay rests unconformably on the Lower Calcareous Grit. The first subdivision to appear beneath it is

[1] P. F. Kendall and H. E. Wroot, 1924, *Geol. Yorkshire*, p. 332.
[2] To end of subsection based on C. Fox Strangways, 1892, *J.R.B.*, pp. 300–70.

the Upper Calcareous Grit, with what is probably the upper part of the Glos Oolite Series represented by an argillaceous rock, to be described in the next section, the North Grimston Cementstone. At Toft House 'a little impure limestone with oolitic grains' makes its appearance and thickens rapidly towards North Grimston, gradually coming to represent the whole of the great limestone divisions of the Yorkshire Corallian.

Thus the country between the Chalk Wolds and the Derwent gives us a fairly accurate idea of what became of the Corallian Beds over the Market Weighton Axis along the line of the present outcrop. It is clear that no part of the formation except the Lower Calcareous Grit reached so far south as the region where the Chalk overstep would have removed it, all the rest of the subdivisions having already disappeared, both by diminution in thickness —'wedging out'—and by being overlapped by the Kimeridge Clay.

V. THE YORKSHIRE BASIN

The thickening of the Corallian Beds, by the incoming of new sub-divisions and the expansion of the old ones, continues north-westward across the Derwent and along the Howardian Hills to beyond the Gilling Gap. In the Hambleton Hills and eastward through the Tubular Range to the sea the formation attains a grander development than in any other part of the British Isles. It wraps around three sides of the Vale of Pickering, its dip-slopes along the north and west margins rising into flat-topped moorland, untamed and heather-covered, and offering as marked a contrast with the gentle hills of the Berkshire and Wiltshire range as do the moors of the Lower Oolites with their contemporary formations in the South of England. The strongest features of all are the lofty escarpments of the Hambleton Hills. For a considerable distance these rise above an average height of 1,000 ft., built of the hard gritstones of the Lower Calcareous Grit, much thicker and more massive than in the South and presenting precipitous faces westward, visible from far away over the Vale of Mowbray, like the Northern Pennines seen from the Vale of Eden.

Between the strata containing the fauna of the Berkshire Oolite Series and the true Lower Calcareous Grit are developed some 120 ft. of limestones and grits, belonging to the upper part of the *cordatum* zone. These beds, which are unknown in the South of England and may be represented in time by the gap below the Pebble Bed, are called the Hambleton Oolite Series, or the Lower Limestones and Passage Beds. Farther down the dip-slopes, forming an inner circle around the Vale of Pickering, follow the Osmington Oolite Series and succeeding beds, with a variable representation of the Upper Calcareous Grit. Coralline formations are widespread in the Osmington Oolites and probably also in the Glos Oolites, on the horizon of the *Trigonia clavellata* Beds. The earliest coral reef in England is found in the Hambleton Oolite Series of Hackness.

To attempt an adequate description of these rocks in any way commensurate with their importance and interest would involve extending this chapter far beyond its already swollen proportions. No other formation in Yorkshire maintains such continuous variety and so rich a fauna, or stimulates inquiry with so many exposures, as the Corallian. Its many-sided interest has attracted also a disproportionate amount of research, and we have exhaustive accounts of the rocks from the stratigrapher's point of view by Fox-Strangways and from the palaeontologist's point of view both by Hudleston and by Blake and Hudleston jointly. Yet even still a vast amount remains to be discovered, as is being proved by the researches of Dr. Vernon Wilson.

It is customary for purposes of description to divide the outcrop into a number of sections, each of which would have an importance equal to that of one of the earlier sections of this chapter. But in view of the essential continuity of the rocks, we will consider the Yorkshire Basin as a whole, confining ourselves to a brief summary of the succession, and noting the localities where each subdivision is best developed.

SUMMARY OF THE SUCCESSION AROUND THE YORKSHIRE BASIN [1]

Upper Calcareous Grit + Upper Glos Oolite Series: 80 ft. east of the Derwent and perhaps in the Howardian Hills; not more than 45 ft. elsewhere.

The second subdivision of the Corallian formation to appear beneath the Kimeridge Clay on the north side of the Market Weighton Axis (the first being the Lower Calcareous Grit) is a thick argillaceous series called the NORTH GRIMSTON CEMENTSTONE. It consists of bands of grey argillaceous limestone, like Blue Lias and used for making 'Blue Lias Cement', separated by and passing down in the lower half of the series into calcareous shales. The total thickness locally may exceed 80 ft.

The Cementstone Series is markedly unconformable with the coral rag and underlying oolite, which it overlaps southward on to the Lower Calcareous Grit, before being in turn overlapped by the Kimeridge Clay.[2] The well-known quarries near North Grimston show the character of the rock well. Fossils are rather scarce and always difficult to extract. The commonest are large specimens of *Gryphæa dilatata* Sow. of several common varieties, including var. *discoidea* Seeley,[3] also *Lucina aspera* Buv. (abundant in the Upper Calcareous Grit) and such undiagnostic forms as *Thracia* and *Goniomya*. Some ammonites have been found, recorded by Blake and Hudleston as '*A biplex-variocostatus*' and '*A.* cf. *alternans* and *A.* cf. *serratus*' (*Prionodoceras* or *Amœboceras*).

The Cementstone Series crosses the Derwent and probably persists throughout the Howardian Hills, but no exposure has been seen for the whole length of 15 miles beyond Hildenley (near Malton). At the extreme west end of the Howardians, at Snape Hill Quarry, the upper part of the series is seen to have become more arenaceous. The total thickness is here still more than 50 ft. At the top are 5–6 ft. (formerly 10 ft.) of ferruginous sandstone—typical Upper Calcareous Grit—overlying $12\frac{1}{2}$–13 ft. of alternating hard,

---

[1] Based on C. Fox-Strangways (1892, *J.R.B.*, pp. 300–70), W. H. Hudleston (1876–8, *P.G.A.*, vol. iv, pp. 353–410, and vol. v, pp. 407–94), and Blake and Hudleston (1877, *Q.J.G.S.*, vol. xxxiii, pp. 315–91), with alterations to nomenclature and some additions from personal observation and the researches of Dr. V. Wilson. See map, fig. 27, p. 138.

[2] Dr. Vernon Wilson informs me that he considers the Cementstone passes gradually up into the Kimeridge Clay.

[3] The *Gryphæa* was given a new name, *G. subgibbosa*, by Blake and Hudleston, but a number which Dr. Wilson has kindly shown me cannot be distinguished from *G. dilatata*. The varieties like var. *discoidea* were recorded by authors as *Ostrea bullata* Sow., the type of which came from the Kimeridge Clay. See Arkell, 'Mon. Corall. Lamell.', Part IV, p. 162.

blue-centred sandstones and softer shaly sandstones, then shales and cement-stones to a thickness of 36 ft., resembling the North Grimston Cementstone but darker and harder; the whole resting on coral rag.[1] This section seems to show that the upper part of the Cementstone Series alternates with and passes laterally into sandstones, which on the north side of the Gilling Gap constitute the Upper Calcareous Grit; though it is not improbable that some of the sandstone here is younger than any of the beds seen about North Grimston.

The UPPER CALCAREOUS GRIT in its more typical development encircles the north-west and north sides of the Vale of Pickering, its outcrop generally in contact with the alluvium, or separated from it only by scattered patches of Kimeridge Clay. The first good section is seen in the railway-cutting midway between Nunnington and Oswaldkirk. Here and all around the west end of the Vale, about Helmsley and onward to Kirkby Moorside, the rock is a massive blue-hearted sandstone, frequently containing lines of large doggers of still harder rock. It is separated from the coral rag by shaly beds which are seldom seen, but they have been reported as full of *Gryphæa dilatata* var. *discoidea* ('*Ostrea bullata*'). The ammonites recorded from the Upper Calcareous Grit include '*A. achilles* d'Orb.', '*A. decipiens* Sow.', '*A. berryeri* Leseur', '*A. serratus* Sow.', '*A. alternans* von Buch.', and '*A. biplex* (small interior whorls)', all suggestive of the *Ringsteadiæ* and associated forms of Wilts. and Dorset. This correlation is confirmed by the presence of *Chlamys midas* (d'Orb.), which in Dorset and the South of England abounds in and is restricted to the Upper Calcareous Grit. The shales below thus fall into line with the Sandsfoot Clay, and so with at least a part of the Ampthill Clay. The southward passage of the Upper Calcareous Grit and underlying shales into the North Grimston Cementstone strengthens our correlation of the Ampthill Clay.

Excepting traces on the Hackness outlier, the Upper Calcareous Grit is not exposed at the surface east of Thornton Dale, 2 miles beyond Pickering. Hereabouts it is usually a soft, brownish-red, ferruginous sandstone, locally crowded with *Lucina aspera*, *Pseudomonotis ovalis*, *Chlamys midas*, &c. Below are shale and a claystone of variable thickness, which in the Pickering quarries is called the Throstler. The Upper Calcareous Grit was proved to extend underground to the sea, however, for it was penetrated by borings at Irton, near Scarborough, and at Filey. The thickness was 45–46½ ft., and most of the underlying shale seems to have passed into sandstone.

Osmington Oolite Series (? + Lower Glos Oolite Series) (CORALLINE OOLITE AND CORAL RAG, OR UPPER LIMESTONE), average 50–60 ft.

The Osmington Oolites and probably also some strata contemporaneous with the *Trigonia clavellata* Beds constitute the main mass of the 'Upper Corallian' of Yorkshire, usually known as the Coralline Oolite and Coral Rag, or Upper Limestone. They comprise a highly variable and complex set of strata, the detailed examination of which is beyond the scope of this work. The stratigraphy is as usual greatly complicated by coral growth and, as was realized by Hudleston, the spreads of coral rag in different places were not

---

[1] C. Fox-Strangways, 1892, p. 367; revised measurements and description from Dr. Wilson, 1931.

all formed at the same time. As a general rule, however, the main mass of rag is found at the top of the division, immediately below the shale or clay just described, and this gave rise to the introduction of the term 'The Coral Rag' as a stratigraphical term, which, as we have seen, is not permissible. Below are many variable beds, the commonest and thickest type being white oolite. The combination of the white oolite and the coralline facies forcibly recalls the Calne district, where the coralline province of Oxon.–Berks.– North Wilts. meets the non-coralline white oolite province of Dorset.

It is highly probable that some of the Yorkshire coral rag is younger than the Osmington Oolites and contemporaneous with the Steeple Ashton and Upware reefs, which we correlated with the *Trigonia clavellata* Beds. The shelly facies of these beds, so conspicuous in Dorset, seems to be wanting entirely in Yorkshire, but locally some of the characteristic shells are found in the highest part of the rag: e.g. *Mytilus pectinatus, Lopha solitaria, Lima spectabilis* Contejean (a species of the Lower Kimeridgian about Mont-béliard), and *Opis virdunensis*, only known in France and in the late rag of Upware.[1] The localities where these fossils were found are all situated around Malton, Hildenley and North Grimston, on the south side of the basin, where also the series is thicker than elsewhere.

At North Grimston is the thickest development of coral rag known in the British Isles, amounting to 37 ft. Blake and Hudleston called it the North Grimston Limestone, and its peculiar characters may be considered to justify a separate name. Besides corals and the usual mollusca and *Cidaris* spines, it contains in places many bands of pale flinty chert. The only other localities where any considerable quantity of chert seems to be known in the coral rag are a quarry beside the road to the south of Helmsley, and at Slingsby, Hildenley and East Ness, near Nunnington.[2] Nowhere else, how-ever, is there more than half this thickness of coral rag, and it would seem that coral growth proceeded here, at the southern edge of the basin, more con-tinuously than anywhere farther north. Within a mile or two towards the south the entire division disappears against the Market Weighton Axis. This suggests that, if the wedging out is even partly original and not due entirely to unconformable overstep of the Cementstone, the water may have been shallow here for a longer period of time than elsewhere, and we may be seeing something of a true fringing reef that grew against the southern margin of the basin.

The strata below the rag are difficult to correlate with certainty with the succession in other parts of Yorkshire. Immediately under the rag are 25 ft. of drab-coloured, marly oolites with an occasional band of hard limestone and corals, called by Blake and Hudleston the MAMILLATED URCHIN SERIES. It is crowded with *Diplopodia hemisphærica, Hemicidaris intermedia, Cidaris florigemma, C. smithii* and *Nucleolites scutatus*. The Urchin Marls of the Wiltshire–Berkshire range are similar, but here the assemblage of urchins is much more varied.

Below the Mamillated Urchin Series Blake and Hudleston described a further 30 ft. of oolites, which have been taken to represent the 'Coralline

---

[1] W. J. Arkell, 1929, 'Mon. Corall. Lamell.', *Pal. Soc.*, pp. 8, 11.
[2] Information from Dr. Wilson; the only place where chert was previously noted was Helmsley, by Blake and Hudleston.

Oolite' below the coral rag of other districts; but if this correlation were altogether correct, the thickness of the series here would be little short of 100 ft., or nearly twice the thickness elsewhere. It seems probable that some of the Lower Limestones are represented, especially as the oolites are said to rest directly upon the Passage Beds, which separate the Hambleton Oolite Series from the Lower Calcareous Grit.

At Malton and the neighbouring village of Hildenley, across the Derwent Valley, the thickness is still great, and there is the same difficulty about the lower beds. The best section is shown by a large quarry at the cross-roads on the north side of Malton, exposing 30–40 ft. of rock. At the top is seen up to 7 ft. of coral rag, resting upon 13 ft. of white oolites which seem to correspond to part of the rag elsewhere, as they have 'a fauna somewhat approaching that of the rag'; 'and', observe Blake and Hudleston, 'this may partly serve to account for the very mixed character of the fossils which come from the Malton district in an undoubtedly oolite matrix'.[1] The more typical Coralline Oolites come in below, to the base of the quarry. Blake and Hudleston named these last the CHEMNITZIA LIMESTONES, owing to the abundance all over the western side of the Yorkshire Basin of *Pseudomelania* [*Chemnitzia*] *heddingtonensis* (Sow.). From them came the bulk of the wonderfully preserved shells for which Malton was famous. The collections of Strickland, Leckenby, Bean and others, which now enrich the museums all over the country, were made at a time when the quarries were more actively worked than now. They contain many perfect specimens of such large shells as *Chlamys splendens, C. intertexta, Lima mutabilis, L. læviuscula, Corbis ampliata* (Phil.), *C. decussata* Buv., *C. umbonata* Buv. Some of these species are not found in any other part of England, while others are known elsewhere only very rarely. The common *Chlamys qualicosta*, however, occurs on this horizon and appears to be restricted to it all over England.

Limestones high in the series, like those immediately under the rag at Malton, and containing some fossils more characteristic of the Upper Calcareous Grit, such as *Lucina aspera*, were also described by Blake and Hudleston at Hildenley. Here they recorded '*Ammonites variocostatus*'.

More or less typical Osmington Oolite, of the type called '*Chemnitzia*' Limestone, with 10–15 ft. of coral rag above, is continuous along the Howardian Hills past Hovingham and Cawton to Gilling, and again on the opposite side of the Coxwold–Gilling Gap in the long spur about Oswaldkirk and Nunnington, and so on to the north side of the basin. The total thickness of the division hereabouts maintains an average of 50–60 ft., which is the same as the thickness of the Osmington Oolite Series in Dorset. The Urchin *Glypticus hieroglyphicus*, usually rare, but common at Calne, is recorded from Wath, near Hovingham. About Oswaldkirk Blake and Hudleston noted a finer profusion of corals than at any other point in Yorkshire, especially of *Montlivaltia dispar* and *Stylina tubulifera*; and here, too, the spines of the sea-urchins are unusually large.

Along the north side of the basin the outcrop is continuous past Helmsley, Kirkby Moorside and Pickering to Thornton Dale, beyond which there is a gap of 5½ miles to the Brompton, Ayton and Seamer district. There are fine sections in the sides of Hutton Beck and also in the gorge of the River

[1] Blake and Hudleston, 1877, loc. cit., p. 365.

Seven at Sinnington. The best of all are afforded by the extensive quarries at Pickering, where the Osmington Oolite is still obtained for a flux in the Cleveland ironworks.

In the country north of Pickering the coral rag dies out locally and gives place to flaggy detrital deposits like the Wheatley Limestones. No true rag is present in the Pickering quarries, which show the complete succession well up into the Upper Calcareous Grit. Instead, the 'Chemnitzia' Limestones are succeeded by 15 ft. of impure, earthy and flaggy limestones without corals. The same type of rock has a considerable extension along the northern outcrop, although locally, and more especially in the lower ground towards the alluvium, the rag is luxuriantly developed. The actual passage from rag into Wheatley Limestone, such as may be seen about Headington and Purton, has been described by Fox-Strangways near the valley of the Hutton Beck, north-west of Sinnington. In a quarry he describes about 12 ft. of 'irregular ragged limestone with Cidaris florigemma and branching corals . . . surrounded, or nearly surrounded, with these impure flaggy limestones; which, as we have suggested, may have been formed from the denudation of the purer beds [of coral]'.[1]

The Sinnington Gorge and the quarries at Pickering (fig. 75) show that the 'Chemnitzia' Limestones portion of the Osmington Oolite Series is hereabouts 18–20 ft. thick; below are 13 ft. of variable limestones with shell-beds and occasional nests of corals, forming a passage downward into beds that are proclaimed by their fauna to belong to the Berkshire Oolite Series.

The isolated area inland of Scarborough, some 6–7 miles in length, from Brompton past East and West Ayton to Seamer, and its continuation in the Hackness outlier, contains beds of coral rag which Hudleston believed to occupy a lower horizon than the rag on the west and north-west sides of the basin.[2] The principal reasons for this opinion were the abundance of certain species of mollusca, such as Nerinea visurgis and Astarte subdepressa Blake and Hudleston (A. duboisiana auct., non d'Orb.) which are characteristic of the base of the Coralline Oolite at Pickering; and the absence of Cidaris florigemma, which abounds in the coral rag of all other parts of Yorkshire, but is here replaced by C. smithii. Dr. Wilson does not agree with Hudleston, however.

The best clean section at present worked is at the eastern extremity of the outcrop, in the Crossgates Quarry, Seamer, which shows 25 ft. of beds. At the top are 12 ft. of alternating oolites, Wheatley Limestones and coral beds, and below 13 ft. of oolites. At the base of the higher division with corals is a 2 ft. band of coral rock crowded with shells, called by Blake and Hudleston the Coral-Shell-Bed. They recognized the same bed, or another identical with it, on the Hackness outlier at Bell Heads Quarry, Silpho.

In Forge Valley, the gorge of the Derwent above Ayton, a complete section shows about 14 ft. of coral rag resting on about 25 ft. of oolites. In the neighbouring Yedmandale the rag is 20 ft. thick. The causes of the variability of these beds was made clear by Fox-Strangways in the following passage:

'Although there may be 20 ft. of coral rag in some of the sections, it is doubtful whether it does not thin out altogether in other places; in fact, the coral rag appears

[1] C. Fox-Strangways, 1892, J.R.B., p. 351.
[2] W. H. Hudleston, 1878, loc. cit., p. 420.

to stand up in irregular lumps or bosses between the oolitic brash and dense pasty limestones which were formed around them, in the same manner that coral reefs at the present day are surrounded by deposits formed from their own destruction.'[1]

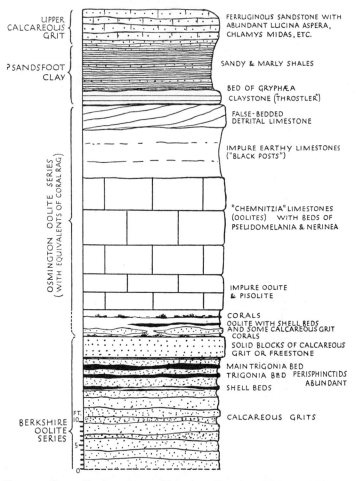

FIG. 75. Generalized section at Pickering. Based on Blake and Hudle-ston, *Q.J.G.S.*, vol. xxxiii, pp. 334–6, and personal observation, with correlation as suggested in the text.

On the coast, only the very base of the Osmington Oolite is seen at the point of Filey Brigg, where it is poorly fossiliferous and is overlain by thick Boulder Clay.

Berkshire Oolite Series (MIDDLE CALCAREOUS GRIT WITH TRIGONIA HUDLE-STONI BEDS), average 10–45 ft., maximum 80 ft.

One of the best datum-lines for correlating the Corallian Beds of York-shire with those of the South of England is the *Trigonia hudlestoni* Beds. Shell-banks, crowded with this highly characteristic species and with many

[1] C. Fox-Strangways, 1892, *J.R.B.*, p. 346.

of its associates farther south, such as *T. reticulata, Cucullæa contracta, Gervillia aviculoides*, and many others, immediately underlie and inosculate with the basal layers of the '*Chemnitzia*' Limestones and the white oolites under them, which we recognize as the Osmington Oolite Series.

The *Trigonia hudlestoni* Beds form part of a predominantly sandy series, called the Middle Calcareous Grit, and mapped separately by the Survey. It extends continuously along the northern and north-western sides of the basin, attaining its greatest development in the north-west, between Helmsley and the Coxwold–Gilling Gap at Ampleforth. Its thickness is as much as 80 ft. in the banks of the Rye between Helmsley and Rievaulx, and it forms a great surface-spread on Wass and Ampleforth Moors, from the Rye to the Coxwold valley.

The clearest sections are in the Pickering Quarries and Hutton Beck, where the thickness is about 20–40 ft.; the rock is exposed in Hutton Beck for nearly a mile. The greater part of the series consists of poorly-fossiliferous sandstones, but locally freestones are developed, and here and there, especially towards the top, some of the slabs are studded with *Trigoniæ* and other shells, forming almost a solid mass. Perisphinctids similar to those in the South are abundant, though they still remain to be figured and identified. A start was made by Buckman, who refigured the type-specimens of three species named by Young and Bird, but unhappily all under redundant new generic names.[1] There are also Cardiocerates of *excavatum* type[2] and Aspidocerates (*A. perarmatum* in the old records), just as in the *T. hudlestoni* Beds in the South of England.

Towards both the east and the south-east the series becomes thinner. On the coast it is represented by 6 ft. of alternating shelly beds and sandstones, overlying 10 ft. of more massive sandstones, called the Filey Brigg Calcareous Grit. In the Howardians it thins down in the same way, until at the eastern end, about Malton, it is doubtfully represented by 10 ft. of gritty limestone containing large lamellibranchs, formerly exposed at Middle Cave. South of Malton it cannot be detected at all.

In view of the unconformable relations of the Berkshire Oolite Series to the beds underneath in so much of the South of England, a quarry described by Fox-Strangways on Wass Moor, north-west of Ampleforth, acquires a special interest.[3] It showed the base of the Middle Calcareous Grit resting on the Hambleton Oolite Series (Lower Limestone), in the top of which a large channel had been excavated and filled with sandstone before the grits were deposited. Erosion therefore preceded the deposition of the *Trigonia hudlestoni* Beds, at least locally, in Yorkshire also.

Hambleton Oolite Series (LOWER LIMESTONE AND PASSAGE BEDS), maximum 90 ft.

The next beds, the Hambleton Oolites of Blake and Hudleston and the Lower Limestone and Passage Beds of the Survey, are altogether peculiar to

---

[1] *Perisphinctes maximus* (Young and Bird) (refigured *T.A.*, 1924, pl. DXII), *P. ingens* (Y. & B.) (*T.A.*, 1920, pl. CLXXXIV) and *P. pickeringius* (Y. & B.) (*T.A.*, 1923, pl. CDXLVIII), all from Pickering.
[2] Including '*Chalcedoniceras*' *chalcedonicum* (Y. & B.) (*T.A.*, 1922, pl. CCXCV, A & B) from Thornton.
[3] C. Fox-Strangways, 1892, *J.R.B.*, p. 341, fig. 17.

Yorkshire, where they are, however, of great stratigraphical importance. Like the Middle Calcareous Grit, the beds are lenticular in shape, the thickest part of the lens being situated over the north-western end of the Vale of Pickering, away from which they thin out in all directions. Fox-Strangways thought that this distribution was determined by the original circumstances of deposition rather than by subsequent erosion, and the same view is taken by Dr. Versey.[1]

Over most of the north-eastern part of the area the Survey were able to map two distinct subdivisions, the Lower Limestone or Hambleton Oolite proper above, and a more gritty series below, called the Passage Beds. The two subdivisions grade imperceptibly one into the other, not only vertically, but also horizontally. On the coast, at Filey, the whole is developed in the gritty facies of Passage Beds, while at the west end of the Vale of Pickering all is limestones.

As remarked above, the lower portion of the great thickness of oolites at North Grimston, usually all grouped as 'Upper Limestone', may probably represent the Hambleton Oolite Series; the Middle Calcareous Grit, which usually separates the two, having wedged out. If this surmise be correct, the Hambleton Oolite Series has not such a localized distribution as has been supposed.

The best section of the Hambleton Oolite Series in the Howardian district is seen in the Brows Quarry, Malton, now unfortunately abandoned. It still shows a mural face of peculiar buff-coloured, fine-grained, gritty, sparsely-oolitic limestone, from which such early forms as *Rhynchonelloidea thurmanni*, *Millericrinus echinatus* and Cardiocerates recorded as '*A. cordatum*' have been obtained. Fossils, however, are always rather scarce in these beds.

Beyond the Coxwold–Gilling Gap the limestone thickens rapidly, and in the country at the west end of the Vale of Pickering and on the dip-slope of the Hambletons it has an average thickness of 50–60 ft., with a maximum of perhaps 90 ft. between Hawnby and Kepwick.[2] Up the dip-slope, however, it thins to 16 ft. on the edge of the western escarpment. In this district the rock is poorly fossiliferous and much silicified. In extensive quarries at Kepwick Blake and Hudleston recorded *Cardioceras cordatum*, *Rhynchonelloidea thurmanni*, *Oxytoma expansa*, *Pseudomonotis ovalis*, *P. lævis* and a few other fossils, all of Lower Calcareous Grit aspect.

'The fact is,' wrote Fox-Strangways, 'in this region the base of the [Lower] Limestone [Hambleton Oolite Series] and the top of the Lower Calcareous Grit are dovetailed together, producing an alternating series of sandstones and limestones, which to the north develop into one thick bed of limestone, while to the south first the lower band of limestone dies out and then the upper, so that in the south-east of this range of hills there is only one thick mass of arenaceous strata to represent the whole of these limestones.'[3]

It is only east of Kirk Dale (north-west of Kirkby Moorside) that the lower part of the Hambleton Oolite Series becomes permanently gritty enough to map separately as the Passage Beds. Along the rest of the Tabular Range the

[1] H. C. Versey, 1929, *Proc. Yorks. Geol. Soc.*, N.S., vol. xxi, p. 210.
[2] Fox-Strangways stated that the maximum was 80 ft., but Dr. Wilson informs me that at least 90 ft. are visible above Arden Hall, near Hawnby.
[3] 1892, *J.R.B.*, p. 331.

PLATE XVIII

Photo. *Godfrey Bingley.*

Gristhorpe Cliff, near Filey, Yorks.

The overhanging bluffs are of Lower Calcareous Grit with some Lower Limestone, capped with Boulder Clay; the cliff below is of Oxford Clay.

Photo. *W. J. A.*

The Carr Naze, north side of Filey Brigg.

Cliffs of Filey Brigg Calcareous Grit, Lower Limestone and Passage Beds, with thick covering of Boulder Clay.

two divisions are then well distinguished, the limestones, 30–50 ft. thick, above, and the Passage Beds, up to 20 ft. thick or more, below. Above Staindale the Passage Beds have been locally cemented by silica, and the hardened portions, having resisted the action of the weather, stand up on the moor as the famous Bride Stones—some as much as 16 ft. high.

The two divisions are still distinguishable on the coast in Scarborough Castle Hill, where the limestone contains an abundance of very large shells of *Gervillia aviculoides*; but farther south, at Filey Brigg, they have coalesced, leaving some 35 ft. of strata, all in the gritty facies like the Passage Beds, but still with the large *Gervilliæ*.

On the Hackness outlier the Hambleton Oolite Series provides evidence of the earliest coral growth of the Corallian period in Britain. At the base of some 35 ft. of limestone are about 6–8 ft. of typical coral rag, which is best seen at Suffield and Silpho, but may extend as far south as Seamer Moor and as far west as Bickley or even Thornton Dale. In this early rag was already gathered together the well-known assemblage of corals, *Isastræa explanata*, *Thamnastræa concinna*, *Thecosmilia annularis* and *Rhabdophyllia phillipsi*, with their inevitable molluscan associates, *Lithophaga inclusa*, *Lima zonata*, *Chlamys nattheimensis*, *Exogyra nana*, *Lopha gregarea* and the ubiquitous *Cidaris smithii* (but not *C. florigemma*); and in addition a rich fauna of sponges and small brachiopods. Immediately above follow oolites with the typical giant *Gervillia aviculoides*, and *Cardioceras* recorded as *C. cordatum*. There can therefore be no doubt, either on stratigraphical or on palaeontological grounds, of the antiquity of this reef—or rather of the reef of which we see the broken fragments, with corals that have taken root amongst the debris. The only feature at all peculiar is the greater abundance than usual of small brachiopods and sponges. Nearly all the Corallian Calcispongiæ figured by Hinde in his monograph were recorded from here.

The determination of the stratigraphical position of the Hambleton Oolite Series in relation to the Corallian rocks of the South of England presents difficulties. The series is evidently to be regarded as a subdivision of the *cordatum* zone, for *Cardioceras cordatum* has been repeatedly recorded from it; and the passage just quoted from Fox-Strangways indicates the manner in which the limestone alternates with the normal sandstones of the Lower Calcareous Grit. On the other hand, such a development of limestones in the *cordatum* zone is known nowhere else in Britain. Some elements in the shell-bed fauna also are unique; in particular the highly characteristic *Gervillia aviculoides* of gigantic dimensions, and two species of *Trigonia*, *T. spinifera* d'Orb. and *T. blakei* Hudleston.

It is suggestive that from Calne, in Wiltshire, throughout North Wilts., Berks., Oxon., Bucks. and Beds., to at least as far as Upware, Cambridgeshire, the Berkshire Oolite Series rests non-sequentially and even unconformably upon the underlying *cordatum* and *præcordatum* zones; and that for forty miles it has at the base a pebble bed giving incontestable proof of erosion. I suggested in 1927 that in the South of England the *cordatum* zone is nowhere complete, and that the Hambleton Oolite Series may represent the missing strata. Until some more satisfactory correlation is put forward, it seems reasonable to suppose that the Yorkshire beds were laid down during the time-interval represented in the South of England by the Pebble Bed.

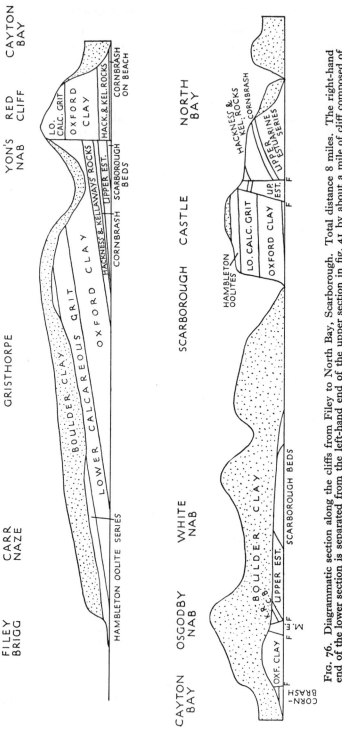

FIG. 76. Diagrammatic section along the cliffs from Filey to North Bay, Scarborough. Total distance 8 miles. The right-hand end of the lower section is separated from the left-hand end of the upper section in fig. 41 by about a mile of cliff composed of Boulder Clay. Vertical scale exaggerated. (After J. F. Blake, 1891, *P.G.A.*, vol. xii, p. 116.) (Included to show the general lie of the strata; the detail requires revision.)

Lower Calcareous Grit, 50–150 ft.

The Lower Calcareous Grit of Yorkshire is officially described as

'a yellow calcareous sandstone with doggers and cherty-bands passing down into softer sandstones, which become more shaly towards the base and gradually pass into the sandy shales of the Oxford Clay. Although the term "Grit" is applied to this rock, there is very little of it that would be called a grit among older formations, in fact many of the sandstones even of the Lower Oolite are harder and firmer and far more gritty.'[1]

This description, although perfectly accurate, conjures up a formation differing little from the Lower Calcareous Grit of the South of England. But in reality the Yorkshire rocks, as may be seen by reference to Pl. XVIII, are on the whole far harder and more massive, almost devoid of loose sand. The strongest bands of blue-hearted or blackish gritstone have few parallels in the South, excepting only in the Lower Calcareous Grit of the Weymouth district, where, however, the formation is so much thinner that the resemblance is not very striking.

In the Yorkshire Basin the Lower Calcareous Grit, as has been stated, is the first part of the Corallian Series to reappear on the north side of the Market Weighton Axis. Thence it is the most persistent member of the formation all round the basin. It attains its greatest development in the Hambleton Hills, where it caps the highest ridge of the escarpment, reaching a maximum height of 1,309 ft. above the sea in Black Hambleton. It also has a wide outcrop through the Tabular Range, extending as long tongues northward along the water-partings between the becks, far into the North York Moors. On the coast it is the only portion of the Corallian Series well exposed in the cliffs, under Scarborough Castle, in Red Cliff, Cayton Bay, and southward from Gristhorpe Cliff to Filey Brigg (fig. 76). Its great hardness and resistance to atmospheric weathering are directly responsible for the magnificent overhanging ramparts of Red Cliff and the jagged and vertical coast fronting the north side of the Carr Naze at Filey—some of the finest coastal scenery in Yorkshire.

Much of the Lower Calcareous Grit in Yorkshire is built up of sponge spicules, which are locally so abundant at certain horizons that they give rise to beds of chert. The species is *Rhaxella perforata* Hinde,[2] the same as that which builds up a large part of the Arngrove Stone and the similar cherty stone of Purton.

Measurements of the thickness vary greatly, from 60 ft. at the south end of the Howardian Hills and on the coast at Filey and Scarborough, to 150 ft. along the Tabular Hills and as much as 200 ft. at the north end of the Hambletons. It will be evident, however, from what has been said of the Passage Beds above and of the Oxford Clay below, that these figures include an indefinite amount of both older and younger strata in different places. In the Hambleton Hills, about Kepwick, as much as 45 ft. of limestones belonging to the Hambleton Oolite Series pass laterally into sandstones, which cannot be distinguished from the Lower Calcareous Grit. At Roulston Scar supposed Lower Calcareous Grit, resting directly on Hackness Rock, has

[1] C. Fox-Strangways, 1892, *J.R.B.*, p. 304.
[2] G. J. Hinde, 1890, *Q.J.G.S.*, vol. xlvi, pp. 54–61.

yielded a *Kosmoceras* of the *athleta* zone (see p. 360). Along most of the inland outcrop the boundaries, both above and below, are quite indefinite. The coast-sections alone present a clearly-defined Lower Calcareous Grit; but even there the collections of the earlier geologists show that at least a part of the grit is in the *præcordatum* zone of the Upper Oxford Clay. *C. præcordatum* identical with specimens found at Studley (Plate XXXVII, fig. 2) are represented in the old collections, in hard gritstone matrix.

In the cliffs below Scarborough Castle and from Cayton Bay to Filey Brigg the thickness is about 60–70 ft., and three fairly-well-marked divisions may be recognized. The highest 10–18 ft. consist of soft sandstone, weathering to a sand, with lines of huge, intensely hard doggers like cannon-balls. These 'Ball Beds' are especially conspicuous in the cliff of Scarborough Castle Hill, and some of the smaller balls closely resemble those in the Lower Calcareous Grit of Redcliff Point, near Weymouth, which Buckman mistook for derived blocks. Inside they sometimes contain nests of fossils, the commonest being *Chlamys fibrosa*, *Rhynchonelloidea thurmanni* and *Pinna lanceolata* Sow. (of which the type-specimen came from Oliver's Mount, Scarborough). On the platforms north-west of Filey Brigg, large double-valved shells of *Trigonia triquetra* Seebach may also be seen.

Under the Ball Beds is a 3–7 ft. block of very hard cherty calcareous gritstone, which is not only a conspicuous feature along the cliffs, jutting out abruptly beyond the softer strata above and below, but also caps many miles of the inland escarpment through the Tabular Hills, forming surface-spreads wide out of all proportion to its thickness.

The main mass of the Lower Calcareous Grit below consists of rough reddish or yellowish-brown grits and gritstones, with siliceous cement concentrated in bands and nodules. The whole weathers with a very irregular and jagged surface.

The fauna of the Lower Calcareous Grit seldom seems rich. *Cardioceras* and *Aspidoceras* are frequently quoted, but it is now difficult to obtain specimens adequate for specific determination. The commonest fossil, besides *Chlamys fibrosa*, is *Rhynchonelloidea thurmanni*, which is crowded in certain bands. It is curious that this Rhynchonellid has been recorded in the South of England at only two places—an old quarry at Catcombe, south of Lyneham, Wilts., now entirely obscured,[1] and in one of the borings in Kent.[2]

VI. EAST SCOTLAND

(a) The Brora District of Sutherland [3]

In Scotland the Corallian rocks are unfortunately not continuously exposed and little is known of any but the lowest portions. In the coastal plain north of Brora, where the bulk of the formation is presumed to occur, there are no exposures and much of the ground is covered with raised beach material and peat. There is an outlier south of the River Brora, in Dunrobin and Uppat Woods, but the surface is densely wooded and largely obscured by Drift (map, fig. 63, p. 366).

Where the low cliffs begin near Kintradwell, 2 miles north of Brora,

[1] Blake and Hudleston, 1877, loc. cit., p. 294.　　　　[2] See p. 439.
[3] Based on G. W. Lee, 1925, 'Geol. Golspie', *Mem. Geol. Surv.*, pp. 95–9, and personal observations.

Kimeridge Clay has already come in, but still farther north a gentle anticline near Loth brings up a series of sandstones which certainly belong mainly to the base of the Kimeridge formation, but may also in part represent the highest beds of the Corallian. Their position has been debated by several geologists. H. B. Woodward originally mapped them as Upper Corallian, but G. W. Lee considered them to be Kimeridgian and his conclusion was supported by Buckman tentatively identifying ammonites in Lee's collection as *Rasenia* and *Pictonia*. More recently, Dr. J. Pringle and Dr. M. Macgregor [1] have found between tide-marks a fossiliferous ledge which was previously overlooked, and from this they have collected a lamellibranch fauna (still unpublished) which they consider to have more affinities with the Corallian than with the Kimeridge Clay. This ledge lies off the mouth of a stream known as Allt na Cuile, about 1 mile south-west of Loth Railway Station, and it is only visible at low tide. Hitherto it is the only known exposure of beds that may represent the higher part of the Corallian formation.

The Lower Calcareous Grit is better known, from several exposures, but although the fauna is fairly abundant near the top, the downward limit within the Brora Sandstone (see p. 367) is purely a matter of conjecture.

The best exposure of the Lower Calcareous Grit in Sutherlandshire is at Ardassie Point, near the golf links, on the north side of the mouth of Brora River. No cliff-section exists, and a geologist first arriving there at high tide is disappointed to see only blown sand. The rocks emerge at low tide as several ledges running out to sea. They consist of seven bands of hard dark limestone separated by beds of carbonaceous shaly sandstone, the whole known as the ARDASSIE LIMESTONES. A total thickness of about 16 ft. is exposed. When the rock is broken up with a heavy hammer it is found to be highly fossiliferous, the most conspicuous members of the fauna being Cardiocerates and lamellibranchs. The Survey have recorded *Cardioceras* cf. *cordatum* (three forms), *C.* cf. *cawtonense*, *C.* cf. *excavatum*, *C.* cf. *maltonense*, *C.* cf. *rouilleri*, *C.* cf. *suessi*, and six unnamed forms, one figured by Buckman as '*Scoticardioceras*' *scoticum*; [2] also *Klematosphinctes vernoni* (?), a Perisphinctid of *biplex* type, and numerous lamellibranchs. The lamellibranchs belong almost entirely to common Corallian species, such as *Chlamys splendens*, *C. fibrosa*, *Camptonectes giganteus*, *Lima mutabilis*, *Oxytoma expansa*, &c., but there are a few unfamiliar elements, especially an abundant small *Grammatodon*.

The Ardassie Limestones strike inland in a westerly direction, and from the fossils found in surface rubble they are believed to cap Braambury Hill (see map, p. 366).

Another fossiliferous section in the Lower Calcareous Grit is afforded by a small road-metal quarry 320 yards south-west of Uppat House, on the wooded outlier south of the Brora River. About 8 ft. of sandstone are exposed, from which numerous impressions of more or less flattened Cardiocerates may be obtained, together with moulds of typical Corallian lamellibranchs. The bed is distinguished as the UPPAT CARDIOCERAS BED, and both by its fossils and by their state of preservation, it bears rather a striking resemblance to the Arngrove Stone. Among the many species common to the two is a small clavellate *Trigonia*.

The species of *Cardioceras*, of which the Survey record eight, are the

[1] 1930, *P.G.A.*, vol. xli, p. 81.   [2] *T.A.* 1925, pl. DXCIX.

same as those in the Ardassie Limestones, and the two deposits cannot be of very different dates. The lamellibranchs constitute a poorer assemblage, however, and the lithology is quite different. Lee considered that the Uppat *Cardioceras* Bed was probably rather earlier in date of formation, and suggested that it might lie on the horizon of a fossil bed in the upper part of the Brora Sandstone east of the bridges at Brora, mentioned by Judd but no longer open to view.

The absence of fossils in the bulk of the Brora Sandstone, below these *Cardioceras* horizons, precludes the drawing of any boundary line between the Lower Calcareous Grit and the Oxford Clay. The sandstone was described in the previous chapter because it forms a part of the Brora Arenaceous Series, of which the lower divisions are certainly of the age of the Oxford Clay; but the upper portions of the Brora Sandstone in the restricted sense are just as likely to correspond with some of the little-known basal portion of the English Lower Calcareous Grit.

A feature of the Sutherland Lower Calcareous Grit which should be mentioned is the lack of exact agreement between the ammonite species and their English prototypes. 'Ardassie', wrote Buckman, 'reveals little correspondence with the fauna of the Yorkshire beds. Its species, with few exceptions, appear to be new to English strata, but they have a likeness to Russian forms figured by Ilovaïsky.'[1]

(b) Ross-shire[2]

At Port-an-Righ, Ross-shire, both of the shore-sections between tide-marks, already referred to in connexion with the Oxford Clay (above, p. 369), are continued up into Corallian Beds, and they show (although obscurely) strata higher in the sequence than any that can be definitely ascribed to the Corallian in Sutherland. The northern section shows the highest beds, those near the top in the southern (Cadh-an-Righ) locality being nearly vertical and unfavourably displayed for fossil-collecting.

The highest Jurassic rocks seen are the PORT-AN-RIGH SANDSTONE, about 60 ft. of dark, shaly-weathering sandstones, with some hard bands and occasional small doggers. Fossils are sparsely distributed, but the Survey succeeded in collecting, besides a few indeterminative lamellibranchs and a Cardiocerate, three highly important Perisphinctids. Buckman identified them as *Perisphinctes antecedens*, *P. martelli* ('cf. *biplex*') and *P.* cf. *stenocycloides* (Neum.),[3] which indicate a date at least as late as the Berkshire Oolite Series.

The beds in which the *Cardioceras cordatum* fauna has been found consist of 7 ft. of soft, dark sandstone with nodules and nodular ribs of ironstone, called by Buckman the PORT-AN-RIGH IRONSTONES. From them the Survey collected the true *Cardioceras cordatum* (Sow.), *C. excavatum* (Sow.), and three other Cardiocerates, with another ammonite referred tentatively to *Klematosphinctes vernoni* (Bean-Phil.). Below are the thick shales of the *præcordatum* zone of the Oxford Clay. It is not justifiable to assume, however, that the Lower Calcareous Grit is reduced to 7 ft. in thickness, for it may also include an indefinite quantity of the Port-an-Righ Sandstones, below the level (not stated) at which the Perisphinctids were found.

[1] S. S. Buckman, 1925, *T.A.*, vol. v, p. 49.   [2] G. W. Lee, 1925, loc. cit., pp. 100-3.
[3] S. S. Buckman, 1925, *T.A.*, vol. v, p. 49.

VII. THE HEBRIDEAN AREA[1]

The impossibility of drawing a definite boundary between the equivalents of the southern Lower Calcareous Grit and Oxford Clay within the Cardioceratan deposits is as manifest in the Inner Hebrides as in Sutherland. The *cordatum* fauna of the Lower Calcareous Grit occurs in three places, in the promontories of Strathaird and Trotternish, Skye, and in the Isle of Eigg. In Strathaird it is succeeded by an assemblage of Perisphinctids correlating with the Berkshire Oolite Series, and the beds containing these are overlain unconformably by Upper Cretaceous sediments and Tertiary basalts; in Trotternish, however, a complete sequence of Corallian and Lower Kimeridgian shales may be expected, even though not fully exposed.[2] Sufficient detailed collecting has not yet been undertaken to show where the *præcordatum* fauna of the Upper Oxford Clay gives place to the true *cordatum* and *excavatum* fauna of the Lower Calcareous Grit, and the Survey have found it impracticable to separate the equivalents of the two formations, either on their maps or in their descriptive memoirs.

The Skye and Eigg occurrences were described in the previous chapter in connexion with the Oxford Clay (p. 371). In Strathaird, Skye, the beds usually called Oxford Clay, which consist of some 200 ft. or more of dark sandy or micaceous shales, passing down into 80 ft. of sandstones, occur in a syncline running along the centre of the promontory. Owing to the overstep of the basalts across the syncline the highest sedimentary horizons preserved come to the surface along the sides of the central valley of Abhuinn Cille Mhairè. Here Wedd found a number of ammonites of obvious Berkshire Oolite affinities, identified as *Perisphinctes ? variocostatus* (Buckl.), *P. ? pickeringius* (Y. & B.), *? P. suevicus* Siem., *P.? mogosensis* Choffat, and *Cardioceras* cf. *quadratum* (Sow.).

At lower levels, in the cliffs north of Elgol, a *Cardioceras* fauna similar to that of the Ardassie Limestones and the Uppat *Cardioceras* Bed is said to occur to the base of the series. The forms recorded by Messrs. Wedd and Kitchin are as follows: *Cardioceras cordatum* (Sow.), *C. cordatum* (d'Orb.) *pars*, *C. excavatum* (Sow.), *C.? funiferum* (Phil.), *C.? nikitinianum* Lahusen, *C. quadratoides* (Nikitin), *C.? quadratum* (Sow.), *C.* cf. *rotundatum* (Nikitin), *C. rouilleri* Lahusen, *C. suessi* Siem., *C. tenuicostatum* (Nikitin), *C. vertebrale* (Nikitin): an assemblage which appears to be a mixture of forms of the *cordatum* and *præcordatum* zones.

At Laig Bay in the Isle of Eigg, in a section exposed on the foreshore at low tide, described in the preceding chapter with the Oxford Clay (p. 372), a definite Lower Calcareous Grit fauna occurs in the dark-grey shales of Oxford Clay facies. Here Barrow found *Cardioceras cordatum* (two forms, coarsely and finely ribbed), *C. excavatum* and *Aspidoceras perarmatum*, associated with some typical Corallian lamellibranchs, the assemblage being very much the same as that of the Ardassie Limestones. No higher strata comparable with the *Perisphinctes* beds of Strathaird appear to be exposed.

An indication of the highest ammonite horizons of the Corallian (Upper

---

[1] Based on C. B. Wedd, 1910, 'Geol. Glenelg, Lochalsh and S.E. Skye', pp. 128–31; and G. Barrow, 1908, 'Geol. Small Isles of Inverness', pp. 26–8; *Mems. Geol. Surv.*
[2] See Mr. M. MacGregor's find of Lower Kimeridge Clay at Kildorais, NW. of Staffin, mentioned in the next chapter, p. 478.

Calcareous Grit), with *Prionodoceras* and *Dichotomoceras*, has recently been found in North Trotternish, Skye, by Mr. Malcolm MacGregor. The downward relations of the beds have not been seen, but upward they are continuous with a series of shales containing *Pictonia* and *Rasenia*, and so the occurrence is described in Chapter XIV, in connexion with the Kimeridge Clay (p. 478).

VIII. KENT[1]

The concealed province of Corallian rocks beneath the Weald is one of the most important in Britain. In it there is an exceptionally thick development of the upper divisions—Upper Calcareous Grit, Glos Oolite and Osmington Oolite Series. The lower divisions, the Berkshire Oolite Series and the Lower Calcareous Grit, on the other hand, are poorly developed and it is difficult to bring them into line with the thick Dorset sequence. This part of the succession is more comparable with the Lower Corallian Beds on the north of the Paris Basin, in Normandy and the Boulonnais, than with those of any other English district.

The complete succession was penetrated in four borings at Dover, at Brabourne, 4 miles east of Ashford, at Elham, and at Folkestone, but of these only the first two yielded complete information. Several borings entered the Corallian Beds but did not reach the base, while a number of others penetrated them in the district where they are incomplete, the upper portions having been removed by the Cretaceous denudations.

The average width of the buried outcrop is 3 miles. It runs from the coast under Dover to Chatham and Rochester, maintaining for the greater part of the distance a nearly straight course. The easterly attenuation against the London landmass is not nearly so marked as in the other formations, the total thickness being over 300 ft. at Brabourne and still 270–80 ft. at Dover.[2] These thicknesses are rivalled nowhere else in England outside the Yorkshire Basin.

The most conspicuous and constant feature of the Kentish Corallian Beds is a 125–35 ft. mass of white coralline limestones. In general it doubtless corresponds with the great central block of white coralline or oolitic limestones elsewhere, the Osmington Oolite Series; but it is twice as thick as the Osmington Oolite Series at its thickest in Dorset or in Yorkshire, excepting only at North Grimston. Other features suggestive of North Grimston are remarkably thick masses of coral rag—one as much as 33 ft. thick at Dover—and the rag facies recurs at all levels throughout.

It is interesting that this exceptional development of coralline limestones should be found close to the margin of the London landmass, against which all the other formations of the Jurassic thin out, while in Dorset, far from any shore-line, the oolites are only half as thick and contain no coral rag. It will be remembered that the great development of rag at North Grimston occupies an analogous position, close to the Market Weighton Axis; and that it and the superjacent portions of the Corallian Series, alone of all the Jurassic rocks, thicken towards the axis before being finally overstepped. The circumstances

[1] Based on G. W. Lamplugh and F. L. Kitchin, 1911, loc. cit., and Lamplugh, Kitchin and Pringle, 1923, loc. cit.

[2] The figures do not include the ironshot marls of the *præcordatum* zone, which the Survey classed as Lower Corallian.

therefore suggest that at both places we are dealing with actual fringing reefs that grew close to the shore.

It was remarked that the expansion of the coralline oolite and rag at North Grimston might be in part accounted for by the Berkshire and Hambleton Oolite Series taking on a coralline facies as they approach the land. The same explanation probably applies in Kent, where there is a poor show of rock that can be definitely assigned to those beds, and nothing at all comparable with the thick Nothe Clay and Bencliff Grit of Dorset.

Another analogy with North Grimston is provided by the thickness of the strata above the main mass of limestones—amounting at North Grimston to 80 ft. At Dover there were 96 ft. of such beds, while at Brabourne they reached the astonishing thickness of 162 ft. No such development of Upper Calcareous Grit and Glos Oolite Series is known elsewhere in the British Isles, but better parallels can be found on the Continent. All the principal types of rock in other parts of England are represented, however, the most characteristic being the ironshot oolite or Westbury Iron Ore, with its fauna of *Ringsteadiæ*.

It will be seen that the detailed correlation of the Kentish Corallian Beds with those of other parts of England presents special difficulties. The difficulties might not be insuperable were the rocks exposed in continuous cliff-sections like those in Dorset, but with no more information than could be obtained from bore-holes to work upon, no very detailed correlation is possible.

SUMMARY OF THE CORALLIAN ROCKS IN KENT

Upper Calcareous Grit and **Glos Oolite Series**, max. 162 ft.

The top of the Corallian Beds at Brabourne was taken above a 4–6 ft. band of 'greenish-grey glauconitic sandy mudstone with black specks', below which were about 14 ft. of 'blue-grey marly clay' and then ironshot oolite like the Westbury Iron Ore. The ore at Brabourne was only 3 ft. thick, all the rest of the succession down to the top of the main coralline limestones consisting of marls and marlstone or smooth argillaceous limestone, like a vastly expanded Sandsfoot Clay. In this were presumably included not only argillaceous equivalents of the Sandsfoot Grits, but also, judging by the fossils, strata of the age of the *Trigonia clavellata* Beds; for although little is known of the fauna of the Sandsfoot Clay, it is certainly not as rich as this would appear from the lists to be. The only possible correlation of the lower part of these upper beds is with the *T. clavellata* Beds, or the slightly later fossiliferous beds of Glos, in Normandy.

At Dover the succession was somewhat more varied and the lower portions were lithologically more suggestive of the *Trigonia clavellata* Beds. Here the Westbury Iron Ore—'millet-seed ore . . . small shining brown globules of iron carbonate crowded in a slightly clayey or loamy matrix'—was 16 ft. thick and yielded a specimen of *Ringsteadia*. Seven feet of clay above it also contained the ironshot grains, and above that were 26 ft. of alternating 'oolitic limestone, calcareous claystone, muddy grit, bands of marly clay and layers of pisolitic rubble'. These highest beds seem to have no counterparts in other regions in England; but the equivalent of the Westbury Iron Ore in the Boulonnais, the Oolithe d'Hesdin-l'Abbé, is 33 ft. thick and is succeeded by some 16 ft. of hard marly limestones called the Caillasses d'Hesdigneul, which in the valley of the Liane are replaced by calcareous sandstone (the Grès de Wirwignes).[1]

Trigonia clavellata Beds? (Pars.), Osmington Oolite Series, Berkshire Oolite Series (Pars.), 125–35 ft.

The highest 20 ft. of the beds grouped together as 'Corallian Limestones' by the Survey were described at Dover as 'creamy or greyish, soft, sandy limestone, with occasional layers of flaggy calcareous sandstone and incoherent sandy shale; and with rubbly bands mainly composed of rolled bits of shell and ooliths and containing many gasteropods, and *Pecten, Lima, Ostrea*, &c.'. Below this came 33 ft. of coral rag in 'irregular tabular masses', then 12 ft. of 'creamy-grey soft calcareous stone of sandy texture, containing few corals, but many shells and *Cidaris* spines', separating the upper coral rag from another 60 ft. of 'coral-limestone in hard tabular masses set in a softer calcareous matrix, with bluish-grey partings of calcareous silt': this last contained abundant Terebratulids.

At Brabourne corals were more or less common through the whole thickness of the limestones—134 ft. As already remarked, no such extensive development of coral rag as this is known elsewhere in England, for it exceeds even that at North Grimston; nor is there anything at all comparable with it either in the Boulonnais or in Normandy. If there were cliff-sections in Kent we might be able to trace out the coral rags at various levels into their contemporaneous shell-beds, or normal, non-coralline deposits, and so date them. But since coral rag of whatever age within the Corallian Series carries with it the same facies-assemblage, it is, as we have seen in Wiltshire and in Yorkshire, impossible to assign a date to any particular coral rag from internal evidence alone. The most we can hope to do is to fix limits to the period during which locally the coral régime endured.

Even if we are right in assigning some of the argillaceous shelly beds above the coral rag to the *Trigonia clavellata* Beds, there is still no reason why a considerable portion of the coral rag should not also be developed on the horizon of part of the *T. clavellata* Beds; for in Dorset they have every appearance of being condensed. It seems improbable that coral growth would have stopped short in Kent while the Steeple Ashton reef was growing in Wiltshire, and when, as we have shown reason to believe, rag was still being formed in the south of the Yorkshire Basin. The less coralliferous uppermost 20 ft. of the limestones at Dover, in fact, seems from the description more like *T. clavellata* Beds than like any part of the Osmington Oolite Series in other parts of England.

Berkshire Oolite Series (Pars.), Lower Calcareous Grit.

The lower limit of the Berkshire Oolite Series (the important plane of transgression at the base of the *martelli* zone) can be fixed with more confidence. This lower part of the succession seems to bear a strong resemblance to that on the coast of Normandy. There the Osmington Oolite Series (Oolithe de Trouville) rests directly on a thin representative of the Berkshire Oolite Series, in the form of 10 ft. of shelly *Trigonia hudlestoni* Beds (lime-

[1] P. Pruvost and J. Pringle, 1924, *P.G.A.*, vol. xxxv, pp. 34–5.

stones and marls) crowded with the usual lamellibranchs and Perisphinctids, and recalling very forcibly the *T. hudlestoni* Beds of England. In addition the shelly limestones are locally ironshot, thus resembling in appearance the Elsworth Rock, on approximately the same horizon.[1]

In Kent, the two most complete records (Dover and Brabourne) are rather obscure in this part of the section, but some of the more recently described borings are highly suggestive. At Dover the coralline oolite was immediately underlain by 11 ft. of 'dark blue clay with massive marly structure, in places indurated to marlstone, with bands of black rather sandy clay; and containing large crushed plicatiloid ammonites, casts of *Trigonia* and other lamellibranchs'. This is suggestive of the Nothe Clay; but 27 ft. of clay and claystone followed below before the ironshot oolite elsewhere at the top of the Oxford Clay was reached, and these beds did not yield sufficient fossils to enable their age to be settled.

At Fredville the Wealden Beds rest directly on strata below the main coralline limestones: 8 ft. of 'irregularly indurated pisolitic marlstone, passing down into 6 ft. of dingy, grey, impure, oolitic limestone with soft rubbly bands'; and below that were 12 ft. of 'compact blue-grey limestone, sparsely oolitic, with marly bands, *Gryphæa*, &c.'. These beds are suggestive of the Urchin Marls and *Trigonia hudlestoni* Limestones of Highworth, which, where they are thickest, do not contain the characteristic *Trigoniæ*; in fact the description of the Fredville strata applies exactly to some of those passed through in depth in the Red Down Boring, Highworth. Unfortunately these beds (like the Nothe Clay) were poorly fossiliferous and they can only be correlated with the Berkshire Oolite Series on stratigraphical and lithological grounds. They rested upon a bed which, by reason of its peculiar appearance, was easily recognized in a number of borings and served as a very useful datum-line—it was encountered at Fredville, Snowdown Colliery, Bere Farm, Brabourne, Chilham, Chilton and probably also at Guildford. This has already been described, in the preceding chapter, as forming the top of the Oxford Clay. It consisted of hard grey marlstone, packed with conspicuous dark-coated fossils, and its rich Cardiocerate and Peltocerate fauna proclaimed it to belong to the *præcordatum* zone. Above it no Cardiocerates of any kind were found, but in it and below it they were abundant.

No trace of the *cordatum* fauna was detected, though the presence of *C.* cf. *excavatum* (Sow.) and *Rhynchonelloidea thurmanni* (Voltz) in the dark-coated-fossil bed, and of *Perisphinctes* cf. *variocostatum* (Buckland) with abundant *Chlamys fibrosa* immediately below it, suggests that it may have been accumulated during a long period of time, perhaps spanning the *præcordatum*, *cordatum* and *martelli* hemeræ. In the absence of ammonites of the *cordatum* zone we must suppose that there is a non-sequence comparable with that at Quainton, Bucks., but no pebble-bed or other signs of erosion were found at the base of the presumed equivalents of the *martelli* zone.

[1] W. J. Arkell, 1930, *P.G.A.*, vol. xli, p. 401.

CHAPTER XIV

KIMERIDGE CLAY

Zones (Plates XXXIX, XL).	Dorset Strata.	South Midlands.
(Zonally unclassified)	Hounstout Marls Hounstout Clay *Rhynchonella* Marls and *Lingula* Shales	absent
Pavlovia pallasioides [1]		Swindon and Hartwell Clays with Lower Lydite Bed at base
Pavlovia rotunda [2]	*Rotunda* Nodules Crushed Amm. Shales	
Pectinatites pectinatus		Shotover Grit Sand
Virgatosphinctoides nodiferus [3]	400 ft. of clays with stone bands; Oil Shale 150 ft. from base	*Wheatleyensis* Nodules
Virgatosphinctoides wheatleyensis [4]		
Subplanites spp.[5]		absent
	YELLOW LEDGE	
Gravesia irius [6]	Lower clays of Hen Cliff 60 ft.	Generally absent
Gravesia gravesiana [7]		
	MAPLE LEDGE	
Aulacostephanus pseudomutabilis	Clays above and below The Flats, 120 ft.	Clays with *Exogyra virgula*[9] and vertebrates
Aulacostephanus yo [8]		
Pararasenia mutabilis	Clays with layers of *Ostrea* *delta* towards base	Clays with layers of *Ostrea* *delta* towards base
Rasenia cymodoce		
Pictonia baylei	*Rh. inconstans* Clay	clay (absent from Oxford)

[1], [2] The *pallasianus* zone of Chatwin and Pringle (1922), divided by Kitchin and Pringle (1923) into the zones of *Pallasiceras pallasianus* above and *P. lomonossovi* below; revised by Neaverson in 1924 and 1925. *Pavlovia* Ilovaïsky, 1917, includes *Pallasiceras* and *Holcosphinctes*, according to Ilovaïsky (1923, *Bull. Soc. Nat. Moscou*, vol. ii, p. 342) and Spath (1931, *Pal. Indica*, N.S., vol. ix, pp. 470–1).

[3], [4] The *Virgatites* zone of Salfeld (1913) and later writers until Neaverson showed (1924–5) that the true Russian Virgatitids are unrepresented at this level. According to Dr. Spath, the *Virgatites* zones are at a higher horizon than the *Pavlovia* zones; and, in fact, *Provirgatites scythicus* (Michalski, plate XL, fig. 3) occurs in the basal Portland Sand of Hounstout Cliff. Dr. Spath informs me that he considers *Virgatosphinctoides* synonymous with *Subplanites*, but until a further study of these extremely perplexing ammonites has been published, I retain the names current hitherto.

[5] *Subplanites* Spath, 1925, *Mon. Hunt. Mus. Glasgow*, No. 1, p. 120; type 'Virgatosphinctes' reisi Schneid (see T. Schneid, 1915, *Geol. u. Pal. Abhandl.*, vol. xvii, pp. 305–414).

[6], [7] Introduced by Salfeld (1913). Involute *Gravesiæ* of the *gravesiana* and *irius* groups appear to be unrepresented in any collections from this country, but crushed forms more evolute than these occur in the lower part of Hen Cliff, Kimeridge (two specimens found, 1932, by Dr. Spath, one in the writer's company; Salfeld stated that he found 'numerous examples').

[8] Apparently never yet found in this country, but retained in accordance with Salfeld's zonal scheme.

[9] Mr. L. R. Cox (1930, *Ann. Mag. Nat. Hist.* [10], vol. vi, p. 298) has pointed out that the name *Exogyra virgula* (Defrance, 1821–31) should lapse as a synonym of *E. striata* (William Smith, 1817); but as there is no other *E. virgula* having prior claim to the name, and as it has become so firmly established in the literature of more than a century, I propose to sacrifice precision for intelligibility and retain it here. The case for *Ostrea delta* Smith is different, because *O. deltoidea* Lamarck is a Cretaceous species.

I. DORSET

(a) The Isle of Purbeck: Chapman's Pool and Kimeridge[1]

THE structural backbone of both the Isle of Wight and the so-called Isle of Purbeck is the anticlinal flexure described in Chapter III, which strikes east and west and plunges gently to the east. In the Isle of Wight the lowest strata are the Wealden Beds of Brixton Bay on the west coast. From beneath these, somewhere under the intervening sea, rise the Purbeck and Portland Beds of Purbeck; and on the west coast of the peninsula is seen the lowest stratum brought to the surface by the anticline, the Kimeridge Clay (map, fig. 77, p. 442).

The Kimeridge Clay of the classic area of Purbeck crops out along the south-west side of the promontory as a narrow strip, measuring about 6 miles in length and never much exceeding 1 mile in width. It gives rise to a small but characteristic oasis of green clay-land, encircled behind by the lofty, steep-fronted escarpment of the Portland Stone, which rises well above the 500 ft. contour and effectively shuts off the coastal lowland from the rest of the 'island'.

The sea frontage from Chapman's Pool in the east to Kimeridge and Brandy Bays in the west consists of low cliffs of grey and black, shaly clay with thin mudstone bands. While in general the erosion of the sea is rapid, keeping the cliffs vertical and free from talus, the mudstone or cementstone bands offer a stubborn resistance, the harder ones, often repeated by small faults, running out to sea as treacherous ledges for more than a mile from the shore. Among mariners, the Kimeridge Ledges have justly deserved an evil reputation.

More famous than the ledges, however, is the 'Kimeridge Coal', a bituminous oil-shale, which for centuries has been dug along the cliffs and put to diverse uses.

In the south-east corner of the semicircular bay of Kimeridge the surface of the fields is scarred by a number of irregular mounds, overgrown with brambles. Near by a curious round building known as Clavell's Tower stands on the summit of Hen Cliff overlooking the eastern entrance to the bay, while on the beach beneath a pile of shaped stones is to be seen at the point marked 'pier' on the map (Pl. XX). These are the last vestiges of the Kimeridge alum industry, started at the beginning of the eighteenth century by Sir William Clavell, who built the neighbouring Smedmore House. Coker records [2] that that gentleman 'being ingenious in diverse faculties, put in tryall the makeing of allom, which hee had noe sooner, by much cost and travell, brought to a reasonable perfection, but the farmers of the allom workes seized to the king's use; and, being not soe skillfull or fortunate as himselfe, were forced with losses to leave it offe, and soe nowe it rests allmost ruined'. But 'Sir William Clavile, who[m] one disaster dismayed not' instead set up

---

[1] The spelling KIMERIDGE was used by H. B. Woodward in *The Jurassic Rocks of Britain*, by Damon in *The Geology of Weymouth and the Coast of Dorset*, and by most earlier authorities. The new form KIMMERIDGE was not heard of before Webster and Buckland introduced it in the nineteenth century and seems to have no justification. According to Hutchins (who in his great work on Dorset never deviated from KIMERIDGE) the spelling was KYMERICH in 1293 and CAMERIC in Domesday Book (Hutchins, *History of Dorset*, 2nd ed., 1774, p. 193).

[2] Rev. Coker, 1732, *A Survey of Dorsetshire*, p. 46, London.

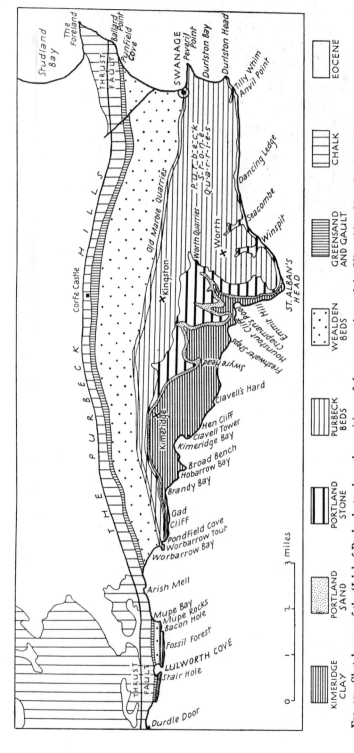

FIG. 77. Sketch map of the 'Isle' of Purbeck, to show the positions of the type-sections of the Kimeridge Clay, Portland Sand and Purbeck Beds.

works for making glass and for boiling the sea-water to extract salt, using the Kimeridge oil-shale as fuel. This earned for it the name 'coal', but Coker records that 'in burning it yields such an offensive savoure and extraordinarie blacknesse, that the people labouring about those fires are more like furies than men'. Notwithstanding, it is still used as a substitute for coal in the cottages round about, constant familiarity apparently enabling the local nose to accustom itself to the strong fumes of sulphuretted hydrogen given off during combustion.

By the middle of the eighteenth century ruins of buildings and heaps of ashes were all that remained of Sir William Clavell's projects, but in the latter half of the nineteenth century the bituminous shale was again exploited for the extraction of oil. The residual matter left after distillation of the oil and gas was sold as a deodorizer, decolorizer, disinfectant and manure. A ton of the shale was supposed to produce $7\frac{1}{2}$ gallons of naphtha, 10 gallons of lubricating oil, 1 cwt. of pitch, and some fine white paraffin wax and gas.[1]

Owing to the high content of sulphur (6–7 per cent.) and to the thinness of the workable seam, however, the cost of extraction has always been too heavy to allow the projects to become an economic success, and no satisfactory method of extracting the sulphur having yet been found, the present century has seen the works totally abandoned.

The coal was described by Sir A. Strahan as 'a highly bituminous layer of shaly stone, about 2 ft. 10 in. thick with its partings, and of a dark brown colour, whence its local name of "blackstone". It breaks with a conchoidal fracture and readily ignites, burning with a bright flame and an offensive smell, and leaving a copious grey ash.' In 1917–18 the Government sank a series of test holes around Kimeridge in order to estimate the total yield of the field should exploitation become necessary owing to the shortage of other oils.

The thickness of Kimeridge Clay visible between Chapman's Pool and Brandy Bay west of Kimeridge is about 700 ft.,[2] although the three lowest zones, those of *Pictonia baylei*, *Rasenia cymodoce* and *Pararasenia mutabilis*, are not exposed. The total thickness of clay present, therefore, may amount to more than 800 ft.

The upper zones rise from beneath the Portland Stone and Sand along the west side of St. Alban's Head, where they are, however, largely concealed by extensive landslips (fig. 80). After a temporary incursion inland, along two deep valleys at Chapman's Pool, they strike the coast once more in the magnificent Hounstout Cliff, which rises 501 ft. above the sea (Pl. XIX). As the Portland Stone forms only a capping, 50 ft. thick, to this cliff, the lower slopes are relatively free from slipped masses, and provide the best section of the uppermost zones of the clay in Dorset.

The level at which the top of the Kimeridge Clay should be drawn is debatable, and a question analogous with the upper boundary of the Oxford Clay. Just as by time-honoured usage the Lower Calcareous Grit is classed as a part of the Corallian, although palaeontologically it is no more than a sandy upper part of the Oxford Clay, and the arenaceous facies undoubtedly begins at different horizons in different places, so the Portland Sands grade down imperceptibly into the Kimeridge Clay, from which there

[1] A. Strahan, 1918, 'Spec. Repts. Min. Resources', vol. vii, p. 27, *Mem. Geol. Surv.*
[2] As much as 930 ft. has been estimated by some authors, but this is certainly excessive.

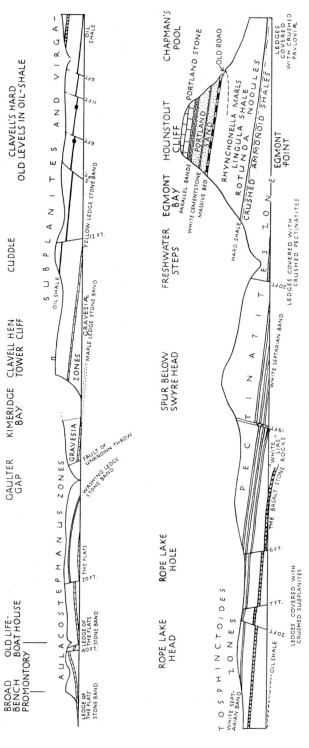

Fig. 78. Section along the cliffs of Kimeridge Clay from Broad Bench Promontory, near Kimeridge, to Chapman's Pool. Total distance 4⅛ miles. Horizontal scale 3·4 in. = 1 mile, vertical scale just under 1 inch = 500 ft. (Outline, stone bands and faults after Strahan, 1898, 'Geol. Isle of Purbeck and Weymouth', *Mem. Geol. Surv.*, pl. x; faults with a throw of less than 5 ft. omitted. The two faults near the lifeboat house have a combined throw of 40 ft.).

is usually little or nothing to distinguish them palaeontologically. The only comparatively constant datum is the base of the Portland Stone.

Fitton originally placed the top of the Kimeridge Clay 120 to 140 ft. below the Portland Stone, assigning the intervening beds to the Portland Sand,[1] and this plan was adopted by Sir A. Strahan in his Memoir on the Geology of the Isle of Purbeck.[2] Blake chose a datum about 100 ft. lower, giving the Portland Sand 244 ft.,[3] while H. B. Woodward preferred 179 ft., grouping the lowest 40–50 ft. as 'Passage Beds'.[4] Then when the deep borings in Kent proved Portland Stone resting directly on clay, it was believed that the Portland Sands of Dorset were there represented by clay. Although no direct palaeontological evidence of this was obtained, the idea gained credence to such an extent that in 1925 Dr. Neaverson wrote:

'There seems to be no doubt that the use of the term "Portland Sands" must be discontinued so far as the sandy beds below the Portland Stone Series of Dorset are concerned, since their faunal affinities are Kimeridgian rather than Portlandian and the sandy facies of Dorset and Bucks is represented by clays in the Kent borings. The ammonites from this horizon are, however, practically unknown . . .'[5]

The following summer, three years before his death, S. S. Buckman visited Chapman's Pool and, with the help of Mr. C. H. Waddington, began a study of Hounstout Cliff. He published a description of the beds below the Portland Stone, recording the fossils obtained *in situ*, and the result was to restore the Portland Sands to their original status, given them by Fitton and the early surveyors.[6]

In a band of white argillaceous cementstone, about 60–65 ft. from the top of the Portland Sands, specimens of *Leucopetrites* were found, a genus characterizing the lower levels of the Portland Stone (Behemothan Age) at Long Crendon, Bucks. Thus at least the upper half of the Sand (the St. Alban's Head Marl) belongs properly with the Portland Beds.

Below the level at which *Leucopetrites* was found are about 30–40 ft. of sandy marls (the Emmit Hill Marls, see below, p. 491), and below them again the most prominent band in Hounstout Cliff—a 5 ft. block of massive yellow and blue-grey sandstone, which Buckman named the Massive Bed. Fallen blocks can be easily recognized, and Buckman found in them ammonites with quadruplicate ribbing which he identified with the Russian *Virgatites scythicus* and *V. pallasi* (Michalski).[7] This was unexpected, since the Russian *Virgatites* zone was originally believed by Salfeld and others to be represented in England by what have been known since Neaverson's revision as the *Virgatosphinctoides* zones. Of the two specimens figured by Buckman, there seems no doubt of the identity of at least '*V.' scythicus* with the Russian form; but this group (as may be seen from Michalski's figures) has a different ontogeny from the true *Virgatites*, and it has for some years been separated under the generic name *Provirgatites*. True *Virgatites* still remains to be found.

[1] W. H. Fitton, 1836, *Trans. Geol. Soc.* [2], vol. iv, p. 212.
[2] A. Strahan, 1898, 'Geol. Isle of Purbeck', p. 60, *Mem. Geol. Surv.*
[3] J. F. Blake, 1880, *Q.J.G.S.*, vol. xxxvi, p. 195.
[4] H. B. Woodward, 1895, *J.R.B.*, p. 192.
[5] E. Neaverson, 1925, *Ammonites of the Upper Kim. Clay*, p. 43.
[6] S. S. Buckman, 1926, *T.A.*, vol. vi, p. 30.
[7] *T.A.*, pl. DCLXXV: and see Michalski, 1890–4, *Mém. Comité géol. Russie*, vol. viii, pl. v, figs. 6, 7.

The Massive Bed was the datum selected by Sir A. Strahan in his Memoir on the Isle of Purbeck at which to draw the arbitrary line of separation between the Portland Sands and Kimeridge Clay, and since it corresponds best with the base of the vague division established by Fitton, the originator of the term Portland Sand, there seem good historical reasons for following suit. Although Buckman and H. B. Woodward preferred a still more arbitrary invisible line 40–50 ft. lower, the Massive Bed has been adopted as the basement bed both by Mr. L. R. Cox in his palaeontological studies[1] and by Mr. M. P. Latter in his petrological account[2] of the Portland Beds, and its claims are accepted here.

SUMMARY OF THE SUCCESSION IN THE CLIFFS FROM
CHAPMAN'S POOL TO KIMERIDGE[3]

Pavlovia Zones (+ ?).

The highest 120–130 ft. of the Kimeridge Clay according to this definition have so far yielded only a few ill-preserved ammonites, badly crushed. Dr. Neaverson recorded 'ammonites similar to *H. pallasioides* (though in a poor state of preservation)',[4] and the lower portions of the beds contain a number of other fossils, especially brachiopods. Buckman recognized four subdivisions (of which he included the highest in the Portland Sand). For the purposes of recording fossils it is essential that these should have names. Since the beds are best exposed in the face of Hounstout Cliff, I name the upper two divisions the Hounstout Marl and Hounstout Clay (see Pl. XIX and fig. 78). The succession is as follows:

[Massive Bed above]

4. HOUNSTOUT MARL, 40–50 ft.: blue sandy marl with thin bands of 'stinkstone' (sandy cementstone giving off a strong odour when struck with a hammer); it extends down to the old road round the face of Hounstout.

3. HOUNSTOUT CLAY: about 30 ft. of dark clay, apparently devoid of fossils.

2. RHYNCHONELLA MARLS, about 20 ft.: grey marls with numerous *Rhynchonella* cf. *subvariabilis* Dav., a small *Oxytoma*, numerous small belemnites, and *Cidaris* spines; well seen at Pier Bottom on the west side of St. Alban's Head, where they first rise from the sea.

1. LINGULA SHALES, 35–40 ft.: dark shales with *Lingula ovalis* auct. non Sow.

[*Rotunda* Nodules below.]

The highest beds yielding well-preserved ammonites belong to the *rotunda* zone. Where they are first seen, at the east end of Chapman's Pool, near the boat-house, they occupy the lowest 10–15 ft. of the vertical cliffs of black shale. The strongly-ribbed ammonites of the type that were long styled '*biplex*' can be seen from a distance, standing out in hundreds as hard claystone casts covered with a white chalky coating. Near the top they can be obtained more or less whole, often embedded in small nodules, and this bed is therefore known as the ROTUNDA NODULES. They have recently been collected carefully by Dr. Pringle and studied by Dr. Neaverson and Buckman, and were referred to the genera *Pallasiceras* and *Lydistratites*, which

---

[1] L. R. Cox, 1925–30, loc. infra cit.   [2] M. P. Latter, 1926, loc. infra cit.
[3] The ledges after A. Strahan, 1898, 'Geol. Isle of Purbeck', *Mem. Geol. Surv.*, pp. 51–6.
[4] E. Neaverson, 1924, *Geol. Mag.*, vol. lxi, p. 149.

PLATE XIX

W. J. A.

Photo.

Hounstout Cliff, Dorset.

The finest section of the Upper Kimeridge Clay and Portland Sand in Europe. The capping is of Portland Stone. The arrow on the right denotes the position of the Massive Bed, here chosen as the base of the Portland Sand. RN= *Rotunda* Nodules, which dip steadily towards sea-level at the right-hand end of the picture.

Dr. Spath regards as synonymous with *Pavlovia*.[1] *P. rotunda* is abundant, and Sowerby's holotype almost certainty came from here. Beneath the Nodules, at the foot of the cliff, the ammonites, though still white, are crushed flat. Most of them belong to the same species, but some are much larger, and Buckman thought that they might belong to *Paravirgatites*.[2] Other abundant fossils flattened in the bedding planes are *Lucina minuscula* and *Orbiculoidea latissima*. From the boat-house to the stream at the head of Chapman's Pool these beds form a hard slippery platform, on which the waves break. Westward they rise gently, providing a conspicuous datum easily followed below Hounstout until they rise to the top of the cliff some distance before reaching Freshwater Steps.

Pectinatites, Virgatosphinctoides and Subplanites Zones.

From the point where the conspicuous, hard, Crushed Ammonoid Shale rises to the top of the cliff in Egmont Bay, east of Freshwater Steps (Plate XIX), the next convenient datum is a thin stone band (1 ft.), which forms the base of the little cascade at Freshwater Steps, at a level estimated at 110 ft. below the top of the Crushed Ammonoid Shale.

In 1931, after heavy gales, I had the good fortune to see a large part of the lowest 40 ft. of these clays exposed as smooth ledges, swept clear of shingle, in Egmont Bay. They were covered with flattened white ammonites, mainly of fine-ribbed forms resembling *Pectinatites*, and some identical with '*Keratinites*' *naso* S. Buck. and '*K.*' *nasutus* S. Buck.,[3] which seem to be, and have been stated by Dr. Spath to be, generically identical with *Pectinatites*. During the winter 1931–2 the shingle returned and completely covered these exposures; but the crumbling cliffs yield numerous fragments of crushed, small, fine-ribbed, involute ammonites, which can only be identified as *Pectinatites*. Specimens may also be seen upon the shale ledges about 100 yds. west of Freshwater Steps, at a level about 6–10 ft. above the Steps stone band.[4]

West of Freshwater Steps a continuous exposure of Kimeridge Clay extends for 1½ miles to Clavell's Hard, where the Blackstone or 'coal' was quarried, and this stretch of cliffs has remained until almost the time of going to press virtually a *terra incognita*. It is inaccessible from either end except at low tide, and the cliffs are unscalable from above.

It was known that immediately above the Blackstone numerous small, crushed, fine-ribbed ammonites, very similar to those in Egmont Bay and at Freshwater Steps, abound on the ledges, and they have been generally identified as *Pectinatites pectinatus* and allied species.[5] The identification was first made by W. H. Hudleston as long ago as 1896, when reporting on an excursion of the Geologists' Association which he directed in that year:

'Pyritized specimens of *Ammonites pectinatus* Phil. are abundant in the Kimeridge Coal—a fact not hitherto noticed, though very important by way of correlation. It serves to show that the Kimeridge Coal is on the horizon of the well-known and richly fossiliferous "Lower Portland Sands" of Swindon. Consequently we have

---

[1] See figures by Neaverson, 1925, loc. cit., pl. I, figs. 6–10, and Buckman, *T.A.*, 1926, pls. DXC (A–C), CCCLIII (C, D), DCXXXIX.
[2] S. S. Buckman, 1926, *T.A.*, vol. vi, p. 33.   [3] *T.A.*, pls. DCLII, DCLXIV.
[4] These I have recently photographed *in situ*.
[5] Two have been so figured by Neaverson, 1925, loc. cit., pl. I, figs. 4, 5.

no difficulty in believing that the Kimeridge Clay of Chapman's Pool is on the horizon of the Hartwell and Swindon clays, at least, approximately.' [1]

For the time at which it was written, this was a remarkable observation, and after having been overlooked for a quarter of a century it seemed to have been vindicated by the independent work of the Survey and Dr. Neaverson.

The bed which yields these (pyritized) specimens identified as *P. pectinatus* and allied species most abundantly is the roof of the 'coal' seam. With them are found little radial plates, also pyritized, of the pelagic crinoid *Saccocoma*,[2] which have such a restricted vertical and such a widespread lateral range (having been encountered on the same horizon in the Kent and Norfolk borings) that Messrs. Kitchin and Pringle have suggested making of them a separate subzone; at Kimeridge they have a total vertical range of 13 ft.,[3] and are most abundant just below the Blackstone.

The ammonites, however, range up very much higher; in fact they may be found at intervals, with little apparent change, almost up to the stone band that sinks to the beach at Freshwater Steps, a measured distance of 155 ft. above the Blackstone.[4] Numbers are well shown upon the ledges on the east side of Rope Lake Head, where I have recently photographed them as they lie, since they invariably break if extraction is attempted. This level, where they seem to attain their acme, is between 20 ft. and 30 ft. above the Blackstone.

To those who have not made a very detailed and comprehensive study of these 'pectiniform' ammonites, the conclusion seems unavoidable that between the Blackstone and the Crushed Ammonoid Shales (*rotunda* zone) all the clays, shales and thin stone bands must be assigned to an extended *Pectinatites* zone, 250 ft. thick. Ammonites which seem indistinguishable in the field from *Pectinatites* abound throughout this great thickness; and I had therefore determined on this course. Dr. L. F. Spath assures me, however, that the only true *Pectinatites* occur above the stone band at Freshwater Steps— namely in the highest 100 ft.—and that all the rest, although they are barely (if at all) distinguishable from *Pectinatites* in the young stages which are by far the most commonly represented, develop later along different lines. These abundant earlier forms he assigns to the genus *Subplanites*, and he considers that the small ammonites found at and about the level of the Blackstone, and identified by Hudleston and by Neaverson as *Pectinatites*, belong to the homœomorphous genus *Lithacoceras* (Hyatt, 1920). It is therefore interesting to note that Buckman also gave them a separate name (*Pectiniformites*, 1925, *T.A.*, pl. DLXVII) and placed *pectinatus* considerably higher.

The 150 ft. of clays below and including the coal seam constitute the *Virgatosphinctoides* zones of Neaverson.

The ammonites of these zones have given rise to much controversy, owing to their general similarity to Russian forms, which Dr. Neaverson showed, however, to be unrelated. Salfeld in 1913 stated that the beds were equivalent to the Russian *Virgatites* zone, although he admitted that he had never found in them a true specimen of *Virgatites*.[5] The Geological Survey at first accepted Salfeld's statement and believed that they had detected the Russian

---

[1] W. H. Hudleston, 1896, *P.G.A.*, vol. xiv, p. 322.
[2] F. A. Bather, 1911, *Sum. Prog. Geol. Surv.* for 1910, pp. 78–9.
[3] Tested in 6 borings (J. Pringle, *in lit.*).
[4] Field-notes and measurements, 1932.   [5] H. Salfeld, 1913, *Q.J.G.S.*, vol. lxix, p. 425.

genus in a number of specimens recovered from the horizon of the oil-shale in borings at Corton, near Abbotsbury. With them were some large ammonites which were identified with another foreign genus, *Pseudovirgatites*, and this they made the index of a new zone to include the oil-shale.[1]

In 1925 Dr. Neaverson published his monograph upon these and other Upper Kimeridge Clay ammonites, showing that the similarity which the English forms bore to the foreign was due to homœomorphy, and the Corton and Kimeridge species previously regarded as *Pseudovirgatites* he named *Virgatosphinctoides grandis*, *V.* cf. *nodiferus* and *V. delicatulus*.[2] The stratigraphical position of the species was first determined inland, at Shotover and Wheatley, where they immediately underlie the thin *pectinatus* zone, but their position in the type-section in Dorset is by no means so certainly fixed. They have been recorded at and below the level of the oil-shale; but if Dr. Spath is right they should be sought at a considerably higher horizon. (For further notes on the ammonites see footnotes on p. 440.)

SUMMARY OF THE SUCCESSION FROM FRESHWATER STEPS TO CLAVELL'S HARD

	Ft.
Clays with crushed *Pectinatites*; some nodules about 25–30 ft. from base .	100
THE THREE STONE BANDS, drawn in Strahan's section (loc. cit.) . . .	35
The highest is the Freshwater Steps Stone Band.	
The lowest is the ill-named White Septarian Band, so called by Strahan in Brandy Bay. It is not septarian, but consists of hard white limestone with numerous paper-thin interlaminated clay-seams, giving it locally the appearance of tourmalinized slate. It forms the conspicuous white rocks east of Rope Lake Head, known by the fishermen as the Lias Rocks. Saurian vertebrae occur.[3]	
Clays and shales with crushed *Subplanites*, &c., more or less hardened . .	35
THE BASALT STONE BAND: hard black stone with conchoidal fracture; looks like a basalt sill; forms ledge bounding east side of Rope Lake Hole, and reaches beach level in W. corner of Brandy Bay	3
Soft shaly clay, many crushed *Subplanites*	70
ROPE LAKE HEAD STONE BAND: forms prominent ledge on W. side of extremity of Rope Lake Head	1½
Clays and shales with crushed *Subplanites*, about	10
THE BLACKSTONE	1
Clays and shales with crushed *Subplanites*, &c., and two conspicuous bands of cementstone, formerly worked for cement	140 to 150
THE YELLOW LEDGE STONE BAND	2

About 140–150 ft. below the 'coal' seam is a prominent band of hard mudstone known as the Yellow Ledge Stone Band. It rises at the ledge from which it takes its name (fig. 78) and, after running along the middle of Hen Cliff, reaches the top a hundred yards west of the Clavell Tower. This marks the line of division between Upper and Lower Kimeridge Clays, and it is the highest level at which *Exogyra virgula* has been found.

[1] Lamplugh, Kitchin and Pringle, 1923, 'Concealed Mesozoic Rocks in Kent', pp. 222–5 and pl. II, *Mem. Geol. Surv.*
[2] E. Neaverson, 1925, *Ammonites from the Upper Kim. Clay*, Liverpool.
[3] No ammonites or traces of ammonites have been found for certain in this band, and it could not possibly have been the source of the *Pavlovia* figured by Damon as 'Am. biplex' and renamed *A. Kimmeridiensis* by von Seebach, as supposed by Buckman.—*T.A.* 1926, vol. VI, p. 38, and pl. DCLXXIII. From its state of preservation that specimen can only have come from the *Rotunda* Nodules.

Gravesia Zones.

In the clays below the Yellow Ledge Stone Band, to the foot of Hen Cliff, the presence of flattened ammonites of the genus *Gravesia* was announced by Salfeld in 1913. He recorded the group of *G. irius* (d'Orb.) in the higher parts and the group of *G. gravesiana* (d'Orb.) below, thus (assuming his identification to be correct) providing an important link with the French and German succession.[1] Repeated search in more recent years, however, has only resulted in the discovery of two badly crushed specimens resembling the more evolute group of *G. gigas*.[2]

The downward limit of the *Gravesia* zones was defined by Salfeld as the Maple Ledge Stone Band,[3] which rises on the north side of the old pier, about 60 ft. below the Yellow Ledge Stone Band.

One hundred yards east of Gaulter Gap the Maple Ledge Stone Band is faulted up above the top of the cliff, and there is no other datum by which to determine the displacement of the fault. Sir A. Strahan wrote:[4] 'There is no reason, however, to suspect a large throw, and we may assume that the strata on the west side of the fault, while certainly below, are not far underneath the lowest seen on its east side.'

Aulacostephanus Zones.

On the other side of the fault, and below the Maple Ledge Stone Band, the ammonite fauna again changes; and, as Salfeld found, down to the lowest levels exposed the clays yield abundant *Aulacostephanus eudoxus*, *A. pseudomutabilis*, and other species of the same genus, with *Aptychi* and small *Aspidoceras*. The total thickness of the *Aulacostephanus* zones exposed is about 120 ft.

The axis of the Purbeck Anticline passes out to sea close to the Broad Bench Promontory, which forms the western horn of Kimeridge Bay. At the point, lying mainly between tide-marks, is a large flat shelf known as the Broad Bench, up which the muddy waves rush with spectacular fury in rough weather. It is formed by the lowest of the stone bands, which on account of repetition by small faults, gives rise to two similar benches nearer Kimeridge called The Flats, and so bears the name of The Flats Stone Band (Pl. XX).

On the west side of Broad Bench Promontory, in Hobarrow and Brandy Bays, the strata dip rather steeply below sea-level once more in the northern limb of the anticline, and all the zones are repeated in the space of less than a mile to the commencement of Gad Cliff. Farther west, beneath the precipitous walls of Portland Stone the highest zones are largely obscured by talus and wild undercliff like that beneath St. Alban's Head.

The lowest beds exposed in the Isle of Purbeck, still belonging to the zone of *Aulacostephanus yo*, are seen for a short distance in Hobarrow Bay, north-west of the axis. Here The Flats Stone Band is thrown about 50 ft. up in the cliff by a fault, so that below it are seen lower beds than are visible in the crest of the anticline on the other side of Broad Bench Promontory.

---

[1] H. Salfeld, 1913, loc. cit., p. 425.
[2] Found by Dr. Spath in 1932.
[3] Inquiries among the fishermen at Kimeridge in 1931 showed that the ledges called Maple and Washing Ledges in Sir A. Strahan's time are no longer known by name.
[4] 1898, loc. cit., p. 55.

PLATE XX

Photo.　　　　　　　　　　　　　　　　　　　　　　　*W. J. A.*

Hen Cliff, Kimeridge.

From the site of the old 'coal' pier. The Yellow Ledge Stone Band can be seen
rather above the middle of the cliff, which is here formed mainly of the
Gravesia zones. Note the muddy sea, and talus slopes continually being
removed by the waves.

Photo.　　　　　　　　　　　　　　　　　　　　　　　*W. J. A.*

Kimeridge Bay from The Flats.

Hen Cliff and Clavell's Tower on right of centre. Smedmore House among
the trees in the distance on left, with Swyre Head behind. In the distance on
right, waves are breaking on the mudstone ledges.

PLATE XXI

Photo. *W. J. A.*

Ringstead Bay and White Nothe Head.

The headland is of Chalk, resting on thin Upper Greensand and Gault, which cut unconformably across the Kimeridge Clay. Holworth House near top of cliff on left, with two masses of Portland Beds and Purbeck Beds below (showing white) brought down by pre-Albian faults.

Photo. *W. J. A.*

Junction of Kimeridge Clay and Corallian Beds, Ringstead Bay.

For explanation see pp. 379 and 451.

It was probably from the *Aulacostephanus* zones that the type specimens of *Steneosaurus manselii* Hulke and *Ichthyosaurus enthekiodon* Hulke were obtained; they were embedded in reefs exposed at low water in Kimeridge Bay. A jaw of a *Teleosaurus* was found by Mansel-Pleydell and figured by Hulke, which had fallen from the cliff during the winter, probably from the same zones.[1]

(b) Northern Limb of the Weymouth Anticline: Ringstead to Abbotsbury

After a gap of 7 miles the Kimeridge Clay again appears in the cliffs of Ringstead Bay, where it is overlain by a small patch of faulted Portland and Purbeck Beds, and all are transgressed unconformably by the Gault (Pl. XXI). The exposures have deteriorated considerably in recent years and fossils are now somewhat scarce. Sir A. Strahan considered it probable that the middle zones are cut out by the faults, but both the highest and the lowest zones are visible. A detailed lithological description was made by Waagen in 1865,[2] and Salfeld in his paper of 1914 identified some of the ammonites collected by Waagen and endeavoured to assign them to their proper beds,[3] but no modern palaeontological work has been published.

The clays at the east end of the bay, which are partly succeeded by the Portland Sand and partly by the Gault, are poorly fossiliferous and all seem to lie above the *rotunda* zone. In the centre of the bay Waagen obtained abundant ammonites of the *Aulacostephanus* zones, below which, and passing westward into the lowest 100–140 ft. of the clay, Salfeld recognized evidences of both the *Rasenia* zones. From the presence of numerous flattened specimens of the typical *Pararasenia mutabilis* (Sow.) he assigned the higher parts of the lowest 140 ft. to the *mutabilis* zone; the *cymodoce* zone he recognized about 10 to 13 ft. from the base. From the *Aulacostephanus* zones of this place came the type-material of *Cardioceras anglicum* Salf. and *C. krausei* Salf.[4]

The *cymodoce* and *Pictonia* zones can be conveniently studied in the low cliffs for about a mile along the west side of the bay (Pl. XXI, lower figure). They contain many layers of large and well-preserved specimens of *Ostrea delta*, which weather out in hundreds on the beach. About $3\frac{1}{2}$ ft. above the base of the Kimeridge Clay is a 6-in. band of marl, partly indurated to form a hard limestone, almost entirely composed of *Exogyra nana* (Sow.), with occasionally *E. prævirgula* Jourdy. The band of pale clay beneath contains abundant *Rhactorhynchia inconstans* and imperfect casts of *Pictoniæ*.[5] Waagen and subsequent writers have taken this by common consent as the basal bed of the Kimeridge Clay, and it is the lowest level at which *Pictoniæ* seem to occur (if not the only level). Immediately below is the Ringstead Coral Bed, the topmost member of the Corallian, in the equivalent of which *Ringsteadiæ* are found farther west.

---

[1] J. W. Hulke, 1869–80; see bibliography.
[2] W. Waagen, 1865, *Versuch einer allgemeinen Classifikation*, p. 4.
[3] H. Salfeld, 1914, loc. cit., pp. 204–6.
[4] H. Salfeld, 1915, 'Monographie der Gattung *Cardioceras* Neum. u. Uhlig, I', *Zeitschr. Deutsch. geol. Gesellsch.*, vol. lxvii, pl. xx, figs 1–9.
[5] Proved by personal observation, which is not in accord with Salfeld's record of *Ringsteadiæ* from the *Rh. inconstans* Bed. Waagen's specimens of *Ringsteadiæ* probably came from the clays with ironstone nodules immediately *below* the Ringstead Coral Bed, where they may still be found *in situ*.

The high cliff west of Holworth House, at the east end of the bay, is still known as Burning Cliff from a spontaneous ignition of the bituminous shale which took place in 1826 and continued for four years. The heat that caused the shale to ignite was supposed to have been produced by the decomposition of iron pyrites. The fire was in progress when Buckland and De la Beche visited the district for the preparation of their paper 'On the Geology of the Neighbourhood of Weymouth', read in 1830. They state that it originally gave off flames for many months, but by the time of their visit it had considerably abated, there being no flame, only 'small fumaroles that exhale bituminous and sulphureous vapours, and some of which are lined with a thin sublimation of sulphur'. 'Much of the shale near the central parts', they wrote, 'has undergone a perfect fusion and is converted to a cellular slag. . . . Where the effect of the fire has been less intense, the shale is simply baked and reduced to the condition of red tiles.' [1]

If the oil-shale is on the same horizon as that at Kimeridge, several other zones must be present that have not been proved palaeontologically.

At Black Head, west of Osmington, faults again bring down the base of the formation nearly to sea-level, and the *Pictonia* zone, with *Rhactorhynchia inconstans* and *Ostrea delta*, and the Ringstead Coral Bed at the bottom, can be traced close above the beach. Most of the Upper Kimeridge Clay is here cut out by the Cretaceous unconformity. Close to Osmington Mills, the Upper Greensand rests directly on the *Rasenia* zones, but higher zones succeed farther west, up to and above the oil-shale. The strata dip into the cliff at 40 to 60°, and where springs are given off at the base of the Chalk the clays have foundered, letting down enormous blocks of chalk, clay and greensand, and forming remarkable mud rivers, which flow slowly from the top of the cliff into the sea, after the manner of lava streams.

Beyond Black Head the Kimeridge Clay strikes inland, passing in a west-north-westerly direction along the north side of the Weymouth Anticline, giving rise to a deep valley parallel to that of the Oxford Clay, but separated from it by the Corallian ridge. It is not cut by the coast at the west end, but terminates at Abbotsbury faulted against Forest Marble.

Little was known of the major portion of the clay along this outcrop before 1917, when the boring of a number of test-holes for oil was commenced simultaneously at Kimeridge and at Corton, between Abbotsbury and Upway. The investigation was undertaken by the Department for the Development of Mineral Resources, oil-shale having been discovered at Portisham in 1856. The borings struck a moderately rich seam of oil-shale up to $2\frac{1}{2}$ ft. thick, but it proved of no commercial value, for it shared the same disadvantage as the Kimeridge shale, namely the excessively high content of sulphur. [2]

Palaeontologically the borings were more profitable, the level of the oil-shale yielding *Saccocoma*, as at Clavell's Hard, associated with crushed ammonites identified as *Virgatosphinctoides* (*Subplanites*?).

The district is chiefly remarkable for another economic product, a thick deposit of iron ore which is locally developed in the *Rasenia* zones at Abbotsbury. The average thickness of the ore is about 20 ft., but in the best section, in the sides of the lane leading from Abbotsbury to Gorwell, the total develop-

[1] Buckland and De la Beche, 1835, *Trans. Geol. Soc.* [2], vol. iv, p. 23.
[2] A. Strahan, 1918, 'Spec. Repts. Min. Resources', vol. vii, pp. 24–40, *Mem. Geol. Surv.*

ment of ferruginous and sandy rock amounts to 45 ft. The main ore consists of 20 ft. of 'crumbling reddish-brown oolitic rock, full of shining pellets of ore in a matrix of fine quartz sand . . . traversed by numerous long thin seams of concretionary iron', with thick beds of sand, more or less ferruginous, above and below. Not much attempt has been made to work the ore, probably on account of its being too siliceous.[1]

A remarkable brachiopod assemblage occurs in the ore bed, the commonest being *Rhynchonella corallina* Leym. and *Ornithella lampas* (Sow.), with which is associated *Rh. inconstans*. From this association Blake and Hudleston inferred the stratigraphical position with tolerable accuracy, giving it as their opinion in 1877 that the beds were 'at least on the horizon of the passage-beds to the Kimmeridge Clay', and definitely above the Sandsfoot Grits.[2] Later Douvillé and then Salfeld showed that the ore contains such ammonites as *Rasenia uralensis* (d'Orb.) and *R. thurmanni* (Oppel), indicative of the *Rasenia* zones of the Kimeridge Clay.

The small size of the area over which the ore and sand deposits are developed is remarkable. Nothing similar is known at this horizon in any other part of England.

(c) Southern Limb of the Weymouth Anticline: Weymouth and Portland

The small area of Kimeridge Clay preserved on the southern limb of the Weymouth Anticline could scarcely be better described than in the words of Buckland and De la Beche, written in 1830:[3]

'The southern belt of Kimeridge Clay near Weymouth occupies a very small portion of the surface, constituting a triangular area, the base of which extends about a mile from Sandsfoot Castle westward to the Chesil Bank, whilst its apex is at Portland Ferry: but although so small a portion of this belt of clay is here visible on the surface, we have evidence of its submarine continuation from hence to Portland Island, in the clay bottom of the excellent anchorage of Portland Road, beyond which also it appears above the level of the sea in the base of the escarpment at the north extremity of the Isle of Portland, and along its west shore almost immediately south of the village of Chesilton. Hence it is clear that the Kimeridge Clay forms the fundamental stratum of the whole island, separated, as we have shown, from the Portland Stone by the Portland Sand and Sandstone. The rapid dip of all these strata towards the south causes the Kimeridge formations to sink below the level of the sea in the southern portion of the island; whilst that part of its western coast, whose base is composed of these perishable sands and clays, is defended from the tremendous south-western waves by a natural breakwater of enormous masses of Portland Stone that have fallen from the summit, and form a barrier against any further encroachments.'

In the top of the Portland Sands on the west side of the island Salfeld found a specimen of '*Perisphinctes gorei*', showing that at least the upper part of the sands here, as at Hounstout, should be classed with the Portland Sand. The same author recorded ammonites of the '*pallasianus*' group (i.e. *pallasioides* or *rotunda* zones) in septaria in the clays below, and in flattened

[1] J. Pringle, 1920, ' Spec. Repts. Min. Resources ', vol. xii, p. 222, *Mem. Geol. Surv.*
[2] J. F. Blake and W. H. Hudleston, 1877, *Q.J.G.S.*, vol. xxxiii, pp. 273-4.
[3] 1836, loc. cit., p. 22.

condition in shaly clays in the railway-cutting.[1] The lowest horizon of the Kimeridge Clay outcropping at the surface on Portland Island is the oil-shale with *Saccocoma*. Between the island and the mainland is a gap, the highest beds exposed in the cliffs south of Sandsfoot Castle belonging to the *Pararasenia mutabilis* zone. An indication of the nature of the intervening clays that occupy this gap is afforded by a number of septaria washed up on the shore of Portland Roads. These have yielded some seven species of *Aula-costephanus*, together with other ammonites, and they indicate a development similar to that at Ringstead Bay.

The low cliffs extending to the north towards Sandsfoot Castle, at the southern extremity of the mainland, begin with the lowest beds of the *mutabilis* zone and show a good section of the *cymodoce* and *baylei* zones. The junction of the latter with the top of the Corallian is not so easy to find here as in Ringstead Bay, but the Ringstead Coral Bed is represented by a thin band of oolitic ironstone, reminiscent of the Westbury Iron Ore, and this should certainly be regarded as the highest bed of the Corallian. The *Exogyra nana* Bed[2] is also conspicuous.

(d) North Dorset and the Vale of Wardour

After reappearing from beneath the Chalk of the Dorset Downs the Kimeridge Clay outcrops for about 20 miles through North Dorset, until it is cut off abruptly by the E.–W. fault along the north side of the Vale of Wardour. On the downthrown (north) side of this fault the Cretaceous rocks still overstep the Kimeridge Clay and Corallian, the edge of the Chalk and Greensand scarp resting upon the Oxford Clay. On the upthrown (south) side of the fault they have been stripped off.[3]

Concerning the Kimeridge Clay of this tract very little is known. The highest zones reach the surface only in a narrow belt within the Vale of Wardour, south of which the Cretaceous scarp, running westward to Shaftes-bury and then south-westward, oversteps on to progressively lower zones. Evidence was obtained by the Survey from a well-shaft between Okeford Fitzpaine and Shillingstone that in the south the Gault rests upon some part of the *Virgatosphinctoides* zones. They also proved the *Aulacostephanus pseudomutabilis* zone near Okeford Fitzpaine Church, in a brookside. The only clear section in the district is a large clay-pit south of the railway station at Gillingham, which exposes 25 ft. of clays falling entirely within the *Rasenia* zones. The clays are black and selenitic, and contain abundant *Exogyra virgula*, with *Lingula ovalis*, *Ostrea delta* and other shells, and bones of *Ophthalmosaurus pleydelli* Lyd. Dr. Pringle has recorded *Rasenia* cf. *stephan-oides* (Oppel), *R.* cf. *trimerus* (Oppel) and numerous specimens of [*Para-rasenia*] *desmonota* (Oppel).[4]

---

[1] H. Salfeld, 1914, loc. cit., p. 203.

[2] Salfeld's Bed 19, loc. cit., p. 202. Salfeld here places *Pictoniæ* below the *Exogyra nana* Bed, just as I have found them (in fragments) at Ringstead (see footnote on p. 451).

[3] Once the hard Chalk and Greensand had been removed, erosion of the Kimeridge Clay proceeded rapidly, so that now the Chalk on the downthrown side of the fault towers several hundred feet above the clay vale on the upthrown side.

[4] J. Pringle, 1923, 'Geol. Shaftesbury', *Mem. Geol. Surv.*, pp. 40–1; and *Sum. Prog.* for 1921, p. 112.

II. WILTSHIRE, BERKSHIRE, OXON. AND BUCKS

(a) Wiltshire

North of the Vale of Wardour the Kimeridge Clay is concealed for 11 miles beneath the extension of Salisbury Plain, formed by the overstepping of the Chalk westward to Mere, Warminster and Westbury. North of this the clay reappears for a few miles at the mouth of the Vale of Pewsey, where the *Pavlovia* zones have been worked in small brickyards, and where there are also (badly exposed) Portland rocks, as in the Vale of Wardour.[1] It is soon overstepped once more by Lower Greensand and Gault for 5 miles nearly to Calne. From Calne onwards the outcrop is uninterrupted until beyond the border of Berkshire; then at Faringdon a narrow tongue of Lower Greensand extends again on to the Corallian.

Throughout most of this tract, excepting the short distance in the Vale of Pewsey, the Kimeridge Clay is incomplete, being unconformably overlain by the Lower Greensand. Swindon, however, lies on a synclinal trough in which are preserved not only the higher Kimeridge zones but also the Portland and Purbeck Beds. The limestones form the capping of the hill on which the Old Town is built, while in the steep slopes below numerous brick-pits, opened to provide building materials during the rapid growth of the new railway town in the plain to the north, have at one time exposed almost the whole succession of the Kimeridge Clay. These pits, some of them of great size, yielded a rich harvest of vertebrate skeletons, belonging to *Ichthyosaurus*, *Ceteosaurus*, *Omosaurus* and rare Chelonia, some of which were described by Owen in his monographs of the Mesozoic Reptilia. Most of the vertebrate remains came out of the lower beds, especially the *Rasenia* and *Aulacostephanus* zones, as at Kimeridge.[2] The pits from which they came have for many years been abandoned.

Evidence has been obtained at Swindon for the presence of all the known zones of the Kimeridge Clay except that of *Subplanites*, up to and including that of *Pavlovia rotunda*, and less certainly that of *P. pallasioides*. Upon this the Portland Beds (*gorei* zone) rest non-sequentially, with a phosphatic pebble-bed at the base, containing derived fossils of *rotunda* and doubtfully *pallasioides* dates (the Upper Lydite Bed). Notwithstanding this apparently almost complete representation of its component parts, the formation reaches only some 300 ft. in thickness, or less than one-third of the total on the Dorset coast. This is in large measure due to the fact that the 400 ft. of the Dorset *pectinatus* and *Subplanites* zones are only represented by some 35–45 ft. of beds.

The most remarkable feature of the Kimeridge Clay at Swindon, a feature common also to the district about Oxford, is the development of the *pectinatus* zone in an arenaceous facies. Although all the other zones are represented by normal clays, the *pectinatus* zone consists of thick beds of sand with large doggers of hard calcareous gritstone. These have been called the Shotover Grit Sands, on account of their good development, with abundant

---

[1] A. J. Jukes-Browne, 1905, 'Geol. Country S. and E. of Devizes', *Mem. Geol. Surv.*, p. 5 (*Ammonites pallasianus* is said to have been 'one of the commonest fossils').

[2] *Tholemys passmorei* Andrews, found by Mr. A. D. Passmore in a temporary excavation and presented to the British Museum, was described as the most perfectly-preserved turtle ever found in the English Kimeridge Clay: C. W. Andrews, 1921, *Ann. Mag. Nat. Hist.* [9], vol. vii, pp. 145–53.

fossils, on Shotover Hill, Oxford. Their equivalence to a part of the Kimeridge Clay as low down as the oil-shale of Kimeridge was suggested in 1896 (see above, p. 447) by W. H. Hudleston, on the ground of the occurrence of *P. pectinatus* (Phil.), but 'the correlation was overlooked for twenty-five years. The true position was misunderstood by Salfeld, who correlated the Shotover Grit Sands at Swindon and Oxford with a part of the Portland Sands of Dorset above the *pallasioides* zone, and at the same time placed them above the Hartwell Clay (and the Atherfield Clay!).[1] Messrs. Chatwin and Pringle, of the Geological Survey, arrived independently, in 1921, at the same conclusion as Hudleston, and they, with the help of Hudleston's collection, gave the first accurate description of the sequence on Swindon hill.

SUMMARY OF THE SEQUENCE AT SWINDON [2]

Pavlovia Zones (SWINDON CLAY AND LOWER LYDITE BED).

On the floor of Okus Quarry, on the top of the west end of the hill, where the Portland Limestones are still actively quarried, a thin band crowded with dark phosphatic pellets and small lydite pebbles separates the glauconitic base of the Portland Stone (or strictly speaking the stony Portland Sand) from the clay below. Among the pebbles are commonly found rolled and phosphatized fragments of small ammonites belonging to the *rotunda* and possibly *pallasioides* zones, and it is concluded that these have been washed out of the

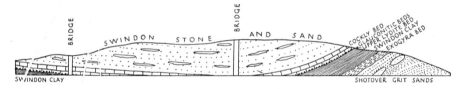

FIG. 79. Section in railway-cutting west of Swindon Town (Old Swindon) Railway Station. (After H. B. Woodward, 1895, *J.R.B.*, vol. v, p. 212, re-drawn.) This cutting, made in the early 'nineties, established finally the inferred succession of the strata at Swindon.

upper part of the Swindon Clay. This Upper Lydite Bed has a wide extent in Oxfordshire and Buckinghamshire, where it always marks the local base of the Portland Series, and was called by Hudleston the 'basal conglomerate of the Portlands'.[3]

The highest surviving Kimeridge stratum is the Swindon Clay, a blue clay, weathering brown, 15–20 ft. thick. The top forms the floor of Okus Quarry, but the lower parts, which are more sandy, are not now exposed, except occasionally in graves in the cemetery. The whole is very poorly fossiliferous, but Dr. Neaverson writes: 'The basal portion . . . consisting of a hard, greenish marl, contains fragments of ammonites which appear to be referable to *Pallasiceras* [= *Pavlovia*], and poorly preserved specimens are occasionally found in the blue clay which forms the mass of the Swindon Clay.'[4] Dr.

---

[1] Loc. cit., 1913, and 1914; see especially, 1914, p. 128, Table I.
[2] Based on C. P. Chatwin and J. Pringle, 1922, *Sum. Prog. Geol. Surv.* for 1921, pp. 162–8 (with emendations).
[3] W. H. Hudleston, 1880, *P.G.A.*, vol. vi, p. 345.
[4] E. Neaverson, 1924, *Geol. Mag.*, vol. lxi, p. 148.

Kitchin has more recently recorded that an excavation lately made in the top of the clay yielded ammonites indistinguishable from *Pavlovia pallasioides*.[1]

At the base is another 8-in. pebble-bed, the Lower Lydite Bed, resting on the

Pectinatus Zone (SHOTOVER GRIT SANDS).

The Shotover Grit Sands, 35–45 ft. thick, consist for the most part of yellow calcareous sands with enormous spherical doggers of hard gritstone, their old name being 'Lower Portland Sands'. Fossils are uncommon in all but the highest 6–8 ft., which consist of green and red marly sandstones, sands and clays, with abundant *Exogyra nana* and a rich fauna of other mollusca, including *Pectinatites pectinatus* (Phil.)[2] and *P. eastlecottensis* (Salf.).[3] Hudleston collected a fine series of fossils from a cutting during the construction of the M.S.W.J.Railway (fig. 79), but little below the Portland Beds can now be seen there. According to Messrs. Chatwin and Pringle:

'These upper beds were at one time (1887) well exposed in the road-cutting below Victoria Street, and the section was recorded by H. B. Woodward. A similar section was visible in May 1921 in a disused clay-pit south of Stafford Street; this pit has now been filled in, but from time to time the beds are exposed when graves are dug in the old cemetery, in the slope overlooking Clifton Street. Many fine uncrushed ammonites have been found in the material thrown out in grave-digging.'[4]

There is now a fine section showing sands with enormous doggers, overlain by the *Exogyra* Bed and Lower Lydite Bed, at the north-west end of the hill, above Hill's brickyard.

Virgatosphinctoides Zones.

The Grit Sands pass down gradually into clays, the junction having been exposed in Turner's upper brick-pit, west of Drove Road, on the north-east side of the hill, and a better section of the clays also in Hill's brickyard, below Okus Quarry. From these beds a number of ammonites with semi-virgatotome ribbing have been collected, which were formerly identified as *Virgatites*, but would now be called *Virgatosphinctoides*. There appears to be no indication of a separate *Subplanites* zone either here or in Oxfordshire. More detailed study of the ammonites is needed, however.

Gravesia Zones.

Messrs. Chatwin and Pringle record that a specimen of *Gravesia* exists in the Hudleston collection from Swindon, but the zone has never been identified *in situ*, and there is no longer any exposure likely to show it.

Aulacostephanus Zones.

In Turner's lower pit, known also as Bazzard's pit (now filled up with

---

[1] F. L. Kitchin, 1926, *Ann. Mag. Nat. Hist.* [9], vol. xviii, p. 452.

[2] A topotype of *A. pectinatus* Phil. (from Shotover Hill, Oxford) in the Oxford University Museum, has been figured by Buckman, *T.A.*, pl. CCCLIV; and another is figured here (pl. XXXIX).

[3] The type, figured *Q.J.G.S.*, 1913, vol. lxix, p. 430, pls. XLI, XLII, was believed by Salfeld to have come from the Upper Lydite Bed, and he made it the index of a separate zone above the Swindon Clay. Chatwin and Pringle proved that it came out of the upper part of the Shotover Grit Sands (loc. cit., p. 166). Buckman assigned it to a new genus, *Wheatleyites*, which Dr. Spath considers synonymous with *Pectinatites* (1931, loc. cit., p. 468).

[4] C. P. Chatwin and J. Pringle, 1922, loc. cit., p. 166.

rubbish and water, and having allotments over a large part of its floor), there were formerly exposed leathery bituminous shales, in which Salfeld recorded numerous crushed specimens of *Aulacostephanus pseudomutabilis* and *A. eudoxus*, together with *Aptychi*.

Rasenia and Pictonia Zones.

The presence of the *Rasenia* zones at Swindon is proved by some specimens collected by Hudleston from The Wharf, while Salfeld recorded that a fragment of *Pictonia baylei* had been found in Turner's lower pit (Bazzard's). At the base of an old clay pit in Telford Road he collected abundant *Priono-doceras serratum* (Sow.), which may have come from the base of the *baylei* zone or from the highest layers of the Corallian, though no *Ringsteadiæ* were found (see p. 395).[1] At the time work in Turner's lower pit was in active progress, a section was probably seen displaying clays of the *Pictonia*, *Rasenia* and *Aulacostephanus* zones, the two last yielding the vertebrate remains. The *Rasenia* and *Pictonia* zones are still visible in the brickyard at Stratton St. Margaret, north-west of the bridge where Ermine St. crosses the railway. They consist of brick-clays with *Ostrea delta* and some lines of septarian crackers containing the zonal ammonites well-preserved. The same zones, and the junction with the Upper Calcareous Grit, were formerly to be seen in an old brickyard at Wootton Bassett.[2]

During the construction of the Great Western Railway some of the longest and finest sections of the Kimeridge Clay in the country must have been seen in the vicinity of Swindon. Deep cuttings pass through the lower beds for several miles between Swindon and Wootton Bassett, and also near Stratton St. Margaret and South Marston; while near Bourton the upper beds were laid bare, and a small portion of them has recently been reopened in making a siding. Unfortunately no accounts of these cuttings were published.

Beneath the small Portlandian outlier of Bourton, near Shrivenham, the Upper Kimeridgian Beds are essentially the same as at Swindon, except that the top part of the Swindon Clay is sandy (see p. 508). A temporary excavation made in 1931 near Bourton End showed 3 ft. of Swindon Clay, greenish and sandy, with the Lower Lydite Bed at the base, resting on blue sands with large doggers, pierced to a depth of 4 ft. The lydite pebbles were exceedingly abundant.[3]

(b) Berkshire Vale and Oxford

Except for a momentary interruption near Faringdon, where a narrow strip of Lower Greensand, marked by high sandy ground, oversteps across the Kimeridge Clay on to the edge of the Corallian escarpment, the outcrop of the clay forms a continuous low-lying vale from Swindon to the Thames at Abingdon. This is the Vale of the White Horse in the narrower sense: the vale of the Gault and Kimeridge Clays, bounded on the north by the Corallian ridge, beyond which lies the wider but separate valley of the Thames.

Along the greater part of the Vale of the White Horse, the Gault and

[1] H. Salfeld, 1914, loc. cit., p. 196.
[2] The holotype of *Pictonia costigera* was collected here; see *T.A.*, 1927, pl. DCCXVI.
[3] Record kindly forwarded by my brother, Mr. J. O. A. Arkell, of Bourton End. I inspected the tip-heap, but found no fossils, only abundant lydite pebbles.

Kimeridge Clays are in contact, possibly because, as explained on p. 77, the outcrop obliquely crosses the continuation of the Birdlip Axis. The relations of the two are well seen at Culham brickyard, near Abingdon, where the *Douvilleiceras mammillatum* zone of the Gault (Middle Albian),[1] with a basal sandy pebble-bed, rests on grey sandy Kimeridge Clay, possibly of *pectinatus* date, or rather earlier. There are now very few other sections, though in days gone by the base of the Kimeridge Clay was to be seen in a brick-pit at Stanford in the Vale and a small exposure of the upper portion is still provided by another brick-pit at Drayton, near Abingdon.

On either side of the Thames south of Oxford, the outcrop extends northward for several miles up the dip-slope of the Corallian limestones, as long spurs with sides rising steeply from the plateau and capped by Lower Cretaceous sands. The western spur forms Boar's Hill and Cumnor Hurst; the eastern spur forms the lofty ridge that runs from Garsington to Shotover, and carries, in addition to the Cretaceous sands, a thick layer of Portland Beds with traces of Purbecks. At the end of each ridge there is a large brick-pit, the one at Shotover giving a nearly complete exposure of the Kimeridge Clay.

Eastward, around the south of Otmoor, the boundaries are obscurely defined owing to the dwindling of the Corallian limestones, but beyond the valley of the River Thame a third spur reaches northward to Brill and Muswell Hills. Here there are again exposures.

Still farther east the relics of a fourth spur, capped by outliers of Portland and Purbeck Beds and Cretaceous sands, survive as the hills at Quainton and Oving. The intervening country is dotted with a broken covering of Portland Beds, dissected by the Thame and its tributaries, and it presents in consequence an undulating scenery quite different from any to be found in other parts of the inland outcrop.

Seven miles north-east of Aylesbury the Kimeridge Clay is completely overstepped by the Gault and Greensand, and disappears under the Dunstable Downs, not to reappear for 32 miles.

Within this tract the best exposures are the brick-pits at the extremities of the two spurs on either side of Oxford, at Cumnor Hurst and Shotover, and a third brickyard on the eastern face of the Shotover ridge, at Littleworth, close to Wheatley. On the eastern side of Brill Hill, Prof. Morley Davies has described a valuable section of the lower beds at Rids Hill Brickyard, while the uppermost beds are or have been seen in other small brickfields at Long Crendon, Thame, Hartwell and Aylesbury. These exposures will be mentioned in connexion with the Hartwell Clay, a consideration of which is best left until the succession near Oxford has been described.

In the vicinity of Oxford the total thickness of the Kimeridge Clay is about 150 ft., although in places, as at Cumnor Hurst, it is still more reduced by pre-Cretaceous denudation.[2] This is only half the thickness at Swindon, but nevertheless the chief features of the Swindon succession, such as the Swindon Clay and the Shotover Grit Sands, are present, and the only zones which appear to be generally missing are those of *Pictonia*, *Gravesiæ*, and *Sub-*

---

[1] See Spath, 1923, 'Mon. Ammon. Gault', *Pal. Soc.*, part i, p. 4.
[2] Such denudation probably accounts for the low records of 94 ft. obtained in borings at Wantage and Culham [see 'Water Supply of Berks.', p. 89, and 'Water Supply of Oxon.', p. 42, *Mems. Geol. Surv.*].

planites. A substantial proportion of the total gap is probably represented, as at Swindon, by the Lower Lydite Bed.

At Chawley Brickworks, Cumnor Hurst, the Lower Greensand is seen resting on deeply channelled clays of the *Virgatosphinctoides* zone, about 20 ft. thick. In the past some relic of the *pectinatus* zone must have been encountered, for specimens of the zonal index fossil from this locality are preserved in the Oxford University Museum. The rest of the pit shows the upper *Aulacostephanus* zone, with *A. eudoxus*, *Exogyra virgula* and Saurian bones.[1]

The sections at Shotover and Wheatley Brickyards are much more complete and either here or at Brill the whole of the Kimeridge Clay has at one time or another been described.

SUMMARY OF THE SEQUENCE NEAR OXFORD, AT SHOTOVER, CUMNOR, WHEATLEY AND BRILL [2]

Pavlovia Zones (SWINDON CLAY AND LOWER LYDITE BED).

During 1930 and 1931 a sand-pit above the Shotover Brickyard has been worked farther back into the hill so that it now exposes not only the Swindon Clay but also the Upper Lydite Bed and the basal portion of the green Glauconitic Beds of the Portland Stone Series. The Upper Lydite Bed is about a foot thick and very conspicuous.

The SWINDON CLAY is a greenish-grey sandy clay and has at the base, as at Swindon and Bourton, the LOWER LYDITE BED. The full thickness of the clay is difficult to estimate, for it has acted as a slide-plane upon which the superincumbent Portland Stone has slipped down the hill-side, and it is in consequence much disturbed. The two lydite beds are only some 5 ft. apart, but an unknown thickness of the clay at the point where it is seen may have been squeezed out, since the beds are inclined at a high angle, in places approaching the vertical.

At Wheatley 12 ft. of the Swindon Clay[3] are exposed, but the top is not seen. Here small brown nodules occur in the clay, not uncommonly yielding fragmentary ammonites of the type of *Pavlovia rotunda* (Sow.). Fossils seem to be no longer found at Shotover, but Dr. Douglas observed in 1910 that 'a number of Kimeridge fossils, including the species usually termed *A. biplex*, were obtained from this clay-band'.[4] The species usually termed *A. biplex* at that time comprised the 'genera' *Pallasiceras*, *Lydistratites*, *Holcosphinctes* and *Aposphinctoceras*, all of which Dr. Spath considers synonymous with *Pavlovia* Ilovaïsky.[5]

Pectinatus Zone (SHOTOVER GRIT SANDS).

At the type-locality, the sand-pit above the brickworks on Shotover Hill, these consist of 16 ft. of yellow and whitish coarse sand with large doggers,

[1] J. Pringle, 1926, 'Geol. Oxford', 2nd ed., *Mem. Geol. Surv.*, p. 68.
[2] Based on J. Pringle, 1926, loc. cit., pp. 66–75.
[3] Buckman's name for the Swindon Clay here was Littleworth Lydite Clay.
[4] J. A. Douglas, 1910, *Geol. in the Field*, p. 206.
[5] The type of *Aposphinctoceras decipiens* Neaverson was figured by Miss M. Healey as '*Olcostephanus pallasianus* (d'Orb.)', *Q.J.G.S.*, 1904, vol. lx, p. 60, pl. XII, figs. 1 and 2, and came from Chippinghurst, Oxon., apparently from the Swindon Clay. This was previously known as *A. biplex*.

some of them resembling enormous cannon-balls. The sands and doggers, especially towards the top, contain abundant small lydite pebbles. The doggers are crowded with shells, of which the commonest are large Pernas (*Isognomon bouchardi* and another species), *Ostrea expansa* Sow., *Exogyra thurmanni*, a *Quenstedtia*, and various other bivalves. Among ammonites it suffices to mention those species of which this is the type-locality: *Pectinatites pectinatus* (Phil.),[1] *Paravirgatites pringlei* Buck.,[2] *Paravirgatites paravirgatus* Buck.,[3] and the horned *Pectinatites* ('*Keratinites*') *naso* Buck.[4] The name Shotover Grit Sands serves to distinguish them from the prior-named Shotover Sands of Lower Cretaceous (Wealden) age, which cap the hill.[5] Buckman also distinguished a Shotover Fine Sand, but this is probably no more than a local variant of the Shotover Grit Sands, which are here understood as including all the sands between the Swindon Clay and the *Virgatosphinctoides* zones.

At Wheatley brick-pit the Shotover Grit Sands are only $9\frac{1}{2}$ ft. thick, consisting of 4 ft. of friable brown sandstone, crowded with shells, overlying $5\frac{1}{2}$ ft. of lilac sandy clay. From the soft sandstone the fossils fall out in perfect condition, making the collection of specimens a much easier task than from the iron-hard doggers at Shotover. Very perfect *Trigoniæ* and other fossils may be obtained, forming a noticeably different assemblage from that at Shotover, but containing the same ammonites with numerous additions. From this bed were obtained the type specimens of seven species of *Pectinatites* (figured by Buckman as '*Wheatleyites*' and '*Keratinites*').

Virgatosphinctoides Zones.

Beneath the sands at Shotover about 10 ft. of blue-black clay is dug for bricks, and out of this come large irregular nodules or crackers, containing white-coated fossils of the *Virgatosphinctoides wheatleyensis* zone.

At Wheatley, the two *Virgatosphinctoides* zones together consist of some 18 ft. of dark clays, with layers of septarian nodules or cementstone crackers, from which Dr. Neaverson has obtained numerous ammonites. These nodules at Wheatley are the source of the types of the following:

Virgatosphinctoides wheatleyensis Neav., 1925, pl. I, fig. 1.	*Allovirgatites tutcheri* Neav., pl. III, fig. 2.
V. delicatulus Neav., pl. I, fig. 2.	*A. robustus* Neav., pl. III, fig. 3.
Sphinctoceras crassum Neav., pl. II, fig. 1.	*A. versicostatus* Neav., pl. III, fig. 4.

while at Shotover were obtained the types of *Virgatosphinctoides nodiferus* Neav. (pl. IV, fig. 1), *Sphinctoceras distans* Neav. (pl. IV, fig. 3), *Allovirgatites woodwardi* Neav. (pl. III, fig. 1). The *V. nodiferus* came from the clay above the nodule band, immediately below the Shotover Grit Sands, and on the strength of this Dr. Neaverson recognizes a separate zone of *V. nodiferus*; but it remains to be shown whether it can be recognized elsewhere.

At Chawley Brickworks, Cumnor Hurst, the clays on which the Lower Greensand is now seen to rest belong to the *Virgatosphinctoides* zones, which may be about 22 ft. thick. They contain the same cementstone crackers,

---

[1-4] See *T.A.*, pls. CCCLIV, DLXII, CCCVIII, DCLII. Dr. Spath considers Buckman's genus *Shotoverites* (type *S. pringlei*) identical with *Paravirgatites* (loc. cit., 1931, p. 472).

[5] The freshwater Wealden sands were so named by Prestwich and other early writers; the alienation of the name for the Kimeridgian dates from Blake (1880, pl. VIII). (See G. W. Lamplugh, 1908, 'Geol. Oxford', *Mem. Geol. Surv*, p. 68; and 2nd ed., p. 87.)

with *V. wheatleyensis*, *Allovirgatites tutcheri*, *Modiola* (*Musculus*) *autissiodorensis*, &c. The zones are also exposed at Rids Hill, Brill, where they are the highest seen in the old brick-pit.

Gravesia Zones.

No evidence for the presence of either fauna or strata of *Gravesiæ* date has been found in the Oxford district, and there seems no doubt that the zones are wanting.

Aulacostephanus Zones.

At Wheatley, Cumnor and Rids Hill Brickyards the *Virgatosphinctoides* clays rest directly upon dark shaly clays of the *Aulacostephanus* zones, with *Exogyra virgula* and *Aptychus latus*. To the base of the pit at Wheatley 8 ft. are seen, most of which belongs to the *pseudomutabilis* subzone, since *A. eudoxus* abounds; but Dr. Pringle has recorded *Physodoceras acanthicum* and *Ph. karpinskii*, from which he concludes that the top of the subzone of *A. yo* is also exposed.

At Cumnor 10 ft. of the same dark, shaly clays with *E. virgula* and *Aptychus* have been seen in the north end of the pit, but they are not visible on the main face to the south, being there thrown down below the floor by a fault. The zone at this place yields numerous Saurian bones, just as at Swindon and at Kimeridge. Remains of *Pliosaurus*, *Plesiosaurus*, *Dacosaurus* (also found at Shotover) and *Ichthyosaurus* have been obtained, and also the type-specimen of *Camptosaurus prestwichi* Hulke. *A. eudoxus* and allied ammonites are tolerably abundant.

At Rids Hill Brickyard, again, in the same clays, Dr. Pringle has obtained the characteristic fossils, *A. eudoxus*, *Aptychus latus* and *Exogyra virgula*.

Rasenia and Pictonia Zones.

The only section of the basal clays below the *Aulacostephanus* zones now open seems to be the old brickyard at Rids Hill, which is becoming obscure since being abandoned in 1911. The section was described by Prof. Morley Davies in 1907 and again by Dr. Pringle in 1924.[1] Immediately below the *Aulacostephanus* zone are two 1 ft. bands of creamy-weathering cement-stone with fragments of *Rasenia*, enclosing 2 ft. of creamy, calcareous clay; below this are 8 ft. of dark grey selenitiferous clay with *Rasenia stephanoides* (Oppel), *Ostrea delta*, and ancestral forms of *Exogyra virgula*.

At the base is what Prof. Morley Davies described as the most fossiliferous bed in the pit—the BRILL SERPULITE BED. It consists of an impersistent band of limestones or flattened doggers, up to 6 in. thick, largely composed of *Serpulæ* and shells. The stone is so tough that while the smaller bivalves such as Cyprinids and *Astarte* usually spring out perfect when it is broken, the larger belemnites and oysters generally shiver to pieces. When the masses are weathered, however, a better idea of the fauna can be obtained. *Serpula tetragona* (Sow.) and the little '*Cyprina*' cf. *cyreniformis* Blake predominate, but *Ostrea delta* and other lamellibranchs also abound,[2] and Buckman has figured from the bed a specimen of *Prionodoceras superstes*.[3] Since this ammonite

---

[1] 1926, loc. cit., p. 74. [2] A. M. Davies, 1907, *Q.J.G.S.*, vol. lxiii, p. 34.
[3] *T.A.*, 1923, pl. CDXXII.

has been found in association with *Pictoniæ* in Scotland,[1] Dr. Pringle suggests that the Brill Serpulite Bed may possibly be a representative of the *Pictonia baylei* zone. He detected the bed *in situ* for the first time in 1924, at the base of the *Rasenia* zones, and he therefore considered the Serpulite Bed the basal bed of the Kimeridge Clay. Below were exposed 14 ft. of black selenitiferous clay probably belonging to the upper part of the Ampthill Clay (Upper Calcareous Grit-equivalent).

In no other exposure near Oxford has so much as a suggestion of the *Pictonia* zone been found. As mentioned in the last chapter, during the reconstruction of the London road in 1925, Dr. Pringle saw a deep trench east of Shotover Lodge, in which clays belonging to the zone of *Rasenia cymodoce* rested directly on a markedly eroded surface of the Corallian limestone. In the old quarry below the brickyard at the foot of Shotover Hill a similar section has only lately become overgrown. Phillips noted that about 15 ft. from the base of the clay there was a band of septaria yielding *Rhactorhynchia inconstans*, while below were layers of *Ostrea delta*, *Exogyra nana* and *E. virgula* (Sowerby's type specimens of the first two oysters came from here). At the base was a layer of coprolites.[2]

The non-sequence in the Oxford district denoted by this hiatus is very considerable, for not only is the *Pictonia baylei* zone missing, but also all the Upper Calcareous Grit and Glos Oolite Series of the Corallian (see p. 403).

The difference between the Shotover Grit Sands or *pectinatus* zone at Shotover and at Wheatley indicates that lateral variation is likely to be considerable. In the Horsepath tunnel, by which the railway passes under Shotover Hill, and in the adjacent cuttings, the sands seem to be much thicker and they have at their base two bands of shelly limestone, $1\frac{1}{2}$ to 2 ft. thick and 4 ft. apart, which run all through the cuttings. No Swindon Clay seems to be present.[3] A comparable section was made out by Prof. Morley Davies in a long field trench dug down the side of the hill near Garsington.[4] The sands were seen to be about 33 ft. thick and to become argillaceous towards the base. They were capped by a well-marked Lydite Bed, which was directly overlain by the green Glauconitic Series of the Portland Beds. The increased thickness of the sands suggests that the Swindon Clay has in this direction passed into sand.

Exactly comparable sections, where the green Glauconitic Beds with their basal Lydite Bed rest directly on some 30 ft. of sands (the Thame Sands of Buckman), had been described by Fitton farther east, at Long Crendon and at Barley Hill, east of Thame[5] (see below).

(c) The Thame and Aylesbury District (Hartwell Clay Area).

In the district about Thame and Aylesbury the best sections to be obtained are afforded by shallow brickyards, most of which are opened in a sandy, greenish clay, usually more or less micaceous and glauconitic—the HARTWELL CLAY.

[1] S. S. Buckman, 1923, *T.A.*, vol. iv, p. 42.
[2] J. Phillips, 1871, *Geol. Oxford*, p. 413.
[3] J. Pringle, 1926, loc. cit., p. 71.
[4] A. M. Davies, 1899, *P.G.A*, vol. xvi, p. 20.
[5] W. H. Fitton, 1836, *Trans. Geol. Soc.* [2], vol. iv, pp. 281–3.

The type-section is Locke's Brickyard, Hartwell, about a mile south-west of Aylesbury, beside the Thame road. Almost a duplicate section was formerly to be seen at Webster and Cannon's (Hill's) Brickyard, on the Bierton Road, north-east of Aylesbury, and in Ward and Cannon's Brickyard at Bierton. At these places the Hartwell Clay is or was exposed to a depth of about 10 ft., and is directly overlain by the Upper Lydite Bed and Glauconitic Series of the Portland Beds.[1] As early as 1880 Hudleston called this lydite bed the 'basal conglomerate of the Portlands' and correlated it with the similar stratum at Swindon; and in the passage already quoted (on p. 448) written in 1896, he definitely correlated the Hartwell Clay with the Swindon Clay. There was at that time no certainty on this point, however, for the fauna had not been properly studied, and the strata below the Hartwell Clay were, and still remain to-day, virtually unknown. Hudleston mentioned a report that a sandy stratum lay between the Hartwell Clay and the lower shaly clays on Aylesbury Hill, but he said that the information was unreliable. There is thus at least an indication that the Shotover Grit Sands may have some representation as far east as Aylesbury.

Concerning the stratigraphical position of the Hartwell Clay there have been widely divergent opinions: Blake in his paper on the Kimeridge Clay of England (1875) first placed it in the Lower Kimeridge, but five years later, probably under the influence of Hudleston, who in that year (1880) conducted an excursion to the district, he admitted his mistake and ascribed it to the Upper Kimeridge. H. B. Woodward in 1895 went to the other extreme and described the Hartwell Clay as part of the Portland Beds. In this he was probably misled by the then recent discovery that the Middle and Upper Kimeridge Clay of England were to be correlated with the Lower and Middle Portlandien of French geologists, but it is plain that he used the term Portland Beds in the English sense, not in the Continental. Apart from matters of nomenclature, he agreed with Hudleston, for he wrote: [2]

'There can be no question that the clay below the Portland Stone at Swindon is homotaxial with the Hartwell Clay, for although the Swindon Clay is not so fossiliferous as that at Hartwell, yet it has yielded some species, and the beds immediately below the Swindon Clay have yielded many fossils identical with those of the Hartwell Clay.'

In recent years Dr. Neaverson has collected many ammonites from the Hartwell Clay and has studied them and the collections in the Aylesbury Museum and elsewhere, with the result that the date of the deposit has been settled beyond much doubt as principally *pallasioides*, the lower parts including also deposits of *rotunda* date. In the monograph, already mentioned, on the ammonites from the Aylesbury–Oxford district (1925), he has described and figured the following forms from the Hartwell Clay:

[3] *'Aposphinctoceras' ailesburiense* Neav., pl. II, fig. 3.
[3] *'A.' hartwellense* Neav., pl. II, fig. 4.
[3] *'A.' variabile* Neav., pl. II, fig. 5.
'Holcosphinctes' pallasioides Neav., pl. III, fig. 5.
'H.' flexicostatus Neav., pl. III, fig. 6.

[1] See W. H. Hudleston, 1880, *P.G.A.*, vol. vi, pp. 344–52; and 1888, ibid., vol. x, pp. 166–72.
[2] 1895, *J.R.B.*, p. 223.
[3] Buckman did not consider these species congeneric with the genotype, which was chosen

while from the lower parts of the clay, obtained at a time when the brickyards were worked to a deeper level, he has described two representatives of the *rotunda* fauna: *P. ultima* (pl. I, fig. 2) and *P. inflata* (pl. II, fig. 2).

Since the discovery, announced by Dr. F. L. Kitchin, of *P. pallasioides* in the upper part of the Swindon Clay,[1] it is evident that the Swindon Clay and Hartwell Clay can be closely correlated, although the greater part of the Swindon Clay is probably of *rotunda* date and represents that portion of the Hartwell Clay lying below the usual floor of the brickyards. At Wheatley and Shotover, on the other hand, the upper or *pallasioides* portion of the clay seems to have been removed in early Portland times, before the formation of the Upper Lydite Bed.

Much additional information of interest might be obtained from a study of the lamellibranch faunas of these uppermost clays; for instance Dr. Kitchin finds that the most characteristic Hartwell species, *Hartwellia hartwellensis* (Sow.), recently described by him as the type of a new genus, is absolutely restricted to this level, although other allied species have been wrongly recorded under the same name from the *pectinatus* and even the *Virgatosphinctoides* zones at Swindon and Shotover.

Before leaving the subject of the Hartwell Clay, reference must be made to an opinion persistently held by Buckman, that the *pallasioides* fauna of the Hartwell Clay should be sought *below* the Shotover Grit Sands of the Oxford district and of Swindon. To the last he maintained that the Survey and Dr. Neaverson were wrong in their correlations.

'There is no objection taken', he wrote in August 1926,[2] 'to the correlation of the *rotunda* zone of Chapman's Pool with the Swindon Clay and with the Littleworth Lydite Clay; but when they also correlate these deposits with the Hartwell Clay of Hartwell, near Aylesbury, serious protest must be made; because the stratigraphical sequence of the beds is quite opposed to it. . . . The likeness of the Ammonoids of the Hartwell Clay and of Chapman's Pool is admitted; but it is a deceptive likeness.'

Again, in December 1926,[3] he wrote:

'There are at Swindon, below the *pectinatus* beds, a thick series of sands (Lower Cemetery Beds) above some feet of clay, whose ammonoid faunas are unknown: these beds are where the Hartwell Clay faunas should be sought. . . . There is little to be gained in reiterating statements of correlation without figuring the ammonoid evidence. I have figured such evidence for the correlations of the beds concerned.'

Any one reading these passages without knowing the history of the controversy might be led to believe that the weight of evidence was on Buckman's side. In 1926, however, at the meeting of the British Association at Oxford, Buckman read a paper upon the position of the Hartwell Clay. The paper was never published and so, in the interests of truth, a word of explanation may be timely.

Cross-questioning by Dr. Pringle elicited the information that the critical ammonites from Long Crendon, upon which the correlations depended, had

by Neaverson as *A. decipiens, nom. nov.* for the specimen figured by Miss Healey as *Olcostephanus pallasianus* in 1904, from the Swindon Clay of Chippinghurst, near Chiselhampton, Oxon. See *T.A.*, vol. vi, 1926, p. 25 (where references are given). Dr. Spath, on the other hand, considers not only all these but also the allied genera *Holcosphinctes* and *Episphinctoceras* Neaverson synonymous with *Pavlovia* (1931, loc. cit., pp. 470–1).

[1] F. L. Kitchin, 1926, *Ann. Mag. Nat. Hist.* [9], vol. xviii, p. 452.
[2] *T.A.*, vol. vi, p. 27.   [3] Ibid., p. 40.

not been dug up either by Buckman or in his sight, but had been purchased long afterwards from workmen, who had also worked in the pits at Hartwell and Brill. They had, therefore, probably never come from Crendon at all, but from the Hartwell Clay of Hartwell.

The sections at Long Crendon, on the hill leading down towards Thame (Barrel Hill), were described by Fitton in his classic memoir on the Strata below the Chalk,[1] and were re-examined in 1898 by Prof. Morley Davies, who confirmed Fitton's sequence.[2] Underlying the typical Portland Limestones are the usual Glauconitic Beds, with the Upper Lydite Bed at the base. Below come 30 ft.[3] of predominantly sandy strata 'with much clayey material in some of the beds' according to Morley Davies, and these rest in turn upon clay, which was formerly dug in the old brickyard near the foot of the hill. Nothing is now visible in the brickyard, the adjoining sand-pit has been filled up with rubbish, and the road-cutting is overgrown.

Similar conditions—Lydite Bed resting directly on sands, as at Garsington —were proved by Morley Davies in drainage excavations in Thame, and Fitton described the Glauconitic Beds resting directly on 30 ft. of sands at Barley Hill, east of Thame. Here the sand (Buckman called it Thame Sands) contained 'nodules of great size scattered in it irregularly', and they reminded Fitton of the doggers in the Shotover Grit Sands at Shotover.[4]

Buckman admitted the approximate contemporaneity of at least a part of the Thame Sands and the Shotover Grit Sands, but his point of divergence from the view accepted by all other recent investigators lay in the belief that the clay below the Thame Sands in the old brickyard at Long Crendon was Hartwell Clay. Blake referred to it as Hartwell Clay when leading an excursion of the Geologists' Association to the brickyard in 1893, but his contribution to Kimeridgian stratigraphy was never profound, and he gave no reasons for the opinion. Prof. Morley Davies also visited the brickyard while it was still in work, but he found no fossils, and a workman 'although he certainly knew what fossils were' informed him that none were found.

To settle the matter an excavation was made in the floor of the old brickyard at Long Crendon, and in the presence of Messrs. Buckman, Pringle and Chatwin, Dr. Pringle informs me, the workmen unearthed some ammonites typical of the *Virgatosphinctoides* zones, but nothing at all suggestive of Hartwell Clay. About 6 ft. below the floor of the pit appeared the *Exogyra virgula* clays.

The true explanation may be that, northward and eastward from Garsington to Shotover Brickyard and Wheatley, the upper part of the sands below the Upper Lydite Bed passes into the sandy Swindon Clay, and again that northward and eastward from Thame and Long Crendon to Brill and Hartwell it passes into the sandy Hartwell Clay. H. B. Woodward appears to have been believed in such a change of facies, for he wrote:[5]

'It is evident that the "sands of the Lower Portland Beds" [i.e. Thame Sands] are gradually replaced by clay as we proceed northwards [from Long Crendon]. At Brill, beneath the lydite-bed, there is 3 ft. of brown and greenish sand, which passes

---

[1] 1836, *Trans. Geol. Soc.* [2], vol. iv, pp. 281–2.
[2] A. M. Davies, 1899, *P.G.A.*, vol. xvi, p. 21.
[3] Buckman's measurements make the sands 80 ft. (*T.A.*, vol. vi, p. 36), but this thickness is improbable.
[4] W. H. Fitton, 1836, loc. cit., p. 283.   [5] 1895, *J.R.B.*, p. 225.

down gradually into stiff blue clay, yielding *Ammonites biplex*, *Thracia*, &c., and is not separable from the Kimeridge Clay.'

Prof. Davies expressed the same view somewhat differently, as follows:

'If we consider the . . . beds below the Pebble Bed [Upper Lydite Bed] we can trace a gradual transition, the clayey facies gradually rising from west to east. From the 39 ft. of pure sands with limestone at Garsington, we pass through the 26 ft. of more or less clayey sands of Long Crendon [Thame Sands] to the sandy clay of Dadbrook Hill and Hartwell [Hartwell Clay].'[1]

An alternative explanation, suggested and preferred by Dr. Kitchin, is that at Garsington, Thame and Long Crendon the non-sequence below the Upper Lydite Bed is greater than at the other places, and the Hartwell Clay has been removed by erosion, leaving the Upper Lydite Bed resting on Shotover Grit Sands.[2] If this view is correct the whole of the Thame Sands should one day be found to contain the *pectinatus* fauna.

III. THE FENS: CAMBRIDGE AND NORFOLK

After a gap of about 32 miles the Kimeridge Clay gradually emerges once more from beneath the Lower Greensand at Great Gransden, 10 miles west of Cambridge.[3] The outcrop remains narrow until due north of Cambridge, where it spreads out to a great width and is lost beneath the Fens. The eastern half of the Fens is floored by the Kimeridge Clay, which rises through the superficial deposits here and there as low, straggling 'islands'. The largest is the Isle of Ely, with its cappings of Lower Greensand, and other large islands have determined the sites of Chatteris and March. In Norman times the artificial causeways were the only means of communication between these isolated tracts and the mainland.

The area was studied principally by T. Roberts of Cambridge, previous to the year 1886,[4] but much additional information has been obtained from borings made during and since the Great War.

The maximum thickness of the Kimeridge Clay was estimated by Roberts to be 142 ft., this figure being arrived at by piecing together a composite section from the principal brick-pits in Cambridgeshire. He was in doubt, however, about 20 ft. of clays, which he considered he might have included twice, at Haddenham and at Littleport. In 1920 three boreholes were drilled in the Fens east of Southery, just over the border of Norfolk, and the results were published by Dr. J. Pringle.[5] The whole of the Kimeridge Clay was pierced in two of the borings, capped by Lower Greensand, and the total thickness was 121 ft. in one boring and 125 ft. in the other.

Pleistocene or Lower Cretaceous deposits rest unconformably on the clay. The Sandringham Sands of the Lower Greensand have at their junction with it a pebble-bed containing phosphatic nodules. The thinness of the Kimeridge Clay, however, is not due primarily to truncation of its upper zones by

---

[1] 1899, *P.G.A.*, vol. xvi, p. 25.   [2] F. L. Kitchin, 1932, *in lit.*

[3] The *Pictonia* zone may appear sporadically much earlier, especially about Ampthill, where *Perisphinctes variocostatus* (Buckland) is said to have been found in clay grouped as Ampthill Clay, and according to Buckman indicates basal Kimeridgian. At Sandy, however, the Kimeridge Clay is entirely overstepped and Lower Greensand rests on basal Corallian.

[4] T. Roberts, 1892, *Jurassic Rocks of Cambridge*, pp. 61–76.

[5] J. Pringle, 1923, *Sum. Prog. Geol. Surv.* for 1922, pp. 126–39.

the pre-Cretaceous denudation, but to piecemeal attenuation of its component parts, as in Oxfordshire. In the Southery borings all the main zones were detected. The formation consists of dark clays with occasional thin bands of cementstone and some brown and greenish-brown oil-shales, just as in the South of England. The thinness would seem, therefore, to be due mainly to slow deposition.

According to Dr. Pringle, the basement-bed of the Sandringham Sands rests upon some part of the *rotunda* zone, for within 4 ft. of the top of the clay fragments of an ammonite were found which compare with some of those abounding in the *rotunda* zone at Chapman's Pool. Some better-preserved examples were found in another boring at King's Lynn.

Only 13 ft. below this horizon at Southery was encountered a shale, 8 in. to 1 ft. thick, crowded with the pyritized radial plates of *Saccocoma*, associated with fish teeth. These crinoid remains occur in Dorset, as has been stated, on the horizon of the Kimeridge Oil-Shale. Twelve feet below the *Saccocoma* band was a cementstone, which, since it marked the lower limit of *Modiola autissiodorensis* and the upper limit of *Exogyra virgula*, Dr. Pringle correlated with the Yellow Ledge Stone Band of Kimeridge.

The *Gravesia* zones were not proved palaeontologically, but there was room for them in the borings, for 30 ft. below the cementstone the *Aulacostephanus pseudomutabilis* zone was proved by a layer of *Amœboceras krausei* (Salf.).

The *Pararasenia mutabilis* fauna was represented by *Pararasenia desmonota* (Oppel), about 18 ft. above the base of the Kimeridge Clay at Southery, while *Rasenia stephanoides* was obtained from several borings near King's Lynn. At the base were 6 ft. of grey limy clay with fragments of smooth-whorled ammonites, identified with some hesitation as *Pictoniæ*.

Thus the bulk of the formation from the *rotunda* zone downwards seems to be represented in the Fenland, and a re-examination of the brickyards of Cambridgeshire on modern palaeontological lines might yield much new information about the several zones, their thicknesses and faunas.

The interpretation of Roberts's descriptions in the light of the borings is a task which could only be undertaken after a study of his fossils and other material, and detailed collecting from the pits.

In the first place it is evident that nearly all the sections described by him fall in the Lower Kimeridge Clay, below the level of the Yellow Ledge Stone Band. The opinion formerly held, that at the Roslyn (or Roswell) Pit, 1 mile north-east of Ely, some 8 ft. of paper-shales and clays with *Orbiculoidea latissima* are above the limit of *Exogyra virgula*, is not sustained by recent work. Dr. Kitchin and Dr. Pringle inform me that they have carefully examined the highest beds there and have found that their position is not above the zone of *Aulacostephanus pseudomutabilis*. The principal *Exogyra virgula* beds, which (Roberts noted) are also characterized by abundance of *Aptychus latus* at a certain level, we may identify with little hesitation as belonging to the *Aulacostephanus* zones. At the Roslyn Pit these beds occupy about 26 ft., and the main *E. virgula* bed is within 3 ft. of the top. Roberts used as zonal index of these *Aptychus* and *virgula* beds the comprehensive '*Ammonites alternans*'.

The lowest 4 ft. of clay and shale exposed from time to time in the Roslyn Pit are marked by a band crowded with *Astarte supracorallina*, and from this level Dr. Pringle has recorded *Pararasenia desmonota* (Oppel), characteristic of the *mutabilis* zone.[1] Normally the floor of the pit is under water.

The downward continuation was believed by Roberts to be represented in three clay-pits at Littleport. If these three pits do not overlap more than Roberts supposed, the *Astarte supracorallina* level is underlain by some 40 ft. of clay; but this needs substantiating by more intensive collecting.

The basement-beds were described in a brick-field half a mile west of Haddenham Railway Station, and in several temporary exposures near the west side of the outcrop. They always contain layers of *Ostrea delta*, and at the base a band of phosphatic nodules, thought to be coprolites. Where the base of the Kimeridge Clay rests upon the coral rag of Upware there is reported to be at the junction a layer of broken and rolled fragments of coral, but the clay is there unfossiliferous. The abundance of *Ostrea delta* and the layer of phosphatic nodules are said to be otherwise of widespread occurrence and to provide an easily recognized guide in mapping the junction of Kimeridge and Ampthill Clays. They recall the Corallian-Kimeridge junction at Shotover, Oxford.

IV. LINCOLNSHIRE: FROM THE WASH TO THE HUMBER

Beneath the Wash and the lowest tract of the surrounding Fens the Kimeridge Clay is totally concealed for 30 miles. On reappearing in Lincolnshire it has more than doubled in thickness, the accepted estimate from a number of borings in the south and middle of the county being 300–320 ft.[2] The most recent boring, at Donnington on Bain, which was undertaken in 1917 in search of oil-shale, stopped short after passing through 245 ft. of Kimeridge Clay, without reaching the bottom.[3] The lowest zone for which any fossil evidence was obtained was that of *Aulacostephanus pseudomutabilis*, at a level 66 ft. above that at which the boring was stopped. In an older boring at the same place the Kimeridge Clay was said to have been penetrated for 309 ft.

In the north of the county the Lower Cretaceous rocks gradually overstep on to successively lower levels within the clay, but the overstep has never been investigated quantitatively. Owing both to the poverty of exposures and to the difficulty of drawing a lower boundary between the Kimeridge and the Ampthill Clays in the north of the county, the extent to which the various zones may individually or collectively attenuate is unknown. All that can at present be said is that the Kimeridge Clay, like the Oxford, becomes much thinner towards the Humber.

The highest beds passed through beneath the Spilsby Sandstone in the Donnington boring were unrepresented by cores; but at 30 ft. from the top Dr. Pringle recognized several specimens of a finely-ribbed Perisphinctid identical with some which characterize the oil-shale at Kimeridge (*Subplanites*?). Two bands of brown bituminous shale were in fact encountered at 21 and 30 ft. from the top. The same forms of ammonite were also recorded at Acre House Mine, north of Claxby.

Perhaps to some part of the Upper Kimeridge, if not to the *Aulacostephanus*

[1] J. Pringle, 1923, loc. cit., p. 135.
[2] H. B. Woodward, 1904, 'Water Supply of Lincs.', p. 10, *Mem. Geol. Surv.*
[3] J. Pringle, 1919, *Sum. Prog. Geol. Surv.* for 1918, pp. 50–2.

zones, also belong about 15 ft. of paper-shales exposed in a brickyard about a mile west of Fulletby.[1] The bedding-planes of the shale are here crowded with white shells of *Orbiculoidea latissima* and *Lucina minuscula*, recalling the paper-shales at Ely, formerly thought to be of Upper Kimeridge date but now assigned to the *Aulacostephanus* zones.

Certainly the majority of the sections in Lincolnshire fall within the Lower Kimeridge Clay. The general succession seems to be normal, consisting of the usual dark clays with some cementstone bands, but much work still remains to be done on the palaeontology. An extremely interesting feature is the representation of the *Gravesia* zones, indicated by the raising of specimens of supposed *Gravesia* in the Donnington boring from a depth of 70 ft. below the top of the Kimeridge Clay. At 179 ft. from the top was found *Aulacostephanus eudoxus*. The absence of *Exogyra virgula* is remarkable, and seems to be a characteristic feature of the tracts north of the Wash. Bands of brown bituminous shale, on the other hand, occur more frequently and at much lower horizons than in any other part of England. They closely resemble the 'Kimeridge Coal' and give a similar yield of oil and other products, all strongly contaminated with sulphur. The lowest bands fall within the zone of *Aulacostephanus pseudomutabilis*.

The beds below the *Aulacostephanus* zones are best known from the renowned brickyards of Market Rasen, the type-locality of the genus *Rasenia* where the ammonites are in beautiful iridiscent and pyritic preservation. The commonest species here is *Rasenia cymodoce*. The zone has also been worked at Brigg and at Horncastle, where in addition *Pictoniæ* are recorded. Roberts noted that *Astarte supracorallina* has a longer range in Lincolnshire than in Cambridgeshire, occurring throughout the Lower Kimeridge; and that '*Ammonites alternans*', which he used as a zone fossil, attains its acme at a lower level. (It would include *C. kitchini*, *C. pingue*, and *C. cricki* of Salfeld.)[2]

The basement-beds of the Kimeridge Clay were recognized by Roberts at Woodhall Spa, West Barkwith and North Kelsey, crowded as usual with *Ostrea delta*.

V. THE MARKET WEIGHTON AXIS

It is certain that the Kimeridge Clay, like the Oxford and Corallian Clays, becomes extremely thin on approaching the Market Weighton Axis, for in South Yorkshire the three together measure little more than 100 ft. in thickness. Owing to the lack of exposures no boundaries have been drawn between the three formations, but 'in a general way, the Oxford Clay may be taken as occupying the flatter part of the ground just above the Kellaways Rock, while the Kimeridge Clay forms the steep slopes beneath the Chalk escarpment'.[3] Definite evidence for the presence of Kimeridge Clay is scanty, and it is improbable that any is present north of Drewton. Between the Humber and Drewton the Cretaceous rocks probably overstep all but the lowest zones,[4] so that the full thickness is nowhere present and the effects of the Market

---

[1] J. F. Blake, 1875, *Q.J.G.S.*, vol. xxxi, p. 201.

[2] Ibid., pp. 206–9; T. Roberts, 1889, *Q.J.G.S.*, vol. xlv, pp. 551–8; H. B. Woodward, 1895, *J.R.B.*, pp. 173–7; H. Salfeld, 1914, loc. cit., Table I, 1915, loc. cit., pl. xix, figs. 1–17.

[3] C. Fox-Strangways, 1892, *J.R.B.*, p. 299.

[4] In a boring at Ferriby, near Hull, Red Chalk was found to rest either on the basal part of the Kimeridge Clay or on clays of Corallian age: J. Pringle, 1921, *Sum. Prog. Geol. Surv.* for 1920, p. 63.

Weighton Axis upon sedimentation in Kimeridgian times will never be known quantitatively.

On the north side of the Chalk Wolds the Kimeridge Clay does not appear at the surface until Acklam and North Grimston, a considerable distance from the axis, and by then the northerly thickening has already set in.

VI. THE YORKSHIRE BASIN

In the Yorkshire Basin the Kimeridge Clay floors the Vale of Pickering, bounded on the north by the rising dip-slope of the Corallian Moors and on the south mainly by the steep escarpment of the Chalk Wolds. Along the eastern part of the Vale, where the southern boundary is so formed by the overstepping Chalk and subjacent Speeton Clay, the structure is in essential features the same as that of the Vale of the White Horse, where the Gault protruding from the foot of the Downs is in contact with the Kimeridge Clay. In the western part of the Vale of Pickering the clay area is bounded on the south by the Corallian rocks of the Howardian Hills, the junction being largely faulted (see map, p. 138).

The greater part of the outcrop of the Kimeridge Clay, forming as it does the lowest tracts of the vale, is thickly covered by the alluvial deposits of the Pleistocene Lake Pickering, and exposures are scarce. Along the south side the alluvium laps up to the foot of the Wolds and the Corallian hills, covering the 'solid' formations to an average thickness of 90 ft. Only for a short distance where the Chalk escarpment turns southward, from Wintringham to Grimston, Birdsall and Acklam, does the Kimeridge Clay break through, and it is repeated in a faulted inlier south-west of Malton. Along the north side of the Vale outcrops are more numerous, but the consequent streams running down from the moors have dissected a once continuous rim of clay into a series of tongues and outliers. East of Brompton the clay becomes entirely hidden beneath the superficial deposits.

The east end of the Vale of Pickering, where good coast-sections might have been expected, is completely blocked with Drift. Under Filey the surface of the Kimeridge Clay is below sea-level, but towards the south end of the bay it rises sporadically above low-tide mark and towards Speeton it forms the base of the cliff, beneath the slipped masses of Cretaceous clays and Chalk. The dip, however, is southerly and soon brings the Cretaceous rocks down to beach-level.

The Speeton cliffs present at first sight a hopeless spectacle, a jumbled mass of slips, largely covered with mud. In the course of a century, however, geologists, chief among whom was the late G. W. Lamplugh, have watched the sections and, by piecing together the information obtained at different times when opportunities were favourable, after the beach has been swept clear by a storm, have succeeded in reconstructing the succession and collecting numerous fossils *in situ*. The state of the cliffs is always slowly but surely changing, and passes through cycles alternately favourable and unfavourable for collecting. The present century has so far been an unfavourable period, but two favourable occasions occurred in the thirties and again in the seventies and eighties of the last century. Of the later of these Lamplugh, who lived near at hand, was able to take full advantage.[1]

[1] G. W. Lamplugh, 1896, *Q.J.G.S.*, vol. lii, pp. 179–220; and 1924, *Proc. Yorks. Geol. Soc.*, N.S., vol. xx, pp. 1–31.

The stratigraphical work accomplished by Lamplugh provided a sure foundation for palaeontological researches, which culminated a few years ago in the application of Dr. L. F. Spath's wide experience of Cretaceous cephalopods to the ammonites collected at Speeton during the past century.[1] He was able to dispel finally the idea, already refuted two years previously by Dr. Kitchin and Dr. Pringle,[2] but still embodied in nearly all text-books, that there are at Speeton argillaceous representatives of the Portland Beds and 'passage-beds' between the Jurassic and Cretaceous. He found the '*Belemnites lateralis* zone', containing the earliest ammonitiferous beds above definite Kimeridge Clay, to be well up in the Valanginian (Lower Neocomian). At the base of the Cretaceous is a layer of phosphatic nodules (the Coprolite Bed), in which are embedded derived Kimeridge ammonites, and this Lamplugh correlated with the similar phosphatic nodule-bed at the base of the Spilsby Sandstone in Lincolnshire; both beds rest non-sequentially on the Kimeridge Clay. Lamplugh's correlation still stands, but over the whole region the strata immediately above the nodule-bed are now known to be of Neocomian date, and to be separated from the Jurassic rocks beneath by a long time-interval. At Speeton, according to Dr. Spath, 'there is a complete absence of the uppermost Kimeridgian, the whole of the Portlandian, the Tithonian (= Purbeckian), the Infra-Valanginian (= 'Upper Berriasian'), and the lowest Valanginian formations'[3]—an enormous non-sequence, but detected solely as the result of modern discrimination in ammonite identification.

The Coprolite Bed, or basal conglomerate of the Cretaceous, contains, among other rolled ammonites, species recorded as '*Olcostephanus*' and '*Virgatites*'[4] (? *Virgatosphinctoides*). Beneath have been seen some 40 ft. of dark shaly clays with compressed fossils and large septaria, but no detailed study of the ammonites has yet been published. Lamplugh recorded *Virgatites* sp. [? *Virgatosphinctoides*]. According to Danford 80 per cent. of the Belemnites belong to *Cylindroteuthis porrectus* (Phil.), as identified by Pavlow (? *C. tornatilis* Phil. sp.).[5] The beds are much contorted, probably as the result of squeezing by large masses of Drift or even by ice. The lowest levels exposed may sometimes be seen at low tide on the shore near Butcher Haven, 1¾ miles south of Filey. They consist of pale blue shaly clay with pyritous nodules, overlain by dark shales with large septaria, in which Lamplugh collected species of *Aulacostephanus*. Salfeld stated that *Aulacostephanus pseudomutabilis*, *A.* ? *yo*, and also *Pictonia* occurred in Filey Bay, but the last was probably from the Drift.[6]

The middle and lower portions of the Kimeridge Clay have been exposed inland in some small brickyards scattered round the rim of the western end of the Vale of Pickering. Since the old records of ammonites when accepted without revision are often misleading, little can usefully be said regarding the

[1] L. F. Spath, 1924, *Geol. Mag.*, vol. lxi, pp. 73–89.
[2] F. L. Kitchin and J. Pringle, 1922, *Geol. Mag.*, vol. lix, p. 197.
[3] L. F. Spath, 1924, loc. cit., p. 80.
[4] H. Salfeld, 1914, loc. cit., Table I.
[5] C. G. Danford, 1906, *Trans. Hull Geol. Soc.*, vol. vi, pt. i, pp. 1–14; and M. Lissajous, 1925, *Répertoire alph. Bélemnites jurass.*
[6] Isolated outcrops of shale appear from time to time in the cliffs nearer Filey, but their fossils prove them to be transported masses of Lias.

representation of the various zones. The lower layers of the shales are usually characterized, as in other parts of England, by large numbers of *Ostrea delta*. In the Coxwold rift valley Fox-Strangways mentions that '*Ammonites biplex* is rather abundant, and a *Discina* [*Orbiculoidea*] occurs'. In a brickyard at Hildenley, Hudleston records the basement-beds with *Ostrea delta*, *Exogyra nana*, &c., and '*Ammonites mutabilis*', suggesting the *Rasenia* zones. The same beds were also to be seen in the brickyard at North Grimston.[1]

The total thickness of the Kimeridge Clay in the Yorkshire basin has never been accurately determined. Borings along the northern and western rims of the Vale of Pickering pass through only the lower beds. On the south a boring at Knapton, close under the Chalk Wolds, penetrated 500 ft. of clay without reaching the bottom. The cores were not critically examined, however, and besides the probability that some of the clays were Cretaceous, there is a possibility that part of the Corallian formation is there represented by an argillaceous development, and that the boring came to an end in a representative of the Ampthill Clay. South of Wass, in the Coxwold fault-valley, a boring proved a thickness of 400 ft. of blue shales, and there are other records from different parts of the county indicating from 200 to 320 ft. A conservative estimate of the maximum thickness could safely be made at 400 ft.

VII. EAST SCOTLAND

(a) The Brora and Helmsdale District, Sutherland

The evidence for another basin of deposition off Eastern Scotland in Kimeridgian times is limited to a strip of rocks along the coast, now nowhere more than half a mile wide. But although forming a mere selvage, they occupy the foreshore and low cliffs beneath the raised beaches for a distance of over 11 miles and provide some of the most interesting and instructive Mesozoic sections in all Scotland (map, p. 151).

The outcrop is widest at the south end, where the basal zones first succeed the Corallian formation near Kintradwell, 2 miles north of Brora, and gradually dwindles northward, finally running out to sea near Dun Glas, north-east of Helmsdale. With the disappearance of the Jurassic rocks near Helmsdale the railway, which has been able to follow the coast from Golspie, is obliged once more to turn inland. With this sign that the coastal platform has come to an end, the last occurrences of Kimeridgian rocks become mere excrescences, hanging, as it were, from the steep wall of the Helmsdale Granite.

Not the least interesting features on this coast are several fine sections across the Ord Fault, by which the Jurassic rocks have been lowered into their present position. On the small headland of Dun Glas the downthrown rock can be seen dipping seaward at a high angle from the fault-plane, and in Navidale Bay a mass of sandstone belonging to the base of the Kimeridge series is caught up in the fault, and deceived Judd into thinking that it was in position above the local Kimeridgian—here probably of *Gravesiæ* date. At this point a throw of at least 700 ft. is therefore proved; but the fact that the Kimeridge Clay is now in contact with the granite suggests that the total throw is somewhere about 2,000 ft.

[1] C. Fox-Strangways, 1892, *J.R.B.*, pp. 375–6.

Owing to almost endless repetition by small fractures and folds, and to the highly variable lithology of the deposits, it is well nigh impossible to describe a continuous section or to arrive at the total thickness by direct measurements. The Survey estimated at least 700 ft., but Prof. Bailey and Dr. Weir measured over 1,000 ft. in unbroken succession and consider the total thickness to be at least 1,500 ft. Even if the whole of the formation were represented, these figures would be great, but modern zonal study by the Survey has proved that the highest strata present cannot be younger than *Gravesiæ* date.[1] The whole of the beds therefore fall within the Lower Kimeridge Clay, below the Yellow Ledge Stone Band of Kimeridge Bay.

The most remarkable feature of the sections is the presence of Boulder Beds intercalated in the series through almost the entire length of the outcrop (Pl. XXII).[2] The boulders are of all sizes up to huge masses many yards in circumference, and most of them are sharply angular, though some are more or less rounded. Here and there they are intermingled with smaller well-waterworn pebbles. Hugh Miller was the first to point out that the boulders contain fish-remains, such as *Gyroptychius* and *Osteolepis*, proving them to have been derived from the Old Red Sandstone.

The boulders usually lie along definite levels, where they protrude as numerous ledges averaging 5 or 6 ft., but sometimes up to 50 ft., in thickness, separated by dark shales. The shales are puckered and squeezed down under them, as if the boulders had fallen on them before they became consolidated. Occasionally, as at Dun Glas, little or no shaly matrix separates the boulders, which then form great thicknesses of breccia.

The largest of all the boulders is a huge mass of Old Red Sandstone popularly known as the 'Fallen Stack', measuring 150 ft. in length, 90 ft. in width and about 30 ft. in height. It lies on the shore, almost covered at high tide, a quarter of a mile south of Portgower, where it is a well-known land-mark from the railway (see Pl. XXII). As it lies on its side the bedding-planes in the sandstone stand vertically, giving the impression very strongly of a fallen erosion-pinnacle or stack, such as are common round the coasts of Caithness at the present day. Yet it is only the largest of many blocks in a boulder-bed 50 ft. thick, underlain by shales and other boulder-beds for hundreds of feet.

Many theories have been put forward to account for these extraordinary beds. Nearly all the explanations suggested during last century involved, in some way or another, glacial action, whether by floating ice, glaciers, rain, snow and frost, or an ice-sheet. But, as Dr. Macgregor has pointed out, a glacial origin is discountenanced by the absence of Drift or striæ, by the fact that the boulders are too angular and too constant in their composition, and finally by the fauna and flora of the intercalated shales; both are of de-cidedly warm-temperate aspect, in no way incongruous with the fauna and flora of the rest of the British province at that time.

Thus we are bound to conclude that the boulders fell from a cliff, and moreover, a cliff composed exclusively of Old Red Sandstone; for the boulder-

[1] Part of a large ammonite found by H. B. Woodward and taken for '*A. giganteus*' and there-fore as indicative of Portland Beds (*Q.J.G.S.*, 1902, vol. lviii, p. 310) is now considered to be a *Gravesia*.

[2] For the best account of these beds see G. W. Lee, 1925, 'Geol. Golspie', pp. 103–13, *Mem. Geol. Surv.*

PLATE XXII

Photo. W. J. A.

'The Fallen Stack', Portgower, near Helmsdale.

The bedding-planes of the Old Red Sandstone stand vertically. The scale is
shown by a cormorant standing on the right hand end.

Photo. Geol. Survey.

Kimeridgian Boulder Bed, near Portgower, Sutherland.

Tumbled blocks of Old Red Sandstone interbedded in Lower Kimeridge
shales of the *Aulacostephanus* zones.

beds contain no other material, although for several miles they now abut against granite. That part of the movements which brought up the granite into its present position along the Ord Fault must, therefore, have been mainly subsequent to the formation of the beds.

The explanation of the boulder-beds as under-water and shore talus fallen from a cliff was first arrived at by H. B. Woodward in about the year 1891, but his conclusions remained in manuscript until 1902,[1] and it was not until 1910 that a condensed account was published in Prof. Seward's monograph on the Jurassic Flora of Sutherland. The theory has been more recently elaborated by Dr. Macgregor in a paper entitled 'A Jurassic Shore Line', in which the history of previous investigation is usefully summarized.[2] It is implicit in this theory that, by an almost miraculous chance, modern coast-erosion and the Ord Fault have interacted in such a way that they have brought about a coincidence of the present shore with a shore of Lower Kimeridge date; and further, that both shores chanced to coincide with the line of the Ord Fault.

To such a chain of coincidences there are obvious objections, and Prof. Bailey and Dr. Weir deny the possibility that the boulders could have fallen from any ordinary sea-cliff.[3] They point out in the first place that there is at the foot of every cliff a platform of erosion upon which the talus falls, but that nowhere along these eleven miles of continuous exposures has any sign of the necessary platform been found; instead, boulder-beds and shales alternate for a thickness of at least 1,000 ft., until the Corallian strata rise conformably from beneath. Even the 'Fallen Stack', they remark, cannot have fallen on its side and remained there ever since, as its popular name implies, for it has no base. Below it are hundreds of feet of shales and other boulder-beds, proving that it must have been transported. In the second place, the cliff would need to have been at least 1,000 ft. high and to have been sinking progressively as the talus collected at its foot, for the highest layers are proved by the associated fossils to be still of submarine origin. Alternatively, they argue, it was much higher and 1,000 ft. of it was all the time below sea-level; but at such a depth, whence came the many large blocks of reef-corals lying loose among the debris?

The facts are ingeniously explained by Prof. Bailey and Dr. Weir in a theory elaborated at the centenary meeting of the British Association in 1931. The proximity of the boulder-beds to the Ord Fault for so many miles they hold to be no chance coincidence, but rather an indication that the cliff from which the blocks fell was none other than the fault-scarp. They explain the absence of a platform of erosion at the foot of the cliff, and the great thickness of the boulder-beds and shales, by supposing that movement was already taking place along the fault, as it was along the Peak Fault. It is visualized as having lain beneath the sea, not far from the shore. On the upthrown side, towards the land, grew the massive corals and shallow-water lamellibranchs; on the downthrown side was tranquil, deep water, in which ammonites and drifted plant-remains became entombed in fine, muddy sediment.

Sporadically, differential movement was renewed and the sea-bed on the outer side of the fault sank a step deeper, causing a seismic wave. At the same

[1] H. B. Woodward, 1902, *Q.J.G.S.*, vol. lviii, p. 205, and p. 310.
[2] M. Macgregor, 1916, *Trans. Geol. Soc. Glasgow*, vol. xvi, pp. 75–85.
[3] E. B. Bailey and J. Weir, 1932, *Rept. Brit. Assoc.* for 1931, pp. 375–6.

time blocks of Old Red Sandstone of all sizes would break away from the submarine cliff and slide down into the muds collecting below, followed by a debris of corals, shells and sand torn from the upper platform by the force of the wave. Then all would lapse into temporary quiescence once more, until the next earthquake.[1]

In recent years the work of the Scottish Geological Survey has thrown much new light on the fauna associated with these remarkable Kimeridgian beds. In places they are very shelly, though usually it is difficult to obtain even passably unbroken specimens. The commonest fossils are oysters and belemnites, while Pectens, Limas, &c., abound at several horizons, with more or less crushed ammonites.

The highest part of the sequence, at Navidale, north of Helmsdale, yielded fragmentary specimens of a large Perisphinctid, which Messrs. Kitchin and Pringle compare with forms occurring in the *Gravesia* zones. The zone having the widest extent of all is that of *A. pseudomutabilis*, and it is from this that most of the fossils have been procured. The commonest forms from this and the underlying zones are *Lima concentrica* (Sow.), *Ctenostreon proboscideum* (Sow.), *Chlamys* cf. *quenstedti* (Blake), *Ostrea delta* Smith, *Exogyra nana*, *E. virgula*, and various belemnites. Fragments of *Ichthyosaurus* and *Pliosaurus* skeletons have also been found.

Among the most interesting components of the fauna are the large loose blocks of reef-coral, *Isastræa oblonga* (Flem.) and rarely *Stylina alveolata* (Goldf.),[2] which are sometimes so numerous after having been torn up during storms that they are collected from the beach by the inhabitants for lime-burning. They appear to be as foreign to the muddy shales in which they are found as the blocks of Old Red Sandstone, and Prof. Bailey's suggestion that they have been swept down from a shallow-water platform on the upthrown side of the fault is the only explanation yet offered to account for them. This was the most northerly station of reef-building corals in Mesozoic Europe.

Other conspicuous fossils, not uncommon in certain places, are *Terebratula joassi* Dav. and *Rhynchonella sutherlandi* Dav., neither of which has been found outside Sutherland. The *Rhynchonella* is one of the largest species known. Its restriction to this locality is reminiscent of the local distribution of several other large brachiopods in the earlier formations in England.

The *Rasenia* zones both rise to the surface in the southern part of the outcrop, about Loth and West Garty. They consist of a hundred feet of shales, interlaminated with thin seams of hard fine-grained sandstone, and Buckman has named them the LOTH RIVER SHALES. They stand in marked contrast with the *Aulacostephanus* shales above, owing to the absence of boulders. The best section is to be seen in the left bank of the Loth River, beneath the railway bridge and in the railway cutting north of it. The Survey have obtained fragmentary ammonites identified by Buckman as *Rasenia* cf. *stephanoides*, *R.* cf. *cymodoce*, *R.* cf. *uralensis*, *R.* cf. *circumplicatus* and *Amœboceras* sp.; *Lima concentrica* is also common.[3]

---

[1] The first statement of this theory, with a fuller discussion of earlier analogues, appeared in E. B. Bailey and others, 1928, *Journ. Geol.*, vol. xxxvi, pp. 577–614.

[2] J. Pringle, &c., 1930, *P.G.A.*, vol. xli, p. 78.

[3] S. S. Buckman, 1923, *T.A.*, vol. iv, pp. 40–4; the ammonites are so fragmentary, however, that the identifications are doubtful.

Under the railway bridge the base of the Loth River Shales is seen resting on thick beds of white sandstone separated by black carbonaceous layers. The sandstone becomes less divided by black partings towards the sea, where it forms conspicuous buttresses of white rock beside the Loth River estuary. Here it seems to be entirely unfossiliferous, but a mile farther south, in the gorge cut by a stream named Allt na Cuile (where nearly 100 ft. are seen), some of the beds of sandstone are crowded with casts of fossils.

The position of the ALLT NA CUILE SANDSTONE in the zonal table still remains to be settled. H. B. Woodward originally mapped it as Corallian, but Buckman identified *Rasenia* and *Pictonia* in it and concluded that it belongs to the Kimeridge Clay, and Lee accepted his conclusion.[1] More recently the Scottish Survey officers have reverted to Woodward's classification.[2] The ammonites are all small and in the form of casts, not fit for specific determination, but if Buckman's identifications are right there can be no question as to the correct position of at any rate the highest part of the sandstone, in which the principal fossil-bed occurs (8 to 12 ft. from the top). Lee considered, indeed, that part of the sandstone replaced part of the Loth River Shales laterally.[3] The principal members of the fauna are Rhynchonellids, which, in the condition of casts showing the muscle-scars, offer a field for minute investigation which might settle the question. The lamellibranchs, though principally of Corallian species, are inconclusive; more Corallian species occur in the Clynelish Quarry Sandstone, which is dated by its ammonites to the Upper Oxford Clay. Moreover, a fragment of a large *Velata* from near the base of the Allt na Cuile section, preserved in the Scottish Geological Survey office, Edinburgh, is definitely not of any Corallian species known in England. The occurrence of spines of *Cidaris smithi* near the base cannot be regarded as of any great significance, for in Dorset the spines abound on the borderland of the highest Upper Calcareous Grit and the lowest Kimeridge Clay, in the Ringstead Coral Bed (see above, p. 379).

Not the least interesting feature of the remarkable Kimeridgian deposits of Sutherland is a rich and varied fossil flora. Petrified wood is found in the Boulder Beds and shales, and leaves and fronds occur in the shales and sandstones. They have been made the subject of two monographs by Prof. Seward, who recorded sixty-five species of plants, comprising Wealden or Upper Jurassic, Middle, and Lower Jurassic species.[4]

The abundance of land plants in the Kimeridge Clay of Sutherland testifies to the proximity of a shore-line during the time they were deposited, but there is no evidence to justify calling the deposits estuarine, as has often been done.

(b) Ross-shire[5]

The sections showing Corallian Beds at Port-an-Righ (above, p. 434) do not extend above that formation, but the presence of submerged Kimeridge Clay off-shore is proved by fossiliferous nodules which are washed up on the beach. From these have been obtained ammonites of several genera

[1] S. S. Buckman, 1923, *T.A.*, vol. iv, p. 40.
[2] 1930, *P.G.A.*, vol. xli, pp. 76–7.
[3] G. W. Lee, 1925, loc. cit., p. 106.
[4] A. C. Seward, 1911–13, see bibliography.
[5] G. W. Lee, 1925, loc. cit., pp. 114–15.

characteristic of the zone of *Pictonia baylei*—*Pictonia*, *Prionodoceras* (several) and *Dichotomoceras*.

Some 6 miles farther south, on the opposite side of the Cromarty Firth, a minute exposure of Kimeridge Clay is visible at low tide on the foreshore at Ethie. The beds, which are much disrupted owing to proximity to the boundary fault, consist of carbonaceous shales, sandstones, grits, bituminous shales and limestones, with fossils of the *Rasenia cymodoce* zone and plant-remains. Some of the fossils are exceptionally well preserved. From this locality were obtained the type-specimens of *Lima concentrica* (Sow.), *Amœbo-ceras kitchini* (Salf.), *A. pingue* (Salf.), *Cylindroteuthis obeliscus* (Phil.)[1] and *C. spicularis* (Phil.),[1] as well as the types of several plants.

VIII. THE HEBRIDEAN AREA

The valuable researches of Mr. Malcolm MacGregor in North-east Skye, already referred to in Chapters XII and XIII, have revealed that Kimeridge Clay is present at the north end of the peninsula of Trotternish. I am in-debted to Mr. MacGregor for very kindly communicating for this book an advance account of his discoveries.

The best section is on the foreshore at Kildorais, 2 miles north-west of Staffin, where about 100 ft. of shales are seen between tide-marks, tilted almost vertically. From the lower part of the shales Mr. MacGregor obtained specimens of *Prionodoceras* and *Dichotomoceras*, and from the upper part numerous *Pictoniæ* and occasional *Raseniæ*; and in addition a form which was identified in London as ? *Aulacostephanus pseudomutabilis*. The lower parts of this succession were again discovered in a quarry in the grounds of Flodi-garry House, and also not far from the Oxford Clay exposure on the foreshore on the north side of Staffin Bay.

Kimeridge Clay has also been detected by Dr. Spath, he kindly allows me to state in advance of publication, in the Island of Mull. He has recognized the presence of numerous *Raseniæ* and Amœbocerates among a collection made there by the Survey and assigned by Buckman to the '*Reineckeia* zone' (= *kœnigi* zone), or Kellaways Beds (see p. 371 for further particulars).[2]

The importance of these discoveries is obviously great, for they prove that marine deposition continued normally in the Hebridean Area as long as in East Scotland, and the community of fossils indicates that the two areas were in connexion.

IX. KENT[3]

The concealed outcrop of the Kimeridge Clay beneath the Cretaceous rocks of Kent is rather wider than that of any of other Jurassic formations. It underlies the coast from a little south of Dover to Folkestone, at which place it is succeeded by the Portland Beds; maintaining a constant width of about 5 miles it then strikes inland in a north-westerly direction, which soon changes to west north-westerly. West of the railway from Canterbury to Ashford its boundaries have not been determined.

The thickness increases from east to west, and the study to which the fossils

---

[1] Not Oxfordian species as generally supposed (e.g. in Lissajous's catalogue).
[2] L. F. Spath, 1932, *Med. om Grønland*, vol. lxxxvii, no. 7, p. 149 (now published).
[3] Based on Lamplugh and Kitchin, 1911, loc. cit.; and Lamplugh, Kitchin and Pringle, 1923, loc. cit.

from the borings were subjected by the Survey led them to the conclusion that the 44 ft. of Kimeridge Clay at Dover might be represented by 138 ft. at Brabourne and by a still greater thickness at Pluckley, west of Ashford. The separation of this thickening from the effects of pre-Cretaceous erosion was only made possible by close attention to palaeontological zones.

At Penshurst 622 ft. of Upper Kimeridge Clay and Portland Sand-equivalents were proved, although the boring was stopped before it reached the base of the *Virgatosphinctoides* zones. At Pluckley the Lower Clays exceeded 300 ft. up to the top of the *Gravesia* zones, and may have been 350 ft. thick, so that the total maximum thickness of the Kimeridge Clay in Kent (including perforce the equivalents of the Portland Sand) must be close on 1,000 ft., or rather more than the development in Dorset.

The highest beds were penetrated in the most westerly boring at Penshurst, south-west of Tonbridge, where about 250 ft. of sandy clay overlies the *pallasioides* zone. Of this about 100 ft. is believed to represent the Portland Sands of Dorset (including the *gorei* zone).[1] Palaeontological evidence as to the exact stratigraphical position of these highest beds is meagre. Some unidentified ammonites were found, likened to some from the beds above the *pallasioides* zone of Hounstout Cliff, and there were also a species of *Grammatodon* and two or three other lamellibranchs of doubtful correlative value, but the exact positions of these in the Hounstout sequence (p. 446) has not been shown. More will be said on this subject in the next chapter on the Portland Beds.

At Ottinge and Brabourne the Portland Stone rests non-sequentially upon clays yielding ammonites believed to indicate an horizon about 300 ft. lower than those immediately below the stone-beds at Penshurst and Pluckley, namely the *pallasioides* or even the *rotunda* zone—an unconformity comparable with that in Wiltshire, Oxfordshire and Bucks.

The *pectinatus* and *Virgatosphinctoides* zones (as then understood, now including the *Subplanites* zones) were proved at Pluckley and were probably penetrated at Brabourne and Ottinge. They consist of shaly clays as in Dorset, with flattened ammonites, which have been discussed at length by Messrs. Kitchin and Pringle, for the most part under the name '*Pseudovirgatites*'. Associated with them, also, were numerous pyritized radial plates of *Saccocoma*, which were taken to correlate with the oil-shale of Kimeridge. According to Messrs. Kitchin and Pringle, *Modiola (Musculus) autissiodorensis* (Cott.) makes its appearance near the base of the zones and ranges up into the Portland Sand, being one of the commonest bivalves of the Upper Clays in Kent. Other common shells are *Protocardia morinica* (de Lor.), *Astarte* cf. *mysis* d'Orb. and *Aporrhais* cf. *piettei* (Buv.).

The Lower Kimeridge Clay was penetrated at Brabourne, Pluckley, Lower Standen, Abbotscliff and Dover. Wherever it was seen, it proved to be more sandy and calcareous than the Upper Clay. Palaeontologically it was separable with equal ease by the presence of *Exogyra virgula* and the absence of *Modiola autissiodorensis*.

Near the top of the Lower Clay a fragment of Perisphinctid was found agreeing with *P. bleicheri* (de Lor.), an ammonite of the *Gravesia irius* zone near Boulogne, and other fragments have been described as possibly repre-

---

[1] Lamplugh, Kitchin and Pringle, 1923, loc. cit., p. 228.

senting species of *Gravesia*. A notable feature of this zone was a *Trigonia* bed containing abundant *T. pellati* Mun.-Chalm., identical with a bed in the *Gravesia irius* zone at Boulogne.

The *Aulacostephanus* zones, although probably developed normally, yielded no ammonites, but their presence was thought to be indicated by the occurrence of *Protocardia morinica* at a considerable distance above beds with *Rasenia*. In Dorset this *Protocardia* makes its appearance in the Upper *Aulacostephanus* zone, that of *A. pseudomutabilis*, below Maple Ledge. Another bivalve of the same zone which occurred abundantly at Brabourne and Abbotscliff was *Astarte ingenua* de Lor.

The *Rasenia* zones proved more fossiliferous, yielding at levels 10 ft.–30 ft. above the base of the Kimeridge Clay such species as *Rasenia* cf. *stephanoides* (Oppel), *R.* (?*Prorasenia*) *witteana* (Oppel), *Pararasenia desmonota* (Oppel), ?*Involuticeras trimerus* (Oppel), *Physodoceras orthocera* (d'Orb.), recalling the upper part of the Abbotsbury Iron Ore and the clay of the type-locality of Market Rasen. At Dover a specimen of *Rasenia* was found only 12 ft. below the Cretaceous.

The *Pictonia baylei* zone cannot be more than 10–20 ft. thick and is probably less, since no ammonites that would prove its presence have been found. As in other parts of England, however, the lowest part of the Kimeridge Clay is full of *Ostrea delta* and *Exogyra nana*.

POSTSCRIPT NOTE:—On the eve of going to press an important paper has come to hand by E. B. Bailey and J. Weir on 'Submarine Faulting in Kimmeridgian Times in East Sutherland', *Trans. Roy. Soc. Edinburgh*, vol. lvii, Dec. 1932, pp. 429–67. The authors elaborate their theory referred to above, pp. 475–6, and claim to have established the following facts: During Kimeridgian times a submarine fault-scarp was maintained by intermittent movement of the sea-floor of the Helmsdale district, while dry land existed a little to the north-west. The aggregate movement of the fault much exceeded 2,000 ft. The fault-scarp separated a comparatively shallow-water facies from a comparatively deep-water facies. Frequent earthquakes caused landslips along the scarp and spread out the debris of Old Red Sandstone in graded boulder-beds in a manner indicating the co-operation of tunamis ('tidal waves'); the movements also opened fissures in the Kimeridgian and produced a chasm-breccia along the fault. Analogues are pointed to in Britain, Switzerland, Canada, and the United States. In addition, the *Gravesia* zones are established at several localities, and the *Subplanites* zone is believed to exist at the northern extremity of the Kimeridgian outcrop.

CHAPTER XV

PORTLAND BEDS

Zones (Plate XL).		Dorset.	Swindon.	Bucks.
		STRATA.		
Titanites titan and T. giganteus [1]	PORTLAND STONE.	Freestone Series 40–50 ft.	Creamy Limestones 10 ft.	Creamy Limestones 7–12 ft.
Kerberites kerberus [2] and K. okusensis [3]		Cherty Series 60–70 ft. Basal Shell Bed 8 ft.	Swindon Sand and Stone 25 ft. Cockly Bed 4 ft.	Crendon Sand 5–6 ft. Rubbly Limestones 6–12 ft.
Glaucolithites gorei [4] and G. glaucolithus [5]	PORTLAND SAND.	Portland Sand 100–20 ft.	Glauconitic Beds 3½ ft. Upper Lydite Bed	Glauconitic Beds 5–10 ft. Upper Lydite Bed.
Provirgatites scythicus [6]			——	——

I. THE DORSET COAST

(a) The Isle of Purbeck

THE Isle of Purbeck owes to the Portland Beds some of the grandest coastal scenery in the South of England. From Durlston Head, south of Swanage, westward for 6 miles to St. Alban's (or more correctly St. Aldhelm's) Head, the Stone Beds ascend sheer from the sea in a perpendicular wall of rock, unscalable, and inaccessible except by boat. Above the cliffs, the home of the sea-aster, the sea-campion, the samphyre and the thrift, the Purbeck Beds rise as grassy slopes to a level of 400 ft.

Here the strike coincides with the coast-line and the rocks appear horizontal, but in reality there is a slight seaward dip from the axial anticline of the Isle of Purbeck. At St. Alban's Head, where the coast-line changes its direction to the north, the Portland Beds rise to 500 ft. and cap the precipices of Emmit Hill and Hounstout, their tumbled blocks forming a wild undercliff upon the Kimeridge Clay beneath. From Hounstout they strike inland round the Golden Bowl, just failing to reach the sea in the prominent hill of Swyre

---

[1] Genotype of *Gigantites* Buckman, but, according to Dr. Spath, Buckman's genera *Gigantites*, *Briareites*, and *Galbanites* are all synonyms of *Titanites* Buckman, 1921. (See Spath, 1931, *Pal. Indica*, N.S., vol. ix, p. 472.)

[2] Formerly known as the zone of *Ammonites pseudogigas* Blake, but replaced by Buckman (1926, *T.A.*, vol. vi, p. 26) owing to his selection of a later species of the *Titanites* zone as neotype (*Trophonites pseudogigas* Blake sp., *T.A.*, 1923, pl. CCCLXXXV).

[3] Named by Salfeld (1914, *Neues Jahrb.*, Beil.–Bd. xxxvii, pp. 130, 198–200) and since retained by all writers as alternative zonal index. Made genotype of *Kerberites* by Buckman (*T.A.*, 1925, pl. DLXX).

[4] Name given by Salfeld (loc. cit.) to *Ammonites biplex* de Loriol non Sow. (1867, pl. 11, figs. 3–4) and subsequently used by all writers as zonal index. (Not placed generically by Buckman; but see Spath 1931, loc. cit., p. 472.)

[5] Used as hemeral index by Buckman (*T.A.*, 1922, vol. iv, p. 26) for part of the Glauconitic Beds of the Thame–Aylesbury district and figured, loc. cit., pl. CCCVI.   [6] See p. 491

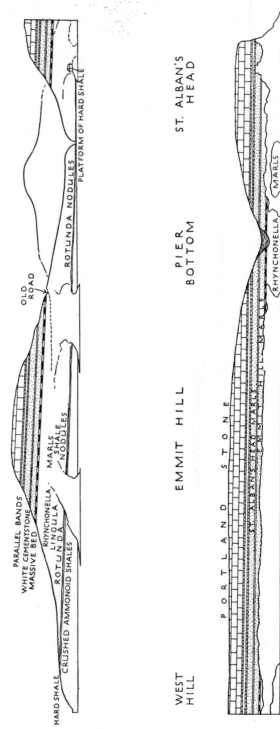

FRESHWATER
STEPS

HOUNSTOUT
CLIFF

CHAPMAN'S
POOL

PARALLEL BANDS
WHITE CEMENTSTONE
MASSIVE BED
RHYNCHONELLA
LINGULA
ROTUNDA

MARLS
SHALE
NODULES

ROTUNDA NODULES

OLD
ROAD

PLATFORM OF HARD SHALE

HARD SHALE
CRUSHED AMMONOID SHALES

WEST
HILL

EMMIT HILL

PIER
BOTTOM

ST. ALBAN'S
HEAD

PORTLAND STONE

ST. ALBAN'S HEAD MARLS

EMMIT HILL MARLS

RHYNCHONELLA

MARLS

Fig. 80. Section along the cliffs from Freshwater Steps, below Encombe, to St. Alban's Head. Total distance 2 miles. Scale, vertical and horizontal, 7·2 in. = 1 mile. (Outline based on Strahan, 1898, 'Geol. Isle of Purbeck and Weymouth', *Mem. Geol. Surv.*, plates X and XI.)

Head (653 ft.), and after encircling the clay tract about Kimeridge with a bold escarpment, fall rapidly in overhanging crags into the sea at Gad Cliff. Here the dip is steeply to the north, towards the thrust-fault in the Chalk, and it is the predominant dip throughout Purbeck (see Pl. XXV and fig. 77, p. 442). Farther west smaller tracts of Portland Beds rise to the surface on either side of Lulworth Cove, forming the Mupe Rocks, the Man o' War in Man o' War Cove, and the base of the promontory and the famous arch at Durdle Door (Pl. XXVIII). These last fragments are close to the thrust-fault and they are tilted almost vertically.

The maximum thickness of the formation along this coast is about 220 ft., consisting of Portland Stone 110–20 ft. and Portland Sand about 100–20 ft. Since the beds are far more continuously and more extensively exposed than in the island from which they take their name, it is preferable to describe Purbeck first.

The Portland freestone was probably quarried in Purbeck long before the fame of the Portland quarries became known, and it is still wrought under the name of Purbeck-Portland. Its powers of resisting the weather are superior to those of the freestone from the island. The many ancient galleries driven into the cliffs east of St. Alban's Head bear witness to the activities of the past. Most of them have been abandoned for centuries and provided safe retreats for smugglers. The old quarries of Tilly Whim (generally called 'caves') are supposed to have supplied the stone for the building of Corfe Castle, and there is no doubt that some of the forgotten shafts and galleries were being worked in the early Middle Ages, when Swanage was exporting Purbeck Marble for churches, not only all over England, but to all parts of Europe.[1]

The Portland Stone is divisible in Dorset into two, the Freestone Series above and the Cherty Series below, corresponding roughly with the *Titanites* and the *Kerberites* zones.

SUMMARY OF THE PORTLAND BEDS OF PURBECK

The cliffs between Durlston and St. Alban's Heads provide a continuous exposure of almost the whole of the Portland Stone Series, and it is again magnificently exposed in Gad Cliff, Tyneham. In spite of this, the vertical (and in Gad Cliff, overhanging) precipices can be studied only at a few points, such as Tilly Whim, Dancing Ledge, Seacombe and Winspit, where quarrying operations have removed the Upper or Freestone Series, leaving the worthless Cherty Series beneath. At the end of the last century the quarries at Winspit (Pl. XXIII) were still in work, but they have now been deserted for some years. After the Great War there was a revival with modern equipment at the old quarries at Seacombe and St. Alban's Head, but about 1930 both of these ventures became derelict. Now the only locality at which work is being actively carried on is inland, at the Sheepsleights Quarry of the Worth Stone Company, in Coombe Bottom, 1 mile north-west of Worth (fig. 82). No waste occurs, for the principal service to which the stone is put is the making of macadam, and the Cherty Series is used as well as the caps and overburden

---

[1] In Grabau's *Textbook of Geology*, 1920, pp. 812–13, two photographs of Tilly Whim are given with the inscription 'Tilly Whim Caves: Elevated sea-caves cut by waves in horizontal (Jurassic) strata'. The engraving here reproduced, from Englefield's *Isle of Wight* (see fig. 81), drawn by Thomas Webster in 1811, showing the 'caves' in the making, may be of interest to users of that excellent text-book.

FIG. 81. Tilly Whim Caves in 1811, drawn by Thomas Webster; showing galleries being driven into the Purbeck-Portland Freestones, with the Cherty Series left as a ledge below. (From Englefield's *Description of the . . . Isle of Wight*, 1816, pl. 33.)

of the freestone. The best freestone is cut and sold for building, but only as a side-line.

This extensive new quarry provides the most valuable exposure of the Freestone Series in Purbeck. Being worked in four stages, it affords abundant access to all the beds *in situ*, a facility never enjoyed by geologists in the old cliffside quarries.[1]

No detailed investigation of the fossils of the numerous individual beds of the Portland Stone has yet been attempted. Mollusca, especially lamelli-branchs, are abundant at almost all horizons, but owing to the hardness of the matrix they make unattractive material for the collector. The best-preserved fossils are the ammonites, but they are so large and heavy that they present difficulties of their own. It is essential for the investigator of the Portland Beds to have sufficient previous knowledge of the palaeontology to enable him to make identifications with accuracy in the field (though with the ammonites this is impossible). In the Freestone Series the ammonites seem almost restricted to a level near the top, where they abound, while the lamellibranchs, at least the common species, range through the whole Freestone Series and Cherty Series in about equal abundance. It is doubtful, therefore, whether satisfactory small palaeontological subdivisions of the Portland Beds will ever be established. For the present it is necessary to retain the terminology of the quarrymen. It has at least the merits of being applicable to all the Purbeck quarries and of being readily understood by all who deal with the stone, while it can also claim the dignity of a high antiquity.

Titanites Zone: Freestone Series, 50 ft.

SHRIMP BED, 8–10 ft. (reaching 16 ft. on St. Alban's Head).

White, fine-grained sublithographic limestone, like the Cream Cheese Bed of the Great Oolite. It derives its name from the carapaces and claws of a small shrimp-like Crustacean, probably of the genus *Callianassa*, found in it.[2] Casts of *Protocardia dissimilis*, *Chlamys* (*Camptochlamys*) *lamellosa* and some other species are also common, and occasionally there appear fragmentary casts of coarsely-ribbed triplicate ammonites (? perhaps *Glottoptychinites* Buckman gen.). The change to the overlying Purbeck Beds is everywhere abrupt, though conformable, but the downward passage to the underlying Spangle is gradual.

TITANITES BED (PERNA BED, BLUE BED or SPANGLE), 10 ft.

Greyish shelly limestone, crowded with *Trigonia gibbosa*, *T. incurva*, *Iso-gnomon* [*Perna*] *bouchardi*, *Chlamys lamellosa*, *Protocardia dissimilis* and many other shells, and the source of all the ammonites obtained in the Worth quarry except the fragments mentioned above, in the Shrimp Bed. The ammonites, which are all giants, were identified with a query by Buckman as *Briareites*,[3]

[1] Through the kindness of the management, I was personally conducted over the excavations by Foreman W. J. Bower, who had spent his life in the Winspit and Seacombe Quarries before the present opening was begun, and whose knowledge of the stone-beds of all the quarries in Purbeck is unrivalled. He was able to give me the correct names for the different beds, together with interesting information as to their qualities, uses, fossil-contents and development in the coastal sections.

[2] Identification kindly supplied by Mr. H. Woods from fragmentary carapace and claw submitted to him from Seacombe and Worth.

[3] S. S. Buckman, 1926, *T.A.*, vol. vi, p. 35.

which Dr. Spath considers synonymous with *Titanites*.[1] The previous naming of this important fossil-bed is so unsatisfactory that I propose to call it the *Titanites* Bed. It was explained to me in the quarry, by the undisputed authority, the foreman mentioned above, that the whole bed may be called Spangle in one place and Blue Bed in another, but that in the Worth quarry the highest 8 ft. is Blue Bed and the rest Spangle; but it was almost immediately admitted that there was little or no perceptible difference. The word Spangle seems to refer to the glint of the fractured surface, due to calcite or shells; it is purely a descriptive lithological term, used equally for the House Cap lower down, and it should not therefore be applied to one particular bed as it has been by Buckman.[2] The foreman at Worth considered this bed the equivalent of the Roach at Portland, in which he was in unconscious agreement with J. F. Blake. Blake split up the bed at St. Alban's Head into 'Roach' above and 'Oolite with *Perna*' below.[3] Hudleston and Monckton called the whole bed the *Perna* Bed at Winspit,[4] but this designation is unsatisfactory for two reasons: first because the genus usually known as *Perna* should correctly be called *Isognomon*, and secondly because the species, *I. bouchardi*, is equally common in the lower bed of Spangle, the House Cap, and also abounds in the Cherty Series at much lower levels. H. B. Woodward[5] called the whole bed the Blue Stone, but this term is again misleading because the stone is never at all blue, becoming at its darkest no more than a mauvish-grey; further, in quarrymen's terminology there is a nice distinction between Blue Stone and Spangle, and the whole course is not always Blue Stone.

In the east, at Tilly Whim, a lenticular oyster-bed develops on this horizon. Locally about 8 ft. of the rock is almost entirely composed of oysters, cemented in an intensely hard matrix but weathering out in perfect condition. The principal constituents are *Exogyra nana* (Sow.), *E. thurmanni* Étall., *Ostrea expansa* Sow. and *Isognomon bouchardi* (Oppel), with a smaller proportion of *Lima rustica* (Sow.) and *Plicatula boisdini* de Lor.

PON OR POND FREESTONE, 7–7$\frac{1}{2}$ ft.

Good oolitic freestone, showing some false-bedding. Fossils rare. The origin of the name seems to have been lost, unless it is an abbreviation of Upon [the Chert Vein].

CHERT VEIN, 2–4 ft.

This, the most easily recognized datum in all the quarries, is usually called the Flint Vein or Flint Stone. Blake raised the boundary of the Cherty Series to include it, an undesirable course which has not been followed by any subsequent writer.

LISTY BED, 6–9 in.

So called 'because it breaks easily'. Present in the cliffs but absent in the Worth quarry. A soft freestone.[6]

---

[1] L. F. Spath, 1931, *Pal. Indica*, N.S., vol. ix, p. 472.
[2] 1926, *T.A.*, vol. vi, p. 35, misspelt Spengel.
[3] J. F. Blake, 1880, *Q.J.G.S.*, vol. xxxvi, p. 194.
[4] 1910, *P.G.A.*, vol. xxi, p. 514 and several earlier excursions.
[5] 1895, *J.R.B.*, p. 190.
[6] Misspelt Lisky Bed by Buckman, loc. cit., p. 35. Apparently the Nist Bed of H. B. Woodward, 1895, *J.R.B.*, p. 190, perhaps so recorded in error, as the Worth foreman and his

HOUSE CAP, 8 ft.

Greyish coarse limestone, like the *Titanites* Bed and also called a Spangle, but on the whole less shelly. A 6-in. shelly vein runs through it at Worth about 2½ ft. from the top, and the lower part is conspicuously shelly at Tilly Whim. The fossils seem to be the same, except that at Worth ammonites are absent; but at Winspit Woodward recorded '*Ammonites giganteus*'. *Trigoniæ* predominate.

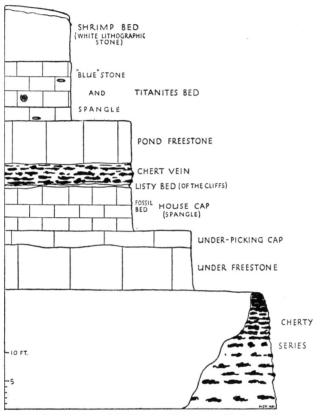

FIG. 82. Section of the Portland Freestone Series at Sheep-sleights Quarry, Worth Stone Quarries, 1 mile north-west of Worth Matravers.

UNDER PICKING CAP, 2–3 ft.

Hard freestone, which had to be blasted in the old cliff quarries, locally called Spangle. Rarely large ammonites were found in it at Winspit, but not at Worth.

UNDER OR BOTTOM FREESTONE, 8–11 ft.

Fine cream-coloured oolite, an easily-worked freestone of excellent quality,

men had never heard the name. At least as early as 1863 the bed immediately underlying the Chert Vein was called the Listy Bed (see a section at Seacombe in Hutchins's *History of Dorset*, 3rd ed., vol. i, p. 687).

with few fossils. This is the bed which was principally mined at all the quarries from Tilly Whim to St. Alban's Head. The numerous 'caves' are adit galleries driven into it, the 'caps' forming the roofs and the Cherty Series the floors (see fig. 81, and footnote on p. 483).

Towards the west most of the Freestone Series dies out. Hudleston pointed out that in Gad Cliff and Worbarrow Tout the Shrimp Bed and *Titanites* Bed are well developed, but that they rest directly upon the Cherty Series.[1] It remains to be proved whether the Pond and Under Freestones and intervening Cap disappear or merely lose their character and become cherty.

Kerberites Zone: Cherty Series, 60–70 ft.

The Cherty Series forms the lower part of the cliffs east of St. Alban's Head and the vertical capping to the higher cliffs to the west, at Emmit Hill and Hounstout. So far it has defied subdivision and it has the distinction of being one of the thickest masses of unsubdivided rock remaining in the English Jurassic System. It consists throughout of intensely hard brown limestone with numerous impersistent veins and nodules of dark chert. It can be easily studied at Tilly Whim, Dancing Ledge, Seacombe, Winspit and St. Alban's Head. At Dancing Ledge one of the thick courses of hard stone forms the ledge up which the waves dance, and the surface is studded with giant ammonites.

The upper part of the series is poorly fossiliferous, but the middle and lower parts are often quite shelly, and some of the courses are largely composed of *Serpula gordialis*. The limestone is so hard that collecting is difficult, but *Isognomon bouchardi*, *Chlamys lamellosa* and some of the other familiar shells may usually be seen. Blake recognized in 1880 that the ammonites differed from those found in the Freestone Series, signifying the difference by calling them '*A. boloniensis* de Lor.' Although they abound on the surface of some of the ledges washed by the sea and upon the fallen blocks beneath the cliffs west of St. Alban's Head and under Hounstout, their characters are still but little known. The reason for this is that the beds have never been quarried and in the natural exposures it is almost impossible to detach specimens from their matrix, while those visible *in situ* are always too much weathered for certain identification. In 1931, however, the Worth Stone Co. started quarrying the Cherty Series at Worth for breaking up to make macadam, and so it may be hoped that some of the ammonites will be obtained. At present our knowledge is restricted almost entirely to a few specimens collected in the Basal Shell Bed of the Isle of Portland (to be described later, below, p. 496).

Although Blake spoke of a Basal Shell Bed in his record of the section of St. Alban's Head,[2] no such bed has been detected by subsequent investigators, or seems to exist at this horizon in Purbeck; the shells are rather distributed through the lower portions of the Cherty Series instead of being concentrated in one particular bed as at Portland. In consequence the boundary between the Cherty Series and the Portland Sands beneath is less easy to define. The best accessible section is at the point of St. Alban's Head: here there is a perfect gradation downward through sandy and cherty

[1] W. H. Hudleston, 1896, *P.G.A.*, vol. xiv, p. 319.
[2] J. F. Blake, 1880, loc. cit., p. 194.

PLATE XXIII

Photo. W. J. A.

Winspit Quarry, near Worth Matravers, Dorset.

Portland Freestones in upper part of cliff quarried back, and Cherty Series left in lower part of cliff. PB = Purbeck Beds; SB = Shrimp Bed; TB = *Titanites* Bed; F = Portland Freestones and Chert Vein. The vertical cliffs of Portland Stone are continued in the distance, surmounted by a slope of Lower and Middle Purbeck Beds.

Photo. W. J. A.

Dancing Ledge Quarry, near Langton Matravers.

Galleries have been driven into the Freestone behind, at a lower level.

PLATE XXIV

Photo. *W. J. A.*

Swyre Head, Hounstout Cliff and Emmit Hill, seen from St. Alban's Head.

The dip in the foreground is Pier Bottom, the next Chapman's Pool, the next
Egmont Bay, below Encombe, with Freshwater Steps at the point where the
grass almost reaches the sea.

Photo. *W. J. A.*

Hounstout Cliff and Emmit Hill.

On Emmit Hill PB = The Parallel Bands; St.AHM = St. Alban's Head
Marls; WCB = The White Cementstone; EHM = Emmit Hill Marls. The
Massive Bed (base of the Portland Sand) crops out at the foot of the bank,
below the photographer.

limestone and calcareous cherty sandstone to sandstone without chert but with honeycomb weathering, denoting that lime, formerly present, has been leached out. To draw the boundary at the downward limit of either lime or chert is unsatisfactory, and at present evidence is insufficient for the marking of any palaeontological division. The difficulty is illustrated by the recent petrological study of the Portland Sands by Mr. M. P. Latter, who fixed no definite upper boundary to the Sands and began his descriptions of sections at different levels in localities so close together as St. Alban's Head and Emmit Hill.[1] In short, wherever we choose our boundary, we cannot avoid grouping either sandy beds with the Cherty Series or cherty beds with the Sands. At the junction are 24 ft. of grey and yellow sandstone, weathering with honeycomb structure, the upper half very cherty and sandy, the lower half free from chert (see fig. 83). I follow Hudleston[2] and Sir A. Strahan[3] in grouping these beds with the Cherty Series of the Portland Stone rather than with the Portland Sand.

Glaucolithites Zone: Portland Sand, 100–20 ft.

The Portland Sand was first so named by Fitton,[4] who described Emmit Hill, the steep cliff on the west side of St. Alban's Head (Pl. XXIV) as the typical section. There is hereabouts a complete and almost continuous exposure from the point of St. Alban's Head to Chapman's Pool and again in Hounstout Cliff (fig. 80). At first sight 'Sand' seems a misnomer, for the series consists rather of grey and black sandy marls, with bands of sandstone and sandy cementstone. The same applies to the Isle of Portland. North of Weymouth, however, the proportion of sand is much higher, and there are in places up to 40 ft. of yellow sand; but even in Purbeck and Portland there is more than is at first sight apparent, for owing to its black colour and the admixture with marl, the sand escapes notice.

In the cliffs of the type-locality the beds fall into several well-defined subdivisions, marked off by prominent bands of cementstone or sandstone, distinctly traceable from end to end of the exposures. To provide a framework for palaeontological investigations in the future it is essential that these thick subdivisions should have names, and accordingly some of them are here named for the first time.

Buckman introduced two major subdivisions, a Cementstone Series above and a Sandy Series below,[5] but his interpretation of the lower boundary of the formation was different from that accepted here, and his Cementstone Series is almost exactly synonymous with the Portland Sand as now understood. All but the highest 5 ft. (the Massive Bed) of his Sandy Series is here classed as Kimeridge Clay, following Strahan, Cox, Latter and probably Fitton (see above, p. 446, where it is named the Hounstout Marl).

THE PARALLEL BANDS, 23 ft.

The most conspicuous strata in Hounstout, Emmit Hill and St. Alban's Head cliffs, remarked on by all observers, are two parallel thick bands of

[1] M. P. Latter, 1926, *P.G.A.*, vol. xxxvii, pp. 76–7.
[2] W. H. Hudleston, 1882, *P.G.A.*, vol. vii, p. 385.
[3] A. Strahan, 1898, 'Geol. Isle of Purbeck', p. 67, *Mem. Geol. Surv.*
[4] W. H. Fitton, 1836, *Trans. Geol. Soc.* [2], vol. iv, pp. 211–12.
[5] S. S. Buckman, 1926, *T.A.*, vol. vi, p. 34.

sandstone or sandy cementstone, which run from one end of the sections to the other without perceptible change. Both Blake and Buckman (and Hudleston and Strahan by implication)[1] took these bands as marking the top of the Portland Sand, and since here, at the type-locality, they provide the only obvious summit to a division admittedly of arbitrary boundaries, it is as well to make use of them. Buckman and Strahan gave no thickness, but according to my own measurements and to those of Mr. Latter they measure in all about 23 ft.[2]—Upper $6\frac{1}{2}$-7 ft.; Lower 10 ft.; intervening beds 5–6 ft.

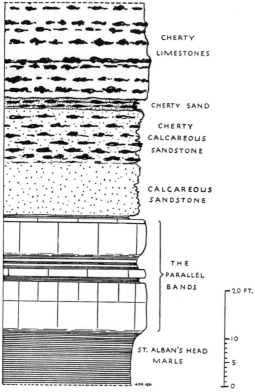

CHERTY LIMESTONES

CHERTY SAND

CHERTY CALCAREOUS SANDSTONE

CALCAREOUS SANDSTONE

THE PARALLEL BANDS

20 FT.

10

ST. ALBAN'S HEAD MARLS

5

0

Under St. Alban's Head, 9 in. below the top of the Upper Parallel Band, there is a noticeably even and persistent parting, which may indicate a physical break. The highest 9 in. form a separate bed of hard sandstone, above which is a dark shaly layer, merging up into the grey sandstones. A convenient position at which to draw the top of the Portland Sand would be at this persistent parting or eroded surface (fig. 83).

The two chief Parallel Bands are separated by light chalky shales with two white layers, and in the lower part of the chalky shales there is a thinner hard band (1 ft. 6 in.) as persistent as the other two and forming a Middle Parallel Band, conspicuous from St. Alban's Head to Hounstout (fig. 83 and Pl. XXIV).

Fig. 83. Section in the cliff under the west side of St. Alban's Head, at the junction of the Cherty Series with the Portland Sand. The arbitrary boundary is taken at the top of the Parallel Bands.

Concerning the palaeontology of the Parallel Bands everything still remains to be discovered. Blake speaks of '*Ammonites biplex*' as abundant, a record that has not so far been interpreted.

[1] 'The base of the Portland Stone, however, can be seen round nearly the whole headland [St. Alban's] and under Emmit Hill also; it contains not only an abundance of chert-nodules, but so much sand that it may be described as a calcareous sandstone' (Strahan, 1898, loc. cit., p. 67). See also Hudleston (1882, *P.G.A.*, vol. vii, p. 385), 'the bare slopes of Portland Sand are seen to be surmounted by the cherty calcgrits which form the base of the Portland Stone capping the precipices of Emmit Hill'. The 'cherty calcgrits' and 'calcareous sandstone' are the sandy base of the Cherty Series immediately above the Parallel Bands, referred to on p. 489.

[2] Blake gave a thickness of 39 ft., but how this was arrived at is not clear. My own measurements and Mr. Latter's, although made quite independently, agreed within 1 ft. See M. P. Latter, 1926, loc. cit., pp. 76–7.

ST. ALBAN'S HEAD MARLS, 40–45 ft.

Below the Parallel Bands are 40–45 ft. of grey sandy marls and marly sands with thin bands and nodules of cementstone, all poorly fossiliferous. These and the underlying marly and sandy beds have not hitherto been named.

THE WHITE CEMENTSTONE BAND, 2 ft.

A conspicuous band of light-grey to whitish cementstone, 2 ft. thick, is persistent in all the cliffs, and Buckman named it the White Cementstone, recording and figuring from it the Behemothan ammonite *Leucopetrites cæmentarius*, found by Mr. C. H. Waddington.[1] The bed breaks with a conchoidal fracture and the broken surfaces often show red staining. It contains casts of *Thracia depressa* (Sow.), *Pleuromya tellina* Ag. and '*Arca*' *fœtida* Cox., in places abundantly, especially in the lower part.[2]

EMMIT HILL MARLS, 30–40 ft.

Below the White Cementstone are three ill-defined bands of partly indurated, blackish, shaly marl or sandy shale, in all 12 ft. thick, containing abundant crushed casts of *Thracia*, *Pleuromya*, &c. Below this again, and distinguished only by being less indurated, are some 25 ft. of blue or blue-black sandy and partly shaly marls, with numerous thin bands of cementstone. Since all these are still in need of a name I propose to call them the Emmit Hill Marls, after the type-locality of the Portland Sand, where they are well exposed and thickest. According to Mr. Latter's measurements they are 39 ft. thick at Emmit Hill, 38 ft. at St. Alban's Head, and 30 ft. at Hounstout.[3] Dr. Spath informs me that he believes a number of the fragments of *Provirgatites* of the *scythicus* group collected by him to have fallen from these beds.

THE MASSIVE BED, 5–6 ft.

The Massive Bed, so named by Buckman,[4] has already been referred to at some length (above, pp. 445–6). It is a prominent band of hard, blue-centred, brown-weathering, shaly to rubbly, calcareous sandstone, which forms an easily-recognized feature about 50 ft. above the old road leading round the face of Hounstout Cliff. Palaeontologically it has proved, owing to the researches of Messrs. Waddington and Buckman, of great interest. It contains in places an abundance of a small rugose variety of *Exogyra nana*, recalling the *Exogyra* Bed of the Isle of Portland, associated with crushed specimens of *Rhynchonella portlandica* Blake. Of paramount importance, however, are the ammonites. Buckman[5] was the first to identify some of these with the Russian *Provirgatites scythicus* (Michalski) (see above, p. 445); and since a fairly large number of fragments have since been found, it seems to be indicated that we place the Massive Bed and Emmit Hill Marls in the *scythicus* zone, already recognized in Germany (Pl. XL, fig. 3).

The Massive Bed was selected by Sir A. Strahan [6] as the arbitrary base of

---

[1] S. S. Buckman, 1926, *T.A.*, p. 35, pl. DCLXXVII.

[2] It seems likely that this is Blake's Bed 13 (cementstone with abundant *Thracia*, 2 ft.) (*Q.J.G.S.*, 1880, p. 195). It is Latter's Bed 6 at St. Alban's Head and Emmit Hill and Bed 4 at Hounstout (*P.G.A.*, 1926, pp. 76–8).

[3] Latter's Beds 3, 4, 5 (loc. cit., pp. 76–7) and 3 (p. 78); Blake's Bed 12 (thickness 30 ft.).

[4] 1926, *T.A.*, pp. 31–3.   [5] Ibid., pp. 32–3, pls. DCLXXV, DCXCIII.

[6] 1898, loc. cit., p. 62.

the Portland Sand, a selection accepted by Messrs. Kitchin and Pringle, by Latter [1] and by Cox.[2] The thickness thus assigned to the Portland Sand, if the Parallel Bands are taken as the top, is about 100 ft., whereas Fitton's original estimate was 120–140 ft. It seems likely, however, that Fitton included in the sands the 24 ft. of sandstone above the Parallel Bands, the upper part of which, as mentioned above (p. 489) contains chert. Sir A. Strahan's estimate of the thickness of the Portland Sand was 100–120 ft., with which the classification here adopted agrees, and he certainly included the sandstones in question with the Cherty Series of the Portland Stone.[3]

The Geological Survey of 1855 mapped an indefinite line in the sandy clays about 40–50 ft. below the Massive Bed, and in this they were followed by H. B. Woodward, who estimated the thickness of the Portland Sand at 170 ft. or 130–70 ft.; [4] and more recently Buckman also has followed suit. Blake preferred a line 70 ft. lower still, giving the Portland Sands 244 ft.,[5] but with this no authors have agreed.

(b) The Isle of Portland

The Portland Beds of the Isle of Portland bear a greater resemblance to those of Eastern Purbeck than to those of Western Purbeck anywhere west of Kimeridge, or to their continuation north of Weymouth. The Stone Beds are divisible into a Freestone Series above and a Cherty Series below, and the Sands consist predominantly of argillaceous sands and cementstones rather than true sands. The details, however, differ considerably, rendering refined correlations uncertain.

The Isle of Portland represents but a small fragment of the southern limb of the Weymouth Anticline. It provides a sample of the great tract of country which formerly lay beyond the present coast-line of Dorset—a limestone upland of grassy downs like the Cotswolds, useful for sheep-grazing and corn-growing had it not nearly all been destroyed by the inroads of the sea. The island is a gently-tilted plateau, completely surrounded by cliffs of Portland Stone, the surface 'so destitute of Wood and Fuell that the inhabitants are glad to burne their Cowe Dung, being first dried against the Stone Walls, with which their Grounds are enclosed altogether', as Coker tells us.[6] At the north end the cliffs rise to 495 ft., above a tumbled undercliff of fallen blocks lodged on the Kimeridge Clay, as at Hounstout and St. Alban's Head; towards the south end or Bill they sink gently into the sea. So extensive have been the quarrying operations of the last two centuries, however, that most of the Freestone Series has been removed from the cliffs, while vast tips of debris have built up a curtain round the greater part of the north, east and west sides of the island. The east side has in addition been much obscured by landslips, due to the heavy stone-beds foundering upon the clay slopes beneath. It is only the magnificent West Weare Cliffs that give tolerably clear and accessible sections of the Portland Sand and Cherty Series at the present day (Frontispiece and Plate XXV, opposite; also map, p. 342).

The quarrying of the Portland Freestone was carried on only for local

[1] 1926, loc. cit., p. 76. [2] 1929, loc. infra cit., p. 133.
[3] A. Strahan, 1898, loc. cit., p. 60; and see footnote above, on p. 490.
[4] 1895, *J.R.B.*, p. 192. [5] 1880, loc. cit., p. 194.
[6] Rev. Coker, 1732, *Survey of Dorsetshire*, p. 38.

PLATE XXV

Photo.   *W. J. A.*

West Weare Cliff, Portland.

C = Cherty Series of Portland Stone, with SB = Basal Shell Bed; WWS = West Weare Sandstones; EB = *Exogyra* Bed; BNB = Black Nore Beds. Chesil Beach, Portland Roads and Weymouth are seen in the distance.

Photo.   *W. J. A.*

Gad Cliff from Worbarrow Tout, Dorset.

Overhanging cliff of Portland Stone and Lower Purbeck Beds. Undercliff of fallen blocks on Kimeridge Clay. Swyre Head in the distance.

PLATE XXVI

Photo.   *W. J. A.*

Photo.   *W. J. A.*

'Divers of those Snake or Snail Stones, as they call them, whereof great varieties are found in Portland, dug out of the very midst of the Quarry, of a prodigious bigness.' (Dr. Robert Hooke, *Lectures and Discourses of Earthquakes*, 1668 (1705), p. 327.)

Specimens of *Titanites* from the Portland Stone and silicified tree-trunks from the Great Dirt Bed of the quarries where the stone was got for rebuilding St. Paul's Cathedral; photographed on Portland Island.

purposes until the seventeenth century. The fame and great reputation of the stone dates mainly from its selection by Sir Christopher Wren for the re-building of St. Paul's Cathedral after the Fire of London, though it was already chosen in 1610 for rebuilding the Banqueting Hall at Whitehall. Since that time it has been used for well-known buildings all over the country, and especially in London, where its resistance to the corroding agencies in the atmosphere renders it particularly valuable. Among the buildings constructed with it are the British Museum (1753), Somerset House (1776–92), the General Post Office (1829), the Horseguards, the Foreign Office, and the Record Office; and the parts of the stone not shipped as a freestone were used in enormous quantities for the great breakwaters of Portland Harbour, one of the largest undertakings of its kind in the world, completed by convict labour in 1864.

As the result of all these activities, most of the north and central parts of the island have been removed bodily, leaving vast expanses of rubbish tips, now overgrown, traversed here and there by walls and strips of unquarried rock. The scene has been graphically described by Sir Frederick Treves:

'Here is a garden of stones with roads deep in dust or deep in mud: the grass is grey with dust, the horses, that in teams of eight or ten drag stone-laden trolleys through the waste, are grey too with the powder of stone. A faded traction-engine rumbles by crunching the stones and blackening the air, driven by earth-coloured men, who are shaken as they pass by the fearful vibrations of the machine. There are deep pits of stone, tanks of oolite, walls of white masonry laid bare by the pick, terrific slopes of loose rubble sliding down into the cool sea.

'Over a wide chasm, with sides as clean-cut as those of a graving-dock, fantastic cranes rise up into the air. They wave titanic arms against the sky, which might be the tentacles of some leviathan insect, or weave threads of wire over the abyss like the strands of some unearthly spider's-web; smoke rises from the gasping engines, while now and then a block of stone glides hissing across the void like a fearsome bird. All round are heaps of litter, piles of wind-blown dust, patches of scarred earth, and deserted pits which are becoming covered with a green mould.' [1]

Since the Great War most of the north end of the island, once the centre of this scene, the nucleus of one of the world's most famous quarrying locali-ties, has been abandoned, and instead smaller pits are being worked farther south. The horse-teams, too, have gone, superseded by files of 'faded traction-engines', which invade the roads, sometimes ten at a time, laden with their huge cubes of stone.

SUMMARY OF THE PORTLAND BEDS OF PORTLAND [2]

Titanites Zone: Freestone Series, 25 ft.

THE ROACH, $1\frac{1}{2}$–4 ft., usually about 3 ft.

White oolitic limestone, a porous mass of hollow moulds and casts from which the shells have been entirely dissolved away. The commonest species are *Aptyxiella* [*Cerithium*] *portlandica* (the 'Portland Screw'), *Trigonia gibbosa* and *Plicatula lamellosa* Cox. The 'Screw' is entirely confined to this horizon. Locally in a thin layer at the top some of the fossils are silicified.

[1] *Highways and Byways in Dorset*, 2nd ed., 1906, p. 229.
[2] The best account of the Portland Stone Series is by H. B. Woodward, 1895, *J.R.B.*, pp. 198–201.

WHIT BED, 7 ft. (in places up to 15 ft.).

Buff to white oolite, with comminuted shells, the best freestone on the island. In Fitton's time this was the only bed worked. Locally an irregular cherty band with 'sand holes' yields, according to H. B. Woodward, a micromorphic fauna of gastropods and marks off an additional 3 ft. to 3 ft. 9 in. of buff oolite below, called the Bottom Whit Bed. The Whit Bed is the principal source of the gigantic ammonites for which Portland is famous. According to Blake '*Ammonites giganteus*' is confined to this horizon, but Woodward also records it from the Curf. Some huge specimens are preserved in the garden of the Portland Stone Firm's Office on the island (see Pl. XXVI). Blake noted a plane of erosion at the base of the Whit Bed, in places cutting out the Curf, which is said by Woodward to be absent south of Weston.

CURF AND CHERT, 3–6½ ft.

Oolitic shelly limestone with numerous chert nodules. According to Woodward '*Ammonites giganteus* is abundant'.

LITTLE ROACH, 0–1½ ft.

Oolitic limestone with hollow moulds and casts of *Trigonia* and other shells, resembling the true Roach above; not always present.

BASE BED, 10 ft. (5–10 ft.).

Fine-grained, buff, oolitic freestone, comparatively unfossiliferous. Some of the old quarries from which the Whit Bed was formerly removed are now being reopened to get at the Base Bed, which may be almost as good a freestone as the Whit Bed.

No detailed correlation of these subdivisions with the thicker Freestone Series of Purbeck has yet been established. It is noteworthy that the peculiar white sublithographic stone, the Shrimp Bed, so constant a feature at the top of the formation in Purbeck, is not to be seen in Portland. Instead the top is formed by the equally peculiar Roach, not found in Purbeck. It is believed by the Worth Foreman, Mr. W. J. Bower (above, p. 485), and was suggested by Blake, that the Roach is equivalent to the *Titanites* Bed of Purbeck. This view is probably based on the tempting supposition that the Whit Bed freestone equals the Pond Freestone and that the cherty Curf is represented by the Chert Vein. But it must not be forgotten that the Curf and the Whit Bed of Portland are the home of the giant ammonites, while in Purbeck ammonites never seem to be found in the Chert Vein or in the Pond Freestone but occur in abundance in the *Titanites* Bed above. Until more detailed work on the specific identity of the ammonites in the two areas has been accomplished, the correlations must remain a matter of surmise.

Kerberites Zone: Cherty Series, 60–70 ft.

The Cherty Series is of about the same thickness in Portland as in Purbeck and of much the same appearance, but there is no proof that its upper limit is on exactly the same horizon in the two localities; the assumption is founded only upon the supposed equivalence of the Under Freestone to the Base Bed Freestone, neither of which has yielded diagnostic fossils. The Cherty Series is well exposed in the high crags forming the summit of West Weare Cliffs and

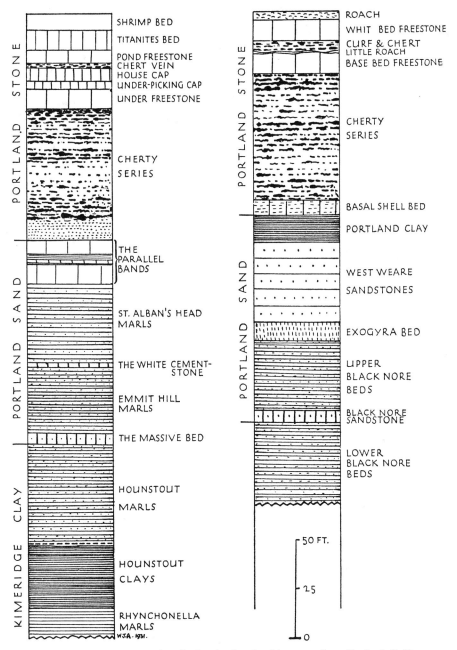

FIG. 84. Tables representing the Portland and subjacent rocks at Purbeck (left) and Portland (right).

M m

Clay Hope, on the north-west side of the island (Pl. XXV). The hard limestones contain bands and nodules of chert throughout, and in places at various levels may be seen lenticles of cellular rock full of hollow moulds of fossils like the Roach. These represent *Trigonia incurva* and *Protocardia dissimilis*, not found in the true Roach, and also the large Behemothan ammonites distinguished collectively as '*Ammonites boloniensis*' by the earlier writers, as in Purbeck. *Serpula gordialis* is in places so abundant that Blake called a band near the top a serpulite.

BASAL SHELL BED, 7–8 ft.

By far the most interesting feature of the series in Portland, and the richest palaeontological horizon in the British Portland strata, is the Basal Shell Bed. It forms a clearly demarcated base to the stone beds, offering a marked contrast to the gradual transition from cherty stone to sand in Purbeck. Until recently the Basal Shell Bed was comparatively little known, but in 1925 an extensive collection of fossils made by Col. R. H. Cunnington was described in a valuable monograph by Mr. L. R. Cox.[1] This has enriched our knowledge of the fauna of the English Portland Beds by nine new species of gastropods, eighteen new species of lamellibranchs, and several new ammonites, echinoids and polyzoa.

The thickness of the bed is 7–8 ft. It consists largely of shells compacted together in a matrix of very hard limestone crowded with the tubes of *Serpula gordialis*. The whole has been almost entirely recrystallized as calcite, which breaks along its characteristic cleavage-planes, irrespective of the boundaries between shells and matrix. It is consequently impossible to separate the shells from the stone, except where they stand out in relief upon the weathered surfaces, whence they can be removed with a hammer and chisel. The fauna consists principally of lamellibranchs, of which Mr. Cox records over 50 species. The commonest forms are those most frequently met with in the middle and lower parts of the Cherty Series in Purbeck, such as *Isognomon bouchardi*, *Chlamys lamellosa*, &c. The ammonites, although scarce, sufficed for Buckman to pronounce them as fairly evidently of late Behemothan date; they include the usual large forms of '*bononiensis*' style, the type-specimen and other material of *Kerberites portlandensis* Cox, and forms resembling *Glaucolithites gorei* (Salf.).

Glaucolithites Zone: Portland Sand, about 100 ft.

The Portland Sand of Portland still remains palaeontologically almost unknown. Its general lithic character is much like that of Eastern Purbeck, but the details are entirely different.

At the north end of the island, under the Verne Fort, the highest member is a 10 ft.–14 ft. band of stiff blue marl, called by Damon the PORTLAND CLAY.[2] This is not present in the other sections; in the clearest of all, at West Weare Cliff, the Basal Shell Bed of the Portland Stone is immediately underlain by 30 ft. of brown and grey marly sandstones and sandy cementstones of varying degrees of hardness. From these Salfeld recorded *Glaucolithites gorei*,[3] and other records by Blake show the desirability of further collecting.

---

[1] L. R. Cox, 1925, *Proc. Dorset N.F.C.*, vol. xlvi, pp. 113–72.
[2] R. Damon, 1860, *Geol. Weymouth*, p. 82.
[3] H. Salfeld, 1914, *Neues Jahrb. für Min.*, B.-B. xxxvii, p. 191.

Towards the north end of the island, according to the measurements of Mr. Latter, these sandstones thicken considerably. It is convenient to distinguish them by the name WEST WEARE SANDSTONES.

Beneath is the most conspicuous band in the series, the EXOGYRA BED. It consists of 6–8 ft. of stiff marl packed with almost a solid mass of *Exogyra nana* (Sow.),[1] and it weathers to a prominent massive band easily recognized in the cliff (Frontispiece and Pl. XXV). According to Latter the top of the *Exogyra* Bed is 31 ft. below the Shell Bed in West Weare Cliff and 57 ft. below it at the north end of the island.

The lower beds are imperfectly exposed, but a considerable thickness of sandy beds is present at West Weare Cliff, below Black Nore Gorge, and they may be appropriately referred to as the BLACK NORE BEDS. The Upper Black Nore Beds consist of black sands with lines of light grey nodules, 35 ft. thick according to Latter.[2] They are separated from the Lower Black Nore Beds by a 6 ft. band of hard, black, argillaceous sandstone with large intensely hard concretions.[3] The Lower Black Nore Beds[4] consist of blue-black sandy clays, extending as far as the foot of the visible section, 40 ft. below the Black Nore Sandstone.

The Portland Sand of Portland may therefore be summarized as follows:

TABLE OF THE PORTLAND SAND OF PORTLAND

		ft.
Portland Clay (north end of island only)		14
West Weare Sandstones . . .	30 ft. to about	40
Exogyra Bed		8
Upper Black Nore Beds		35
Black Nore Sandstone		6
Lower Black Nore Beds	. seen to	40

Total about 140

In the present state of palaeontological knowledge it would be rash to attempt any correlations with Purbeck. If we assume that the upper limit is drawn on the same horizon as in Purbeck (itself an unsafe assumption) then there is a close agreement in aggregate thickness if we take the base at the bottom of the Black Nore Sandstone, and moreover the Black Nore Sandstone falls into line with the Massive Bed of Purbeck. The presence in the Massive Bed in one locality at Chapman's Pool of a considerable quantity of *Exogyra nana* suggests correlation rather with the *Exogyra* Bed; but *Rhynchonella portlandica*, which is associated with it, has not been found in the *Exogyra* Bed, and the futility of placing any reliance on the oyster is demonstrated by the much later oyster-bed near the top of the Freestone Series at Tilly Whim (above, p. 486). Aggregate thicknesses are here likely to be a safer guide for a working hypothesis, and it seems likely that the sandy clays constituting the Lower Black Nore Beds will one day prove to be the Hounstout Marl, regarded here as belonging to the Kimeridge Clay (above, p. 446).

---

[1] Usually known by the synonym, *E. bruntrutana* Thurm., see Jourdy, 1924, 'Hist. nat. des Exogyres', *Ann. Pal.*, vol. xiii.
[2] Latter's Bed 3 at Black Nore Gorge, 1926, loc. cit., p. 82. Latter calls all the nodules and concretions, both here and in Purbeck, septaria, but very few are septarian.
[3] Latter's Bed 2, loc. cit., p. 82.   [4] Latter's Bed 1.

(c) The Northern Limb of the Weymouth Anticline

The Portland Beds in the northern limb of the Weymouth Anticline are distant from Portland Island only 5 miles across the bay, but they differ profoundly from their equivalents on the island. They also differ as much from the Portland Beds of Purbeck, and the change from the Purbeck and Portland type of development can be traced north-westwards through the numerous small connecting outcrops along the sea-coast, at Lulworth, Durdle Door and the Cow and Calf Rocks. Beyond the last outlying rock there is a gap of 2 miles to the faulted mass below Holworth House, Ringstead Bay, whence the outcrop is continuous, except for minor interruptions, to Portisham. The Portland Beds, capped by Purbecks, form a lower escarpment and spurs jutting out below the main cuesta of the Chalk Downs. Where they enter into the Chaldon Anticline they surround and floor an elongate inlier, giving rise to a striking amphitheatre known as Poxwell Circus.

It was remarked that even so far east as Gad Cliff great changes were to be seen in both the Portland Stone and the Portland Sand. There Hudleston noted that the Shrimp Bed and *Titanites* Bed seem to rest directly on the Cherty Series, while the Portland Sands are both sandier and thinner, and the divisions established farther east cannot be recognized. In the main outcrop from Ringstead Bay westward the differences are intensified, the whole formation dwindling to less than 80 ft. at Portisham.

The **Titanites Zone** or **Freestone Series** presents a curious blend of the various beds developed in Portland and Purbeck. A close study of their local changes might provide a key to the correlation of the two better-known areas, but no such investigation has yet been undertaken. In Ringstead Bay the series is represented by variable, oolitic, shelly and chalky limestones, 12 to 18 ft. thick, containing Roach as at Portland, in the upper and lower parts. Roach is also present at Sutton Poyntz, in the neighbourhood of which '*Ammonites giganteus*' is described as abundant. At Greenhill Barton there is an oyster-bed like that at Tilly Whim.[1]

The greatest attenuation seems to take place in the middle of the outcrop, at Upway, where only 3 ft. of Roach intervenes between the Purbeck Caps and the Cherty Series. The thickness at Portisham is 5 ft., but the roach has there given place to a hard limestone. We are warned by the presence of the Chert Vein in the centre of the Freestone Series of Purbeck against assuming that the chert nodules terminate everywhere on one and the same horizon.

The **Kerberites Zone** or **Cherty Series,** however, is hardly likely to embrace representatives of much of the overlying strata, for it is itself greatly reduced in thickness, measuring only 30 ft. at Portisham. Its most noticeable feature is an extreme whiteness of the limestone, which renders the series hard to distinguish from Chalk-with-flints. The Basal Shell Bed is represented at Portisham, where the lowest 6 ft. of the series consists of limestone crowded with *Serpula gordialis* and shells. Farther east, however, the bed does not seem to have been recognized, and there is nothing that could be said to represent it in the cliff-section below Holworth House in Ringstead Bay.

The **Portland Sand** has recently been the object of petrological investigation by Mr. M. P. Latter, who emphasized the much higher proportion of

[1] A. Strahan, 1898, loc. cit., p. 69.

true sand here as compared with Purbeck or Portland. Commensurate with the increase of sand, an increase in the quantities of heavy minerals was noticed, especially garnet, tourmaline, zircon and rutile, accompanied by kyanite, staurolite, muscovite and glauconite.[1] These facts he rightly considered to indicate that the shore-line lay to the north-west. Concerning the origin of the heavy minerals, however, his results are diametrically opposed to those of Dr. A. W. Groves. While Mr. Latter concluded that the heavy minerals 'point to the granite massifs of Devon and Cornwall as being the main source of origin for the Portland Sand', Dr. Groves asserts 'No indication whatever has been found of direct derivation of detritus from Dartmoor in the Jurassic rocks of Dorset, nor inland as far as the Oxford district . . . in the Portland Sand detritus resembling Dartmoor material is referred with more certainty to the granites of Normandy and Brittany.'[2]

The best section of the Portland Sand is to be seen in the faulted block below Holworth House, in Ringstead Bay. It has been described by Woodward, Strahan, Latter and Buckman, but the descriptions vary so greatly that it is almost impossible to discern a single feature in common. Woodward's total thickness adds up to 69 ft., Strahan's to 57 ft., Latter's to 121 ft., and Buckman's to 48 ft. 2 in.! Buckman, however, stated that his thicknesses were 'only guess-work',[3] while Latter evidently included in the Portland Sand most, if not the whole, of the Cherty Series. The downward passage from Cherty Series to Sands is, in fact, quite gradual, as at St. Alban's Head, but here there are no Parallel Bands to guide us in our arbitrary selection of a junction.

Buckman found, somewhere near the base of the section, a 2-in. seam of marly clay containing *Rhynchonella portlandica* Blake, *Lingula ovalis* auctt., and two minute species of *Orbiculoidea* no bigger than a large pin's head. On account of the *Rhynchonella* he was inclined to correlate this seam with the Massive Bed of Purbeck.[4]

Woodward noted, about 20 ft. from the top selected by him, 10 ft. of calcareous sandstone with *Exogyra*, *Ostrea* and *Trigonia*, which suggests a possible correlation with the *Exogyra* Bed of Portland; he also recorded *Exogyra* in about the same position at Portisham.[5]

The presence of glauconite grains in the Portland Sands of this district is of great interest in view of the important part that they play in rocks placed on palaeontological grounds on the same horizon farther inland. The chalky condition of the Cherty Series is also of considerable assistance in correlating with the next district to the north—the Vale of Wardour.

II. THE WESTERN RIM OF THE WILTSHIRE DOWNS

Owing to the unconformable overstep of the Cretaceous rocks, the Portland Beds are almost entirely concealed inland. Only two glimpses of the main outcrop are obtained under the western rim of the Wiltshire Downs, where deep notches have been cut back into the Chalk to form the Vales of Wardour and of Pewsey; and farther north two small outliers have been preserved in synclines on the Kimeridge Clay at Swindon and Bourton. The fact that the

---

[1] M. P. Latter, 1926, loc. cit., pp. 89–90.
[2] A. W. Groves, 1930, *Q.J.G.S.*, vol. lxxxvii, pp. 86, 70.
[3] *T.A.*, vol. vi, p. 37.   [4] Ibid.
[5] H. B. Woodward, 1895, *J.R.B.*, p. 194.

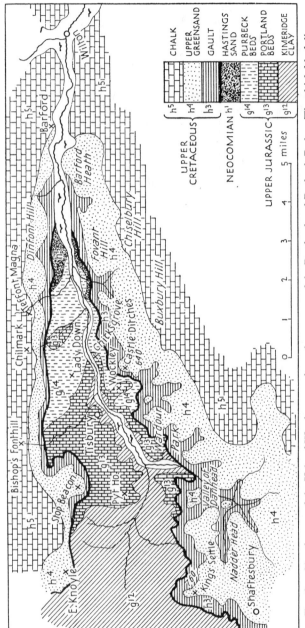

FIG. 85. Sketch-map of the Vale of Wardour, to show the outcrop of the Portland and Purbeck Beds. The thick black line represents the unconformity at the base of the Upper Cretaceous (Albian). (Based on W. H. Hudleston, 1881, *Proc. Geol. Assoc.*, vol. vii, p. 163; re-drawn.)

only two recessions in the Downs disclose Portland Beds in normal super-position above the Kimeridge Clay, while the larger, the Vale of Wardour, shows Purbeck Beds and Wealden Sands also, suggests strongly that the out-crop of these beds is continuous beneath the Cretaceous covering. The cover-ing is so thick that it has never been penetrated by wells.

In a series so variable, the discontinuity of the visible outcrops renders detailed correlation from place to place exceedingly difficult. The total area of Portland rocks exposed at the surface in Wiltshire does not exceed 4 miles square. On the other hand, the distance of the Vale of Wardour from the nearest outcrop to the south, the Isle of Purbeck, is 30 miles; while another 14 miles intervenes between the Isle of Wardour and the Vale of Pewsey, which is separated by a further 18 miles from the outlier at Swindon. From such fragmentary evidence we cannot hope to piece together more than a very incomplete picture of the inland Portland Beds.

(a) The Vale of Wardour

By far the largest and most important surface-outcrop is that of the Vale of Wardour, lying in the extreme south-west corner of Wiltshire, 10 miles west of Salisbury. The remarkable geological features of the Vale have been described by many writers. The River Nadder gathers the surface water from the Kimeridge Clay north of Shaftesbury and carries it eastward across the strike of both the Jurassic and the Cretaceous formations into the heart of the Chalk Downs, to join the Avon at Salisbury. In the apex of the narrowing valley, before the river passes on to the Chalk, the Portland, Purbeck and Wealden Beds are laid bare in a trough 8 miles long and 2 miles wide at the widest end. They strike N.–S. across the floor of the valley and disappear on both sides beneath the walls of Gault, Upper Greensand and Chalk (map, opposite).

The main outcrop of the Portland Beds occupies a dissected upland plateau within the Vale, surrounding the town of Tisbury. A number of quarries have been exploited for building-stone in the past, and one or two are still working. The principal exposures, however, lie about a mile to the east of the main outcrop, in the sides of a valley known as the Chilmark Ravine, where a roll brings the Portland Beds to the surface through the Purbeck Beds in an outcrop about a mile long and a few hundred yards wide. This pic-turesque ravine, thickly wooded, and scarred beneath the vegetation by numberless ancient quarries and tip-heaps, has for hundreds of years yielded the Chilmark Stone, which was well known long before the Portland Stone was heard of. It was employed in many important buildings in the South of England, among them the cathedrals of Salisbury, Chichester and Rochester, the Chapter Houses of Westminster Abbey, Christchurch Priory, Wardour Castle, Lulworth Castle, and Fonthill Abbey.

The Chilmark Stone is at present worked only in one locality, by means of an underground gallery in the west side of the ravine. In the past it was obtained in large open quarries as well as underground galleries, and some still afford good exposures, especially the largest quarry at the north-west end of the ravine.

The main building-stones are about 18 ft. in thickness and consist of peculiar glauconitic and sandy limestones, very different in appearance from the Portland and Purbeck-Portland freestones. Moreover, they form the base of the local Portland Stone, while above them follow thick chalky limestones with veins and nodules of chert, identical with the Cherty Series of the Upway-Portisham district; and these are in turn succeeded by an upper series of freestones more like those of Dorset.

There seems little doubt on stratigraphical grounds that the main building-stones of the Vale of Wardour correspond with the upper part of the Port-land Sands of Dorset, which become glauconitic in the northern limb of the Weymouth Anticline (p. 499). They are probably to be correlated with the West Weare Sandstones of Portland, and so with the upper part of the *Glaucolithites* or '*gorei*' zone. Beneath are some 38 ft. of sandy strata corre-sponding with the rest of the zone. So far as I am aware this correlation has not been put forward previously, and the study of the ammonites is not yet sufficiently advanced to establish it on a palaeontological basis. But since it seems to be the only one by which the Portland Beds of the Vale of Wardour can be brought into line with those in Dorset, it is tentatively adopted here.

The succession has been very thoroughly elucidated by Hudleston, Blake, and H. B. Woodward. They built on foundations laid by Fitton, W. R. Andrews of Teffont Evias, and Miss Etheldred Benett, one of the earliest of lady geologists, who in the early part of the nineteenth century resided at the stately Pyt House, west of Tisbury. In the following summary the succession established by these geologists is fitted into the classification used throughout this chapter.

SUMMARY OF THE PORTLAND BEDS OF THE VALE OF WARDOUR [1]

Titanites Zone: Upper Building Stones, 0–16 ft.

The Upper Building Stones consist of fine-grained, white or buff, siliceous, oolitic limestones, locally passing into Roach, full of cavities whence the fossils have been dissolved out. The better quality stone takes fine carving and, being very durable, was used in the elaborately carved west front of Salisbury Cathedral. The principal fossils are *Aptyxiella portlandica* (con-fined to this horizon as at Portland), *Trigonia gibbosa*, *Neomiodon cuneatum* (Sow.) [= *Cyrena* or *Cytherea rugosa* auctt.], *Eodonax dukei* (Mor. and Lyc.), *Protocardia dissimilis*, *Chlamys lamellosa*, *Neritoma sinuosa*, *Ampullina ceres*, and *Lucina portlandica*. There are occasional chert veins in the lower part of the series.

Locally in the south of the area, as at Chicksgrove Mill Quarry, the Upper Building Stones are absent, owing either to attenuation or to the unconform-able overlap of the Purbeck Beds, thus recalling the wedging out of the Free-stone Series in Gad Cliff.

Kerberites Zone: Cherty Series and Ragstone Beds, 30–5 ft.

CHALKY OR CHERTY SERIES, 24 ft.

The Cherty Series is identical with that in the northern limb of the Wey-mouth Anticline, consisting of soft, white, chalky limestone, burnt for lime,

---

[1] Based on J. F. Blake, 1880, loc. cit., pp. 199–203; W. H. Hudleston, 1883, 'The Geology of the Vale of Wardour', *P.G.A.*, vol. vii, pp. 161–85; H. B. Woodward, 1895, *J.R.B.*, pp. 203–9.

with many veins and nodules of black flinty chert. The fossils are mainly of large size and include many big ammonites recorded as '*A. boloniensis*', like those abounding on the slabs at Dancing Ledge and Winspit, but as yet unstudied (Buckman figured one specimen from Chilmark as '*Galbanites' cretarius*').[1] The other fossils are principally *Chlamys lamellosa, Ostrea expansa, Protocardia dissimilis* and *Trigonia gibbosa*; the last forms locally a bed of Roach at the base, comparable with the roach-beds in the Cherty Series of Portland. According to Hudleston the upper part of the series is missing at Chicksgrove Mill. A local chert vein near Tisbury formerly yielded silicified specimens of the coral, *Isastræa oblonga*, which were cut and polished and were famed as long ago as 1729 under the name of 'Starr'd Agates'.[2]

THE RAGSTONE BEDS, 8–10 ft. [= BASAL SHELL BED?].

At the base of the Cherty Series are 8–10 ft. of strongly-bedded, very shelly limestone, containing quartz grains but no glauconite, and divided by marly partings. In the state of preservation of their fossils as well as in their stratigraphical position these Ragstones resemble the Basal Shell Bed of the Isle of Portland. Before the wonderful fauna of the Portland stratum was made known by Mr. Cox, Hudleston declared that no Portland rock in England contained such well-preserved fossils as the Ragstones of Chilmark. The fauna, however, has received relatively little study, and no ammonites are known. The most abundant species is *Neomiodon cuneatum* (Sow.), a small lamellibranch that has given rise to more discussion and misconceptions than any other in the Upper Jurassic.[3] It is in appearance like a strongly-ribbed *Astarte*, but the hinge bears the dentition of *Neomiodon*, to which belongs the freshwater Cyprinid shells common in the Purbeck Beds and formerly known as *Cyrena*. On this account Hudleston called the Ragstones the *Cyrena* Beds and the Upper Building Stones the Upper *Cyrena* Beds, even going so far as to say that they had an estuarine or fluvio-marine origin. But of this *Neomiodon cuneatum* is no evidence, for although it has not been found in the Basal Shell Bed (or any other part of the Portland Series) of Portland, it ranges on the Continent throughout the Kimeridgian and Portlandian stages, where it certainly lived under purely marine conditions. It differs from the Purbeck species in being strongly ribbed like an *Astarte*, whereas they are smooth, and its generic affinities are still somewhat doubtful.

Other common fossils in the Ragstones are '*Cerithium*' *concavum* Sow., a number of other gastropods, and such familiar marine forms as *Protocardia dissimilis* and *Lucina portlandica*.

Glaucolithites Zone: Portland Sand, 50–60 ft.

MAIN BUILDING STONES, 18 ft.

The main building stones of Chilmark and Tisbury consist of glauconitic and sandy limestones varying from green to brownish-buff in colour. They are of generally uniform appearance, but the quarrymen detect slight differences and give to each bed a distinguishing name. The names vary in different

[1] *T.A.*, 1925, pl. DCXXI.
[2] H. B. Woodward, 1895, *J.R.B.*, p. 208.
[3] L. R. Cox, 1929, *Proc. Dorset N.F.C.*, vol. 1, pp. 175–6.

parts of the district and they are now for the most part dying out; Woodward listed the following:

	ft.	in.	
Trough Bed with *T. gibbosa*	2	8	
Green Bed	5	0	
Slant Bed	1	0	
Pinney Bed	2	0	18 ft. 0 in.
Cleaving or Hard Bed	1	0	
Fretting Bed	3	4	
Under Beds	3	0	

Fossils are generally scarce, but *Trigonia gibbosa* is not uncommon in the Trough Bed, and the Pinney Bed receives its name from an abundance of *Serpulæ*. Some ammonites occur, recorded by Blake as '*A. boloniensis*' and '*A. biplex*'. The latter is suggestive of *Glaucolithites*, but the forms have not been worked out according to modern standards. One, however, has been figured by Buckman from the Green Bed as *Gyromegalites polygyralis*.[1]

BASEMENT BEDS WITH UPPER LYDITE BED, 30–40 ft.

Woodward records that a well was sunk at the base of the Chilmark (Teffont) Quarry beneath the Building Stones to a depth of 38 ft., through clays and calcareous sandy bands to black Kimeridge Clay. These sandy strata, called the Basement Beds by Hudleston, seem to have been seen only twice in temporary excavations. In the railway-cutting (now overgrown) west of Chicksgrove Mill, east of Tisbury, Woodward described 15–20 ft. of 'brown and greenish-brown sand with clay seams and bands of indurated sand, containing casts of shells here and there and thin beds of stone near the top'. These beds seemed to merge up gradually into sandy, shelly and glauconitic limestone, presumably representing the base of the Building Stones.[2]

A more interesting section was described by Blake and by Hudleston, exposed in a road-cutting near Hazelton, between Tisbury and Wardour.[3] Beneath the Building Stones were 7 ft. of loose sands with doggers, underlain by 3 ft. of greenish concretionary gritstone, and below all, sands and clays seen to 21 ft. The chief centre of interest was the 3 ft. of gritstone, which contained not only badly-preserved *Trigoniæ, Isognomon bouchardi, Chlamys lamellosa, Protocardia dissimilis, Exogyra nana*, &c., but also occasional lydite pebbles.

This is the most southerly known occurrence of lydite pebbles. They become a constant feature in the more northerly areas, at Swindon and in Oxfordshire and Buckinghamshire, where they occur always on the same horizon and form the Upper Lydite Bed.

(b) The Vale of Pewsey

A small area of Portland Beds peeps out from beneath the Cretaceous rocks at the mouth of the Vale of Pewsey, between Potterne and Coulston, near Devizes, but exposures are too meagre for the succession to be made out. Woodward described diminutive exposures of several types of rock: cherty

---

[1] *T.A.*, pl. DCXX, c, d.　　　　[2] H. B. Woodward, 1895, *J.R.B.*, pp. 206, 204.
[3] W. H. Hudleston, 1883, *P.G.A.*, vol. vii, p. 172.

beds, fossiliferous gritty limestone, gritty and glauconitic limestone with an occasional lydite pebble, and glauconitic sands with concretions.[1] Not for another 18 miles, until Swindon is reached, 32 miles from the Vale of Wardour, do the Portland Beds again come to the surface.

(c) The Swindon Outlier

The hill upon which Old Swindon is built is capped with Portland and Purbeck Beds. They probably owe their preservation to being situated in a synclinal depression on the outskirts of the area where the Cretaceous rocks normally would overstep on to Kimeridge Clay, for they seem to be an outlier: no trace of them was found in a boring carried through the Cretaceous rocks at Burdrop, between Wroughton and Chiseldon, on the south.

Thanks to two extensive quarries on the summit of the hill, described in detail by Blake, and to a railway-cutting opened subsequently to Blake's work and described by H. B. Woodward (fig. 79, p. 456), the succession is known in considerable detail. The zonal positions of the subdivisions were established by the palaeontological work of Dr. Salfeld, whose results were later corrected and amplified by Messrs. Pringle and Chatwin.

The highest part of the Portland Stone (the *Titanites* zone) and the Purbeck Beds are seen only in the Town Gardens Quarry, long ago abandoned and now rapidly deteriorating. The lower zones and the junction with the Kimeridge Clay, which were also exposed in the railway-cutting near Old Town Railway Station, are well displayed in Okus Quarry, on the western end of the hill. Here work is still carried on apace and many gigantic ammonites may be obtained. Although some of them reach almost the dimensions of the giants of Portland and Purbeck, they are of different genera, characteristic of the earlier Behemothan Age, and they are associated with species of *Kerberites*.

There are several new features not met with farther south, but perhaps the most important is the extremely thin representation of the Portland Sand, which is reduced, both here and in the more north-easterly areas, to a few feet of Glauconitic Beds with the Upper Lydite Bed at the base.

SUMMARY OF THE PORTLAND BEDS OF SWINDON [2] (fig. 86)

Titanites Zone: The Creamy Limestones, about 10 ft.

As elsewhere, the beds of the *Titanites* zone are highly variable in short distances, a fact which has given rise to considerable discrepancies in the accounts of the succession seen in the old Town Gardens Quarry. In places the whole zone is absent and the beds beneath are overlain directly by Purbeck Beds.

Where they are preserved, there is at the top up to 6 ft. of pale, cream-coloured marly limestone, suggestive of some of the Purbeck Beds, but containing, rarely, shells of *Trigonia*. Beneath comes 4 or 5 ft. of Roach like that at Portland, riddled with hollow moulds of *Aptyxiella portlandica*, here associated with an equal abundance of *Neomiodon cuneatum*, and the usual *Trigonia*

---

[1] H. B. Woodward, 1895, *J.R.B.*, p. 210.

[2] J. F. Blake 1880, loc. cit., pp. 203–13; H. B. Woodward, 1895, *J.R.B.*, pp. 210–15; Chatwin and Pringle, 1922, *Sum. Prog. Geol. Surv.* for 1921, pp. 162–8 (see also 1922, *P.G.A.*, vol. xxxiii, pp. 152–5).

gibbosa, Chlamys lamellosa and *Protocardia dissimilis*. Apparently no ammonites have been found in these beds.

According to Blake there is at the base a distinct discordance.

Kerberites Zone: Swindon Sand and Cockly Bed, 30 ft.

THE SWINDON SAND AND STONE.

This is the thickest part of the Portland Beds of Swindon, consisting of

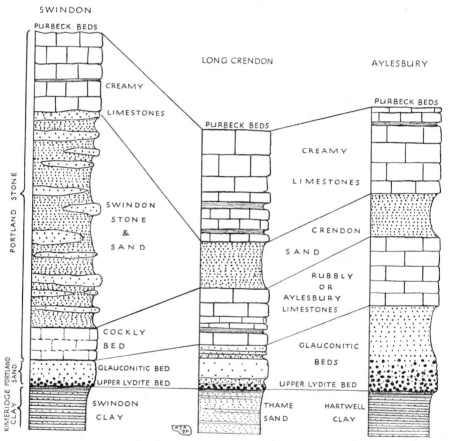

FIG. 86. Comparative sections showing the Portland Beds of Swindon, Long Crendon (after Buckman) and Aylesbury (after Woodward and Hudleston).

25 ft. of buff and white false-bedded sands with bands and lenticles of calcareous sandstone. Locally the sands are largely made up of comminuted oysters and Pectens, and they also contain some lignite; but on the whole they are very poorly fossiliferous, and ammonites seem to be unknown. The Swindon Sand forms the lower part of the section in the Town Gardens Quarry and the upper part of that in Okus Quarry, and it is still well exposed in the railway-cutting (fig. 79). Blake correlated the beds with the Cherty Series of Portland, a view which still seems probably correct.

THE COCKLY BED.

At the base of the Swindon Sand in Okus Quarry is a 4 ft. bed of creamy limestone crowded with fossils, for the most part in the form of moulds and casts, named the Cockly Bed to distinguish it from the Roach of higher horizons. The Cockly Bed is the source of nearly all the ammonites found in the Portland Stone of Swindon, and some are of gigantic size, but they still remain to be adequately described and figured. Salfeld and subsequent writers identified the commonest form as '*Perisphinctes*' *pseudogigas* Blake, but more recently Buckman has chosen a neotype of Blake's species from the *Titanites* zone of Buckinghamshire. A number of the Swindon forms will probably fall into the genus *Behemoth* of Buckman. The other common type of ammonite is the genus with heavy triplicate ribbing now known as *Kerberites*, of which the Cockly Bed of Okus Quarry is the source of the types of *K. trikranus* Buck.[1] and *K. okusensis* (Salfeld).[2] *K. kerberus* Buck. is also common; the holotype came from Chicksgrove near Tisbury, but from exactly which horizon is uncertain.[3]

Other common fossils of the Cockly Bed, approximately in order of abundance, are *Protocardia dissimilis*, *Trigonia incurva*, *T. gibbosa*, *Isognomon bouchardi*, *Pleurotomaria rugata*, *Pleuromya tellina*, *Mytilus suprajurensis* Cox. The bed is simulated in appearance by the lenticles of Roach in the Cherty Series of Portland.

Glaucolithites Zone: Glauconitic Beds and Upper Lydite Bed, $3\frac{1}{2}$ ft.

The Portland Sand is very greatly reduced, consisting of only $3\frac{1}{2}$ ft. of glauconitic, sandy limestone, with at the base a well-marked layer of lydite and phosphatic pebbles—the Upper Lydite Bed, so called to distinguish it from the Lower Lydite Bed at the base of the Swindon Clay (see p. 457).

The glauconitic limestone contains *Glaucolithites gorei* (Salf.) which is found in the upper part of the Portland Sand of Dorset (the West Weare Sandstone), and a number of similar forms, some of which have been figured and named by Buckman; e.g. *Polymegalites polypreon*[4] and *Gyromegalites polygyralis*.[5] The last was also figured from the Green Bed of the Main Building Stones of Chilmark.[6] The Upper Lydite Bed, aptly styled many years ago by Hudleston 'the basal conglomerate of the Portlands', contains rolled and phosphatized fragments of small ammonites derived from the *Pavlovia* zones of the Kimeridge Clay, upon which it reposes non-sequentially. If we correlate it with the Lydite Bed in the Vale of Wardour (as seems reasonable) it is evident that the bulk of the Portland Sands of that locality are missing at Swindon. Assuming that the portion of the Portland Sands above the Lydite Bed in the Vale of Wardour (the Main Building Stones and 7 ft. of sands) is approximately equivalent to the West Weare Sandstone of Portland and the St. Alban's Head Marl of Purbeck, then it emerges also that some 130 ft. of Dorset strata are wanting in this position at Swindon. The *scythicus* zone has nowhere been detected inland.

[1] *T.A.*, 1924, pl. DXXXV.
[2] *T.A.*, 1925, pl. DLXX.
[3] *T.A.*, 1924, pl. DXX and 1926, pl. DXX, a–b.
[4] *T.A.*, 1925, pl. DXCI.
[5] *T.A.*, 1925, pl. DCXX, a and b.
[6] *T.A.*, 1927, pl. DCXX, c and d.

(d) The Bourton Outlier

Five miles east of Swindon, just over the Berkshire border, a tiny outlier of Portland Beds, only half a mile in diameter, forms the hill on which the village of Bourton stands. The stone was formerly quarried in a large pit on the south of the village, at Bourton End, but the sides are now sloped and overgrown, and the excavation is full of water. The complete succession on the hill was made out by Godwin-Austen in 1850,[1] and the quarry was again described by Sir Andrew Ramsay in 1858.[2] The succession is essentially the same as at Swindon, except that there are a few feet (7 ft.?) of sands below the Lydite Bed. It is possible that this represents some portion of the Portland Sand, preserved here below the Lydite Bed, but it is more probable that it is a sandy development of the upper part of the Swindon Clay, equivalent to the Thame Sand of Oxfordshire (see above, p. 458). A recent excavation already referred to (p. 458), lower down the hill south-east of the old quarry, has proved that the lower part of the Swindon Clay is present in normal development, with the Lower Lydite Bed at the base, resting on the Shotover Grit Sands or *pectinatus* zone.

Combining Godwin-Austen's and Ramsay's description with this new information, it is possible to piece together the full succession at Bourton, as follows:

SUCCESSION AT BOURTON

Interpretation.	Record (Godwin–Austen).	Ft.
CREAMY LIMESTONE	Stratified earthy oolite: Ammonites and casts of *Trigonia*	8?
SWINDON SAND	Buff and yellow sand; no fossils . . .	12
	Flat-bedded white oolitic sand . . .	8
COCKLY BED	Rubbly oolite: large *Pleurotomariæ* . .	1
	Ostrea and *Perna* bed	?
GLAUCONITIC BEDS AND UPPER LYDITE BED	Pebbles in calcareous beds: fossils numerous (Ramsay calls the base of this 'a bed of hard bluish fossiliferous limestone with pebbles of Lydian stone and white quartz')	10
? THAME SAND SWINDON CLAY	Fine sands . . . . . . Kimeridge Clay below. (Swindon Clay, base greenish, sandy)	7?
LOWER LYDITE BED	(Lower Lydite Bed, lydites abundant, ½ ft.)	Seen in 1930
SHOTOVER GRIT SANDS	(Sands with large doggers, seen to 4 ft.)	

(Left bracket label: KIM. CLAY)

III. THE OXFORDSHIRE AND BUCKINGHAMSHIRE AREA

The most northerly area of Portland Beds exposed at the surface in Britain, and also the largest, extends from Nuneham Courtenay, on the east bank of the Thames, in a north-easterly direction for 25 miles past Thame and Aylesbury to Whitchurch and Stewkley, on the watershed of the Ouse. It is now dissected by the River Thame and its tributaries into scattered groups of outliers, but the original area covered by the Portland Beds now exposed must

[1] R. A. C. Godwin-Austen, 1850, *Q.J.G.S.*, vol. vi, p. 462.
[2] A. Ramsay, 1858, 'Geol. Parts of Wilts. & Gloster', p. 27, *Mem. Geol. Surv.*

have been at least 25 miles long and more than 8 miles wide. The south-western edge of the area is 21 miles from Bourton.

The principal outlier, usually designated the main outcrop, lies between Thame and Aylesbury, and, with the many others to the north and west, forms a pleasant country of small hills, often capped by Purbeck Beds, Lower Greensand or Gault, and separated by valleys of Kimeridge Clay.

The Cretaceous rocks overstep from the south and south-east in such a way that the Portland Beds seem to have been isolated like those at Swindon. In consequence also of this overstep, especially on the southern and south-eastern margins of the area, the sequence is often incomplete. The effect is most marked in Oxfordshire, about Haseley, the Baldons, the Miltons, and the long ridge of Shotover Hill, running from Garsington and Cuddesdon to Headington, where the *Titanites* zone is often absent. In the parallel ridge running from Thame to Long Crendon and Chilton and terminating in the outliers of Brill and Muswell Hills, and also in the central area about Hadden-ham, Cuddington and Hartwell, higher beds are present. From here many giant ammonites of the *Titanites* zone, like those at Portland, have been obtained. Down the dip-slope, about Thame, on the contrary, the Portland Stone Series seems to have been in places entirely removed by the Cretaceous denudation.

In the most northerly outliers, about Quainton, Oving, Whitchurch and Stewkley, exposures are now meagre, but the succession was accurately described by Fitton, and it is evident that it is complete. North-east of Ayles-bury, the Gault cuts out the Portland Beds entirely, coming to rest on the Kimeridge Clay and finally on the Oxford Clay.

Although this area is so large in comparison with the Wiltshire outcrops, it will suffice for our present purpose to summarize its features as a whole, in a single generalized succession. The same subdivisions as at Swindon may be recognized; in particular there is almost the same reduction of the Portland Sands to a few feet of Glauconitic Beds, with the Upper Lydite Bed at the base.

The sequence has been elucidated by a number of distinguished geologists. The groundwork was as usual laid by Fitton,[1] who described in detail many sections long since abandoned and too often completely obliterated. His careful work, which has preserved for our benefit many valuable and irretriev-able records, provides an outstanding example of the importance of recording every exposure before economic vicissitudes render it too late and the infor-mation is lost. Blake in 1880 summarized the succession,[2] introducing the four main subdivisions used at the present day. At the top he divided off beds of 'compacted shell brash' and 'creamy limestones' (now grouped together as the Creamy Limestones), separated by sands (the Crendon Sands of Buckman forty-five years later) from a lower set of highly fossiliferous 'Rubbly Limestones'. Below these he recognized the Glauconitic Beds and the Lydite Bed at the base. Blake's correlation of these subdivisions with those at Swindon, however, was somewhat wild.

Between 1880 and 1890 Hudleston led several excursions of the Geologists' Association over the ground in the neighbourhood of Aylesbury, publishing detailed accounts of the succession at the Bugle Pit, Hartwell, and at Aylesbury

[1] W. H. Fitton, 1836, *Trans. Geol. Soc.* [2], vol. iv, pp. 269–95.
[2] J. F. Blake, 1880, loc. cit., pp. 213–21.

and Bierton.[1] The whole area was revised by H. B. Woodward in 1895 and Blake's correlation was criticized. In particular Woodward emphasized the stratigraphical continuity of the Lydite Bed with that at Swindon.[2]

By far the most detailed investigation was made by Prof. A. Morley Davies and published in 1899.[3] He described a number of new exposures and above all was able to establish three complete vertical sections in the west and central parts of the area, at Garsington, Long Crendon and Haddenham, to compare with that published by Hudleston for the Aylesbury district in the east. The stratal succession was thus established on a sure foundation, and in certain localities it was now known in considerable detail.

Finally S. S. Buckman settled in the district. He made his home at South-field, on the road between Thame and Long Crendon, and so began in earnest his incursions into Upper Jurassic stratigraphy. He soon collected a number of large ammonites from the pits at Long Crendon, especially one now already nearly filled up at Barrel Hill,[4] south of the village, and others north-west of the village beside the road to Oakley. In 1922 he published a table of hemeræ, which was corrected and enlarged in 1926.[5] The quarrymen's terms for the beds were listed and almost every bed was assigned a zonal index; in all, the Portland Beds of Long Crendon were divided into 23 beds, for which 16 zonal indices were provided, covering his two Behemothan and Gigantitan Ages. At the same time photographs of the ammonites were published in *Type Ammonites*, most of them under not only new specific but also new generic names. Buckman explained that this was to be considered, not so much a splitting up of previously-existing zones, as an amplification of the known zonal sequence.

In his treatment of the Portland Beds of Long Crendon Buckman displayed as clearly as in any part of his work the limitations of the principles which he followed, and the illogical results inevitably obtained (both in systematy and in stratigraphy) by pursuing them beyond their reasonable capacity. In his more cautious youth he would never have pressed sound principles to so outrageous a conclusion, but in his old age he seemed to become reckless. Every bed that yielded an ammonite became for him a zone, every new thread discerned in the mighty tangle of ammonite phylogeny, however slightly different it might be from the rest, was for him a new genus. The ordinary cautions proclaimed by ecology went unheeded, and the absence of an expected ammonite was hailed as infallibly indicating stratal failure.

A different view has been taken by the most recent reviser of the families of ammonites to which those from Long Crendon belong. Dr. L. F. Spath, in his monumental work on the Indian and English Cephalopods, states that most of Buckman's Portlandian genera are quite unjustifiable, and he believes that the diversity of types upon which they are founded constitute a purely local English fauna without chronological value.[6] Nevertheless this fauna,

[1] W. H. Hudleston, see bibliography.
[2] H. B. Woodward, 1895, *J.R.B.*, pp. 216–28.
[3] A. M. Davies, 1899, *P.G.A.*, vol. xvi, pp. 15–58.
[4] Not to be confused with Barley Hill, east of Thame, one of Fitton's localities (1836, p. 282) now entirely obliterated, which Prof. Davies believes to be located at the rising ground north of Kingsey Road, Thame (1899, p. 31).
[5] S. S. Buckman, 1922, *T.A.*, vol. iv, p. 26; and 1926, vol. vi, p. 35.
[6] L. F. Spath, 1931, *Pal. Indica*, N.S., vol. ix, p. 472.

local or not, would well repay detailed examination. An interesting work awaits any palaeontologist who will undertake to describe and figure it systematically (the Portland and Purbeck specimens as well as those from Buckinghamshire), for it not only represents the expiring effort in the evolution of that almost incredibly diverse family, the Perisphinctidæ, but also their acme in diversity of form and immensity of size.

Moreover it must be recognized that Buckman, by his work at Long Crendon and by means of his refinement in ammonite identification, achieved the correlation of the Oxon.-Bucks. Portland Beds with those of Wilts. and Dorset. He showed that the Creamy Limestones contain the same assemblage of *Titanites* ammonites as the Freestone Series of the Isle of Portland; that the Rubbly Limestones contain *Kerberites kerberus*, which characterizes the Cockly Bed at Swindon; and that the Glauconitic Beds contain the *gorei* forms (*Glaucolithites*, &c.) found on the same horizon at Swindon and in the top of the Portland Sands of Dorset.

Thus we are enabled for the first time to give a summary of the Portland Beds of Oxfordshire and Bucks. fitted into the classification adopted in the type-locality.

SUMMARY OF THE PORTLAND BEDS IN OXON. AND BUCKS.[1] (fig. 86, p. 506)

Titanites Zone: Creamy Limestones, 7–12 ft.

The highest rock in all the quarries was described by Blake as several beds of a compacted shell brash, 3–4 ft. thick, having at the bottom of the topmost block a number of *Trigoniæ* (*T. gibbosa*). From these beds Buckman figured some peculiar ammonites which he did not find at lower horizons. The highest 3 ft. at Barrel Hill, Long Crendon, in the quarry now in process of being filled up, consisted of four beds with Cadicone Gigantids and was called by the quarrymen the Upper Witchett. Beneath was a 2 ft. 6 in. shell-bed with massive Gigantids called the Osses Ed.[2] From these beds Buckman figured the genotypes and holotypes of the following ammonites: *Hippostratites hippocephaliticus*,[3] *H. rhedarius*,[4] *Glottoptychinites glottodes*,[5] *G. audax*,[6] the first from Scotsgrove near Haddenham, the second and third from Barrel Hill, and the last from Coney Hill near Over Winchendon. He remarked that he had seen nothing comparable with these ammonites among the giants from Portland and suggested that the beds yielding them were younger than any Portland Beds preserved in Dorset. Perhaps they correspond with the Shrimp Bed of Purbeck, in which unidentified ammonites of similar style occur (see p. 485).

The true Creamy Limestones, as originally so-called by Blake, consist of some 5 ft.–9 ft. of white, chalky, non-oolitic and highly fossiliferous limestone, crowded with *Ostrea expansa*, *Trigonia gibbosa*, *Protocardia dissimilis* *Ampullina ceres*, &c., and with the usual giant ammonites of the *Titanites* zone. Blake stated that he found *Aptyxiella portlandica* at one locality, but he omitted to mention where.[7]

---

[1] Based on Blake, Hudleston, Woodward, Morley Davies and Buckman, loc. cit.

[2] Horses' Heads is the name given by quarrymen in many parts of England to the casts of *Trigoniæ*. Plot adopted the name as *Hippocephaloides* (*Nat. Hist. Oxfordshire*, 1676, p. 128, pl. VII, fig. 1).

[3] *T.A.*, 1924, pl. CDXCV.   [4] *T.A.*, 1924, pl. DXIV.   [5] *T.A.*, 1923, pl. CDIII.
[6] *T.A.*, 1927, pl. DCCXVII.   [7] 1880, loc. cit., p. 217.

According to Buckman certain forms of ammonites at Barrel Hill, Long Crendon, are restricted to definite beds, and he lists the workmen's terms for the beds and the ammonites derived from them as follows (the Roman numerals refer to plates in *Type Ammonites* in which the forms are figured):

Sandstone	1 ft. 7 in. with *Briareites polymeles* (CCLVII).
[Supposed position of bed at Haddenham with *Titanites titan*] (CCXXXI).	
Hard Lime	1 ft. 3 in. with *Gigantites giganteus* (CCLVI).
Soft Rock	9 in. with *Trophonites trophon* (CCCXXV and CCCXLIII).
Lower Witchett . . .	1 ft. 0 in. with *Galbanites fasciger* (CDLI) and *Pleuromegalites forticosta* (DXIII).
Hard Stone . . .	1 ft. 3 in.
Waste	7 in.
Bottom Bed or Hard Brown .	1 ft. 2 in. with *Vaumegalites vau* (DXXXVI).

It was from this district that the types of the familiar species *A. giganteus* Sowerby and *A. pseudogigas* Blake came, and Buckman has figured chorotypes.[1] The new genus *Gigantites*, which he created for *A. giganteus*, is according to Dr. Spath synonymous with the earlier-named *Titanites*, and the genus *Trophonites*, to which he assigned *A. pseudogigas*, Dr. Spath thinks may be the same as the earlier-named *Kerberites*. It is to be noted that between 1922 and 1925 Buckman reversed the order of the *Titanites* and *Briareites* hemeræ without explanation.[2] Dr. Spath considers that *Galbanites* as well as the other three genera can be considered synonymous with *Titanites*.[3]

Kerberites Zone: Crendon Sand and Aylesbury Limestone, *c.* 12–15 ft.

THE CRENDON SAND, 5–17 ft.

At the base of the Barrel Hill Quarry, Long Crendon, were 5 ft. of non-glauconitic, unfossiliferous, yellowish-brown sands, called by Buckman the Crendon Sand. Blake noticed that they are continuous throughout the district and occur 'within the limits which we must assign to the Portland Stone as distinguished from the Portland Sand'. Hudleston and Woodward noted them, still 5 ft. thick, at Aylesbury, at the other end of the district (see fig. 86), and Fitton described them on the outlier of Stewkley (7 ft. thick). In a recent excavation for a new reservoir at the northern extremity of Shotover Hill they were considerably thicker and were directly overlain by the Shotover Ironsands (Cretaceous). Here they also contained bands of hard calcareous, unfossiliferous sandstone, recalling unmistakably the Swindon Sand and Stone, with which they are beyond much doubt to be correlated. Dr. Pringle noted a similar section in a well at Coombe Wood, on the Shotover ridge, and here the thickness of the sand and stone bands (Swindon Stone) was 17 ft. He noted them also at the City Farm, where their thickness was 12–15 ft.[4]

---

[1] *T.A.*, 1921, pl. CCLVI, and 1923, pl. CCCLXXXV.
[2] Compare *T.A.*, iv, 1922, p. 26; vol. v, 1925, p. 71; and vol. vi, 1926, p. 35.
[3] L. F. Spath, 1931, loc. cit., p. 472.
[4] J. Pringle, 1926, 'Geol. Oxford', 2nd ed., *Mem. Geol. Surv.*, pp. 80–1.

THE RUBBLY LIMESTONES OR AYLESBURY LIMESTONE,[1] 6–12 ft.

Blake described the Rubbly Limestones as well displayed in several excavations now almost disappeared, at Coney Hill, Quainton, Brill, Lodge Hill and Aylesbury railway-cutting. They are still to be seen in the small quarries north-west of Long Crendon, beside the road to Oakley, where they were described under the quarrymen's terms by Buckman; the thickness here is about 7 ft. In 1922 he recorded '*P. gorei*' from about the middle, but later (1925–6) he omitted this record and inserted *Kerberites kerberus* at the base and '*Crendonites*' *leptolobatus* near the top. But Dr. Spath thinks '*Crendonites*' is probably synonymous with *Glaucolithites*, and Buckman's specimen was picked up on a heap of quarried stone and has a slightly glauconitic matrix, and so it may have come from the Glauconitic Beds.[2]

In the recent excavation for the reservoir on Shotover Hill a large quantity of the Rubbly Limestones was thrown out. The material, in the identity and state of preservation of its fossils as well as in its lithic appearance, bore a striking resemblance to the Cockly Bed of Swindon. Dr. J. A. Douglas and Mr. C. J. Bayzand obtained a number of the large ammonites comparable with those from the Cockly Bed, and it is hoped that a publication will soon appear on the subject. A similar fauna was collected by Hudleston from material thrown up during drainage operations near the George Hotel, Aylesbury.

Blake noticed that at Coney Hill many of the fossils seemed to have been derived, for their matrix differed from that of the surrounding Rubbly Limestone.

Glaucolithites Zone: Glauconitic Beds and Upper Lydite Bed, 5–10 ft.

Over the whole area the base of the Portland Beds consists of 5–10 ft. of highly glauconitic, green, rubbly, sandy limestone and sand, with the Upper Lydite Bed at the bottom, resting upon the *Pavlovia* zones of the Kimeridge Clay (or the Thame Sands; see p. 466). Two of the best sections, described by Hudleston, were the pits in the Hartwell Clay at Hartwell and Bierton, near Aylesbury, where the Lydite Bed is unusually thick and fossiliferous. A cutting on the Metropolitan Railway at Walton proved the sand and Lydite Bed to be 10 ft. thick.[3] Another good section may now be seen on Shotover Hill, since the large quarry in the Shotover Grit Sands has recently been enlarged farther into the hill and displays the Swindon Clay with the lower Lydite Bed below and the Upper Lydite Bed above, overlain by the Glauconitic Beds.

Buckman has investigated the ammonite fauna of the Glauconitic Beds in the quarries north-west of Long Crendon and he records the following succession, using again the quarrymen's terms and introducing a number of new genera. As these have not yet been revised, I give the original names.

[1] The name Aylesbury Limestone was introduced by Farey in 1811 (*teste* H. B. Woodward, 1895, *J.R.B.*, p. 224). It was also used by Conybeare and Phillips in their *Geology of England and Wales*, 1822 (map and sections).

[2] *T.A.*, 1923, pl. CDI; and 1926, vol. vi, p. 35.

[3] C. P. Chatwin and J. Pringle, 1922, 'Geol. Aylesbury', *Mem. Geol. Surv.*, p. 6, Bed 1. (Presumably Bed 2 = Rubbly Limestone, 8 ft.; Bed 3 = Crendon Sands, 5 ft.; Bed 4 = Creamy Limestones, 5 ft. 8 in., overlain by Gault.)

(Rubbly Limestones above.)

Green Specked Bed . .	9 in.	with *Polymegalites polypreon* (cf. DXCI).
Brown layer . . .	3 in.	
Green marl (ammonites in white matrix)	8 in.	with *Leucopetrites leucus* (CCCVII a–c).
Building Stone (glauconitic) .	3 ft. 0 in.	with *Glaucolithites glaucolithus* (CCCVI).
Waterstone . . .	10 in.	with *Aquistratites aquator* (DXXXIV) and *Hydrostratites bifurcus* (DCLXXVI).
Lydite Bed	5 in.	with '*Lydistratites*' *lyditicus* (= derived Pavlovids).

(Thame Sands below.)

In addition to these ammonites he figured from the Glauconitic Beds of Haddenham the genoholotype of *Behemoth megasthenes*[1] and from Thame *B. lapideus*.[2]

The Upper Lydite Bed provides one of the most constant and most useful datum lines for the correlation of the Portland Beds from Aylesbury to the Vale of Wardour. Its essential contemporaneity was insisted upon first by Hudleston, then by Woodward, and more recently by Prof. Davies.

'There are certain general reasons', wrote the last, 'why a pebble-bed of this kind should be a more trustworthy indication of a definite horizon than a line of lithological change. . . . Such a bed, especially when continuous over a large area, betokens most probably an interruption of normal sedimentation for a time by the action of a current—some of the fine material previously deposited being probably swept away at the beginning of the current-phase, while at the end of that phase the first slackening in the current would cause a deposition of pebbles previously swept to and fro, and then the deposit of fine material would be resumed.'[3]

Prof. Davies found among the usual fragments of black chert constituting the bulk of the 'lydites' at Garsington several pebbles of spherulitic felsite and one of phosphatized bone. Later he recorded also a fragment of a phosphatized ammonite at Dadbrook Hill near Haddenham.[4] More recently a detailed study has been devoted to the bed by Dr. Neaverson.[5] He divides the pebbles into two sorts, those of local origin, and those which are far-travelled. The pebbles of the first category are highly phosphatic, containing as much as 56 per cent. of calcium phosphate; they comprise phosphatic casts of lamellibranchs and Pavlovid ammonites, dull black in colour, composed of clear quartz grains, with occasional grains of glauconite, set in a brown matrix. The ammonites have for the most part been washed out of the underlying Hartwell or Swindon Clay and are thus derived, though not in the sense of Dr. Neaverson's more distantly derived pebbles of the other category.

The pebbles brought into the district from a distant source consist mainly of black chert, and in some of them Dr. Neaverson found spicules of *Hyalostelia*, a sponge found in the Lower Carboniferous Beds of Scotland, Yorkshire, North Wales and Ireland. Besides the cherts, there are a number of fragments of vein quartz and some, of especial interest, of silicified oolite. These may also be regarded as a chert, with ooliths which have been nearly

[1] *T.A.*, 1922, p. CCCV.   [2] *T.A.*, 1922, pl. CCCXLII a–c.
[3] A. M. Davies, 1899, loc. cit., p. 25.   [4] Ibid., pp. 20, 158.
[5] E. Neaverson, 1925, *P.G.A.*, vol. xxxvi, pp. 245–6.

obliterated during silicification. Similar pebbles occur in the Lower Green-sand of Faringdon, and in boulders derived from the Lower Greensand in Glacial deposits near Quainton. Their origin is problematic, but again the existing occurrences of silicified oolite, like those of spicular chert, are in the North-West Highlands of Scotland (Torridonian and Cambrian) and South Wales (Carboniferous).

These considerations, taken in conjunction with the failure of the lydites in Dorset, lead Dr. Neaverson to postulate a northerly origin for the pebbles. That they were derived originally from some part of the Caledonian mountain chain seems highly probable, but they may have come from a part of the chain lying to the east of the British Islands; and they may have travelled south in late Palaeozoic or Triassic times and have been rederived during the Jurassic period from the London–Ardennes landmass or its extension under the North Sea.[1]

IV. KENT[2]

No Portland Beds are known to exist in the British Isles north of the last faulted outlier at Stewkley, near Leighton Buzzard. An idea formerly prevalent (introduced by Pavlow), and reiterated even in certain recent text-books, that equivalents of the Portland Beds exist in the Speeton Clay of Yorkshire, has been dispelled by recent work (see above, p. 472). Another basin of deposition, however, or more probably a continuation of the Dorset trough, has been proved by the borings beneath the Weald of Kent. The Portland Beds were penetrated in at least six places, four of them in a straight line running nearly west to east from Tonbridge towards Dover, and the rest near Battle, on the south of the Weald.

Lithologically the rocks are markedly different from those in Dorset and accurate correlation of the subdivisions, with the limited palaeontological evidence supplied by the cores, was impossible. The Portland Stone was found to be much more sandy than in Dorset and there was no suggestion of a Freestone Series or a Cherty Series. The Portland Sand, on the other hand, was more argillaceous and graded down so imperceptibly into the Kimeridge Clay that for convenience it had to be classed with the Upper Kimeridge Clay, no separation being practicable. For these reasons the over-all thickness of the Portland Beds as defined in this chapter cannot be ascertained from the records.

It was found that the stone-beds were thickest in the west, at Penshurst boring, south-west of Tonbridge, where they attained a thickness of 131 ft., and at Battle, where supposedly the same beds reached 141 ft. Eastward they thinned rapidly. At Pluckley and Brabourne, two borings nine miles apart, west and east of Ashford, proved thicknesses of 70 ft. and 30 ft., while at Ottinge, midway between Ashford and Dover, only 17 ft. were left. At the first four places Purbeck Beds capped the Portlands, but at Ottinge the Purbecks had been cut out by an unconformable overstep of the Wealden Beds, and an unknown thickness of Portland Stone may also have been missing. In a boring still farther east, at Dover, the Wealden Sands had crossed both

[1] See W. J. Arkell, 1927, *Phil. Trans. Roy. Soc.*, vol. ccxvi B, pp. 80–2, where this problem is discussed in connexion with the similar lydite pebbles in the Lower Calcareous Grit.

[2] Based on Lamplugh and Kitchin, 1911, loc. cit.; and Lamplugh, Kitchin and Pringle, 1923, loc. cit.

Purbeck and Portland Beds and had come to rest on the Lower Kimeridge Clay.

The borings at Penshurst and Battle, which penetrated the fullest sequence, provide the best illustration of the development in Kent. The Lower Purbeck Beds at Penshurst, consisting here of gypsiferous marlstone, were separated from the Portland Beds, the top composed of a calcareous sandstone full of marine fossils, by a sharp line of demarcation, wanting, however, in any signs of erosion. Only the highest 6 ft. of the Portland Stone contained sufficient calcareous matter to be entitled a sandy limestone. It abounded in such fossils as *Chlamys lamellosa*, *Trigonia gibbosa* and *Isognomon bouchardi*. The sandstone below this, according to Lamplugh,

'for the next 60 ft. .... was much less limy and yielded comparatively few fossils, being a massive homogeneous rock with the bedding planes very feebly developed. At lower depths the bedding became more pronounced again, being often accentuated by clayey streaks; and the sandstone also (as at Pluckley) included hard calcareous bands containing many fossils. These beds merged gradually downward into sandy shale, which formed an unbroken passage into the Kimmeridge Clay.'[1]

The lowest 6 ft. of strata grouped with the Portland Beds were described as 'dark grey, sandy, calcareous shale with a few obscure fossils, chiefly *Ostrea*, *Pecten*, &c., passing down gradually into dark blue clay of Kimmeridge type'.[2] Beneath this were 185 ft. of dark blue and paler grey clay with bands of calcareous claystone at intervals, some of them nodular, and containing *Lingula ovalis*, *Modiola autissiodorensis*, &c., before really shaly and black clays were reached. Evidently a great part of this, if not all of it (Messrs. Kitchin and Pringle have suggested 100 ft.), corresponds to the Portland Sand of Dorset as defined above, but it is quite impossible to say where the line of separation should be drawn. Ammonites suggestive of the *Pavlovia* zones were met with about 215 ft. below the base of the stone.

In the Pluckley boring, about 5 miles west of Ashford, the succession was identical, except that the Portland Stone was only 70 ft. thick, and an ammonite, which may indicate the *Pavlovia* zones, was met with only 131 ft. below the stone.

At Brabourne, about the same distance east of Ashford, great changes had set in. Here the stone-beds were only 31 ft. thick, although the Purbeck Beds were still present above. They could be roughly divided into three blocks. At the top were 11 ft. of greyish-yellow sandy limestone with many *Serpulæ* and casts of bivalves, and a fragment of a large ammonite. The fossils were the same as in the upper portion at Penshurst and Pluckley. Below came 16 ft. of greenish-grey sandy mudstone and semi-indurated calcareous sandstone, again with *Serpulæ* and casts of bivalves. At the base, 4 ft. of hard conglomeratic rock rested abruptly on dark, highly fossiliferous shaly clay with ammonites of the *Pavlovia* zones.

At Ottinge, still farther east, where the Wealden Beds had cut down unconformably upon them, only 17 ft. of the Portland Beds were left. The whole of this consisted of glauconitic sand and sandstone, and it rested abruptly on hard dark Kimeridge Clay, evidently with an even more pronounced nonsequence than that revealed at Brabourne.

---

[1] G. W. Lamplugh, 1911, loc. cit., p. 76. [2] Ibid., p. 75.

Messrs. Lamplugh, Kitchin and Pringle considered that the non-sequence here excluded not only the greater part of the *pallasioides* zone and the overlying 'passage-beds' but also the whole of the equivalents of the Portland Sand of Dorset, and in addition some basal portion of the Portland Stone (about 300 ft. of strata at Penshurst). No lydite pebbles were reported from any of the borings, but in view of the 'conglomeratic or brecciated appearance'[1] of the basement-bed at Brabourne and the presence of glauconite, so conspicuous at Ottinge, it is tempting to suppose that the transgressive stratum is on the horizon of the Upper Lydite Bed and Glauconitic Beds of Wiltshire, Oxon. and Bucks. If this is so, then there was on both sides of the London–Ardennes ridge an overstep at about the same time, and a simultaneous appearance of glauconite in the sediment. By such a correlation, the magnitude of the non-sequence in Kent is somewhat reduced, for the Glauconitic Beds and Upper Lydite Bed of the Midlands undoubtedly represent an upper portion of the Portland Sand of Dorset.

A feature of the Portland Stone in Kent, possibly of secondary origin, but none the less remarkable, was the impregnation of the stone with bitumen. In some of the borings it was so abundant that it had collected in the joints and other crevices in thick drops. In North-West Germany, also, the presumed equivalent of the Portland Beds, the 100 ft.-thick Eimbeckhäuser Plattenkalk Series, is highly bituminous. Although shales are often bituminous, it is a rare feature in limestones and sandstones.

[1] Ibid., p. 44.

CHAPTER XVI

PURBECK BEDS

Divisions.	Ostracod Zones (Pl. XLI).	Strata in Durlston Bay.	Thickness.	
UPPER	Cypridea punctata and C. ventrosa	Viviparus Clays Marble Beds and Ostracod Shales Unio Beds Broken Shell Limestone	11′ (+ ?) 46′ 5′ 10′	72′(+)
MIDDLE	Subzone. Cypridea granulosa var. fasciculata and Metacypris forbesii	Chief Beef Beds Corbula Beds Upper Building Stones Cinder Bed Lower Building Stones Mammal Bed	30′ 34′ 50′ 8½′ 34′ 1′	157′
LOWER	Cypris purbeckensis and Candona ansata.	Marls with Gypsum and Insect Beds Broken Beds Caps and Dirt Beds	135′ 15′ 19′	170′

THE last phase of the Jurassic period, represented by the Tithonian Stage in Southern Europe where normal marine sedimentation continued, saw a profound change over the British area. The sea retreated, leaving swamps and lakes, in which a freshwater fauna flourished from Dorset to Buckinghamshire and Kent. Ammonites became extinct in the area, not to reappear until brought in with the Cretaceous transgressions.

Minor oscillations lifted the surface temporarily into dry land, on which Coniferous and Cycadean forests took root and flourished, providing a habitat for insects and, most important of all, for small mammals and reptiles, whose bones became interred in the land soils and lacustrine muds. The sea remained not far distant, however, for periodically the fresh and brackish-water fauna or land flora was replaced by a marine assemblage comprising such familiar genera as *Ostrea*, *Trigonia*, *Modiola*, *Thracia*, *Isognomon* and *Hemicidaris*. The most important of these transgressions left evidence that it affected Dorset, Sussex, Kent and South Wiltshire. In Dorset and in the Vale of Wardour was formed an oyster-bank like those in the Portland Beds, locally 8¼ ft. thick, composed of *Ostrea distorta* Sow. The other marine invasions were relatively minor episodes in a long period of lacustrine or mud-flat and swamp conditions, under which great accumulations of banded shales, clays and thin-bedded limestones were built up, often crowded with Ostracods and small freshwater molluscs, such as *Viviparus* and *Unio*.

These deposits together form the Purbeck Beds, which in the type-locality in Dorset attain a maximum thickness of 400 ft. and under the Weald of Kent reach no less than 560 ft. Superficially, as noted by several geologists, they bear an unmistakable resemblance to the Rhætic Series: banded clays and shales, fine-grained white limestone like White Lias, laminated insect-beds and even arborescent stone resembling Cotham Marble, all are found in the Purbeck Beds. It would thus appear that the sedimentary conditions under which the Jurassic period in the British area opened were to some extent repeated in its closing phase.

It is not easy to subdivide the Purbeck Beds, or to correlate the subdivisions in different parts of England. Lithologically there is little by which to be guided, for the many types of rock alternate rapidly in short distances. This is not surprising in view of the probable mode of formation of the deposits, in changing and often separate basins covered by shallow water, now open to the sea, now cut off from it by shifting sand or mud bars. The palaeontology at first sight likewise affords few criteria, for the freshwater mollusca are the same from base to summit, the assemblages at successive levels differing only in the relative abundance of certain species—differences of no use in making correlations over any but small areas.

The deficiency left by the macroscopic fauna, however, is made good by the Ostracoda. The study of these minute Crustacea *in situ* in the cliffs of Dorset was begun by Edward Forbes, who based on them a tripartite subdivision of the beds, which he published in 1850.[1] The subdivisions were at the same time accurately defined by Bristow, who mapped the Isle of Purbeck and measured the sections in great detail.[2] Unfortunately Forbes did not live to fulfil his intention of monographing the Purbeck invertebrate fauna, but the study of the Ostracods was taken up, with important results, by the late Prof. T. Rupert Jones.

Prof. Jones described and figured all the known Ostracods from the English Purbecks and determined as accurately as possible their range. He concluded as follows:

'There are fourteen species of Ostracoda in E. Forbes' three divisions of the Purbeck series of deposits. Five of them occur only in the Lower Purbeck. Of the others, six occur in both the Middle and the Upper. . . . *Cypridea punctata* for the Upper, *C. granulosa* (*fasciculata*) for the Middle, and *Cypris purbeckensis* for the Lower Purbeck, are especially characteristic.'[3]

Since this passage was written some authors have promoted several of Jones's varieties to new species, but his general conclusions as to their range have been found to hold, not only for Dorset, but also for the inland localities. Andrews and Jukes-Browne early applied Forbes's and Jones's classification to the Vale of Wardour Purbecks, with marked success, the Ostracods enabling them to prove that all three subdivisions are there represented. Detailed collecting by Prof. Morley Davies and still more recently by Mr. E. A. Merrett has shown that two zones of Ostracods can be recognized also in Buckinghamshire, although the strata are greatly reduced in thickness; the lower zone there represents the Lower Purbeck Beds and the higher zone the Middle and perhaps also the Upper Purbeck of Dorset. In N.W. Germany, 550 miles away, Koert has proved the same succession and established all three divisions (see next chapter, p. 551).

So much work has now been successfully done on the Purbeck Ostracods

[1] E. Forbes, 1850, *Edinburgh Phil. Journ.*, vol. xlix, pp. 311–13, 391; and 1851, *Rept. Brit. Assoc.* for 1850, pp. 79–81.
[2] H. W. Bristow, Vertical Sections Geol. Survey, sheet 22, No. 1.
[3] T. R. Jones, 1885, *Q.J.G.S.*, vol. xli, p. 331.

that it seems justifiable to accept their chronological value as established. It does not seem practicable, however, to distinguish more than two major zones. The upper corresponds with the range of *Cypridea punctata* (Forbes) and *C. ventrosa* Jones, embracing the Upper and Middle Purbeck of Forbes.[1] The lower zone is defined by the range of the five species enumerated by Jones, and of them may be chosen as zonal indices the two most typical and abundant, *Cypris purbeckensis* Forbes and one of the two common species of *Candona*, *C. ansata* Jones or *C. bononiensis* Jones (see Pl. XLI).

Two species seem to be confined to the Middle Purbeck, *Cypridea granulosa* (Sow.) (especially its more common variety, var. *fasciculata* Forbes) and *Metacypris forbesii* Jones. But since both of these occur within the range of *Cypridea punctata* and *C. ventrosa*, they do not indicate more than a subzone.

In contrast to the advanced state of our knowledge of the Ostracoda, the rest of the invertebrate palaeontology remains in much the same position as seventy years ago. A revision of the mollusca is badly needed, but the continued neglect of the subject is fostered by the unattractive condition of the material. Much patience would be required to procure specimens of all the species fit for photography and description.

The vertebrate fauna has fared better, as its extreme importance to palaeontology (but not to stratigraphy) merited. The mammals, reptiles and fishes are known the world over, and a voluminous literature has grown up about them. Even since the Great War revisionary monographs have appeared upon the fishes by Sir Arthur Smith Woodward and upon the mammals by Dr. G. G. Simpson. The vertebrate fossils will be noticed in the following pages when the type localities are described.

I. DORSET

(a) The Isle of Purbeck

The Purbeck Beds of the Isle of Purbeck give rise to a grassy upland, partitioned by stone walls, and simulating an outlying fragment of the Cotswold Hills. The outcrop is widest in the east, about the quarrying centres of Swanage, Langton and Worth, where the Portland Stone lies deep underground, only appearing above the surface along the south coast of the peninsula. Above the vertical cliffs of Portland Stone described in the last chapter the softer Purbecks rise in steep grassy and flowery slopes to a general height of 400 ft., clad during the summer in scabius, rest-harrow, pyramid orchis and bee orchis, and the closer turf at the top carpeted with thyme and blue, pink, and white veronica.

The hills are scarred with innumerable old quarries and the tip-heaps from abandoned shafts, which, with the remains of ancient quarrymen's shelters overgrown with brambles, suggest to a new-comer to the district the site of some ruined town of remote antiquity.

These relics scattered all over Purbeck mark the site of one of the oldest stone-quarrying industries in the country. Purbeck Stone, together with the Portland Stone mined in the cliffs and now called Purbeck-Portland, were transported far and wide as building materials long before the stone of Port-

---

[1] *C. dunkeri* Jones is unsuitable as a zonal index, for it ranges up into and abounds in the Wealden Beds in Kent.

land Island had acquired fame through Wren's selecting it for the rebuilding of St. Paul's. Purbeck Marble also, which is a freshwater limestone made up of myriads of *Viviparus* [*Paludina*] shells, was transported to Scotland, Ireland and the Continent for church decoration in the Middle Ages. The causes of its demand were its green (and occasionally red) mottled colouring and its property of taking a high polish. The most striking and extensive use to which it has been put is probably in the interior decoration of Salisbury Cathedral, where the dark green, slender detached shafts of the columns and windows are of marble, contrasting strongly with the white background of freestone. This building testifies to an active quarrying industry in Purbeck at least as early as 1258. After a life of some 700 years, or more, the industry seems at last to have died out, killed by the importation of more varied and more durable true marbles from Italy.[1]

The quarrying of the Purbeck Stone still continues, and is still largely in the control of the closed guild who have handed down the tradition from father to son from time immemorial. To the stone industry has been directly due the growth of Swanage. More recently, however, since the exploits of George the Third at Weymouth introduced the cult of sea bathing, the little stone-mining village has found in its golden sands a new and more valuable asset, and the austerities of its earlier development are becoming so thoroughly masked that they will soon be forgotten.

Several complete sections of the Purbeck Beds are afforded by the cliffs from Swanage to beyond Lulworth. The finest is that in Durlston Bay (fig. 87) to the south of Swanage, where the formation attains its maximum thickness in Dorset of 400 ft.; most of the sequence is repeated by a fault in the centre of the bay. Westward from Durlston Head, past Dancing Ledge, Seacombe and Winspit to St. Alban's Head, the basement-beds and the abrupt junction with the Portland Stone are seen discontinuously in the brow of the cliffs, while inland along the same length of outcrop many small sections of the middle and upper parts of the series are afforded by the quarries and stone mines. The next complete section is seen on the south side of Worbarrow Bay, in the small peninsula of Worbarrow Tout and the adjoining Pondfield Cove (Pl. XXVII); here the total thickness has shrunk by more than a quarter to 290 ft. Towards the west the strata become more and more steeply inclined as they approach nearer to the Purbeck Thrust Fault, and this contributes even more than the diminution in thickness to the narrowing of the outcrop.

West of Worbarrow Tout the outcrop lies for the most part beneath the sea, but several disconnected stretches remain on either side of Lulworth, pro-tected on their seaward side by bulwarks of Portland Stone. The longest runs from the south side of Mupe Bay, where outliers cap the stacks of Portland Stone called the Mupe Rocks, to Lulworth Cove. Here are some of the best of all the sections of the Purbeck Beds, especially in Bacon Hole, a cove immediately west of the Mupe Rocks, and at the famous Fossil Forest between Bacon Hole and Lulworth (Pl. XXVII). The total thickness still further diminishes, from 250 ft. at Mupe Bay to 176 ft. at Lulworth Cove (map, p. 442).

[1] The last building in which any extensive use was made of the marble was the Eldon Memorial Church at Kingston, near Corfe Castle, opened in 1880; large quantities were employed for facing the columns of the colonnades in the nave, for the chancel screen and the window pillars, and the result is extremely pleasing. The pits were opened in a field at Blashenwell Farm, north of the village.

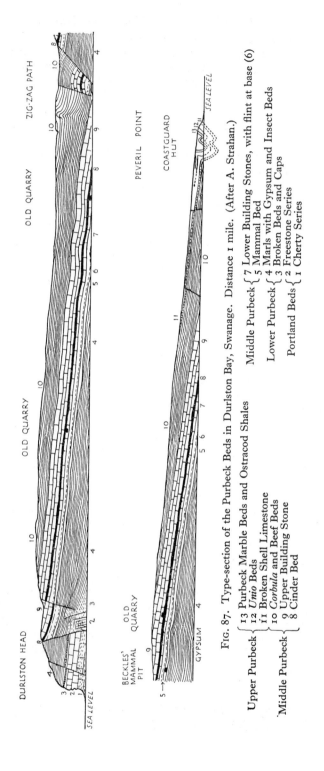

Fig. 87. Type-section of the Purbeck Beds in Durlston Bay, Swanage. Distance 1 mile. (After A. Strahan.)

Upper Purbeck {
13 Purbeck Marble Beds and Ostracod Shales
12 *Unio* Beds
11 Broken Shell Limestone
10 *Corbula* and Beef Beds
9 Upper Building Stone
8 Cinder Bed
}

Middle Purbeck {
7 Lower Building Stones, with flint at base (6)
5 Mammal Bed
4 Marls with Gypsum and Insect Beds
3 Broken Beds and Caps
}

Middle Purbeck { 8 Cinder Bed

Lower Purbeck {

Portland Beds {
2 Freestone Series
1 Cherty Series
}

West of Lulworth Cove the beds are much contorted, as is shown in the well-known section in Stair Hole, so often portrayed by pen and camera. The last remaining fragment forms the neck of the Promontory of Durdle Door, where the beds, being close beneath the Thrust Fault, seen in the Chalk cliff of the adjoining mainland, stand vertically (Pl. XXVIII). The extreme thinness of the formation here, and to a lesser extent also at Stair Hole, is due to the squeezing out of the clays, for much of the Gault also is missing.

SUMMARY OF THE PURBECK BEDS OF PURBECK [1]

Zone of Cypridea punctata and C. ventrosa: Upper Purbeck, 27–70 ft.+

VIVIPARUS CLAYS, 1 ft.–11 ft.?

The highest beds of all consist of blue, green, grey and purple marls and sandy shale with fish-remains, *Viviparus* and Ostracods. These beds are not seen at Swanage, where the junction with the Wealden Series is concealed by the harbour and town, as well as by alluvium. At Worbarrow also it is obscured by the stream that enters the sea near the boat-house. At Mupe Bay it can be clearly studied, however, and the passage to the Wealden is perfectly conformable and gradual; the thickness of the *Viviparus* Clays there is 11 ft.

MARBLE BEDS AND OSTRACOD SHALES [Upper *Cypris* Shales], 17–46 ft.

Near the top of this division are the two chief bands of Purbeck Marble, one red and the other green, composed of *Viviparus cariniferus* and *V. elongatus*. A short distance below is a tough bed of greenish sandy limestone with *Unio* (the *Unio* Bed). These hard bands crop out at Peveril Point, where they run for a mile out to sea and form some of the ledges upon which the Danish fleet is supposed to have been wrecked in the year 877. Inland they run along the High St., and outside the town they can be traced at the junction of the vale and the hills by a line of old marble workings, overgrown with woods and thickets. Below and between the limestones are clays, shales with 'beef', red and green marls, and thin seams of limestone, some of which are made up almost entirely of the minute valves of Ostracods. A piece of one of the Ostracod-limestones has been figured by Chapman,[2] who states that the species, in order of abundance, are *Cypridea punctata* (Forbes), *C. posticalis* Jones and *C. ventrosa* Jones; *Darwinula leguminella*, fish-remains and coprolites are also found.

UNIO BEDS, 5 ft. 9 in.–4 ft. 9 in.

The name *Unio* Beds is given to a thin series of clays with occasional beef and bands of greenish limestone containing *Unio*, but not so conspicuously as in the *Unio* Bed higher up. One band of carbonaceous limestone has yielded crocodilian remains in Durlston Bay and has been called the Crocodile Bed. Other fossils are remains of fish and turtles, coprolites, *Cypridea punctata*, *Viviparus*, and the freshwater bivalve *Neomiodon* ['*Cyrena*' auct.]. One of the best sections of the *Unio* Beds is to be seen on the north side of Worbarrow Tout, close to the beach, near the boat-house.

---

[1] Based on H. B. Woodward, 1895, *J.R.B.*, pp. 243–56; and A. Strahan, 1898, 'Geol. Isle of Purbeck and Weymouth', pp. 91–103, *Mem. Geol. Surv.*; thicknesses by H. W. Bristow. Alterations have been introduced in nomenclature.

[2] F. Chapman, 1906, *P.G.A.*, vol. xix, pl. v, p. 283.

BROKEN SHELL LIMESTONE OR BURR, 3 ft. 9 in.–10 ft.

This limestone, largely made up of comminuted shells of *Neomiodon* and *Unio*, with *Limnæa*, *Viviparus* and fish-remains, is 10 ft. thick at Peveril Point, where it forms the most northerly of the reefs. It also forms the shelving platform along the shore below the Grosvenor Hotel, where it contains curious round depressions noticed a hundred years ago by Fitton, but not yet satisfactorily explained.[1]

Subzone of Cypridea granulosa and Metacypris forbesii: Middle Purbeck, 55–157 ft.

CHIEF BEEF BEDS, 8–30 ft.

Dark shales with much 'beef' and selenite, with beds of limestone and layers of perished shells of *Neomiodon*; also *Cypridea*.

CORBULA BEDS, 16–34 ft.

Layers of shelly limestone, shale and marl, with beef and selenite, and marine molluscs: *Corbula* spp., *Modiola*, *Ostrea*, '*Pecten*', *Isognomon*, *Thracia*, *Protocardia*; also *Melanopsis harpæformis*, *Neomiodon*, turtle- and fish-remains, insects and Ostracods. The *Corbula* Beds contain more limestone than the Chief Beef Beds, but otherwise there is no essential difference in the lithology. The species recorded from the Middle Purbeck of Swanage by Koert are *Corbula alata* Sow., *C. forbesi* De Lor., *C. inflexa* Roemer and *C. durlstonensis* Maillard.[2]

UPPER BUILDING STONES, 8½–50 ft.

The Purbeck building-stones are not a strictly-defined stratigraphical unit, for towards the west end of the promontory they are in part replaced by marl and clay; this may be seen at Worbarrow Tout, where the great reduction in their thickness is not wholly due to the general westerly attenuation.

The stone mines lie entirely east of Kingston, and are thickest in the district between Worth, Langton and Durlston, where the downs are almost covered with tip-heaps and depressions. Many of the mines are still worked by the ancient methods, an inclined shaft following the chosen rock-band underground, the dip being generally steep.

The stone is a more or less shelly, pale limestone, occurring in rather thin beds separated by shale partings with *Cypridea punctata*. The highest 4½ ft. at Durlston, called the White Roach, and by Forbes the Pecten Bed, contains a marine fauna, including '*Pecten*', *Ostrea*, *Gervillia obtusa* Roem., *Corbula* spp. and *Protocardia*. The same assemblage (less the unidentified '*Pecten*') is met with at several horizons, separated by beds in which only freshwater or estuarine forms are found—*Viviparus*, *Limnæa*, *Hydrobia chopardi* de Lor., *Valvata helicoides* Forbes, &c.[3]

Of the greatest importance, however, is the vertebrate fauna of the Building Stones, which is large and varied, and has provided science with a number of unique forms of fishes, turtles and crocodilians. Even on a casual visit to

[1] See A. Strahan, 1898, loc. cit., p. 91.
[2] W. Koert, 1898, Inaugural-Dissertation, Göttingen, pp. 52–3.
[3] Koert, 1898, loc. cit., pp. 52–3, gives a list of mollusca from the Middle Purbeck of Swanage which he says are preserved in Göttingen Museum and are identical with forms from North-West Germany.

a mine it is usually possible to obtain from the quarrymen fin-spines of sharks ('ichthyodorulites') or bones or plates of the curious turtle, *Pleurosternum*, intermediate in structure between the later Cryptodira and Pleurodira, and so relegated to a separate sub-order, the Amphichelydia. The rich fish fauna of the Purbeck Beds, which has been monographed by Sir Arthur Smith Woodward,[1] comes almost entirely from the Upper and Lower Building Stones; the list is now as follows (compiled from his monograph):

Hybodus strictus Ag.
Asteracanthus verrucosus Eg.
A. semiverrucosus Eg.
Undina purbeckensis Smith Woodw.
Lepidotus minor Ag.
L. notopterus Ag.
Mesodon daviesi Smith Woodw.
Eomesodon depressus Smith Woodw.
Microdon radiatus Ag.
Cœlodus lævidens Smith Woodw.
C. arcuatus Smith Woodw.
Ophiopsis dorsalis Ag.

Histionotus angularis Eg.
Caturus purbeckensis Smith Woodw.
C. tenuidens Smith Woodw.
Amiopsis austeni (Eg.)
Aspidorhynchus fischeri Eg.
Pholidophorus ornatus Ag.
P. granulatus Eg.
P. purbeckensis Davies
Pleuropholis crassicauda Eg.
P. longicauda Eg.
Pachythrissops lævis Smith Woodw.
Thrissops molossus Smith Woodw.

CINDER BED, 4 ft.–8 ft. 6 in.

The Cinder Bed derives its name from its black scoriaceous appearance where weathered on the shore. It is the most conspicuous and the most constant bed in the whole Purbeck Series and marks a marine invasion of widespread importance. The mass of the rock is made up of the compacted shells of *Ostrea distorta* Sow., but in addition there are occasionally found other marine forms, such as *Trigonia, Isognomon, Protocardia purbeckensis* (de Loriol) *Serpula coacervata* Blumb., and above all the significant echinoid, *Hemicidaris purbeckensis* Forbes.[2] Unlike the oyster-banks in the Portland Series, with their restricted distribution, this bed is found on the same horizon all along the outcrop as far west as Portisham, and even in the Vale of Wardour. The marine episode to which it bears witness seems also to be marked by the presence of the same molluscan assemblage in the Weald. *Cypridea granulosa* var. *fasciculata* is recorded in the Cinder Bed by Jones.

LOWER BUILDING STONES, 5 ft. 9 in.–34 ft.[3]

The fauna of the Lower Building Stones is mainly freshwater. The top course (2 ft. 2 in. thick) immediately below the Cinder Bed is known to the quarrymen as the Feather Bed. From it was obtained the first jaw (the type-specimen) of the mammal *Trioracodon major* (Owen) and another jaw of *T. ferox* (Owen) came from immediately below.[4]

Sixteen feet below the top of the Feather Bed is a conspicuous 3 ft. band of white limestone with nodules of black chert. From this band Bristow recorded *Viviparus, Valvata, Physa, Planorbis* and *Neomiodon*.

From the chert vein to the Mammal Bed at the base of the Middle Purbeck,

---

[1] 1916–19, *Pal. Soc.*
[2] Usually rare, but Prof. Hawkins found 35 tests at Durlston: *Q.J.G.S.*, vol. lxxxi, p. cxxviii.
[3] Bristow's measurement, quoted by Woodward (1895, p. 244), was 43 ft., but according to Strahan's figures it is only 34 ft. (1898, p. 94).
[4] *Teste* Simpson. It should be noted, however, that Owen sometimes referred to the Mammal Bed as the Feather Bed.

shales predominate. Mr. F. W. Anderson has recently made a microscopic study of this part of the sequence and has discovered three rhythmic or phasal changes in the salt-content of the shales and in the Ostracod fauna. He attributes these changes to the periodic influx of fresh water into basins which were otherwise becoming steadily more saline.

'The increase in salinity had a marked influence on the Ostracod fauna. The fresh-water deposits at the commencement of each phase contain *Darwinula leguminella* (Forbes) and *Cyprione bristovii* (Jones); these are followed by an early form of *Cypridea dunkeri* (Jones) and *Cypridea punctata* (Forbes), which are later in the phase replaced by *Cypridea granulosa* (Sow.), these latter becoming more abundant as the water became more brackish.'

The third phase, according to Mr. Anderson, was terminated by the marine invasion marked by the Cinder Bed.[1]

MAMMAL BED, 0–1 ft.

The bed chosen as the base of the Middle Purbeck division is a brown or grey-brown, carbonaceous, earthy layer or 'dirt bed', suggesting an ancient soil, and filling inequalities in the hard marl below. The wonderful vertebrate fauna obtained from this bed in the nineteenth century has made Durlston Bay world-famous. Except for the few mammalian remains found in the Stonesfield Slate and the Rhætic Beds, the fauna is unique in Europe; it correlates, however, with an assemblage entombed in somewhat similar beds in the Morrison formation of North America. Most of the Purbeck specimens were found in a special excavation known as Beckles's Mammal Pit, near the top of the cliff a short distance north of Belle Vue restaurant. An exceptionally rich pocket was struck, for a large excavation made subsequently led to the discovery of only a single additional specimen. Beckles's collection, described by Owen, was purchased by the British Museum, where it can now be seen.

The revision completed in 1928 by G. G. Simpson[2] has made it clear that the time-honoured belief that these early mammals belonged to the marsupials has no foundation in fact. Simpson ranges them in four orders, *Multituberculata, Triconodonta, Symmetrodonta* and *Pantotheria*, and he regards these as derived independently from the Cynodont reptiles. The last two orders may, he thinks, have diverged from the same stock as all the living mammals, but they branched off long before the marsupials and the placentals became differentiated.

Since this bed is to the vertebrate palaeontologist perhaps the most important in the Jurassic System, it may be useful to give the complete list of the mammals found in it, as revised by Simpson—in all 19 species.

LIST OF MAMMALIA FROM THE MAMMAL BED OF DURLSTON BAY

MULTITUBERCULATA; PLAGIAULACOIDEA.

Plagiaulax becklesii Falconer.
Ctenacodon minor (Falconer).
C. falconeri (Owen).
Bolodon crassidens Owen.
B. osborni Simpson.
B. elongatus Simpson.

TRICONODONTA; TRICONODONTIDÆ.

Triconodon mordax Owen.
Trioracodon ferox (Owen).
T. oweni Simpson.
T. major (Owen).

[1] F. W. Anderson, 1932, *Rept. Brit. Assoc.* for 1931, p. 380.
[2] G. G. Simpson, 1928, *Cat. Mesozoic Mammalia Brit. Mus.*

SYMMETRODONTA; SPALACOTHERIIDÆ.
Spalacotherium tricuspidens Owen.
Peralestes longirostris Owen.

PANTOTHERIA; PAURODONTIDÆ.
Peramus tenuirostris Owen.

DICROCYNODONTIDÆ.
Peraiocynodon inexpectatus Simpson.

PANTOTHERIA (contd.)

DRYOLESTIDÆ.
Amblotherium pusillum (Owen).
A. nanum (Owen).
Kurtodon pusillus Osborn.
Peraspalax talpoides Owen.
Phascolestes mustelula (Owen).

Only second in interest to the mammals are the unique dwarf crocodiles found in this bed and at some other horizons. Most of them were described by Owen, and they have been more recently revised by Sir Arthur Smith Woodward.[1] The Teleosauridæ are represented by the genus *Petrosuchus*; but the majority belong to the Goniopholidæ, a family much more akin to the marsh-loving forms of the present day. The genus *Goniopholis* is full-sized; the rest, *Nannosuchus*, *Theriasuchus* and *Oweniasuchus*, are dwarfed.

Zone of Cypris purbeckensis and Candona ansata: Lower Purbeck, $93\frac{1}{2}$–170 ft.

MARLS WITH GYPSUM AND INSECT BEDS, 87–135 ft.

The thick marls and clays forming the greater part of the Lower Purbeck Beds contain irregular masses of gypsum, some so large that they were formerly worked in the cliff of Durlston Bay for making Plaster of Paris. In the upper part there are also pseudomorphs in mud of crystals of rock-salt, the cavities filled with carbonate of lime, and other salts also have been detected: the deposits seem to have formed in muddy evaporating basins or salterns. Mollusca are rare, but insects were apparently blown or washed into the water in abundance. Two levels of insect-beds were recognized by Fisher and Bristow, who divided the series into six subdivisions, but the details are so inconstant that they are not worth noting; Bristow's sequence (quoted by H. B. Woodward) does not tally with Fisher's and was not adopted by Strahan. The insects include butterflies, beetles, dragon flies, locusts, grasshoppers, ants and aphides, an assemblage indicative of a temperate climate.

Ostracods are, as usual, abundant at certain levels, but the species are not the same as those found in the higher divisions; the most characteristic are *Candona ansata* Jones, *Candona bononiensis* Jones and *Cypris purbeckensis* Forbes. The marine bivalve *Protocardia purbeckensis* is not uncommon but badly preserved, and gave rise to the somewhat inappropriate name Cockle Beds for the bulk of the Lower Purbecks. Occasionally *Neomiodon*, *Planorbis*, *Corbula*, &c. may be found, and also the Isopod, *Archæoniscus brodiei*. At the base are numerous bands of freestone interbedded with the marls and shales to a thickness of 36 ft. in Durlston Bay and 22 ft. at Lulworth. Bristow called this basal portion the *Cypris* Freestone, from the occurrence of *Cypris purbeckensis*.

BROKEN BEDS, 10–15 ft.

Towards the base of the Lower Purbeck Beds limestones predominate, the lowest of all, the Caps, simulating at a distance the Portland Stone. Above the Caps all over Purbeck (perhaps the best section being at the Fossil Forest,

---

[1] A. Smith Woodward, 1886, *P.G.A.*, vol. ix, pp. 318–26.

O o

near Lulworth) are remarkable shattered limestones known as the Broken Beds, the formation of which has given rise to a controversy of long standing. They consist of fragments of limestone set at all angles in an earthy rubble, like a crush-breccia. The Rev. O. Fisher first suggested in 1854[1] that the brecciation might be due to the sediments having been deposited over a mass of vegetation, which subsequently decayed and allowed the covering rock to fall in. But H. B. Woodward considered that the fact that the disturbance affects higher beds in some places than in others proved the shattering to have been subsequent to Purbeck times. He believed that the Broken Beds represent a gigantic slide-plane, upon which the whole of the Purbeck Beds above have been moved forward by pressures set up during the formation of the Purbeck Anticline and the Thrust Fault.[2] Sir A. Strahan, however, reverted to Fisher's view. He pointed out that the Broken Beds follow approximately the same stratigraphical horizon, not only from Durlston Head to the Isle of Portland, but also in the Vale of Wardour, and that they everywhere overlie a forest-grown land surface; moreover, that the nearest approach to the Broken Beds in other formations is a brecciated clay found above the Cromer Forest Bed.[3]

CAPS AND DIRT BEDS, with FOSSIL FOREST, 9–19 ft.

The junction with the Portland Stone is obscured by faulting at the south end of the Durlston Bay section, in Durlston Head, but numerous exposures of the junction and the basal Purbeck Caps are to be seen in the quarries along the cliffs westward to St. Alban's Head. The finest sections of all are at the Fossil Forest near Lulworth, and in Pondfield Cove, Worbarrow. Everywhere the junction of the two formations, though sharply defined, is perfectly conformable.

THE CAPS consist of two bands, a Soft Cap above ($4\frac{1}{2}$–$7\frac{1}{2}$ ft.) and a Hard Cap below ($4\frac{1}{2}$–11 ft.), with thin impersistent fossil soils and limestone gravel, called DIRT BEDS, at the base of either or both. The stone composing the Caps is a peculiar, porous, tufaceous and botryoidal limestone, occasionally enclosing such mixed fossils as fish-remains, *Viviparus*, Ostracods and *Archæoniscus*.

The section of these beds at the Fossil Forest, midway between Bacon Hole and Lulworth, is a highly instructive and vivid sight. The cliff there rises to a height of about 100 ft. above the sea, and the lower part, formed by the Portland Stone, is vertical or even overhanging; but the upper part, formed by the softer Purbecks, is weathered back in a gentler slope down which it is easy to scramble. Midway down, the Caps form a broad ledge upon which the famous 'forest' is displayed (Pl. XXVII).

By looking over the edge of the ledge it is possible to see that the Hard Cap, which is very thick, rests directly on the Portland Stone with the usual sharp but conformable junction. Between the Caps is the principal Dirt Bed of this part of the district, a seam of black earth 6–18 in. thick, so full of pebbles of limestone that it resembles a gravel. The Soft Cap above, 2–4 ft. thick, here also a hard tufaceous limestone, encloses the silicified trees, and the standing

[1] O. Fisher, 1856, *Trans. Cambr. Phil. Soc.*, vol. ix, p. 566.
[2] H. B. Woodward, 1895, *J.R.B.*, p. 248.
[3] A. Strahan, 1898, loc. cit., pp. 80–2.

PLATE XXVII

Photo.   *W. J. A.*

Bacon Hole and the Mupe Rocks, looking across to Worbarrow Bay
and Gad Cliff.

Rocks of Portland Stone and cliff of Purbeck Beds, tilted in the Purbeck
Anticline. Chalk cliffs of Arish Mell behind on left.

Photo.   *W. J. A.*

The Fossil Forest, Lulworth.

Silicified boles of conifers encased in 'burrs' of tufa, in position of growth in the
dirt bed, basal Purbeck Beds.

PLATE XXVIII

Photo. *W. J. A.*

Durdle Door, near Lulworth.

The arch is of Portland Stone and the highest part of the promontory is of Purbeck Beds. The soft Wealden Beds form the neck as far as the conspicuous band of dark Gault Clay, in front of which is the Chalk.

Photo. *W. J. A.*

Vertical Purbeck Beds, Durdle Cove, Lulworth.

The whole Purbeck formation is seen, with the Wealden Beds on left, all greatly reduced in thickness (partly by squeezing out of the clays).

stumps, enveloped in tufa, protrude on the upper surface as huge round bosses (called 'burrs'), their centres sometimes hollowed out to form cup-shaped depressions. Beside them lie the prostrate trunks, broken off once close to the root and again some 8–10 ft. higher up, and likewise enclosed in sheaths of tufa, although the trees within are silicified.

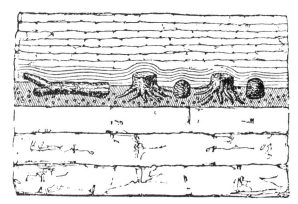

FIG. 88. The Great Dirt Bed of the Isle of Portland, show-ing boles of trees with roots, cycads, and pebbles of Portland Limestone. The Dirt Bed rests on the tufaceous Top Cap limestone and is overlain by the Aish. (After Buckland and De la Beche, 1836, *Trans. Geol. Soc.* [2], vol. iv, p. 13.)

(b) The Isle of Portland

Only Lower Purbeck Beds remain on the Isle of Portland. The greatest thickness survives north and west of Southwell, where little less than 100 ft. are exposed in the cliff. An old geological map by Buckland and De la Beche[1] shows that a century ago Purbeck Beds also overspread much of the northern part of the island, whence they have since been completely quarried away.

The essential features of the basement-beds, the tufaceous Caps with silicified trees and Dirt Beds, are present, but the details differ considerably in adjoining parts of the island and there is nowhere an exact replica of the development on the mainland. Little is to be gained by following out the changes of detail, and a typical section will suffice to illustrate the succession (fig. 89, p. 530).

The bulk of the beds consists of marls with bands of white fissile limestone, up to 3 ft. thick but splitting into thin laminæ; these bands are called by the quarrymen 'Slatt' (slate). The chief interest, however, is centred in the base-ment-beds. Below the lowest slatt is a hard streaky limestone with layers of sand, graphically called the Bacon Tier (about 2 ft.), below which, and often joined on to it, is another 2 ft. band of soft argillaceous limestone called the Aish. This rests on a tufaceous limestone called the Burr Bed or Soft Burr, which probably corresponds with the Soft Cap of Purbeck. It contains silicified tree-stumps and trunks enveloped in tufa (the 'burrs' of the quarry-men); Woodward records that one was found so large that people came from

---

[1] *Trans. Geol. Soc.* [2], vol. iv, pl. 1.

far and wide to view the 'fossil elephant'. The Soft Burr rests on the Upper or Great Dirt Bed, a blackish layer in places up to 1 ft. thick, with rolled limestone pebbles and well-preserved silicified trees and *Cycadeoidæ*. The trees are much more perfectly preserved than in Purbeck, some having been

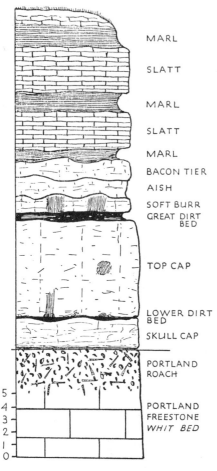

obtained as much as 23 ft. in length and from 2 to 4 ft. in diameter (see fig. 90 and Pl. XXVI). At the base are two massive blocks of tufaceous limestone called the Top Cap (10 ft.) and the Skull Cap (2–3 ft.), easily mistaken at first sight for Portland Stone. Between them, however, is in places an impersistent Dirt Bed, thinner than that above, and in the Top Cap may be seen many hollow casts of tree-trunks. The lower Dirt Bed has yielded some of the best specimens of Cycads known.

At the south end of the island the Top Cap, Upper Dirt Bed and Soft Burr are not represented.

The basement-beds of the Purbeck rest everywhere upon the Roach of the Portland Stone and the junction is perfectly even and conformable. Nevertheless, some of the pebbles in the Dirt Bed have been identified as derived from the Portland Stone,[1] thus proving that there was erosion somewhere in the vicinity, possibly over the crest of the Weymouth Anticline.

(c) **The Northern Limb of the Weymouth Anticline from Ringstead to Portisham, and the Chaldon Anticline.**

The Purbeck Beds follow the Portland feature along the outcrop in the northern limb of the Weymouth Anticline and round the Poxwell Circus, but there are now only small sections. About 12 ft. of the Lower Division is also seen in the faulted block below Holworth House, in Ringstead Bay. The details again differ from those noted elsewhere.

A complete section (total thickness 190 ft.) was formerly seen in the railway-cutting near the entrance to the Bincombe Tunnel,[2] and building-stones in the Lower Purbeck, on the horizon of the *Cypris* Freestone, are still

Fig. 89. Section of the basement-beds of the Lower Purbeck in the quarries at the south end of the West Weare Cliff, north of Black Nore, Isle of Portland.

---

[1] H. B. Woodward, 1895, *J.R.B.*, p. 265.
[2] Described by C. H. Weston, 1852, *Q.J.G.S.*, vol. viii, p. 116.

from time to time quarried near by, close to Upway. These exposures were studied in great detail by the Rev. O. Fisher, who described Upper and Lower Insect Beds and obtained many specimens from them, and also noted the chert-beds in the same position as at Durlston, and the Cinder Bed.[1] The Lower Purbeck has yielded two fish not known at Swanage—*Ophiopsis penicillata* Ag. and *Amiopsis damoni* (Eg.).[2] It is interesting to notice that the Upper Purbeck Beds here were said to be coarser and sandier than in the Isle of Purbeck and to contain a larger proportion of the spoils of the land, such as lignite and plants. This evidence agrees with that provided by the Portland Sand in pointing to a land-area in the west.

II. THE VALE OF WARDOUR

In the Vale of Wardour a considerable area of Purbeck Beds occupies the surface on the east of the Portlandian outcrop about Tisbury and surrounds the inlier in the Chilmark Ravine (see map, p. 500). The total maximum thickness is not accurately known, but it was estimated by Woodward to be about 85 ft., and, by Andrews and Jukes-Browne, 110 ft.

The general aspect of the beds and the fossils are the same as on the Dorset coast, and all three divisions have been recognized by their Ostracod faunas. The Middle Division bears witness, as in Dorset, to an important marine invasion, in which the Cinder Bed was formed, composed of *Ostrea distorta* and its associated marine molluscs and Hemicidarid.

Numerous sections, especially of the Lower Division, have been described in detail by Fitton, the Rev. P. B. Brodie, the Rev. O. Fisher, the Rev. J. F. Blake, the Rev. W. R. Andrews, A. J. Jukes-Browne, W. H. Hudleston, and H. B. Woodward. To Messrs. Andrews and

[1] O. Fisher, 1856, loc. cit.
[2] A. Smith Woodward, 1916–19, loc. cit.

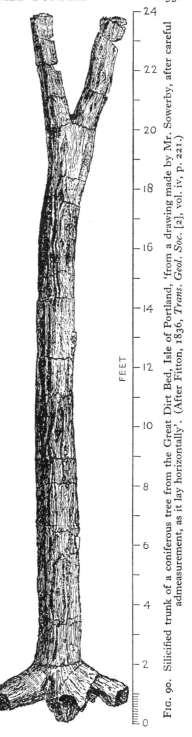

FEET

FIG. 90. Silicified trunk of a coniferous tree from the Great Dirt Bed, Isle of Portland, 'from a drawing made by Mr. Sowerby, after careful admeasurement, as it lay horizontally'. (After Fitton, 1836, *Trans. Geol. Soc.* [2], vol. iv, p. 221.)

Jukes-Browne, however, belongs the credit for deciphering the Ostracod succession and thus establishing the correlation of the beds on a palaeontological basis. They were able to show, by careful collecting *in situ*, that the attenuated Purbeck Beds of the Vale of Wardour contain representatives of all the three divisions of Dorset, and at the same time to set bounds to the divisions. It was thus established that the feeble representation of the formation in the Vale was not due to removal of the major portion of the beds by pre-Cretaceous denudation, but to a piecemeal thinning out of its component parts.

SUMMARY OF THE PURBECK BEDS OF THE VALE OF WARDOUR[1]

Zone of Cypridea punctata and C. ventrosa: Upper Purbeck, 20 ft.

Only one good section of the Upper Purbeck Beds has been described, a complete one in the railway-cuttings west of Dinton Railway Station, first studied by Messrs. Andrews and Jukes-Browne, and later by H. B. Woodward in company with Sir A. Strahan. There was some difference of opinion regarding the horizon at which the boundary between the Wealden and the Purbeck Beds should be drawn, and in spite of the opinion of the earlier authors that there were signs of discordance between the two formations, H. B. Woodward and Sir A. Strahan showed that the passage was in reality perfectly conformable. At the same time they assigned to the base of the Wealden some considerable thickness of white clays and sands regarded by the earlier writers as part of the Purbeck.

The Upper Purbeck Beds at Dinton as so restricted comprise 20 ft. of highly variable strata, ranging from blue clay and white or blue marl to brown sandstone and blue-hearted shelly limestone—all types of rock common in the same division in Dorset. Woodward regarded the highest 'lithological Purbeck' bed, a band of stiff white marl, as the top of the formation. At the base were about 6 ft. of blue clay. Messrs. Andrews and Jukes-Browne collected *Cypridea punctata* from three levels and at one horizon they also obtained *Cyprione bristovii*. About the middle of the division *Unio, Neomiodon* and *Viviparus* were collected from a shelly limestone which they considered to be probably equivalent to the *Unio* Beds of Durlston Bay.

Subzone of Cypridea granulosa and Metacypris forbesii: Middle Purbeck, 22 ft.[2] (20–5 ft.).

The Middle Purbeck Beds also were completely exposed in the Dinton railway-cuttings, as well as in several quarries, especially at Teffont Evias, described by Andrews and Jukes-Browne.

The topmost bed grouped with this division was a band of hard white marl, its upper surface eroded and the hollows filled with clay. Below this were about 6 ft. of shales and sandstones with occasional *Neomiodon*, and then a remarkable 3-in. layer of hard marly limestone in places crowded with the Isopod *Archæoniscus brodiei*.

About 2 ft. below the *Archæoniscus* Bed is the most easily recognized horizon of all, the Cinder Bed, 1 ft. 3 in. thick at Dinton, 2 ft. 6 in. thick at

---

[1] W. R. Andrews and A. J. Jukes-Browne, 1894, *Q.J.G.S.*, vol. l, pp. 44–69; H. B. Woodward, 1895, *J.R.B.*, pp. 267–75.

[2] Woodward's estimate plus a portion of his Lower Purbeck which contains Ostracods belonging to the Middle Division. Andrews and Jukes-Browne estimated at least 32 ft.

Teffont Evias railway-cutting, and 1 ft. 6 in. thick at Teffont quarry (near the church); as in Dorset, this bed is largely made up of *Ostrea distorta* Sow. Although it would usually be rash to consider oyster-beds contemporaneous at places so distant as the Vale of Wardour and the Dorset coast, the correlation in this instance is supported by several other marine forms—occasional spines of *Hemicidaris purbeckensis* and two species of *Trigonia*, one likened to the Portlandian *T. gibbosa* Sow., the other a new species, *T. densinoda* Etheridge.

Below the Cinder Bed are more alternations of compact marlstone, shelly limestone and shale, for the most part indistinguishable lithologically from the local Lower Purbecks. The best section of these beds was described by Andrews and Jukes-Browne in the large quarry west of Teffont Evias Church. Here they found, by carefully collecting the Ostracoda *in situ*, that *Cypridea punctata* extended to 7 ft. 10 in. and *C. granulosa* var. *fasciculata* to 11 ft. below the Cinder Bed, while at lower levels only *Cypris purbeckensis* was found; they therefore drew the line between the Middle and Lower Purbecks 11 ft. below the Cinder Bed. In the lowest bed of their Middle Purbeck *Cypris purbeckensis* and *Cypridea granulosa* both occurred plentifully, but they justifiably selected 'the point where *Cypridea fasciculata* first makes its appearance'. 'It is true', they wrote, 'that *Cypris purbeckensis* is still the most abundant form, so that the bed might be grouped with either division, but we prefer to regard the incoming of *Cypridea fasciculata* as marking the base of the Middle Purbeck Beds.'[1]

Although this careful work showed that the classification of the Purbeck Beds by means of the Ostracods could be applied in the Vale of Wardour as well as in Dorset, H. B. Woodward remained unconvinced of the zonal value of Ostracods, and having declared his scepticism in the discussion following the reading of Andrews's and Jukes-Browne's paper, he ignored their results when compiling *The Jurassic Rocks of Britain*. In describing the Teffont Evias quarry he was guided by the lithology, and in order to group all the main bands of limestone together, he carried the top of the Lower Purbeck up to within 4 ft. 7 in. of the Cinder Bed (and to within 1 ft. 5 in. of it in the Dinton railway-cutting). At the same time, Andrews's and Jukes-Browne's records of Ostracods appear in his account merely as 'Cyprides'.[2]

Zone of Cypris purbeckensis and Candona ansata: Lower Purbeck, about 45 ft.[3]

The Lower Purbeck Beds have been exposed in numerous quarries. The highest portions, best seen in the base of the quarry south-west of Teffont Evias Church, contain white limestone bands locally known as Lias and bearing a resemblance to true Lower Lias limestones. Lithologically they are indistinguishable from the basal portion of the Middle Purbeck, but they may be recognized by the characteristic *Cypris purbeckensis*. Fish-remains also are numerous, some as fine as any obtained in Dorset in the Middle Division. From the Lower or Middle Purbeck of Teffont came the types of *Coccolepis*

---

[1] R. W. Andrews and A. J. Jukes-Browne, 1894, loc. cit., p. 54.
[2] 1895, *J.R.B.*, pp. 271, 274.
[3] Woodward's estimate, less that portion of his Lower Purbeck which is here assigned to the Middle Division. See, however, Andrews and Jukes-Browne, 1894, p. 63, who estimate the thickness of the division to be 70 ft.

andrewsi Smith Woodw., *Mesodon parvus* Smith Woodw., *Pholidophorus pur-beckensis* Davies and *Pleuropholis formosa* Smith Woodw.; while from the Vale of Wardour, but less localized, came the types of *Ophiopsis breviceps* Egerton (a common species), *Enchelyolepis andrewsi* Smith Woodw., *Cera-murus macrocephalus* Egerton and *Leptolepis brodiei* Agassiz.[1]

The basement-beds have been described in the Chilmark Ravine and at Chicksgrove, Wockley and other places near Tisbury, the details differing in every exposure; on the whole, however, they are strikingly like their counter-parts in Dorset. At Chilmark the base is formed by a 6 ft. block of tufaceous 'cap' with seams of chert, and a little higher up there are two Dirt Beds with gravelly stones, fossil wood and *Cycadeoidea*. In the basal bed at Wockley Messrs. Andrews and Jukes-Browne found the characteristic Ostracods *Candona ansata* and *C. bononiensis*, but as these were supposed to be estuarine forms, they were led to include the bed with the Portland Stone. Woodward, however, maintained that it unquestionably belonged to the Purbeck and that there was nowhere in the Vale of Wardour any difficulty in determining the junction between the two formations. Subsequent work on the Ostracods has brought the palaeontology into line with Woodward's opinion, for *Candona ansata* and *C. bononiensis* have now been listed as definitely freshwater forms.[2]

Hudleston considered, from the way in which the Purbeck Beds rest upon different subdivisions of the Portland Stone, that there is undoubtedly an unconformity between the two.[3] But this view was contested by Woodward.[4]

'No doubt', he wrote, 'there are abrupt changes here and there between the for-mations, as there sometimes are between individual beds in the Portland series. There is, however, no discordance such as would imply upheaval and denudation of the strata. The phenomena may be attributed in part to contemporaneous erosion, in part to the attenuation and local deposition of certain [Portlandian] sediments; while again the variations in the lithological characters of different layers serve to render the results of minute correlation very difficult and uncertain.'

III. THE SWINDON OUTLIER

A small patch of both zones of the Purbeck Beds, with a maximum thick-ness of 19 or 20 ft., caps the Portlandian outlier on Swindon hill. The only exposure described is the old Town Gardens Quarry, but the condition of the section is now so bad that we have to rely almost entirely on the published accounts of the last century.

The beds comprise a varied series of marls, clays and hard white marly limestone, from which *Viviparus*, *Unio*, *Planorbis* and the Ostracods *Candona ansata*, *C. bononiensis*, *Cypris purbeckensis*, *Cypridea dunkeri*, *C. punctata* and *Cythere retirugata* have been recorded. But of any Dirt Beds or tufaceous limestones suggestive of the Caps farther south, there is no sign.

At Swindon there is no doubt about the relations of the Purbeck to the Portland Beds: they are distinctly unconformable. The uneven base of the Purbecks was seen in the west face of the quarry to cut down across the *Titanites* zone or Creamy Limestones on to the Swindon Sand and Stone

---

[1] A. Smith Woodward, 1916–19, op. cit.
[2] F. Chapman, 1899, *P.G.A.*, vol. xvi, p. 43.
[3] W. H. Hudleston, 1883, *P.G.A.*, vol. vii, pp. 170–4.
[4] H. B. Woodward, 1895, *J.R.B.*, p. 208.

(*Kerberites* zone), and in the base of the Purbeck both Blake and Woodward found derived blocks of Swindon Stone and pebbles of Creamy Limestone containing *Aptyxiella portlandica*. To Woodward it appeared also that there was a certain amount of overlap of the lower layers of the Purbeck Beds by the higher.[1]

Blake somehow became obsessed with a belief that the beds here were older than those farther south, and were to be correlated with a part of the Portland Beds of Dorset.[2] But Woodward remarked 'such a view seems to me purely hypothetical'.[3] It would seem more probable, if the overlap suggested by Woodward be a fact, that the beds seen at Swindon are not even so old as the lowest Purbeck Beds in the type-locality. In this way might be explained the absence of tufaceous Caps or Dirt Beds and the unconformity. The records of *Cypridea punctata* and *C. dunkeri* point to the presence of Middle or Upper as well as Lower Purbeck Beds. Although the summit is not seen at Swindon, these Ostracods give an indication of great attenuation in the formation between South and North Wiltshire.

IV. THE OXON.-BUCKS. AREA

No Purbeck Beds are known on the Portlandian outlier at Bourton, but in the large tract in Bucks. and on the borders of Oxford they are scattered over almost as wide an area as the Portland Beds. They are reduced to a number of small outliers capping hills of Portland Stone and inliers showing through the Gault, or to small outcrops protruding from beneath Lower Cretaceous sands. At the best these are merely fragments, left over by the accidents of denudations, not only of the subaerial denudation of the recent period which has dissected the country into scattered hills, but also of three subaqueous denudations in Lower and Upper Cretaceous times. As explained in the last chapter, the Portland sequence is not everywhere complete, owing to the overstep of the Cretaceous rocks, and from this it follows that in places the Purbeck Beds were either entirely removed or never deposited.

In the ridge of Shotover Hill, near Oxford, a small patch of Purbeck Beds occurs at the south end, about Garsington, but towards the north the Lower Cretaceous sands rest on the *Kerberites* zone of the Portland Stone. This is of especial interest because the sands, called the Shotover Sands or Shotover Ironsands, have from time to time yielded freshwater fossils (of which a collection is preserved in the Oxford University Museum) denoting that they are of Wealden date. They seem to be overlapped or overstepped in turn by the marine Lower Greensand. Lamplugh investigated the age of the Shotover Sands (including those of Brill); and after revising the stratigraphy of the district of their occurrence and summarizing the palaeontological evidence, he agreed with the opinion reached by Strickland, and later by Phillips and by Prestwich, that they are entirely of freshwater origin and that 'they represent the lowermost part of the Wealden Series of Kent and Sussex, and may be correlated with the Hastings Beds'.[4]

Thus there is in the Oxford district the first unequivocal evidence of overlap

[1] J. F. Blake, 1880, *Q.J.G.S.*, vol. xxxvi, pp. 203, 207; H. B. Woodward, 1895, *J.R.B.*, p. 277.
[2] 1880, loc. cit., pp. 203–13, and repeated in 1885, *Q.J.G.S.*, vol. xli, p. 352.
[3] 1895, *J.R.B.*, p. 278.
[4] G. W. Lamplugh, 1908, in Pocock, 'Geol. Oxford', p. 66, *Mem. Geol. Surv.*

of the Purbeck by the Wealden Beds. No such overlap is apparent in the west; but under the north and east of Kent, where the formations approach nearer to the London–Ardennes landmass than at Oxford, it is proportionately more strongly marked.

The relics of Purbeck Beds that survive in Oxon. and Bucks. are now only poorly exposed, but Fitton described numerous sections, since for the most part obscured.[1] A general survey of the area was published in 1899 by Prof. A. Morley Davies, who described most of the exposures then extant.[2] He enumerated eighteen patches of Purbeck Beds, at the following localities: Garsington, Long Crendon, Brill, Towersey, Haddenham and district, Stone, Bishopstone, Aylesbury, Coney Hill, Quainton Hill, Oving Hill, Weedon and Stewkley. Since then the part of the area around Aylesbury has been again covered by Mr. E. A. Merrett, who searched especially for Ostracoda *in situ*.[3]

The most complete sections that have been described are the well-known Bugle Pit at Stone (Hartwell), where 12 ft. of Purbeck Beds can still be seen, and a quarry west of the cross-roads (King's Cross) north of Haddenham. This quarry, which was described by Prof. Davies and yielded interesting Ostracods, showed 8 ft. of Purbeck Beds without reaching the top, but it is now filled up with tins. The other exposures are all considerably smaller.

Lithologically the beds are highly variable, as in the other localities, consisting chiefly of green or grey clays and marls, with thin seams of hard whitish limestone, occasionally botryoidal or showing obscure oolitic structure. Some of the bands are shelly, and the genera *Viviparus*, *Planorbis*, *Neomiodon*, *Modiola* and *Mytilus* have been recorded, but nearly always too badly preserved for specific identification. Fish teeth and scales, too, are not uncommon: *Pleuropholis serrata* Egerton and *Arthrodon intermedius* Smith Woodw. have been collected from the Bugle Pit, Hartwell, and near Aylesbury.

Of greatest importance, however, are the Ostracoda, which Messrs. Chapman and Merrett have turned to good account in correlating with other areas. Mr. Merrett's collecting in those pits which are still open, and Prof. Davies's in some others now closed, have established that Middle or possibly even Upper Purbeck Beds are present in some of the localities in the Aylesbury district. At the Bugle Pit near Hartwell, at Haddenham near Ford, and at Creslow Farm near Whitchurch, they have found that the uppermost portion of the sequence contains only the Middle or Upper Purbeck forms *Cypridea punctata*, *C. dunkeri*, *C. granulosa* and *Cyprione bristovii*, while the rest of the sequence is characterized by the Lower Purbeck forms *Candona ansata*, *C. bononiensis*, *Cypris purbeckensis*, *Cythere transiens* and *C. retirugata*. In the Bugle Pit, Mr. Merrett describes the junction of the two divisions as clearly visible in the form of an uneven line running through a thick bed of unstratified marl, from 2 ft. 10 in. to 3 ft. 3 in. below the top of the pit. He speaks of the relations of the two divisions, from the upper of which he obtained only the Middle or Upper Purbeck Ostracods, as being unconformable and proclaiming erosion within the Purbeck formation. At this pit and

[1] W. H. Fitton, 1836, *Trans. Geol. Soc.* [2], vol. iv, pp. 269–92.
[2] A. M. Davies, 1899, *P.G.A.*, vol. xvi, pp. 15–58.
[3] E. A. Merrett, 1924, *Geol. Mag.*, vol. lxi, pp. 233–8.

at two other places in the district, namely Brill and the railway-cutting ¼ mile west of Haddenham, seams of peculiar earthy breccia have been noticed, consisting of angular fragments of pale mudstone and bits of lignite scattered in a muddy loam. Similar rock was found in Kent.[1]

At some localities, such as the pit now filled in, at King's Cross, north of Haddenham, and again between Bishopstone and Walton, Prof. Davies collected from the Lower Purbeck Beds a mixture of species of different habitat, apparently indicating a gradual transition from marine to freshwater conditions, but probably only of minor and local importance, like the phases in the Middle Purbecks of Durlston Bay, detected by Mr. Anderson. There is never any difficulty in separating the Purbeck Beds from the purely marine Portland Stone with its highly characteristic fauna of large mollusca. The dividing line is as usual sharp, although there are no signs of unconformity.

The types of basement-beds so characteristic of Dorset and the Vale of Wardour are generally entirely absent. Silicified wood has been noted at Garsington, however, and Buckland recorded traces of a dirt-bed above the Portland Stone about 2 miles north of Thame—probably at Long Crendon.[2]

The recent studies in the Ostracods enable us to obtain a far truer conception of the meaning of the Purbeck Beds of this area than was possible before. We may now visualize a thin formation made up of both the Lower and the Middle and perhaps also the Upper Divisions of Dorset, and therefore as truly representative of the fully developed sequence as are the greatly reduced but composite Portland Beds beneath.

V. KENT AND SUSSEX[3]

The part of the trough in which the Purbeck Beds attain their greatest thickness in Britain is that underlying the Weald of Kent. It may be presumed that the underground extension is continuous from Dorset, but on this point there is no evidence. The beds crop out at the surface over a small area in the crest of the Wealden Anticline, north and north-west of Battle. The total length of the outcrop does not exceed 10 miles and the breadth is nowhere so much as 1 mile; and it is separated into three portions, of which the two easterly are due to faulting.

The beds have been worked at many places on the outcrop. The general development is closely comparable with that in Dorset, the succession consisting of shales and clays with subordinate thin limestones, wrought in open quarries and in shafts, and the fauna comprising the familiar molluscs *Viviparus*, *Unio*, *Corbula*, *Neomiodon*, *Melanopsis* and *Ostrea distorta*, with insect remains, Purbeck Ostracods and fish and reptile bones. Conybeare and Phillips in 1822 and Webster and Fitton in 1824 recognized the similarity between these oldest strata exposed in the Weald and the Purbeck Beds. From surface indications the thickness was estimated at about 400 ft., as in Dorset.

In the present century a flood of new light has been thrown on these beds by the borings for coal. The formation was penetrated from top to bottom

[1] G. W. Lamplugh, 1908, in Pocock, 'Geol. Oxford', p. 63; and 1911, 'Mesoz. Rocks in Coal Expl. in Kent', p. 40 (with photograph); *Mems. Geol. Surv.*

[2] H. B. Woodward, 1895, *J.R.B.*, p. 279.

[3] Based on Lamplugh and Kitchin, 1911, loc. cit.; and Lamplugh, Kitchin and Pringle, 1923, loc. cit.; also H. B. Woodward, 1895, *J.R.B.*, pp. 280–6.

by five borings: one was on the south side of the Weald at Battle, near the outcrop but started on Wealden Beds; the others farther north, at Penshurst south-west of Tonbridge, and at Pluckley, Hothfield and Brabourne, in the Ashford district.

The Battle site was 2¾ miles east-south-east of the old sub-Wealden borings of 1874–5, but they had started on Purbeck Beds and therefore did not penetrate the full thickness. The new sinking of 1907–09 first passed through 424 ft. of Wealden Beds and proved the total thickness of the Purbecks (according to the grouping adopted by Lamplugh) to be 387 ft. The greatest thickness known in Britain was proved at Penshurst, the most westerly of the borings, where 562 ft. of strata were grouped with the Purbeck Beds.

The more easterly borings showed great attenuation: at Pluckley, 23 miles east of Penshurst, the thickness was 100 ft., at Hothfield it was 68 ft., and at Brabourne only 60 ft. Finally, in the numerous borings in the country farther east and north, the formation was altogether absent.

An interesting fact in connexion with this easterly and northerly attenuation is the overlap of the Purbeck Beds by the Wealden, already noticed at Oxford. In the westerly borings, those in which Purbecks were present, the boundary could only be fixed with difficulty and arbitrarily, owing to the perfect gradation from the one formation to the other. Farther east, however, the Wealden Beds were proved to pass on after the disappearance of the Purbecks and they were met with in numerous other borings, resting upon the lower formations. At Dover and Folkstone the Wealden Beds (themselves attenuated) rested upon an eroded surface of the Kimeridge Clay; at Guildford, Chilton, Fredville and Harmansole they had descended on to Corallian Beds; at Oxney on to Kellaways Beds; and at several other more remote places they were in contact with Great Oolite or the Palaeozoic platform.

In the Penshurst section, where the beds were most fully developed, the succession consisted almost entirely of clays and shales with subordinate mudstones or thin sandstones, with only very occasionally minor bands of limestone; in the more easterly sections limestones became somewhat more prominent. In all the borings the palaeontological succession was much the same. The fauna was predominantly a freshwater one, except in a band about the middle, which denoted a marine invasion probably to be correlated with that of the Middle Purbeck Beds of Dorset and the Vale of Wardour.

The commonest fossil in the highest 130 ft. of shales at Penshurst was *Viviparus*, recalling the *Viviparus* Clays and Purbeck Marble of Dorset. The marine invasion, with the usual *Ostrea distorta*, *Protocardia* and *Isognomon*, was detected between 268 ft. and 278 ft. from the top. The same assemblage occurred also about the middle of the section at Brabourne, where some more specific identifications were possible: Dr. Kitchin was able to recognize *Protocardia purbeckensis* (de Lor.), also an undescribed species of *Protocardia* and *Corbula alata* Sow. Throughout most of the rest of the succession the fauna was found to be essentially freshwater. In the upper part, but below the maximum of *Viviparus*, bivalves, principally *Neomiodon* and *Unio*, were so numerous that at intervals they formed flaggy layers of shelly limestone.

In the Lower Division fossils were as usual rare, consisting principally of fish-remains and Ostracods. On the whole the use that could be made of the

Ostracods in correlating was disappointing when compared with the results obtained in Buckinghamshire and abroad. However, the two zones are definitely recognizable from the records: *Cypris purbeckensis* was restricted to the lower beds, below the marine horizon, and *Cypridea punctata* was recorded from a number of levels in the upper part of the series, the lowest record being 17 ft. below the marine band. Several additional Ostracods were found in both Upper and Lower Divisions but could not be identified.

The lowest 200 ft. of the beds at Penshurst (and lesser thicknesses at the base of all the other sections) were found to be highly charged with gypsum. The mineral is also present at the outcrop in Sussex, where it is about 60 ft. thick, and it is extensively mined near Netherfield. This provides another link with the Purbecks of Dorset, where it will be remembered gypsum was formerly worked for making plaster of Paris from the Lower Purbeck Beds of Durlston Bay.

The junction with the Portland Beds wherever seen in the Kent borings was abrupt but devoid of any signs of erosion or unconformity.

CHAPTER XVII

THE END OF JURASSIC TIME AND THE CRETACEOUS BOUNDARY

I. THE TRANSITION FROM JURASSIC TO CRETACEOUS IN THE BRITISH ISLES

OVER all the northern part of the British Isles evidence bearing on the events in the closing phases of the Jurassic period is entirely lacking. In Scotland and in Northern and Central England as far south as Bedfordshire the record is cut off abruptly in the Kimeridge Clay, and it is unprofitable to speculate on what may have taken place in the interval between the deposition of the Kimeridge Clay and its truncation by the transgressive Lower Cretaceous (Neocomian and Aptian) seas. Some rolled pebbles and fossils incorporated in the basement-beds of the Cretaceous in this area have been identified with Portlandian forms and have been taken to indicate a former extension of Portland Beds far beyond the present outcrop; but the identifications have been seriously questioned, and it seems probable that all the species not derived from the Kimeridge Clay are Cretaceous.[1]

We will therefore abandon the north for a time and concentrate our attention on the tract south of Leighton Buzzard in an endeavour to trace out the concluding events of Jurassic time in Britain. The quest is not a new one, but we can embark on it equipped with a great deal more accurate information than was at the disposal of our predecessors.

We now know, thanks to Buckman's researches on the ammonites, that the marine Upper Portland Beds of Oxfordshire and Buckinghamshire are complete, and that there are therefore no grounds for a supposition at one time entertained, that the Purbeck Beds of that area and of Swindon were laid down while Portland Stone was forming in Dorset.[2] Moreover, a knowledge of the Ostracods enables us to state that both of the zones recognizable in the Purbeck Beds of Dorset are present in these northerly outcrops. We must therefore picture the peculiar conditions of the Purbeck period, the low-lying swamps and lakes, now elevated into dry land, now depressed beneath the sea, as extending from the English Channel between East Kent and West Dorset at least as far inland as Swindon, Oxford, Aylesbury and Leighton Buzzard. A number of facts point to these most northerly existing outcrops having lain near the fringe of the area of deposition: the extreme thinness of the beds, the overlapping of successive members one over another at Swindon, and their local unconformity to the Portland Beds, the discordance between the two Purbeck zones at Hartwell, and the fact that the marine advance of the Middle Purbeck left no trace and so presumably did not extend so far north. Similarly the rapid attenuation of the Dorset Purbecks towards the west, commensurate with an increase in the coarseness of the sediment, points to permanent land not far away in that direction. In North and East Kent we

[1] G. W. Lamplugh, 1896, *Q.J.G.S.*, vol. lii, pp. 195–8.
[2] J. F. Blake, 1880, loc. cit., and 1885, loc. cit.; and A. J. Jukes-Browne, 1892, *Build. Brit. Isles*, 2nd ed., p. 240.

have seen that the attenuation is even more rapid against the London–Ardennes landmass.

The area from South-West Kent and Sussex to East Dorset experienced the maximum amount of subsidence and deposition, through the Purbeck as through the preceding Portland and Kimeridge periods. Here too, in the central region of the depression, was subsequently accumulated the greatest thickness (more than 2,000 ft.) of materials, still perfectly conformable with the Purbecks and of freshwater origin—the Wealden Formation.

Like the Purbecks, the Wealden Beds thin out rapidly towards the east, west and north. Although more than 2,000 ft. thick at Swanage, they are reduced at Mupe Bay to 750 ft. and north of Weymouth to little more than 350 ft.—a rate of diminution at which they would disappear when they reached Bridport. At the same time there is a marked westerly increase in the coarseness of the sediments, sands replacing clays and pebbly grits taking the place of sands.

Similarly, in the Weald, although the beds in Sussex are as thick as at Swanage, the numerous borings in Kent have shown that between Dover and Canterbury they nowhere exceed 85 ft., and beyond this, towards the north-east and north, soon after overlapping the Purbecks on to the Palaeozoic platform, they die out altogether. In the Vale of Wardour the thickness is much reduced, and in the most northerly outcrops of all, on the hills of Shotover and Brill, east of Oxford, it does not exceed 50–60 ft.

It will thus be seen that, although the Wealden Beds are everywhere four or five times as thick as the Purbeck Beds, the two behave as one in regard to their changes of thickness and their distribution.

The Wealden Beds are too well known to need describing in detail. In the region of greatest thickness, from West Kent and Sussex to East Purbeck, they comprise two main divisions, a predominantly sandy series below, called the Hastings Beds (maximum thickness about 1,000 ft.) and a predominantly argillaceous series above, known as the Weald Clay (maximum thickness 1,200 ft.). At the outcrop in the Weald a number of subdivisions have been established; in particular the Hastings Beds are divided into Ashdown Sand below and Tunbridge Wells Sand above, separated by the Wadhurst Clay (with a maximum thickness of 180 ft.). But these subdivisions are not constant even at the outcrop, both of the sandy divisions being liable to replacement by clay (the Fairlight Clays, up to 350 ft. thick, at the base of the Ashdown Sand near Eastbourne, and the Grinstead Clay, up to 50 ft. thick, at any horizon in the Tunbridge Wells Sand). Away from the outcrop, as in the borings in North and East Kent and in the equally-well-developed succession at Swanage, the subdivisions cannot be identified, and only a broad grouping of the series into Weald Clay above and Hastings Beds below is possible, the details varying considerably from place to place.[1]

In view of this inconstancy, the lithology of the Wealden Beds can only be described in general terms. The coarser sediments vary from fine sand to coarse quartz sand with pebbly seams, and wedge-bedding is frequent, testifying to the action of changing currents. Some of the pebbles in the Weald have been identified as derived from the Palaeozoic rocks of the

[1] G. W. Lamplugh, 1911, 'Mesoz. Rocks in Coal Expl. in Kent', p. 85; and A. Strahan, 1898, 'Geol. Purbeck and Weymouth', pp. 122–32; *Mems. Geol. Surv.*

London–Ardennes landmass, but in Purbeck the quartz pebbles increase westward and obviously came from the west. In the Upper Wealden of Purbeck and the Weald have been found the first heavy minerals derived from the Dartmoor granite, now freshly laid bare of its thick sedimentary covering.[1]

The clays are sometimes soft and soapy, sometimes shaly and interbedded with lenticular and irregular limestone bands, crowded with freshwater shells and closely resembling some of the Purbeck Beds. Locally there are deposits of drift-wood, of which the classic example is the well-known 'Pine Raft' on the west coast of the Isle of Wight.

'The character of the greater part of the formation', wrote Sir A. Strahan, 'suggests the action of a river distributing clay, sand and gravel irregularly and locally on a subsiding area.'[2] The nature of the subsiding area has long been visualized as a great lake-basin: 'The essentially lacustrine character of the Wealden deposits is now generally admitted,' writes Jukes-Browne.[3]

'By some geologists they have been regarded as the delta of a great river, comparable to that of the Nile or the Mississippi; but as a matter of fact the deposits do not resemble those of a single delta, but rather those of several rivers or streams pouring their sediment from different directions into a large freshwater lake. This view was held by Searles V. Wood and by Godwin-Austen, and was strongly advocated by Meyer in 1872.'[4]

De Lapparent rightly remarks that the area drained by these rivers must have been far greater than all the existing mountainous regions of the British Isles.

The conception of a lake filled by rivers is not radically different from the picture of the flat and swampy landscape in Purbeck times. All the differences would be accounted for by increased subsidence of the floor of the basin, compensated, as is usual in similar changes of level, by increased elevation of the surrounding land. This would set in motion the normal forces of denudation accompanying a cycle of activity: revival of the rivers and accelerated sedimentation; in fact a repetition of the processes involved in the filling up of the Old Red Sandstone lakes, but on a modified scale. From time to time in the Wealden lake, conditions locally simulated those in which the freshwater portions of the Purbecks were formed, as may be seen by the repetition even in late Wealden times of Ostracod-shales interbedded with *Viviparus*-limestones ('Sussex Marble') in Sussex and Kent.

In view of this essential continuity of conditions between the Purbeck and Wealden periods, it is interesting to examine more closely the junction of the two formations—'the convenient plane for the base of the Cretaceous' which, according to the most recent text-book, is 'marked by the lithological change from limestones and marls to sands and clays'.[5]

The best sections are undoubtedly those exposed on the coast of the Isle of Purbeck, especially at Worbarrow and Mupe Bays, and at Lulworth and Durdle Door. No one knew these sections better than Sir A. Strahan, who of all men might be expected to provide us with an exact level at which to draw so important a boundary-line. Instead, he summarized his conclusions by

---

[1] A. W. Groves, 1931, *Q.J.G.S.*, vol. lxxxvii, pp. 70–2.
[2] A. Strahan, 1898, loc. cit., p. 122.   [3] 1911, *Building Brit. Isles*, 3rd ed., p. 302.
[4] 1872, *Q.J.G.S.*, vol. xxviii, p. 247.
[5] 1929, *Handbook to the Geol. of Gr. Britain*, p. 388.

saying that 'no line can be drawn which does not either include beds of Purbeck type in the Wealden or beds of Wealden type in the Purbeck, the two formations being absolutely inseparable'.[1] In the most favoured locality for studying the junction of the Jurassic and Cretaceous Systems, then, there is clearly no 'convenient plane' of separation at all, and any one familiar with the sections knows that any line is purely arbitrary.

In the Vale of Wardour, again, the passage is so gradual that many feet of argillaceous strata have been grouped with the Wealden by some authorities (Woodward and Strahan)[2] and with the Purbeck by others (Andrews and Jukes-Browne).[3]

The same difficulty was met with in the more complete of the borings in Kent and Sussex, where it had to be admitted that, when Purbeck and Wealden Beds were both present, the position at which the boundary between the two should be placed was a matter of opinion.

'At Penshurst the lowest beds [of the part of the core grouped as Wealden] were dark laminated shales and greenish clays with bands of ironstone and hard sandstone, hardly differing from the underlying Purbecks and containing the same abundance of fresh-water fossils. By gradual upward transition and alternation these beds became more sandy and less fossiliferous, containing for the most part the remains of plants and fish only, until in the upper 225 feet of the boring there was a predominance of soft yellowish sandstone, with subsidiary beds of clay.'[4]

Upon this evidence it may be considered established that in the central parts of the trough, where the Purbeck and Wealden Beds are most fully developed, the formations grade imperceptibly one into the other and form a perfectly continuous and conformable series.

But, as might be expected, the earth-movements which failed to interrupt the essential continuity of sedimentation in the central parts of the trough made themselves felt at the margins, especially in the vicinity of the London–Ardennes landmass, a traditionally unstable region. The Kentish borings proved that the greatly attenuated Wealden formation oversteps all the Jurassic formations and comes to rest on the Palaeozoic platform, before being in turn overstepped by the Lower Greensand. Close to the same Palaeozoic platform in the Boulonnais, also, it can be seen overstepping on to rocks as old as the Bathonian.[5] There is nothing by which to correlate these reduced representatives of the Wealden with the fuller sequence that is conformable with the Purbecks, and they may represent only the uppermost part of the formation. The highest beds would be expected to cover a wider area than the lowest and to overlap them as the result of the continued subsidence. This was appreciated by the authors of the Memoir on the 'Concealed Mesozoic Rocks in Kent', who, although they noted that even where the Wealden Beds are thinnest they are still argillaceous and shaly in the upper portion and sandy and silty in the lower, considered that these subdivisions do not truly represent the thick Weald Clay and Hastings Beds of the fuller sequence.[6]

Again, in the northerly outcrop on Shotover Hill, east of Oxford, the thin

[1] A. Strahan, 1898, loc. cit., p. 126.
[2] H. B. Woodward, 1895, *J.R.B.*, pp. 267, 273.
[3] 1894, *Q.J.G.S.*, vol. l, pp. 59–63 and 66–8.
[4] G. W. Lamplugh, in Lamplugh and Kitchin, 1911, loc. cit., p. 85.
[5] M. Parent, *Ann. Soc. géol. Nord*, vol. xxxii, p. 17.
[6] Lamplugh, Kitchin and Pringle, 1923, loc. cit., p. 17.

representatives of the Wealden Beds (the Shotover Sands) overlap the Purbecks and come to rest on the Portland Stone; but on the neighbouring Brill Hill, a few miles to the east, there seems to be no such overlap (fig. 91).[1]

The evidence of the palaeontology in England is entirely in favour of assigning the Wealden as well as the Purbeck formation to the Jurassic, and

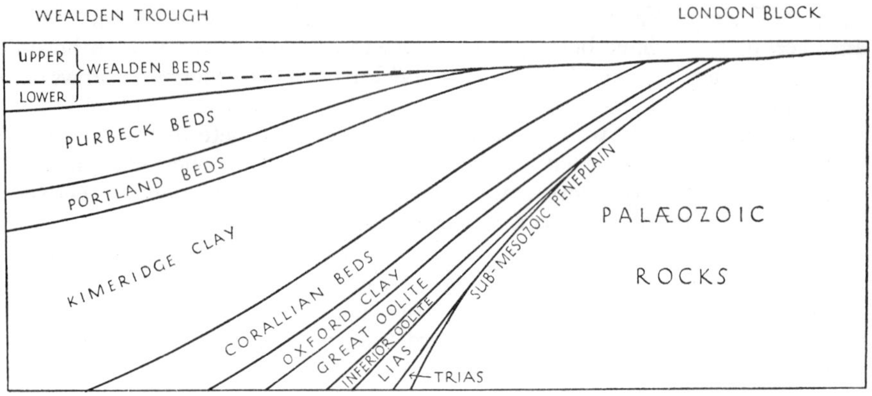

FIG. 91. Diagrammatic section across the margin of the Wealden trough of deposition, where the strata abut against the London landmass; the top of the Wealden Beds restored to horizontality. Note that in the centre of the trough the whole series from Trias to Wealden is perfectly conformable but that towards the edge overlaps and oversteps develop. In particular the Upper Wealden overlaps the Lower, and oversteps the whole Jurassic System unconformably. Towards the centre of the trough, however, it is progressively difficult to distinguish overlap from overstep. (Relative thicknesses approximately correct, but vertical scale greatly exaggerated for the sake of clearness.)

that they should be divided between the two systems seems wholly irrational if we consider only the English fauna and flora.

The most recent revisers only endorse opinions arrived at before the middle of the last century.

'So far as known', writes Sir Arthur Smith Woodward in 1919, 'the fishes of the Wealden and Purbeck formations are essentially Jurassic, and not mingled with any typically Cretaceous forms. Most of them are, indeed, the specialised and evidently final representatives of the Jurassic families to which they belong, and very few can be regarded as possible ancestors of fishes which followed in Cretaceous and later times.'[2]

Sir Richard Owen stated the same thing of the Wealden Reptilia in 1850, when he brought it forward as an objection to Forbes's proposal to sever the Wealden from the Purbeck Beds and to place one in the Cretaceous and the other in the Jurassic System. Now lastly comes Dr. Simpson's verdict upon the Purbeck mammals, some of which have been found also in the Wealden Beds. He finds (1928) that they are unrelated to any Cretaceous or more recent stocks, but that some belong to the same orders as those represented in the Stonesfield Slate.[3]

Among invertebrates the testimony of the Ostracods is the most important.

[1] See above, p. 535.
[2] 1919, 'Mon. Purbeck and Wealden Fishes', *Pal. Soc.*, p. 141.
[3] 1928, *Cat. Mesozoic Mammals in Brit. Mus.* See above, pp. 526–7.

T. R. Jones showed that of the six species of Ostracods which he recognized in the Middle and Upper Purbeck Beds, five range up into the Wealden, where some of them become more abundant.[1] The Middle and Upper Purbecks are therefore much more closely linked by their Ostracod fauna to the Wealden than they are to the Lower Purbecks.

The freshwater molluscs of the two formations are largely identical, but like all freshwater molluscan faunas, they are somewhat featureless for dating purposes. It has often been remarked that, taken from their context, they have a Tertiary or even Recent appearance. The marine mollusca (which occur only in the Purbecks) are all of typical Jurassic genera; but of even more significance than these is the Jurassic echinoid, *Hemicidaris purbeckensis*, found in the Cinder Bed, and also in the Portland Beds in the North of France.

The plant life of the period tells the same story. The greatest authority, Prof. Seward, wrote in 1895, 'The evidence of palaeo-botany certainly favours the inclusion of the Wealden rocks in the Jurassic series.'[2] In 1931 he is still of the same opinion: 'Whether or not we include the Wealden series in the Jurassic period is a matter of secondary importance,' he writes, but, 'of greater importance from our point of view is the general agreement of the plants preserved in Wealden deposits with those of the older Jurassic floras.'[3]

In face of this consensus of opinion that the Wealden and Purbeck Beds cannot be separated on palaeontological or on stratigraphical grounds, it is important to understand the stratigraphical relations of the Wealden Beds to the strata of undisputed Cretaceous date in other countries, as well as in England.

So far as the South of England is concerned, the facts are once more abundantly clear in the coast-sections of the Isle of Wight and the Isle of Purbeck, especially at Atherfield in the Isle of Wight and at Punfield in Swanage Bay. At Atherfield the basement-bed of the marine Lower Greensand (the base of the Atherfield Clay) is the *Perna* Bed (so called from the occurrence of *Isognomon [Perna] mulleti*). It consists of

'a brown calcareous grit, highly fossiliferous, and always contains scattered pebbles, some of which are rolled ammonites or other marine fossils; and it rests upon a slightly eroded surface of the Wealden Shales. At Punfield a highly fossiliferous grit, displaying the conglomeratic character of the *Perna* Bed, occupies a corresponding position at the base of the Atherfield Clay, and also rests upon a disturbed surface of Wealden Shales.'[4]

The Atherfield Clay yields ammonites (*Parahoplitoides*) which belong to an horizon 'well above the base of the Aptian', and the ferruginous sands immediately above at Punfield (formerly known as the 'Punfield Beds') have yielded forms of the same genus pointing to a slightly later date, but still to the Lower Aptian or Bedoulian.[5]

Westward through Purbeck the Lower Greensand (Aptian) oversteps steadily on to lower horizons in the Wealden Beds.[6] Similarly, in the exposures about Oxford the two formations are unconformable: the marine Aptian

[1] T. R. Jones, 1885, *Q.J.G.S.*, vol. xli, p. 331.
[2] 1895, *Cat. Plants Brit. Mus., Wealden Flora*, part 2, p. 240.
[3] 1931, *Plant Life through the Ages*, p. 340.   [4] A. Strahan, 1898, loc. cit., p. 136.
[5] L. F. Spath, 1923, *Sum. Prog. Geol. Surv.* for 1922, p. 148.
[6] A. Strahan, 1898, loc. cit., p. 122.

TABLE ILLUSTRATING THE RELATIONS OF THE STRATA ON THE JURASSIC-CRETACEOUS BORDERLINE

Ages of Neocomian and Tithonian and correlations of Neocomian after Spath, 1923 and 1924 (ages of the Tithonian and their correspondence with the three divisions of the Purbeck rather hypothetical)

Stages.	Ages. (Marine Sequence)	S. France and Jura.	NW. Germany.	Speeton.	Lincs. and Norfolk.	S. England.
APTIAN		APTIAN	APTIAN	APTIAN	APTIAN (CARSTONE)	APTIAN (ATHERFIELD CLAY)
BARREMIAN	Heteroceratan			Cement Beds	Snettisham Clay Tealby Limestone Tealby Clay	(Aptian pebble-bed rests on Wealden)
	Paracrioceratan			Zone of B. brunsvicensis (B)		
	Hoplocrioceratan				(gap)	
HAUTERIVIAN	Simbirskitan	MARINE NEOCOMIAN	MARINE NEOCOMIAN	Zone of B. jaculum (C)	Claxby Ironstone	
	Crioceratan			Zone of B. lateris pars. (D)	(gap)	
	Lyticoceratan			(gap)	Hundleby Clay	
	Hoplitidan			Zone of B. lateris pars. (D) (?)	(gap)	
VALANGINIAN	Polyptychian		Wealden	Coprolite Bed on Kim. Clay	Spilsby Sandstone Phosphatic Pebble Bed on Kim. Clay	Wealden
	Platylenticeratan					
INFRAVALANGINIAN	Subcraspeditan					
	Spiticeratan					
TITHONIAN or PURBECKIAN	Berriasellidan	Absent or much reduced and inosculating with above.	Freshwater Limestones Zone of Cypridea punctata			Upper Purbeck Zone of Cypridea punctata
	Kossmatian	? Middle Purbeck	Serpulite Sub-Zone of Metacypris forbesii			Middle Purbeck Sub-Zone of Metacypris forbesii
	Aulacosphinctoidean	Lower Purbeck Zone of Cypris purbeckensis	Munder Marls Zone of Cypris purbeckensis			Lower Purbeck Zone of Cypris purbeckensis
PORTLANDIAN	Titanitan	Marine Portland (chiefly dolomites)	Partly represented by the Plattenkalk (mainly marine)			Marine Portland
	Behemothan					

(SPEETON CLAY spans the Speeton column. Right-margin brackets: CRETACEOUS, NEOCOMIAN, JURASSIC.)

sands come to rest at numerous places inland upon Kimeridge Clay, and over the Charnwood and Nuneaton Axes they extend on to Oxford Clay. In the Weald the discordance is less marked, but even there the Aptian eventually oversteps the Wealden and passes on to the Palaeozoic platform.[1]

And so, as Sir A. Strahan wrote in 1898, 'there is not only in the Isle of Wight and in the Isle of Purbeck, but in the South of England generally, a well-defined base to the Lower Greensand, above which, and nowhere below it, occur the characteristic and purely marine species of that formation, while the brackish and freshwater fauna is confined . . . to the Wealden group'.[2]

Were we dealing only with the South of England there could be no question of drawing the dividing line between the Jurassic and Cretaceous Systems anywhere but at this major transgression, which marked the invasion of the area covered by the freshwater lake-deposits, containing a purely Jurassic fauna and flora, by the Aptian sea with its purely Cretaceous and marine assemblage. But when we take a wider area into account the matter assumes an entirely different aspect. For in Yorkshire, Lincolnshire and North Germany there were two Lower Cretaceous transgressions, the Aptian invasion being preceded by a Neocomian one, which left no traces in, and so probably did not extend over, the country occupied by our Wealden lake.

The Neocomian deposits of Yorkshire, Lincolnshire and Norfolk were fruitfully studied for many years by Lamplugh, but it is only recently that Dr. Spath's elucidation of the ammonites has enabled them to be effectively brought into line with the Continental sequence. Everywhere the Neocomian deposits rest upon an eroded surface of the Kimeridge Clay and contain at their base a bed of phosphatic pebbles, among which are rolled Kimeridgian ammonites and other fossils. In Yorkshire the phosphatic pebble-bed (called the Coprolite Bed) is succeeded by the Speeton Clay, containing ammonites belonging to all the three subdivisions of the Neocomian—the Valanginian, Hauterivian and Barremian (though with a number of gaps in their faunas)— and overlain by the Aptian. In Lincolnshire and Norfolk the sequence is somewhat different, but a similar general succession of ammonite faunas has been made out, ascending from the Spilsby Sandstone through the Hundleby Clay, Claxby Ironstone, Tealby Clay, Tealby Limestone and Snettisham Clay to the Aptian Carstone. Here the earliest ammonites found, in the Spilsby Sandstone, are of slightly earlier date than those in the lowest ammonite-bearing horizon of the Speeton Clay (the Upper part of the Infravalanginian or Upper Berriasian);[3] but, as pointed out by Lamplugh, the phosphatic pebble-bed at the base of the Spilsby Sandstone can without much doubt be correlated with the so-called Coprolite Bed at the base of the Speeton Clay.[4] The two denote but a single transgression of the Jurassic plains by the Neocomian sea.

This northern Neocomian sea was completely excluded from the area of the Wealden lake by the London–Ardennes landmass: no sign of marine Neocomian deposits exists anywhere on the south of the barrier in England or in Northern France. If, however, we follow along the north of the land-barrier

[1] Lamplugh, Kitchin and Pringle, 1923, loc. cit., p. 16.
[2] A. Strahan, 1898, loc. cit., p. 135.
[3] L. F. Spath, 1924, *Geol. Mag.*, vol. lxi, p. 80, Table.
[4] G. W. Lamplugh, 1896, *Q.J.G.S.*, vol. lii, p. 159 and Table.

eastward and cross into Germany, we find to the south of Hanover an extensive tract of richly ammonitiferous Neocomian clays, evidently a continuation of those in Yorkshire and Lincolnshire, and beneath them other freshwater beds resembling our Wealden and Purbeck formations.

Before we can profitably discuss any further the vexed question of the upper limit of the Jurassic System, therefore, we must abandon the insular outlook which we have maintained hitherto in this book and consider the relations of the formations in the classic region around Hanover and Hildesheim.

II. THE TRANSITION FROM JURASSIC TO CRETACEOUS TIMES IN NORTH-WEST GERMANY, AND ON THE JURASSIC-CRETACEOUS BOUNDARY IN GENERAL

The Jurassic and Lower Cretaceous rocks of the neighbourhood of Hanover were long ago described in memoirs by Struckmann, Roemer, Brauns and von Seebach, and more recently they have received attention at the hands of Koert, von Koenen, Salfeld and other geologists. By far the most important modern contribution to an understanding of the uppermost part of the Jurassic series was made by Koert, who carefully collected the Ostracods from the various levels.

The general succession from the Neocomian to the Kimeridgian is as follows:

SUMMARY OF THE SUCCESSION NEAR HANOVER

Hils Clays. The Hils Clays, studied with minute care by von Koenen in numerous brickyards, especially in the Hilsmulde, contain a full suite of Neocomian ammonite faunas, with which Dr. Spath has been able to correlate the ammonite succession of the Speeton Clay and its equivalents in Lincolnshire. Since the clays are undisputedly of Cretaceous date they only concern us here in so far as they provide us with a datum-line.

Unlike the Neocomian of Lincolnshire and Yorkshire, the Hils Clays do not rest non-sequentially upon much older Jurassic beds, with a basal pebblebed, but pass down conformably into a thick series of strata like our Wealden. The lowest ammonite-bearing marine horizon is the *Platylenticeras* zone of the Lower Valanginian, which is unrepresented in England, but which Dr. Spath places between the faunas of the Hundleby Clay and the Spilsby Sandstone (see table on p. 546). Immediately above it is the *Polyptychinites* zone or Middle Valanginian, equivalent to the Hundleby Clay.[1] At Hils the *Platylenticeras* fauna has not been found, and the *Polyptychinites* zone seems to rest directly on the Wealden Beds. At other places the basal portion of the *Platylenticeras* zone or marine Lower Valanginian inosculates with the freshwater Wealden, so that there are several alternations of marine and freshwater strata.[2] On the evidence of this inosculation of the uppermost beds of the Wealden with strata slightly older than the Hundleby Clay (which rests on the Spilsby Sandstone), Dr. Spath correlates the Wealden of North-West Germany with the Spilsby Sandstone and the Infravalanginian of other regions.[3]

Wealden. The German Wealden, like the English, is a highly variable

[1] L. F. Spath, 1924, loc. cit., p. 80, Table.
[2] C. Gagel, 1893, *Jahrb. Preuss. Geol. Landesanst.*, vol. xiv, pp. 158–79.
[3] L. F. Spath, 1924, loc. cit.

series of freshwater, lacustrine and deltaic deposits, and reaches a maximum thickness not far short of 1,000 ft. Some authors have divided it into Wealdenthon above and Wealdensandstein below, suggesting our Weald Clay and Hastings Sands, while others have considered it divisible into three, a predominantly sandstone division between two clays—the differences depending on the place. The sandstones make good building-stones and have been used in Cologne Cathedral. Interbedded with them are bituminous, pyritous or carbonaceous shales and seams of coal. The flora, like that of our Wealden, is purely Jurassic, and so is the vertebrate fauna. The clays are grey or black, rarely sandy, with thin-bedded limestones made up of *Neomiodon, Unio waldensis, Melania strombiformis, Viviparus fluviorum, Cypris waldensis,* &c. The same fossils occur more rarely in the sandstones; and saurian footprints have also been found.

Freshwater Limestones. Locally at Hilsmulde, Little Deister, Osterwald and Nesselberg there are up to 185 ft. of marly, freshwater limestones and marls; but they seem to fail near Hanover, at Deister, at Süntel and in the Teutoburger Wald, where various writers have described the Wealden sandstones resting directly on the next stratum beneath, the Serpulite. Koert showed that these limestones yield *Cypridea punctata* Forbes in almost every exposure, and also *Physa bristovii* Forbes, *Hydrobia, Bythinia, Valvata, Planorbis* and *Chara*. Some of the beds are described by Koert as crowded with *C. punctata,* like the Ostracod Shales of the Upper Purbeck, and the species is found throughout the sections, especially with *Planorbis* and *Chara*-remains.[1]

Serpulite. The freshwater limestones pass down, without any sharp line of demarcation, into a series of some 50–150 ft. of mixed marine and freshwater beds, mainly massive grey limestones in the upper part, and grey clays with subordinate limestones in the lower part. This series has long borne the name of Serpulite on account of the abundance of *Serpula coacervata* Blumb., which largely composes some of the limestones. The marine fauna predominates, but there are many freshwater fossils, including *Physa bristovii* Forbes, *Neomiodon angulatam* (Sow.), *Valvata helicoides* Forbes. Of greatest significance are the Ostracoda, carefully collected by Koert, which include *Metacypris forbesii* Jones, *Cypridea punctata* and *C. dunkeri* Jones,[2] clearly a Middle Purbeck assemblage. The marine genera recorded are *Ostrea, Exogyra, 'Pecten', Gervillia, Corbula* and *Actæonina.* Other fossils stated by Koert to be represented in the collection at Göttingen from both the Serpulite and the Middle Purbeck of Swanage are *Melanopsis harpæformis* Dunk. & K., *M. attenuata* Sow., *Nerita valdensis* Roem., *Litorinella elongata* (Sow.), *Turritella minuta* Dunk. & K., *Corbula durlstonensis* Maillard and other *Corbulæ,* &c.

The Serpulite transgresses the underlying beds down to the Lias and Trias, but where the Munder Marls are developed the passage is gradual and conformable.

Munder Marls. The Munder Mergel, often called Purbeck Mergel, consists of up to 1,000 ft. of red and green marls, containing much gypsum and

---

[1] W. Koert, 1898, *Geol. und Pal. Untersuch. der Grenzschichten zwischen Jura und Kreide auf der Südwestseite des Selter',* Inaug.-Dissertation, Göttingen, pp. 23–30.

[2] Ibid., pp. 17–23.

other salts. Fossils, except for *Cypris purbeckensis*, are rare, but certain beds yield such marine species as *Gervillia arenaria* Roem., *Corbula* and *Serpula coacervata*; while other layers, especially towards the top, contain some of the same lacustrine forms as the Serpulite—*Valvata helicoides* Forbes and *V. sabaudiensis* Maillard. The marls are in many respects comparable with the Lower Purbeck, with which they have long been correlated by authors, and with which Koert's discovery of abundant *Cypris purbeckensis* accords.[1] They have been likened to the marls of the Keuper, and seem to have had a similar mode of origin.

Eimbeckhäuser Plattenkalk. The Munder Marls pass down gradually and conformably into the Plattenkalk, which is a very widespread formation in North-West Germany. It consists of a variable series of peculiar platy, bituminous, marly or shaly limestones, some oolitic, separated by marly clays, in all up to more than 300 ft. thick. Some of the layers are highly impregnated with bitumen. The fossils recorded by various authors include *Eodonax pellati*, *Protocardia* cf. *dissimilis*, *Corbula alata*, *C. inflexa*, *C. autissiodorensis* and *Modiola autissiodorensis*, all rather abundant, and *Isognomon bouchardi*, *Trigonia variegata*, *Neomiodon cuneatum* and *N. angulatum*. With these are still occasionally found *Viviparus*, *Bythinia*, *Melania* and *Carychium*.

Gigas Beds. Below the Plattenkalk, and resting on Kimeridge Clay with *Exogyra virgula* (Lower Kimeridge Clay), are the fossiliferous *Gigas* Beds with large ammonites previously known as *Ammonites gigas* and *A. giganteus*, and supposed to represent the Portland Stone. Salfeld, however, recognized these forms as the Kimeridgian genus *Gravesia*. The beds consist of marls with bands of limestone. They also contain *Exogyra virgula*.

[Lower] **Kimeridge Clay**, with *Exogyra virgula*, below.

Until recently the correlation of this succession with that in Dorset was considered straightforward. The view held by the pioneers, Roemer, Struckmann and Seebach, and greatly strengthened by Koert, was that the *Gigas* Beds, with or without the Plattenkalk, represent the Portland Stone, and that the remaining strata up to the base of the Wealden are equivalent to the Purbeck Beds. This is the correlation set forth in all the text-books, such as the *Traités* of De Lapparent and Haug, and, so far as the strata above the *Gigas* Beds are concerned, there is much to be said in its defence.

In 1914 Dr. H. Salfeld introduced a revolutionary classification, based on his discovery that the large ammonites of the *Gigas* Beds belonged, not to the species of the Portland Stone, but to the superficially similar genus which he called *Gravesia*, characteristic of the Lower Kimeridge Clay in England. As a result of this correction, he sought to discover representatives of the higher Kimeridgian and Portlandian Beds amongst the overlying strata in Germany usually assigned to the Purbeck. Von Koenen had already suggested that the Serpulite might perhaps be correlated with the Portland Stone,[2] and Salfeld now took up the idea and proclaimed it as his opinion that the underlying Munder Marls and Plattenkalk probably bridged the zones of '*pallasianus*' [*Pavloviœ*], *pectinatus* and *gorei*.[3] As the solitary representative of the Purbeck

---

[1] W. Koert, 1898, loc. cit., pp. 16 and 17.
[2] Von Koenen, 1900, 'Ueber das Alter des norddeutschen Wälderthon's', *Nachrichten des K. Gesell. Göttingen*, aus 1899, pp. 311–14.   [3] H. Salfeld, 1914, loc. cit., p. 173.

Beds, he left the impersistent Freshwater Limestones (Koert's Upper Purbeck).

Salfeld's correlation seems to be based solely on the three following points:[1]

(1) in one place, in the conglomeratic and highly transgressive Serpulite, von Koenen found a single specimen of *Belemnites* cf. *absolutus* Fischer, a species which is also found in the *Titanites* (*giganteus*) zone (see von Koenen, 1900, p. 313).

(2) the abundant *Serpulæ* in the Serpulite recalled to Salfeld portions of the Cherty Series of the Isle of Portland, likewise crowded with *Serpulæ*.

(3) the Serpulite marks a period of marine transgression, whereas in England and North France, he states, the sea withdrew after *Titanites* (*giganteus*) times; therefore the Serpulite is unlikely to be later than the *Titanites* zone.

Against these considerations I would draw attention to the following: (1) The single belemnite found in the Serpulite (even if its identification were not beyond suspicion, but it is only *compared* with *B. absolutus* Fischer) could never be relied upon for dating purposes, for it might have been derived from an older formation; the associated pebbles are derived from all parts of the Jurassic and even the Trias.[2] Andrée has pointed out in another connexion that not much value can be attached to this single find.[3] (2) The *Serpula* of the Portland Cherty Series is *S. gordialis*, whereas that of the Hanoverian Serpulite is *S. coacervata*, a species which Koert identified in the Middle Purbeck of Swanage.[4] (3) The statement that the sea withdrew altogether after *Titanites* times is incorrect, for there was a widespread marine invasion in the Middle Purbeck (see above, pp. 518, 531, 538).

Against the slender arguments adduced by Salfeld must be set the resemblance of the whole of the fauna and most of the lithology of the Munder Marls, Serpulite, and Freshwater Limestones to the Purbecks. There is, in fact, definite evidence for believing that these three formations represent the Lower, Middle and Upper Divisions of the Purbeck Beds: (i) In the Munder Marls there is the zone fossil *Cypris purbeckensis*, which Koert says is abundant at certain horizons, as well as the general lithology and the masses of gypsum to suggest the Marls-with-Gypsum of the Lower Purbeck; the gypsum and other salts are in evidence not only in Dorset and the Weald[5] but also in the Charente and the Jura Mountains.[6] (2) In the Serpulite, Koert's discovery of the Ostracods *Metacypris forbesii*, *Cypridea punctata* and *C. dunkeri*, taken with the predominance of limestones and the evidence of a marine transgression, calls for correlation with the Middle Purbeck. (3) The Freshwater Limestones correspond well with the Upper Purbeck, even to the abundance of the zone-fossil, *Cypridea punctata*, to the exclusion of other Ostracods.

Von Seebach classed the Plattenkalk also with the Purbecks,[7] but the marine lamellibranch fauna is that of the Portland Beds. The abundance of

---

[1] H. Salfeld, 1914, loc. cit., pp. 158–9.
[2] H. Stille, 1905, 'Muschelkalkgerölle im Serpulit des nördlichen Teutoburger Waldes', *Zeitschr. deutsch. geol. Gesell.*, vol. lvii, p. 168.
[3] K. Andrée, 1908, *Neues Jahrb.*, Beilage Band, xxv, p. 389.
[4] W. Koert, 1898, loc. cit., p. 53. [5] See above, p. 539.
[6] At la Rivière, south-west of Pontarlier, there is a gypsum works, where the Lower Purbeck gypsum is exploited as in the Weald and formerly in Durlston Bay. (Maillard, 1884, *Mém. Soc. pal. Suisse*, vol. xi, pp. 9–26 and map.)
[7] K. von Seebach, 1864, *Hannoversche Jura*, p. 59.

bitumen, also, it may be remembered, was one of the characters of the Portland Beds in the Kentish borings, where, especially at Brabourne, 'both the
limestone and the muddy sandstone were strongly impregnated with bitumen,
which had frequently exuded in brown pitchy drops and flakes along joints
and other crevices'.[1]

When all things are considered, therefore, there seems good reason for
supposing that, however imperfectly the Portland Beds and Upper Kimeridge
Clay may be represented in Hanover (the latter perhaps cut out entirely by
a major non-sequence represented in part by the very considerable ones below
our two Midland Lydite Beds), the Purbeck and Wealden Beds are well
developed, the Purbecks more than twice as thick and the Wealden not quite
half as thick as in Southern England. So great a thickening of the Purbeck
Beds would probably cause no surprise if the full sequence of marine equivalents in other parts of Europe were more properly understood. Dr. Spath,
who has recently studied the question, states that 'recent researches indicate
that there is room for a very long and important epoch between the Cretaceous
and the Portlandian, an epoch for the widely-scattered marine equivalents of
which the terms "Purbeckian" or "Aquilonian" seem inappropriate'.[2]

The reason for this long digression to inquire whether the North German
Purbeck-Wealden sequence is homotaxial or contemporaneous with that of
Southern England, although accumulated in a separate basin, will be evident:
it provides us with a common factor for linking up the Wealden lake-deposits of
Southern England with the marine Neocomian in Lincolnshire and Yorkshire.
If the German Purbeck-Wealden can be correlated with the English, then at
least the upper part of the Wealden Beds south of the London–Ardennes ridge
is contemporaneous with the Spilsby Sandstone to the north of the ridge.

In view of this correlation a new interest attaches to the transgression of the
attenuated upper part of the Wealden Beds as it approaches the London
Palaeozoic rocks under the Weald of Kent, and again in the Boulonnais and
in Belgium (where the famous Iguanodon deposits of Bernissart, near Mons,
rest on a channelled surface of the older rocks). The same transgression of
the upper part of the Wealden has been reported in Germany. The Weald
Clay rests on Bajocian at Sehnde, near Hanover, and a boring at Borgloh-
Ösede has been said to show it in contact with Sequanian; borings on the
borders of Westphalia and Holland proved it again, much diminished in
thickness, resting on Trias.[3] If these records have been rightly interpreted,
the German Upper Wealden reaches out, as it were, to meet the Anglo-
Franco-Belgian on the other side of the London–Ardennes ridge. Such a
transgression on both sides of the ridge speaks eloquently in support of the
palaeontological correlation with the Spilsby Sandstone and other basement-
deposits of the highly transgressive Neocomian in Lincolnshire and Yorkshire. That some of the transgressive deposits are marine and others lacustrine
is unimportant, since a general subsidence of the ridge would have the same

[1] G. W. Lamplugh, 1911, loc. cit., p. 43, and see other borings.
[2] L. F. Spath, 1923, Q.J.G.S., vol. lxxix, p. 304. For the stage-name he uses Oppel's term
Tithonian.
[3] De Lapparent, 1906, Traité, p. 1310, where references are given. See, however, K.
Andrée, 1908, Neues Jahrb. für Min. &c., Beilage Band, xxv, pp. 389–90, who throws doubt
on some of these interpretations. He believes the marls with gypsum and anhydrite at
Borgloh-Ösede, which Gagel called Weald Clay, to be the Munder Marls.

effect on both sides whether the water were fresh or salt; moreover the subsidence led, as we have seen, to a mingling of the fresh water and the sea in the neighbourhood of Hanover.

If this subsidence was not purely a local one affecting only the London–Ardennes ridge, then we begin to see some justification for De Lapparent's proposal to divide up the Wealden Beds so as to assign the Weald Clay to the Cretaceous and the Hastings Beds to the Jurassic.[1] But such a partition of a perfectly continuous series is a highly unnatural one, and all the arguments used against the separation of the Wealden from the Purbeck Beds apply again with equal force. Since the lack of fossils of any chronological value makes it impossible to determine which portions of the fully-developed sequence are represented in the attenuated and overstepping strata at the margins of the basin, no line could be drawn with any certainty in the Wealden Beds of the South of England. Salfeld has protested also against the attempt to draw any such line through the Hanoverian Wealden.[2] Further, it may be questioned whether any separation of the transgressive from the non-transgressive portions of the formation is justifiable on theoretical grounds; for the process of subsidence of the land was probably continuous and the overlap progressive throughout the life of the Wealden lakes. The accumulation of over 2,000 ft. of sediments proves the bed of the English lake to have sunk at least that amount during the Wealden period.

Once again we seem driven to look farther afield for relevant facts indicating where the dividing line between Jurassic and Cretaceous is to be drawn in the South of England. Comparison with Germany has shown us that our Wealden Beds are Neocomian, and therefore Cretaceous. It now remains to see what are the relations of the Neocomian formation, when fully developed in its marine facies, to the Purbeck or Portland Beds. For this we have to pass to the South of Europe.

In the Jura Mountains and the surrounding districts to the west and north-west, ammonite-bearing deposits, earlier than any found in the Neocomian of North-West Europe, and belonging to the Lower Infravalanginian, rest directly upon a thin but widespread representative of the freshwater Purbeck Beds. Locally they even inosculate with the upper portion of the Purbecks just as do the *Platylenticeras* deposits of the Lower Valanginian with the top of the Wealden in Germany.[3] So closely do the Purbeck Beds here resemble, in their lithic characters and their fauna, those of England, that the conclusion that they are at least homotaxial is unavoidable. Moreover, they rest upon apparently authentic Portland Beds (correlated by several authors with the Plattenkalk). Maillard, who monographed the freshwater fauna, concluded his résumé by stating emphatically 'the Purbeckian of the Jura, reduced to two subdivisions, corresponds exactly with the Purbeckian of Hanover as restricted to the Munder Marls and the Serpulite, and probably with the Lower and Middle Purbeckian of England. . . . There is no gap between the Purbeckian and the [Infra]valang[in]ian, which is the time-equivalent of the Wealden.'[4] In support of his conclusions he figured from the Jura a number of English and Hanoverian Purbeck fossils, including *Cypris purbeckensis* Forbes.

[1] De Lapparent, loc. cit., p. 1262.   [2] H. Salfeld, 1914, loc. cit., p. 161.
[3] G. Maillard, 1884, loc. cit., p. 136; stated to have been observed also by Benoît, Bertrand and Gilliéron.   [4] G. Maillard, 1884, p. 144.

Koert endorsed Maillard's statements after his careful study of the Purbeckian of the Selter in North-West Germany, and the table which he published, indicating the numerous species in common, leaves little doubt as to the equivalence of the Purbeck Beds in the three areas.[1]

In the Jura Mountains these beds are so insignificant in thickness that Maillard regarded them as merely a freshwater facies of the upper part of the Portland Stone, and this doubtless influenced Haug, who classed them in this text-book as a substage of the Portlandian.[2] The inosculation of the top of the Purbeck Beds with the base of the marine Neocomian then led to the fallacious conclusion that part of the Portlandian in some places might be contemporaneous with part of the Neocomian in others.

What really takes place is that the Purbeck Beds, when followed towards the Alps, pass laterally into marine strata, resembling and perfectly conformable with the Portland Beds—the Tithonian formation. As its name, derived from Greek mythology, implies,[3] this formation is not only conformable with, but also in some ways foreshadows the Cretaceous, although it is always classed with the Jurassic. Here in the deeper parts of Tethys we have, in fact, a continuous sequence of marine deposits linking the Jurassic System with the Cretaceous on the one hand and with the Triassic on the other. Here no earth-movements such as disturbed the shallower sea-bed of North-Western Europe were profound enough to cause great breaks in the sedimentation or to give rise to important unconformities. This is no region in which to establish convenient stratigraphical subdivisions or to set other than palaeontological boundaries to the formations.

The Tithonian is comparable (on a smaller scale) with the Tethyan Trias, in that both formations pass laterally over wide areas in North-Western Europe into freshwater and terrestrial formations. The continental episode represented by the Permo-Trias (or 'Epiric') formation marks one of the major interludes in the sedimentary history of North-Western Europe and is made use of to separate the Primary from the Secondary rocks.[4] The difficulties of correlating it with the marine strata and of deciding to which formation it should rightly be assigned are illustrated by the still lively controversy. The one point upon which all now seem agreed is that the Jurassic System should begin with the advance of the Rhætic sea from the shrunken Tethys across the desert plains and salt lakes of North Germany and Britain. This advance, which, as we have seen, reached to the North of Scotland, brought in a new era and with it we began this book. If we are to be consistent we must make use of the corresponding datum in choosing the base of the Cretaceous System. The succession in the Jura and the Rhone Basin in general shows us that the continental Purbeck episode, of which the lacustrine and terrestrial deposits extend from the borders of the Alps to North Germany and Central England, was brought to an end by the marine invasion of the Infravalanginian. This, and no other, should be the basal member of the Cretaceous.[5]

---

[1] W. Koert, 1898, loc. cit., pp. 52–3.
[2] E. Haug, 1911, *Traité de Géologie*, vol. ii, pp. 1075 et seq.
[3] Tithon was the spouse of Eos (Aurora), goddess of the dawn.
[4] See R. L. Sherlock, 'Correlation of the British Permo-Triassic Rocks', *P.G.A.*, 1926, vol. xxxvii, pp. 1–72, and 1928, vol. xxxix, pp. 49–94.
[5] For a convincing exposition of this principle of using widespread gaps in sedimentation

Complication arises from the fact that the Infravalanginian transgression did not affect the whole of the area covered by the Purbeck lakes and swamps. The regions where the Purbeck Beds were thickest—North Germany and Southern England—became isolated by land barriers, and, although they too were depressed at the same time as the sea found its way north of the London landmass and over the plains of Lincolnshire and Yorkshire, they did not sink beneath the sea but formed vast lakes. The lacustrine fauna of the Purbeck era there survived with but little modification, for it found itself in an environment differing little from that to which it was already adapted. The land, too, which supplied the flora washed into the Wealden lakes, was the same as that which surrounded the Purbeck swamps and lakes; and this in turn was nothing but an extension of the old Jurassic land-surface. Consequently the terrestrial fauna and flora still remained essentially Jurassic.

Thus we see that in the South of England (and in North Germany) we are in an abnormal area. We cannot judge of the events at the close of Jurassic and the dawn of Cretaceous times without travelling far afield, for we have none of the right kind of evidence, and the little that we have is misleading. Although palaeontologists and palaeobotanists affirm that the land and freshwater fauna and the flora of the Wealden Beds cannot be separated from those of the Purbecks, the Wealden Beds are contemporaneous with Cretaceous marine strata on the other side of the London landmass and in the Jura Mountains, while the Middle Purbeck incursion of marine animals proves that the Purbeck Beds were laid down at a time when the fauna in the neighbouring sea was still purely Jurassic.

Lacustrine mollusca are traditionally resistant to the forces of evolution, and their assemblages stable, while terrestrial faunas and floras also may be expected to survive unless wiped out by wholesale marine transgression or revolutionary changes of climate. If consideration of them, on their rare occurrences in the stratigraphical record, were allowed to outweigh the evidence of the normal marine succession, chaos would result. When Messrs. Topley and Jukes-Browne reported to the International Geological Congress in 1855 that 'the separation of Purbeck from Wealden is a mistake due only to the occurrence in Purbeck of one or two Oolitic forms of marine life, . . . is unnatural',[1] they were displaying an outlook anything but international. Further, they were disregarding the principles of correlation, for the 'one or two Oolitic forms of marine life' are worthy of more regard than all the rest of the fauna and flora together. No matter if Jurassic mammals and reptiles still peopled a Jurassic land-area, about lakes where an imprisoned assemblage of Jurassic fishes and freshwater mollusca still survived; once the Cretaceous sea with its teeming new population was already encroaching over the Continent, then Cretaceous times had begun. When Fitton and Webster, in the eighteen-twenties, united the Purbeck and Wealden series[2] they knew nothing of the marine provinces beyond our own islands, and more than twenty years were still to elapse before the discovery by Forbes of the few precious marine species in the Middle Purbecks.

over continental areas as guiding lines in stratigraphical classification see C. Diener, 1925, *Grundzüge der Biostratigraphie*, pp. 160–1.

[1] *Compte Rendu, Congrès géol. int.*, 3me Session, Berlin, 1885, p. 453.

[2] T. Webster, 1826, *Trans. Geol. Soc.* [2], vol. ii, p. 44; W. H. Fitton, 1827, ibid., vol. iv, pp. 105, 159.

The conclusion, therefore, seems to be that, since the *relatively* sudden change of sedimentation at the commencement of the Wealden Series may be due to the same earth movements as those which elsewhere caused the Neocomian transgression to begin, however unsatisfactory it may be in the abnormal Southern English area, it is the only available datum-line at which to draw the important boundary between the Jurassic and Cretaceous Systems.

PART IV

CHAPTER XVIII

PALAEOGEOGRAPHICAL CONCLUSIONS

HAVING now reviewed briefly the principal geological data, we are better qualified to approach the most fascinating but at the same time the most dangerous part of our subject, the palaeogeography. This entails leaving the comfortable society of ascertained facts and venturing into the alluring but treacherous realms of deduction, inference and speculation. It is to the guiding of such speculations, however, that the data set forth in the stratigraphical part of this book are directed, and to leave the mass of facts as they stand would be to abandon them as meaningless. As Prof. Watts has said: 'It is to the elucidation of Earth History that all branches of geology are contributory, and from a knowledge and interpretation of the details of this history that the applications of geology proceed. . . . We must be accurate in our geological facts; but we may, as Darwin advised us, speculate freely.'[1]

To do justice to the palaeogeography of the Jurassic period would require a second volume almost as large as this; for the subject is no longer a mere constructing of palaeogeographical maps, but a full-blown science, containing within it many -ologies. If we were to pursue it thoroughly, we should have to proceed systematically from the points of view of palaeorography, palaeohydrography, palaeoceanography, palaeobiogeography, palaeoclimatology, palaeoastronomy, and many others.[2] No such attempt is possible here, at the end of a single volume already swollen beyond its intended dimensions; and consequently the following brief remarks may appear superficial. But it should be observed that certain fundamental aspects of palaeogeography have already been dealt with at greater length in Part II. We have already pictured the trough sea that stretched across the British area between the western and the eastern landmasses, from Kent and Dorset north-eastward to Yorkshire and north-westward by way of Shropshire, the Irish Sea and Antrim to the Hebrides; and we have examined the stages by which the bed of the sea sank progressively deeper, keeping pace with the accumulation of sediment. Now it remains to review briefly certain other matters upon which light is shed by the stratigraphy: (1) the distribution of the more important marine organisms, which give some indication of the conditions of life in the seas—especially the depth (this comes under 'palaeobiogeography' or 'palaeoecology'); and (2) the probable extent of the Jurassic seas over regions where no trace of sediments is to be found; the height of the surrounding land; and the connexions between the British seas and those on the Continent or elsewhere (palaeogeography proper, or 'palaeocartography').

I. SOME ASPECTS OF PALAEOECOLOGY

(a) The Distribution of Corals

The reef-building corals are probably the most important of all marine organisms for the palaeogeographer. At the present day, although there is an

---

[1] W. W. Watts, 1911, Address to the Geological Society, *Q.J.G.S.*, vol. lxvii, p. lxiii.
[2] E. Dacqué, 1915, *Grundlagen und Methoden der Paläogeographie*, pp. 1–499.

almost boundless diversity of types and forms, they all live under the same conditions, within definite limits of depth, temperature, salinity and clarity of water; it is therefore reasonable to infer that where reef-building corals flourished in the past, similar conditions obtained. Since sheets of coral are intercalated at a number of horizons in the Jurassic System in various parts of the country, we are provided with a number of valuable (because definite) controls. We may not postulate abysmal depths for the deposition of sediments in which any coral growth took place. Further, it is improbable that there was a great depth of water during the formation of any part of a series in which coralline intercalations are so frequent as in the English Jurassic; otherwise we have to imagine an unlikely degree of activity of the sea-bed, alternately rising and sinking through prodigious distances, every time the bottom was brought within the necessary reach of the surface to enable coral growth to become established. Moreover, since thin and irregular sheets of coral sometimes completely overspread the troughs of deposition from side to side, we know that at those times the depth of water *even in the centres of the troughs* cannot have exceeded 120–50 ft. (see p. 53).

It is difficult to classify the Jurassic coral sheets or to find analogues in modern seas. They are neither barrier reefs nor atolls, for they seldom exceed 20 ft. in thickness, and no suggestion of atoll shape has been detected. On the other hand, although many of them evidently grew far from any shore, and so are hardly comparable with the best known and most typical fringing reefs (such as those along the steeply-shelving coasts of oceanic islands or the Red Sea), it would seem that they may be grouped with Fringing Reefs as understood in Darwin's classification; for the width of a fringing reef, as Darwin pointed out, depends entirely on the degree of inclination of the submarine slope, combined with the fact that the corals cannot grow below a certain depth. Thus when the sea is very shallow and the coast gently-shelving, there is nothing to prevent the fringing reefs spreading almost indefinitely. Then, as Darwin found in the shallows of the Persian Gulf and the East Indian Archipelago, 'the reefs lose their fringing character and appear as separate and irregularly scattered patches, often of considerable area' (see p. 394).

It is noteworthy that the only two localities where uninterrupted coral rock exceeding 20 ft. in thickness is known are at North Grimston (p. 423) and in Kent (especially in the borings at Brabourne and Dover; p. 438). North Grimston lies on the southern margin of the Yorkshire Basin, against the flank of the relatively-stable (non-subsiding) axis of Market Weighton; the Kent localities are on the margin of the Wealden trough, close to the flank of the London landmass. As these are the only truly fringing (marginal) reefs known, the idea suggests itself that the prevailing thinness of the coral sheets elsewhere may be due to frequent periodic subsidences having interrupted their growth, perhaps carrying them below the 150 ft. limit, or admitting muddy or sandy currents, before they had time to attain any greater thickness; and that it was only at the margins of the troughs and near the stable axes that coral growth was able to keep pace with subsidence. The borings through the Corallian rocks in Kent lead to the surmise that if the marginal Inferior Oolite deposits in the west had been preserved, between the Cotswold escarpment and the Welsh mainland, the thin and scattered bands of coral in the Cotswolds would have been found to join up as they were traced westward over the Vale of Gloucester and the Vale of Berkeley, until they passed into thick fringing reefs along the Palaeozoic land margin. There is, indeed, evidence of a thickening of the existing coral bands in this direction.

The history of coral growth in the British Jurassic seas is interesting. Very rarely, isolated fragments of coral are met with in the Rhætic Beds (e.g. p. 108), and scattered examples and seams, chiefly of *Montlivaltia*, occur in many places in the Lower Lias—even commonly enough to form an ornamental stone in one place in the *davœi* zone (Banbury Marble, p. 134)—but in general the Liassic sea was too muddy for coral growth to flourish. In two widely separated localities, however, in the neighbourhood of the Cowbridge Island, Glamorgan, and in Skye, corals became well established in early Lower Lias times (pp.127, 146). Here there was apparently unusually clear water. It is possible, too, that the building of calcareous structures was facilitated in the neighbourhood of the Mendip Archipelago by the solution of the Carboniferous Limestone of which the islands were composed, and in Skye by the solution of the neighbouring outcrops of Durness Limestone. The coral fauna of the Sutton Stone in Glamorgan is a rich one, but the area over which it occurs is small.

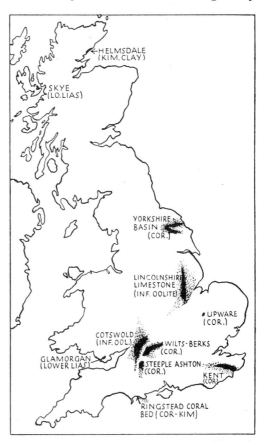

FIG. 92. Map showing the distribution of Jurassic coral reefs in Britain. Existing fossil coral reefs black; extensions for which there is more or less indirect or strong presumptive evidence dotted.

It was not until the Inferior Oolite period that coral growth covered large areas in England. It then became the most characteristic feature of the Cotswold and the Lincolnshire Limestone provinces. Although, as just mentioned, most of the Inferior Oolite reefs, of which so much of the Cotswold Hills are the detritus, probably lay in the main west of the present outcrop and were destroyed by the excavation of the Severn Valley, some true coral reefs remain. The earliest and perhaps the best-developed (15–20 ft. thick) is found between the Pea Grit and the Lower Freestone (*murchisonæ* zone) in the neighbourhood of Stroud, where the Oolite escarpment draws close to the Palaeozoic tract; and thence it can be traced for some miles along the escarpment, as far as Crickley Hill, Coberley and Cowley, but it dies out towards the north-east.[1] In the same district a Middle Coral Bed appears in the Oolite Marl (*bradfordensis* zone), and this can be traced to Leckhampton Hill;[2] but neither this nor the last extends south of the South Cotswolds, the region of the Mendips and the rest of Somerset and Dorset being a non-coralline province. Finally, in Upper Inferior Oolite times a new sheet of corals appears all along the South Cotswolds, comprising the same species as the two earlier reefs, but spreading over a wider area (p. 240). This, the Upper Coral Bed (*truellei* zone), for the first time encroaches on part of the previously non-coralline province to the south, for it continues through the Bath district and on to Dundry Hill, and corals were rolled round the east end of and beyond the Mendips, being found as far away as Bruton (p. 237).

In the eastern counties patches and straggling sheets of coral were formed all over the basin of deposition of the Lincolnshire Limestone, probably at several different dates or more or less continuously in different places (p. 210), but the stratigraphy of this district still remains to be worked out. It may be presumed that the two basins of Lincolnshire and the Cotswolds were in open communication north of the present outcrop, divided only by a belt of shallows over the Vale of Moreton Axis (p. 67).

In the Great Oolite period conditions in the Cotswolds–Lincolnshire province were similar to those of Lower Inferior Oolite times, at least during the deposition of the White Limestone and (in the Cotswolds) earlier. Corals are abundant in the White Limestone from the neighbourhood of Bath to mid-Lincolnshire; but there is nothing entitled to be called a reef, the coral beds rarely exceeding 2 or 3 ft. in thickness. There is no doubt, however, that the corals in these beds are often in the positions where they grew, for the beds are lenticular in shape and contain perfectly-preserved specimens of delicate branching varieties, standing erect. The coralline province was then coextensive with the area where the Great Oolite limestones were forming. In the Mendip district and farther south, as in Lower Inferior Oolite times, there were no corals; while north of the Market Weighton Axis, as before, deltaic conditions held sway. The species in the Great Oolite are for the most part new, but perhaps 10 per cent. are survivals from the Inferior Oolite. Although the coral beds are so thin, some of them can be traced intermittently over considerable areas on the same horizons, and, since there are a number of horizons, the Great Oolite limestones are on the average probably as richly coralliferous as those of the Inferior Oolite. Coral growth seems to have been extinguished by the disturbed conditions with strong currents which gave rise to the Forest Marble, when this facies, which may be regarded as a modified extension of the deltaic ('estuarine') type of Yorkshire and Scotland, spread south over both coralline and non-coralline provinces alike.

The next burst of coral growth was in the Corallian period. It was the greatest of all, and all the species were new. After the general shallowing and clarifying of the sea which heralded the second phase of the Oxford Clay cycle and is attested by the piling up of the Lower Calcareous Grit sandbanks, coral growth sprang up in widely separated parts of England. The earliest reef appeared in the Hackness Hills on the north side of the Yorkshire Basin (Hambleton Oolite Series) (p. 429). South of the Market Weighton Axis, the

---

[1] T. Wright, 1868, 'On Coral Reefs, Present and Past,' *Proc. Cots. N.F.C.*, vol. iv, p. 149.
[2] Ibid., p. 151.

earliest reefs preserved are in the lower part of the Berkshire Oolite Series in Wiltshire. They consist of small patches of true reef-coral, which were soon smothered by deposits of Nothe (or Highworth) Clay and Bencliff (or Highworth) Grit (p. 400 and fig. 69). The main burst of coral growth in this part of England came later, in the Osmington Oolite period, when more or less continuous reefs stretched for forty miles from the neighbourhood of Oxford to Calne, in Wiltshire (fig. 66) and again over a large area in Kent, where they probably continued into the succeeding period. At the same time an isolated reef appeared at Upware near Cambridge; but south of Calne, in Dorset, and in most parts of Yorkshire, oolites were deposited at this time.

Towards the end of the Osmington Oolite period, and in the succeeding Glos Oolite period, coral growth attained its maximum in Yorkshire and at Upware, while in Wiltshire it extended southward to Steeple Ashton, and coral debris even reached North Dorset (p. 388). At all of these places the reefs and coral debris overlie the main mass of the Osmington Oolite, into which those of the Oxford–Calne area pass laterally.

With the incoming of the Sandsfoot Clay (and in Yorkshire the Grimston Cementstone) at the beginning of the next cycle (p. 54) coral growth in England was almost extinguished for good, although on the Continent it continued through the Kimeridgian period. The thin Ringstead Coral Bed, at the top of the Upper Calcareous Grit, testifies to a recrudescence of coral growth somewhere in the vicinity, either off the Purbeck coast or under the Chalk Downs; but the exact location of the reef, if one existed, is unknown (p. 379). Still later, in Sutherland, considerable masses of reef-building coral grew in the peculiar conditions prevailing there during the time of the Lower Kimeridge Clay (*Aulacostephanus* zones) (p. 476).

The Portland Beds contain thin bands of corals in one or two localities, but usually they are wholly absent, and there is no sign of anything like true reef-formation having occurred at this period in any part of Britain.

One of the most important and suggestive aspects of coral reefs is their influence upon the rest of the fauna.

It has been shown how, in the only reefs that have been studied from this point of view—those in the Corallian Beds—a small assemblage of mollusca, comprising certain lamellibranchs and gastropods, not closely related to one another, are always found in association with the corals and seldom if ever away from them (p. 397). More important than this, however, are the much larger assemblages of mollusca which appear to have found the proximity of coral reefs uncongenial, since they have been shown to be seldom found in any quantity in association with them. Among the lamellibranchia this mutual exclusiveness between the coral-dwelling and other forms is especially noticeable in the genera *Trigonia, Lima, Gervillia, Isognomon, Pinna*, &c., of which the large and fragile species occur in great profusion in shell-banks in the non-coralline area, but are only found occasionally (and then usually in fragments) among the corals. There may or may not be a causal connexion, but it is evident that these large and brittle lamellibranchs are mechanically unsuited to a coral habitat, where they would have to withstand the pounding of surf and strong currents. The same applies still more forcibly to the floating ammonites, whose shells would be doomed to destruction if they wandered or were driven by winds or currents too close to the reefs. It is, therefore, interesting to notice that ammonites are not only scarcely ever found actually among the reefs, but that they are always rare in coralline provinces. This mutual exclusiveness on the part of ammonites and corals has hitherto been little appreciated; but the fact that either the conditions which favoured coral growth were unfavourable to ammonites, or else that the conditions produced in the vicinity of coral reefs provided an unfavourable habitat, explains many of the anomalies in the distribution of ammonites upon which altogether different constructions have been put.

(b) The Distribution of Ammonites

Buckman drew attention to the fact that in certain formations ammonites are abundant all over the British area, from Dorset to the North of Scotland; but that in other formations, notably the Oolites, they abound and are well preserved only in certain districts, while in other districts only a few miles away they may be unaccountably scarce. Moreover, he pointed out that in the districts in which ammonites are scarce, such few specimens as are found are more or less broken or worn and frequently have *Serpulæ* or oysters attached; in fact they bear signs of having been drifted into their present positions after death.[1] The abundant and well-preserved ammonites he called autochthonous,[2] or autochthones, as distinct from those which give evidence of having been drifted away from some district where they were autochthonous.

The changing history of ammonite distribution may be summarized as follows (the Ages and their equivalents are shown in Table IV, p. 24):

Ages.	*Ammonite Distribution.*[3]
Pre-Psiloceratan	During the deposition of the Rhætic and Pre-*planorbis* Beds ammonites seem to have been completely excluded from the British area.
Psiloceratan-Cana-varinan	Autochthonous ammonites all over Britain.
Ludwigian-Sonninian	Autochthonous ammonites abundant in the Dorset–Somerset area, including Dundry Hill (north of the Mendip Axis), and also in the Hebrides; drifted in the Cotswolds and Lincs. areas; none in Yorkshire.
Stepheoceratan	As before, but in addition ammonites spread over the Yorkshire Basin (Scarborough Limestone).
Parkinsonian	Ammonites still common in the Hebrides and in Dorset, but retreated from Yorkshire, and from Somerset, as far south as the Yeovil–Sherborne district.
Zigzagiceratan and Tulitan	Ammonites common in the Normandy–Dorset–Somerset area; north of this, in the Cotswolds and Lincs. areas, rare and mostly drifted. None in the Hebrides or Yorkshire.
Early Clydoniceratan	Ammonites retreat altogether from the British area, while deposits somewhat similar to the 'estuarines', previously confined to the North, spread all over England (Forest Marble).

[1] S. S. Buckman, 1922, *T.A.*, vol. iv, pp. 19–20.
[2] *Oxford Dict.*: 'sprung from that land itself'; i.e. aboriginal.
[3] Compare a table by Buckman, 1922, loc. cit., p. 19, drawn up, however, to illustrate a

Ages.	Ammonite Distribution.
Late Clydoniceratan	Autochthonous ammonites again abundant, in the Lower Cornbrash, from Normandy as far north as the Market Weighton Axis; north of this, deltaic conditions or no deposition at all.
Macrocephalitan–Early Perisphinctean	Autochthonous ammonites all over Britain.
Late Perisphinctean	Autochthonous ammonites retreat from all parts of England with the general spread of coral reefs (Osmington and Glos Oolite Series); strata not exposed in Scotland.
Ringsteadian (or Prionodoceratan?)–Gigantitan	Autochthonous ammonites all over Britain (except when the sea retreated from the North in Portlandian times).
Post-Gigantitan	During the Purbeck and Wealden periods ammonites appear to have been once more completely excluded from the British area.

Buckman explained these facts by postulating movements along axes of uplift, which he supposed cut off certain areas from free communication with the open ocean. Thus, to account for the abundance of autochthonous ammonites in the Dorset and Somerset area and on Dundry Hill in the Lower Inferior Oolite, and their absence from the Cotswolds, he wrote: 'The Mendip axis divided in the main, but it was breached between Somerset and South Wales, so that Dundry had autochthones like Dorset. Dundry was cut off from the non-autochthonous area of the Cotswolds by an elevation of the Malvern Axis.' Then, to account for the retreat of ammonites towards the south in the Upper Inferior Oolite period, he supposed 'the North Devon Axis divided'.[1]

This explanation led Buckman to some very extraordinary palaeogeographical restorations. Having assumed that communication between the Dorset-Somerset area and the Cotswolds was severed, he was faced with finding some other connexion between the southern area and the Hebrides, where the same ammonites swarmed as in Dorset and Somerset; and he provided it by means of a connecting channel round the west of Ireland.[2] On this view, which is opposed to all previous palaeogeographical restorations, Ireland, Wales, Lyonesse and Brittany were an island, which Buckman named Juroceltia. Both to the north-west and to the south of it (in the Paris Basin) were supposed to be freely-communicating reservoirs of autochthonous ammonites, from which the seas over England were supplied; and Buckman considered that during the time when the Mendip and Malvern Axes were supposed to have prevented communication across England, the drifted specimens in the Cotswolds and Midlands came from the northern reservoir, via the Hebrides.

A further difficulty was the presence of relatively numerous and well-preserved ammonites in the Yorkshire Basin in the Stepheoceratan Age (during the deposition of the Scarborough Limestone). To account for this

tectonic theory; the present table contains many amplifications and emendations and the tectonic bias is eliminated.

[1] S. S. Buckman, 1922, loc. cit., p. 19.
[2] 1922, loc. cit., p. 19; and 1923, map, p. 52.

another sea had to be postulated to the north-east, for in Buckman's view communication with the south and west was severed at that time by the Market Weighton Axis.[1]

To say the least of these palaeogeographical restorations, they seem extremely far-fetched. The Mendip and Malvern Axes, as was shown in Chapter III of this book, were probably regions of shallows and interrupted deposition and they certainly marked off two well-defined areas—a coralline and a non-coralline province—but there is nothing to warrant the supposition that throughout their whole length they stood above water, interrupting communication completely. There seems no reason to suppose that ammonites could not have floated from south to north across parts of the axes visible at the present outcrop, and concerning the much wider belt to eastward, hidden under the later rocks, there is almost no evidence at all. Deep borings farther east might reveal Inferior Oolite transitional between the two types—in fact the Westbury boring of 1921 yielded fragments of ammonites which Buckman himself identified as probably belonging to the *murchisonæ* zone (p. 200). But even if no such rocks were found, this would be no proof of interrupted communications; for over the axis region there was probably little deposition and there were certainly frequent erosions. It would be as legitimate to base a similar assertion on the absence of connecting deposits between the Cole Syncline and the Sherborne District.

Buckman's theory takes no account of the facts that of the lamellibranchs, gastropods, brachiopods and echinoderms of the Cotswolds, while a number of species are not found in the Dorset and Somerset province (and on these emphasis has been laid), yet a much larger number are common to the two provinces; and that they constitute an incomparably richer fauna than is found in the Hebrides. For these animals, migration via the Hebrides seems out of the question.

If the rest of the fauna was able to migrate across the axes of uplift, we cannot suppose that the ammonites were unable to do so. The only conclusion seems to be that the ammonites were able to enter the Cotswold province from the south, and did so, *but that they did not colonize it because it was an uncongenial or unsuitable habitat.*

A few that did enter died and were entombed in the rapidly-accumulating coral debris of the Cotswold basin; some floated in *discites* times as far as Lincoln, there to become interred in the Silver Bed; but the majority of those that crossed the axis passed on until they reached Western Scotland. There they found a favourable environment similar to that in the Dorset–Somerset area, and in it they settled and became truly autochthonous. The fact that the epiboles and faunizones succeed one another in the Hebrides almost exactly as in Dorset proves that migration was continuous, and moreover that, as Buckman frequently taught, the time taken to perform such a journey was negligible in relation to species-duration. Since the rare and non-autochthonous individuals in the Cotswolds compose an identical, though incomplete and not easily recognized, zonal succession, it seems reasonable to regard them as the chance casualties in a non-favourable area, through which the species migrated without staying to colonize it.

In Yorkshire, throughout most of Lower Oolite times, deltaic deposits were

[1] S. S. Buckman, 1923, loc. cit., p. 52.

accumulating and all forms of marine life, including ammonites, were absent. On several occasions, however (pp. 217–22), the deltaic type of deposition was temporarily arrested and a marine fauna entered the district, and on the last of these occasions—during the Stepheoceratan Age—ammonites tarried in fair numbers, their shells becoming embedded in the Scarborough Limestone.

In the Cotswold and Lincolnshire provinces it was not deltaic conditions that produced an inimical or unsuitable environment and prevented colonization by ammonites: it was either the concomitants of coral reefs—the heavy breakers and strong currents—or else something about the peculiar conditions that favoured the growth of coral reefs, but that was repellent to ammonites.

In the Upper Inferior Oolite period (Parkinsonian Age), Buckman notes, autochthonous ammonites retreated south from Dundry and most of Somerset, at the time when the Mendip Axis and the associated shallows to the north (the Malvern Axis?) became submerged. Buckman now shifts the role of barrier on to the North Devon Axis; it was this, he says, that in the Parkinsonian Age confined ammonites to the Sherborne and Yeovil district and the area to the south (as quoted above, p. 563). But if the Mendip and Malvern Axes previously prevented ammonites from advancing northward into the Cotswolds, as Buckman maintained, why should the breakdown of those axes cause the ammonites to retreat still farther to the south? With the removal of the barrier, if there were no other impediment, they might be expected to advance northward. Since they did not, Buckman can only have wished us to suppose that the North Devon Axis rose above water before the Mendips sank, so that the advance of the main body of autochthones was cut off before the northern barrier was removed, and then those that had been isolated between the two barriers were starved or otherwise extinguished before the Mendips sank.

This elaborate explanation (which Buckman did not put forward, but which seems to be necessarily implicit in his hypothesis) does not survive examination of the stratigraphy of the North Devon Axis. From the careful work of Richardson it is apparent that the Upper Inferior Oolite passes across the axis without interruption (p. 236), and the North Devon Axis therefore offered no resistance to the free passage of ammonites during the Parkinsonian Age.

Rather it would seem that some force, liberated by the further subsidence of the Mendip and Malvern Axes, repelled the ammonites southward at the very time when, on Buckman's hypothesis, they should have been free to spread northward. It can hardly be mere coincidence that the retreat of the ammonites corresponds both in time and in place with the southerly advance of corals from the Cotswold province. The spread of the coralline type of deposition was in turn probably conditioned by the subsidence of the Mendip Axis, which had previously been effective in separating two different basins of deposition, but it was only in this indirect manner that the axis influenced the ammonites. There is a very great difference between a barrier rising above water without a break, in such a way as to bar the passage of floating and swimming cephalopods (such as Buckman visualized), and a belt of shallows and strong currents, offering no bar to migration, but nevertheless sufficient to provide a boundary between two basins which, for some other reason, were

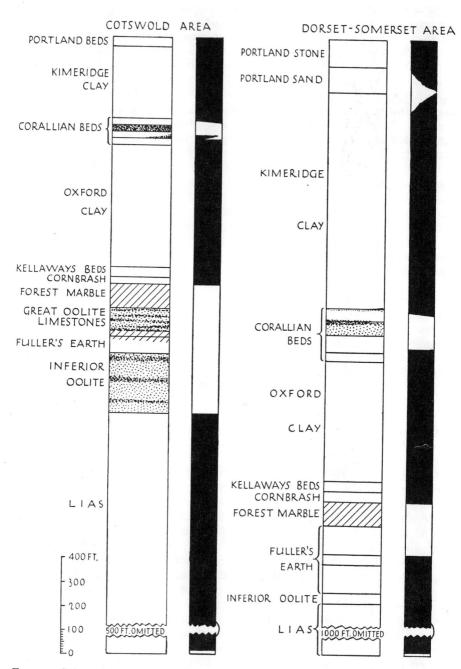

FIG. 93. Columns showing the relation between coralline deposits and autochthonous ammonites. In left column in each area, dotted = coralline deposits (coralline oolites and detrital limestones, &c.); thick dots = coral-beds; cross-hatched = 'Forest Marble' facies. In right columns, black denotes autochthonous ammonites.

characterized by different types of sedimentation and different environmental conditions.

Whenever we come to compare the distribution of ammonites and coral reefs the interdependence of the two is striking (fig. 93). Where the deltaic (or what is generally known as estuarine) type of deposition prevailed, neither corals nor ammonites existed; but in the districts where other types of deposit were being formed, either ammonites were autochthonous or corals, but never both. Moreover, at the phase of maximum growth of coral reefs (the Osmington–early Glos Oolite periods), when almost the whole of England was covered by coral seas, autochthonous ammonites disappeared altogether, their scattered shells being only very rarely found even in the areas where oolite or debris was accumulating although no actual coral growth was taking place (e.g. in South Dorset). The presence of even a few drifted ammonites or fragments of ammonites in such deposits, however, precludes their being placed in the same category as the Rhætic and Purbeck Beds, which were apparently formed in water definitely cut off by land barriers from all connexion with seas inhabited by ammonites, since no trace whatever of cephalopods is to be found in them.

It is just possible that there may be some more subtle reason why corals and ammonites should not have flourished in the same waters—perhaps connected with the food-supply of the adult ammonites or their free-swimming larvae. But since ammonites are extinct, this will always remain a matter of surmise.

II. PALAEOCARTOGRAPHY

Whether the resulting conclusions be accepted or rejected, the foregoing remarks will at least have served to illustrate the necessity for ecological studies as a prelude to attempting palaeocartographical reconstructions. Unhampered by the awkward land-barriers or isthmuses which Buckman thought it necessary to erect, we may more readily tackle the wider problems: the outer boundaries of the Jurassic seas, where they abutted on the ancient landmasses, and where they were connected with the continental seas and the Tethys. We now enter more deeply into the realms of surmise than at any previous stage of this book.

(a) The General Boundaries of the British Troughs in the West and North

All (with the exception of Buckman) who have attempted to construct palaeogeographical maps of the British region in Mesozoic times have agreed in one respect, namely, in the existence of a vast landmass, North Atlantis, to the west of these islands, over the site of the present Atlantic. Of this landmass the last relics are supposed to be the mountains of Wales and the West of Ireland, the peninsulas of Devon and Cornwall and Brittany, and perhaps the Outer Hebrides. The rest has foundered in the bed of the Atlantic or been eaten away by the waves.

How wide the landmass is likely to have been we do not know; but for all we can tell it may have been continuous with the Mesozoic continent which Canadian and United States geologists are accustomed to picture as they look eastward over the eastern part of North America and the Western

Atlantic—their Jurolaurentia.[1] Further speculations will be idle until the Wegener hypothesis that the continents have drifted apart has been either proved or disproved. Of more immediate concern is the extent to which the Jurassic seas overspread North Atlantis from the east, and upon this question many views have been expressed, supported by many different classes of argument. The relevant facts are briefly as follows:

At the present time three tongues of Jurassic rocks extend more or less intermittently westward into the Palaeozoic region and separate the ancient highlands of Brittany, Devon and Cornwall, Wales, and Scotland (see fig. 2, p. 38):

(i) The most southerly tongue occupies the site of the English Channel; the evidence for it being the dredging up of fragments of Liassic limestone 30–34 miles south-east and 45 miles south-south-east of the Lizard,[2] and of *Psiloceras planorbis* or an allied species 30 miles south of the Eddystone Lighthouse (above, p. 121).

(ii) The second coincides with the Bristol Channel: it is denoted by the westerly extensions of the Lower Lias and Rhætic Beds in South Glamorgan and near Minehead, and these unquestionably extended at one time still farther. That there has been Lower Lias in the vicinity of the Gower peninsula seems to be indicated by the presence of oysters allied to *O. irregularis* in stalactites in the Carboniferous Limestone at Mumbles near Swansea.[3]

(iii) The northern tongue, by far the most important of all, is denoted by the Liassic outliers at Prees in Shropshire (p. 135) and near Carlisle (p. 142), by the Lias and Rhætic Beds preserved beneath the basalt plateau of Antrim (pp. 112, 142), and in the Tertiary volcano of Arran (p. 115), and by the disconnected but important relics of Jurassic rocks in the Inner Hebrides (ranging in age, as we have seen, from Rhætic to Kimeridgian). A further outlier connecting the Shropshire and the Antrim areas is believed, from indirect evidence, to exist upon the floor of the Irish Sea, where Mr. E. Greenly maps an elliptical area of Jurassic rocks covered by Chalk, as shown in fig. 2 (p. 38). The evidence for this is the presence of pieces of a crumbly ferruginous oolite like that in the *jurensis* and *opalinum* zones and of 'Jurassic-looking boulders, chiefly of calcareous shale' in the Drift on the east side of Anglesey, associated with numerous Chalk flints, all of which seem to have come from the bed of the Irish Sea to the north-east.[4] Liassic boulders containing fossils have also been found in the Drift on the adjoining north coast of the mainland, at Penmaenmawr, Carnarvonshire.[5]

Upon the intervening projections of Palaeozoic rocks, the supposed salients of Atlantis, there are no traces of Mesozoic sediments, but there are conspicuous elevated peneplains at various levels and of uncertain dates.

The crux of the matter is to decide how far the present arrangement of the Jurassic rocks is original: whether they were laid down as they are found, in low-lying troughs separated by projecting masses of highland, or whether they owe their preservation in those areas to subsequent depression, while

[1] See a series of palaeogeographical maps of N. America by C. H. Crickmay, 1931, 'Jurassic History of North America', *Proc. Amer. Phil. Soc.*, vol. lxx, pp. 80–93.
[2] A. J. Jukes-Browne, 1911, *Building of the British Isles*, 3rd ed., p. 260.
[3] A. E. Trueman, 1922, *P.G.A.*, vol. xxxiii, p. 278.
[4] E. Greenly, 1919, 'Geol. Anglesey', vol. ii, pp. 777–8, *Mem. Geol. Surv.*
[5] *Summ. Prog. Geol. Surv.* for 1931, p. 38.

erosion has caused the removal of the rest of the strata in the intervening areas. Upon this question opinion has always been divided; but it is now becoming increasingly probable that neither view is altogether right: that some of the high Palaeozoic tracts were submerged and some were not, and that often they formed shallows—areas of retarded deposition alternating with contemporaneous erosion—rather than dry land. The two schools of thought are exemplified by Jukes-Browne and Judd, whose conflicting views we will have occasion to examine in the following pages.

(b) The Boundaries of the Jurassic Sea in Wales and the Bristol Channel

It was a favourite line of reasoning with Jukes-Browne that, although the Jurassic formations, if produced, would in places pass over tracts occupied by Palaeozoic rocks, the present relief of those tracts is only due to Tertiary and Quaternary denudation; therefore that they probably stood a great deal higher in Jurassic times, and the Jurassic sea would not have been able to cover them. The complementary inference, that if Tertiary and Quaternary denudation has been so potent, it has probably destroyed Jurassic rocks that may formerly have existed over those self-same areas, received scant consideration.

Thus Jukes-Browne wrote:

'This [1,010 ft.] is about the height of the escarpment [on the east side of the Forest of Dean] at the present day, and as it must have been very much higher in Jurassic times, the sea of the Upper Lias is not likely to have passed over the Forest of Dean.'[1] Again: 'We may assume that during the formation of the Lias the coast-line ran from the eastern border of Dean Forest to Malvern, the gaps which now intervene between the higher elevations being due to post-Jurassic erosion. Thence it probably trended north-westward through Shropshire and Denbighshire, and across the Irish Sea to the west coast of Ireland. . . . From Ireland the sea extended over the site of the western Scottish [Keuper] lake, and thence probably up the Great Glen to the north-eastern basin.'[2]

For this conservative view there is much to be said, and the distribution of land and water outlined by Jukes-Browne in the latter of these passages has been generally adopted in text-books and is implicitly accepted by most research workers in Jurassic stratigraphy. The great stumbling-block to the acceptance of the Malvern and Abberley Hills as the actual shore-line, however, has always been the Malvern Fault. This everywhere forms the western boundary of the Trias, which it throws down against the Palaeozoic and Archaean rocks of Wales. It is hinged in such a way that, although the throw is only perhaps 100 ft. near Newnham in the south, it increases to a maximum of 1,000 ft. near Malvern. It seems to be entirely post-Triassic, for the Trias adjoining it is of the thickness normal for the district and appears not to be banked up against it; moreover there seems to be no marginal change of facies. From these facts it is inferred that the Trias 'must have formerly extended well over the site of the Malvern and Abberley Hills on to the area of the Old Red Sandstone'.[3] How far the Trias may have extended no one has ventured to calculate; and the question has no certain bearing on the former extension of the Jurassic formations, since movement may have

[1] A. J. Jukes-Browne, 1911, loc. cit., p. 265.   [2] Ibid., p. 266.
[3] T. Groom, 1910, Geol. in the Field, pp. 717 and 726.

begun along the fault immediately after the end of Triassic times. As we have seen, the Rhætic Beds of the Berrow Hill outlier and the western side of the Worcestershire outcrop, as well as the Inferior Oolite of the Cotswolds, show signs of a shore-line at no great distance (p. 108). It would not be surprising if the fault were a marginal fracture caused by unequal subsidence of the Jurassic trough, for it adjoins the deepest part of the trough, where the Lias and the Lower Oolites attain thicknesses far exceeding any developed in other parts of the South of England.

A reasoned attempt to ascertain to what extent the Mesozoic rocks, especially the Trias, may have encroached upon the Palaeozoic tract of Wales was made by Prof. O. T. Jones in 1930, in the course of a suggestive study of the Bristol Channel region.[1] His most important innovation is an endeavour to trace Tertiary folding from the Mesozoic region on the east into the Palaeozoic tract of the west, and he comes to some novel conclusions. He believes that Central and North Wales have been uplifted by a broad anticlinal axis extending east and west, probably a prolongation of that which gave rise to the major deflexion of the Chalk outcrop and strike in Fenland. Similarly, he considers the high elevation of Exmoor to be due, not so much to the superior hardness of the rocks of which it is composed, as to its being on the axis of uplift between the syncline of Central Devon and another centred along the Bristol Channel. The Bristol Channel he considers to be a broad E.–W. Miocene downfold, analogous with the syncline of the London Basin, but resembling it as a mirror image and pitching to the west.

These conclusions are reached along two lines of reasoning. The first is the drainage. As Strahan pointed out,[2] the rivers of South Wales bear no relation to the structures resulting from the Armorican and Charnian orogenic movements, but disregard them entirely, cutting through the highest escarpments by deep gorges. South Wales has, in fact, a superimposed drainage, which can only be explained by supposing it to have originated upon a blanket of Mesozoic strata sloping gently south-eastward, but now entirely removed. Strahan supposed that this blanket was composed of sediments of Upper Cretaceous date—an idea which was acclaimed by Jukes-Browne, for it fitted in with his view of the former wide extension of the Upper Chalk sea over all except the highest districts of North and Central Wales and the Brecon Beacons.[3] Quite apart from the question whether the sedimentary blanket was composed of Jurassic or Cretaceous strata, Strahan's theory, if correct, is held to involve a post-Cretaceous (presumably Miocene) uplift of North Wales relative to South Wales. Actually, however, it is conceivable that the drainage may have started upon an original slope of deposition, and so it does not seem absolutely necessary to invoke tilting.

The other line of reasoning by which Prof. Jones was led to the belief in a Miocene uplift of North Wales was a consideration of the slope of the sub-Mesozoic platform when traced beyond the region where Mesozoic deposits remain upon it. This subject is of such importance that we will proceed to a short independent examination of the data; for Prof. Jones considers it legitimate to assume that marine Rhætic and Liassic (if not later) strata were

---

[1] O. T. Jones, 1931, *Rept. Brit. Assoc.* for 1930, Presid. Add. Sect. C, pp. 57–82.
[2] A. Strahan, 1902, *Q.J.G.S.*, vol. lviii, pp. 207–25.
[3] A. J. Jukes-Browne, 1911, loc. cit., fig. 53, and p. 335.

deposited, not only over all parts of the platform where Trias is now found, but also over all those parts from which it has been removed by subsequent erosion.[1] But apart from this assumption being invalidated by the existence of a fault such as that at Malvern, which although post-Triassic, may be pre-Jurassic or at least intra-Jurassic, it is important that there should be absolute certainty that the sub-Mesozoic platform can be correctly identified. It is therefore to the identification of the platform that we will pay special attention.

Within the British Isles the ideal district for studying the sub-Mesozoic platform where it first begins to emerge from beneath its sedimentary covering is the area surrounding the head of the Bristol Channel, in Somerset, Monmouth and Glamorgan. Here recent denudation has so far stripped off the covering that it is possible to obtain a clear idea of the surface that lies beneath, while at the same time a sufficient amount remains still buried to leave no doubt as to the age of the feature.

A more or less unevenly sculptured surface of Palaeozoic rocks lies revealed, evidently an old landscape, formed by the wearing down of the Armorican mountains under subaerial agencies—probably in a desert climate. All the original tectonic features had been planed down, so that the Triassic deposits overstep rapidly across the basset edges of the various Palaeozoic formations; and in places, such as near Cardiff, where the Keuper rests on Silurian, at least 7,000–8,000 ft. of Upper Palaeozoic rocks were removed. Here and there, however, residual hills or 'monadnocks' were left standing above the general level of the undulating plain, and later these formed islands, first in the Keuper lake and then in the Rhætic and Liassic seas (as described on pp. 103–5 above). The most important of the smaller monadnocks were the Mendips, Quantocks, and Cowbridge Island (see fig. 17, p. 104). Beyond these the scarp of the South Wales Coalfield certainly rose to even higher elevations; but whether it merely formed the edge of a much larger manadnock, or was the fringe of a great tract of ground lying to westward at a higher level and worthy of the name of mainland, usually bestowed upon it, is a difficult problem.

The configuration of the sub-Mesozoic platform in the Bristol Channel district and the gradual submergence of the monadnocks by the marine transgressions of the Rhætic and Lias have been vividly described in the following passage by Strahan,[2] who elucidated the sequence of events in Glamorgan:

'The features of the old landscape were due primarily to the effects of denudation upon the sharply folded Palaeozoic rocks. Then, as now, the Carboniferous Limestone formed scarps and the Old Red Sandstone stood up as rounded hills. Around and between these features the marl was spread out, mixed at its margin with the debris that fell from their sides. As the subsidence continued the marl and breccias extended further up the slopes, but, though they had covered the small crags, they had failed to surmount the higher tracts when they were succeeded by the Tea Green Marls. Gradually diminishing in size, these same tracts can be easily recognized through the period of the Tea Green Marls, of the Rhætic, and of part of the Lias, but they all finally disappeared during the deposition of the Lower Lias.'

It might be thought that such a well-marked peneplain, with its residual monadnocks, would be easy to follow over wide areas in other parts of Britain.

[1] O. T. Jones, 1931, loc. cit., p. 64.
[2] A. Strahan, 1904, 'Geol. South Wales Coalfield, Part VI, Bridgend', p. 22, *Mem. Geol. Surv.*

But directly we leave the fringe of the surviving Triassic deposits we are faced with two difficulties: (i) uncertainty (or, rather, complete ignorance) of the level at which to look for it, and (ii) interference by other peneplains of much later date.

The first difficulty will be readily appreciated by considering the differences in level of the sub-Mesozoic peneplain or Palaeozoic platform beneath those parts of England that are still under the Mesozoic covering. As pointed out in Chapter II, it has been continually depressed beneath the Jurassic troughs of deposition, and to an increasing extent towards the centres of those troughs, throughout the Jurassic period; and as remarked on p. 50, it may have been still further depressed in the same places by the very Miocene movements that gave rise to anticlines in the Chalk covering above (for example under the Weald). The Palaeozoic platform is two or three thousand feet lower under Hampshire and Sussex than in Somerset, where it comes to the surface (see figs. 3–6). It plunges rapidly below sea-level again beneath the Bristol Channel. At what level, therefore, are we to look for it over Wales, and so how is it to be distinguished from other peneplains?

This brings us to the second difficulty. The sub-Mesozoic peneplain has been identified with the 400-ft. coastal platform of Carmarthen and Pembrokeshire. No Mesozoic strata are found *in situ* so far west, but the presence of red marls of Triassic appearance in gash-breccias, or the filling of collapsed caverns in the Carboniferous Limestone, and the local red staining of the limestone, have been taken to indicate the former presence of the Trias. Prof. Jones therefore suggests that the conspicuous monadnocks, such as the Prescelly Mountains and Carn Llidi and others, which dominate the Pembroke peninsula, are survivals of the Triassic landscape just as are those in Glamorgan and Somerset; in fact, that the coastal plain of Pembrokeshire is substantially the sub-Mesozoic peneplain.[1] Mr. E. E. L. Dixon, who has studied the peninsula in detail, also considers that 'though the Triassic floor has undergone some later planation, this has merely touched up the work of the earlier erosion'. Nevertheless he admits that 'parts of it show little trace of Triassic staining and none [shows any trace] of Triassic deposits; on the contrary it supports relics of post-Triassic sediments':[2] 'as though the surface of the original Triassic landscape had been considerably pared down', as he wrote a few years earlier.[3]

This coastal platform compares perfectly, in fact, with the Pliocene platform of Cornwall. From the Cornish platform rise the monadnocks of Carn Menellis, Carn Brea, Carn Marth, St. Agnes' Beacon, &c., for all the world like Carn Llidi and its companions across the Bristol Channel; and upon its surface, as proof of its age, lie the relict patches of Pliocene deposits at St. Erth and St. Agnes. There are upon parts of the Pembrokeshire platform also the same type of gravels as in Cornwall, probably dating from the slightly earlier period at which the peneplain began to be formed by subaerial agencies. In one place, too, not far from Pembroke, is a considerable patch of pipe-clay of uncertain Upper Tertiary age, and at least 45 ft. thick.[4]

---

[1] O. T. Jones, 1931, loc. cit., pp. 62–4.

[2] E. E. L. Dixon, 1921, 'Geol. South Wales Coalfield, Part XIII, Pembroke and Tenby', p. 162, *Mem. Geol. Surv.*

[3] E. E. L. Dixon, 1914, 'Geol. South Wales Coalfield, Part XI, Haverfordwest', p. 199, *Mem. Geol. Surv.*   [4] E. E. L. Dixon, 1921, loc. cit., pp. 166 et seq.

The Pliocene coastal platform occurs also in North Wales and Anglesey, varying within a hundred feet or more in height, and often separable into two or more stages. In Anglesey and the adjoining mainland Greenly has described three such platforms, at 550 ft., 430 ft., and 275 ft., all of which he considers probably of approximately Pliocene date; and since the highest has monadnocks rising out of it as in Pembroke and Cornwall, he calls it the Monadnocks Platform.[1] The point of interest about these platforms in our present context is not their precise heights, but their late Tertiary (conveniently generalized as Pliocene) date, and their widespread distribution. They enable it to be said with confidence that the existence of a platform at about 400 ft. with monadnocks in Pembrokeshire, even though there may be some places farther east along the coast where it coincides with the sub-Mesozoic peneplain, is no proof of the former existence of the Mesozoic base at anything like that level over the Pembrokeshire peninsula. The point is one of some importance, for if the Mesozoic base did extend at so low a level across the peninsula, it would seem to be shaping to pass round behind the highlands of Wales.

The question is precisely similar to that arising in the promontory formed farther north by the Isle of Anglesey. Near at hand under the Irish Sea, as under the Bristol Channel, the base of the Trias is well below sea-level (p. 568 and fig. 2, p. 38). The adjoining land is peneplaned, and from the peneplain rise monadnocks. But Greenly does not attribute the peneplain and the monadnocks to the Trias. Although he asserts that 'it is, of course, evident that the Mesozoic rocks must have extended over the tract that is now the Isle of Anglesey', he considers that the Mesozoic base over the island was probably more than 700 ft. above sea-level.[2]

Prof. Jones has endeavoured to trace the sub-Mesozoic peneplain farther into Central Wales. He describes how in North Pembrokeshire, Carmarthenshire and South Cardiganshire the coastal plateau at 400 ft. slopes upward inland and merges imperceptibly into the high plateau of Central and North Wales, at 1,900 or 2,000 ft.[3] The high plateau was noticed as early as 1846 by Ramsay (who attributed it, however, to marine erosion)[4] and it has been described also by Prof. Fearnsides[5] and others. It is a well-marked feature from which the highest mountains of the Snowdon group, Cader Idris, the Arenigs and Arans, Plynlimon, &c., stand out as monadnocks, like those on the lower plateau. Prof. Fearnsides also spoke of it as 'sloping gently away to the south-east across Merioneth and Cardiganshire', and Prof. Jones in an earlier paper described how it 'extends to the foot of the Upper Old Red Sandstone escarpment of Breconshire and Carmarthenshire, which overlooks it as a line of high cliffs overlooks a level foreshore'.[6] The continuous slope of the platform is taken by Prof. Jones to indicate that the whole of it is the sub-Mesozoic peneplain which has been warped up in Central Wales, as explained on p. 570. The logical conclusion at which he arrives is that 'it would be rash to assert that the Lias did not extend over the Palaeozoic

[1] E. Greenly, 1919, 'Geol. Anglesey', vol. ii, p. 783, *Mem. Geol. Surv.*; see also W. G. Fearnsides, 1916, *Q.J.G.S.*, vol. lxxii, p. 76 (platforms in Carnarvonshire).

[2] E. Greenly, 1919, loc. cit., p. 778.

[3] O. T. Jones, 1931, loc. cit.

[4] A. Ramsay, 1846, *Mem. Geol. Surv.*, vol. i, pp. 331, 333; and 1866, ibid., vol. iii, p. 236.

[5] W. G. Fearnsides, 1910, *Geol. in the Field*, pp. 820–1.

[6] O. T. Jones, 1924, *Q.J.G.S.*, vol. lxxx, p. 568.

area of Wales and the Welsh Borders; the formation may, indeed, have attained a thickness of several hundred feet over that area'.[1]

Before accepting so far-reaching a conclusion as this it is well to examine for comparison the ancient Palaeozoic tract under London, which, as already remarked (p. 52), offers unique opportunities for studying the relations of the Mesozoic rocks to the platform. Here all the evidence has been preserved and can be studied by means of borings, whereas around the western and northern landmasses the critical strata have been completely destroyed by erosion.

The first fact to strike the eye on referring to Lamplugh's sections is that, although the surface of the platform has a continuous slope, it is not all of the same age, in the sense that neither Trias nor Lias nor any other single formation of the Jurassic stretches or apparently ever stretched all over it. Owing to the peculiar combination of oversteps and overlaps occurring at the margins of the trough of deposition (as shown in fig. 91, p. 544) progressively later rocks transgress on to it at progressively higher levels. Therefore, although the lower part of the peneplain may have lain beneath the Liassic water-level, the higher parts were not inundated until the Upper Cretaceous. Yet, if the Mesozoic rocks were to be stripped entirely away, the peneplain would presumably look much the same as that of Wales. The points that must be emphasized are (1) that the slope of the peneplain was not imposed upon it by the Miocene earth-movement, but was slowly acquired, stage by stage, through the subsidence of the Jurassic trough around it; and (2) that the land-area was exposed to prolonged subaerial denudation, the highest parts for the longest time, before finally sinking beneath the sea and perhaps having a submarine peneplain superimposed upon the subaerial one.

By analogy, it is not unreasonable to suppose that by the end of Jurassic times the slope of Palaeozoic rocks in Central Wales was already tilted southward and eastward so steeply that, if there has been any Miocene uplift, its effects have been relatively trivial.[2]

The Miocene uplift of Exmoor, which, as Prof. Jones points out, produced a water-parting from which the streams flow southward into the syncline of Central Devon and northward into that of the Bristol Channel, was merely a revival of an earlier uplift. Prof. Jones calls Exmoor an analogue of the Wealden anticlinorium and the Bristol Channel an analogue of the London Basin synclinorium.[3] The Miocene movements, which conditioned the modern drainage, may, it is true, have had similar effects in the two areas; but it is the fundamental structure that concerns us here, and in this there is no analogy. Exmoor, as we saw in Chapter III (pp. 71–3), lies upon the North Devon Axis of uplift, which was a line of retarded deposition, indicating non-subsidence or relative elevation, in the Jurassic period. The Weald, on the contrary, was a region of depression and heavy sedimentation throughout the Jurassic period. Further, the Bristol Channel, as is shown by the surviving outcrops of Lias, was depressed early in Jurassic history, while the London Basin probably remained above water for most of the Jurassic period —at least it is the very region where the Palaeozoic platform rises nearest the surface and attains its maximum convexity. So far from being analogous,

---

[1] O. T. Jones, 1931, loc. cit., p. 66.

[2] I say 'if' because the drainage may have been initiated upon an original slope of deposition of the Cretaceous covering.　　　　[3] O. T. Jones, 1931, loc. cit., p. 81.

therefore, the Weald and Exmoor, in their fundamental structure, are opposites. Yet both appear to have been elevated during the Miocene orogeny, while both the London Basin and the Bristol Channel, equally opposite, were depressed.

This is only one of several instances illustrating the apparent contradiction between Godwin-Austen's 'law' (p. 59) that the Miocene anticlines arose along the lines of earlier anticlines, traceable in the Jurassic and earlier rocks, and Lamplugh's 'law' (p. 47) that the greatest uplifts that occurred in the Miocene period were superimposed upon broad synclines in the Jurassic rocks beneath. At first sight the contradiction is puzzling, and seems to render both laws worthless. But the explanation is, I think, that Godwin-Austen and Lamplugh made their observations on two different types of uplift, both of Miocene date, but arising upon different foundations from different immediate causes (though both were due ultimately to pressure); and so both laws are true when applied to the type of phenomenon from which they were first deduced.

Godwin-Austen had in mind the type of anticline exemplified by the Vale of Pewsey and the Hog's Back—a long but relatively narrow fold or ripple in the strata, often consisting of strings of periclines. Such folds, of which the east–west axes are the best examples, are all based upon older folds of Armorican date, and their Miocene revival was only the last of many revivals of activity spanning all Mesozoic time (see Chapter III).

Lamplugh, on the other hand, studied the Weald, which is not a single anticline but an anticlinorium. Folds which replace those of the Vales of Pewsey and Wardour eastward pass on over the Wealden anticlinorium merely as ripples upon its surface. It is not one of the separate anticlines that is based upon a syncline, but the whole anticlinorium that is based upon a synclinorium in the Jurassic rocks underneath; and the synclinorium is the Jurassic trough of deposition. The supposed cause of the upfold of the Cretaceous rocks here, by compression of the Jurassic trough already filled with sediment, was explained on p. 50.

It so happens that in the Weald the Jurassic trough coincides in direction with the east–west or Armorican folds. But elsewhere (fig. 7, p. 48) the folds cut across it at all angles, and so give rise to 'axes of uplift' across the troughs. The Mendip and the North Devon Axes intersect the trough of deposition in Somerset and Wilts. roughly at right-angles (see map, fig. 15, p. 86). Therefore the elevations along these axes acted at right-angles to the broader bulging up of the Chalk over the deepest parts of the trough. But over the axes the trough was shallow and there was correspondingly little bulging up; therefore an anticline of east–west directrix was always the dominant result of the Miocene earth-pressures over the ancient axes of uplift, no matter what the direction of the trough thereabouts might be.

The bearing of this upon our present problem will be obvious. It explains why the uplift of the North Devon Axis was dominant, although the trough of deposition crossed it at right angles, and therefore how the east–west elevation of Exmoor came to condition the drainage, rather than any bulging up of the Cretaceous rocks over the Jurassic trough in a contrary direction. In this way came about the analogy between the drainage of Exmoor and that of the Weald, which proves so deceptive when used as the starting-point for inquiries into the tectonic history of Mesozoic times.

To summarize the conclusions to which a study of the ancient landmass under London leads us:

(1) The surface of the London landmass is a peneplain, which slopes gently under the Mesozoic covering. The covering is still intact owing to its having been carried for the most part below the reach of the forces of erosion by the south-easterly tilt to which the British Isles were subjected early in the Tertiary period. On the other side of the Jurassic trough the Welsh highlands rise similarly with a peneplaned surface, but they have been stripped bare in the Tertiary and Quaternary periods.

(2) If, as seems probable, the two landmasses on opposite sides of the Jurassic trough were analogous, none of the Triassic or Jurassic rocks is likely to have passed right over the top of the Welsh highland, although they may have encroached to an unknown extent round its margins, beyond their present outcrop.

(3) The slope of the Welsh peneplain was not acquired as the result of the Miocene orogeny, although it is possible that it was steepened at that time. The surface sagged gradually throughout the Jurassic period as the trough subsided. It is probable that the south-easterly slope would be considerably greater had not the stresses found relief in the Malvern Fault, whereby the subsidence of the trough was probably enabled to continue without dragging the margin of the adjoining highland with it.

(4) The depression of the Bristol Channel, where Trias and Lias still exist below present sea-level, represents a branch of the Jurassic trough of deposition, analogous with the Kentish Weald. The floors of these depressions sagged to their present low relative levels during the Jurassic period, not as the result of the Miocene folding.

(5) The highland of Exmoor lies upon an Armorican anticline, along which activity was repeatedly revived during the Mesozoic period, when it and its continuation eastward formed the North Devon Axis of uplift, crossing the Jurassic trough of deposition at right-angles. It was finally re-elevated during the Miocene orogeny, and this revival gave rise to the Vale of Wardour fold and initiated the present drainage of Exmoor.

(c) The Dartmoor Highland

Concerning the elevation of the Dartmoor highland in the Mesozoic period an altogether new line of inquiry has been opened up by the study of the mineral assemblage of the Dartmoor granite. Dr. A. W. Groves has carried out a series of investigations in the sediments of all ages in the South of England with the express object of detecting the earliest appearance of this assemblage of heavy minerals.[1]

Assuming that the batholith was intruded in Permo-Carboniferous times, and that its granitic structure proves it to have crystallized slowly under a thick covering (probably 5,000 ft.) of Palaeozoic rocks, the problem was to find at what period the covering first began to be removed so far as to expose the granite. Dr. Groves finds that the sought-for detritals first appear in the Wealden Beds, while the many Triassic and Jurassic sands contain not a trace. The inference is, therefore, that throughout Triassic and Jurassic times the

[1] A. W. Groves, 1931, 'The Unroofing of the Dartmoor Granite', *Q.J.G.S.*, vol. lxxxvii, pp. 62–96.

Dartmoor highlands were some thousands of feet higher than at present, relative to the sub-Mesozoic floor, the granite still lying buried beneath its covering of Palaeozoic strata and remaining untouched by denudation until the period of the Wealden lakes. This seems to vindicate Jukes-Browne's line of reasoning that the Palaeozoic land areas were much higher in the Jurassic period than now.

Striking evidence is available of the height and steepness of the mountain tract of Dartmoor during the formation of the earlier New Red rocks. The Triassic peneplain in Devonshire comes to an end abruptly against the high ground in a manner suggestive of its termination against the escarpment of the South Wales Coalfield. Beyond, buried valleys, filled with New Red rocks, run for considerable distances westward into the folded Carboniferous tract, out of which they were carved during the Epiric or New Red continental period. Although their excavation would date from an earlier stage of the erosion of the Armorican mountains than the formation of the desert peneplain of the Trias, the fact that they were not destroyed by the production of the peneplain testifies to the survival of high ground into the Jurassic period. The steep gradient of the valleys is remarkable. In the largest, which runs from Exeter westwards through Crediton to Hatherleigh, the floor lies more than 2,000 ft. lower than the top of the Dartmoor granite only 4 miles away; yet apparently erosion in the upper course of the valley had not reached the granite, for no detritus from it is found in the New Red filling below.[1]

In dealing with Wales we are hampered by the fact that not only has all trace of Jurassic rocks been swept away from the neighbourhood of the existing high ground, but all trace of the Cretaceous rocks also. It is therefore particularly instructive that in Devonshire, in one restricted area, we are afforded an opportunity of observing the relations of Cretaceous and Eocene strata to the Palaeozoics, within a few miles of the highland of Dartmoor.

The Upper Cretaceous rocks of the Blackdown Hills and the dissected plateau to the south, which themselves project many miles farther west than any other Cretaceous outcrops in England, are continued after a gap of about 15 miles by the diminutive outliers of the Haldon Hills, south of Exeter. Already at Seaton on the coast the cliff-sections show the Albian overstepping the Lias and Rhætic Beds and coming to rest on the Keuper Marls, and the discordance can be followed in the cliffs as far as the Peak Hills west of Sidmouth. The inland outcrops continue the overstep until on the Haldon Hills the Albian rests upon the Permian. But there is also an overlap within the Albian, for the lower part of the Upper Gault (in the arenaceous facies known as Blackdown Beds), present in the Blackdown Hills, is absent at Haldon.

'Not only is there this overlap westward, but the sands at Haldon have become in places coarse-grained, granitic, or even pebbly. The fauna they yield is also highly suggestive of shoal water, or even of a shore-line within a few miles. Compound corals resembling littoral forms, absent from Blackdown, are here abundant and varied; the mollusca are thick-shelled strong species, such as could stand much knocking about.'[2]

Within 3 miles to the west, in fact, is the present edge of the Dartmoor

---

[1] A. W. Groves, 1931, loc. cit., p. 69.
[2] C. Reid, 1913, 'Geol. Newton Abbot', p. 93, *Mem. Geol. Surv.*

granite, beyond which heights rise rapidly to 1,200 ft. and 1,500 ft. above sea-level, and eventually to about 2,000 ft. on the highest points of the moor. Since the base of the Cretaceous rocks on Haldon is only about 700–50 ft. above sea-level and there is no sign of folding or faulting on a comparable scale, it is evident that they abutted against a shore-line somewhere in the next few miles to the west, where a deep valley has since been excavated by the River Teign. The total thickness of the Albian rocks is about 65 ft. on Great Haldon and 90 ft. on Little Haldon. Upon them lies a sheet of gravel of presumably Eocene age, composed mainly of Chalk flints mixed with finer-grained detritus from Dartmoor. The fossils in the flints indicate the former existence of nearly all the zones of the English Chalk.

If a section be drawn from Dartmoor through the Haldon Hills to the main Mesozoic outcrop of Dorset, filling in the gaps caused by subsequent erosion (fig. 94), certain marked resemblances to the section through the eastern margin of the Mesozoic trough emerge. In the first place the Cretaceous rocks appear as a transgressive sheet, overstepping the Jurassic and Triassic strata outwards; secondly, the Cretaceous rocks show overlap within the series, the higher overlapping the lower. This is in principle exactly what occurs on the margin of the London landmass. But there are important differences. The principal are, that in Devon (1) the New Red rocks are vastly thicker and are not overlapped by the Jurassics before those are in turn overstepped by the Cretaceous; and (2) the Cretaceous oversteps regularly on to progressively older members of the Jurassic System westward, the last being the Lower Lias and Rhætic; whereas against the London landmass it jumps from Lower Oolites on to Palaeozoics, the Lias having been previously overlapped and overstepped by the Oolites (see fig. 8, p. 52). The second difference is accounted for by the fact that the Cretaceous overstep against Dartmoor is considerably steeper (previous tilting was greater) than against the London landmass, so that the edges of the higher Jurassic formations, which may once have spread beyond the boundaries of the Lias, would have been removed prior to the Cretaceous transgression (this is suggested diagram-matically in fig. 94). A comparison between the feather-edges of the Jurassic rocks in Devonshire with those around the London landmass is therefore legitimate. We may presume that a similar combination of overlaps and oversteps and paucity of sedimentation within the Jurassic Series took place towards the shore-line of Dartmoor, but we must imagine all the evidence as having lain at some unknown (but probably not very great) height above the present plane of discordance at the base of the Cretaceous. The marked westerly attenuation of the Green Ammonite Beds of the Lower Lias between Golden Cap and Black Ven may be significant in this connexion (p. 119).

The alternative explanation to the one here offered, namely that the Jurassic System (some 3,000 ft. thick in Dorset), or at least a substantial part of it, was laid down over the top of the Dartmoor highland, but has since been removed owing to Tertiary uplift of the highland, as Prof. Jones would have us believe of Wales and Exmoor, is here discountenanced by the existence of the Albian patches on the Haldon Hills. These patches prove that the necessary uplift and erosion must have taken place *before the Albian period* instead of in the Miocene as postulated for Wales and Exmoor; and although there cer-tainly was pre-Albian and post-Jurassic uplift of the landmass relative to the adjoining trough, such colossal upheaval and stupendous erosion between the Jurassic and Cretaceous periods as would be required on this view seems altogether improbable.

Incidentally the low level of the Haldon Cretaceous and Eocene outliers relative to the adjacent granite mass throws interesting light on the age of the erosion features on the top of Dartmoor. If the granite was not submerged by the Cretaceous sea, it has been undergoing subaerial denudation for a very long period. Any one who is acquainted with the vast sheets of fresh and coarse granitic detritus, largely of Dartmoor origin, spread out over the Dorset–Hampshire Basin and constituting the Bagshot Beds, will readily ascribe the main roughing out of the existing features on the highest part of the moor to the Eocene period. Hence if there is anything in the suggestion of Travis[1] that the highest points on Dartmoor[2] and Exmoor[3] are part of a subaerial peneplain and continuous with the 1,900–2,000 ft. plain of North and Central Wales, sloping towards a base-level of erosion to the south or south-east, then the Eocene is at least a likely period for the formation of this peneplain. The Eocene gravels, granitic sand and pipe-clay of the Dorset and Hampshire heaths can only have reached their present positions across a continuous sheet of Chalk, stretching up to the flanks of Dartmoor over the sites of the modern valleys of the Teign, Exe and Otter. It is highly probable, and in accordance with Jukes-Browne's restoration of the Chalk sea, that the highest parts of the peneplain in North Wales, with the largest monadnocks rising from it, may likewise never have been submerged beneath any Mesozoic sea

[1] C. B. Travis, 1914, Pres. Add. *Liverpool Geol. Soc.*, vol. xii, p. 26.
[2] High Willhays, 2039 ft.; Cut Hill, 1981 ft.; White Horse Hill, 1,974 ft.; Lints Tor, 1,908 ft.   [3] Dunkery Beacon, 1,707 ft.

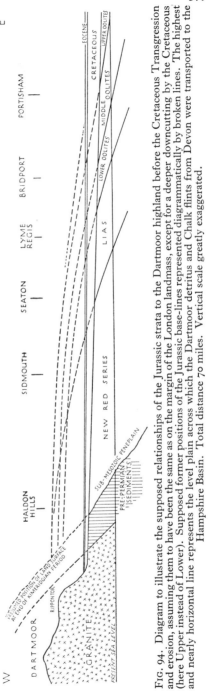

Fig. 94. Diagram to illustrate the supposed relationships of the Jurassic strata to the Dartmoor highland before the Cretaceous Transgression and erosion, assuming them to have been the same as on the margin of the London landmass, except for a deeper downcutting by the Cretaceous (here Upper instead of Lower). Supposed former positions of the Jurassic base-lines represented diagrammatically by broken lines. The highest and nearly horizontal line represents the level plain across which the Dartmoor detritus and Chalk flints from Devon were transported to the Hampshire Basin. Total distance 70 miles. Vertical scale greatly exaggerated.

but may have been still undergoing subaerial erosion in the Eocene, when the Mesozoic rocks themselves formed the continuation of the same plain to eastward. This seems to have been the view of Prof. Fearnsides, who was the first in modern times to give a description of the feature in North Wales; he wrote: 'As to the age of the peneplain we have no evidence. . . . Probably it is an early Tertiary surface of subaerial denudation which, reduced to a condition of low relief, had its drainage rejuvenated by the further uplift of the Miocene.'[1]

(d) The Pennine Range and the Lake District

In considering the question of the possible extension of the Jurassic sea over the high Palaeozoic tracts of the North of England, and in particular the Pennine Range and the mountains of the Lake District, there are two lines of approach: (1) direct internal evidence for or against land to the west or north-west in the Jurassic sediments of Leicestershire and Lincolnshire, the Yorkshire Basin, and the Carlisle outlier; (2) evidence bearing on the date of upheaval of the high features beyond the present confines of the Jurassic outcrop, so that it may be judged whether it is probable that the Jurassic sea ever overspread them. These two classes of data will now be briefly considered.

INTERNAL EVIDENCE OF PROXIMITY TO LAND IN THE JURASSIC ROCKS OF
NORTHERN ENGLAND AND THE MIDLANDS

Until very recently the microscope was not used for examining the sands of the Yorkshire Jurassic, and the data considered in inferring their source of origin were entirely macroscopic—false-bedding, thickening and increase in coarseness. On these grounds it was generally believed that the materials of the thick deltaic deposits of the Middle and Upper Jurassic were brought into Yorkshire from the north, or from rather west of north. Eighty years ago Sorby came to the conclusion, on taking the average of a large number of measurements, that the false-bedding indicates currents from about WNW.,[2] and more recently Mr. Black has ascertained that the washout channels also indicate a flow of water from north to south (p. 315). As a whole the 'Estuarine Series' thickens towards the north of the Yorkshire Basin and, as Fox-Strangways pointed out, the sand-grains become coarser towards the west, thus suggesting a source of supply in the north-west.[3]

Most of the other Jurassic formations of the Yorkshire Basin were considered by Fox-Strangways to show signs of having been formed off land lying in approximately the same direction, ranging from north to west, on account of their either thickening in that direction or becoming more arenaceous, or both. Thus the Middle Lias changes in facies towards the north-west from a deeper-water argillaceous deposit to a series of calcareo-arenaceous ironstones (pp. 160–1), suggesting to Fox-Strangways that land lay not far north-west of Eston Moor near Middlesborough.[4] The Scarborough Beds or Grey Limestone Series likewise become excessively coarse and arenaceous at Eston Moor, so that Fox-Strangways says: 'It is therefore probable that

---

[1] W. G. Fearnsides, 1910, *Geol. in the Field*, p. 821.
[2] H. C. Sorby, 1852, *Proc. Yorks. Phil. Soc.*, vol. i, pp. 111–13.
[3] C. Fox-Strangways, 1892, *J.R.B.*, p. 391.   [4] Ibid., pp. 387, 389.

a shore-line existed somewhere to the west, and not far from the present north-west outcrop; arenaceous beds were formed in the west, while calcareo-argillaceous strata were deposited in the east.'[1] The Kellaways Beds thicken in general towards the north, like the Estuarine Series, the constituents of which the sands resemble; and Fox-Strangways considers this 'incontestable proof that the sandy sediment must have been derived from the north'.[2] Mere increase of thickness, however, is no proof of approach towards a shore-line, since it may be caused by thinning in the opposite direction against an axis of uplift, where sedimentation was retarded. Fox-Strangways himself records on the next page that the Oxford Clay thickens to a maximum in the south-east, but he does not infer from this that land lay in that direction, but rather the deepest part of the sea, while uplift was going on in the north-west. The lenticular shape of the Hambleton Oolite Series and the westerly thickening of the Middle Calcareous Grit were likewise taken by the same author to indicate land in the west, and he suggested that the sub-oolitic limestone of the Hambleton Oolite might be composed of detritus from a coral reef 'extending along the flank of the old Palaeozoic land to the west, which we have every reason for believing existed during the whole of the period'.[3] Hudleston was equally convinced that land existed not far to the north-west of the Yorkshire Basin in the Oxfordian period.[4]

We cannot accuse Fox-Strangways of being biased and of interpreting all evidence in favour of a preconceived notion that there was land to the north and west; for he records of one formation—the Dogger—that a more calcareous development along the western escarpment seems to him to indicate 'somewhat deeper water' in that direction, and he is led to suggest that the channels of the Dogger (p. 224) may be 'a series of narrow inlets connected with a larger sea to the west'.[5] He makes no comment, however, on the inconsistency of this with the evidence derived from the other formations, and the reader is left to reconcile the opinions as best he may. The suggestion as to the Dogger channels has, in fact, little to recommend it. The increase in the calcareousness of the Middle Lias towards the NW. is taken by Fox-Strangways on another page to indicate approach to a shore, where presumably Carboniferous Limestone was undergoing solution.

All these inferences drawn by Fox-Strangways and his predecessors from the macroscopic data bear solely on the direction from which the sands were brought into Yorkshire; they suggest nothing as to the actual source of the materials. Upon this subject the microscopic investigation of the heavy minerals is now being brought to bear by Dr. Rastall, but the work is still in its infancy and unfortunately no very definite conclusions are yet available. It appears certain, however, from information which Dr. Rastall has kindly communicated to me, that the sands could not have come from any adjoining British source. The inference is that they were derived from the Fennoscandian mainland or Atlantis, probably from some part of the site of the present North Sea. They therefore have no bearing on the existence of land over the Pennines.

The clay formations, such as the Lias and the Kimeridge Clay, give no

[1] Ibid., p. 394.   [2] Ibid., p. 396.   [3] Ibid., pp. 399, 401.
[4] W. H. Hudleston, 1876, *P.G.A.*, vol. iv, p. 370.
[5] C. Fox-Strangways, 1892, loc. cit., p. 390.

indications of the source of their materials. It was suggested by Goodchild that the dark muddy sediment of the Lias might have been derived from the disintegration of Coal Measures, but it is stated that the mica flakes in the Lias are larger than those in the Coal Measures.[1]

South of the Market Weighton Axis all the formations thicken steadily and, as has been emphasized on earlier pages, the outcrops of the Jurassic formations cut obliquely across a trough of deposition, of which the opposite margin lay to the south-east. Only one definite instance is known of materials in the Jurassic strata of this trough having been derived from the north-west: at Leicester the base of the Rhætic Beds contains small pebbles of igneous rocks from the neighbouring Charnwood Forest (p. 111). On the Forest the Keuper Marls overlap the earlier New Red rocks and rest directly upon the Archæan, and the pebbles at Leicester indicate that, as with some of the islands in the Mendip Archipelago, the surface was not submerged until during or after Rhætic times. Of the later deposits, only the Northampton Sands have been investigated, and Mr. Skerl is of the opinion that their materials were derived from the south-east (p. 209).

The Lower Lias and Rhætic Beds of the Carlisle outlier show no abnormalities that could be ascribed to deposition particularly near to a shore-line; nor do the Liassic strata of the Shropshire outlier.

THE AGE OF THE PRESENT HIGH FEATURES OF THE PENNINE RANGE AND THE
LAKE DISTRICT AND THE POSSIBILITIES OF THEIR HAVING
BEEN COVERED BY THE JURASSIC SEA.

When we examine the surface upon which the New Red rocks rest as they approach the Pennines we find a peneplain similar to the sub-Mesozoic peneplain of the South-West of England. The Trias and the Permian together form a conformable series, which passes across the denuded edges of the structures formed during the Armorican orogeny and itself takes no part in them. There is the same proof of enormous erosion after or during the uplift of the Armorican mountains, before the New Red rocks were laid down. Along the flanks of the Pennines the coal-basins were already determined in their long N.–S. synclines and the Coal Measures had been entirely removed from the intervening anticlines before the earliest Permian strata were laid across their surfaces.[2] On the west side of the Pennines, in the Ribble Valley at Clitheroe, one of the most easterly patches of Permian rests upon Viséan limestone. This indicates that all the Coal Measures, the Millstone Grit and the Yoredale–Pendleside Series had previously been removed—amounting to 15,000 ft. of strata.[3]

Nevertheless, the central anticline of the South and Mid Pennines (the northern part of the range will be considered separately) still rises to an average height of 1,000–1,500 ft. above sea-level, with points over 2,000 ft. in Derbyshire, many hundreds of feet above any surviving New Red rocks. The patch at Clitheroe lies at an elevation of only some 200 ft., in a valley

[1] P. F. Kendall and H. E. Wroot, 1924, *Geol. Yorkshire*, p. 310.

[2] J. J. H. Teall and E. Wilson, 1880, *Geol. Mag.* [2], vol. vii, pp. 92–5; E. Wilson, ibid., 1879, vol. vi, pp. 500–4; G. V. Wilson, 1926, 'Concealed Coalfield of Yorks. and Notts.', 2nd ed., *Mem. Geol. Surv.*, p. 54.

[3] R. H. Tiddeman, 1875, in 'Geol. Burnley Coalfield', pp. 121–2, *Mem. Geol. Surv.*; and Kendall and Wroot, 1924, loc. cit., p. 262.

hollowed out of the crest of an anticline, and is surrounded by hills of Carboniferous strata ranging from 1,000 to nearly 2,000 ft. in height;[1] it is thus analogous with the Permian valleys which run westward into the Culm Measures towards the Dartmoor highland. This suggests strongly that the South and Central Pennines stood high above the Permian sea and have not been first removed by the peneplanation and later re-elevated to their present height. Moreover, the difference in the fauna of the marine Permian on the two sides demands a barrier with only a roundabout connexion.

That there has been re-elevation along the Pennine Axis from time to time can hardly be doubted. But, by analogy with comparable axes under the Jurassic covering, the movement is more likely to have been gradual, prolonged through the Mesozoic period and mostly relative to the sagging troughs of deposition on either side, rather than positive, paroxysmal, and restricted to the Tertiary period, as commonly stated. The line of the Derbyshire Pennines is continued accurately by the Vale of Moreton Axis, and if the two are truly connected, as seems probable from the relation of the Mendip, Malvern, North Devon and other axes to corresponding uplifts in the Mesozoic covering, then the Vale of Moreton Axis is a key to the history of the Pennine Chain.

The movements along the Vale of Moreton Axis were probably only faint tremors of the main disturbances farther north, for the Vale of Moreton lay near the extreme south end of the anticline, where it pitched underground and disappeared or joined the opposite side of the trough. Yet we have seen (Chapter III) that the movements were enough to control deposition through much of the Lower and Middle Jurassic period. If a continuation so remote, although below water, was sufficiently near the surface to cause retarded deposition and repeated contemporaneous erosions, it is difficult to believe that the main Pennine Axis was not in part above the sea like the Mendips in Jurassic times. The least we can suppose is that it constituted a broad hog's-back of shallows, sometimes above water, sometimes submerged, separating the eastern from the western trough of deposition. We must also not forget the evidence of an island at least as late as Rhætic times at Charwood Forest. The Charnwood range is only a small branch of the Pennines, and it is hardly likely to have stood above water while they were submerged, an island by itself in the middle of an extensive sea.

The many eminent geologists who have discussed the evolution, tectonic and physiographical, of the Pennines and Northern England have usually evaded the question of the former extent of the Jurassic sea, or have dismissed it in a few lines, including it, if at all, simply to complete their histories.

Goodchild wrote: ' I have no doubt in my own mind that all the rocks up to the Lias, and even all the Jurassic rocks, once overspread the whole district and were continuous with those of other parts of the kingdom.'[2] He gave no reasons, however. Prof. Marr reached the same conclusion at the end of an ingenious and interesting chain of reasoning.[3] Jukes-Browne wrote:'When discussing the extension of the Keuper Marls across England we came to the conclusion that the whole of the southern and central parts of the Pennine

[1] Kendall and Wroot, 1924, loc. cit., p. 263.
[2] J. G. Goodchild, 1889, *P.G.A.*, vol. xi, p. 277.
[3] J. E. Marr, 1906, *Q.J.G.S.*, vol. xlii, pp. lxxxvii–lxxxviii.

Chain were submerged beneath the great inland sea of later Triassic time. From this it follows that the Liassic sea also spread over this part of England from Cheshire, Lancashire, and Westmoreland on the west to Yorkshire and Lincolnshire on the east, and thence across the whole breadth of the North Sea area'; and 'I strongly suspect that the Oxfordian sea spread completely over the greater part of the Pennine area, as the Liassic sea had done'.[1] Messrs. Kendall and Wroot 'are driven to the conclusion that . . . the general lowering of level had been carried sufficiently far to bring the waters of the Jurassic sea over all that remained of at any rate the South Pennines. In that case probably all the Jurassic series was deposited over the Pennine area, except perhaps the heights of the northern fault blocks.'[2] Lastly, Mr. J. S. Turner has published a similar opinion, though adding no new evidence.[3]

I should be placing myself in an extremely unenviable position if I openly expressed disagreement with such a consensus of opinion. Nevertheless no harm can be done by reviewing the evidence brought forward in support of these statements, calling attention to some of the objections. The arguments, so far as I have been able to discover them, may be ranged under the following headings:

(1) The base-lines of the Trias and the Rhætic, when produced, pass over the top of the Pennines and Lake District (Jukes-Browne and Marr).

(2) If the uplift of the present elevations was post-Triassic, as shown by (1), then it was probably Tertiary, because in the Mesozoic period 'there is no evidence of great movements in the British Area', such as would suffice for the task (Marr; and implied by others).

(3) If the amount of Permo-Trias present in West Lancashire were restored over the relict outlier of Permian in the Ingleton Coalfield, the top of the Keuper would be considerably higher than the Carboniferous rocks of the adjoining Pennines (Jukes-Browne).

(4) Denudation during and after the Armorican uplift was so drastic that all features were probably lowered to such an extent as to bring them below the Jurassic sea-level (Kendall and Wroot).

(5) To supply the Jurassic sediments of the Yorkshire Basin 'it seems likely that some land exposure of much greater area must have been drawn upon. The thick beds of grit at the Peak [of Yorkshire; i.e. Ravenscar] must have been brought by a larger river than could have been formed on so narrow a belt as the Pennine uplift' (Kendall and Wroot).

(6) 'If the Skye uplift can have been sculptured into its present condition in late Tertiary times, the same explanation applies to Lakeland, and if we suppose Lakeland to have existed as land since, say, the end of New Red Sandstone times, it would long ago have been reduced to a nearly level surface by denuding agencies.' On the contrary, the district is physiographically young, and erosion is still in progress (Marr).

(7) The south-easterly tilt is probably alone responsible for the preservation of Mesozoic rocks in the South-East of the British Isles; hence their absence from the North-West of England and Scotland is probably due to their having been removed by erosion (Marr).

[1] A. J. Jukes-Browne, 1911, *Building of the Brit. Isles*, pp. 266, 277.
[2] P. F. Kendall and H. E. Wroot, 1924, *Geol. Yorkshire*, p. 310.
[3] J. S. Turner, 1927, *P.G.A.*, vol. xxxviii, p. 372.

When we come to examine these arguments, we find that each one harbours a fallacy, which I will endeavour to the best of my ability to show up, since they are dangerous when allowed to remain latent.

(1) and (2) The production of base-lines was a favourite line of reasoning with Jukes-Browne, who employed it throughout his book. The assumption is that if the base-line of the Keuper or Rhætic passes over the top of any high ground separating two outcrops, the folding is subsequent to deposition, and therefore the sheet of sediment was originally continuous—was not banked up against the barrier. A good instance of the application of this argument is the following:[1] 'The broad Derbyshire anticline lies between two equally deep and broad synclines, and the manner in which both Trias and Lias come into these basins makes it almost certain that they both dipped off the intervening anticline; in other words, the final arching of the anticline was completed by *post-Jurassic* flexuring.' The two words that I have placed in italics imply the application of argument No. (2) above, and this leads to such statements as the following: 'if the Keuper passed across the Palaeozoic ridges [of the North Pennines] then the greater part of the Jurassic series must have done so, and the Palaeozoic floor would have been covered by some 3,000 to 4,000 feet of Trias, Lias and Oolites'.[2]

Now this line of argument is plausible, but it rests upon a misconception of the method of accumulation of the Jurassic sediments and upon a lack of appreciation of the tectonic movements which, as we have seen, were proceeding all through the Jurassic period. It is true there was no major orogeny in the Jurassic, comparable with the Alpine and capable of upheaving the Pennines and Lake District all in one spasm; but there was continual subsidence of the troughs of deposition, amounting altogether to 3,000–4,000 ft., compensated to an unknown extent by elevation of the adjacent more stable blocks. This subsidence was greatest in the centres of the troughs and decreased or changed into elevation towards the margins. Consequently the earlier strata to be formed (i.e. the Rhætics and Lias) were by the end of the epoch depressed into deep synclinals, although the latest stratum was still being laid down horizontally. When the later strata were then removed from the edges of the troughs (because that was where they were highest and thinnest), the earlier layers revealed underneath would be disposed in synclinal fashion without having undergone any 'post-Jurassic flexuring'. The two completely different interpretations that can therefore be put upon the same facts are illustrated by fig. 95, p. 586.[3]

(3) The third argument depends upon the first; for it assumes that the same thickness of Permo-Trias was deposited at Ingleton, on the flank of the Pennines, as farther west, over the syncline of West Lancashire. On the view put forward here, an equal thickness of Permo-Trias would certainly not have been formed at the edge of the trough in such a position, and still less an equal amount of Jurassics. There would, in fact, probably have been slow sedimentation interrupted by erosions. It is curious that Jukes-Browne himself should note, on the next page, that 'round Derby and Ashbourne, where the Pennine axis passes below the Trias, the Bunter is reduced to a small thickness (about 100 ft.), and the Keuper sandstones thin out entirely'. If there is this

---

[1] A. J. Jukes-Browne, 1911, loc. cit., p. 245.　　　　　　　　　　[2] Ibid., p. 244.
[3] For a further discussion of these principles see Chapter III.

attenuation of the Trias across the axis so far south, where it was pitching and dying out, how much more might both Trias and Jurassic have thinned out over and towards the principal part of the uplift farther north? Even at the Vale of Moreton much of the Jurassic is missing.

(4) and (5) Prof. Kendall's and Mr. Wroot's first suggestion seems to me to be robbed of much of its weight by the fact that the denudation did not remove the high Carboniferous ground surrounding the Permian valley at Clitheroe. Moreover, drastic as this denudation certainly was everywhere, it left high features in the South-West, such as Dartmoor and the South Wales Coalfield, across which the Jurassic sea was unable to spread. Their other suggestion (5) can be readily agreed to without prejudicing the existence of a Pennine ridge. It seems certain that whether the Pennines were high above the water or deep beneath the sea, a large tract of land lay to the north and north-east (Fennoscandia), and that this supplied the sediments, not the small and narrow Pennine ridge. Any such ridge, even if it was above water, was too small to provide a catchment area adequate to supply the huge mass of deltaic sediments in Yorkshire; moreover, modern work on the heavy minerals being begun by Dr. R. H. Rastall shows that the sediments are of the wrong sort to have come from the rocks composing the Pennines. Yet the Pennines may have been land

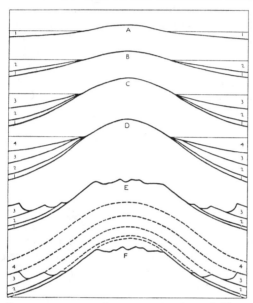

FIG. 95. A, B, C, D show stages in the subsidence and filling of two troughs of deposition, compensated by uprise of the intervening anticline. E shows the same after epeirogenic uplift of the whole area and consequent subaerial denudation, with relics of strata 1, 2, 3 dipping off the anticline. F represents the reconstruction inferred by producing the base-lines of the strata 1–4.

nevertheless, just as Cowbridge Island was above water while the thick mass of sandstones derived from the mainland were accumulating to northward of it in the Rhætic Period (p. 103; and compare fig. 97, p. 597).

(6) Prof. Marr's comparison of the Lake District with the Skye uplift, where Tertiary rocks of an equal or greater hardness have been sculptured to about an equal extent, is a singularly apt reminder of the extreme length of geological time. No one could deny that the Lake District would have been converted to a nearly level plain had it been upheaved to its present position in relation to sea-level in New Red Sandstone times and had those relations remained the same ever since. But the present rate of erosion depends entirely on the relation of the ground to sea-level, a factor which cannot possibly have remained even approximately constant for so great a length of time. That the present state of erosion of the Lake District is the same as that of Skye only proves that the last major adjustment of sea-level took place at about the same

time (late Tertiary or early Quaternary) in the two regions. This has no bearing on the relations of the ancient rocks of the Lake District to the Jurassic sea. That there was upheaval over the site of the Lake District during the Armorican movements is proved by the overstep of the New Red rocks across Lower Palaeozoics along the present seaboard, between Whitehaven and Millom. And, as will be mentioned presently, there seems good reason for supposing that high ground survived and that the Carboniferous covering continued to be stripped off during the formation of the Permian of the Vale of Eden. Whether this high ground remained above water in Jurassic times or was only a region of shallows it is impossible to say; but the synclinal arrangement of the Lower Lias and Rhætic Beds of the Carlisle outlier does not imply that the Jurassic rocks were deposited in force all over the Lake District, but may even indicate the reverse (see (1) above, and fig. 95).

(7) It is in general true that the south-easterly tilt of the Tertiary epoch was responsible for the preservation of the Mesozoic rocks in the South-East of the British Isles; and that they have been removed from much of the North-West is a legitimate corollary. But Jurassic rocks were not deposited everywhere in the South-East (e.g. they are absent over the London–Ardennes landmass) and consequently they need not have been deposited everywhere in the North-West.

There is a direct bearing upon these problems in the new picture of the structure of the Northern Pennines that has begun to emerge as the result of recent work. The stratigraphical results achieved by Prof. Garwood and the illuminating investigations by Profs. Marr and Jones, Dr. Rastall, Mr. Turner, and Messrs. Trotter and Hollingworth depict the northern end of the Pennine Range east of the Craven, Dent and Pennine Faults as a stable block with a rigid foundation of Archæan rocks, more or less surrounded by geosynclinals filled with Lower Palaeozoic sediments.[1] In Lower Carboniferous times sedimentation began early over the geosynclinal troughs, but it was not until late in the S_2–D period that the Rigid Block was submerged.

The previous view has been that the great faults were initiated at latest in Carboniferous or Permo-Carboniferous times and had arrived at essentially their present positions by the Permian period; that the Rigid Block stood high above the subsided country of the Vale of Eden, where fans of torrent-breccia were spread out from the foot of the escarpment to form the Permian brockrams. Prof. Kendall explained the introduction of Lower Palaeozoic pebbles in the Upper Brockram and their absence from the Lower by supposing that in the interval between the formation of the two layers of breccia a movement took place along the faults, which brought up lower rocks within the influence of erosion. This view of the physiography of the Northern Pennines in New Red Sandstone times, of which Prof. Kendall was the champion,[2] seemed to be corroborated by Prof. Marr's conclusions as to the Tertiary date of upheaval of the Lake District dome.[3]

[1] J. E. Marr, 1921, 'The Rigidity of North-West Yorkshire', *The Naturalist*, pp. 63–72; and O. T. Jones, 1927, 'The Foundations of the Pennines', *Journ. Manchester Geol. Assoc.*, vol. i, part 1, pp. 5–14.
[2] P. F. Kendall, 1902, 'The Brockrams of the Vale of Eden . . .', *Geol. Mag.* [4], vol. ix, pp. 510–13; and 1924, *Geol. Yorkshire*, p. 261.
[3] J. E. Marr, 1906, loc. cit.; and 1919, *Geol. of the Lake District*, pp. 122–3.

Certain facts, however, make it seem possible that a complete revision of this view will be necessary. Messrs. Turner[1] and Trotter and Hollingworth[2] have brought forward evidence that in New Red times the Rigid Block actually lay at a lower level than its surroundings. Turner has observed that the Permian brockrams do not thicken and become coarser and more angular towards the Pennine Faults, but away from them, as if the materials had been derived from the south. He interprets the appearance of Lower Palaeozoic rocks in the Upper Brockram as a sign that between the formation of the two layers of brockram the Carboniferous covering over the high ground of the Howgill Fells and the Lake District, from which the materials were being derived, had begun to be worn through, so that Lower Palaeozoic rocks were exposed to erosion. Confirmation of this view is adduced by Trotter and Hollingworth. They point out that north of Armathwaite (9 miles SE. of Carlisle), where the adjoining Carboniferous rocks of the escarpment that overlooks the Vale of Eden are composed of sandstones of the Yoredale facies, there are lenticular beds of conglomerate in the Penrith (New Red) Sandstone made up almost entirely of limestone pebbles. If any pebbles in this position had been derived from the escarpment on the north-east they would have consisted mainly of sandstone: the inference is, therefore, that these came from high ground in the opposite direction, where the Carboniferous Limestone was being stripped off the Lake District.

Turner has shown, moreover, that the westerly downthrows along the Outer and Middle Pennine Faults, which lowered the present Vale of Eden, are post-Triassic, while along the Dent and Inner Pennine Faults, which downthrow eastwards, the movements were late Carboniferous and pre- or early-Permian, since they do not affect the New Red rocks. He also draws interesting corroborative conclusions from a study of the dolomitization, to which the Carboniferous Limestone was subjected below the floor of the magnesian sea or salt lake of the Permian period. The limestone of the Rigid Block, east of the faults, has not been dolomitized like that on the west, and from this he infers that in Permian times it was protected by a thick covering of Yoredales, Millstone Grit and Coal Measures, through which the dolomitizing waters were unable to percolate. This points again to considerable pre-Permian depression of the Rigid Block relative to its surroundings. The depression was apparently brought about by movements along the Dent and Inner Pennine Faults; the subsequent post-Triassic movements, which reversed the relative levels of the Rigid Block and the Vale of Eden, occurred along the parallel Outer and Middle Pennine Faults.

Along the E.–W. line of the Stublick Fault, which forms the abrupt northern edge of the Rigid Block, some of the protective covering of higher Yoredales, Millstone Grit and Coal Measures still survives as a string of faulted outliers extending from Midgeholme eastward to south of Hexham. North of the Stublick Fault region there ran in Lower Palaeozoic and Lower Carboniferous times a geosyncline, striking E.-W., where sedimentation began at a much earlier stage of the Carboniferous period than over the Rigid Block to the south of the fault. Over the site of this geosyncline arose during the Armorican orogeny the Bewcastle Anticline, the crest of which was

[1] J. S. Turner, 1927, *P.G.A.*, vol. xxxviii, pp. 358–72.
[2] F. M. Trotter and S. E. Hollingworth, 1928, *Geol. Mag.*, vol. lxv, pp. 433–47.

denuded down until in places (as at Kirkcambeck, 12 miles north-east of Carlisle) the Trias came to rest directly upon the Cementstones at the base of the Lower Carboniferous. Here we have proof that before the deposition of the Trias some 10,000 ft. of Carboniferous strata had been removed and therefore that the anticline had been elevated by that amount relative to the Rigid Block.[1]

Another geosynclinal trough of Carboniferous beds, with a pre-New Red anticline superimposed upon it, is traced by Messrs. Trotter and Hollingworth from east to west across the Rigid Block. Starting south of St. Bees Head and pitching east like the Bewcastle Anticline, it passes through the Howgill Fells and the Pass of Stainmore until lost beneath the main outcrop of the Magnesian Limestone at Middleton Tyas, between Richmond and Darlington. On this line there is an enormous overstep of the Carboniferous strata by the New Red rocks on both sides of the Pennines. The feature severs the northern end of the Rigid Block from the rest and makes of it a separate structural unit, which Trotter and Hollingworth term the Alston Block.

The Alston Block, it is interesting to note, is surrounded on three sides (north, south, and west) by broad anticlines, elevated during the Armorican movements over the sites of previous geosynclinal troughs, much as after the Alpine movements the London landmass was surrounded on the same three sides by upraised Chalk, arched over geosynclinal troughs filled with Jurassic rocks. The Carboniferous and the Cretaceous strata seem to have played an analogous role in the two places. We are reminded of the conclusion that the relatively low level of the Chalk under the London Basin is due not so much to its having been folded down as to the bulging up of the surroundings owing to the compression of the buried geosynclinal troughs beneath (p. 50).

The question of importance in our present context is: At what period was the Alston Block re-elevated relative to its surroundings, and when was the thick canopy of Millstone Grit, Coal Measures and Permo-Trias removed from its top? There is a certain amount of indirect evidence. Many years ago Goodchild[2] pointed out the existence of relics of a peneplain, passing from the level tops of the Lake District mountains and the Howgill Fells across the New Red country eastward, until it sloped down to the foot of the Pennine escarpment. He believed that this peneplain was formerly continuous with that which is a marked feature at about 2,000 ft. at the top of the Pennine Range, from which Cross Fell (2,930 ft.) and a few other high points protrude as monadnocks; and on his view its age was Cretaceous. To Goodchild, therefore, belongs the credit for noticing a peneplain of post-Triassic date *which has been displaced along the Pennine Faults.*

Goodchild's views, because often imaginatively expressed, have frequently been rejected or even treated with contempt. But other investigators, when they have become as familiar with the district as he was, have also come to recognize this peneplain.[3] It has now been accepted whole-heartedly by Trotter, who considers that it has not only been faulted along the Pennine and

---

[1] Trotter and Hollingworth, loc. cit.
[2] J. G. Goodchild, 1889, *P.G.A.*, vol. xi, p. 268 and fig. 8; and *Trans. Cumb. and West. Assoc.*, No. XIV, pp. 73–90.
[3] e. g. T. McK. Hughes, 1901, *Proc. Yorks. Geol. Soc.*, vol. xiv, p. 129.

Stublick Faults but also warped or gently folded, as well as having participated in the general south-easterly tilt of England.[1] He believes that the same peneplain is continued by the level top of the Chalk Wolds of East Yorkshire (at about 800 ft.) and therefore that the date is not Cretaceous as Goodchild supposed, but Tertiary—that it is, in fact, the Early Tertiary peneplain recognized in the Mesozoic areas of England by W. M. Davis.

By an able study of the drainage of the region, Trotter shows that the faulting and gentle folding of the peneplain are events 'so recent that their impress is still boldly stamped on the topography'; and while they are pre-Glacial, he considers that they are probably not earlier than Pliocene. This latest movement along the faults was only enough to give rise to the present main topographic features—to allow the Vale of Eden and Tyne Gap to subside to their present low levels relative to the top of the Alston Block. The amount of the throw is measurable by the present height of the escarpment—a maximum of about 1,500–2,000 ft. If we restore the two sides of the fault as they were before the displacement, so that the peneplain is continuous once more, it becomes evident that there was a much greater displacement before the formation of the peneplain; for by joining the faulted edges of the peneplain we only bring Trias against Lower Carboniferous. This means that if the Permian and Trias and all the intervening Carboniferous Beds originally passed across the Alston Block, some 8,000 ft. of strata had been removed from the block by the time the peneplain was completed. Had this mass of strata been already carried away by Jurassic times, or did it remain throughout the Jurassic period and act as a barrier separating the east from the west?

Mr. Trotter considers that, since the idea of the Northern Pennines towering above the surrounding country in Permian times has to be replaced by a conception of the Rigid Block depressed relatively to its surroundings by the Armorican movements, the idea that the Mesozoic formations may not have extended across the area from east to west must inevitably collapse. This conclusion, however, seems too hasty. There are still many uncertain factors. It cannot be assumed that because the Alston Block was depressed during the Armorican orogeny it remained a depression all through the Mesozoic era. We know, in fact, that it rose to such an extent that all the extra mass of sediments was removed by the time the Tertiary peneplanation was completed. The vital factor is the date (or dates) at which this elevation took place.

According to Turner, part of the post-Triassic movements along the great faults consisted of thrusting from the south or south-west. During this episode dolomitized limestone was brought into direct contact with un-dolomitized limestone and Silurian rocks were thrust over Carboniferous. Such pressures suggest the Miocene orogeny and were perhaps a prelude to the final (Miocene or Pliocene?) subsidence along the faults. But the main trend of the movement was in the opposite direction, resulting in elevation of the Alston Block or lowering of the surrounding country. Both Turner and Trotter assume that this elevation took place in early Tertiary times. The amount of the movement is so great, however, that it seems likely that it was spread out over a very long time—perhaps through the whole Jurassic and Lower Cretaceous periods—denudation perhaps keeping pace with elevation

[1] F. M. Trotter, 1929, *Proc. Yorks. Geol. Soc.*, vol. xxi, pp. 161–80.

as movement continued along the faults. There is an example of an intra-Jurassic fault close at hand, at Peak (Ravenscar) (p. 180). Moreover, we know that the work was brought to an end by the completion of the great peneplain; and by comparison with other areas the date of the peneplain is Eocene.

Another line of speculation as to the state of the Lake District in the Jurassic period is worth mentioning. We have seen that, even if Prof. Marr's evidence of the radial drainage demands a final doming up of the district in the Tertiary period, there was certainly an elevation as early as the end of Carboniferous times. Mr. Green considers that this earlier uplift took place along an anticline of NW.–SE. trend and pitch, which can be traced through the Ingleton and Horton-in-Ribblesdale inliers until, bending round W.–E., it is lost in the Pennines beyond Pateley Bridge in Nidderdale.[1] On this view the Lake District is nothing more than a periclinal enlargement of an anticline pitching to the SE. Now it is noteworthy that where this anticline loses itself in the centre of the Pennines it is replaced by several others having a W.–E. or NW.–SE. strike, which pass on eastward until they disappear beneath the Mesozoic covering. The most important is the Wharfe Anticline, which continues the earlier line of the Lake District—Horton Anticline; and this has been identified by Prof. Kendall with the Market Weighton Axis (p. 62). Therefore, if Green's view of the structure is correct, it is not unreasonable to suppose that in Mesozoic times the Lake District had a history similar to that of the Market Weighton Axis.

(e) The Scottish Highlands

It has been emphasized that the present occurrences of Jurassic rocks in the Hebrides bear no relation whatever to the original distribution of the sea which deposited them, but that their preservation has been determined entirely by faulting combined with the protective action of coverings of Tertiary lava. The surviving relics, in fact, represent fragments that were faulted down into safe positions in the solid sub-Mesozoic platform, out of reach of denuding agencies.[2]

Thus it would seem that, but for the faulting and the outpourings of basalt, there would be no trace at all of the Jurassic sediments. The most impressive demonstration is the occurrence of Rhætic rocks and fossils caught up in the neck of the Tertiary volcano on the Isle of Arran, many miles from any other vestige of Mesozoic strata. Yet the Arran and other Hebridean Jurassic rocks show no signs of having been deposited in a very restricted area; on the contrary, most of them closely resemble their counterparts in England.

Consequently, when in imagination we restore the landscape as it stood before the denudation, it seems that we cannot argue as Jukes-Browne was wont to do on the Welsh Border, that the surrounding ancient rocks were formerly so much higher that the Jurassics cannot have spread far beyond their present limits. Instead, we are bound to infer that from areas where mountainous tracts now form the surface, great thicknesses of Jurassic strata have been removed.

---

[1] J. F. N. Green, 1920, 'Geol. Structure of the Lake District', *P.G.A.*, vol. xxxi, p. 126.
[2] J. W. Judd, 1878, *Q.J.G.S.*, vol. xxxiv, p. 669; and G. W. Lee, 1920, 'Mesozoic Rocks of Applecross, Raasay and N.E. Skye', p. 1, *Mem. Geol. Surv.*

Judd, who made the first comprehensive study of the Jurassic rocks of Scotland, was so impressed by these indications that he found it

'impossible to avoid the conviction that these patches of Secondary strata, although now so minute in dimensions and isolated in position, once formed portions of a great series of connected deposits which covered the greater part of the vast area [of Scotland] and attained in places a thickness of from 4,000 to 5,000 ft. . . . that the whole of the north and north-western portions of the British archipelago—now sculptured by denudation into a rugged mountain land—were, like the south and south-eastern parts of the same islands, covered by sedimentary deposits, ranging in age from Carboniferous to Cretaceous inclusive.'[1]

This was an extreme view, and it was opposed by such authorities as Sir Andrew Ramsay and Professor Sollas. The latter justly remarked that 'to say from the present absence of Oolitic beds over large areas that such beds had never been was more philosophical than to deduce from it their former existence', and upheld the view that also 'all Wales, Devonshire, Cornwall and the Pennine chain have been above water since the time of the Lower Lias'.[2] Nevertheless, it cannot be denied that there has been colossal loss by denudation, and that a very considerable area of what is now high mountainous country in the Hebrides and along the west coast must have been at one time covered by Jurassic and Cretaceous strata.

Unfortunately there is no hope of gaining any idea of the former extent of these sediments by looking for the sub-Mesozoic peneplain, for all traces of it have been obliterated by two or more phases of Tertiary peneplanation.

The loftiest mountains in Scotland, such as Ben Nevis and the Cairngorms, rise as monadnocks from a High Plateau at 2,000–3,000 ft., which is the most conspicuous physiographical feature of the Highlands. W. M. Davis ascribed this plateau to subaerial denudation, and he considered that, by analogy with other peneplains, it can hardly have been exposed to dissection longer than since late Mesozoic times, and that the work of dissection probably commenced in the Tertiary era.[3] According to Peach and Horne, the arrangement of the Mesozoic rocks shows that 'they must have entered, largely into the structure of the High Plateau'[4], and if that is so the peneplanation cannot be earlier than Tertiary. This conclusion falls into line with Trotter's views as to the Tertiary date of the peneplain at 2,000 ft. in the North Pennines, and with Fearnsides's conjecture regarding the one at the same height in North and Central Wales; hence it becomes extremely unlikely that any information on Mesozoic water-levels is to be gained from the peneplains in any of the highlands of Western and Northern Britain.

Along most of the west coast of Scotland a still later plateau is conspicuous, with an upper limit at about 1,000 ft. This is partly cut in the Eocene volcanic and plutonic rocks and is presumably a shelf of marine erosion, like the 1,000 ft. and lower platforms in Cornwall. Where it is found there is still less possibility of tracing any sub-Mesozoic peneplain.[5] Clearly, if we are to

---

[1] J. W. Judd, 1878, loc. cit., pp. 668–9.
[2] W. J. Sollas, in discussion of Judd's paper, ibid., p. 743.
[3] W. M. Davis, 1896, Geol. Mag. [4], vol. iii, p. 526.
[4] B. N. Peach and J. Horne, 1930, Chapters on the Geology of Scotland, p. 7.
[5] Peach and Horne call the 1,000 ft. the Intermediate Plateau, the lowest being the Continental Shelf; 1930, loc. cit., p. 2.

arrive at any decision as to how much of the Highlands was covered by the Jurassic sea, we must approach the question from other directions.

When Judd came to the conclusion that all Scotland was submerged and had the full thickness of Jurassic sediments deposited over it, he seems to have been influenced by two principal considerations: (1) the fact that the rocks in the Hebrides are not of littoral facies any more than those in England; and (2) the fact that it is only owing to faulting and lava-flows that any fragments have been preserved at all; which prompts the question: 'How much more may have been lost entirely, leaving no trace behind?' Let us now examine the value of these considerations as arguments.

The first may be dismissed, since all the Jurassic rocks of the Hebrides could have been deposited sufficiently far from a shore-line for them to contain no littoral deposits, without the whole of Scotland being submerged. The shore may have lain 10 or even 20 miles east of the existing deposits.

The second line of argument implies that the faulting and the igneous activity to which the sediments owe their preservation were restricted to the area where the sediments are found. For if the sediments previously existed everywhere and their survival anywhere is directly due to these processes, then they should be preserved as a natural consequence wherever the faulting and the igneous activity have occurred.

This raises an interesting point, because to a certain extent it is true that the relics of Mesozoic strata are centred precisely in the part of Scotland where Tertiary igneous activity and faulting were most thickly concentrated. The igneous activity, indeed, *was* virtually confined to the western coast and islands. The faulting seems to have been at least most frequent in the same area. Prof. Gregory believed that the majority of the pre-Glacial fiord-like lochs and valleys were due primarily to faults,[1] and, although this view has been questioned, it is remarkable that the faulted relics of Mesozoic rocks survive just in the area in which most of the fiords occur.

But it is possible that all three phenomena—igneous activity, faulting, and the occurrence of Mesozoic sediments—are merely effects of some common cause. If the area was one of subsidence in Jurassic times, it is not improbable that the same crustal weakness that gave rise to the subsidence and consequent sedimentation was later responsible for the localization of the faulting and igneous outbreaks of the Eocene period.

At least it would be very rash to assert, as Judd seems to imply, that the absence of Mesozoic sediments in other parts of Scotland can be attributed solely to the absence of those causes which conditioned their survival in the Hebrides—namely, lava-sheets and faulting.

Some of the largest faults of all, which have lowered the whole surface of the country thousands of feet, cross Scotland from sea to sea. The most important examples are the great trough faults of the Central Lowland or Midland Valley and the Great Glen (Glenmore). It is said that there are traces still extant of the river-system which drained the country now occupied by the Central Lowland and the surrounding country before the faults came into

---

[1] J. W. Gregory, 1913, *Nature and Origin of Fiords*, pp. 172–5, and map on p. 144. All agree that faults often determined the sites of fiords even though the fiord form was not directly due to faulting; for even if we suppose fiords to be ice-deepened valleys, the pre-Glacial rivers which originally formed the valleys frequently selected the lines of faults as paths of least resistance.

existence, and that the faults are therefore (at least in the main) of Tertiary date; and other faults which belong to the same group cut some of the Eocene igneous dykes.[1] Consequently, if thick Mesozoic strata had covered the area, they would have been lowered by the faults and some would surely have survived—at least until the Pleistocene Ice Age, when the Boulder Clay would have been filled with their fragments. In the Great Glen it is still more noteworthy that Mesozoic rocks do survive also at the eastern end of the troughed belt (in the Brora and Ethie district), but they are to be found *at the ends only*. There is nothing to show that the faulting which gave rise to the Great Glen was more intense at the extremities than in the middle, and indeed such an idea is on theoretical grounds highly improbable. Why, then, should there be no traces of Mesozoic rocks throughout the central and longest part of the trough?

The only logical conclusion seems to be that the Jurassic strata were only deposited, at least in anything like their full thickness, on either side of the Highlands; that Scotland formed an island, a non-subsiding block, between two sinking troughs or geosynclinals. Hence, even though the Jurassic seas may have transgressed to an unknown extent around its edges, the general tendency of the area would have been one of elevation, and no great thickness of Jurassic sediments is likely to have been deposited upon it.

On the other hand, the Jurassic rocks and their faunas, from the Lower Lias to the Kimeridge Clay, are so nearly identical on the two sides of Scotland[2] that a direct sea-connexion is demanded. De Lapparent's maps show continuous sea across the North-West Highlands[3] from the time of the Lower Lias,[4] and Prof. Wills also submerges the North-West Highlands during the Oxfordian period.[5] But the present heights of the North-West Highlands are little short of those of the Grampians: there are large areas over 1,500 ft. high, and many points exceed 3,000 ft.; on the same latitude as the Jurassic area of Ross-shire Ben Dearg rises to 3,547 ft.; and there is a point of 3,040 ft. (Ben Hope) within a few miles of the north coast. All this country consists of ancient rocks, the relics of the Caledonian mountains, and there is no more reason for supposing that it was ever submerged beneath the Jurassic sea than for supposing that the Grampian Highlands or Snowdonia were submerged.

Another form of connexion that is often portrayed on palaeogeographical maps is a narrow strait through the Great Glen, where Jukes-Browne imagined a river to have existed as early as the Triassic period, linking up the main Keuper lake with an isolated lake over the Golspie district and the Moray Firth.[6] But apart from the fact that this unjustifiably presupposes the existence of a Triassic and Jurassic Great Glen, as we have already remarked, if Jurassic sediments had been deposited all along it, there seems no reason why none should have been preserved in the middle. The central reaches of

---

[1] E. B. Bailey, 1910, 'Geol. East Lothian', p. 9, *Mem. Geol. Surv.*; and J. W. Gregory, 1913, loc. cit., p. 175.

[2] Excepting local accidents, such as the boulder-beds in the Kimeridgian of Sutherland.

[3] The term North-West Highlands is here used as defined by Peach and Horne, to denote the part of Scotland north and west of the Great Glen (Glenmore); the mountain mass between the Great Glen and the Central Lowland or Midland Valley being the Grampian Highlands.

[4] De Lapparent, 1906, *Traité*, p. 1126.

[5] L. J. Wills, 1929, *Physiog. Evolution of Britain*, p. 150.

[6] A. J. Jukes-Browne, 1911, loc. cit., pp. 248, 274; and Wills, loc. cit., p. 134.

the Great Glen would have provided an ideal place for the survival of some fragments of faulted Mesozoic strata had any existed there.

In view of these difficulties in the way of supposing the Highlands or Low-lands to have been submerged in the Jurassic period, the simplest course seems to be to make the necessary connexion round the north of the North-West Highlands, over the low-lying tract mainly occupied by sea at the present day. From the lie of the Trias at Elgin, on the south side of the Moray

FIG. 96. Sketch-map to show the supposed distribution of land and sea over the British Isles during the deposition of the Lower Lias. Sea stippled; some special shallows dotted.

Firth, it is well-nigh certain that the Mesozoic rocks lapped round the Grampians across the low-lying promontory of North-East Aberdeenshire and Banff into the bay now occupied by the Moray, Cromarty and Dornoch Firths and the Coastal Lowland of Sutherland and Ross. It imposes no strain on probability if we imagine that they similarly lapped round the North-West Highlands, perhaps across part of Caithness and the Orkneys, into the Hebridean sea. The connexion that is demanded is one, not only from the Oxfordian period onward, but from the time of the Lower Lias and probably the Trias.

On this view the Grampians and the North-West Highlands, which, like the Northern Pennines, were considerably higher before the Tertiary pene-planation, would have stood above water in Mesozoic times. By analogy with other such massifs of elevated ancient rocks in Britain and on the Continent, upward movements would have been in excess of downward during Jurassic times, and there would never have accumulated those thousands of feet of sediments which Judd conjured up, but which only characterize geosyn-clinal regions.

(f) The North Sea Region and the Connexions with the Continental Troughs

De Lapparent[1] and most other palaeogeographers depict the greater part of the North Sea as submerged in Jurassic times beneath a broad channel leading to the Arctic Ocean. Fossils brought back in recent years from Spitsbergen and other Arctic islands prove that after the Triassic period there was a marine transgression in those regions at least as early as the time of the Upper Lias, and that shallow epeiric seas survived with sporadic interruptions until late in the Cretaceous period. The faunas of these far northern seas are of the same general facies as those which inhabited Europe, and they prove that the two areas lay within one and the same zoological province. There are many minor differences—in fact there is seldom specific identity—but this may be accounted for partly by the colder climate and partly by the fact that there was probably often emergence and loss of deposit in one area while there was submergence and deposition in the other, with the result that the faunas which appear most nearly comparable are not strictly synchronous.[2] How-ever this may be, the correspondences are so close that there can be no doubt of the existence of a direct sea-connexion with North-Western Europe through-out most of the Mesozoic era.

A number of the Spitsbergen fossils are more closely allied to forms found only on the Continent than to British species. There must, therefore, have been some more direct connexion between the Continent and the Arctic than across England by way of the Hebrides.

These considerations make it probable that a channel or trough lay over at least a part of the North Sea, between the highlands of Scotland and Scandinavia (fig. 96). It is not by any means certain, however, that the channel remained always open. In the Bathonian period, for instance, when enormous quantities of sand were brought into the north of the British area, presumably by a large river or rivers from the Scandinavian highlands, so that deltaic and estuarine conditions prevailed on both sides of Scotland and in Yorkshire and Lincolnshire, it is difficult to see how any passage could have remained open. Probably deltaic material such as that which was carried southward on both sides of Scotland and all down Eastern England completely filled the narrow trough sea, cutting off communication with the Arctic (fig. 97).

Interesting light has been thrown on the south-westward extension of the ancient landmass of Fennoscandia, and the extent to which it supplied sedi-ment to the adjoining Jurassic seas, by petrographical and stratigraphical

---

[1] 1906, *Traité*, maps on pp. 1126, 1189, 1211, 1225, 1239, 1245.

[2] H. Frebold, 1928, 'Strat. u. Paläog. des Jura u. der Kreide Spitzbergens', *Centralblatt für Min.* &c., Abt. B, p. 629, and *Skrifter om Svalbard og Ishavet*, 1929, no. 20, and 1930, no. 31, with numerous palaeogeographical maps. Spath rejects Frebold's explanation in favour of climatic races.

investigations in North Germany. Brinkmann[1] and Schott[2] have shown that at least from Vesulian to Kimeridgian times the bulk of the clastic sediment deposited in the North German sea was derived from the north. Every formation when followed from north to south passes from sand into clay. On the strength of a great deal of evidence brought together from an examination of outcrops, borings and Drift pebbles, it is concluded that at least from the

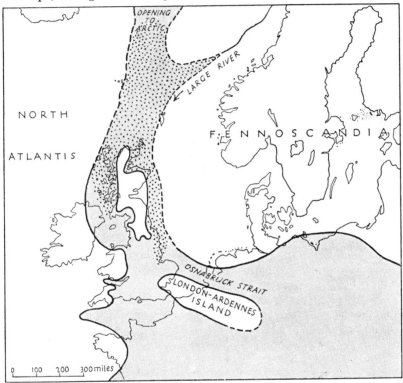

FIG. 97. Sketch-map to show the supposed distribution of land and sea during the Bathonian period (Tulitan Age). Deltaic deposits dotted. The boundary of Fennoscandia over Germany according to Schott.

Parkinsonian to the Perisphinctean Ages the coast-line ran roughly E.–W. across the South Baltic region from Riga towards Copenhagen, and thence, after encircling a gulf that lay over most of Denmark, on by way of Hamburg and the Zuider Zee in the direction of The Wash. On this view most of the North Sea was land, an extension of the continent of Fennoscandia.[3]

Somewhat farther south, as we have seen in Chapters II and III, lay the London–Ardennes landmass, and although this was negligible as a source of sediment in Germany according to Schott, its existence as a large island is

[1] R. Brinkmann, 1923, 'Dogger u. Oxford des Südbaltikums', *Jahrb. Preuss. Geol. Landesanst.*, vol. xliv, pp. 477–513.

[2] W. Schott, 1930, 'Paläogeographische Untersuchungen über den Oberon Braunen und Unteren Weissen Jura Nordwestdeutschlands': Inaugural-Dissert., Göttingen, *Abh. Preuss. Geol. Landesanst.*, N.F., Heft 133, pp. 1–51.

[3] Schott uses the name Cimbria for the south-western peninsula separated by the Danish Gulf (which only came into existence with the Callovian (Macrocephalitan) transgression).

manifest in the thinning out of the formations towards its shores both in England and in North-West Germany, and in its contributions of sediment in England. Its northern shore-line, striking WNW.–ESE., is found by Schott to be constantly in evidence owing to the attenuation of the formations south-west of the Teutoburger Wald, which bounds the Jurassic region of Hanover, Bielefeld and Osnabrück.

Between the London–Ardennes landmass and the south-westerly extension of Fennoscandia the Jurassic rocks are arranged in a typical trough of deposition, which Schott has called the Osnabrück Strait. It strikes approximately SE.–NW., in the direction of East Anglia, and is crossed, like its English analogues, by axes of uplift running at right angles to the direction of the trough and causing local attenuation of the rocks.

There can be no doubt that the Osnabrück Strait made connexion with the English Midland trough in Lincolnshire and the district of The Wash. This connexion accounts for several facts, notably the occurrence in the Coral Rag of Upware and Yorkshire of a group of ammonites and lamellibranchs which have never been found in the South of England, but which were described on the Continent, especially in the Department of the Meuse (p. 416). It may also explain the resemblance of the Cardiocerates of Eastern Scotland to Russian species (p. 434); though, if a more critical study of the German Cardiocerates fails to reveal these species, it will have to be supposed that they migrated round the north of Fennoscandia by way of the Arctic. It was undoubtedly through the Osnabrück Strait that the Jurassic sea first invaded the British Isles from the Continent. The Rhætic Beds of Germany and their fauna are almost identical with our own, whereas the correspondence with those of France is much less close. In the Rhætic and Liassic periods the strait was probably wider than in Upper Jurassic times and may have covered a considerable area which later became part of Fennoscandia. Northward, as already mentioned, it seems highly probable that it was continued east of Scotland to the Arctic Sea.

The direct connexion between England and the Paris Basin seems to have been first opened at the time of the Lower Lias. Thenceforward there was always a wide sea channel between Normandy and the Boulonnais, perfectly continuous with that between Dorset and Kent.[1] M. Lemoine has shown that the deepest part of the trough lay along its north-eastern side, close to the London–Ardennes island, from the Pays de Bray eastward along the basins of the Marne and the Aisne, across that of the Meuse, towards the Vosges. Here in a broad belt the Lias is more than 1,000 ft. thick and the Palaeozoic floor reaches its greatest depths below the surface.[2] This deep zone is separated from that in Kent by a ridge of shallows which runs approximately under the French coast. It appears to be a typical 'axis of uplift' with Caledonian strike, like those which Schott has detected crossing the Osnabrück Strait, and upon it the whole of the Lias is exceptionally thin (105 ft. at Abbeville, probably less at Rouen, and said to be nil at Havre). Little has been learnt about this ridge, however, for the Cretaceous covering is thick and there

---

[1] See W. J. Arkell, 1930, 'A Comparison between the Jurassic Rocks of the Calvados Coast and those of Southern England', *P.G.A.*, vol. xli, pp. 396–411, esp. 396–7.
[2] P. Lemoine, 1930, 'Structure d'ensemble du bassin de Paris', *Livre jubilaire, Soc. géol. France*, vol. ii, pl. XLIX and L.

have been few borings. The attenuation of the Jurassic rocks westward in Normandy is due to their approaching the margin of the trough of deposition and has no bearing on the subject.

Back at the shores of the Celtic highland, we must bring our brief survey of British Jurassic palaeocartography to an end.

The picture of contemporary geography which we are led to draw differs from those extreme reconstructions which either conjure up land almost wherever no Jurassic sediments are found at the present day or extend the sea across the whole British area from horizon to horizon. It enables us to recognize the significance of the outliers in Ireland and the Hebrides without submerging the Scottish and Welsh Highlands beneath a sea with thousands of feet of sediment accumulating on its bed; it admits of a strait up the western part of the North Sea, connecting Germany with Eastern Britain and the Arctic, although most of the area was occupied by the Fennoscandian mainland. Our picture is of a landscape diversified with rather narrow and strait-like seas, occupying a subsiding cleft between the mainlands of Fennoscandia on the one side and North Atlantis on the other, and broken by two elongated islands—the London–Ardennes Island in the south-east, the Scottish–Pennine Island in the north.

Around the margins of the land the seas transgressed and regressed with an unceasing ebb and flow, as in one age subsidence exceeded elevation and in another elevation overcame subsidence. On the whole, however, the upward tendency prevailed on the land and the downward in the troughs, until in the end the centres of the troughs had subsided between 3,000 and 4,000 feet, and they had become filled with the richly-fossiliferous and inexhaustibly interesting series of varied sediments which we call the Jurassic System.

ILLUSTRATIONS OF THE PRINCIPAL SPECIES OF AMMONITES EMPLOYED AS ZONAL INDICES IN THE BRITISH JURASSIC

with the characteristic lamellibranchs of the Rhætic and Pre-*planorbis* Beds and the zonal ostracods of the Purbeck Beds.
Mainly from photographs by J. W. TUTCHER, Esq., M.Sc.

EXPLANATORY NOTES

DIMENSIONS.—After the name and particulars of each specimen the dimensions are given, following S. S. Buckman. The first figure represents the maximum diameter in millimetres, the second the height of the last whorl, the third the thickness of the last whorl, the fourth the width of the umbilicus. The last three dimensions are expressed as percentages of the diameter. At the end comes the amount of the reduction or enlargement of the figure.

TYPE SPECIMENS.—Types are designated as follows:

Holotype: the original specimen on which a species was founded.

Syntypes: the original specimens on which a species was founded, when two or more specimens were figured or described and no holotype was designated.

Paratypes: any other specimens figured or referred to by the author in his original description as belonging to his species in addition to a specified holotype.

Lectotype: a specimen selected as type by a subsequent author from among the original syntypes.

Topotypes: specimens from the same horizon and locality as the original types.

Chorotypes: specimens from the same district and horizon as the original types, but not from exactly the same locality.

Neotype: a specimen designated as type by a subsequent author when the original types are lost or destroyed or too fragmentary for recognition. A neotype may only be chosen from among paratypes, topotypes, or chorotypes.

Genotype: the original specimen on which a genus was founded. And so on for genosyntypes, genolectotype, &c.

PLATE XXIX
RHÆTIC BEDS AND PRE-PLANORBIS BEDS

PRE-PLANORBIS BEDS

1. *Oxytoma longicostata* (Strickland),
Montpelier, Bristol. J.W.T. coll. fig. ×0·6.
2. *Ostrea liassica* Strickland (= *O. irregularis* Schlotheim?),
Ostrea Beds, Stoke Gifford, nr. Bristol. J.W.T. coll. fig. ×1·0.
3. *Lima terquemi* Tate,
Shepton Mallet, Somerset. J.W.T. coll. fig. ×1·0.
4. *Modiola minima* Sowerby,
Ashley Hill, Bristol. J.W.T. coll. fig. ×1·5.
8. *Pleuromya tatei* Richardson & Tutcher,
Stapleton, Bristol. J.W.T. coll. fig. ×1·0.

LANGPORT BEDS OR WHITE LIAS

5. *Dimyodon intus-striatus* (Emmrich),
Cold Knap, Barry, Glamorgan. L. Richardson coll. fig. ×2·0.
6. *Modiola langportensis* (Richardson & Tutcher),
Langport, Somerset. J.W.T. coll. fig. ×1·0.

COTHAM BEDS

7. *Pseudomonotis fallax* (Pflücker),
Kilmersden, Somerset. J.W.T. coll. fig. ×1·0.

WESTBURY BEDS OR CONTORTA SHALES

9. *Pteromya crowcombeia* Moore,
Blue Anchor, nr. Watchet. L. Richardson coll. fig. ×2·0
(from Richardson & Tutcher, 1916, Proc. Yorks. G.S., xix, pl. VIII, fig. 1).
10. *'Schizodus' ewaldi* (Bornemann) (= *'Axinus' cloacinus* Quenst. sp. et
Moore; = *'Isocyprina' ewaldi* Healey). Genus indet. Pending determination of the genus, *Schizodus* is retained here, as in the bulk of the
previous literature, although it and all other generic determinations
made hitherto are certainly wrong.
Beer Crowcombe, Somerset. J.W.T. coll., ex Moore coll. fig. ×2·0.
11. *Chlamys valoniensis* (Defrance),
Stoke Gifford, Glos. J.W.T. coll. fig. ×1·0.
12. *Protocardia rhætica* (Mérian),
Charlton, near Bristol. J.W.T. coll. fig. ×1·75.
13. *Mytilus cloacinus* Tutcher,
Bone Bed, Aust Cliff. J.W.T. coll. fig. ×1·0.
14. *'Cardium' cloacinum* Quenstedt,
Blue Anchor, nr. Watchet. J.W.T. coll. fig. ×1·0.
15. *Pleurophorus elongatus* Moore,
Aust Cliff. J.W.T. coll. fig. ×1·0.
16. *Pteria contorta* (Portlock),
Charlton, near Bristol. J.W.T. coll. fig. ×1·3.

SULLY BEDS

17. *Ostrea bristovi* Etheridge,
St. Mary's Well Bay, Glamorgan. J.W.T. coll. fig. ×0·8.

PLATE XXIX

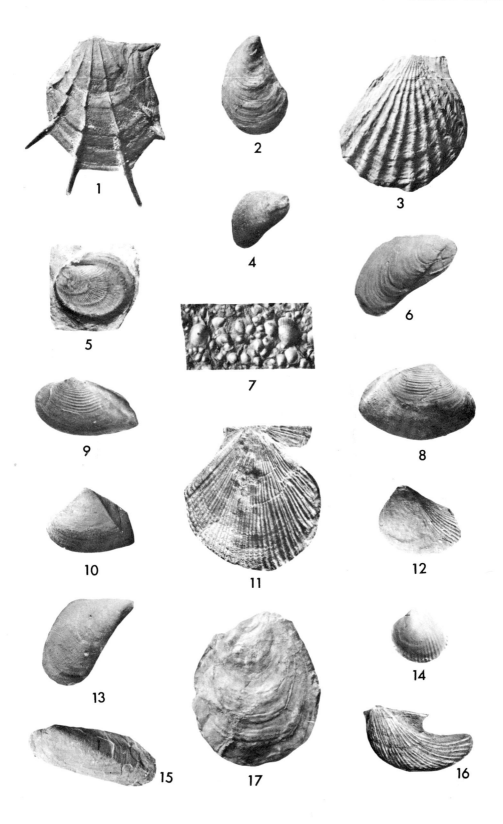

PLATE XXX

1

2

3

4

5

6

7

PLATE XXX

LOWER LIAS

1. *Oxynoticeras oxynotum* (Quenstedt),
Stonehouse, near Stroud, Glos. J.W.T. coll.
48. 50.15.20. fig. × 1·29.

2. *Asteroceras obtusum* (Sowerby),
Charmouth, Dorset. Chorotype. J.W.T. coll.
70. 37.31/34.40. fig. × 0·8.

3, 4. *Arnioceras semicostatum* (Young & Bird),
3. Semur, Côte d'Or, France; from Hyatt, *Genesis of the Arietidae*,
1889, pl. 2, fig. 10. fig. × 1·0.

4. The holotype, Robin Hood's Bay, Yorkshire. Whitby Museum;
from Buckman, *Type Ammonites*, vol. 2, pl. cxii.
38. 29.23.52. fig. × 1·0.

5. *Coroniceras bucklandi* (Sowerby),
Keynsham, Somerset. Chorotype. J.W.T. coll. (The whorl-section is
drawn from another figure and is too quadrate.)
493. 26.32.55. fig. × 0·15.

6. *Scamnoceras angulatum* (Schlotheim),
Hanham, near Bristol. J.W.T. coll.
48. 35.25.34. fig. × 1·0.

7. *Psiloceras planorbis* (Sowerby),
Watchet, Somerset. Topotype. J.W.T. coll. (The crushed condition
is typical.)
68. 32.?.42. fig. × 0·75.

PLATE XXXI
LOWER AND MIDDLE LIAS

1. *Paltopleuroceras spinatum* (Bruguière),
Stoke-sub-Hamdon, near Ilchester, Somerset. J.W.T. coll.
48. 34.30.40. fig. ×1·0.

2. *Amaltheus margaritatus* (Montfort),
South Petherton, Somerset. J.W.T. coll.
66. 43.20.28. fig. ×0·8.

3. *Prodactylioceras davœi* (Sowerby),
Charmouth, Dorset. Topotype. J.W.T. coll.
66. 25.27.56. fig. ×0·8.

4. *Tragophylloceras ibex* (d'Orbigny),
Branch Huish, near Radstock, Somerset. J.W.T. coll.
43. 43.23.26. fig. ×1·0.

5. *Echioceras raricostatum* auct. pars, *E. raricostatoides* Vadasz (= *E. raricostatum* Bayle non Zieten).
The true *E. raricostatum* Zieten, which is rare in Britain, has a somewhat less rounded whorl and rather more widely-spaced ribbing on the outer whorls: see Trueman & Williams, 1925, *Trans. Roy. Soc. Edinburgh*, vol. liii: there are, however, many closely comparable forms, doubtfully true species.
Grove Quarry, Radstock, Somerset. J.W.T. coll.
60. 20.19/23.62. fig. ×0·9.

6. *Uptonia jamesoni* (Sowerby),
Paulton, Somerset. J.W.T. coll.
96. 33.22.51. fig. ×0·78.

PLATE XXXI

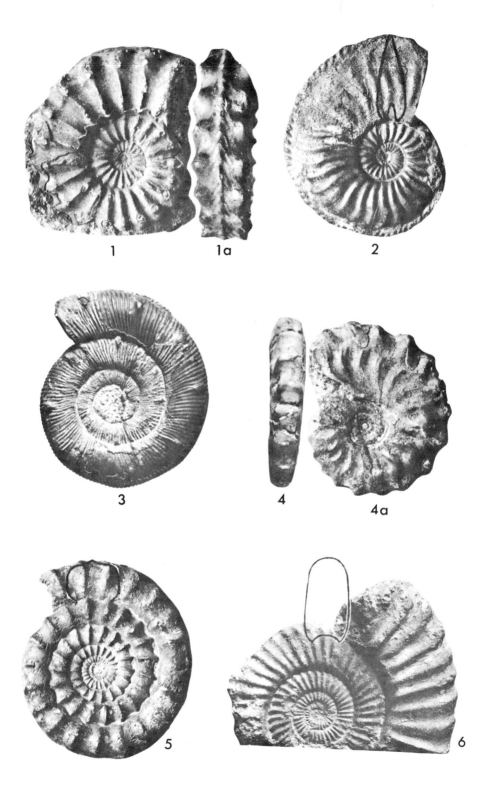

1   1a   2

3   4   4a

5   6

PLATE XXXII

PLATE XXXII

UPPER LIAS

1. *Lioceras opalinum* (Reinecke),
Near Bridport, Dorset. J.W.T. coll.
55. 50.20.16. fig. ×1·0.

2. *Lytoceras jurensis* (Zieten),
Frocester Hill, Glos. J.W.T. coll.
101. 42.36.30. fig. ×0·65.

3. *Hildoceras bifrons* (Bruguière),
Whitby, Yorks. J.W.T. coll.
60. 32.30.41. fig. ×1·0.

4. *Dactylioceras commune* (Sowerby),
Whitby, Yorks. Topotype. J.W.T. coll.
53. 25.24.60. fig. ×1·0.

5. *Harpoceras falcifer* (Sowerby),
Moolham, near Ilminster, Somerset. J.W.T. coll.
107. 40.21½.31. fig. ×0·6.

6. *Dactylioceras tenuicostatum* (Young & Bird),
Whitby, Yorks. Chorotype. J.W.T. coll.
66. 22½.20.56. fig. ×0·8.

PLATE XXXIII
LOWER AND MIDDLE INFERIOR OOLITE

1. *Shirbuirnia stephani* Buckman,
Sandford Lane, near Sherborne, Dorset. Chorotype. J.W.T. coll.
117. 51.31.18. fig. ×0·55.

2. *Hyperlioceras discites* (Waagen) pars., *H. discitiforme* Buckman. The holotype of *H. discites* (Waagen) in the Berlin Museum has been figured by Buckman (1907, 'Mon. Ammonites Inf. Ool.', *Pal. Soc.*, p. cxxii, fig. 88), who says that all the British specimens differ slightly in either ribbing or proportions or both. He therefore regards them as distinct species.
Bradford Abbas, Dorset. Topotype. J.W.T. coll.
65. 50.20.13. fig. ×1·0.

3. *Ludwigella concava* (Sowerby),
Wyke, near Sherborne, Dorset. J.W.T. coll.
71. 52.21.15. fig. ×0·8.

4. *Tmetoceras scissum* (Benecke),
Burton Bradstock, Dorset. J.W.T. coll.
35. 33.28.45. fig. ×1·1.

5. *Brasilia bradfordensis* Buckman,
Barrofield, Beaminster, Dorset. J.W.T. coll.
123. 48.21.21. fig. ×0·45.

6. *Ludwigia murchisonæ* (Sowerby),
Bradford Abbas, Dorset. From Buckman, 'Mon. Ammonites Inf. Ool.'
Pl. III, fig. 1.
114. 41.26.32. fig. ×0·6.

PLATE XXXIII

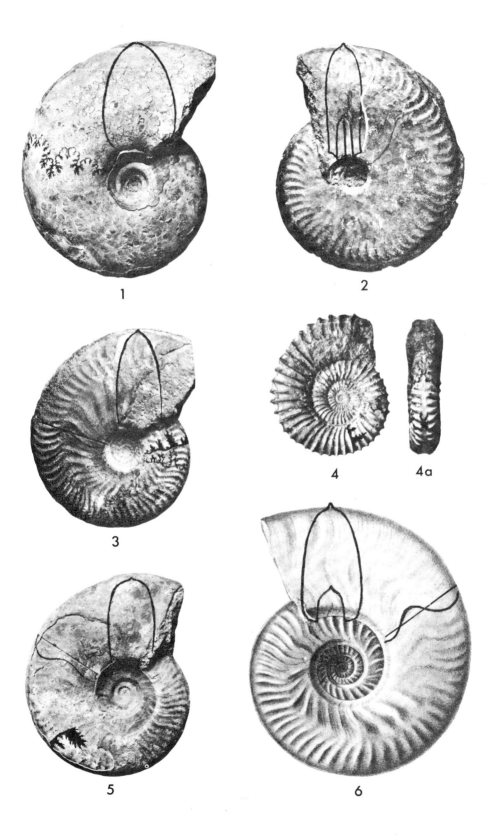

1

2

3

4

4a

5

6

PLATE XXXIV

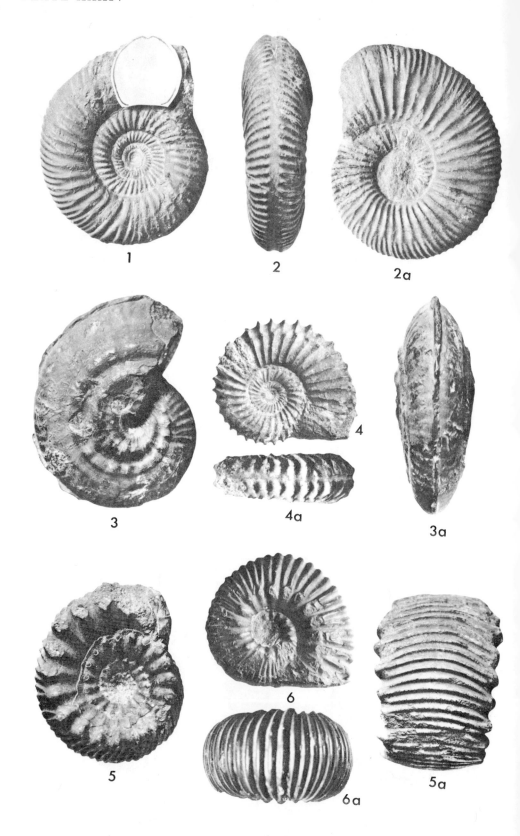

1

2

2a

3

4

4a

3a

5

6

6a

5a

PLATE XXXIV

MIDDLE AND UPPER INFERIOR OOLITE

1. *Parkinsonia schlœnbachi* Schlippe,
Burton Bradstock, Dorset. G. A. Kellaway coll.
75. 34.33.44. fig. ×0·7.
2. *Garantiana garantiana* (d'Orbigny),
Near Sherborne, Dorset. J.W.T. coll.
76. 38.34.36. fig. ×0·75.
3. *Strigoceras truellei* (d'Orbigny),
Burton Bradstock, Dorset. J.W.T. coll.
79. 56.36.7. fig. ×0·7.
4. *Strenoceras niortensis* (d'Orbigny),
Oborne, Dorset. J.W.T. coll.
40. 35.30.41. fig. ×1·0.
5. *Teloceras blagdeni* (Sowerby),
Near Sherborne, Dorset. J.W.T. coll.
41. 29.71.47. fig. ×1·2.
6. *Otoites sauzei* (d'Orbigny),
Dundry Hill, Somerset. J.W.T. coll.
36. 42.50.23. fig. ×1·2.

PLATE XXXV
GREAT OOLITE SERIES AND CORNBRASH

1. *Macrocephalites macrocephalus* (Schlotheim),
   Upper Cornbrash, Shorncote, Glos. W.J.A. coll.
   Identified by L. F. Spath and mentioned Douglas & Arkell, *Q.J.G.S.*,
   vol. lxxxiv, p. 135.
   151. 53.52½.12. fig. ×0·33.

2. *Clydoniceras discus* (Sowerby),
   Lower Cornbrash, South Brewham, Somerset. J.A.D. & W.J.A. coll.
   Identified by S. S. Buckman and figured *T.A.*, plate DVI B; mentioned
   Douglas & Arkell, *Q.J.G.S.*, vol. lxxxiv, p. 146.
   60. 60.25.0 fig.×1·0.

3. *Tulites subcontractus* (Morris & Lycett),
   Great Oolite, Minchinhampton, Glos. Lectotype. Lycett coll.
   Geol. Survey Engl., No. 25610. Figd. Buckman, *T.A.*, vol. 4, pl. CCLXX.
   86. 35.51.35. fig. ×0·72.

4. *Morrisiceras morrisi* (Lycett),
   Great Oolite, Minchinhampton. Topotype. J.W.T. coll.
   68. 50.64.15. fig. ×0·85.

5. *Oppelia fusca* (Quenstedt),
   Basal Fuller's Earth, Broad Windsor, Dorset. J.W.T. coll.
   Identified by S. S. Buckman.
   38. 55.18.9. fig. ×1·24.

6. *Zigzagiceras zigzag* (d'Orbigny),
   Zigzag Bed, Crewkerne, Somerset. J.W.T. coll.
   37. 33.32.43. fig. ×1·0.

PLATE XXXV

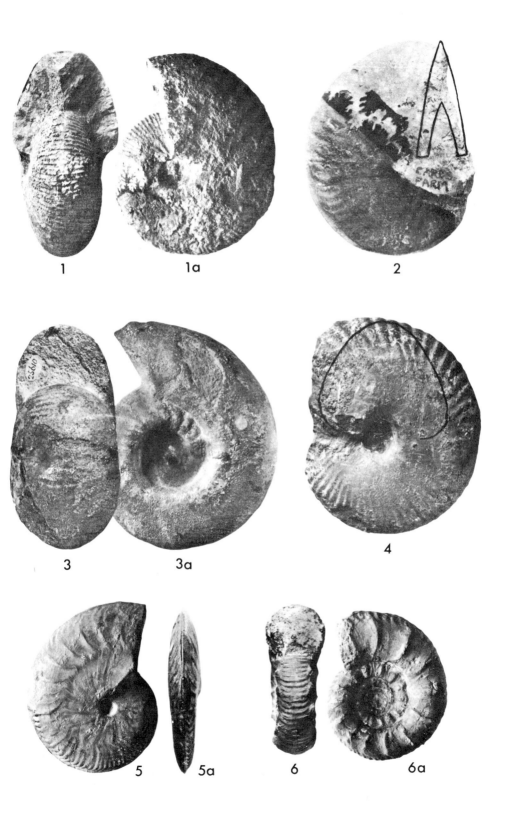

1 1a 2

3 3a 4

5 5a 6 6a

PLATE XXXVI

1   1a   2

3   3a   4   4a

5   5a   6

PLATE XXXVI
KELLAWAYS BEDS AND LOWER OXFORD CLAY

1. *Erymnoceras reginaldi* (Morris),
Trowbridge, Wilts. Topotype. British Museum.
430. 38.36.36. fig. ×0·16.

2. *Kosmoceras* (*Spinikosmoceras*) *pollux* (Reinecke),
Christian Malford, Wilts. Brit. Mus., No. 62229. (The flattening is highly characteristic in England.)
60. 38½.—.32. fig. ×0·9.

3. *Kosmoceras* (*Gulielmites*) *jason* (Reinecke),
Gammelshäusen, Württemberg. Chorotype. Alte Akademie, Munich. After S. S. Buckman, *T.A.*, 1924, plate DIII.
24. 44.26.29. fig. ×1·0.

4. *Cadoceras sublæve* (Sowerby),
Kellaways Rock, Kellaways, Wilts. Topotype. J.W.T. coll. Figd. Buckman, *T.A.*, 1922, pl. CCLXXV.
65. 39.80.25½. fig. ×0·76.

5. *Sigaloceras calloviensis* (Sowerby),
Kellaways Rock, Kellaways, Wilts. Holotype, Sowerby coll., Brit. Mus. Photographed from plaster cast; suture from another specimen (chorotype, ?topotype) superimposed.
94. 47.41.22. fig. ×0·63.

6. *Proplanulites kœnigi* (Sowerby),
Kellaways Clay, near Chippenham, Wilts. Chorotype. J.W.T. coll. Suture from another specimen.
70. 40.30.29. fig. 1×·0.

PLATE XXXVII

MIDDLE AND UPPER OXFORD CLAY

1, 2. *Cardioceras præcordatum* R. Douvillé,
   1. Marnes à *Creniceras renggeri*, Jura Mountains. Paratype; from R. Douvillé, 1912, *Mém. Soc. géol. France*, vol. xix, p. 62, pl. IV, fig. 21.
   28. 45.33.29. fig. ×1·0.
   2. Horton-cum-Studley, near Oxford. W.J.A. coll.
   38. 42.32.31. fig. ×1·0.

3. *Quenstedtoceras lamberti* (Sowerby),
   Tidmoor Point, Weymouth. Topotype. J.W.T. coll.
   63. 40.25.32. fig. ×1·0.

4. *Creniceras renggeri* (Oppel) (*Ammonites cristatus* Sowerby non Defrance),
   Ludgershall, Bucks. Geol Survey coll. No. 27791.
   15. 49.22.24. fig. ×1·5.

5. *Kosmoceras proniæ* Teisseyre,
   Summertown Brickyard, Oxford. J.W.T. coll.
   41. 45.30.25. fig. ×1·0.

6. *Kosmoceras duncani* (Sowerby),[1]
   Summertown Brickyard, Oxford. J.W.T. coll.
   33. 42.29.29. fig. ×1·0.

7. *Peltoceras athleta* (Phillips),
   Hackness Rock, Scarborough, Yorks. Topotype, Neotype. Brit. Mus., No. 89052. Identified and figured by L. F. Spath, 1931, *Mem. Geol. Survey India*, vol. ix, pl. CVI, fig. 3, and CVII, fig. 5.
   70. 28.46.49. fig. ×1·0.

[1] This specimen compares as nearly as possible with Sowerby's figure of *Ammonites duncani*; the original, which came from the Oxford Clay of St. Neots, is lost. Brinkmann's so-called neotype, from Etrochey, Côte d'Or (*Abh. Ges. Wiss. Göttingen*, 1929, M-P. Kl., N.S., vol. xiii, pl. I, fig. 7), should not be accepted, since it bears no resemblance to Sowerby's figure and was not chosen from among topotypes or even chorotypes.

PLATE XXXVII

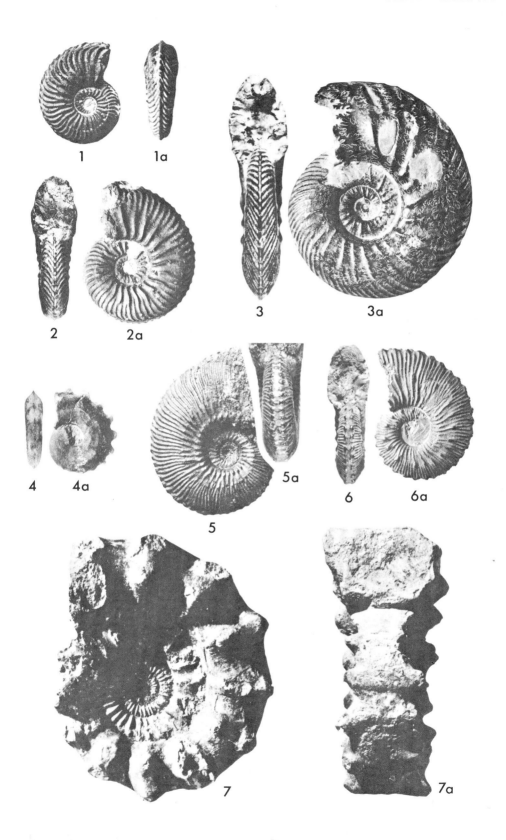

PLATE XXXVIII

1

2   2a

3   4   4a

5   5a   6   6a   7

PLATE XXXVIII

CORALLIAN BEDS

1. *Ringsteadia anglica* Salfeld,
Iron Ore, Westbury, Wilts. Topotype. Devizes Museum.
263. 34.19½.37. fig. ×0·25.

2. *Perisphinctes* (*Dichotomosphinctes*) *wartæ* Bukowski,
Weisser Oxfordkalk, Czenstochau, Poland. Copy of original figure,
Beitr. z. Geol. u. Pal. Österr.-Ungarns, vol. v, 1887, pl. XXXVII, fig. 1.
156. 27.17.51. fig. ×0·35.

3. *Perisphinctes* (*Dichotomosphinctes*) *antecedens* Salfeld,
Junction of Heersumer Schichten and Korallenoolith, near Hildesheim,
North-West Germany. Copy of original figure, 7e *Jahresber. d.
Niedersäch. geol. Vereins z. Hannover*, 1914, fig. 3.
133. 27.24.55. fig. ×0·42.

4. *Perisphinctes* (*Perisphinctes*) *martelli* (Oppel),
Holotype (*Amm. biplex* d'Orbigny non Sow.); locality unknown.
From *Palaeontologia Universalis*, no. 51.
268. 23.30.60. fig. ×0·2.

5, 6. *Cardioceras cordatum* (Sowerby),
5, 5 a. Lower Calcareous Grit, Marcham, Berks. Chorotype. Brit.
Mus. No. 88978.
60. 40.28.33½. fig. ×1·0.
6, 6 a. Lower Calcareous Grit, Shotover, Oxon. Holotype. Brit. Mus.
Sowerby coll.
20. 35.25.40. fig. ×1·0.

7. *Aspidoceras* (*Euaspidoceras*) *catena* (Sowerby),
Lower Calcareous Grit, Marcham, Berks. Topotype. Oxford University
Museum.
272. 28.27.51½. fig. ×0·2.

PLATE XXXIX

KIMERIDGE CLAY

1. *Virgatosphinctoides wheatleyensis* Neaverson,
Kimeridge Clay, Wheatley, Oxon. Holotype. Brit. Mus. No. C. 26897.
From Neaverson, *Amm. Up. Kim. Clay*, pl. 1, fig. 1.
122. 33.28.48. fig. ×0·6.

2. *Gravesia gravesiana* (d'Orbigny),
Near Auxerre, Yonne, France. Syntype, Geno-syntype. From
Palaeontologia Universalis, No. 178, figs. T 1, T 1 *a*.
85. 34.76.42. fig. ×0·68.

3. *Aulacostephanus pseudomutabilis* (de Loriol),
Speeton, Yorkshire. Geol. Survey coll., No. 18150.
Specimen identified by Dr. F. L. Kitchin.
73. 41.26.27½. fig. ×0·93.

4. *Rasenia cymodoce* (d'Orbigny),
Lectotype, Genotype. Locality unknown. From *Palaeontologia Universalis*, No. 55.
81. 33.26½.45. fig. ×0·6.

5. *Pararasenia mutabilis* (Sowerby),
Horncastle, Lincolnshire. Holotype. British Museum, Sowerby coll.
70. 38½.28½.36. fig. ×1·0.

6. *Pictonia baylei* Salfeld,
Wootton Bassett, Wilts. J.W.T. coll.
135. 31.22.48. fig. ×0·5.

PLATE XXXIX

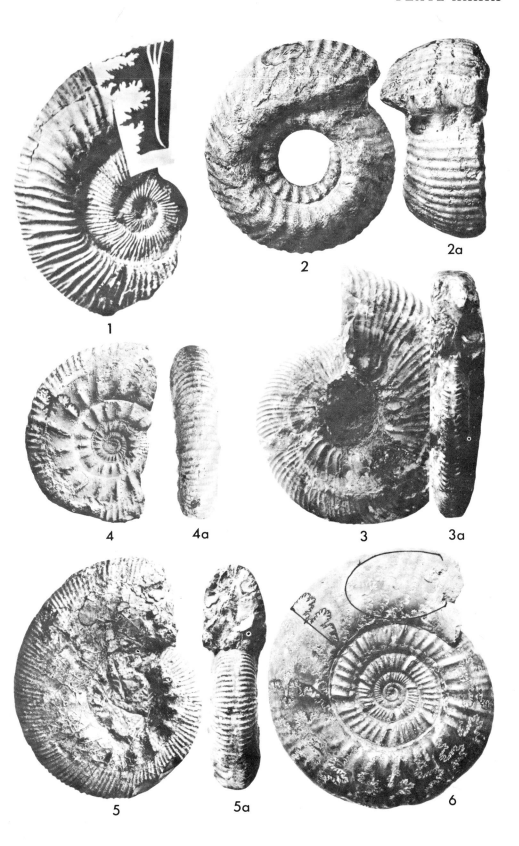

PLATE XL

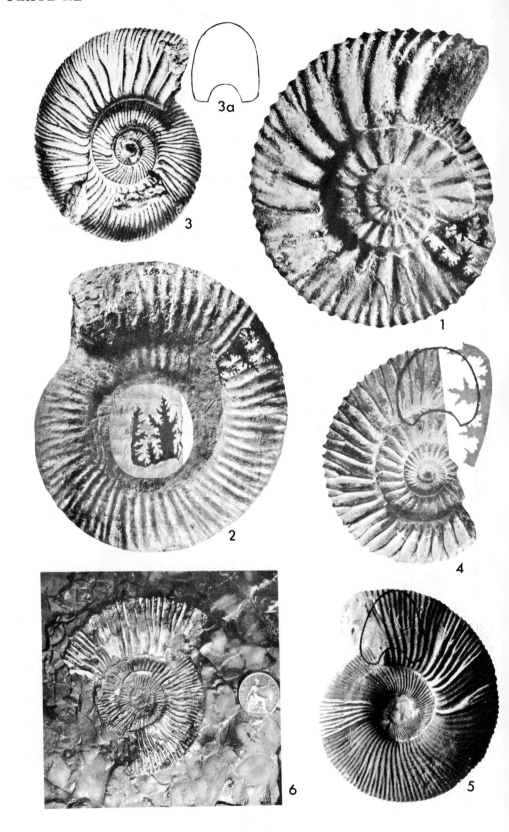

PLATE XL
PORTLAND BEDS AND UPPER KIMERIDGE CLAY

1. *Kerberites kerberus* Buckman,
Portland Stone, Chicksgrove, Tisbury, Wilts. Genotype, Holotype.
J.W.T. coll. Figd. Buckman, *T.A.*, pl. DXX.
105. 33.41.40. fig. ×0·76.

2. *Glaucolithites glaucolithus* Buckman,
Glauconitic Beds, Portland Sand, Long Crendon, Bucks. Genotype,
Holotype. S.S.B. coll. From *T.A.*, pl. CCCVI *a, b.*
271. 33.35.48. fig. ×0·3.

3. *Provirgatites scythicus* (Michalski),
'Kimeridge Clay', near Moscow. Original figure, from Michalski,
Mém. Comité géol. Russie, 1894, vol. viii, pl. v, figs. 6*a* and *b.*
58. 34½.31.39. fig. × 1·0.
(Note that the ribbing on the early whorls is not virgatotome as in
Virgatites.)

4. *Pavlovia rotunda* (Sowerby),
Rotunda Nodules, Chapman's Pool, Dorset. Topotype. Brit. Mus.,
No. C. 26903. From Neaverson, *Amm. Upp. Kim. Clay*, pl. I, fig. 6.
96. 27.31.48. fig. ×0·6.

5. *Pectinatites pectinatus* (Phillips),
Shotover Grit Sands, Shotover Hill, Oxford. Topotype. Oxford
University Museum.
114. 38.33½.35. fig. ×0·5.

6. *Subplanites* sp.,
Typical specimen in characteristic state of preservation, photographed
by the author *in situ* on the foreshore, at a level 14 ft. above the Rope
Lake Head Stone Band and about 25 ft. above the Blackstone, on the
east side of Rope Lake Head, Dorset. (Identified by Dr. L. F. Spath.)
129. 33⅓.—.40. fig. ×0·37.

Specimens of *Titanites*, from the highest zone of the Portland Stone,
are figured on pl. XXVI, p. 493.

PLATE XLI

PURBECK BEDS

All the figures are reproduced from T. Rupert Jones's paper 'On the Ostracoda of the Purbeck Formation', *Q.J.G.S.*, vol. xli, 1885, pp. 311–53, pls. VIII and IX.

1–8. *Cypridea punctata* (Forbes) and varieties,
Middle and Upper Purbeck: 1–3, Mupes Bay; 4, 5, Durlston Bay; 6–8, Ridgeway.

9, 10. *Cypridea ventrosa* Jones,
Upper Purbeck, Durlston Bay.

11–14. *Cypridea granulosa* (Sowerby), var. *paucigranulata* Jones,
Middle Purbeck, Whitchurch?

15–20. *Metacypris forbesii* Jones,
Middle Purbeck, Ridgeway.

21–6. *Cypris purbeckensis* Forbes,
Lower Purbeck: 23, 27, Hartwell; 25, Whitchurch; the rest from the Vale of Wardour.

27–8. *Candona bononiensis* Jones,
Lower Purbeck, Hartwell.

29–32. *Candona ansata* Jones,
Lower Purbeck, Hartwell; and Oving (30 only).

PLATE XLI

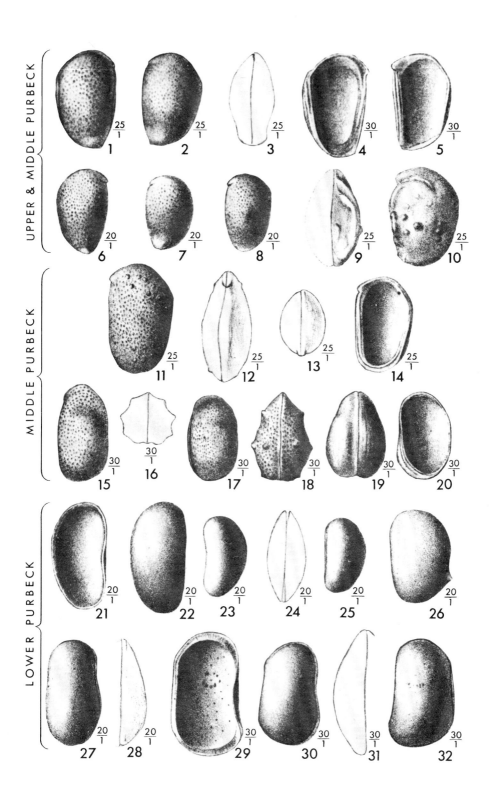

APPENDIX I

NOTES ON THE TERMINOLOGY OF FOSSIL CORAL REEFS, GEOSYNCLINES AND PENEPLAINS

(a) Coral Reefs

WHEN the proofs of this book were nearly completed, a paper appeared advocating greater precision in the terminology of fossil coral reefs and cognate structures (E. R. Cumings, 1932, 'Reefs or Bioherms?', *Bull. Geol. Soc. America*, vol. xl, pp. 331–52).

The author urges that coral reefs and all 'reeflike, moundlike, lenslike or otherwise circumscribed structures of strictly organic origin, embedded in rocks of different lithology' should in future be called 'bioherms'; while 'purely bedded structures, such as shell beds, crinoid beds, coral beds, &c., consisting of and built mainly by sedentary organisms and not swelling into moundlike or lenslike forms' should be known as 'biostromes'. He then goes on to such suggestions as 'hermatolith' or 'hermatobiolith' for reef rock, and 'hermatopelago' for a group of coral islands. The reason adduced for these innovations is alleged ambiguity in existing usage of the terms coral reef, coral bed, shell bed, &c.; the definition of 'reef' quoted from Murray's Dictionary is 'A narrow ridge or chain of rocks, shingle or sand, lying at or near the surface of the water'.

Admittedly the word reef carries this meaning in nautical language; but as such it is purely a generic term. The definition quoted fits it only when it is used in its generic sense. It is difficult to see how there can be any confusion when it is accompanied by the customary specific qualification: e.g. coral reef, reef of rocks, shingle reef. Even without such qualifying words, the context usually renders the meaning unmistakable in scientific literature; moreover, no English scientist in the midst of a discourse on coral structures would court confusion by referring to a 'reef' if he meant a reef of rocks or shingle: he would surely call it a rock ledge (cf. the Kimeridge Ledges, above, p. 441) or a shingle spit or a sand bar. As to 'purely bedded structures, such as shell beds, crinoid beds, coral beds, &c.', any author who referred to these as reefs would not be competent to contribute to scientific knowledge and might safely be ignored.

The true test of such innovations in our terminology is their descriptive power: whether they make not only for the greater precision that they claim, but also for increased lucidity and vividness. The two following passages may be compared from this point of view:

1. *Ordinary English:*

The sea floor was dotted with a small archipelago of coral islets, in which quarries show typical reef rock made up of corals in position of growth. When traced outwards from the reefs, the rock is seen to pass first into beds of rolled corals, then into bedded detrital deposits almost entirely made up of coral debris, and finally into limestones with shell beds.

2. *The same passage with the so-called ambiguous terms superseded:*

The sea floor carried a hermatopelago, in which quarries show typical hermatolith. When traced outwards from the bioherms, the rock is seen to pass first into coral-biostromes and then into limestones with shell-biostromes.

It should be noted that, in order to express the grading of the detrital beds from rolled corals to coral debris by the method advocated, it would be necessary to add so much qualifying or descriptive matter to the word 'biostrome' that the word itself would become redundant.

I venture to suggest that Professor Cumings has fallen into the same error as did S. S. Buckman when he imagined that he was making himself clearer by referring to a 'colomorphic nomomorph' when he meant an ammonite of normal size reduced by the removal of the outer whorls; or by calling a deformity a 'plagiomorphic kakomorph'.[1]

As Dr. Seitz has remarked: 'The art of description does not consist in stringing a greater or lesser number of terms disjointedly together, but in bringing out in clear language the features that are essential and distinctive.' [2]

(b) Geosynclines

Followers of the Haug and Kober schools will take exception to the free use made in Chapter XVIII of the word geosyncline. I have not used it in this way in ignorance of what has been said in favour of restricting the term to the much deeper troughs out of which the folded mountains arose; but because, having read and studied *Der Bau der Erde*, I remain unconvinced. To those who are familiar with the sedimentary troughs of the extra-Alpine region of North-West Europe, their secular downsinking and the deformation which they have undergone during orogenic times, Kober's rigid distinction between orogen and kratogen seems artificial. The difference between the deformation of the kratogen and the deformation of the orogen (which Kober would have us believe is something fundamental) seems to me to be purely one of degree. Not only do the elongate troughs of deposition on the kratogen answer to the American definition of geosynclines—'all the greater long-continued down-flexured parts of the lithosphere'—but the folding to which they were subjected generally follows the lines laid down during past orogenies (as pointed out in Chapter III).

Admittedly there are different classes of geosynclines, and those which were deepest and subsequently broke forth into mountain ranges (probably as the direct result of their excessive depth) should rightly be placed in a different sub-class (the Alpine type) from those with less violent histories (as Stille calls them, the Germano-type). But there is nevertheless a gradation in depth which links them all together as fundamentally one class of phenomenon.[3] A consideration of the sedimentary systems of the world reveals a network of subsiding troughs of all different degrees of size and depth, and periodically the deepest buckled, bringing on an orogenic convulsion with the formation of a range of folded mountains. As Stille has aptly said, if we are to restrict the term geosyncline to those troughs which have given rise to mountains 'we might with as much justice deny the title of wife to a wife because she has not brought forth a family'.[4]

(c) Peneplains

The spelling of this word as 'peneplain' (almost a plain) or 'peneplane' (almost a plane surface) has now become nearly an infallible distinction between English and

[1] *T.A.*, vol. iii, pp. 6, 32.

[2] O. Seitz, 1929, *Jahrb. Preuss. Geol. Landesanst.*, vol. i, p. 154.

[3] The significant depth is, of course, not the depth of water, but the depth of water plus sediment. This gives a measure of the subsidence of the bottom.

[4] *Grundfragen*, p. 8. The conception of the geosyncline here given conforms with that in the two principal works on geosynclines by American authors—and it was in America that the expression was first used: see Ch. Schuchert, 1923, 'Sites and Nature of the North American Geosynclines', *Bull. Geol. Soc. America*, vol. xxxiv, pp. 151–230, and A. W. Grabau, 1924, 'Migration of Geosynclines', *Bull. Geol. Soc. China*, vol. iii, pp. 208–347. Grabau gives a map showing Asia covered with a network of geosynclinal troughs of different ages. It is very doubtful whether his distinction between geosynclines (in this broad sense) and oceanic deeps can be maintained. He calls the latter fore-deeps; but the Indo-Gangetic plain is a typical fore-deep in a tectonic sense, and because this is filled with sediment he calls it a geosyncline. The presence or absence of a plentiful supply of sediment is surely an extraneous factor.

American writings. It is not on this account, however, that I have adopted peneplain.

The question has been fully discussed by Professor D. W. Johnson in his *Shore Processes and Shoreline Development* (1919, pp. 164–9). He defines his use of certain terms as follows: '(1) The level erosion surface produced in the ultimate stage of any cycle will be called a *plane*. (2) The undulating erosion surface of moderate relief produced in the penultimate stage of any cycle will be called a *peneplane*. (3) A low-relief region of horizontal rocks will be called a *plain*.' And later he states: 'A peneplain is not "almost a plain" of horizontal sediments, but is almost a plane surface in the mathematical sense of the term; therefore, "peneplane" more nearly expresses the true meaning of the term than does the older and commoner spelling.'

The British reader will at once detect in these passages an unfamiliar element in the definition of a 'plain', which seems unwarrantably to be made to depend on the horizontality of the strata. In English the expression carries no such implications regarding the strata. The prototypes of plains, such as Salisbury Plain, the Plain of York, the Plain of Esdraelon, &c., were so called centuries before any idea of the stratification can have been present in the minds of those who conferred the names. Plain in English, in fact, simply means 'a level tract of country'[1] and nothing more. Therefore the peneplains referred to in Chapter XVIII, whether of marine or sub-aerial origin, are best described as 'almost plains'. It was certainly this meaning, rather than almost mathematically plane surfaces, that W. M. Davis expressed when he first introduced the word as 'peneplain'.

APPENDIX II

LIST OF 120 STAGE-NAMES HITHERTO PROPOSED FOR PARTS OF THE JURASSIC SYSTEM IN EUROPE, WITH REFERENCES.

Arranged approximately in descending stratigraphical order, synonyms in order of date.

May.-Eym., 1864, 1881, 1884, 1888, refer to the four works by C. D. W. Mayer-Eymar, enumerated in the bibliography, p. 624.

CLAVINIAN. Jukes-Browne, 1884. Geol. Mag. [3], vol. i, p. 526. Clavinium = Weymouth, Dorset. = Middle and Upper Oolites.

GLEVONIAN. Jukes-Browne, 1884. Geol. Mag. [3], vol. i, p. 526. Glevonium = Gloucester. = Lower Oolites up to Cornbrash incl.

PURBECKIEN. Brongniart, 1829. Tabl. terr. séd. = Purbeck Beds.

DUBISIEN. Desor, 1859. Étude Jura neuchat., p. 45. = Purbeckian of Doubs and the Jura Mts.

TITHONIAN. Oppel, 1865. Zeitschr. Deutsch Geol. Gesell., vol. xvii, p. 535 = Marine Purbeckian of the Alps. Tithon was spouse of Eos (Aurora), Goddess of the Dawn (ref. dawn of Cretaceous). Discussed Haug, 1898, Bull. Soc. géol. France, vol. xxvi, pp. 197–228. Restr. Spath, 1923, Q.J.G.S., vol. lxxix, p. 304.

AQUILONIAN. Pavlow, 1891. Bull. Soc. natur. Moscou, N.S., vol. v, p. 550; & 1896, Q.J.G.S., vol. lii, p. 542. Aquilo, Nord, France. = Marine Portland–Purbeck pars. Discussed Spath, 1923, Q.J.G.S., vol. lxxix, p. 304.

ALLOBROGIEN. Rollier, 1909. Compte rendu, 9ᵉ Congrès Assoc. franc-comtoise à

Pontarlier, pp. 13–30; and see Mém. Soc. pal. Suisse, 1917, vol. xlii, p. 624. = Purbeck and Portland Beds of the 'Rhodano-Swabian Province'.

PORTLANDIEN. Brongniart, 1829. Tabl. terr. séd. = Portland Beds.

FREIXIALIN. Choffat, 1887. Rech. terr. second. au sud du Sado, Secção dos Trabalhos Geologicos, vol. i, p. 307. Freixial, Portugal. = Portlandian of Portugal.

KIMMÉRIDIEN. Thurmann, 1833. Essai sur les soulèvemens jurassiques du Porrentruy. Mém. Soc. Hist. nat. Strasbourg, vol. i, 2 livr., p. 12.

KIMÉRIDGIEN. d'Orbigny, 1846–9. Pal. franç. terr. jurass., vol. i, p. 610. = Lower Kimeridge Clay.

BOULOGNIAN. Blake, 1880. Q.J.G.S., vol. xxxvi, p. 196. See Bolonian.

BOLONIAN. Blake, 1881. Q.J.G.S., vol. xxxvii, p. 581. Boulogne-sur-Mer, France. = Upper Kimeridge Clay, between the clays with *Exogyra virgula* and the Portland Beds. = Portlandien inférieur, *sensu gallico*.

BOLONIN. May.-Eym., 1888.

BOULONIEN. Pavlow, 1891. Bull. Soc. natur. Moscou, N.S., vol. v, p. 550.

BONONIAN. Pavlow, 1896, Q.J.G.S., vol. lii, p. 542. Bononia = Boulogne-sur-Mer. = Portland Beds *sensu stricto* by definition, though apparently only intended as a different spelling of Bolonian, Blake.

VOLGIEN. Nikitin? Defined as the 'Wolgaformation' by Nikitin, 1881, Jura-Ablag. zw. Rybinsk, Mologa u. Myschkin, Mem. Acad. Imp. Sci. St. Petersburg [7], vol. xxviii, p. 36. = Upper Kimeridgian of Russia (River Volga). Discussed Spath, 1923, Q.J.G.S., vol. lxxix, p. 304; and Haug, 1898, Bull. Soc. géol. France [3], vol. xxvi, pp. 197–228.

HÂVRIEN. Brongniart, 1829. Tabl. terr. séd. = Kimeridgian of Hâvre. = Lower Kimeridgian.

VIRGULIEN. Thurmann, 1852. Mitth. Bern. Naturf. Gesell., p. 217. = Lower Kimeridge Clay with *Exogyra virgula*.

SOLENHOFIN. May.-Eym., 1881. Solenhofen, Bavaria. = *nom. nov.* for Virgulien. = Lower Kimeridgian pars, above the Ptérocerian (Bannéin).

SALINIEN. Rollier, 1909. Compte rendu, 9ᵉ Congrès Assoc. franc-comtoise à Pontarlier, pp. 13–30; and see Mém. Soc. pal. Suisse, 1917, vol. xlii, p. 624. Salins, near Besançon, Jura. = Virgulien of France, = Solenhofin.

DANUBIEN. Rollier, 1909. Compte rendu 9ᵉ Congrès Assoc. franc-comtoise à Pontarlier, pp. 13–30; and see Mém. Soc. pal. Suisse, 1917, vol. xlii, p. 624. = Salinien of 'Rhodano-Swabian Province'; all between Elsgovian and Bononian.

PTÉROCERIEN. Thurmann, 1852. Mitth. Bern. Naturf. Gesell., p. 217. = Lowest Kimeridgian, zone of *Pteroceras oceani*, below the range of *Exogyra virgula*.

BANNÉIN. May.-Eym., 1881. = Marls of Banné, Jura salinois; see Marcou, 1848, Jura salinois, p. 104. = Lo. Kimeridgian (Ptérocerian) of Porrentruy, Jura bernois.

ELSGOVIEN. Rollier, 1909. Compte rendu, 9ᵉ Congrès Assoc. franc-comtoise à Pontarlier, pp. 13–30; and see Mém. Soc. pal. Suisse, 1917, vol. xlii, p. 624.= Marls of Banné. = Ptérocerian. = Bannéin.

CRUSSOLIEN. Rollier, 1909. idem; *Tenuilobatus* and *Aulacostephanus* zones and coral rag of Nattheim. = Elsgovien. = Bannéin of 'Rhodano-Swabian Province'.

SEQUANIEN. (Thurmann MS.), Marcou, 1848. Rech. géol. sur le Jura salinois; Mém. Soc. géol. France [2], iii, pp. 96, 102–3. Sequania = Franche-Comté (Jura). = *Astarte supracorallina* Beds = Astartien = Lower Kimeridgian, Thurmann. As used by De Loriol in the Boulonnais and thence by Arkell for Dorset, this includes the Upper Corallian, an extension that does not seem legitimate.

ASTARTIEN. Thurmann, 1852. Mitth. Bern. Naturf. Gesell., p. 217. = *Astarte supracorallina* Beds of Porrentruy, immediately below the Ptérocerian. = Sequanian.

DICÉRATIEN. Étallon, 1857. Esquisse Descr. géol. Haut-Jura; Annales Soc. agric. et indust. Lyon, vol. i, pp. 253, 292. = Reef facies of Sequanian with *Diceras arietina*.

RANDENIEN. Rollier, 1897. Le Malm du Jura et du Randen, Compte rendu, Congrès géol. internat., Sess. vi, Zurich, 1894 (1897), p. 337. Randen, Switzerland. = Ammonitiferous and spongiferous facies of the Sequanian.

CORALLINIEN. Étallon, 1861. Mém. Soc. Emul. Doubs, vol. vi, p. 53. = Reef facies of Upper Sequanian of the Jura.

VERDUNIN. May.-Eym., 1881. = Upper Sequanian of Verdun, Lorraine.

FRINGELIN. May.-Eym., 1881. = Lower Sequanian of Fringeli, Jura bernois.

CORALLIEN. Thurmann, 1833. Mém. Soc. Hist. nat. Strasbourg, vol. i, p. 14. D'Orbigny, 1845, Pal. franç. terr. jurass., vol. i, p. 609. = Sequanian (*not* Corallian *sensu anglico*, which is defined above, Chapter XIII, p. 376).

ARGOVIEN. Marcou, 1848. Rech. géol. sur le Jura salinois; Mém. Soc. géol. France [2], vol. iii, p. 88. Canton of Argovie, Switzerland. = Zones of *Pelt. transversarium* (? and *Cardioceras cordatum*; mentioned by Marcou). = Lower Corallian, *sensu anglico*.

LUSITANIEN. Choffat, 1885. Descr. Faune jurass. Portugal, Moll. Lamell.; Sect. Trav. géol. Portugal, p. 17. Lusitania = Portugal. = Argovian of Portugal. Adopted for whole Corallian by Haug.

RAURACIEN. Gressly, 1867. Essai sur le Jura, p. 72; and 1870, Mat. Carte géol. Suisse, 8ᵉ livr., p. 75. Rauracia = district about Bâle, Jura Mts. = Corallian of the Swiss Jura. = Reef-facies of the Argovian, *teste* Rollier, 1897, Compte rendu, Congrès géol. int., Session vi (1894), p. 335.

GEISSBERGIN. May.-Eym., 1881. (?Moesch, 1857.) Geissberg, Argovian Jura. = Upper Argovien = Pholadomyen = Oltenin, Mayer-Eymar, 1888.

OLTENIN. May.-Eym., 1888. Olten, Soleure. = Upper Argovian = Pholadomyen = Geissbergin, May-Eym., 1881.

ZOANTHARIEN. Étallon, 1861. Mém. Soc. Emul. Doubs, vol. vi, p. 53. = Reef-facies of Upper Argovian of the Jura.

GLYPTICIEN. Étallon, 1861. Mém. Soc. Emul. Doubs, vol. vi, p. 53. = Argovien? or Lower Sequanien? with *Glypticus hieroglyphicus*.

PHOLADOMYEN. Étallon, 1861. Mém. Soc. Emul. Doubs, vol. vi, p. 53. = Upper Argovian with *Pholadomya*. = Geissbergin May.-Eym., 1881. = Oltenin, May.-Eym., 1888.

MIHIÉLIN. May.-Eym., 1881. St. Mihiel, Meuse. = Reef-facies of the Argovien.

EFFINGIN. May.-Eym., 1881. (?Moesch, 1857.) Effingen, Switzerland. = Middle Argovian, Effinger Schichten.

SPONGITIEN. Étallon, 1857. Esquisse Descr. géol. Haut-Jura, Ann. Soc. agric. et indust. Lyon, vol. i, pp. 253, 275. = Spongiferous Lower Argovian of the Jura.

BIRMENSTORFIN. May.-Eym., 1881. (?Stutz, 1859.) Birmenstorf, Switzerland. = Lower Argovian.

OXFORDIEN. Brongniart, 1829. = Middle Jurassic (so used also by Marcou, &c.). Emend. d'Orbigny, 1846–9, = Argovian + Divésian pars.

DIVÉSIEN. Renevier, 1874. Tabl. terr. séd., ed. 1. Dives, Normandy. = Middle Oxfordian, zones of *Kosmoceras ornatum* to *Cardioceras cordatum* inclusive.

NEUVISYEN. Lapparent, 1893. Traité de Géol., ed. 3, p. 1032. Neuvisy, Ardennes. = Upper Divésian, zone of *Cardioceras cordatum*.

VILLERSIN. May.-Eym., 1881. Villers-sur-Mer, Normandy. = Upper Callovian of Oppel, zone of *Peltoceras athleta*.

VILLERSIEN. Lapparent, 1893. Traité, ed. 3; emend. = Upper Divésian, = Upper Oxford Clay+Lower Calc. Grit.

KELLOVIEN. d'Orbigny, 1844.

CALLOVIEN. d'Orbigny, 1846. Pal. franç. terr. jurass., vol. i. Callovium = Kellaways, Wilts. = Kellaways Beds+Lower Oxford Clay, zones of *Proplanulites kœnigi* to *Kosmoceras jason*. Emend. Oppel, 1858, Die Juraformation, p. 506.

CHANASIAN. Parona and Bonarelli, 1895. Mém. Acad. Sci., &c., Savoie [4], vol. vi, p. 30. Chanaz, Savoy. = Kellaways Beds.

NIORTIN. May.-Eym., 1881. Niort, Deux-Sèvres, France. = Lower Callovian *sensu lato* (Upper Cornbrash), zone of *Macrocephalites macrocephalus*.

BEDFORDIN. May.-Eym., 1881. Bedford, England. = Cornbrash.

HANDTHORPIAN.
STALBRIDGIAN.   }Buckman, 1927. Q.J.G.S., vol. lxxxiii, p. 7.
CLOSEWORTHIAN.

Handthorpe, Lincs.; Stalbridge and Closeworth, Somerset. = Cornbrash. For criticism see Douglas and Arkell, 1928, Q.J.G.S., vol. xxxiv, pp. 125–7.

HINTONIAN.
CORSHAMIAN.   }Buckman, 1927. Q.J.G.S., vol. lxxxiii, p. 7.
WYCHWOODIAN.

Hinton Charterhouse and Corsham, Wilts.; Wychwood Forest, Oxon.; = Forest Marble. See Arkell, 1931, Q.J.G.S., vol. lxxxvii, p. 595.

BRADFORDIEN. Desor, 1859. Étude Jura neuchat., p. 85. = Bradford Clay of Bradford-on-Avon, Wilts.+Forest Marble.

BATHONIAN. D'Omalius d'Halloy, 1843. Précis élémentaire de Géologie, p. 470. = Dogger. Restr. d'Orb., Pal. franç. terr. jurass., vol. i, p. 607, to Great Oolite Series (+Vesulian pars).

BATHIEN. May.-Eym., 1864. Bathonian+Bajocian. Restr. May.-Eym., 1884, to Cornbrash+Bradford Clay.

MANDUBIEN. Marcou, 1860. Lettres du Jura, p. 344. Mandubii = people of the Jura Mts. = Bathonian of the Jura.

FALAISIN. May.-Eym., 1881. Falaise, Normandy. = Middle Bathonian; Great Oolite with *Nerinea voltzi*.

VÉSULIEN. Marcou, 1848. Mém. Soc. géol. France [2], vol. iii, p. 73. Vesulum = Vésoul, Haute-Saône. Marls of Vésoul, with *Ostrea acuminata* and *O. knorri*. = Fuller's Earth+Bath Oolite (*pro parte*) by original definition. Emend. May.-Eym., 1881 = Cadomin+Stonesfieldin+Falaisin (i.e. Great Oolite Series less the Bradford Clay and higher beds).

STONESFIELDIN. May.-Eym., 1881. Stonesfield Slate Beds+Oolithe miliaire. Later (1884) included by Mayer-Eymar in his Cadomin.

CADOMIN. May.-Eym., 1881. Cadomum = Caen, Normandy. = Fuller's Earth, Lower Vesulian, marls with *Ostrea acuminata*+Caen Stone.

FULLONIAN. Woodward, 1894. Jurassic Rocks of Britain, vol. iv, p. 229. = Fuller's Earth of England.

EHNINGIN. May.-Eym., 1881. Ehningen, Württemberg. = Upper Inferior Oolite, zone of *Parkinsonia parkinsoni* (*sensu lato*).

BAJOCIAN. d'Orbigny, 1847. Pal. franç. terr. jurass., vol. i, p. 606. Bayeux, Normandy. = Inferior Oolite Series, from the zone of *Ludwigia murchisonœ* to that of *Zigzagiceras zigzag*. Restricted by the founding of the Aalenian.

LAEDONIEN. Marcou, 1848. Rech. géol. Jura salinois, Mém. Soc. géol. France [2], vol. iii, p. 70. Laedo = Lons-le-Saunier, Switzerland. = Bajocian, pars.

TORGONIEN. Rutot and Van den Broëck, 1885. Sketch Geol. Belgium; Geol. Assoc., London, p. 38 = Bajocian of the Ardennes.

SCARBOROUGHIN. May.-Eym., 1881. Scarborough, Yorks. = Bajocian, pars, zone of *Stephanoceras humphriesianum* (*sensu lato*).

MÂCONIN. May.-Eym., 1881. Mâcon, Saône et Loire. = Bajocian, pars, zone of *Sonninia sowerbyi.*

AALENIAN. May.-Eym., 1864. Aalen, Württemberg. = Zones of *Lioceras opalinum* and *Ludwigia murchisonæ* (*sensu lato*).

CHELTENHAMIN. May.-Eym., 1881. = Upper Aalenian. = Lower Inferior Oolite of Cheltenham.

GUNDERSHOFIN. May.-Eym., 1881. Gundershofen, Alsace. = Lower or Middle Aalenian = Gundershofen Schichten, limestones with *Trigonia navis.*

BOLLIN. May.-Eym., 1881. Boll, Baden, Germany. = Lower Aalenian.

OPALINIEN. Renevier, 1874. Tabl. terr. séd., ed. 1. = Lower Aalenian, zone of *Lioceras opalinum.*

TOARCIEN. d'Orbigny, 1849. Pal. franç. terr. jurass., vol. i, p. 106. Toarcium = Thouars, Deux-Sèvres, France. = Upper Lias.

THOUARSIEN. May.-Eym., 1864.

MUSSONIEN. Rutot and Van den Broëck, 1885. Sketch Geol. Belgium; Geol. Assoc., London, p. 38. Musson, Belgium. = Toarcian of the Ardennes.

ALFELDIN. May.-Eym., 1881. Alfeld, near Hanover, Prussia. = Upper Toarcian = Yeovilian.

YEOVILIAN. Buckman, 1910. Q.J.G.S., vol. lxvi, p. 88. Yeovil, Somerset. = Upper Toarcian, subzones of *Grammoceras striatulum* to *Dumortieria moorei.*

ALDORFIN. May.-Eym., 1881.

ALTDORFON. May.-Eym., 1888. Altdorf, nr. Nürnberg, Bavaria. = Lower Toarcian = Whitbian.

WHITBIAN. Buckman, 1910. Q.J.G.S., vol. lxvi, p. 88. Whitby, Yorkshire. = Lower Toarcian, subzones of *Dactylioceras tenuicostatum* to *Haugia variabilis.*

LIASIEN. d'Orbigny, 1849. Pal. franç. terr. jurass., vol. i, p. 605. = Middle and part of Lower Lias. Adopted for whole of the Lias by Jukes-Browne, 1884, Geol. Mag. [3], vol. i, p. 526.

PLIENSBACHIEN. Oppel, 1858. Die Juraformation, p. 815. Pliensbach, near Boll, Württemberg. = Liasien d'Orb.

CHARMOUTHIAN. May.-Eym., 1864. Charmouth, Dorset. = Pliensbachien.

VIRTONIEN. Rutot and Van den Broëck, 1885. Sketch Geol. Belgium; Geol. Assoc., London, p. 38. Sandstone of Virton, Belgium. = Pliensbachian.

CYMBIEN. Leymerie, 1872. Bull. Soc. géol. France, vol. xxix, p. 168. = Middle Lias with *Gryphæa cymbium.* = Upper Pliensbachian.

BANZIN. May.-Eym., 1881. Banz, near Coburg, Bavaria. = Upper Pliensbachian.

DOMERIEN. Bonarelli, 1894. Atti della R. Accad. delle Scienze di Torino, vol. xxx, p. 85; and 1895, Rendiconti Reale Istituto Lombardo, Ser. 2, vol. xvxiii, pp. 326, 415. Monte Domero (Domaro), Lombardy Alps. = Middle Lias = Upper Pliensbachian.

MENDIN. May.-Eym., 1881. Mende, Lozère, France. = Middle Pliensbachian, with *Lytoceras fimbriatum.*

ROTTORFIN. May.-Eym., 1881. Rottorf, Brunswick. = Lower Pliensbachien, zones of *Tragophylloceras ibex* and *Uptonia jamesoni.* = Carixian, pars.

CARIXIAN. Lang, 1913. Geol. Mag. [5], vol. x, p. 401. Carixa = Charmouth. = Lower Pliensbachian, zones of *Prodact. davœi, Trag. ibex* and *Uptonia jamesoni.*

ROBINIEN. Rollier, 1915. Eclogæ Geologicæ Helvetiæ, vol. xiii, p. 374. Robin Hood's Bay, Yorkshire. = Zones of *Prodactylioceras davœi*, *Uptonia jamesoni* and *Deroceras armatum*.

SINÉMURIEN. d'Orbigny, 1849. Pal. franç. terr. jurass., vol. i, p. 604. Sinemurum = Sémur, Côte d'Or. = Lower Lias pars.

SÉMURIEN. May.-Eym., 1864.

ARLONIAN. Rutot and Van den Broëck, 1885. Sketch Geol. Belgium; Geol. Assoc., London, p. 38. Marls of Arlon, Belgium = Lower Lias of Belgium.

LOTHARINGIEN. Haug, 1911. Traité de Géologie, p. 961. Lothringen = Lorraine. = Lower Lias above Hettangian.

BALINGIN. May.-Eym., 1881. Balingen, Württemberg. = Upper Sinemurian, zones of *Echioceras raricostatum*, *Oxynoticeras oxynotum* and *Asteroceras obtusum*.

LOURNANDIN. May.-Eym., 1888. Lournand, Saône-et-Loire. = Upper Sinemurian. = Balingin, May.-Eym., 1881.

FILDERIN. May.-Eym., 1881. Filder, near Stuttgart, Germany. = Lower Sinemurian, zones of *Arnioceras semicostatum* and *Coroniceras bucklandi*.

SUÉVIEN. Rollier, 1915. Eclog. Geol. Helvet., vol. xiii, p. 374. Suévie = fr. Swabia. = Zones of *Arnioceras semicostatum*, *Coroniceras bucklandi*, *Scamnoceras angulatum*, *Psiloceras planorbis* and *Pteria contorta* (outside the Mediterranean Province).

HWICCIAN. Hwiccas = people of Gloucester, Worcester and part of Warwick.

WESSEXIAN. Wessex = the kingdom of the West Saxons.

RAASAYAN. Raasay = island in the Hebrides.

DEIRAN. Deira = the kingdom embracing Yorkshire.

MERCIAN. Mercia = the kingdom embracing the Midlands.

LYMIAN. Lyme Regis, Dorset.

Buckman, 1917. Q.J.G.S., vol. lxxiii, pp. 263–77. = Sinemurian+Carixian.

HETTANGIEN. Renevier, 1864. Not. Alp. vaud., vol. i, p. 51. Hettange, Lorraine. = Sandstone with *Psiloceras planorbis*.

RHÉTIEN (RHÆTIEN). Guembel, 1861. Bay. Alp., p. 122; and Renevier, tabl. terr., ed. 1. Rhætia = Rhætic Alps (Grisons). Rhætic Beds.

SOMERSETIAN. Richardson, 1911. Q.J.G.S., vol. lxvii, p. 73. = Cotham Beds+ Langport Beds+Watchet Beds of Somerset, England. = Upper Rhætic.

KŒSSENIN. May.-Eym., 1881. (?Suess, 1852.) Kœssen or Kössen, Tyrol. = *Pteria contorta* zone and Bone Bed. = Lower Rhætic.

BIBLIOGRAPHY

Note: District and Sheet Memoirs, sections and maps of the Geological Survey are not included, because a complete list is published and may be obtained cheaply from the Geological Survey Museum, London.

ABBREVIATIONS

The following contractions are used throughout the text, in footnotes, and in the bibliography:

Q.J.G.S. *Quarterly Journal of the Geological Society*, London.
P.G.A. *Proceedings of the Geologists' Association*, London.
J.R.B., 1892, 1893, 1894, 1895. *Jurassic Rocks of Britain*, vols. i–v, by C. Fox-Strangways (vol. i, 1892) and H. B. Woodward (vols. ii–v); *Memoirs of the Geological Survey of Great Britain*.
T.A. *Type Ammonites* (including *Yorkshire Type Ammonites*), by S. S. Buckman.
Proc. Cots. N.F.C. *Proceedings of the Cotteswold Naturalists' Field-Club*.
The other abbreviations used are self-explanatory.

THEORY AND PRACTICE OF CORRELATION AND CLASSIFICATION IN THE JURASSIC SYSTEM

ARKELL, W. J., 1930. A Comparison between the Jurassic Rocks of the Calvados Coast and those of Southern England. P.G.A., vol. xli, pp. 396–411.

BLAKE, J. F., 1881. On the Correlation of the Upper Jurassic Rocks of England with those of the Continent. Part I, The Paris Basin. Q.J.G.S., vol. xxxvii, pp. 497–587.

BLANFORD, W. T., 1884. On the Classification of Sedimentary Strata. Geol. Mag. [3], vol. i, pp. 318–21 and Comptes rendus, Congrès géol. int., Session iii, Berlin. Criticism by Jukes-Browne, Geol. Mag. [3], vol. i, pp. 525–6, and reply, ibid., vol. ii, pp. 239–40.

BORN, A., 1920. Die Bedeutung der Meeresströmungen für die geologische Zeitrechnung. Ber. Senckenb. Naturf. Gesell., 1920, pp. 207–17.

BOSWELL, P. G. H., 1921. Sedimentation, Environment and Evolution in Past Ages. Proc. and Trans. Liverpool Biol. Soc., Presidential Address, vol. xxxv, pp. 5–28.

BUCKMAN, S. S., 1891. The Ammonite Zones of Dorset and Somerset. Rept. Brit. Assoc. (Cardiff), p. 655; and Geol. Mag. [3], vol. viii, pp. 502–4.

——, 1898. On the Grouping of some Divisions of so-called 'Jurassic' Time. Q.J.G.S., vol. liv, pp. 442–62.

——, 1902–3. The Term Hemera. Geol. Mag. [4], vol. ix, pp. 554–7 and vol. x, pp. 95–6.

——, 1917–20. Jurassic Chronology. Q.J.G.S., vol. lxxiii, pp. 257–327, and vol. lxxvi, p. 62.

——, 1922–5. Chronology [Jurassic]. T.A., vol. iv, pp. 5–54; vol. v, pp. 34–55, 71–8.

CHAMBERLIN, T. C., 1909. Diastrophism as the Ultimate Basis of Correlation. Journ. Geol. Chicago, vol. xvii, pp. 685–93.

CORROY, G., 1927. Synchronisme des horizons jurassiques de l'est du bassin de Paris. Bull. Soc. géol. France [4], vol. xxvii, pp. 95–113.

DAVIES, A. M., 1930. The Geological Life Work of Sidney Savoury Buckman. P.G.A., vol. xli, pp. 221–40.

DIENER, C., 1918. Die Bedeutung der Zonengliederung für die Frage der Zeitmessung in der Erdgeschichte. Neues Jahrb. für Min., &c., Beil.-Bd. xlii, pp. 65–172.

——, 1925. Grundzüge der Biostratigraphie.

ELLES, G. L., 1924. Evolutional Palaeontology in Relation to the Lower Palaeozoic Rocks. Rept. Brit. Assoc. for 1923, pp. 83–107. [For discussions of theory.]

ETHERIDGE, R., 1881. On the Analysis and Distribution of the British Jurassic Fossils. Q.J.G.S., vol. xxxviii, pp. 59–236.

FIEGE, K., 1927. Die paläontologischen Grundlagen der geologischen Zeitmessung.

FRAAS, O., 1851. On the Comparison of the German Jura Formations with those of France and England. Q.J.G.S., vol. vii, pp. 42–83.

FREBOLD, H., 1924. Ammonitenzonen und Sedimentationszyklen in ihrer Beziehung zueinander. Centralblatt für Min., &c., 1924, pp. 313–20.

GOODCHILD, J. G., 1904. On Unconformities and Palaeontological Breaks in Relation to Geological Time. Trans. Edinburgh Geol. Soc., vol. viii, pp. 275–314.

HÉBERT, E., 1875. Les mers anciennes et leur rivages dans le bassin de Paris, ou classification des terrains par les oscillations du sol. 1re partie, Terrain jurassique, Paris.

HUDLESTON, W. H., & BLAKE, J. F., 1888–91. Report of Sub-Committee No. 3 on Classification and Nomenclature. A, Jurassic, a. The Oolites, b. Lias and Rhætic. Compte rendu, Congrès géol. int., Session iv, London, 1888, Appendix, pp. B 81–B 118.

JUKES-BROWNE, A. J., 1903. The Term 'Hemera'. Geol. Mag. [4], vol. x, pp. 36–8 [Term Secule introduced]. (See reply by Buckman, pp. 95–6.)

KITCHIN, F. L., 1919. The Faunal Characters and Correlation of the Concealed Mesozoic Rocks in Kent. Sum. Prog. Geol. Surv. for 1918, pp. 37–45.

LANGE, W., 1931. Die biostratigraphischen Zonen des Lias a und Vollraths petrographische Leithorizonte. Centralblatt für Min., &c. B., 1931, pp. 349–72.

LAUX, N., 1924. La Méthode analytique Buckman et son application a l'étude du système jurassique. Bull. Soc. géol. France [4], vol. xxiv, pp. 198–212.

MARCOU, J., 1857–60. Lettres sur les Roches du Jura et leur distribution géographique dans les deux Hémisphères. Paris.

MARR, J. E., 1905. The Classification of the Sedimentary Rocks. Q.J.G.S., vol. lxi, pp. lxi–lxxxvi.

MAYER-EYMAR, C. D. W.,[1] 1864. Tableau synchronistique des terrains jurassiques. Zurich. (Folding sheet.)

——, 1881. Classification internationale, naturelle, uniforme, homophone et pratique des terrains de sédiment, suivie dans un cours de stratigraphie. Zurich. 4to.

——, 1884. Classification et terminologie internationales des étages naturels des terrains de sédiment. Zurich. 4to.

——, 1888. Tableau des terrains de sédiment. Zurich. 4to.

MUNIER-CHALMAS, & DE LAPPARENT, A., 1894. Note sur la nomenclature des terrains sédimentaires. Bull. Soc. géol. France [3], vol. xxi, pp. 438–88.

NEAVERSON, E., 1928. Stratigraphical Palaeontology. Chapters xv–xvi.

NEUMAYR, M., 1870–1. Jurastudien. Jahrb. d.k.k. Reichsanst., vol. xx, pp. 549–58, and vol. xxi, pp. 297–378.

ODLING, M., 1915. Correlation and Facies of the Upper and Middle Oolites in England and North-West France. Proc. Yorks. Geol. Soc., N.S., vol. xix, pp. 272–87.

OPPEL, A., 1856–8. Die Juraformation Englands, Frankreichs und des Südwestlichen Deutschlands.

ORBIGNY, A. D'., 1842–9. Paléontologie française; Terrains jurassiques, vol. i.

---

[1] I am indebted to Dr. C. D. Sherborne for the information that K. (or C.) Mayer and C. D. W. Mayer-Eymar, often indexed separately, were one and the same man.

PHILLIPS, J., 1844. Memoirs of William Smith, LL.D. London.

POMPECKJ, J. F., 1914. Die Bedeutung des Swäbischen Jura für die Erdgeschichte. Stuttgart.

PRUVOST, P., & PRINGLE, J., 1924. A Synopsis of the Geology of the Boulonnais, including a Correlation of the Mesozoic Rocks with those of England. P.G.A., vol. xxxv, pp. 29–56.

QUENSTEDT, F. A., 1856–8. Der Jura. Tübingen.

RAMSAY, A. C., 1864. The Breaks in Succession of the British Mesozoic Strata. Q.J.G.S., vol. xx, pp. 40–60.

RENEVIER, E., 1897. Chronographe géologique: IIe édit. du Tableau des terrains sédimentaires, suivi d'un Repertoire stratigraphique polyglotte. Comptes rendus, Congrès géol. int., Session iv (1894), Zurich, 1897, pp. 523–695.

ROBERTS, T., 1887. On the Correlation of the Upper Jurassic Rocks of the Swiss Jura with those of England. Q.J.G.S., vol. xliii, pp. 229–69.

SALFELD, H., 1913. Certain Upper Jurassic Strata of England. Q.J.G.S., vol. lxix, pp. 423–32.

——, 1914. Die Gliederung des Oberen Jura in Nordwest-Europa. Neues Jahrb. für Min., &c., Beil.-Bd. xxxvii, pp. 125–246.

——, 1922. Formänderung und Vererbung bei fossilen Evertebraten [and application to zonal work]. Palaeontologische Zeitschrift, vol. iv, pp. 107–12.

SCHUCHERT, C., 1916. Correlation and Chronology in Geology on the Basis of Paleogeography. Bull. Geol. Soc. America, vol. xxvii, pp. 491–514.

SCUPIN, H., 1923. Der chronologische Wert der Leitfossilien. Centralblatt für Min., &c., 1923, pp. 433–47.

SEDGWICK, A., 1826. On the Classification of the Strata which appear on the Yorkshire Coast. Ann. Phil. [2], vol. xi, pp. 339–62.

SHEPPARD, T., 1915. William Smith: His Maps and Memoirs. Proc. Yorks. Geol. Soc., N.S., vol. xix, pp. 75–254.

SMITH, J. PERRIN, 1895. Geologic Study of Migration of Marine Invertebrates. Journ. Geol. Chicago, vol. iii, pp. 481–95.

——, 1900. Principles of Palaeontologic Correlation. Journ. Geol. Chicago, vol. viii, pp. 673–97.

SMITH, W., 1816–19. Strata identified by Organized Fossils. London.

——, 1817. Stratigraphical System of Organized Fossils. London.

SPATH, L. F., 1923. Note on the Upper Limit of the Jurassic; in Ammonites from New Zealand; Appendix to Trechmann. Q.J.G.S., vol. lxxix, pp. 303–8.

——, 1931. On the Contemporaneity of Certain Ammonite Beds [in the Lower Lias] in England and France. Geol. Mag., vol. lxviii, pp. 182–6.

STAMP, L. D., 1925. Some Practical Aspects of Correlation: a Criticism. P.G.A., vol. xxxvi, pp. 11–25.

STRUCKMANN, C., 1881. Ueber den Parallelismus der hannoverschen und der englischen oberen Jurabildungen. Neues Jahrb. für Min., &c., vol. ii, pp. 77–192. Translation by W. S. Dallas, Geol. Mag. [2], vol. viii, pp. 546–57.

TRUEMAN, A. E., 1923. Some Theoretical Aspects of Correlation. P.G.A., vol. xxxiv, pp. 193–206.

WAAGEN, W., 1865. Versuch einer allgemeinen Classifikation der Schichten des oberen Jura. München.

WEDEKIND, R., 1916. Über die Grundlagen und Methoden der Biostratigraphie.

——, 1918. Über Zonenfolge und Schichtenfolge. Centralblatt für Min., &c., 1918, pp. 268–84.

WETZEL, W., 1924. Beiträge zur Stratigraphie des Mittleren Doggers von Nordwesteuropa. Palaeontographica, vol. lxv, pp. 155–246.

WILLIAMS, H. S., 1893. The Making and Elements of the Geological Time-Scale. Journ. Geol., Chicago, vol. i, pp. 180–96, and 283–95.

——, 1894. Dual Nomenclature in Geological Classification. Journ. Geol., Chicago, vol. ii, pp. 145–60.

——, 1901. The Discrimination of Time-Values in Geology. Journ. Geol., Chicago, vol. ix, pp. 570–85.

——, 1905. Bearing of some New Paleontologic Facts on Nomenclature and Classification of Sedimentary Formations. Bull. Geol. Soc. Amer., vol. xvi, pp. 137–50.

WILLIAMS, H. S., 1913. Correlation Problems suggested by a Study of the Faunas of the Eastport Quadrangle, Maine. Bull. Geol. Soc. America, vol. xxiv, pp. 377–98.

WILLIAMSON, W. C., 1837. On the Distribution of Fossil Remains on the Yorkshire Coast, from the Lower Lias to the Bath Oolite inclusive. Trans. Geol. Soc. [2], vol. v, pp. 223–42, and Proc. Geol. Soc., vol. ii, pp. 82–3, 429–32.

——, 1842. On the Distribution of Fossil Remains on the Yorkshire Coast, from the Upper Sandstone to the Oxford Clay, inclusive. Trans. Geol. Soc. [2], vol. vi, pp. 143–52, and Proc. Geol. Soc., vol. ii, pp. 671–2.

WOODWARD, H. B., 1892. On Geological Zones. P.G.A., vol. xii, pp. 295–315.

WRIGHT, T., 1872. On the Correlation of the Jurassic Rocks in the Department of the Côte-d'Or, France, with the Oolitic Formations in the Counties of Gloucester and Wilts. Proc. Cots. N.F.C., vol. v, pp. 143–238.

UNDERGROUND STRUCTURE AND DEEP BORINGS

ARBER, E. A. N., 1916. The Concealed Oxfordshire Coalfield. Trans. Inst. Mining Eng., vol. l, pp. 373–84.

BURR, M., 1913. Ten Deep Borings in East Kent. Colliery Guardian, vol. xvi, pp. 731–4.

CANTRILL, T. C., & PRINGLE, J., 1914. On a Boring for Coal at Hemington [near Radstock], Somerset. Sum. Prog. Geol. Surv. for 1913, pp. 98–101.

COX, A. H., 1919. Report on Magnetic Disturbances in Northants and Leicestershire, and their Relations to Geological Structure. Phil. Trans. Royal Soc., vol. ccxix A, pp. 73–135.

DAVIES, A. M., & PRINGLE, J., 1913. On two Deep Borings at Calvert Station (North Bucks.) and on the Palaeozoic Floor north of the Thames. Q.J.G.S., vol. lxix, pp. 308–42.

DAWKINS, W. BOYD, 1894. The Probable Range of the Coal Measures under the Newer Rocks of Oxfordshire and the Adjoining Counties. Geol. Mag. [4], vol. i, pp. 459–62; Rept. Brit. Assoc. for 1894, pp. 646–7.

——, 1913. The South-Eastern Coalfield, the Associated Rocks, and the Buried Plateau. Trans. Inst. Mining Eng., vol. xliv, pp. 350–78.

DEWEY, H., PRINGLE, J., & CHATWIN, C. P., 1925. Some Recent Deep Borings in the London Basin. Sum. Prog. Geol. Surv. for 1924, pp. 127–37.

EDWARDS, W., & PRINGLE, J., 1926. On a Borehole in Lower Oolitic Rocks at Wincanton, Somerset [Bridport Sands to Kellaways Beds]. Sum. Prog. Geol. Surv. for 1925, pp. 183–8.

EUNSON, H. J., 1886. Notes on the Deep Boring at Orton, near Kettering, Northants. Journ. Northants N.H.S., vol. iv, pp. 57–68.

GIBSON, W., 1918. On a Deep Boring near Market Weighton. Sum. Prog. Geol. Surv. for 1917, pp. 42–5.

GODWIN-AUSTEN, R. A. C., 1856. On the Possible Extension of the Coal-Measures beneath the South-Eastern Part of England. Q.J.G.S., vol. xii, pp. 38–73.

GOSSELET, J., 1898. Étude préliminaire des récents sondages faits dans le Nord de la France pour la recherche du bassin houiller. Ann. Soc. géol. Nord, vol. xxvii, pp. 139–49. English transl. Trans. Inst. Mining Eng., vol. xviii, 1899–1900, pp. 317–25.

HULL, E., 1860. On the South-Easterly Attenuation of the Lower Secondary Formations of England; and the probable Depth of the Coal Measures under Oxfordshire and Northamptonshire. Q.J.G.S., vol. xvi, pp. 63–81.

JUDD, J. W., 1884. On the Nature and Relations of the Jurassic Deposits which underlie London; with an Introductory Note on a Deep Boring at Richmond, Surrey, by C. Homersham. Q.J.G.S., vol. xl, pp. 724–64, and Nature, vol. xxix, p. 329.

JUKES-BROWNE, A. J., 1889. The Occurrence of Granite in a Boring at Bletchley [supposedly in the Kellaways Rock, more probably at the base of the Lias]. Geol. Mag. [3], vol. vi, pp. 356–61. [For more recent interpretation see Morley Davies, Q.J.G.S., 1913, pp. 332–3.]

——, 1891. On a Boring at Shillingford near Wallingford. Midland Naturalist, vol. xiv, pp. 201–8.

JUKES-BROWNE, A. J., 1893. On Some Recent Borings through the Lower Cretaceous Strata in East Lincolnshire. Q.J.G.S., vol. xlix, pp. 467–77.

LAMPLUGH, G. W., 1917. The Underground Range of the Jurassic and Lower Cretaceous Rocks in East Kent. Sum. Prog. Geol. Surv. for 1916, pp. 45–52.

——, 1917. On a Deep Boring made in 1907–9 at Battle [Purbeck and Portland Beds and Kimeridge Clay]. Sum. Prog. Geol. Surv. for 1916, pp. 40–4.

LAMPLUGH, G. W., & KITCHIN, F. L., 1911. The Mesozoic Rocks in some of the Coal Explorations in Kent. Mem. Geol. Surv.

LAMPLUGH, G. W., KITCHIN, F. L., & PRINGLE, J., 1923. The Concealed Mesozoic Rocks in Kent. Mem. Geol. Surv.

LEMOINE, P., 1910. Résultats géologiques des sondages profonds du bassin de Paris. Bull. Soc. industr. minér., May 1910, p. 367.

——, 1930. Considérations sur la structure d'ensemble du bassin de Paris. Livre Jubilaire, Cent. Soc. géol. France, vol. ii, pp. 481–98.

PRESTWICH, J., 1856. On a Boring through the Chalk at Kentish Town, London. Q.J.G.S., vol. xii, pp. 6–14.

PRESTWICH, J., & MOORE, C., 1878. On the Section of Messrs. Meux & Co.'s Artesian Well in the Tottenham Court Road . . . and on the Probable Range of the . . . [Palaeozoic Rocks under London]. Q.J.G.S., vol. xxxiv, pp. 902–23.

PRINGLE, J., 1917. On Deep Borings for Coal and Ironstone at Bere Farm, Elham and Folkestone, Kent. Palaeozoic Rocks only. Sum. Prog. Geol. Surv. for 1916, pp. 34–40.

——, 1922. On a Boring for Coal at Westbury, Wiltshire. Sum. Prog. Geol. Surv. for 1921, pp. 146–53.

——, 1923. On the Concealed Mesozoic Rocks in South-West Norfolk. Sum. Prog. Geol. Surv. for 1922, pp. 126–39.

——, 1929. Note on a Deep Boring (No. 4) at the Waterworks, Tetbury, Gloucestershire. Proc. Cots. N.F.C., vol. xxiii, pp. 187–90. [Upper Lias–Great Oolite.]

PRUVOST, P., 1928. Le sondage de Ferrières-en-Bray. Ann. de l'Office nation. des Combustibles liquides, 3ᵉ année, 3ᵉ livr., pp. 429–57; and Comptes rendus Acad. Sciences, vol. clxxxci, 1928, pp. 242, 386.

RASTALL, R. H., 1926. Note on the Geology of the Bath Springs. Geol. Mag., vol. lxiii, pp. 98–104.

——, 1927. The Underground Structure of Eastern England. Geol. Mag., vol. lxiv, pp. 10–26.

RICHARDSON, L., 1914–21. Deep Borings at: Kemble, Chavenage near Tetbury, Wallsworth Hall near Gloucester, at Gloucester, Shipton Moyne near Tetbury, and Tetbury Waterworks. Proc. Cots. N.F.C., vol. xviii, pp. 185, 243, 247, 250, and vol. xix, pp. 49, 57, and vol. xx, p. 151.

STRAHAN, A., 1912. On a Deep Boring for Coal at Ebbsfleet, near Ramsgate [Palaeozoic Rocks]. Sum. Prog. Geol. Surv. for 1911, pp. 34–40.

——, 1913. The Form and Structure of the Palæozoic Platform on which the Secondary Rocks of England rest [with complete list of deep borings]. Q.J.G.S., vol. lxix, pp. lxx–xci.

——, 1913. Boring at the East Anglian Ice Co.'s Works, Lowestoft. Sum. Prog. Geol. Surv. for 1912, pp. 87–8.

——, 1913. Batsford (or Lower Lemington) Boring, near Moreton in Marsh. Sum. Prog. Geol. Surv. for 1912, pp. 90–1.

——, 1916. On a Deep Boring for Coal near Little Missenden, in Buckinghamshire. Sum. Prog. Geol. Surv. for 1915, pp. 43–6.

THOMPSON, B., 1890. The Bletchley Boring. Journ. Northants N.H.S., vol. v, pp. 20–5.

WHITAKER, W., 1889. On the Extension of the Bath Oolite under London, as shown by a Deep Boring at Streatham. Rept. Brit. Assoc. for 1888, p. 656.

WOODWARD, H. B., 1886. Account of a Well-Sinking made by the Great Western Railway Co. at Swindon, with lists of Fossils by E. T. Newton [Kim. Clay–Forest Marble]. Q.J.G.S., vol. xlii, pp. 287–308.

CONTEMPORANEOUS TECTONICS AND SEDIMENTATION

ANDRÉE, K., 1908. Über stetige und unterbrochene Meeressedimentation, ihre Ursachen, sowie über deren Bedeutung für die Stratigraphie. Neues Jahrb. für Min., &c., B.-B. xxv, pp. 366–421.

BEILBY, E. M., 1930. The Market Weighton Axis in Middle Jurassic Times. Trans. Leeds Geol. Assoc., xx, pp. 10–12.

BERTRAND, M., 1893. Sur la continuité du phénomène de plissement dans le bassin de Paris. Bull. Soc. géol. France [3], vol. xx, pp. 118–65.

BEURLEN, K., 1927. Stratigraphische Untersuchungen in Weissen Jura Schwabens, Neues Jahrb. für Min., &c., B.-B. lvii, pp. 78–124, 161–230.

BRINKMANN, R., 1925. Über die sedimentäre Abbildung epirogener Bewegungen, sowie über das Schichtungsproblem. Nachr. Gesell. Wissensch. Göttingen, Math.-nat. Klas., pp. 202–28.

COX, A. H., & TRUEMAN, A. E., 1920. Intra-Jurassic Movements and the Underground Structure of the Southern Midlands. Geol. Mag., vol. lvii, pp. 198–208.

DAHLGRÜN, F., 1921. Tektonische, insbesondere kimmerische Vorgänge im mittleren Leinegebiet. Jahrb. Preuss. Geol. Landesanst., vol. xlii, pp. 723–64.

DAVIES, A. M., 1929. Morphology; England and Wales. Handbook of the Geology of Great Britain, ch. i, pp. 1–4.

FREBOLD, H., 1925. Über cyclische Meeressedimentation. Leipzig.

——, 1926. Zur Frage der epirogenen Bewegungen im unteren Jura Mesoeuropas und der für ihre Ableitung wichtigen stratigraphischen Unterlagen. Centralblatt für Min., &c., B., 1926, pp. 73–8.

——, 1927. Die paläogeographische Analyse der epirogenen Bewegungen und ihre Bedeutung für die Stratigraphie. Geol. Arch., 4 Jahrg., pp. 223–40.

GRACHT, W. A. M. VAN W. VAN DER, 1913. Proeve eener Tektonische Schetskaart van het Belgisch-Nederlandsch-Westfaalsche Kolenveld en het aangrenzende noordelijke Gebied tot aan de breedte van Amsterdam. Jaarverslag der Rijkopsporung van Delfstoffen over 1913. Amsterdam.

GRUPE, O., 1926. Die Einzelphasen der saxonischen Gebirgsbildung am Deister. Jahrb. Preuss. Geol. Landesanst., vol. xlvii, pp. 357–82.

HAUG, E., 1900. Les géosynclinaux et les aires continentales. Bull. Soc. géol. France [3], vol. xxviii, pp. 617–711.

HAWKINS, H. L., 1918. Notes on the Geological Structure of the Vale of Kingsclere. Proc. Hampshire F.C., vol. viii, part ii, pp. 191–212.

KENDALL, P. F., 1905. Sub-Report on the Concealed Portion of the Coalfield of Yorkshire, Derbyshire and Nottinghamshire [with table of borings]. Final Report of the Royal Commission on Coal Supplies, Part IX, Appendix 3, pp. 188–205 (18–35).

KLÜPFEL, W., 1916. Über die Sedimente der Flachsee im Lothringer Jura. Geol. Rundschau, vol. vii, pp. 97–109.

——, 1926. Beziehungen zwischen Tektonik, Sedimentation und Paläogeographie in der Weser-Erzformation des Ober–Oxford. Zeitschr. Deutsch. Geol. Gesell., vol. lxxviii, pp. 178–92.

LAMPLUGH, G. W., 1919. The Structure of the Weald and Analogous Tracts. Q.J.G.S., vol. lxxv, pp. lxxiii–xcv.

——, 1920. On Differential Earth Movement in North-East Yorkshire during the Jurassic Period. Proc. Yorks. Geol. Soc., N.S., vol. xix, pp. 383–94.

MOLENGRAAFF, G. A. F., & GRACHT, W. A. M. VAN W. VAN DER, 1913. Tektonik und Bau des posttriadischen Mesozoikums in den Niederlanden. In 'Niederlande'; Handb. Reg. Geol., vol. i, Abt. 3, pp. 33–6.

PHILIPPI, E., 1908. Über das Problem der Schichtung und über Schichtbildung am Boden der heutigen Meere. Zeitschr. Deutsch. Geol. Gesell., vol. lx, pp. 346–77.

PRUVOST, P., 1930. Sédimentation et Subsidence. Livre jubilaire, Soc. géol. France, vol. ii, pp. 545–64.

RASTALL, R. H., 1925. On the Tectonics of the Southern Midlands. Geol. Mag., vol. lxii, pp. 193–222.

STILLE, H., 1902. Über präcretacische Schichtenverschiebungen im älteren Mesozoikum des Egge-Gebirges. Jahrb. Preuss. Geol. Landesanst., vol. xxiii, pp. 296–322.

STILLE, H., 1905. Über Strandverschiebungen in hannoverschen oberen Jura. Zeitschr. Deutsch. Geol. Gesell., vol. lvii, Monatsber., pp. 515–34.

——, 1905. Zur Kenntniss der Dislokationen, Schichtenabtragungen und Transgressionen im jüngsten Jura und der Kreide Westfalens. Jahrb. Preuss. Geol. Landesanst., vol. xxvi, pp. 103–25.

——, 1910. Senkungs-, Sedimentations- und Faltungsräume. Compte rendu, Congrès géol. int., Session ix, Stockholm, pp. 819–36.

——, 1913. Die saxonische 'Faltung'. Zeitschr. Deutsch. Geol. Gesell., vol. lxv, Monatsber., pp. 575–93.

——, 1913. Die Kimmerische (vorcretacische) Phase der saxonischen Faltung des deutschen Bodens. Geol. Rundschau, vol. iv, pp. 362–83.

——, 1922. Studien über Meeres- und Bodenschwankungen. Nachrichten kgl. Gesell. Wiss. Göttingen, M.P. Klasse, pp. 83–95.

——, 1924. Grundfragen der vergleichenden Tektonik. Berlin.

STRAHAN, A., 1904. [Discrimination of Periods of Earth-Movement in the British Isles.] Rept. Brit. Ass. for 1904, Pres. Address Section C; reprinted Geol. Mag. [5], vol. i, 1904, pp. 449–62.

TOPLEY, W., 1874. On the Correspondence between some Areas of Apparent Upheaval and the Thickening of Subjacent Beds. Q.J.G.S., vol. xxx, pp. 186–95.

TWENHOFEL, W. H., 1926. Treatise on Sedimentation. London. [See review, Geol. Mag. 1927, vol. lxiv, pp. 180–4.]

——, 1931. Environment in Sedimentation and Stratigraphy. Bull. Amer. Geol. Soc., vol. xlii, pp. 407–24.

VAUGHAN, T. W., SCHUCHERT, C., and others, 1920. Researches on Sedimentation. Bull. Amer. Geol. Soc., vol. xxxi, pp. 401–32.

VERSEY, H. C., 1929. The Tectonic Structure of the Howardian Hills and Adjacent Areas. Proc. Yorks. Geol. Soc., N.S., vol. xxi, pp. 197–227.

——, 1931. Saxonian Movements in East Yorkshire and Lincolnshire. Proc. Yorks. Geol. Soc., N.S., vol. xxii, pp. 52–8.

GENERAL WORKS, ON BRITAIN AS A WHOLE AND ON CERTAIN DISTRICTS

ARKELL, W. J., 1929. Jurassic; Encyclopædia Britannica, ed. xiv, vol. xiii, pp. 193–6.

BARRETT, C., 1878. The Geology of Swyre, Puncknowle, Burton Bradstock, Loders, Shipton Gorge, Litton Cheney, Longbredy, Littlebredy and Abbotsbury, Dorset. 8vo, Bridport.

BEESLEY, T., 1877. The Geology of the Eastern Portion of the Banbury and Cheltenham Direct Railway [Upper Lias to Great Oolite]. P.G.A., vol. v, pp. 165–85.

BONNEY, T. G., 1872. Cambridgeshire Geology. Cambridge.

BOSWELL, P. G. H., 1917. Rhætic and Jurassic of Scotland. Handb. Reg. Geol., vol. iii, pp. 211–12, 233–4.

BRYCE, J., & ARMSTRONG, J., 1876. The Jurassic Strata of Skye and Raasay: with List of the Jurassic Fossils of Skye, Raasay and Mull. In 'Catalogue of the Western Scottish Fossils'. Glasgow.

BRYCE, J., & TATE, R., 1873. On the Jurassic Rocks and Palaeontology of Skye and Raasay. Q.J.G.S., vol. xxix, pp. 317–51.

BUCKLAND, W., & DE LA BECHE, H. T., 1835. On the Geology of the Neighbourhood of Weymouth and the adjacent parts of the Coast of Dorset. Trans. Geol. Soc. [2], vol. iv, pp. 1–46.

BUCKMAN, J., 1843. Geological Chart of the Oolitic Strata of the Cotteswold Hills, and the Lias of the Vale of Gloucester. Folio, Cheltenham.

——, 1858. On the Oolite Rocks of Gloucestershire and North Wilts. Q.J.G.S., vol. xiv, pp. 98–103.

CROSS, J. E., 1875. The Geology of North-West Lincolnshire. Q.J.G.S., vol. xxxi, pp. 115–30.

DAMON, R., 1860. Geology of Weymouth and the coast of Dorset, and Supplement (with 11 plates of fossils).

DAVIES, A. M., 1909. Buckinghamshire: in Geology in the Field, vol. i, ch. vi, pp. 179–91.

——, 1917–29. Jurassic; England and Wales, Handb. Reg. Geol., vol. iii, pp. 213–45. Re-written in Handbook to Geol. Great Britain, 1929.

DAVIS, J. W., 1892. On Some Sections in the Liassic and Oolitic Rocks of Yorkshire. Proc. Yorks. Soc., vol. xii, Lias, pp. 170–89; Oolites (Lower–Upper), pp. 189–214.

DOUGLAS, J. A., 1909. The Oxford and Banbury District: in Geology in the Field, vol. i, ch. vii, pp. 192–209.

FEARNSIDES, W. G., 1904. The Geology of Cambridgeshire. In Marr and Shipley's Handbook to the Natural History of Cambridgeshire, pp. 9–50. Cambridge.

FOX-STRANGWAYS, C., 1892. The Jurassic Rocks of Britain, vol. i, Yorkshire. Mem. Geol. Survey.

GEIKIE, A., 1858. On the Geology of Strath, Skye; and Notes on the Fossils from the Lias of the Isles of Pabba, Scalpa and Skye, by T. Wright. Q.J.G.S., vol. xiv, pp. 1–36.

HERRIES, R. S., 1906. The Geology of the Yorkshire Coast between Redcar and Robin Hood's Bay. P.G.A., vol. xix, pp. 410–45.

——, 1910. East Yorkshire: in Geology in the Field, vol. iii, pp. 592–623.

HUDLESTON, W. H., & MONCKTON, H. W., 1910. Dorset: in Geology in the Field, vol. ii, pp. 365–413.

IBBETSON, L. L. B., 1848. Notice of the Geology of the Neighbourhood of Stamford and Peterborough [Upper Lias–Oxford Clay]. Rept. Brit. Assoc. for 1847, pp. 127–31.

JUDD, J. W., 1873–8. The Secondary Rocks of Scotland.
  I. [Introduction and East Coast]. Q.J.G.S., vol. xxix, pp. 97–197 (1873).
  II. On the Ancient Volcanoes of the Highlands and the Relations of their Products to the Mesozoic Strata. Ibid., vol. xxx, pp. 220–301 (1874).
  III. The Strata of the Western Coast and Islands. Ibid., vol. xxxiv, pp. 660–743 (1878).

JUKES-BROWNE, A. J., 1910. Lincolnshire: in Geology in the Field, vol. iii, pp. 488–517.

KENDALL, P. F., & WROOT, H. E., 1924. Geology of Yorkshire. Printed for the authors, pp. 1–995.

LEE, G. W., & BUCKMAN, S. S., 1913. [Palaeontological notes on the Lias, Inferior Oolite and Oxford Clay of the Isle of Mull.] Sum. Prog. Geol. Surv. for 1912, pp. 74–5, and 35.

LEE, G. W., 1914. Mesozoic Rocks: West Highland District [Notes on Rhætic, Upper Lias and Great Estuarine Strata of Raasay and Skye]. Sum. Prog. Geol. Surv. for 1913, pp. 39–43.

LEE, G. W., & PRINGLE, J., 1932. A Synopsis of the Mesozoic Rocks of Scotland. Trans. Geol. Soc., Glasgow, vol. xix, pp. 158–224.

LLOYD MORGAN, C., & REYNOLDS, S. H., 1909. Sketch of the Geology of the Bristol District. Proc. Bristol Nat. Soc. [4], vol. ii, pp. 5–26.

LONSDALE, W., 1832. On the Oolitic District of Bath. Trans. Geol. Soc. [2], vol. iii, pp. 241–76.

LYCETT, J., 1857. The Cotteswold Hills. Handbook introductory to their Geology and Palaeontology, London. [Inferior Oolite–Cornbrash.]

MANSEL-PLEYDELL, J. C., 1873. A Brief Memoir on the Geology of Dorset. Geol. Mag., vol. x, pp. 402–13, 438–47.

MORRIS, J., 1853. On Some Sections in the Oolitic District of Lincolnshire. Q.J.G.S., vol. ix, pp. 317–44. [Inferior Oolite–Oxford Clay.]

——, 1869. Geological Notes on Parts of Northampton and Lincolnshire. Geol. Mag., vol. vi, pp. 99–105. [Upper Lias–Great Oolite.]

MURCHISON, R. I, 1827–8. On the Coalfield of Brora, Sutherlandshire, and some other Stratified Deposits in the North of Scotland. Trans. Geol. Soc. [2], vol. ii, pp. 293–326, and Supplementary Remarks on the Strata of the Oolitic Series ... in the Counties of Sutherland and Ross, and in the Hebrides, Ibid., pp. 353–68.

——, 1834–45. Outline of the Geology of the Neighbourhood of Cheltenham, 1834. 2nd edition, with palaeontological appendix by J. Buckman, 1845.

PHILLIPS, J., 1829. Illustrations of the Geology of Yorkshire, part i: The Yorkshire Coast.

PHILLIPS, J., 1855. The Neighbourhood of Oxford and its Geology. Oxford Essays for 1855. 8vo, Oxford.

——, 1858. On Some Comparative Sections in the Oolitic and Ironstone Series of Yorkshire [Mid. Lias–Kim. Clay]. Q.J.G.S., vol. xiv, pp. 84–98.

——, 1871. Geology of Oxford and the Valley of the Thames. Oxford.

PORTER, H., 1861. The Geology of Peterborough and its Vicinity. Peterborough.

RASTALL, R. H., 1909. Cambridgeshire, Bedfordshire and West Norfolk: in Geology in the Field, vol. i, ch. v, pp. 124–78.

REYNOLDS, S. H., & VAUGHAN, A., 1902. On the Jurassic Strata cut through by the South Wales Direct Line between Filton and Wootton Bassett [Rhætic–Corallian]. Q.J.G.S., vol. lviii, pp. 719–52.

RICHARDSON, L., 1910. The Neozoic Rocks of Gloucestershire and Somerset: in Geology in the Field, vol. ii, pp. 329–64.

SCROPE, G. P., 1859. Geology of Wiltshire. Wilts. Arch. N.H. Mag., vol. v, pp. 89–113.

SEDGWICK, A., 1846. On the Geology of the Neighbourhood of Cambridge, including the Formations between the Chalk and the Great Bedford Level. Rept. Brit. Assoc. for 1845, Sections, p. 40.

SOLLAS, W. J., 1926. The Geology of the County round Oxford, in 'The Natural History of the Oxford District', edited by J. J. Walker, pp. 32–59. Oxford.

THOMPSON, B., 1910. Northamptonshire and Parts of Rutland and Warwickshire: in Geology in the Field, vol. iii, pp. 450–87.

VERNON, W., 1826. An Account of the Strata North of the Humber, near Cave. Ann. Phil. [2], vol. xi, pp. 435–9.

WALCOTT, J., 1799. Descriptions and Figures of Petrifications found in the Quarries, Gravel-Pits, &c. near Bath. Hazard, Bath.

WEBSTER, T., 1816. Geological Observations on the Isle of Wight and Adjacent Coast of Dorsetshire [written 1811–12]; in Englefield, H., 1816. Description of the Principal Picturesque Beauties, &c., of the Isle of Wight, pp. 117–238. [Kimeridge Clay–Purbeck Beds.]

WEDD, C. B., 1902. [Notes on the Jurassic Rocks—Lower Lias–Corallian—of Skye.] Sum. Prog. Geol. Surv. for 1901, pp. 142–4.

WHITE, H. J. O., 1910. Berkshire and Part of the Thames Valley: in Geology in the Field, vol. ii, pp. 210–35.

WITCHELL, E., 1882. The Geology of Stroud and the Area Drained by the Frome. Stroud [Lower Lias–Forest Marble.]

WOODWARD, H. B., 1893–5. The Jurassic Rocks of Britain (Yorkshire excepted): vol. iii, The Lias of England and Wales (1893); vol. iv, Lower Oolitic Rocks of England (1894); vol. v, Middle and Upper Oolitic Rocks (1895). Mems. Geol. Surv.

——, 1897. Geology of the London Extension of the Manchester, Sheffield, and Lincolnshire Railway. Part II: Rugby to Quainton Road, near Aylesbury. Geol. Mag. [4], vol. iv, pp. 97–105.

——, 1899. South Wales Direct Railway [Notes on Rhætic–Corallian, Filton to Wootton Bassett]. Sum. Prog. Geol. Surv. for 1898, pp. 188–92.

——, 1910. Wiltshire: in Geology in the Field, vol. ii, pp. 293–307.

YOUNG, G., & BIRD, J., 1822. A Geological Survey of the Yorkshire Coast: describing the strata and fossils occurring between the Humber and the Tees, from the German Ocean to the Plain of York. Whitby, 1822; 2nd ed. Whitby, 1828.

RHÆTIC BEDS

ADAMS, W., & HARRISON, W. J., 1877. On the Penarth or Rhætic Beds of Glamorganshire, Leicestershire, &c. Trans. Cardiff. Nat. Soc., vol. viii, pp. 96–106 [largely reprinted from Harrison, Q.J.G.S., vol. xxxii, 1876, pp. 212–18].

BATES, E. F., & HODGES, L., 1886. Notes on a recent Exposure of the Lower Lias and Rhætics in the Spinney Hills, Leicester. Trans. Lit. and Phil. Soc. Leicester [2], part i, pp. 22–3.

BATHER, F. A., 1909–10. Fossil Representatives of the Lithodomous Worm, *Polydora* [at the base of the Rhætics]; and Some Fossil Annelid Burrows. Geol. Mag. [5], vol. vi, pp. 108–10, and vol. vii, pp. 114–16.

BRISTOW, H. B., 1864. On the Rhætic or Penarth Beds of the neighbourhood of Bristol and the South-West of England. Geol. Mag., vol. i, p. 236; and Rept. Brit. Assoc. for 1864 (1865), p. 50.

BRODIE, P. B., 1874. Notes on a Railway-section of the Lower Lias and Rhætics between Stratford-on-Avon and Fenny Compton, &c. Q.J.G.S., vol. xxx, pp. 746–9.

——, 1886. On two Rhætic Sections in Warwickshire. Q.J.G.S., vol. xlii, pp. 272–5.

——, 1888. On the Range, Extent and Fossils of the Rhætic Formation in Warwickshire. Proc. Warwicks. N.F.C. for 1887, p. 19.

BROWNE, M., 1893–6. Vertebrate Remains from the Rhætic Beds of Britain. Rept. Brit. Assoc. 1893, pp. 748–9; 1894, pp. 657–8; 1894 (and Rhætic Bone Bed of Aust Cliff, &c.), pp. 804–5.

BURTON, F. M., 1867. On the Rhætic Beds near Gainsborough. Q.J.G.S., vol. xxiii, pp. 315–22.

CALLAWAY, C., 1901. The Pre-Rhætic Denudation of the Bristol Area. Proc. Cots. N.F.C., vol. xiv, pp. 47–58.

COLE, G. A. J., 1917. Rhætic; Ireland. In Handb. Reg. Geol., British Isles, p. 212.

DAVIS, J. W., 1881. Notes on the Fish-Remains of the Bone Bed at Aust. Q.J.G.S., vol. xxxvii, pp. 414–25.

DAWKINS, W. B., 1864. On the Rhætic Beds and White Lias of Western and Central Somerset, &c. Q.J.G.S., vol. xx, pp. 396–412.

ETHERIDGE, R., 1865. On the Rhætic or *Avicula contorta* Beds at Garden Cliff, Westbury-upon-Severn. Proc. Cots. N.F.C., vol. iii, pp. 218–34.

——, 1872. On the Physical Structure and Organic Remains of the Penarth (Rhætic) Beds of Penarth and Lavernock; with description of the Westbury-on-Severn Section (with Plates). Trans. Cardiff Nat. Soc., vol. iii, pp. 39–64.

GRÖNWALL, K. A., 1906. On the Occurrence of the [Lamellibranch] Genus *Dimyodon* Mun. Chalm. in the Mesozoic [Rhætic] Rocks of Great Britain. Geol. Mag. [5], vol. iii, pp. 202–5.

HARRIS, T. M., 1931. Rhætic Floras. Biological Reviews and Proc. Cambridge Phil. Soc., vol. vi, pp. 133–62.

HARRISON, W. J., 1876. On the Occurrence of the Rhætic Beds in Leicestershire. Q.J.G.S., vol. xxxii, pp. 212–18.

HOWARD, F. T., 1895. The Geology of Barry Island. Trans. Cardiff Nat. Soc., vol. xxvii, pp. 42–55.

——, 1897. Notes on the Base of the Rhætic Series at Lavernock Point. Trans. Cardiff Nat. Soc., vol. xxix, pp. 64–6.

——, 1899. The Geology of the Cowbridge District. Trans. Cardiff Nat. Soc., vol. xxx, pp. 36–47.

JONES, T. R., 1894. On the Rhætic and some Liassic Ostracoda of Britain. Q.J.G.S., vol. l, pp. 156–68.

MISKIN, F. F., 1922. The Triassic Rocks [including Rhætic] of South Glamorgan. Trans. Cardiff Nat. Soc., vol. lii, pp. 17–25 and plates.

NEWTON, E. T., 1899. On a Megalosauroid Jaw from Rhætic Beds near Bridgend, Glamorganshire. Q.J.G.S., vol. lv, pp. 89–96.

PARIS, E. T., 1906. Additional Notes on the Denny-Hill [Rhætic] Section near Minsterworth, Gloucestershire. Proc. Cots. N.F.C., vol. xv, pp. 263–6.

PEACH, B. N., GUNN, W., & NEWTON, E. T., 1901. On a remarkable Volcanic Vent of Tertiary Age in the Island of Arran, enclosing Mesozoic Fossiliferous Rocks [Rhætic and Lower Lias]. Q.J.G.S., vol. lvii, pp. 226–43.

RAMSAY, A. C., 1871. On the Physical Relations of the New Red Marl, Rhætic Beds and Lower Lias. Q.J.G.S., vol. xxvii, pp. 189–99.

REYNOLDS, S. H., & VAUGHAN, A., 1904. The Rhætic Beds of the South-Wales Direct Line. Q.J.G.S., vol. lx, pp. 194–214.

RICHARDSON, L., 1903. The Rhætic Rocks of North-West Gloucestershire. Proc. Cots. N.F.C., vol. xiv, pp. 127–74, and Supplement, pp. 251–7.

——, 1903. Two Sections of the Rhætic Rocks in Worcestershire. Geol. Mag. [4], vol. x, pp. 80–2.

——, 1903. The Rhætic and Lower Lias of Sedbury Cliff, near Chepstow, Monmouthshire. Q.J.G.S., vol. lix, pp. 390–5.

RICHARDSON, L., 1903. Observations on the Rhætic Rocks of Worcestershire. Trans. Worcester Nat. C., vol. iii, part 2, pp. 92–101.

——, 1904. The Evidence for a Non-Sequence between the Keuper and Rhætic Series in North-West Gloucestershire and Worcestershire. Q.J.G.S., vol. lx, pp. 349–58.

——, 1905. The Rhætic Rocks of Monmouthshire. Q.J.G.S., vol. lxi, pp. 374–84.

——, 1905. The Rhætic and Contiguous Deposits of Glamorganshire. Q.J.G.S., vol. lxi, pp. 385–424.

——, 1905. On the Occurrence of Rhætic Rocks at Berrow Hill, near Tewkesbury. Q.J.G.S., vol. lxi, pp. 425–30.

——, 1906. On the Rhætic and Contiguous Deposits of Devon and Dorset. P.G.A., vol. xix, pp. 401–9.

——, 1906. On the Occurrence of *Ceratodus* in the Rhætic at Garden Cliff, Westbury-on-Severn, Gloucestershire, with some remarks upon its distribution in British Formations. Proc. Cots. N.F.C., vol. xv, pp. 267–70.

——, 1909. The Rhætic Section at Wigston, Leicestershire. Geol. Mag. [5], vol. vi, pp. 366–70. [See also A. R. Horwood, Rept. Brit. Assoc. 1911, Portsmouth, p. 388.]

——, 1911. The Rhætic and Contiguous Deposits of West, Mid, and part of East Somerset. Q.J.G.S., vol. lxvii, pp. 1–74.

——, 1912. The Rhætic Rocks of Warwickshire. Geol. Mag. [5], vol. ix, pp. 24–33.

——, 1917–29. Rhaetic: England and Wales. Handb. Reg. Geol., vol. iii, 1917, pp. 207–11. And in Handb. Geol. Great Britain, 1929, pp. 341–6.

RICHARDSON, L., & TUTCHER, J. W., 1914. On *Pteromya crowcombeia* Moore, and some species of *Pleuromya* and *Volsella* from the Rhætic and Lower Lias. Proc. Yorks. Geol. Soc., N.S., vol. xix, pp. 51–8.

SHORT, A. R., 1903. On the Cotham Marble. Proc. Bristol Nat. Soc., N.S., vol. x, pp. 3–54 and Plates.

——, 1904. A Description of some Rhætic Sections in the Bristol District, with considerations on the Mode of Deposition of the Rhætic Series. Q.J.G.S., vol. lx, pp. 170–93.

SOLLAS, I. B. J., 1901. Structure and Affinities of the Rhætic Plant, *Naiadita*. Q.J.G.S., vol. lvii, pp. 307–12.

STORRIE, J., 1895. Notes on the Tooth of a Species of *Mastodonsaurus*, found with some other bones near Lavernock. Trans. Cardiff Nat. Soc., vol. xxvi, pp. 105–6.

STRAHAN, A., 1903–4. Cutting on the South Wales Direct Railway, near Chipping Sodbury [Rhætic Beds]. Sum. Prog. Geol. Surv. for 1902, pp. 192–4, and for 1903, pp. 171–2.

TATE, R., 1867. On the Liassic affinities of the *Avicula contorta* Series. Geol. and Nat. Hist. Repertory, vol. i, pp. 364–9.

TAWNEY, E. B., 1866. On the Western Limit of the Rhætic Beds in South Wales and on the position of the Sutton Stone, with a note on the Corals by P. M. Duncan. Q.J.G.S., vol. xxii, pp. 68–93.

THOMPSON, B., 1894. Landscape Marble. Q.J.G.S., vol. l, pp. 393–410.

TOMES, R. F., 1884. A Comparative and Critical Revision of the Madreporaria of the White Lias of the Middle and Western Counties of England, and of those of the Conglomerate at the Base of the South Wales Lias. Q.J.G.S., vol. xl, pp. 353–75.

——, 1903. Description of a Species of *Heterastræa* from the Lower Rhætic of Gloucestershire. Q.J.G.S., vol. lix, pp. 403–7.

WICKES, W. H., 1900. A New Rhætic Section at [Redland], Bristol. P.G.A., vol. xvi, pp. 421–3. Also Proc. Bristol Nat. Soc., N.S., vol. ix, 1901, pp. 99–103; and Additional Observations by J. Parsons, ibid., pp. 104–8.

WILSON, E., 1882. The Rhætics of Nottinghamshire. Q.J.G.S., vol. xxxviii, pp. 451–6.

——, 1891–4. Section of the Rhætic Rocks at Pylle Hill (Totterdown). Q.J.G.S., 1891, vol. xlvii, pp. 545–9; and Proc. Bristol Nat. Soc., N.S., vol. vii, 1894, pp. 213–31.

WINWOOD, H. H., 1876. Notes on a Rhætic and Lower Lias Section on the Bath and Evercreech Line near Chilcompton. Proc. Bath F.C., vol. iii, pp. 300–4.

WINWOOD, H. H., & RICHARDSON, L., 1910. Notes on a White Lias Section at Saltford near Bath. Proc. Cots. N.F.C., vol. xvii, pp. 45–50.

WOODWARD, H. B., 1889. Notes on the Rhætic Beds and Lias of Glamorganshire. P.G.A., vol. x, pp. 529–38.

——, 1892. Remarks on the Formation of Landscape Marble. Geol. Mag. [3], vol. ix, pp. 110–14.

WOODWARD, H. B., & BLAKE, J. H., 1872. Notes on the Relations of the Rhætic Beds to the Lower Lias and Keuper Formations in Somersetshire. Geol. Mag., vol. ix, pp. 196–202.

WRIGHT, T., 1860. On the Zone of *Avicula contorta* and the Lower Lias of the South of England. Q.J.G.S., vol. xvi, pp. 374–411.

——, 1864. On the White Lias of Dorsetshire. Geol. Mag., vol. i, pp. 290–2; and Rept. Brit. Assoc. for 1864, p. 75.

LOWER LIAS

BECHE, H. T. DE LA, 1822. Remarks on the Geology of the South Coast of England, from Bridport Harbour, Dorset, to Babbacombe Bay, Devon [Lower Lias and Keuper]. Trans. Geol. Soc. [2], vol. i, pp. 40–7.

——, 1826. On the Lias of the Coast in the Vicinity of Lyme Regis. Trans. Geol. Soc. [2], vol. ii, pp. 21–30.

BEESLEY, T., 1877. The Lias of Fenny Compton, Warwickshire, Proc. Warwicks. N.F.C. for 1877, pp. 1–22; and 8vo, Banbury.

BLAKE, J. F., 1871. Foraminifera from the Lias of Cliffe, near Market Weighton. Proc. Yorks. Nat. Club for 1871, p. 13.

——, 1872. On the Infralias in Yorkshire. Q.J.G.S., vol. xxviii, pp. 132–46.

——, 1887–8. On a Starfish [*Solaster*] from the Yorkshire Lias. Rept. Brit. Assoc. for 1887, p. 716; Nature, vol. xxxvi, p. 591; Geol. Mag. [3], vol. iv, pp. 529–31.

——, 1891. The Geology of the Country between Redcar and Bridlington. P.G.A., vol. xii, pp. 115–44.

BRISTOW, H. W., 1867. On the Lower Lias or Lias-Conglomerate of a part of Glamorganshire. Q.J.G.S., vol. xxiii, pp. 199–207.

BRODIE, P. B., 1857. On Some Species of Corals in the Lias of Gloucestershire, Worcestershire, Warwickshire, and Scotland. Edinburgh New Phil. Journ., N.S., vol. v, pp. 260–4.

——, 1860. Remarks on the Lias of Barrow, in Leicestershire, compared with the Lower Part of that formation in Gloucestershire, Worcestershire and Warwickshire. Proc. Cots. N.F.C., vol. ii, pp. 139–41.

——, 1861. On the Distribution of Corals in the Lias. Q.J.G.S., vol. xvii, pp. 151–2, and Rept. Brit. Assoc. for 1860, pp. 73–4.

——, 1864–5. On the Lias Outliers at Knowle and Wootton Wawen in S. Warwickshire, and on the Presence of the Lias or Rhætic Bone Bed at Copt Heath. Q.J.G.S., vol. xxi, pp. 159–61; and Rept. Brit. Assoc. for 1864, p. 52.

——, 1866. Notes on a Section of Lower Lias and Rhætic Beds near Wells, Somerset. Q.J.G.S., vol. xxii, pp. 93–5.

——, 1867. On the Correlation of the Lower Lias at Barrow-on-Soar in Leicestershire with the same strata in Warwickshire, Worcestershire and Gloucestershire. Ann. Mag. Nat. Hist. [3], vol. xix, pp. 31–4.

BROWN, T., 1843. Description of some New Species of the Genus *Pachyodon* [= *Cardinia*]. Ann. Mag. Nat. Hist., N.S., vol. xii, pp. 390–6.

BUCKLAND, W., 1829. On the Discovery of Coprolites, or Fossil Faeces, in the Lias at Lyme Regis, and in other Formations. P.G. Soc., vol. i, pp. 96–8. Trans. Geol. Soc. [2], vol. iii, pp. 217–22.

BUCKMAN, J., 1842. On the Lias Beds near Cheltenham. Geologist (Moxon's), pp. 14–20.

——, 1879. On *Belemnoteuthis montefiorei* [from the Lower Lias between Charmouth and Lyme Regis]. Proc. Dorset N.F.C., vol. iii, pp. 141–3.

BUCKMAN, S. S., 1906. Some Lias Ammonites: *Schlotheimia* and Species of other Genera. Proc. Cots. N.F.C., vol. xv, pp. 231–54.

——, 1907. Some Species of the Genus *Cincta*. Proc. Cots. N.F.C., vol. xvi, pp. 41–63.

——, 1915. A Palaeontological Classification of the Jurassic Rocks of the Whitby

District, with a Zonal Table of Lias Ammonites. In Fox-Strangways and Barrow, Geol. Whitby and Scarboro., Mem. Geol. Surv., 1915 (2nd edition).

BUCKMAN, S. S., 1917–20. Jurassic Chronology; I. Lias. Q.J.G.S., vol. lxxiii, pp. 257–327; and Supplement: West of England Strata. Q.J.G.S., vol. lxxvi, pp. 62–103.

COLE, G. A. J., 1917. Jurassic [Lower Lias]; Ireland. In Handb. Reg. Geol., British Isles, pp. 245–6.

COX, L. R., 1928. Gastropods and Lamellibranchs from the Belemnite Marls. Appendix to paper by W. D. Lang. Q.J.G.S., vol. lxxxiv, pp. 233–45.

COYSH, A. W., 1931. U-shaped Burrows in the Lower Lias of Somerset and Dorset. Geol. Mag., vol. lxviii, pp. 13–15.

CRICK, G. C., 1920. On *Nautilus pseudotruncatus* n.sp. from the Liassic Rocks of England. Proc. Cots. N.F.C., vol. xx, p. 245.

CRICK, W. D., & SHERBORN, C. D., 1891–2. On Some Liassic Foraminifera from Northamptonshire. Journ. Northants. N.H.S., vol. vi, pp. 208–14, and vol. vii, pp. 67–73.

DAY, E. C. H., 1865. On the Lower Lias of Lyme Regis. Geol. Mag., vol. ii, pp. 518–19.

DUNCAN, P. M., 1867. On the Madreporaria of the Infra-Lias of South Wales [chiefly from the Sutton and Southerndown Series]. Q.J.G.S., vol. xxiii, pp. 12–28.

——, 1886. On the Structure and Classificatory Position of some Madreporaria from the Secondary Strata of England and South Wales. Q.J.G.S., vol. xlii, pp. 113–42.

——, 1886. On the Astrocœniæ of the Sutton Stone and other Deposits of the Infra-Lias of South Wales. Q.J.G.S., vol. xlii, pp. 101–12.

ETHERIDGE, R., 1864. Description of New Species of Mollusca, &c. [from the Lower Lias near Belfast, Ireland]. Q.J.G.S., vol. xx, pp. 112–14.

GAVEY, G. E., 1853. On the Railway Cuttings [in Lower Lias] at the Mickleton Tunnel and at Aston Magna, Gloucestershire. Q.J.G.S., vol. ix, pp. 29–37.

GROOM-NAPIER, C. O., 1868. On the Lower Lias Beds occurring at Cotham, Bedminster and Keynsham, near Bristol. Q.J.G.S., vol. xxiv, pp. 204–5.

HOLMES, T. V., 1881. The Permian, Triassic and Liassic Rocks of the Carlisle Basin. Q.J.G.S., vol. xxxvii, pp. 286–98.

JONES, J., & TOMES, R. F., 1865. The Sutton Beds of Glamorganshire. Proc. Cots. N.F.C., vol. iii, p. 191.

JUKES-BROWNE, A. J., 1908. The Burning Cliff and the Landslip at Lyme Regis. Proc. Dorset N.F.C., vol. xxix, pp. 153–60.

LANG, W. D., 1913. The Lower Pliensbachian—'Carixian'—of Charmouth. Geol. Mag. [v], vol. x, pp. 401–12.

——, 1914. The Geology of the Charmouth Cliffs, Beach and Foreshore. P.G.A., vol. xxv, pp. 293–360.

——, 1917. The Ibex-Zone at Charmouth, and its Relation to the Zones near it. P.G.A., vol. xxviii, pp. 30–6.

LANG, W. D., SPATH, L. F., & RICHARDSON, W. A., 1923. Shales-with-'Beef': a Sequence in the Lower Lias of the Dorset Coast. Q.J.G.S., vol. lxxix, pp. 47–99.

LANG, W. D., 1924. The Blue Lias of the Devon and Dorset Coasts. P.G.A., vol. xxxv, pp. 169–85.

LANG, W. D., SPATH, L. F., COX, L. R., & MUIR-WOOD, H. M., 1926. The Black Marl of Black Ven and Stonebarrow in the Lias of the Dorset Coast. Q.J.G.S., vol. lxxxii, pp. 144–87.

——, 1928. The Belemnite Marls of Charmouth, a Series in the Lias of the Dorset Coast. Q.J.G.S., vol. lxxxiv, pp. 179–257.

LANG, W. D., 1932. The Lower Lias of Charmouth and the Vale of Marshwood. P.G.A., vol. xliii, pp. 97–126.

LUCAS, S., 1862. Section of the Lias Clay in a railway-cutting near Stow on the Wold. Geologist, vol. v, pp. 127–8.

LUCY, W. C., 1886. On the Southerndown, Dunraven, and Bridgend Beds. Proc. Cots. N.F.C., vol. viii, pp. 254–64.

LYCETT, J., 1863. On *Gryphæa incurva* and its varieties. Proc. Cots. N.F.C., vol. iii, pp. 81–96; and Correspondence as to Geological Position, by J. Jones and R. F. Tomes, ibid., pp. 191–4.

LYDEKKER, R., 1891. Note on a nearly perfect Skeleton of *Ichthyosaurus tenuirostris* from the Lower Lias of Street, Somerset. Geol. Mag. [3], vol. viii, pp. 289–90.

McDonald, A. I., & Trueman, A. E., 1921. The Evolution of Certain Liassic Gastropods, with Special Reference to their Use in Stratigraphy. Q.J.G.S., vol. lxxvii, pp. 297–344.

Moore, C., 1867. On Abnormal Conditions of Secondary Deposits when connected with the Somersetshire and South Wales Coal-basins, and on the Age of the Sutton and Southerndown Series. Q.J.G.S., vol. xxiii, pp. 449–568.

——, 1877. The Liassic and other Secondary Deposits of the Southerndown Series. Trans. Cardiff Nat. Soc., vol. viii, pp. 53–60.

Morton, G. H., 1864. On the Lias Formation as developed in Shropshire. Proc. Liverpool Geol. Soc., Session 5, pp. 2–6.

Paris, E. T., 1908. Notes on some Echinoids from the Lias of Worcestershire, Gloucestershire and Somerset. Proc. Cots. N.F.C., vol. xvi, pp. 143–50.

Portlock, J. E., 1843. Report on the Geology of the County of Londonderry and of parts of Tyrone and Fermanagh [Lower Lias]. Dublin.

Quilter, H. E., 1881. Exposure of the Middle Series of the *Bucklandi* Beds in Leicestershire. Mid. Nat., vol. iv, p. 265.

——, 1886. The Lower Lias of Leicestershire. Geol. Mag. [3], vol. iii, pp. 59–65.

Reade, T. M., 1876. The Lower Lias of Street, Somerset. Proc. Liverpool Geol. Soc., vol. iii, pp. 97–9.

Richardson, L., 1905. The Lias of Worcestershire [almost entirely Lower Lias]. Trans. Worcestershire Nat. C., vol. iii, part iv, pp. 188–209.

——, 1906. On a Section of Lower Lias at Maisemore, near Gloucester. Proc. Cots. N.F.C., vol. xv, pp. 259–62.

——, 1906. Liassic Dentaliidæ. Q.J.G.S., vol. lxii, p. 573.

——, 1908. On a Section of Lower Lias at Hock Cliff, Fretherne, Gloucestershire. Proc. Cots. N.F.C., vol. xvi, pp. 135–42.

——, 1918. The Geology [Lias and Superficial Deposits] of the Cheltenham–Stratford on Avon Railway (G.W.R.). Trans. Woolhope N.F.C. for 1916 (1918), pp. 137–53.

——, 1926. A Deep Boring at Rooksbridge, East Brent, Somerset [Alluvium and Lower Lias]. Proc. Somerset Arch. N.H. Soc. [4], vol. lxxii, pp. 73–5.

Richardson, L., Walters, R. C. S., & Trueman, A. E., 1922. Lower Lias Section at Hayden, near Cheltenham, with notes on the Ammonites. Proc. Cots. N.F.C., vol. xxi, pp. 167–76.

Seeley, H. G., 1870. On *Zoocapsia dolichorhamphia*, a Sessile Cirripede from the Lias of Lyme Regis. Ann. Mag. Nat. Hist. [4], vol. v, p. 283.

Simpson, M., 1855. The Fossils of the Yorkshire Lias. Whitby; 2nd edition, 1884.

Sollas, W. J., 1881. On a Species of *Plesiosaurus* (*P. conybeari*) from the Lower Lias of Charmouth; with Observations on *P. megacephalus* Stutchbury and *P. brachycephalus* Owen. Q.J.G.S., vol. xxxvii, pp. 440–80.

——, 1882. On a Rare Plesiosaur from the Lias at Bridport. Proc. Bristol Nat. Soc. [2], vol. iii, pp. 322–3.

Spath, L. F., 1914. The Development of *Tragophylloceras loscombi* [Lower Lias]. Q.J.G.S., vol. lxx, pp. 336–62.

——, 1922. On Lower Lias Ammonites from Skye. Geol. Mag., vol. lix, pp. 170–6.

——, 1922. On the [Lower] Liassic Succession of Pabay, Inner Hebrides. Geol. Mag., vol. lix, pp. 548–51.

——, 1923. Correlation of the *Ibex* and *Jamesoni* Zones of the Lower Lias. Geol. Mag., vol. lx, pp. 6–11.

——, 1924. The Ammonites of the Blue Lias. P.G.A., vol. xxxv, pp. 186–211.

——, 1925. Notes on Yorkshire Ammonites:
  I. On the genus *Oxynoticeras* Hyatt, Naturalist, 1925, pp. 107–12.
  II. On a Deroceratid, ibid., 1925, pp. 137–41.
  III. On the '*Armatus* Zone', ibid., 1925, pp. 167–72.
  IV. On Some Schlotheimidæ, ibid., 1925, pp. 201–6.
  V. *Arietites*, *Asteroceras*, and allied Genera, ibid., 1925, pp. 263–9.
  VI. On *Ammonites planicosta* J. Sow., ibid., 1925, pp. 299–306.
  VII. On *Ammonites semicostatus* Y. & Bird, ibid., 1925, pp. 327–31.
  VIII. More Lower Liassic Forms, ibid., 1925, pp. 359–64; 1926, pp. 45–9, 137–40, 169–71.
  IX. On Recent Criticisms, ibid., 1926, pp. 265–8.

SPATH, L. F., 1929. Corrections of Cephalopod Nomenclature [includes *Deroceras, Microceras, Palæoechioceras (Protechioceras)*, &c.] Naturalist, 1929, pp. 269–71.

STODDART, W. W., 1868. Notes on the Lower Lias-Beds of Bristol. Q.J.G.S., vol. xxiv, pp. 199–204.

STRICKLAND, H. E., 1842–4. On the Genus *Cardinia* Agassiz, as characteristic of the Lias Formation. Rept. Brit. Assoc. for 1841, pp. 65–6, and Ann. Mag. Nat. Hist., vol. xiv, 1844, pp. 100–8.

STUTCHBURY, S., 1839. Description of a new fossil [*Oxytoma longicostata*] from the [Lower] Lias Shale of [Saltford], Somersetshire. Ann. Mag. Nat. Hist. [2], vol. iii, pp. 163–4.

STUTCHBURY, S., 1842. On a New Genus of Fossil Bivalve Shells [*Pachyodon* Stutchbury = *Cardinia* Agass.] Ann. Mag. Nat. Hist., N.S., vol. viii, pp. 481–5.

SWINTON, W. E., 1930. Preliminary Account of a New Genus and Species of Plesiosaur [*Macropterus tenuiceps*, Lower Lias, Harbury]. Ann. Mag. Nat. Hist. [10], vol. vi, pp. 206–9.

TATE, R., 1864. On the [Rhætic and] Liassic Strata of the Neighbourhood of Belfast. Q.J.G.S., vol. xx, pp. 103–11.

——, 1867. On the Lower Lias of the North-East of Ireland. Q.J.G.S., vol. xxiii, pp. 297–305.

——, 1867. On the Fossiliferous development of the zone of *Ammonites angulatus* Schloth. in Great Britain. Q.J.G.S., vol. xxiii, pp. 305–14.

——, 1871. A Census of the Marine Invertebrate Fauna of the Lias. Geol. Mag., vol. viii, pp. 4–11.

——, 1875. On the Lias about Radstock. Q.J.G.S., vol. xxxi, pp. 493–510.

TATE, R., & BLAKE, J. F., 1876. The Yorkshire Lias. London.

THOMPSON, B., 1899. The Geology [Lower Lias and Superficial] of the Great Central Railway: Rugby to Catesby. Q.J.G.S., vol. lv, pp. 65–88.

THOMPSON, C., 1925. An Addition to the Ammonites of the Yorkshire [Lower] Lias [*Wæhnoceras* sp.], Naturalist, 1925, pp. 43–4.

THORNTON, W., 1881. The Liassic Strata in the Western Highlands of Scotland. Journ. Northants. N.H.S., vol. i, pp. 170–8.

TOMES, R. F., 1888. On *Heterastræa*, a New Genus of Madreporaria from the Lower Lias. Geol. Mag. [3], vol. v, pp. 207–18.

——, 1893. Description of a new Genus of Madreporaria from the Sutton Stone of South Wales. Q.J.G.S., vol. xlix, pp. 574–8.

——, 1878. On the Stratigraphical Position of the Corals of the Lias of the Midland and Western Counties of England and of South Wales. Q.J.G.S., vol. xxxiv, pp. 179–95.

TRUEMAN, A. E., 1915. The Fauna of the Hydraulic Limestones in South Notts. Geol. Mag. [6], vol. ii, pp. 150–2.

——, 1918. The [Lower] Lias of South Lincolnshire. Geol. Mag. [6], vol. v, pp. 64–73, 101–3.

——, 1918. The Evolution of the Liparoceratidæ. Q.J.G.S., vol. lxxiv, pp. 247–98.

——, 1920. The Liassic Rocks of the Cardiff District. P.G.A., vol. xxxi, pp. 93–107.

——, 1922. The Liassic Rocks of Glamorgan. P.G.A., vol. xxxiii, pp. 245–84.

——, 1922. The Use of *Gryphæa* in the Correlation of the Lower Lias. Geol. Mag., vol. lix, pp. 256–68.

——, 1930. The Lower Lias (*Bucklandi* Zone) of Nash Point, Glamorgan. P.G.A., vol. xli, pp. 148–59.

——, 1930. Records of some Ammonites from the Lower Lias of Gloucestershire and Worcestershire. Proc. Cots. N.F.C., vol. xxiii, pp. 245–51.

TRUEMAN, A. E., & WILLIAMS, D. M., 1925. Studies in the Ammonites of the Family Echioceratidæ. Trans. Roy. Soc. Edinburgh, vol. liii, pp. 699–739.

——, 1927. Notes on some [Lower] Lias Ammonites from the Cheltenham District [Cheltenham-Honeybourne Railway]. Proc. Cots. N.F.C., vol. xxii, pp. 239–53; also in Geol. Moreton in Marsh, by L. Richardson, Mem. Geol. Surv., 1929.

TUTCHER, J. W., 1917. The Zonal Sequence of the Lower Lias (Lower Part). Q.J.G.S., vol. lxxiii, pp. 278–81.

TUTCHER, J. W., & TRUEMAN, A. E., 1925. The Liassic Rocks of the Radstock District, Somerset. Q.J.G.S., vol. lxxxi, pp. 595–666.

UPTON, C., 1913. On the abundant occurrence of *Involutina liassica* (Jones) in the Lower Lias at Gloucester. Proc. Cots. N.F.C., vol. xviii, p. 72.

VAUGHAN, A., 1903. The Lowest Beds of the Lower Lias at Sedbury Cliff. Q.J.G.S., vol. lix, pp. 396–402.

VAUGHAN, A., & TUTCHER, J. W., 1903. The Lias of the Neighbourhood of Keynsham. Proc. Bristol Nat. Soc., N.S., vol. x, p. 22.

VEITCH, W., 1886. Three new Species observed in the Yorkshire Lias. [*Chonetes clevelandicus* Veitch, *Pleuromya navicula* Veitch and the coral *Isis liasica* Veitch.] Proc. Yorks. Geol. Soc., vol. ix, p. 54.

WEDD, C. B., 1920. Frodingham [Iron Ore] (Lower Lias); in Lamplugh, Wedd and Pringle, Spec. Repts. Min. Resources Gt. Brit., vol. xii—Bedded Ores, ch. iv, pp. 71–105. Mem. Geol. Survey.

WHIDBORNE, G. F., 1881. On a new Species of *Plesiosaurus* (*P. Conybeari*) from the Lower Lias of Charmouth, with observations on other species. Q.J.G.S., vol. xxxvii, pp. 440–81.

WILSON, E., 1887. British Liassic Gasteropoda. Geol. Mag. [3], vol. iv, pp. 193–202, 258–62.

WITHERS, T. H., 1920. The Cirrepede Subgenus *Scillælepas*; its Probable Occurrence in the Jurassic [Lower Lias] Rocks (*S. gaveyi* sp. n.). Ann. Mag. Nat. Hist. [9], vol. v, pp. 258–64.

WOODWARD, A. S., 1890. Notes on some Ganoid Fishes from the English Lower Lias. Ann. Mag. Nat. Hist. [6], vol. v, pp. 430–6.

——, 1895. On the Liassic Fish *Osteorachis macrocephalus*. Geol. Mag. [4], vol. ii, pp. 204–6.

WOODWARD, H., 1866. On a new Crustacean (*Aeger marderi*) from the Lias of Lyme Regis, Dorset. Geol. Mag., vol. iii, pp. 10–13.

——, 1888. On a New Species of *Aeger* from the Lower Lias of Wilmcote, Warwickshire. Geol. Mag. [3], vol. v, pp. 385–7.

——, 1888. On *Eryon antiquus* Broderip sp., from the Lower Lias of Lyme Regis, Dorset. Geol. Mag. [3], vol. v, pp. 433–41.

——, 1892. On a Neuropterous Insect from the Lower Lias, Barrow-on-Soar, Leicestershire. Geol. Mag. [3], vol. ix, pp. 193–8.

WOODWARD, H. B., 1905. Notes on the Railway-cuttings [in Lower Lias and some Rhætic Beds] between Castle Cary and Langport, in Somerset. Sum. Prog. Geol. Surv. for 1904, pp. 163–9.

MIDDLE AND UPPER LIAS

BATHER, F. A., 1886. Notes on some Recent Openings in the Liassic and Oolitic Rocks of Fawler in Oxfordshire, and the Arrangement of those Rocks near Charlbury. Q.J.G.S., vol. xlii, pp. 143–6.

BEESLEY, T., 1872. A Sketch of the Geology of the Neighbourhood of Banbury. Proc. Warwicks. N.F.C., 1872, pp. 11–34, and Geol. Mag., vol. ix, pp. 279–82.

BLAKE, J. F., 1887. On a new Specimen of *Solaster murchisoni* from the Yorkshire Lias. Geol. Mag. [3], vol. iv, p. 529.

BOSWELL, P. G. H., 1924. The Petrography of the Sands of the Upper Lias and Lower Inferior Oolite in the West of England. Geol. Mag., vol. lxi, pp. 246–64.

BRADY, H. B., 1865. On the Foraminifera of the Middle and Upper Lias of Somersetshire. Rept. Brit. Assoc. for 1864, p. 50.

BRODIE, P. B., 1843. Notice on the Discovery of the Remains of Insects in the Lias of Gloucestershire. Proc. Geol. Soc., vol. iv, pp. 14–16, and Rept. Brit. Assoc. for 1842, p. 58.

——, 1849. Notice on the Discovery of a Dragon-fly and a new Species of *Leptolepis* in the Upper Lias near Cheltenham, with a few remarks on that Formation in Gloucestershire. Q.J.G.S., vol. v, pp. 31–7.

——, 1860. Remarks on the Inferior Oolite and Lias in parts of Northamptonshire compared with the same formations in Gloucestershire. Proc. Cots. N.F.C., vol. ii, pp. 132–4.

BUCKMAN, J., 1843. On the Occurrence of the Remains of Insects in the Upper Lias of the County of Gloucester. Proc. Geol. Soc., vol. iv, pp. 211–12.

BUCKMAN, J., 1853. Remarks on *Libellula Brodiei* (Buckman), a Fossil Insect from the Upper Lias of Dumbleton. Proc. Cots. N.F.C., vol. i, pp. 268–70.

——, 1875. On the Cephalopoda-Bed and the Oolite Sands of Dorset and Part of Somerset. Proc. Somerset Arch. Soc., vol. xx, pp. 140–64.

——, 1877. The Cephalopoda-Bed of Gloucester, Dorset and Somerset. Q.J.G.S., vol. xxxiii, pp. 1–9.

——, 1879. On the so-called Midford Sands. Q.J.G.S., vol. xxxv, pp. 736–43.

BUCKMAN, S. S., 1887. On *Ammonites serpentinus, Am. falcifer, Am. elegans*, &c. Geol. Mag. [3], vol. iv, pp. 396–400.

——, 1889. On the Cotteswold, Midford, and Yeovil Sands, and the Division between the Lias and Oolite. Q.J.G.S., vol. xlv, pp. 440–73.

——, 1890. On the so-called 'Upper Lias' Clay of Down Cliffs. Q.J.G.S., vol. xlvi, pp. 518–21.

——, 1890. On the *Jurense*-Zone. Journ. Northants. N.H.S., vol. vi, pp. 76–80.

——, 1892. The Reported Occurrence of *Ammonites jurensis* in the Northampton Sands. Geol. Mag. [3], vol. ix, pp. 258–60.

——, 1903. The Toarcian of Bredon Hill, and a Comparison with Deposits elsewhere. Q.J G.S., vol. lix, pp. 445–64.

——, 1905. On Certain Genera and Species of *Lytoceratidæ*. Q.J.G.S., vol. lxi, pp. 142–54.

——, 1910. Certain Jurassic (Lias–Oolite) Strata of South Dorset; and their Correlation. Q.J.G.S., vol. lxvi (Upper Lias), pp. 80–9.

——, 1911. Comparison of the Upper Toarcian Beds in Yorkshire and the Cotteswold Hills. Appendix to paper by L. Richardson. Proc. Yorks. Geol. Soc., N.S., vol. xvii, pp. 209–12.

——, 1922. Jurassic Chronology; II. Preliminary Studies. Certain Jurassic Strata near Eypesmouth (Dorset); the Junction Bed of Watton Cliff and Associated Rocks. Q.J.G.S., vol. lxxviii, pp. 378–475.

BURTON, J. J., 1913. The Cleveland Ironstone. Naturalist, 1913, pp. 161–8, 185–94.

CARR, W. D., 1883. The Lincoln Lias [Middle and Upper]. Geol. Mag. [2], vol. x, pp. 164–9.

COOKE, J. H., 1897. On a New Section in the Middle Lias of Lincoln. Geol. Mag. [4], vol. iv, pp. 253–9.

CRICK, G. C., 1920. On some Dibranchiate Cephalopoda from the Upper Lias of Gloucestershire. Proc. Cots. N.F.C., vol. xx, pp. 249–56.

DAY, E. C. H., 1863. On the Middle and Upper Lias of the Dorsetshire Coast. Q.J.G.S., vol. xix, pp. 278–97.

DUNN, J., 1831. On a Large Species of *Plesiosaurus* in the Scarborough Museum [from the Upper Lias near Whitby]. P.G.A., vol. i, pp. 336–7.

GORHAM, A., 1930. The Upper and Lower Junctions of the Midford Sands at Limpley Stoke, south-east of Bath. Geol. Mag., vol. lxvii, pp. 289–97.

FAWELL, J., 1897. Description of two New Species of Gasteropoda from the Upper Lias of Yorkshire. Proc. Yorks. Geol. Soc., vol. xiii, pp. 199, 200.

HINDE, G. J., 1889. On a true Leuconid Calcisponge from the Middle Lias of Northamptonshire. Ann. Mag. Nat. Hist. [6], vol. iv, pp. 352–7.

HULL, E., 1860–1. On the Blenheim Iron-ore and the thickness of the formations below the Great Oolite at Stonesfield. Geologist, vol. iii, pp. 303–5, and Rept. Brit. Assoc. for 1860, pp. 81–3.

HUNTON, L., 1836. Remarks on a Section of the Upper Lias and Marlstone of Yorkshire, showing the limited vertical range of the Species of Ammonites and other Testacea, with their value as Geological Tests. Trans. Geol. Soc. [2], vol. v, pp. 215–21.

JACKSON, J. F., 1922. Sections of the Junction-Bed and Contiguous Deposits. Appendix to Jurassic Chronology, II, by S. S. Buckman. Q.J.G.S., vol. lxxviii, pp. 436–48.

——, 1926. The Junction-Bed of the Middle and Upper Lias on the Dorset Coast. Q.J.G.S., vol. lxxxii, pp. 490–525.

LAMPLUGH, G. W., 1920. Cleveland District (Lias and Oolite) [Iron Ores]; in Lamplugh, Wedd, and Pringle. Spec. Repts. Min. Resources Gt. Brit., vol. xii—Bedded Ores, chs. i and ii, pp. 1–64. Mem. Geol. Surv.

640   BIBLIOGRAPHY

LYCETT, J., 1862. On some Sections of the Upper Lias recently exposed at Nailsworth,
Gloucestershire. Proc. Cots. N.F.C., vol. ii, pp. 155–63.
——, 1862. On the Sands intermediate the Inferior Oolite and Lias of the Cotteswold
Hills compared with a similar Deposit upon the Coast of Yorkshire. Proc. Cots.
N.F.C., vol. ii, pp. 142–9; and Notes on the Ammonites of the Sands. Vol. iii,
pp. 3–10.
MACMILLAN, W. E. F., 1925–31. Yeovilian Ammonites in the Inland Area of the
Yorkshire Moors. Naturalist, 1925, pp. 236, 316; 1926, pp. 51–3; and 1931, p. 345.
MELMORE, S., 1931. A Reptilian Egg from the [?Upper] Lias of Whitby. Proc.
Yorks. Phil. Soc. for 1930, pp. 1–3.
MOORE, C., 1853. On the Palaeontology of the Middle and Upper Lias. Proc. Somer-
set Arch. Soc., vol. iii, pp. 61–76.
——, 1867. On the Middle and Upper Lias of the South-West of England. Proc.
Somerset Arch. Soc., vol. xiii, pp. 119–244.
NEWTON, E. T., 1888. On the Skull, Brain and Auditory Organ of a new Species of
Pterosaurian (Scaphognathus purdoni) from the Upper Lias near Whitby. Proc. Roy.
Soc., vol. xliii, pp. 436–40; Phil. Trans. Roy. Soc., vol. clxxix B, pp. 503–37.
PARKIN, C., 1882. On Jet Mining. Trans. N. Eng. Inst. Mining Eng., vol. xxxi,
pp. 51–8.
PRESTON, H., 1903. On a new Boring at Caythorpe, Lincolnshire [mainly in Middle
and Upper Lias]. Q.J.G.S., vol. lix, pp. 29–32.
PRESTON, H., & TRUEMAN, A. E., 1917. Oolite Grains in the Upper Lias of Grantham.
Naturalist, 1917, p. 217.
RASTALL, R. H., 1905. The Blea Wyke Beds and the Dogger in North-East Yorkshire.
Q.J.G.S., vol. lxi, pp. 441–60.
RICHARDSON, L., 1902. Notes on the Geology of Bredon Hill. Trans. Woolhope
N.F.C. for 1902, p. 65.
——, 1906. On a Section of Middle and Upper Lias Rocks near Evercreech, Somerset.
Geol. Mag. [5], vol. iii, pp. 368–9.
——, 1906. On a Well-Sinking in the Upper Lias at Painswick, near Stroud. Proc.
Cots. N.F.C., vol. xv, p. 208.
——, 1909. On some Middle and Upper Lias Sections near Batcombe, Somerset.
Geol. Mag. [5], vol. vi, pp. 540–2.
——, 1921. Ammonites from the Upper Lias, Railway-cutting, Bloxham, Oxon.
Geol. Mag., vol. lviii, pp. 426–8.
SCHINDEWOLF, O. H., 1928–31. Über Farbstreifen bei Amaltheus (Paltopleuroceras)
spinatus (Brug.). Pal. Zeitschr., vol. x, pp. 136–43; and vol. xiii, pp. 284–7.
SEELEY, H. G., 1865. On Plesiosaurus macropterus, a New Species from the Lias of
Whitby. Ann. Mag. Nat. Hist. [3], vol. xv, pp. 49–53, 232.
——, 1880. Notes on the Skulls of a large Teleosaurus and an Ichthyosaurus from the
Whitby Lias, preserved in the Museum of the University of Cambridge. Q.J.G.S.,
vol. xxxvi, pp. 627–47.
SEWARD, A. C., 1902. The Structure and Origin of Jet. Rept. Brit. Assoc. for 1901,
Glasgow, p. 856.
SMITHE, F., 1865. Geology of Churchdown Hill. Proc. Cots. N.F.C., vol. iii,
pp. 40–9.
——, 1877. On the Middle Lias of North Gloucestershire. The Spinatus Zone. And
on the Occurrence of Plicatula lævigata d'Orb. in the Middle Lias. Proc. Cots.
N.F.C., vol. vi, pp. 341–404.
——, 1895. On the [Middle and Upper] Liassic Zones and Structure of Churchdown
Hill, Gloucester. Proc. Cots. N.F.C., vol. xi, pp. 247–56.
SMITHE, F., & LUCY, W. C., 1892. Some Remarks on the [Middle and Upper] Lias
of Alderton, Gretton and Ashton-under-Hill. Proc. Cots. N.F.C., vol. x, pp. 202–12.
SORBY, H. C., 1857. On the Origin of the Cleveland Hill Ironstone. Proc. Yorks.
Geol. Soc., vol. iii, pp. 457–61.
SPATH, L. F., PRINGLE, J., TEMPLEMAN, A., & BUCKMAN, S. S., 1922. The Upper
Lias Succession near Ilminster, Somerset, and Correlation. In Buckman, Q.J.G.S.,
vol. lxxviii, pp. 449–55.
STRICKLAND, H. E., 1840. On the Occurrence of a Fossil Dragon-fly, from the Lias
of Warwickshire [Æschna liassina]. Ann. Mag. Nat. Hist. [2], vol. iv, pp. 301–3.

TATE, R., 1870. Note on the Middle Lias [fossils from the Drift] in the North-East of Ireland. Q.J.G.S., vol. xxvi, pp. 324–5.

——, 1870. On the Palaeontology of the Junction Beds of the Lower and Middle Lias in Gloucestershire. Q.J.G.S., vol. xxvi, pp. 394–408.

——, 1875. On some New Liassic Fossils [*Ammonites acutus* and gastropods from Middle Lias]. Geol. Mag. [2], vol. ii, pp. 203–6.

TATE, R., BLAKE, J. F., & LECKENBY, J., 1872–3. Notes on the Discovery of the oldest known *Trigonia* (*T. lingonensis* Dum.) in Britain [*Spinatum* zone, Yorks.]. Geol. Mag., vol. ix, p. 306, and vol. x, pp. 135, 186.

THOMPSON, B., 1889. The Middle Lias of Northamptonshire, considered 1, stratigraphically; 2, palaeontologically; 3, economically; 4, as a source of water-supply; 5, as a mitigator of floods. (Reprinted from the 'Midland Naturalist'.) London.

——, 1892. Report of the Committee to work the very Fossiliferous Transition Bed between the Middle and Upper Lias in Northamptonshire. Rept. Brit. Assoc. for 1891, pp. 334–51.

THOMPSON, C., 1909. The Ammonites called *A. serpentinus* [with descriptions and figures of allied forms]. Naturalist, 1909, pp. 214–19.

THORNEYCROFT, W., 1914. Note on the Upper Lias of the Western Islands in Reference to the Iron-ore Deposit therein. Trans. Geol. Soc. Edin., vol. x, p. 196.

TOMES, R. F., 1882. Description of a New Species of Coral from the Middle Lias of Oxfordshire. Q.J.G.S., vol. xxxviii, pp. 95–6.

——, 1886. On the Occurrence of Two Species of Madreporaria in the Upper Lias of Gloucestershire. Geol. Mag. [3], vol. iii, pp. 107–11.

TRUEMAN, A. E., 1918. The [Middle and Upper] Lias of South Lincolnshire. Geol. Mag. [6], vol. v, pp. 103–11.

UPTON, C., 1906. On a Section of Upper Lias at Stroud. Proc. Cots. N.F.C., vol. xv, pp. 201–7.

WALFORD, E. A., 1879. On some Upper and Middle Lias Beds in the Neighbourhood of Banbury. Proc. Warwicks. N.F.C., Supplt. for 1878, pp. 1–23.

——, 1887. Notes on some Polyzoa from the [Middle] Lias. Q.J.G.S., vol. xliii, pp. 632–6.

——, 1894. On Cheilostomatous Bryozoa from the Middle Lias. Q.J.G.S., vol. l, pp. 79–84.

——, 1889. The Lias Ironstone of North Oxfordshire. Banbury.

——, 1902. On some Gaps in the Lias. Q.J.G.S., vol. lviii, pp. 267–78.

WALKER, J. F., 1892. On Liassic Sections near Bridport, Dorsetshire. Geol. Mag. [3], vol. ix, pp. 437–43.

WATSON, D. M. S., 1909. A Preliminary Note on two new Genera of Upper Liassic Plesiosaurs [*Microcleidus* and *Sthenarosaurus*, from Whitby]. Mem. Manchester Lit. Phil. Soc., vol. liv, no. 4, pp. 1–28.

——, 1910. Upper Liassic Reptilia: the Sauropterygia of the Whitby Museum. Mem. Manch. Lit. Phil. Soc., vol. liv, no. 11, pp. 1–13.

WEDD, C. B., & PRINGLE, J., 1920. Marlstone (Middle Lias) Ores of Lincolnshire, Leicestershire, Northamptonshire, Oxfordshire and Warwickshire; in Spec. Repts. Min. Resources Gt. Brit., vol. xii—Bedded Ores, ch. v, pp. 106–40. Mem. Geol. Surv.

WILSON, E., & CRICK, W. D., 1889. The Lias Marlstone of Tilton, Leicestershire, with Palaeontological Notes. Geol. Mag. [3], vol. vi, pp. 296–305 and pp. 337–42.

WITCHELL, E., 1865. On some Sections of the Lias and Sands exposed in the Sewerage Works at Stroud. Proc. Cots. N.F.C., vol. iii, pp. 11–14.

WOODWARD, A. S., 1891. *Pholidophorus germanicus*: An Addition to the Fish Fauna of the Upper Lias of Whitby. Geol. Mag. [3], vol. viii, pp. 545–6.

——, 1896–9. On the Fossil Fishes of the Upper Lias of Whitby. Proc. Yorks. Geol Soc., part 1, vol. xiii, pp. 25–42, 155–70, 325–37, 455–72.

——, 1911. *Euthynotus*: a Fossil Fish from the Upper Lias of Dumbleton, Glos. Proc. Cheltenham Nat. Sci. Soc., N.S., vol. i, part v, pp. 322–3.

——, 1911. On a New Species of *Eryon* from the Upper Lias, Dumbleton Hill. Geol. Mag. [5], vol. viii, pp. 307–11.

——, 1929. Note on a Specimen of the Ganoid Fish, *Lepidotus elvensis* (Blainville), from the Upper Lias near Ilminster. Proc. Somerset Arch. Soc. [4], vol. xv, pp. 91–3, and plate.

WOODWARD, H., 1878. On *Penæus Sharpii*, a Macrurous Decapod Crustacean, from the Upper Lias, Kingsthorpe, near Northampton. Geol. Mag. [2], vol. v, pp. 164–5.

WOODWARD, H. B., 1872–88. Notes on the Midford Sands. Geol. Mag., vol. ix, 1872, pp. 513–15; and [3], vol. v, 1888, pp. 650–1.

——, 1887. Notes on the Geology of Brent Knoll, Somerset. Proc. Bath Nat. Hist. Club, vol. vi, pp. 125–30.

——, 1887. Notes on the Ham Hill Stone. Ibid., pp. 182–4.

——, 1893. On a Bed of Oolitic Iron-ore in the Lias of Raasay. Geol. Mag. [3], vol. x, p. 493.

——, 1914. Notes on the Geology of Raasay. Trans. Geol. Soc. Edin., vol. x, part 2, p. 164.

WRIGHT, T., 1856. The Palaeontological and Stratigraphical Relations of the so-called 'Sands of the Inferior Oolite'. Q.J.G.S., vol. xii, pp. 292–325.

INFERIOR OOLITE SERIES

BARROW, G., 1877. On a new Marine Bed [Eller Beck Bed] in the Lower Oolites of East Yorkshire. Geol. Mag. [2], vol. iv, pp. 552–5.

BATHER, F. A., & STATHER, J. W., 1925. U-shaped Burrows [of *Arenicolites*] on [Lower] Estuarine Sandstone near Blea Wyke. Proc. Yorks. Geol. Soc., vol. xx, pp. 182–99.

BRODIE, P. B., 1851. The Basement Beds of the Inferior Oolite in Gloucestershire. Q.J.G.S., vol. vii, pp. 208–12.

BRODIE, P. B., & STRICKLAND, H. E., 1850. On Certain Beds in the Inferior Oolite near Cheltenham (Brodie) with notes on a section at Leckhampton Hill (Strickland). Q.J.G.S., vol. vi, pp. 239–51.

BRODRICK, H., 1909. Notes on Footprint Casts from the Inferior Oolite near Whitby, Yorkshire. Proc. Liverpool Geol. Soc., vol. x, pp. 327–35; and Rept. Brit. Assoc. for 1908 (Dublin), pp. 707–8.

BUCKMAN, J., 1877. On the Fossil Beds of Bradford Abbas and its Vicinity. Proc. Dorset N.F.C., vol. i, pp. 64–72.

——, 1879. On a Series of Sinistral Gastropods [*Cirrus* spp. from the Inferior Oolite]. Proc. Dorset N.F.C., vol. iii, pp. 135–40.

——, 1880. On the New Genus of Bivalve, *Curvirostrum striatum* [= *Isoarca*, from the *Sowerbyi* zone of the Inferior Oolite, Halfway House, near Yeovil]. Proc. Dorset N.F.C., vol. iv, pp. 102–3.

——, 1881. On the *Trigonia bella* from [the Inferior Oolite of] Eype, near Bridport, Dorset. Proc. Dorset N.F.C., vol. v, pp. 154–6, and plate.

BUCKMAN, S. S., 1878. On the Species of *Astarte* from the Inferior Oolite of the Sherborne District. Proc. Dorset N.F.C., vol. ii, pp. 81–92.

——, 1880. Some New Species of Ammonites from the Inferior Oolite. Proc. Dorset N.F.C., vol. iv, pp. 137–46.

——, 1880. The Brachiopoda from the Inferior Oolite of Dorset and a portion of Somerset. Proc. Dorset N.F.C., vol. iv, pp. 1–52.

——, 1881. A Descriptive Catalogue of some of the Species of Ammonites from the Inferior Oolite of Dorset. Q.J.G.S., vol. xxxvii, pp. 588–608.

——, 1886. Notes on Jurassic [Inferior Oolite] Brachiopoda [*Rhynchonella liostraca* Buck. and *Terebratula euides* Buck.]. Geol. Mag. [3], vol. iii, pp. 217–19.

——, 1886–1907, 1908. A Monograph of the Ammonites of the 'Inferior Oolite Series'; and illustrations of Type Specimens of Inferior Oolite Ammonites (edited by the Secretary; 1908).

——, 1887. The Inferior Oolite between Andoversford and Bourton on the Water. Proc. Cots. N.F.C., vol. ix, pp. 108–35.

——, 1887. Some New Species of Brachiopoda from the Inferior Oolite of the Cotteswolds. Proc. Cots. N.F.C., vol. ix, pp. 38–43.

——, 1889. The Descent of *Sonninia* and *Hammatoceras*. Q.J.G.S., vol. xlv, pp. 651–63.

——, 1890. The Relations of Dundry with the Dorset–Somerset and Cotteswold Areas during part of the Jurassic Period. Proc. Cots. N.F.C., vol. ix, pp. 374–87.

——, 1890. Some New Species of Brachiopoda from the Inferior Oolite of the Cotteswolds. Proc. Cots. N.F.C., vol. ix, pp. 38–43.

BUCKMAN, S. S., 1892. The Sections [of Inferior Oolite] exposed between Andoversford and Chedworth: a Comparison with Similar Srata upon the Banbury Line. Proc. Cots. N.F.C., vol. x, pp. 94–100.

——, 1892. The Morphology of '*Stephanoceras*' *zigzag*. Q.J.G.S., vol. xlviii, pp. 447–52.

——, 1893. The Bajocian of the Sherborne District. Q.J.G.S., vol. xlix, pp. 479–522.

——, 1893. 'The Top of the Inferior Oolite' and a Correlation of 'Inferior Oolite' Deposits. Proc. Dorset N.F.C., vol. xiv, pp. 37–43.

——, 1895. The Bajocian of the Mid-Cotteswolds. Q.J.G.S., vol. li, pp. 388–462.

——, 1897. Deposits of the Bajocian Age in the Northern Cotteswolds; the Cleeve Hill Plateau. Q.J.G.S., vol. liii, pp. 607–29.

——, 1901. The Bajocian and Contiguous Deposits in the North Cotteswolds: the Main Hill-Mass. Q.J.G.S., vol. lvii, pp. 126–55.

——, 1906. A Cotteswold Brachiopod [*Rhynchonella acutiplicata* Brown]. Proc. Cots. N.F.C., vol. xv, pp. 209–14.

——, 1910. Certain Jurassic (Lias–[Inferior] Oolite) Strata of South Dorset, and their Correlation. Q.J.G.S., vol. lxvi, pp. 52–89.

——, 1910. Certain Jurassic (Inferior Oolite) Species of Ammonites and Brachiopoda. Q.J.G.S., vol. lxvi, pp. 90–110.

——, 1912. Ammonites from the Scarborough Limestone. Appendix to paper by L. Richardson, Proc. Yorks. Geol. Soc., N.S., vol. xvii, pp. 205–8.

BUCKMAN, S. S., & WILSON, E., 1896. Dundry Hill: its Upper Portion, or the Beds Marked as Inferior Oolite (g⁵) in the Maps of the Geological Survey. Q.J.G.S., vol. lii, pp. 669–720.

BUNBURY, C. J. F., 1851. On some Fossil Plants from the Jurassic Strata [Lower and Middle Estuarines] of the Yorkshire Coast. Q.J.G.S., vol. vii, pp. 179–94.

CRICK, G. C., 1898. Descriptions of new or imperfectly known Species of *Nautilus* from the Inferior Oolite, preserved in the British Museum. Proc. Malacol. Soc., vol. iii, pp. 117–39.

DAVIDSON, T., 1877. On the Species of Brachiopoda that occur in the Inferior Oolite at Bradford Abbas and its Vicinity. Proc. Dorset N.F.C., vol. i, pp. 73–88.

DUNCAN, P. M., 1886. On a New Species of *Axosmilia* (*A. elongata*) from the Pea Grit of the Inferior Oolite. Geol. Mag. [3], vol. iii, pp. 340–2.

EVANS, J., 1879. On some Fossils from the Northampton Sands. Rept. Brit. Assoc. for 1878, pp. 534–5.

GRAY, J. W., 1925. Leckhampton Hill, Gloucestershire: The Wellington Quarry and other Sections. Proc. Cots. N.F.C., vol. xxii, pp. 33–42.

HALLE, T. G., 1912. On the Fructification of Jurassic Fern Leaves of the *Cladophlebis denticulata*; and *Cloughtonia*, a Problematic Fossil Plant from the Yorkshire [Inferior Oolite]. Arkiv für Botanik, vol. x, nos. 14, 15. Stockholm.

——, 1913. On Upright *Equisetites* Stems in the Oolitic [Lower Estuarine] Sandstone in Yorkshire [Peak]. Geol. Mag. [5], vol. x, pp. 3–7, with note on Stratigraphical Position by P. F. Kendall.

HAWELL, J., 1903. On an Oolitic Plant Bed in North Cleveland. Naturalist, p. 312.

——, 1904. Bajocian Plant Beds of Yorkshire. Proc. Cleveland N.F.C. for 1902, pp. 229–34.

HEPWORTH, E., 1923. The [Lower] Estuarine Series of the Yorkshire Coast. Trans. Leeds Geol. Soc., part xix, pp. 24–8.

HOLL, H. B., 1863. On the Correlation of the several Subdivisions of the Inferior Oolite in the Middle and South of England. Q.J.G.S., vol. xix, pp. 306–17.

HUDLESTON, W. H., 1874. The Yorkshire Oolites. Part I, The Lower Oolites: P.G.A., vol. iii, pp. 283–333.

——, 1887–96. A Monograph of the Inferior Oolite Gasteropoda. Pal. Soc.

IBBETSON, L. L. B., 1847. On three Sections of the Oolitic Formations [chiefly Inferior Oolite] on the Great Western Railway at the West End of Sapperton Tunnel. Rept. Brit. Assoc. for 1846, p. 61.

JACKSON, J. W., 1911. A New Species of *Unio* [*U. kendalli*] from the Yorkshire [Lower] Estuarine Series. Naturalist, 1911, pp. 211–14.

JONES, J., 1860. On *Rhynchonella acuta* and its Affinities. Proc. Cots. N.F.C., vol. ii, pp. 1–8.

KENDALL, P. F., 1908. Reptilian Footprints in the Lower Oolites at Saltwick. Naturalist, p. 384.

KENT, A. U., & WALKER, J. F., 1879. The Finding of *Terebratula morieri* [in the Inferior Oolite] at Bradford Abbas, and on its Occurrence in England. Proc. Dorset N.F.C., vol. iii, pp. 39–47.

LUCY, W. C., 1889. Remarks on the Dapple Bed of the Inferior Oolite at the Horsepools. Proc. Cots. N.F.C., vol. ix, pp. 388–92.

LANE, G. J., 1909. Jurassic Plants from the Cleveland Hills. Proc. Cleveland N.F.C., vol. ii, pp. 172–3, also Naturalist, 1909, p. 81. [Lists from Carlton Bank, Marske and Upleatham.]

——, 1910. Notes on the Jurassic Flora of Cleveland. Proc. Cleveland N.F.C., vol. ii, p. 206.

LANE, G. J., & SAUNDERS, T. W., 1910. Oolitic Plant Remains in Yorkshire. Naturalist, 1910, p. 15.

LUCY, W. C., & TOMES, R. F., 1889. Notes on the Jurassic Rocks at Crickley Hill, with an Amended List of the Madreporaria. Proc. Cots. N.F.C., vol. ix, pp. 289–308.

LYCETT, J., 1850. Tabular View of Fossil Shells from the Middle Division of the Inferior Oolite in Gloucestershire. Ann. Mag. Nat. Hist. [2], vol. vi, and Proc. Cots. N.F.C., vol. i, pp. 62–86.

——, 1853. On some New Species of *Trigonia* from the Inferior Oolite of the Cotteswolds, with Preliminary Remarks on that Genus. Proc. Cots. N.F.C., vol. i, pp. 246–61.

——, 1853. Note on the *Gryphæa* of the Bed called Gryphite Grit in the Cotteswolds. Proc. Cots. N.F.C., vol. i, pp. 235–6.

MANSEL-PLEYDELL, J. C., 1881. Notes on a Cone [*Araucaria cleminshawii*] from the Inferior Oolite Beds of Sherborne. Proc. Dorset, N.F.C., vol. v, pp. 141–3.

MOORE, C., 1855. On New Brachiopoda from the Inferior Oolite of Dundry, &c. Proc. Somerset Arch. Soc., vol. v, pp. 107–28.

MURCHISON, R. I., 1832. On the Occurrence of Stems of Fossil Plants in Vertical Positions in the Sandstone of the Inferior Oolite in the Cleveland Hills. Proc. Geol. Soc., vol. i, p. 391.

MURRAY, P., 1828. Account of a Deposit of Fossil Plants discovered in the Coal Formation near Scarborough [Gristhorpe Plant Bed]. Edin. New. Phil. Journ., vol. v, pp. 311–17.

NEWBITT, T., 1911. Catalogue of Oolite Fossil Plants found in the Whitby Cliffs, together with the Museum Numbers of the Specimens exhibited in the Whitby Museum. Rept. Whitby Lit. Phil. Soc., no. 88, for 1910, pp. 23–6.

PARIS, E. T., 1911. Notes on some Yorkshire Echinoids [from the Bajocian]. Appendix to paper by L. Richardson, Proc. Yorks. Geol. Soc., N.S., vol. xvii, pp. 213–15.

PARIS, E. T., & RICHARDSON, L., 1915. Some Inferior Oolite Pectinidæ. Q.J.G.S., vol. lxxi, pp. 521–35.

RICHARDSON, L., 1902. On the Sequence of the Inferior Oolite Deposits at Bredon Hill, Worcestershire. Geol. Mag. [4], vol. ix, pp. 513–14.

——, 1903. On a Section at Cowley, near Cheltenham, and its Bearing upon the Interpretation of the Bajocian Denudation. Q.J.G.S., vol. lix, pp. 382–9.

——, 1905. Excursion to Leckhampton Hill. Proc. Cots. N.F.C., vol. xv, pp. 182–9.

——, 1907. On the 'Top-Beds' of the Inferior Oolite at Rodborough Hill, near Stroud. Proc. Cots. N.F.C., vol. xvi, pp. 71–80.

——, 1907. The Inferior Oolite and Contiguous Deposits of the Bath–Doulting District; and appendix on their correlation with those of Dorset by S. S. Buckman. Q.J.G.S., vol. lxiii, pp. 383–426.

——, 1907. The Inferior Oolite and Contiguous Deposits of the District between Rissington and Burford. Q.J.G.S., vol. lxiii, pp. 437–44.

——, 1907. On New Species of *Amberleya* and of *Spirorbis* (from the Upper Inferior Oolite). Q.J.G.S., vol. lxiii, pp. 434–6.

——, 1907. On the Stratigraphical Position of the Beds [Inferior Oolite] from which *Prosopon richardsoni* H. Woodward was obtained. Geol. Mag. [5], vol. iv, pp. 82–4.

——, 1908. On the Phyllis Collection of Inferior Oolite Fossils from Doulting. Geol. Mag. [5], vol. v, pp. 509–17.

RICHARDSON, L., 1908. The Geology of Ebrington Hill, North Cotteswolds. Proc. Cots. N.F.C., vol. xvi, pp. 129–34.

——, 1909. Note on *Pollicipes aalensis* Richardson. Proc. Cots. N.F.C., vol. xvi, p. 265.

——, 1910. The Inferior Oolite and Contiguous Deposits of the South Cotteswolds. Proc. Cots. N.F.C., vol. xvii, pp. 63–136.

——, 1910. On the Sections of Inferior Oolite on the Midford-Camerton Section of the Limpley Stoke Railway, Somerset. P.G.A., vol. xxi, pp. 97–100.

——, 1910. On the Stratigraphical Distribution of the Inferior Oolite Vertebrates of the Cotteswold Hills and the Bath–Doulting District. Geol. Mag. [5], vol. vii, pp. 272–4.

——, 1910. Excursion to Stonesfield and Fawler. [Description of Inferior Oolite and Upper and Middle Lias.] Proc. Cots. N.F.C., vol. xvii, pp. 28–31.

——, 1911. The Inferior Oolite and Contiguous Deposits of the Chipping Norton District, Oxfordshire. Proc. Cots. N.F.C., vol. xvii, pp. 195–231.

——, 1911. The Lower Oolitic Rocks of Yorkshire. Proc. Yorks. Geol. Soc., N.S., vol. xvii, pp. 184–204.

——, 1914. The Chronological Succession of the Echinoid Faunas in the Inferior Oolite Rocks of the Stonesfield–Burton Bradstock District. Proc. Cots. N.F.C., vol. xviii, pp. 251–6.

——, 1915. The Inferior Oolite and Contiguous Deposits of the Doulting–Milborne-Port District (Somerset). Q.J.G.S., vol. lxxi, pp. 473–520.

——, 1916. On the Stratigraphical Distribution of the Inferior Oolite Vertebrates of the Cotteswold Hills and the Bath–Burton Bradstock District. Proc. Dorset N.F.C., vol. xxxvii, pp. 48–55.

——, 1918. The Inferior Oolite and Contiguous Deposits of the Crewkerne District, Somerset. Q.J.G.S., vol. lxxiv, pp. 145–73.

——, 1920. Remains of Macrurous and Brachyurous Crustacea in the Inferior Oolite of the Stonesfield (Oxon.)–Burton Bradstock (Dorset) District. Proc. Cots. N.F.C., vol. xx, pp. 243–4.

——, 1922. Certain Aalenian–Vesulian Strata of the Banbury District, Oxfordshire. Proc. Cots. N.F.C., vol. xxi, pp. 109–32.

——, 1923. Certain Jurassic (Aalenian-Vesulian) Strata of Southern Northampton-shire. P.G.A., vol. xxxiv, pp. 97–113.

——, 1926. Certain Jurassic Strata of the Duston Area, Northamptonshire. Proc. Cots. N.F.C., vol. xxii, pp. 137–52.

——, 1928–30. The Inferior Oolite and Contiguous Deposits of the Burton Bradstock –Broad Windsor District, Dorset. Proc. Cots. N.F.C., vol. xxiii, pp. 35–68, 149–86, and 253–64.

——, 1932. The Inferior Oolite and Contiguous Deposits of the Sherborne District, Dorset. Proc. Cots. N.F.C., vol. xxiv, pp. 35–85.

RICHARDSON, L., & PARIS, E. T., 1908–13. On the Stratigraphical and Geographical Distribution of the Inferior Oolite Echinoids of the West of England. Proc. Cots. N.F.C., 1908, vol. xvi, pp. 151–89, and Supplement, ibid. 1913, vol. xviii, pp. 73–82.

RICHARDSON, L., & THACKER, A. G., 1920. On the Stratigraphical and Geographical Distribution of the Sponges of the Inferior Oolite of the West of England. P.G.A., vol. xxxi, pp. 161–86.

RICHARDSON, L., & UPTON, C., 1913. Some Inferior Oolite Brachiopoda. Proc. Cots. N.F.C., vol. xviii, pp. 47–58.

SEWARD, A. C., 1903. On the Occurrence of *Dictyozamites* in [the Lower Estuarine Series of Yorks.], with remarks on European Mesozoic Floras. Q.J.G.S., vol. lix, pp. 217–33.

——, 1906. Jurassic Plants from the [Inferior Oolite] Rocks of East Yorkshire. Rept. Brit. Assoc. for 1905, p. 568; and Geol. Mag. [5], vol. iii, p. 518.

SHARP, S., 1869. Notes on the Northampton Oolites. Geol. Mag., vol. vi, pp. 446–8.

——, 1870–3. The Oolites of Northamptonshire. Q.J.G.S., vol. xxvi, pp. 354–93 (and appendix on a starfish by T. Wright); and ibid., vol. xxix, pp. 225–302.

——, 1874. Sketch of the Geology of Northamptonshire. P.G.A., vol. iii, pp. 243–52.

SKERL, J. G. A., 1926–8. The Petrography of some Jurassic 'Sands' from the North Cotteswolds. Proc. Cots. N.F.C., vol. xxii, pp. 153–60; and 'Rocks', ibid., pp. 281–6, and vol. xxiii, pp. 25–30.

SKERL, J. G. A., 1927. Notes on the Petrography of the Northampton Ironstone. P.G.A., vol. xxxviii, pp. 375–94.

SOLLAS, W. J., 1883. Descriptions of Fossil Sponges from the Inferior Oolite, with a Notice of some from the Great Oolite. Q.J.G.S., vol. xxxix, pp. 541–54.

STODDART, W., 1867. Geology of Dundry Hill. Proc. Bristol Nat. Soc. [2], vol. ii, pp. 29–33.

STODDART, W., 1877. List of Characteristic Fossils of the Dundry Oolite. Proc. Cots. N.F.C., vol. vi, pp. 297–300.

TAWNEY, E. B., 1874. Museum Notes: Dundry Gasteropoda. Proc. Bristol Nat. Soc. [2], vol. i, p. 2.

THOMAS, H. D., 1930. On a New Sponge [*Stelletta*] from the Inferior Oolite of Gloucestershire. Proc. Cots. N.F.C., vol. xxiii, pp. 265–7.

THOMAS, H. H., 1911. On the Spores of some Jurassic Ferns [in the Lower Estuarine Series of Yorkshire]. Proc. Camb. Phil. Soc., vol. xvi, pp. 384–8. Abstracts in Nature, p. 472, and Naturalist, p. 347.

——, 1911–13. Recent Researches on the Jurassic Plants of East Yorkshire [Mainly Inferior Oolite]. Naturalist, 1911, pp. 409–10, and Rept. Brit. Assoc. for 1911 (Portsmouth), pp. 569–70; and ibid. for 1912 (Dundee), pp. 294–5.

——, 1911–22. On some New and Rare Jurassic Plants from [the Lower and Middle Estuarine Series of] Yorkshire:
  (1) *Eretmophyllum*, a new type of Ginkgoalean leaf. Proc. Camb. Phil. Soc., vol. xvii, pp. 256–62.
  (2) The male flower of *Williamsonia gigas*; ibid., vol. xviii, pp. 105–10.
  (3) Fertile specimens of *Dictyophyllum rugosum*; ibid., vol. xxi, pp. 111–16.

——, 1912. *Stachypteris hallei*; a new Jurassic Fern [from the Lower Estuarine Series of Yorkshire]. Proc. Camb. Phil. Soc., vol. xvi, pp. 610–14.

——, 1913. The Jurassic Plant-Beds of Roseberry Topping [Yorkshire]. Naturalist, 1913, p. 198.

——, 1913. The Fossil Flora of the Cleveland District of Yorkshire: (1) The Flora of the Marske Quarry [Lower Estuarine Series]; with notes on the Stratigraphy by G. J. Lane. Q.J.G.S., vol. lxix, pp. 223–51.

——, 1915. The *Thinnfeldia* Leaf-Bed of Roseberry Topping. Naturalist, 1915, p. 7.

——, 1915. On *Williamsoniella*, a New Type of Bennettitalean Flower [from the Middle Estuarine Plant-Bed of Gristhorpe]. Phil. Trans. Roy. Soc., vol. ccvii B, pp. 113–48.

——, 1925. The Caytoniales, a New Group of Angiospermous Plants from the [Middle Estuarine Series] of Yorkshire. Phil. Trans. Roy. Soc., vol. ccxiii B, pp. 299–363.

THOMPSON, B., 1921–8. The Northampton Sand of Northamptonshire. Journ. Northants. N.H.S., 1921–8. Reprinted in book form 1928; Dulau, London.

——, 1924. The Inferior Oolite Sequence in Northamptonshire and Parts of Oxfordshire. P.G.A., vol. xxxv, pp. 67–76.

TOMES, R. F., 1878. A List of the Madreporaria of Crickley Hill, Gloucestershire, with Descriptions of some New Species. Geol. Mag. [2], vol. v, pp. 297–305.

——, 1879. On the Fossil Corals obtained from the [Inferior] Oolite of the Railway Cuttings near Hook Norton, Oxfordshire. P.G.A., vol. vi, pp. 152–65.

——, 1882. On the Madreporaria of the Inferior Oolite of the Neighbourhood of Cheltenham and Gloucester. Q.J.G.S., vol. xxxviii, pp. 409–50.

——, 1886. On some New or Imperfectly Known Madreporaria from the Inferior Oolite of Oxfordshire, Gloucestershire and Dorsetshire. Geol. Mag. [3], vol. iii, pp. 385–98, 443–52.

——, 1889. Notes on an Amended List of Madreporaria of Crickley Hill. Proc. Cots. N.F.C., vol. ix, pp. 300–7.

TONKS, L. H., 1923. Recent Notes on the Dogger Sandstone of the Yorkshire Coast. Trans. Leeds Geol. Soc., part xix, pp. 29–33.

UPTON, C., 1899–1905. Some Cotteswold Brachiopoda. Proc. Cots. N.F.C., vol. xiii, pp. 121–32, and vol. xv, pp. 82–92.

WALFORD, E. A., 1883. On the Relation of the so-called 'Northampton Sand' of North Oxon. to the *Clypeus* Grit. Q.J.G.S., vol. xxxix, pp. 224–45.

WALFORD, E. A., 1889–94. On some Bryozoa from the Inferior Oolite of Shipton Gorge, Dorset. Q.J.G.S., vol. xlv, pp. 561–74 (1889), and vol. l, pp. 72–8 (1894).

WALKER, J. F., 1878. On the Occurrence of *Terebratula Morieri* in England. Geol. Mag. [4], vol. v, p. 552.

WALKER, J. F., & BUCKMAN, S. S., 1889. On the Spinose *Rhynchonellæ* (Genus *Acanthothyris* d'Orbigny) found in England. Rept. Yorks. Phil. Soc. for 1888, pp. 41–57.

WEDD, C. B., 1920. Northampton (Inferior Oolite) Ironstone of Mid and South Lincolnshire, Rutland and Northamptonshire; in Lamplugh, Wedd & Pringle. Spec. Repts. Min. Resources Gt. Brit., vol. xii—Bedded Iron Ores, ch. vi, pp. 141–207. Mem. Geol. Surv.

WETHERED, E. B., 1891. The Inferior Oolite of the Cotteswold Hills, with Special Reference to its Microscopical Structure. Q.J.G.S., vol. xlvii, pp. 550–70.

WETHERED, E. B., & WITCHELL, E., 1886–7. The Pea-Grit of Leckhampton Hill. Geol. Mag. [3], vol. iii, p. 525, and vol. iv, pp. 46–7.

WHIDBORNE, G. F., 1883. Notes on some Fossils, chiefly Mollusca, from the Inferior Oolite. Q.J.G.S., vol. xxxix, pp. 487–540.

WILLIAMSON, W. C., 1855–70. On the Restoration of *Zamites gigas* from the Lower Sandstone and Shale of the Yorkshire Coast. Rept. Brit. Assoc. for 1854, p. 103, and Trans. Linn. Soc., vol. xxvi, 1870, p. 663.

WITCHELL, E., 1880. On a Section of Stroud Hill, and the Upper Ragstone Beds of the Cotteswolds. Proc. Cots. N.F.C., vol. vii, pp. 118–35.

——, 1886. On the Basement Beds of the Inferior Oolite of Gloucestershire. Q.J.G.S., vol. xlii, pp. 264–71.

——, 1886. On the Pisolite and the Basement Beds of the Inferior Oolite of the Cotteswolds. Proc. Cots. N.F.C., vol. viii, pp. 35–49.

——, 1887. The Pea Grit of Leckhampton Hill. Geol. Mag. [3], vol. iv, pp. 46–7.

——, 1890. On the Genus *Nerinæa* and its Stratigraphical Distribution in the Cotteswolds. Proc. Cots. N.F.C., vol. ix, pp. 21–37.

WITHERS, T. H., 1911. On the Occurrence of *Pollicipes* [a Cirripede] in the Inferior Oolite. Proc. Cots. N.F.C., vol. xvii, p. 275 (with list of references).

WOODWARD, H. B., 1908. The New Great Western Branch Railway from Camerton to Limpley Stoke, Somerset. Sum. Prog. Geol. Surv. for 1907, pp. 155–7.

WRIGHT, T., 1860. On the Subdivisions of the Inferior Oolite in the South of England, compared with the Equivalent Beds of that Formation in Yorkshire. Q.J.G.S., vol. xvi, pp. 1–48.

WRIGHT, T., & LYCETT, J., 1853. Contributions to the Palaeontology of Gloucestershire:—On the *Strombidæ* of the Oolites, with the Description of a New and Remarkable *Pteroceras*. Proc. Cots. N.F.C., vol. i, pp. 115–19.

GREAT OOLITE SERIES AND CORNBRASH

ARKELL, W. J., 1931. The Upper Great Oolite, Bradford Beds and Forest Marble of South Oxfordshire, and the Succession of Gastropod Faunas in the Great Oolite. Q.J.G.S., vol. lxxxvii, pp. 563–629.

——, 1933. The Geological Results of a Boring into the Great Oolite at Latton, near Cricklade, Wilts. Proc. Cots. N.F.C., vol. xxv (in the press).

——, 1933. A New Section of the Upper Great Oolite at Bladon, near Woodstock, Oxon. P.G.A., vol. xliv (in the press).

ARKELL, W. J., RICHARDSON, L., & PRINGLE, J., 1933. The Lower Oolites exposed in the Ardley and Fritwell Railway-cuttings, Oxon. P.G.A., vol. xliv (in the press).

BARROW, G., 1908–9. The New Great Western Railway from Ashendon to Aynho, near Banbury. Sum. Prog. Geol. Surv. for 1907, pp. 141–54, and P.G.A., vol. xxi, 1909, pp. 35–45.

BEAN, W., 1836. Description and Figures of *Unio distortus* and *Cypris concentrica* from the Upper Sandstone and Shale of Scarborough. Ann. Mag. Nat. Hist., vol. ix, pp. 376–7.

——, 1839. Catalogue of Fossils in the Cornbrash of Scarborough. Ann. Mag. Nat. Hist., N.S., vol. iii, pp. 57–62.

BLACK, M., 1923. Fossil Plants from the Upper Estuarine Series. Naturalist, p. 57.

BLACK, M., 1928. 'Washouts' in the Estuarine Series of Yorkshire. Geol. Mag., vol. lxv, pp. 301–7.

——, 1929. Drifted Plant Beds of the Upper Estuarine Series of Yorkshire. Q.J.G.S., vol. lxxxv, pp. 389–437.

BLAKE, C. C., 1863. On Chelonian Scutes from the Stonesfield Slate. Geologist, vol. vi, pp. 183–4.

BLAKE, J. F., 1905–7. A Monograph of the Fauna of the Cornbrash. Pal. Soc. [incomplete].

BRODERIP, W. J., & FITTON, W. H., 1828. Observations on the Jaw of a Mammiferous Animal found in the Stonesfield Slate, and on the Strata from whence the Fossil . . . was obtained. Zoological Journal, vol. iii, pp. 408–18.

BRODIE, P. B., & BUCKMAN, J., 1845. On the Stonesfield Slate of the Cotteswold Hills. Q.J.G.S., vol. i, pp. 220–5; and Proc. Geol. Soc., vol. iv, pp. 437–42.

BROWETT, A., 1889. The Bath Oolite and Method of Quarrying it. Midland Naturalist, vol. xii, pp. 187–90.

BUCKLAND, W., 1824. Notice on the *Megalosaurus*, or Great Fossil Lizard of Stonesfield. Trans. Geol. Soc. [2], vol. i, pp. 390–6.

——, 1838. On the Discovery of a Fossil Wing of a Neuropterous Insect in the Stonesfield Slate. Proc. Geol. Soc., vol. ii, p. 688.

BUCKMAN, J., 1853. On the Cornbrash of the Neighbourhood of Cirencester. Proc. Cots. N.F.C., vol. i, pp. 262–7, and Ann. Mag. Nat. Hist. [2], vol. xii, pp. 324–9.

——, 1854. On the Cornbrash of Gloucestershire and part of Wilts. Rept. Brit. Assoc. for 1853, pp. 50–1.

——, 1860. On some Fossil Reptilian Eggs from the Great Oolite of Cirencester. Q.J.G.S., vol. xvi, pp. 107–10.

BUCKMAN, S. S., 1922. Jurassic Chronology, II: Watton Cliff, between Eypesmouth and West Bay. Q.J.G.S., vol. lxxviii, pp. 380–9.

——, 1927. Jurassic Chronology, III: Some Faunal Horizons in Cornbrash. Q.J.G.S., vol. lxxxiii, pp. 1–37.

——, 1927. Troll Quarry, Thornford, Dorset [Fuller's Earth Rock]. T.A., vol. vi, pp. 50–1.

BUTLER, A. G., 1873. On a Fossil Butterfly belonging to the Family *Nymphalidæ*, from the Stonesfield Slate near Oxford. Geol. Mag., vol. x, pp. 2, 3.

CARPENTER, P. H., 1882. On some New or Little-Known Jurassic Crinoids [Great Oolite and Kellaways Rock]. Q.J.G.S., vol. xxxviii, pp. 29–43.

CARRUTHERS, W., 1867. On an Aroideous Fruit from the Stonesfield Slate. Geol. Mag., vol. iv, pp. 146–7.

——, 1871. On some Supposed Vegetable Fossils [Reptilian eggs from Stonesfield]. Q.J.G.S., vol. xxvii, pp. 443–9.

COOKE, J. H., 1896. A Section in the Lower Oolites [Cornbrash and Upper Estuarine Series] of Scarborough. Naturalist, pp. 289–92.

COTTERELL, T. S., 1905. Bath Stone. Journ. Brit. Arch. Assoc., N.S., vol. xi, pp. 49–58.

CUNNINGTON, W., 1860. On the Bradford Clay and its Fossils. Wilts. Arch. N.H. Mag., vol. vi, pp. 1–10.

DOUGLAS, J. A., & ARKELL, W. J., 1928–32. The Stratigraphical Distribution of the Cornbrash. I. The South-Western Area. Q.J.G.S., vol. lxxxiv, pp. 117–78. II. The North-Eastern Area, ibid., vol. lxxxviii, pp. 112–70.

DRAKE, H. C., 1910. *Asteracanthus* in the Yorkshire Cornbrash. Naturalist, pp. 141–2.

FORBES, E., 1851. On the Estuary Beds and the Oxford Clay at Loch Staffin in Skye. Q.J.G.S., vol. vii, pp. 104–13.

GAUDRY, A., 1853. Note sur Stonesfield, près Oxford (Angleterre). Bull. Soc. géol. France [2], vol. x, pp. 591–6.

GOODRICH, E. S., 1894. On the Fossil Mammalia of the Stonesfield Slate. Quart. Journ. Micros. Sci., vol. xxxv, pp. 407–31.

GREGSON, C. S., 1863. On a Fossil Elytron from the Stonesfield Slate. Proc. Liverpool Geol. Soc., Session 3, p. 8.

HARKER, A., 1890. On the Sections in the Forest Marble and Great Oolite formations exposed by the New Railway from Cirencester to Chedworth. Proc. Cots. N.F.C., vol. x, pp. 82–94.

HORTON, W. S., 1860. On the Geology of the Stonesfield Slate and its Associate Formations. The Geologist, vol. iii, pp. 249–58.

HUXLEY, T. H., 1859. On *Rhamphorhynchus bucklandi*, a Pterosaurian from the Stonesfield Slate. Q.J.G.S., vol. xv, pp. 658–70.

JACKSON, J. W., 1909. On the Type-Specimen of *Pseudomelania vittata* (Phillips) [Cornbrash of Scarborough; now at Manchester]. Geol. Mag. [5], vol. vi, pp. 542–3.

——, 1911. On *Unio distortus* Bean and *Alasmodon vetustus* Brown, from the Upper Estuarine Beds of Gristhorpe, Yorks. Naturalist, 1911, pp. 104–7, 119–22.

JONES, T. R., & SHERBORN, C. D., 1888. On some Ostracoda from the Fuller's Earth Oolite and Bradford Clay. Proc. Bath N.H. and Antiq. F.C., vol. vi, pp. 249–78.

JUDD, J. W., 1871. On the Anomalous Mode of Growth of Certain Fossil Oysters [*Ostrea undosa* (Phil.) from the Cornbrash]. Geol. Mag., vol. viii, pp. 355–9.

LANG, W. D., 1907. The Evolution of *Estomatopora dichotomoides* (d'Orbigny) [from the Cornbrash of Scarborough]. Geol. Mag. [5], vol. iv, pp. 20–4.

LECKENBY, J., 1864. On the Sandstones and Shales of the Oolites of Scarborough, with Descriptions of some New Species of Fossil Plants. Q.J.G.S., vol. xx, pp. 74–82.

LEE, G. W., 1913. The Occurrence of Oil-Shale among the Jurassic Rocks [Great Estuarine Series] of Raasay and Skye. Nature, 1913, p. 169.

LYCETT, J., 1848. On the Mineral Character and Fossil Conchology of the Great Oolite, as it occurs in the Neighbourhood of Minchinhampton. Q.J.G.S., vol. iv, pp. 181–91.

——, 1863. A Supplementary Monograph on the Mollusca from the Stonesfield Slate, Great Oolite, Forest Marble, and Cornbrash. Pal. Soc.

MACKIE, S. J., 1863. Turtles in the Stonesfield Slate. Geologist, vol. vi, pp. 41–3.

MANSEL-PLEYDELL, J. C., 1877. Notes on a Gavial Skull [*Steneosaurus stephani*] from the Cornbrash of Closworth. Proc. Dorset N.F.C., vol. i, pp. 28–32.

MORRIS, J., 1857. Descriptions of the Figured Species of [Great Oolite] Fossils; in E. Hull's 'Geology of the Country around Cheltenham', Mem. Geol. Surv., pp. 103–4, and plate.

MORRIS, J., & LYCETT, J., 1850–3. A Monograph of the Mollusca from the Great Oolite, chiefly from Minchinhampton and the Coast of Yorkshire. Pal. Soc.

MURCHISON, R. I., 1843. Observations on the Occurrence of Freshwater Beds in the Oolitic Deposits of Brora, Sutherland, &c. Proc. Geol. Soc., vol. iv, pp. 173–6.

ODLING, M., 1913. The Bathonian Rocks of the Oxford District. Q.J.G.S., vol. lxix, pp. 484–513.

OWEN, R., 1838. Description of the Remains of Marsupial Mammalia from the Stonesfield Slate. Proc. Geol. Soc., vol. iii, pp. 17–21.

——, 1841. A Description of a Portion of the Skeleton of the *Cetiosaurus*, a gigantic extinct Saurian Reptile. Proc. Geol. Soc., vol. iii, pp. 457–62.

——, 1883. On the Skull of *Megalosaurus*. Q.J.G.S., vol. xxxix, pp. 334–47.

PARIS, E. T., 1911. New Lamellibranchs (*Gervillia* and *Perna*) from the Neæran Beds. Proc. Cots. N.F.C., vol. xvii, pp. 233–5.

——, 1911. Notes on some Species of *Gervillia* from the Lower and Middle Jurassic Rocks of Gloucestershire, and from the Scarborough Limestone. Proc. Cots. N.F.C., vol. xvii, pp. 237–56.

PHILLIPS, J., 1860. On some Sections of the Strata near Oxford. I. The Great Oolite in the Valley of the Cherwell. Q.J.G.S., vol. xvi, pp. 115–19.

——, 1866. Oxford Fossils [Dragon Flies from Stonesfield]. Geol. Mag., vol. iii, pp. 97–9.

PLATNAUER, H. M., 1887. On the Occurrence of *Strophodus rigauxi* Sauv. in the Yorkshire Cornbrash. Rept. Yorks. Phil. Soc. for 1886, pp. 36–40.

PRÉVOST, C., 1825. Observations sur les schistes calcaires oolitiques de Stonesfield en Angleterre, dans lesquels ont été trouvés plusieurs ossemens fossiles de mammifères. Ann. Sci. Nat. [1], vol. iv, p. 389 and plates.

PRINGLE, J., 1909–10. On a Boring in the Fullonian and Inferior Oolite at Stowell, Somerset. Sum. Prog. Geol. Surv. for 1908, pp. 83–6, and for 1909, pp. 68–70.

RICHARDSON, L., 1903. Notes on a Section of Great Oolite Beds [Chipping Norton Limestone] at Condicote, near Stow on the Wold. Geol. Mag. [4], vol. x, p. 404.

RICHARDSON, L., 1910. On a Fuller's Earth Section at Combe Hay, near Bath. P.G.A., vol. xxi, pp. 425–8.

——, 1910. The Great Oolite Section at Groves' Quarry, Milton-under-Wychwood. Geol. Mag. [4], vol. vii, pp. 537–42.

——, 1911. On the Sections of Forest Marble and Great Oolite on the Railway between Cirencester and Chedworth, Gloucestershire. P.G.A., vol. xxii, pp. 95–115.

——, 1925. A Boring at Lewis Lane, Cirencester [Great Oolite]. P.G.A., vol. xxxvi, pp. 93–9; and Proc. Cots. N.F.C., vol. xii, p. 53.

——, 1925. Excursion to the North Cotteswolds [Stonesfield Slate, Neæran Beds and Chipping Norton Limestone]. Proc. Cots. N.F.C., vol. xxii, pp. 67–73.

——, 1930. Excursion to the Mid Cotteswolds [Great Oolite and Forest Marble]. Proc. Cots. N.F.C., vol. xxiii, pp. 209–13.

——, 1930. The Fosse Lime and Limestone Quarry and Works [in Great Oolite]. Proc. Cots. N.F.C., vol. xxiii, pp. 269–72.

RICHARDSON, L., & ARKELL, W. J. The Fuller's Earth of England [in preparation].

RICHARDSON, L., & WALKER, J. F., 1907. Remarks on the Brachiopoda from the Fuller's Earth. Q.J.G.S., vol. lxiii, pp. 426–34.

SEELEY, H. G., 1880. On *Rhamphocephalus prestwichi* Seeley, an Ornithosaurian from the Stonesfield Slate of Kineton, Glos. Q.J.G.S., vol. xxxvi, pp. 27–35.

STOPES, M. C., 1907. The Flora of the Inferior [Great] Oolite of Brora, Sutherland. Q.J.G.S., vol. lxiii, pp. 375–82.

TATE, R., 1870. The Fuller's Earth in the South-West of England. Quart. Journ. Sci., vol. vii, pp. 68–71.

TAUNTON, J. A., 1872. Sapperton Tunnel on the Thames and Severn Canal. Proc. Cots. N.F.C., vol. v, pp. 255–70.

THOMPSON, B., 1891. The Oolitic Rocks at Stowe-Nine-Churches. Journ. Northants N.H.S., vol. vi, pp. 295–310.

——, 1924. The Geology of Roade Cutting [near Blisworth, Northants]. Geol. Mag., vol. lxi, pp. 210–18.

——, 1930. The Upper Estuarine Series of Northamptonshire and Northern Oxfordshire. Q.J.G.S., vol. lxxxvi, pp. 430–62.

TOMES, R. F., 1883. On the Fossil Madreporaria of the Great Oolite of the Counties of Gloucester and Oxford. Q.J.G.S., vol. xxxix, pp. 168–96.

——, 1885. On some new or imperfectly known Madreporaria from the Great Oolite of the Counties of Oxford, Gloucester and Somerset. Q.J.G.S., vol. xli, pp. 170–90.

VERSEY, H. C., 1928. Cornbrash at Kepwick, N.E. Yorkshire. Naturalist, pp. 117–18.

VINE, G. R., 1892. Notes on the Polyzoa, *Stromatopora* and *Proboscina* Groups, from the Cornbrash of Thrapston, Northants. Proc. Yorks. Geol. Soc., N.S., vol. xii, pp. 247–58.

WALFORD, E. A., 1883. On some Crinoidal and other Beds in the Great Oolite of Gloucestershire, and their probable equivalents in North Oxfordshire. Proc. Warwicks. N.F.C. for 1882, pp. 20–7.

——, 1895–7. Stonesfield Slate. Report of the Committee appointed to open further sections in the Neighbourhood of Stonesfield in order to show the relationship of the Stonesfield Slate to the underlying and overlying strata. Rept. Brit. Assoc. for 1894 (Oxford), pp. 304–6; 1895 (Ipswich), p. 415; 1896 (Liverpool), p. 356.

——, 1906. On some New Oolitic Strata in North Oxfordshire. Buckingham (and abstract in Q.J.G.S., vol. lxi, 1905, p. 440).

——, 1917. The Lower Oolite of North Oxfordshire. Banbury.

WELCH, F. B. A., 1926. On the Further Extension of the Great Oolite in the North Cotteswolds. Proc. Cots. N.F.C., vol. xxii, pp. 219–38.

WESTWOOD, J. O., 1854. Fossil Beetle and Dragon-fly from the Stonesfield Slate; in 'Contributions to Fossil Entomology', Q.J.G.S., vol. x, pp. 379–81.

WHITEAVES, J. F., 1861. On the Invertebrate Fauna of the Lower Oolites of Oxfordshire. Rept. Brit. Assoc. for 1860 (Oxford). Trans. Sect., pp. 104–8.

WINWOOD, H. H., 1913. Well Sinkings on Lansdown, Bath [in Great Oolite]. Proc. Cots. N.F.C., vol. xviii, pp. 83–7.

WITCHELL, E., 1875. On a Bed of Fuller's Earth at Whiteshill, near Stroud. Proc. Cots. N.F.C., vol. vi, pp. 144–6.

WITCHELL, E., 1886. On the Forest Marble and Upper Beds of the Great Oolite between Nailsworth and Wotton-under-Edge. Proc. Cots. N.F.C., vol. viii, pp. 265-80.

WOOD, H. H., 1877. Notes on some Cornbrash Sections in Dorset. Proc. Cots. N.F.C., vol. i, pp. 22-7.

WOODWARD, A. S., 1910. On a Skull of *Megalosaurus* from the Great Oolite of Minchinhampton (Gloucestershire). Q.J.G.S., vol. lxvi, pp. 111-15.

WOODWARD, H., 1866. On the oldest known British Crab (*Palæinachus longipes*) from the Forest Marble, Malmesbury, Wilts. Q.J.G.S., vol. xxii, pp. 493-4.

——, 1868. On a New Brachyurous Crustacean (*Prosopon mammillatum*) from the Great Oolite, Stonesfield. Geol. Mag., vol. v, pp. 3-5.

——, 1890. On a New British Isopod (*Cyclosphæroma trilobatum*) from the Great Oolite of Northampton. Geol. Mag. [3], vol. vii, pp. 529-33.

WOODWARD, H. B., 1887. Note on some Pits near Chipping Norton, Oxon. Essex Naturalist, vol. i, pp. 265-6.

——, 1888-9. The Relations of the Great Oolite to the Forest Marble and Fuller's Earth in the South-West of England. Geol. Mag. [3], vol. v, pp. 467-8, and Rept. Brit. Assoc. for 1888, pp. 651-2.

——, 1902. Further Notes on the Cuttings along the South Wales Direct Railway [Forest Marble]. Sum. Prog. Geol. Surv. for 1901, pp. 59-60.

WRIGHT, T., 1882. On a New Species of Star Fish from the Forest Marble of Road, Wilts. (*Uraster spinigera* Wright). Proc. Cots. N.F.C., vol. viii, pp. 50-2.

OXFORD CLAY AND CORALLIAN BEDS

ANDREWS, C. W., 1895. Note on a Skeleton of a young Plesiosaur from the Oxford Clay of Peterborough. Geol. Mag. [4], vol. ii, pp. 241-3.

——, 1895. On the Development of the Shoulder-Girdle of a Plesiosaur (*Cryptoclidus oxoniensis* Phil.) from the Oxford Clay. Ann. Mag. Nat. Hist. [6], vol. xv, pp. 333-46.

——, 1909. On some New Steneosaurs from the Oxford Clay of Peterborough. Ann. Mag. Nat. Hist. [8], vol. iii, pp. 299-308.

——, 1910-13. A Descriptive Catalogue of the Marine Reptiles of the Oxford Clay, parts 1 and 2. Brit. Mus., London.

——, 1915. Note on a Mounted Skeleton of *Opthalmosaurus icenicus* Seeley [ex Leeds coll., Oxford Clay, Peterborough]. Geol. Mag. [6], vol. ii, pp. 145-6.

ARKELL, W. J., 1926. Studies in the Corallian Lamellibranch Fauna of Oxford, Berks. and Wilts. Part I—Limidæ; Part II—Pectinidæ. Geol. Mag., vol. lxiii, pp. 193-210 and 534-55.

——, 1926. The Corallian Period. Appendix to Chapter by W. J. Sollas in 'The Natural History of the Oxford District', edited by J. J. Walker. Oxford.

——, 1927. The Corallian Rocks of Oxford, Berks., and North Wilts. Phil. Trans. Roy. Soc., vol. ccxvi B, pp. 67-181.

——, 1927. The Red Down Boring, Highworth, and its Geological Significance; with notes on Neighbouring Wells. Wilts. Arch. Nat. Hist. Mag., vol. xliv, pp. 43-8.

——, 1928. Aspects of the Ecology of Certain Fossil Coral Reefs. Journ. of Ecology, vol. xvi, pp. 134-49.

——, 1929- . A Monograph of the British Corallian Lamellibranchia. Pal. Soc.

——, 1931. The Age of the *Natica* Band in the Corallian of Cumnor, near Oxford; and Descriptions of two New Sections at Cothill, Berks. P.G.A., vol. xlii, pp. 44-9.

——, 1932. An Unknown Kellaways Locality in Dorset? [The West Fleet near Langton Herring]. Geol. Mag., vol. lxix, pp. 44-5.

BAILY, W. H., 1860. Description of a New Pentacrinite from the Oxford Clay of Weymouth, Dorset. Ann. Mag. Nat. Hist. [3], vol. vi, pp. 25, 152.

BLAKE, J. F., & HUDLESTON, W. H., 1877. The Corallian Rocks of England. Q.J.G.S., vol. xxxiii, pp. 260-405.

BRINKMANN, R., 1929. Statistisch-biostratigraphische Untersuchungen an Mitteljurassischen Ammoniten über Artbegriff und Stammesentwicklung. [Kosmoceratids from the Oxford Clay of Peterborough.] Abh. Gesell. Wissensch. zu Göttingen, M.P. Klasse, N.F., vol. xiii, 3, pp. 1-250. [Review in Geol. Mag., vol. lxviii, 1931, pp. 373-6.]

BRINKMANN, R., 1929. Monographie der Gattung *Kosmoceras*. Abh. Gesell. Wissensch. Göttingen, M.P. Klasse, N.F., vol. xiii, 4, pp. 1–123.

BUCKMAN, J., 1878. On some Slabs of *Trigonia clavellata* from Osmington Mills, Dorset. Proc. Dorset N.F C., vol. ii, pp. 19–20 and plate.

BUCKMAN, S. S., 1913. The Kelloway Rock of Scarborough. Q.J.G.S., vol. lxix, pp. 152–68.

——, 1924–7. Oxford Clay and Corallian Chronology. T.A., vol. v, 1924–5, pp. 34–55, 62–73; and vol. vi, 1927, p. 49.

CAMERON, A. C. G., 1888. The Clays of Bedfordshire. P.G.A., vol. x, pp. 446–54.

——, 1890. Note on the Recent Exposures of Kellaways Rock at Bedford. Rept. Brit. Assoc. for 1889, pp. 577–8.

——, 1892. On the Continuity of the Kellaways Beds over extended Areas near Bedford. Geol. Mag. [3], vol. ix, pp. 66–71.

CARTER, J., 1886. On the Decapod Crustaceans of the Oxford Clay. Q.J.G.S., vol. xlii, pp. 542–59.

CHADWICK, S., 1886. *Asteracanthus ornatissimus* in the Corallian Oolite near Malton. Naturalist, 1886, p. 102.

COBBOLD, E. S., 1880. Notes on Strata exposed in laying out the Oxford Sewage Farm at Sandford-on-Thames. Q.J.G.S., vol. xxxvi, pp. 314–20.

CRICK, W. D., 1887. Note on some Foraminifera from the Oxford Clay at Keyston, near Thrapston. Journ. Northants N.H.S., vol. iv, p. 232.

CUNNINGTON, W., 1847. On the Fossil Cephalopoda from the Oxford Clay constituting the genus *Belemnoteuthis* (Pearce). Lond. Geol. Journ., pp. 97–9.

DAVIES, A. M., 1907. The Kimeridge Clay and Corallian Rocks of the Neighbourhood of Brill (Buckinghamshire). Q.J.G.S., vol lxiii, pp. 29–49.

——, 1916. The Zones of the Oxford and Ampthill Clays in Buckinghamshire and Bedfordshire. Geol. Mag., vol. liii, pp. 395–400.

——, 1926. An Oolitic Rhaxella Chert from Little Hayes, South-East Essex. Geol. Mag., vol. lxiii, pp. 273–4.

DOUVILLÉ, R., 1912. Études sur les Cardioceratidés de Dives, Villers-sur-Mer et quelques autres gisements. Mem. Soc. géol. France, vol. xix, no. 45, pp. 1–77.

DRAKE, H C., 1907. Remains of *Gyrodus* from the Coral Rag of East Yorkshire. Trans. Hull Field Nat. Club, vol. iii, p. 290.

——, 1909. Remains of a Chimæroid Fish from the Coral Rag of North Grimston. Naturalist, 1909, p. 196.

——, 1911. *Asteracanthus* in the Coralline Oolite [at Seamer, Yorks.]. Naturalist, 1911, p. 130.

FOX-STRANGWAYS, C., 1897. Filey Bay and Brigg. Proc. Yorks. Geol. Soc., N.S., vol. xiii, pp. 338–45.

HARKER, A., 1884. On a Remarkable Exposure of the Kellaways Rock in a Cutting [at South Cerney] near Cirencester. Proc. Cots. N.F.C., vol. viii, pp. 176–87.

HEALEY, M., 1904. Notes on Upper Jurassic Ammonites, with Special Reference to Specimens in the University Museum, Oxford. [Mainly Corallian, 1 Kimeridge.] Q.J.G.S., vol. lx, pp. 54–64.

HINDE, G. J., 1890. On a new Genus of Siliceous Sponges from the Lower Calcareous Grit of Yorkshire [*Rhaxella perforata*]. Q.J.G.S., vol. xlvi, pp. 54–61.

HUDLESTON, W. H., 1876–8. The Yorkshire Oolites [Middle Oolites: Oxford Clay and Corallian]. P.G.A., vol. iv, pp. 353–410 (1876), vol. v, pp. 407–94 (1878).

——, 1880–1. Contributions to the Palaeontology of the Yorkshire Oolites; Corallian Gastropoda. Geol. Mag. [2], vol. vii, pp. 241–8, 289–98, 391–404, 481–8, 529–38, and vol. viii, pp. 49–59, 119–31.

——, 1885. The Geology of Malton and Neighbourhood. Ann. Rept. Malton Nat. Soc., no. 2, for 1884–5, pp. 5–30.

JACKSON, J. W., 1911. *Strophodus* teeth in the Corallian Beds of Malton. Naturalist, 1911, p. 151.

KEEPING, W., 1878. On *Pelanechinus*, a new Genus of Sea-Urchins from the Coral Rag. Q.J.G.S., vol. xxxiv, pp. 924–30.

KEEPING, W., & MIDDLEMISS, C. S., 1883. On some New Railway Sections and other Rock Exposures in the District of Cave, Yorkshire. Geol. Mag. [2], vol. x, pp. 215–21.

KENDALL, P. F., 1915. On an Artesian Well at Oswaldkirk [in Corallian and Oxford Clay]. Proc. Yorks. Geol. Soc., N.S., vol. xix, pp. 284–5.

LECKENBY, J., 1859. On the Kelloway Rock of the Yorkshire Coast. Q.J.G.S., vol. xv, pp. 4–15.

LEEDS, E. T., 1908. On *Metriorhynchus brachyrhynchus* (Deslong.) from the Oxford Clay near Peterborough. Q.J.G.S., vol. lxiv, pp. 345–57.

LYDEKKER, R., 1888. On the Skeleton of a Sauropterygian from the Oxford Clay near Bedford; Q.J.G.S., vol. xliv, pp. 89, 90; and on others: Geol. Mag. [3], vol. v, pp. 350–6.

——, 1890. On a Crocodilian Jaw from Oxford Clay of Peterborough; and on Ornithosaurian Remains from the same. Q.J.G.S., vol. xlvi, pp. 284–8, 429–31.

——, 1893. On the Jaw of a New Carnivorous Dinosaur from [the Oxford Clay of] Peterborough. Q.J.G.S., vol. xlix, pp. 284–7.

——, 1899. Note on a Fossil Crocodile [*Steneosaurus*] from [the Oxford Clay of] Chickerell, Dorset. Proc. Dorset N.F.C., vol. xx, pp. 171–3.

MANTELL, G. A., 1848. Observations on some Belemnites and other Fossil Remains of Cephalopoda, discovered by Mr. R. N. Mantell in the Oxford Clay near Trowbridge, Wiltshire. Phil. Trans. Roy. Soc., vol. cxxxviii B, pp. 171–83.

MANTELL, R. N., 1850. An Account of the Strata and Organic Remains exposed in the Cuttings of the Branch Railway from the G.W. line near Chippenham, through Trowbridge to Westbury in Wiltshire. Q.J.G.S., vol. vi, pp. 310–15.

MARCH, M. C., 1911. Studies in the Morphogenesis of Certain Pelecypoda: (3) The Ornament of *Trigonia clavellata* and some of its Derivatives. Mem. Manchester Lit. Phil. Soc., vol. lv, no. 15, pp. 1–13.

M'COY, F., 1848. On some new Mesozoic Radiata [from the Corallian]. Ann. Mag. Nat. Hist. [2], vol. ii, pp. 397–420.

MORRIS, J., 1845. On the Occurrence of the Genus *Pollicipes* in the Oxford Clay [of Christian Malford]. Ann. Mag. Nat. Hist., vol. xv, pp. 30–1.

——, 1850. List of Organic Remains [and descriptions of new Oxford Clay and Kellaways Mollusca] obtained by R. N. Mantell, Esq., from the railway cuttings near Chippenham, through Trowbridge to Westbury, in Wiltshire. Q.J.G.S., vol. vi, pp. 315–19.

NEAVERSON, E., 1925. The Zones of the Oxford Clay near Peterborough. P.G.A., vol. xxxvi, pp. 27–37.

NEWTON, E. T., 1878. Notes on a Crocodilian Jaw from the Corallian Rocks [Sandsfoot Grit] of Weymouth. Q.J.G.S., vol. xxxiv, pp. 398–400.

——, 1907. Note on Specimens of 'Rhaxella Chert' or 'Arngrove Stone' from Dartford Heath. P.G.A., vol. xx, p. 127.

OWEN, R., 1844. A Description of certain Belemnites, preserved with a great proportion of their soft parts in the Oxford Clay at Christian Malford, Wilts. Phil. Trans. Roy. Soc., vol. cxxxiv B, pp. 65–85.

PRATT, S. P., 1841. Description of some New Species of Ammonites found in the Oxford Clay on the Line of the Great Western Railway, near Christian Malford. Ann. Mag. Nat. Hist., N.S., vol. viii, pp. 161–5.

PRESTWICH, J., 1876. The Thickness of the Oxford Clay [at Oxford]. Geol. Mag. [2], vol. iii, pp. 237–9.

PRINGLE, J., 1920. Wiltshire [Westbury Iron Ore]; in Lamplugh, Wedd and Pringle, Spec. Repts. Min. Resources Gt. Brit., vol. xii—Bedded Iron Ores, pp. 216–20. Mem. Geol. Surv.

RICHARDSON, L., 1922. A Boring [in Oxford Clay and Kellaways Beds] at Calcutt, near Cricklade, Wiltshire. Geol. Mag., vol. lix, pp. 354–5.

ROBERTS, T., 1889. The Upper Jurassic Clays of Lincolnshire. Q.J.G.S., vol. xlv, pp. 545–60.

——, 1892. The Jurassic Rocks of the Neighbourhood of Cambridge. Sedgwick Prize Essay for 1886, Cambridge.

SALFELD, H., 1909. Die Beziehung zwischen Oxford Clay and Kellaway Beds (Rocks). Geol. Abt. Naturhistor. Gesellsch. Hannover, vol. ii, pp. 65–8.

——, 1914. Über einige stratigraphisch wichtige und einige seltene Arten der Gattung *Perisphinctes* aus dem oberen Jura Nordwestdeutschlands [*P. antecedens*

654   BIBLIOGRAPHY

&c. figured]. Jahresbericht des Niedersächsischen geologischen Vereins zu Hannover (Geol. Abt. Naturhistor. Gesellsch. Hannover), 1914.

SALFELD, H., 1917. Monographie der Gattung *Ringsteadia* (gen. nov.) Palaeontographica, vol. lxiii, pp. 69–84.

SEELEY, H. G., 1861. Notice of the Elsworth and other New Rocks in the Oxford Clay, and of the Bluntisham Clay above them. The Geologist, vol. iv, pp. 460–1.

——, 1861. On the Fen Clay Formation. Ann. Mag. Nat. Hist. [3], vol. viii, pp. 503–4.

——, 1862. Notes on Cambridge Geology: 1. Preliminary Notice of the Elsworth Rock and Associated Strata. Ann. Mag. Nat. Hist. [3], vol. x, pp. 97–110.

——, 1874. On *Murænosaurus leedsii* and on *Opthalmosaurus*, from the Oxford Clay. Q.J.G.S., vol. xxx, pp. 197–208, 696–707.

——, 1875. On the Femur of *Cryptosaurus eumerus* Seeley, a Dinosaur from the Oxford Clay of Great Gransden. Q.J.G.S., vol. xxxi, pp. 149–51.

——, 1893. On *Omosaurus phillipsi* (Seeley) [from the Corallian near Slingsby, Yorks.]. Ann. Rept. Yorks. Phil. Soc. for 1892, pp. 52–7.

SHEPPARD, G., 1913. The Kellaways Rock of South Cave, East Yorkshire. Naturalist, 1913, pp. 359–61.

SHEPPARD, T., 1900. Notes on some Remains of *Cryptocleidus* from the Kellaways Rock of East Yorkshire. Geol. Mag. [4], vol. vii, pp. 535–8, and Trans. Hull. Geol. Soc., vol. v, pp. 23–4.

——, 1914. Teeth of *Diplopodia versipora* [from the Corallian of Seamer, Yorks.]. Naturalist, 1914, p. 144.

SHERBORN, C. D., 1888. Note on *Webbina irregularis* from the Oxford Clay of Weymouth. Proc. Bath N.H. and A.F.C., vol. vi, pp. 332–3 and plate.

SORBY, H. C., 1851. On the Microscopical Structure of the Calcareous Grit of the Yorkshire Coast. Q.J.G.S., vol. vii, pp. 1–6.

SPATH, L. F., 1926. Notes on Yorkshire Ammonites, no. x: On some Post-Liassic Ammonites and a New Species of *Bonarellia* [principally Kellaways Beds and Oxford Clay]. Naturalist, 1926, pp. 321–6.

——, 1929. *Ammonites* [*Peltomorphites*] *williamsoni* Phillips, and some Allied Forms [from the Yorkshire Corallian]. Naturalist, 1929, pp. 293–8.

TEISSEYRE, L., 1889. Über *Proplanulites* n. g. Neues Jahrb. für Min., &c., B.-B. vi, pp. 148–76.

TILLYARD, R. J., 1923. *Tarsophlebiopsis mayi*, a Dragon-fly, found in the Body-chamber of a Corallian Ammonite. Geol. Mag., vol. lx, pp. 146–52 and plate.

TOMES, R. F., 1883. On some new or imperfectly known Madreporaria from the Coral Rag and Portland Oolite of the Counties of Wilts., Oxford, Cambridge, and York. Q.J.G.S., vol. xxxix, pp. 555–65.

——, 1900. Contributions to a History of the Mesozoic Corals of the County of York. Proc. Yorks. Geol. Soc., xiv, pp. 72–85.

WALKER, J. F., 1888. On the Occurrence of *Terebratula gesneri* in [the Corallian of] Yorkshire. Ann. Rept. Yorks. Phil. Soc. for 1887, pp. 33–4, and plate.

WEDD, C. B., 1898. On the Corallian Rocks of Upware, Cambridgeshire. Q.J.G.S., vol. liv, pp. 601–19.

——, 1901. On the Corallian Rocks of St. Ives (Huntingdonshire) and Elsworth. Q.J.G.S., vol. lvii, pp. 73–85.

WHITEAVES, J. F., 1861. On the Palaeontology of the Coralline Oolites of the Neighbourhood of Oxford. Ann. Mag. Nat. Hist. [3], vol. viii, pp. 142–7.

WILSON, V., 1932. A Borehole Section in the Upper Jurassic [mainly Corallian] at Irton, near Scarborough. Trans. Leeds Geol. Assoc., vol. v, part 1, pp. 20–2.

WOODWARD, A. S., 1897. On a New Specimen of the Mesozoic Ganoid Fish, *Pholidophorus*, from the Oxford Clay of Weymouth. Proc. Dorset F.C., vol. xviii, p. 150.

WRIGHT, T., 1882–7. On a New Species of Brittle Star from the [Sandsfoot Grit] of Weymouth [*Ophiurella nereida* Wright]. 1882, Proc. Cots. N.F.C., vol. viii, pp. 53–5; and 1887, Geol. Mag. [3], vol. iv, pp. 97–8.

KIMERIDGE CLAY AND PORTLAND BEDS

ANDREWS, C. W., 1921. On a New Chelonian [*Tholemys passmorei*] from the Kimmeridge Clay of Swindon. Ann. Mag. Nat. Hist. [9], vol. vii, pp. 145–53.

ANDREWS, W. R., 1881. Outline of the Geology [Portland and Purbeck Beds] of the Vale of Wardour. Proc. Dorset N.F.C., vol. v, pp. 57–68.

BAILEY, E. B., & WEIR, J., 1932. Submarine Faults as Boundaries of Facies [Kimeridgian Boulder Beds of Sutherland]. Rept. Brit. Assoc. for 1931 (London), pp. 375–6.

BATHER, F. A., 1911. Note on Crinoid Plates from the Penshurst Boring [*Saccocoma* from the Kimeridge Clay]. Sum. Prog. Geol. Surv. for 1910, pp. 78–9.

——, 1916. A Cidarid from the Hartwell Clay. Geol. Mag. [6], vol. iii, pp. 302–4.

BLAKE, J. F., 1875. On the Kimmeridge Clay of England. Q.J.G.S., vol. xxxi, pp. 196–237.

——, 1880. On the Portland Rocks of England. Q.J.G.S., vol. xxxvi, pp. 189–236.

——, 1880. The Portland Building Stone. Pop. Sci. Rev., N.S., vol. iv, pp. 205–12.

——, 1902. On a Remarkable Inlier in the [Kimeridgian] Rocks of Sutherland, and its Bearing on the Origin of the Breccia-Beds. [The 'Fallen Stack'.] Q.J.G.S., vol. lviii, pp. 290–312.

BRODIE, P. B., 1853. Notice of the Occurrence of an Elytron of a Coleopterous Insect in the Kimmeridge Clay at Ringstead Bay, Dorset. Q.J.G.S., vol. ix, pp. 51–2.

BRODIE, W. R., 1876. Notes on the Kimmeridge Clay of the Isle of Purbeck. P.G.A., vol. iv, pp. 517–18.

BUCKMAN, S. S., 1922–6. Kimeridge-Portland Chronology. T.A., vol. iv, 1922–3, pp. 26–45; vol. vi, 1926, pp. 9–16, 24–40.

CHATWIN, C. P., & PRINGLE, J., 1922. The Zones of the Kimmeridge and Portland Rocks at Swindon. Sum. Prog. Geol. Surv. for 1921, pp. 162–8.

COX, L. R., 1925. The Fauna of the Basal Shell-Bed of the Portland Stone, Isle of Portland. Proc. Dorset N.F.C., vol. xlvi, pp. 113–72. (See also Q.J.G.S., 1925, pp. cxxvii–viii.)

——, 1929–30. A Synopsis of the Lamellibranchia and Gastropoda of the Portland Beds of England. Proc. Dorset N.F.C., vol. l, pp. 131–202.

DANFORD, C. G., 1906. Notes on the Belemnites of the Speeton Clays. Trans. Hull Geol. Soc., vol. vi, part i, pp. 1–14.

——, 1907. Notes on Speeton Ammonites. Proc. Yorks. Geol. Soc., vol. xvi, pp. 101–14.

DAVIES, A. M., 1899. Contributions to the Geology of the Thame Valley [Kimeridge Clay, Portland and Purbeck Beds]. P.G.A., vol. xvi, pp. 15–58.

DAVIES, W., 1876. On the . . . large Reptile, *Omosaurus armatus* Owen, from the Kimmeridge Clay of Swindon, Wilts. Geol. Mag. [2], vol. iii, pp. 193–7.

DOUVILLÉ, R., 1909. Sur l'*Ammonites mutabilis* Sow. et sur les genres *Proplanulites* Teisseyre et *Pictonia* Bayle. Bull. Soc. géol. France [4], vol. ix, pp. 234–48.

FITTON, W. H., 1836. Observations on some of the Strata between the Chalk and the Oxford Oolite in the South-East of England. Trans. Geol. Soc. [2], vol. iv, pp. 103–388.

GODWIN-AUSTEN, R. A. C., 1850. On the Age and Position of the Fossiliferous Sands and Gravels of Farringdon. Q.J.G.S., vol. vi, pp. 454–78. [Includes descriptions of the Portland Beds of Swindon and Bourton.]

GRAY, W., 1862. On the Geology of the Isle of Portland. P.G.A., vol. i, pp. 128–47.

GRIFFITHS, J., 1921. Description of Portland Stone Quarries. The Quarry, 1921, pp 217–22.

GROVES, T. B., 1887. The Abbotsbury Iron Deposits. Proc. Dorset N.F.C., vol. viii, pp. 64–6.

HUDLESTON, W. H., 1881. Note on some Gastropoda from the Portland Rocks of the Vale of Wardour and of Bucks. Geol. Mag. [2], vol. viii, pp. 385–95.

——, 1883. On the Geology of the Vale of Wardour. [Portland and Purbeck Beds.] P.G.A., vol. vii, pp. 161–85.

HULKE, J. W., 1869–72. Notes on Saurians from Kimeridge Bay. Q.J.G.S., vol. xxv, pp. 386–400, vol. xxvi, pp. 167–74, 611–22, vol. xxvii, pp. 440–3, vol. xxviii, pp. 34–5.

HULKE, J. W., 1880. *Iguanodon Prestwichii*, a New Species from the Kimeridge Clay . . . at Cumnor. Q.J.G.S., vol. xxxvi, pp. 433–56 [and see Seeley, 1888, Rept. Brit. Assoc. for 1887, p. 698].

ILOVAÏSKY, D., 1923–4. *Pavlovia*, un nouveau genre d'Ammonites. Bull. Soc. des Naturalistes de Moscou, sect. géol., vol. ii (no. 4), pp. 329–60.

JUDD, J. W., 1868. On the Speeton Clay. Q.J.G.S., vol. xxiv, pp. 250–5.

KITCHIN, F. L., 1926. A new Genus of Lamellibranchs (*Hartwellia*, gen. nov.) from the Upper Kimmeridge Clay of England, with a Note on the Position of the Hartwell Clay. Ann. Mag. Nat. Hist. [9], vol. xviii, pp. 433–55.

LAMPLUGH, G. W., 1896. On the Speeton Series in Yorkshire and Lincolnshire. Q.J.G.S., vol. lii, pp. 179–220.

——, 1924. A Review of the Speeton Clays. Proc. Yorks. Geol. Soc., N.S., vol. xx, pp. 1–31.

LATTER, M. P., 1926. The Petrology of the Portland Sand of Dorset. P.G.A., vol. xxxvii, pp. 73–91.

MACGREGOR, M., 1916. A Jurassic Shore-line [in the Kimeridgian of Sutherland]. Trans. Geol. Soc. Glasgow, vol. xvi, pp. 75–85.

MANSEL-PLEYDELL, J. C., 1890. Memoir upon a New Ichthyopterygian from the Kimmeridge Clay of Gillingham, Dorset, *Opthalmosaurus pleydelli*. Proc. Dorset N.F.C., vol. xi, pp. 7–15.

——, 1892. Kimmeridge Coal-Money and other Manufactured Articles from the Kimmeridge Shale. Proc. Dorset N.F.C., vol. xiii, pp. 178–90.

——, 1894. Kimmeridge Shale. Proc. Dorset N.F.C., vol. xv, pp. 172–83.

MASON, J. W., 1869. On *Dakosaurus* from the Kimmeridge Clay of Shotover Hill. Q.J.G.S., vol. xxv, pp. 218–20.

MELMORE, S., 1931. Plesiosauridæ from the Kimeridge Clay of Yorkshire. Naturalist, 1931, pp. 337–8.

MORRIS, J., 1849. On *Neritoma*, a Fossil Genus of Gasteropodous Mollusks allied to *Nerita* [from the Portland Limestone of Swindon]. Q.J.G.S., vol. v, pp. 332–5.

NEAVERSON, E., 1924. The Zonal Nomenclature of the Upper Kimmeridge Clay. Geol. Mag., vol. lxi, pp. 145–51.

——, 1925. The Petrography of the Upper Kimmeridge Clay and Portland Sand in Dorset, Wiltshire, Oxfordshire and Buckinghamshire. P.G.A., vol. xxxvi, pp. 240–56.

——, 1925. Ammonites from the Upper Kimmeridge Clay. University Press, Liverpool.

NEWTON, E. T., 1881. Notes on the Mandible of an *Ischyodus townesendii* found at Upwey, Dorset, in the Portland Oolite. P.G.A., vol. vii, pp. 116–19.

OWEN, R., 1861–89. Monograph on the Reptilia of the Kimmeridge Clay and Portland Stone. Pal. Soc.

PAVLOW, A., 1889. Études sur les couches jurass. et crétacées de la Russie: Jurassique sup. et crét. inf. de la Russie et d'Angleterre; et suppl. Bull. Soc. Nat. Moscou, N.S., vol. iii.

——, 1896. On the Classification of the Strata between the Kimeridgian and Aptian. Q.J.G.S., vol. lii, pp. 542–55.

PAVLOW, A., & LAMPLUGH, G. W., 1895. Argiles de Speeton et leurs équivalents. Bull. Soc. Nat. Moscou, N.S., vol. v, p. 181.

PHILLIPS, J., 1860. Notice on some Sections of the Strata near Oxford. 2: Sections South of Oxford. Q.J.G.S., vol. xvi, pp. 307–11.

PRESTWICH, J., 1879–80. Note on the Occurrence of a New Species of Iguanodon in a brickpit of the Kimeridge Clay at Cumnor Hurst, three miles W.S.W. of Oxford. Q.J.G.S., vol. xxxvi, pp. 430–2, and Geol. Mag. [2], vol. vi, pp. 193–5.

PRINGLE, J., 1919. Palaeontological Notes on the Donnington Borehole [Lincs.] of 1917. [Kimeridge Clay.] Sum. Prog. Geol. Surv. for 1918, pp. 50–2.

——, 1920. Dorset [Abbotsbury Iron Ore]; in Lamplugh, Wedd and Pringle, Spec. Repts. Min. Resources Gt. Brit., vol. xii—Bedded Iron Ores, pp. 221–3.

PRUVOST, P., 1925. Les subdivisions du Portlandian boulonnais d'après les Ammonites. Ann. Soc. géol. du Nord, vol. xliv, pp. 187–215.

RAISIN, C. A., 1903. The Formation of Chert and its Micro-Structures in some Jurassic Strata. P.G.A., vol. xviii, pp. 71–82.

SALFELD, H., 1915. Monographie der Gattung *Cardioceras*, Neumayr und Uhlig, Teil 1. Die Cardioceraten des oberen Oxford und Kimmeridge. Zeitschr. Deutsch. Geol. Gesell., vol. lxvii, pp. 149–204.

SEELEY, H. G., 1871. On a New Species of *Plesiosaurus* from the Portland Limestone. Ann. Mag. Nat. Hist. [4], vol. viii, p. 181.

——, 1888. On *Cumnoria*, an Iguanodont Genus founded upon the *Iguanodon prestwichi* Hulke [from the Kimeridge Clay of Cumnor]. Rept. Brit. Assoc. for 1887, p. 698.

SEWARD, A. C., 1911. The Jurassic [mainly Kimmeridgian] Flora of Sutherland. Trans. Roy. Soc. Edinburgh, vol. xlvii, pp. 643–709.

——, 1912. A Petrified *Williamsonia* from [the Kimeridgian of Ethie], Scotland. Phil. Trans. Roy. Soc., vol. cciii B, pp. 101–26.

SEWARD, A. C., & BANCROFT, T. N., 1913. Jurassic [Kimeridgian] Plants from Cromarty and Sutherland, Scotland. Trans. Roy. Soc. Edin., vol. xlviii, pp. 867–88.

SHEPPARD, T., 1902. Note on a Nearly Complete Specimen of *Ichthyosaurus thyreospondylus* from the Kimeridge Clay of Speeton. Geol. Mag. [4], vol. ix, p. 427, and Hull Museum Publications, no. 10, pp. 1–7.

SPATH, L. F., 1924. On the Ammonites of the Speeton Clay and the Subdivisions of the Neocomian. Geol. Mag., vol. lxi, pp. 73–89.

——, 1925. Ammonites and Aptychi [Kimeridgian] from Somaliland [with a zonal table of the Kimeridgian and Portlandian]. Mon. Hunt. Mus. Glasgow, vol. i, part vii, pp. 111–64.

STRAHAN, A., 1918. Kimmeridge Oil-Shale; Spec. Repts. Min. Resources Gt. Brit., vol. vii, ch. 2, pp. 24–40. Mem. Geol. Surv.

WOODWARD, A. S., 1890. On a Head of *Eurycormus* from the Kimmeridge Clay of Ely. Geol. Mag. [3], vol. vii, pp. 289–92.

——, 1890. On a New Species of Pycnodont Fish (*Mesodon damoni*) from the Portland Oolite. Geol. Mag. [3], vol. vii, pp. 158–9.

——, 1892. On some Teeth of New Chimæroid Fishes from the Oxford and Kimmeridge Clays of England. Ann. Mag. Nat. Hist. [6], vol. x, pp. 13–16 and 94–6.

——, 1895. Note on Megalosaurian Teeth discovered by Mr. J. Alstone in the Portlandian of Aylesbury. P.G.A., vol. xiv, pp. 31–2.

——, 1906. On a New Chimæroid Fin-Spine from the Portland Stone. Proc. Dorset N.F.C., vol. xxvii, pp. 181–2 and plate.

——, 1906. On a Pycnodont Fish of the Genus *Mesodon* from the Portland Stone. Proc. Dorset N.F.C., vol. xxvii, pp. 183–7 and plate.

WOODWARD, H., 1876. On some New Macrurous Crustacea from the Kimmeridge Clay of the Sub-Wealden Boring, Sussex. Q.J.G.S., vol. xxxii, pp. 47–50.

WOODWARD, H. B., 1888. Note on the Portland Sands of Swindon and elsewhere. Geol. Mag. [3], vol. v, pp. 469–70.

PURBECK BEDS

ANDERSON, F. W., 1932. Phasal Deposition in the Middle Purbeck Beds of Dorset. Rept. Brit. Assoc. for 1931, pp. 379–80.

ANDREWS, W. R., & JUKES-BROWNE, A. J., 1894. The Purbeck Beds of the Vale of Wardour. Q.J.G.S., vol. l, pp. 44–69.

BRODIE, P. B., 1839. A Notice on the Discovery of the Remains of Insects and a new genus of Isopodous Crustacea . . . in the [Purbeck] Formation of the Vale of Wardour, Wilts. Proc. Geol. Soc., vol. iii, pp. 134–5.

——, 1847. Notice of the Existence of Purbeck Strata with Remains of Insects and other Fossils, at Swindon, Wilts. Q.J.G.S., vol. iii, pp. 53–4.

——, 1854. On the Insect Beds of the Purbeck Formation in Wilts. and Dorset. Q.J.G.S., vol. x, pp. 475–82.

——, 1867. On the Presence of the Purbeck Beds at Brill, in Buckinghamshire, &c. Q.J.G.S., vol. xxiii, pp. 197–9.

CHAPMAN, F., 1896. Notes on the Microzoa from the Jurassic Beds [Kimeridge Clay and Purbeck Beds] at Hartwell. P.G.A., vol. xv, pp. 96–7.

——, 1900. Remarks upon the Ostracoda [chiefly Lower Purbeck forms from the Thame Valley]. P.G.A., vol. xvi, p. 58.

CHAPMAN, F., 1906. Note on an Ostracodal Limestone from Durlston Bay, Dorset. P.G.A., vol. xix, pp. 283–5.

DAVIES, W., 1887. On New Species of *Pholidophorus* from the Purbeck Beds of Dorset. Geol. Mag. [3], vol. iv, pp. 337–9.

ETHERIDGE, R., 1881. On a New Species of *Trigonia* from the Purbeck Beds of the Vale of Wardour. Q.J.G.S., vol. xxxvii, pp. 246–53.

FALCONER, H., 1857. Description of two Species of the Fossil Mammalian Genus *Plagiaulax* from Purbeck. Q.J.G.S., vol. xiii, pp. 261–82.

——, 1862. On the Disputed Affinity of the Mammalian Genus *Plagiaulax*. Q.J.G.S., vol. xviii, pp. 348–69.

FISHER, O., 1856. On the Purbeck Strata of Dorsetshire. Trans. Cambridge Phil. Soc., vol. ix, pp. 555–81.

FITTON, W. H., 1835. Notice on the Junction of the Portland and Purbeck Strata on the Coast of Dorsetshire. Proc. Geol. Soc., vol. ii, pp. 185–7.

FORBES, E., 1851. On the Succession of Strata and Distribution of Organic Remains in the Dorsetshire Purbecks. Rept. Brit. Assoc. for 1850, pp. 79–81, and Edinburgh Phil. Journ., vol. xlix, pp. 311–13, 391.

HAWKINS, H. L., 1925. [Note on the finding of thirty-eight tests of *Hemicidaris purbeckensis* in Durlston Bay.] Q.J.G.S., vol. lxxxi, pp. cxxviii–ix.

HULKE, J. W., 1878. Note on two Skulls from the Wealden and Purbeck Formations indicating a new Sub-group of Crocodilia. Q.J.G.S., vol. xxxiv, pp. 377–82.

JONES, T. R., 1878. Notes on some Fossil Bivalved Entomostraca: iv [Purbeck and Wealden Ostracoda]. Geol. Mag. [2], vol. v, pp. 103–10, 277–8.

——, 1885. On the Ostracoda of the Purbeck Formation. Q.J.G.S., vol. xli, pp. 311–53.

——, 1890. On some Fossil *Estheriæ*: B. Purbeck *Estheriæ*. Geol. Mag. [3], vol. vii, pp. 385–90.

KOERT, W., 1898. Geol. u. pal. Untersuch. der Grenzschichten zwischen Jura u. Kreide auf der Südwestseite des Selter. [Successful application of the Ostracod zones to the N.W. German Purbecks and comparison with England, with list of fossils common to the two.] Inaug.-Dissert. Göttingen, pp. 1–57.

LYDEKKER, R., 1889. On Certain Chelonian Remains from the Wealden and Purbeck. Q.J.G.S., vol. xlv, pp. 511–18.

LYDEKKER, R., & BOULENGER, G. A., 1887. Notes on Chelonia from the Purbeck, Wealden and London Clay. Geol. Mag. [3], vol. iv, pp. 270–5.

MAILLARD, G., 1884. Invertébrés du Purbeckien du Jura. [Descriptions of many English Purbeck fossils and detailed comparisons of the strata.] Mém. Soc. pal. Suisse, vol. xi.

MANSEL-PLEYDELL, J. C., 1885. A Fossil Chelonian Reptile [*Pleurosternon ovatum* Owen] from the Middle Purbecks. Proc. Dorset N.F.C., vol. vi, pp. 66–9 and plate.

——, 1896. On the Footprints of a Dinosaur (*Iguanodon?*) from the Purbeck Beds of Swanage. Proc. Dorset N.F.C., vol. xvii, p. 115.

MERRETT, E. A., 1924. Ostracoda found in the Purbeck Beds of the Vale of Aylesbury (Fossil Ostracoda in Stratigraphy). Geol. Mag., vol. lxi, pp. 233–8.

OWEN, R., 1854. On some Fossil Reptilian and Mammalian Remains from the Purbecks. Q.J.G.S., vol. x, pp. 420–33.

——, 1879. On the Association of Dwarf Crocodiles with the Diminutive Mammals of the Purbeck Shales. Q.J.G.S., vol. xxxv, pp. 148–55.

PHILLIPS, J., 1858. On the Estuary Sands in the Upper Part of Shotover Hill. Q.J.G.S., vol. xiv, pp. 236–41.

SEELEY, H. G., 1869. Note on the *Pterodactylus macrurus* (Seeley) a New Species from the Purbeck Limestone. . . . Proc. Camb. Phil. Soc., parts vii–x, p. 130.

——, 1875. On an Ornithosaurian (*Doratorhynchus validus*) from the Purbeck Limestone of Langton, near Swanage. Q.J.G.S., vol. xxxi, pp. 465–8.

——, 1893. On a Reptilian Tooth with two Roots [*Nuthetes*, from the Purbeck Beds]. Ann. Mag. Nat. Hist. [6], vol. xii, pp. 227–30, 274–6.

SEWARD, A. C., 1897. On *Cycadeoidea gigantea*, a New Cycadean Stem from the Purbeck Beds of Portland. Q.J.G.S., vol. liii, pp. 22–39.

WEBSTER, T., 1826. Observations on the Purbeck and Portland Beds. Trans. Geol. Soc. [2], vol. ii, pp. 37–44.

WESTON, C. H., 1852. On the Sub-escarpments of the Ridgway Range, and their Contemporaneous Deposits in the Isle of Portland. Q.J.G.S., vol. viii, pp. 110–20.

WESTWOOD, J. O., 1854. Fossil Insects from the Middle and Lower Purbeck Beds of Dorset; in 'Contributions to Fossil Entomology'. Q.J.G.S., vol. x, pp. 382–96.

WETHERED, E., 1892. On the Occurrence of Fossil Forms of the Genus *Chara* in the Middle Purbeck Strata of Lulworth, Dorset. Proc. Cots. N.F.C., vol. x, pp. 101–4.

WILLETT, H., & E. W., 1881. Notes on a Mammalian Jaw from the Purbeck Beds at Swanage, Dorset. Q.J.G.S., vol. xxxvii, pp. 376–80.

WOOD, S. V., 1863. On the Events which produced and terminated the Purbeck and Wealden Deposits of England and France, and on the Geographical Conditions of the Basin in which they were accumulated. Phil. Mag. [4], vol. xxv, pp. 268–89.

WOODWARD, A. S., 1890. On some New Fishes from the English Wealden and Purbeck Beds, referable to the Genera *Oligopleurus*, *Strobilodus* and *Mesodon*. Proc. Zool. Soc., pp. 346–53.

——, 1895. A Contribution to Knowledge of the Fossil Fish Fauna of the English Purbeck Beds. Geol. Mag. [4], vol. ii, pp. 145–52, 401–2.

——, 1909. Note on a Chelonian Skull from the Purbeck Beds of Swanage. Proc. Dorset N.F.C., vol. xxx, pp. 143–4, and plate.

——, 1912. Notes on a Maxilla of *Triconodon* from the Middle Purbeck Beds of Swanage. P.G.A., vol. xxiii, pp. 100–1 and plate.

——, 1916–19. The Fossil Fishes of the English Wealden and Purbeck Formations. Pal. Soc.

YOUNG, J. T., 1878. On the Occurrence of a Freshwater Sponge in the Purbeck Limestone. Geol. Mag. [2], vol. v, pp. 220–1.

EXCURSIONS OF THE GEOLOGISTS' ASSOCIATION

in the course of which Jurassic rocks were studied; arranged geographically.
Those especially important marked *.

Isle of Purbeck: Portland–Purbeck Beds, HUDLESTON, W. H., 1882, vol. vii, pp. 377–90.

*Swanage, Corfe Castle, Kimeridge, &c. Kimeridge Clay—Purbeck Beds, HUDLES-- TON, W. H., MANSEL, O. L., and MONCKTON, H. W., 1896, vol. xiv, pp. 307–24.

*Swanage and Lulworth Cove. Kimeridge Clay–Purbeck Beds, HOVENDEN, F., MONCKTON, H. W., and WOODWARD, A. S., 1910, vol. xxi, pp. 510–21.

Weymouth. Oxford Clay–Purbeck Beds, HUDLESTON, W. H., 1889, vol. xi, pp. xlix–lvii.

Weymouth and Portland: Corallian–Portlandian, BLAKE, J. F., and HUDLESTON, W. H., 1879, vol. vi, pp. 172–4.

Bridport and Weymouth: Middle Lias–Portland Beds, BLAKE, J. F., HUDLESTON, W. H., BUCKMAN, S. S., and MONCKTON, H. W., 1898, vol. xv, pp. 293–304.

Bridport, Bothenhampton, Burton Bradstock and Eype: Middle Lias–Forest Marble, WOODWARD, H. B., 1885, vol. ix, pp. 200–9.

*Sherborne and Bridport: Upper Lias–Inferior Oolite, HUDLESTON, W. H., 1885, vol. ix, pp. 187–99.

*Bridport, Beaminster and Crewkerne: Upper Lias–Inferior Oolite, RICHARDSON, L., 1915, vol. xxvi, pp. 47–78.

Lyme Regis: Rhætics–Upper Lias, WOODWARD, H. B., 1889, vol. xi, pp. xxvi–xlix.

Lyme Regis: Rhætic Beds and Lower Lias, WOODWARD, H. B., 1906, vol. xix, pp. 320–40.

*Charmouth and Lyme Regis: Lower and Middle Lias, LANG, W. D., 1915, vol. xxvi, pp. 111–18.

West Somerset and North Devon: Rhætics at Watchet, HERRIES, R. S., 1896, vol. xiv, pp. 433–40.

*West Somerset: Rhætics and Lower Lias of Minehead coast, RICHARDSON, L., 1914, vol. xxv, pp. 97–102.

Yeovil District: Middle Lias to Inferior Oolite, LOBLEY, J. L., 1871, vol. ii, pp. 247–50.

*Dunball, Burlescombe, Ilminster, Chard, Ham Hill and Bradford Abbas; Rhætic Beds–Fuller's Earth, RICHARDSON, L., USSHER, W. A. E., and others, 1911, vol. xxii, pp. 246–63.

*Frome District, Somerset: Lower Lias–Fuller's Earth, RICHARDSON, L., 1909, vol. xxi, pp. 209–28.

Dundry Hill: Lower Lias–Inferior Oolite, WILSON, E., 1893, vol. xiii, pp. 128–32.

Dundry Hill: Lower Lias–Inferior Oolite, BUCKMAN, S. S., 1901, vol. xvii pp. 152–8.

Bristol District: Rhætics and Lower Lias of Radstock, TUTCHER, J. W., 1919, vol. xxx, pp. 116–18.

Aust Cliff: Rhætic Beds, WINWOOD, H. H., 1901, vol. xvii, pp. 163–6.

Bath: Rhætic–Upper Lias, WINWOOD, H. H., 1872, vol. iii, pp. 89–92.

Bath: General, MOORE, C., 1879, vol. vi, pp. 196–201.

Bath and Midford: Inferior Oolite–Fuller's Earth, WOODWARD, H. B., 1893, vol. xiii, pp. 125–8.

*Bradford-on-Avon: Great Oolite–Forest Marble, WINWOOD, H. H., and WICKES, W. H., 1893, vol. xiii, pp. 132–7.

*Chippenham, Calne, Kellaways and Corsham: Great Oolite–Corallian, WOODWARD, H. B., and WINWOOD, H. H., 1896, vol. xiv, pp. 339–54.

Westbury, Wilts.: Corallian and Kimeridge Clay, WOODWARD, H. B., 1893, vol. xiii, pp. 137–40.

Mere and Maiden Bradley in Wiltshire: Corallian of the Vale of Wardour, BARTLETT, B. POPE, 1916, vol. xxvii, pp. 127–32.

Salisbury, Stonehenge and Vale of Wardour: Portland–Purbeck, HUDLESTON, W. H., 1881, vol. vii, pp. 134–42.

Teffont Evias, Chilmark and Tisbury (Vale of Wardour): Portland–Purbeck Beds, ANDREWS, W. R., 1903, vol. xviii, pp. 149–58.

New G.W.R. Line from Wootton Bassett to Filton: Rhætics–Oxford Clay, WINWOOD, H. H., 1901, vol. xvii, pp. 144–50.

*Mid and South Cotswolds: Lower Lias–Inferior Oolite, RICHARDSON, L., 1908, vol. xx, pp. 514–29.

Cheltenham District: Rhætic (also Inferior Oolite mentioned), LOBLEY, J. L., 1874, vol. iv, pp. 167–74.

Cheltenham and Stroud: Lower Lias–Inferior Oolite, BUCKMAN, S. S., 1897, vol. xv, pp. 175–82.

*Vale of Evesham and the North Cotswolds: Lower Lias–Inferior Oolite, RICHARDSON, L., 1904, vol. xviii, pp. 391–408.

Cirencester and Minchinhampton: Middle Lias–Kellaways Beds, HARKER, A., 1888, vol. x, pp. 157–63.

Cirencester and District: Great Oolite, RICHARDSON, L., 1925, vol. xxxvi, pp. 80–92.

Swindon: Portland Beds, BLAKE, J. F., 1892, vol. xii, pp. 326–7.

*Swindon: Kimeridge Clay and Portland Beds, GORE, C. H., CHATWIN, C. P., and PRINGLE, J., 1922, vol. xxxiii, pp. 152–5.

Swindon and the Vale of White Horse: Corallian–Portland Beds, HAWKINS, H. L., and PRINGLE, J., 1923, vol. xxxiv, pp. 233–41.

*Swindon and Faringdon: Kimeridge Clay–Purbeck Beds, HUDLESTON, W. H., 1876, vol. iv, pp. 543–54.

Faringdon: Corallian, TREACHER, L., 1907, vol. xx, pp. 115–21.

Faringdon, [Marcham] and Abingdon: Corallian, HINDE, G. J., and WOODWARD, H. B., 1892, vol. xii, pp. 327–33.

Culham and Abingdon: Corallian–Kimeridge Clay, TREACHER, L., 1908, vol. xx, pp. 548–52.

Culham and Wallingford: Kimeridge Clay, WHITE, H. J. O., 1903, vol. xviii, pp. 300–6.

Corallian Rocks South-West of Oxford: Corallian, ARKELL, W. J., 1931, vol. xlii, pp. 50–2.

Oxford: Corallian–Kimeridge Clay, BADCOCK, P., 1862, vol. i, pp. 155–7.

Oxford: Middle Lias–Portland, HUDLESTON, W. H., and PARKER, J., 1874, vol. iv, pp. 91–7.

*Oxford: Corallian, HUDLESTON, W. H., 1880, vol. vi, pp. 338–44.

Oxford District: Middle Lias–Portland Beds, PRINGLE, J., 1926, vol. xxxvii, pp. 447–58.

Headington, Shotover and Wheatley: Corallian–Portland Beds, BLAKE, J. F., 1902, vol. xvii, pp. 383–5.

Wheatley and Arngrove: Oxford Clay, Corallian and Ampthill Clay, DAVIES, A. M., 1909, vol. xxi, pp. 234–6.

*Dorton, Brill and Arngrove: Oxford, Ampthill and Kimeridge Clays, DAVIES, A. M., 1907, vol. xx, pp. 183–6.

Brill [and Blackthorn Hill]: Forest Marble–Purbeck Beds, BLAKE, J. F., 1893, vol. xiii, pp. 71–4.

Ashendon and Dorton Railway Cuttings: Oxford and Ampthill Clays, DAVIES, A. M., 1909, vol. xxi, pp. 394–5.

New Railway from Bicester to Aynho: Inferior Oolite–Cornbrash, BARROW, G., 1908, vol. xxi, pp. 36–45.

Thame District: Portland–Purbeck, DAVIES, A. M., 1899, vol. xvi, pp. 157–9.

Aylesbury: Kimeridge Clay–Purbeck Beds, MORRIS, J., 1873, vol. iii, pp. 210–11.

*Aylesbury: Kimeridge Clay–Purbeck Beds, HUDLESTON, W. H., 1880, vol. vi, pp. 344–52.

Aylesbury: Kimeridge Clay–Purbeck Beds, HUDLESTON, W. H., 1888, vol. x, pp. 166–72.

Aylesbury, Hartwell and Stone: Kimeridge Clay–Purbeck Beds, DAVIES, A. M., and EMARY, P., 1897, vol. xv, pp. 90–4.

Aylesbury, Hartwell and Stone: Kimeridge Clay and Portland Beds, DAVIES, A. M., 1912, vol. xxiii, pp. 254–7.

Whitchurch, Oving and Quainton: Oxford Clay–Purbeck Beds, DAVIES, A. M., 1897, vol. xv, pp. 207–9.

Bedford and Clapham: Cornbrash–Oxford Clay, CAMERON, A. C. G., 1889, vol. x, pp. 504–10.

Bedford: Great Oolite–Kellaways Beds, WOODWARD, H. B., 1905, vol. xix, pp. 142–6.

Cambridge and Ely: Kimeridge Clay, MARR, J. E., and LEIGHTON, T., 1894, vol. xiii, pp. 292–5.

*Peterborough: Oxford Clay (with list and bibliography of Oxford Clay Vertebrata in the Leeds collection), LEEDS, A. N., and WOODWARD, A. S., 1897, vol. xv, pp. 188–93.

Oxford University Museum, Enslow Bridge, Kirtlington and Woodstock: Great Oolite, ALLORGE, M. M., and BAYZAND, C. J., 1910, vol. xxii, pp. 1–5.

Oxford, Stonesfield and Fawler: Middle Lias–Oxford Clay, BAYZAND, C. J., 1908, vol. xxi, pp. 25–9.

*Chipping Norton: Lower Lias to Great Oolite, HUDLESTON, W. H., 1878, vol. v, pp. 378–89.

*Banbury: Middle Lias–Great Oolite, BEESLEY, T., 1873, vol. iii, pp. 197–204.

Banbury: Middle Lias–Great Oolite, WALFORD, E. A., 1914, vol. xxv, pp. 71–7.

Banbury, Bloxham, Edge Hill and Hook Norton: Middle Lias–Inferior Oolite, WALFORD, E. A., 1895, vol. xiv, pp. 177–85.

*Banbury and Towcester Districts: Upper Lias–Great Oolite, RICHARDSON, L., 1921, vol. xxxii, pp. 109–22.

*Northants: Marlstone–Cornbrash (especially Great Oolite): THOMPSON, B., and CRICK, W. D., 1891, vol. xii, pp. 172–90.

*Northamptonshire: Upper Lias–Great Oolite, THOMPSON, B., 1921, vol. xxxii, pp. 219–26.

*New Railway at Catesby, Northants: Lower–Upper Lias, THOMPSON, B., 1896, vol. xiv, pp. 421–30.

Kettering and Thrapston: (Brief general remarks), BLAKE, J. F., and THOMPSON, B., 1900, vol. xvi, pp. 516–17.

*Wellingborough: Upper Lias–Great Oolite, THOMPSON, B., and CRICK, W. D., 1894, vol. xiii, pp. 283–91.

Stamford, Collyweston and Ketton: Inferior Oolite, THOMPSON, B., 1906, vol. xix, pp. 367–70.

Weldon, Dene and Gretton: Inferior Oolite, THOMPSON, B., 1899, vol. xvi, pp. 226–31.

Leicestershire: Lower Lias–Lincolnshire Limestone, HARRISON, W. J., 1877, vol. v, pp. 142–6.
Nottingham District: The Vale of Belvoir: Rhætics–Upper Lias, SWINNERTON, H. H., 1914, vol. xxv, pp. 84–5.
Mid-Lincolnshire: Middle Lias–Inferior Oolite, also Kimeridge Clay, PRESTON, H., KENDALL, P. F., and CARTER, W. L., 1905, vol. xix, pp. 114–30.
Lincoln: Lower, Middle and Upper Lias, CARR, W. D., 1884, vol. viii, pp. 383–5.
Lincoln: Middle and Upper Lias and Kimeridge Clay, DALTON, W. H., and others, 1885, vol. viii, pp. 383–9.
East Coast of Yorkshire: General, BLAKE, J. F., and others, 1891, vol. xii, pp. 207–22.
Yorkshire Coast: Lower, Middle and Upper Lias (higher beds briefly mentioned), HERRIES, R. S., 1906, vol. xix, pp. 464–77.
*East Sutherland: Lower Lias–Kimeridgian, MACGREGOR, M., READ, H. H., MANSON, W., and PRINGLE, J., 1930, vol. xli, pp. 63–86.

PALAEONTOLOGY

Some palaeontological works not classifiable in the stratigraphical section.

[*Note.*—This selection is added for the sake of usefulness and not with any idea of completeness.]

VERTEBRATA

WOODWARD, A. S., & SHERBORN, C. D., 1890. A Catalogue of British Fossil Vertebrata. London.

MAMMALIA

LYDEKKER, R., 1885–7. Catalogue of the Fossil Mammalia in the British Museum. 5 parts. London.
OSBORN, H. F., 1888. On the Structure and Classification of the Mesozoic Mammalia. Journ. Acad. Nat. Sci. Philad., vol. ix, p. 282.
OWEN, R., 1871. Monograph of the Fossil Mammalia of the Mesozoic Formations. Pal. Soc.
SIMPSON, G. G., 1928. A Catalogue [Monograph] of the Mesozoic Mammalia in the Geological Dept. of the British Museum.

REPTILIA AND AMPHIBIA

BOULENGER, G. A., 1891. On British Remains of *Homœosaurus*, with Remarks on the Classification of the Rhynchocephalia. Proc. Zool. Soc., pp. 167–72.
LYDEKKER, R., 1888. Note on the Classification of the Ichthyopterygia, with a Notice of Two New Species, Geol. Mag. [3], vol. v, pp. 309–14.
——, 1888–90. Catalogue of the Fossil Reptilia and Amphibia in the British Museum. 8vo. London.
——, 1889. On the Remains and Affinities of five Genera of Mesozoic Reptiles. Q.J.G.S., vol. xlv, pp. 41–58.
——, 1890. Contributions to our Knowledge of the Dinosaurs of the Wealden and the Sauropterygians of the Purbeck and Oxford Clay. Q.J.G.S., vol. xlvi, pp. 37–53.
——, 1904. Vertebrate Palaeontology of Cambridgeshire; In Marr and Shipley's Handbook to the Natural History of Cambridgeshire. Pp. 51–70. Cambridge.
MANSEL-PLEYDELL, J. C., 1888. Fossil Reptiles of Dorset. Proc. Dorset N.F.C., vol. ix, pp. 1–40.
NEWTON, E. T., 1888. Notes on Pterodactyls. P.G.A., vol. x, pp. 406–24.
OWEN, R., 1849–84. British Fossil Reptiles. London.
——, 1861–81. A Monograph of the Fossil Reptilia of the Liassic Formations. Pal. Soc.
——, 1874–89. A Monograph on the Fossil Reptilia of the Mesozoic Formations. Pal. Soc.
SEELEY, H. G., 1869. Index to the Fossil Remains of Aves, Ornithosauria and Reptilia from the Secondary System of Strata, arranged in the Woodwardian Museum at Cambridge. Cambridge.

WOODWARD, A. S., 1885. On the Literature and Nomenclature of British Fossil Crocodilia. Geol. Mag. [3], vol. ii, pp. 496–510.
——, 1886. The History of Fossil Crocodiles. P.G.A., vol. ix, pp. 288–344.

PISCES

WOODWARD, A. S., 1886. On the Palaeontology of the Selachian Genus *Notidanus*. Geol. Mag. [3], vol. iii, pp. 205–17, 253–9.
——, 1889–1901. Catalogue of Fossil Fishes in the British Museum. Parts i–iv. [With full bibliography.] London.
——, 1890. On the Palaeontology of the Sturgeons. P.G.A., vol. xi, pp. 24–44; and The Fossil Sturgeon of the Whitby Lias. Naturalist, 1890, pp. 101–7.
——, 1890–2. A Synopsis of the Fossil Fishes of the English Lower Oolites, and Supplementary Observations; P.G.A., vol. xi, pp. 285–306, and vol. xii, pp. 238–41.
——, 1893. On some British Upper Jurassic Fish-remains, of the genera *Caturus*, *Gyrodus* and *Notidanus* [Oxford Clay, Corallian and Portland Stone]. Ann. Mag. Nat. Hist. [6], vol. xii, pp. 398–462.
——, 1906. The Study of Fossil Fishes. P.G.A., vol. xix, pp. 266–82.

INVERTEBRATA
ARTHROPODA
CRUSTACEA

DARWIN, C., 1851. A Monograph on the Fossil Lepadidæ or Pedunculated Cirripedes of Great Britain. Pal. Soc.
JONES, T. R., 1862. A Monograph of the Fossil Estheriæ [Phyllopoda]. Pal. Soc.
——, 1884. Notes on the Foraminifera and Ostracoda from the Deep Boring at Richmond. Q.J.G.S., vol. xl, pp. 765–83.
MERRETT, E. A., 1924. Fossil Ostracoda and their use in Stratigraphical Research. Geol. Mag., vol. lxi, pp. 228–38.
WITHERS, T. H., 1929. Catalogue of the Fossil Cirripedia in the Dept. of Geology; vol. i, Triassic and Jurassic. Brit. Mus. Cat. [with full bibliography].
WOODS, H., 1925–31. A Monograph of the Fossil Macrurous Crustacea of England. Pal. Soc. [with full bibliography].
WOODWARD, H., 1868. Contributions to British Fossil Crustacea. Geol. Mag., vol. v, pp. 353–7.
——, 1877. A Catalogue of British Fossil Crustacea. London.

INSECTA

BRODIE, P. B., 1845. A History of the Fossil Insects in the Secondary Rocks of England. London.
——, 1874–5. On the Correlation of Fossil Insects. Proc. Warwicks. N.F.C. for 1873, pp. 12–28, and for 1874, pp. 16–38.
COCKERELL, T. D. A., 1915. British Fossil Insects. Proc. U.S. Nat. Mus., vol. xlix (1916), pp. 469–99. [Some of Brodie's material re-described and figured.]
GOSS, H., 1878. The Insect Fauna of the Mesozoic Period. P.G.A., vol. vi, pp. 116–50.
HANDLIRSCH, A., 1906–8. Die Fossilen Insecten und die Phylogenie der Rezenten Formen. Leipzig. Jurassic, vol. i, pp. 411–660. [A complete inventory and revision.]
TILLYARD, R. J., 1925. Fossil Insects, No. 1: The British Liassic Dragon-flies (Odonata). Brit. Mus. Nat. Hist., London.
——, 1930. The Evolution of the Class Insecta. Papers and Proc. Royal Soc. Tasmania, 1930; and Nature, vol. cxxvi, pp. 996–8.
WESTWOOD, J. O., 1854. Contributions to Fossil Entomology. Q.J.G.S., vol. x, pp. 378–96. [First descriptions and figures of many insects from the Purbecks, Stonesfield Slate, Lias and Wealden; revised in Handlirsch, above.]

MOLLUSCA

LAMELLIBRANCHIA

AGASSIZ, L., 1841–5. Études critiques sur les Mollusques fossiles: (1) Mémoire sur les Trigonies, 1841; (2) Monographie des Myes, 1842–5.

ALLEN, H. A., 1905–6. Catalogue of Types and Figured Specimens of British Lamellibranchiata from the Rhætic Beds, Lias, Lower, Middle and Upper Oolites, preserved in the Museum of Practical Geology, London. Summ. Prog. Geol. Surv. for 1904, pp. 172–7, and for 1905, pp. 175–95.

ARKELL, W. J., 1930. The Generic Position of Phylogeny of some Jurassic Arcidæ. Geol. Mag., vol. lxvii, pp. 297–310 and pp. 337–52.

——, 1931. Berichtigungen der Identität gewisser jurassischer Pecten-Arten. Centralblatt für Min., &c., Abt. B., pp. 430–43.

BIGOT, A., 1893–4. Contributions à l'étude de la faune jurassique de Normandie: (1) Mémoire sur les Trigonies. (2) Mémoire sur les Opis. Mém. Soc. Linn. Normandie, vols. xvii and xviii.

COX, L. R., 1929. Notes on the Mesozoic Family Tancrediidæ, with Descriptions of several British Upper Jurassic Species, and of a New Genus, *Eodonax*. Ann. Mag. Nat. Hist. [10], vol. iii, pp. 569–94.

DEECKE, W., 1925. Fossilium Catalogus. 1. Animalia; Pars 30. Trigoniidæ Mesozoicæ (*Myophoria* exclusa).

EUDES-DESLONGCHAMPS, J. A., 1858. Essai sur les Plicatules fossiles des terrains de Calvados. Mém. Soc. Linn. Normandie, vol. vi, p. 220.

GERBER, E., 1918. Beiträge zur Kenntnis der Gattungen *Ceromya* und *Ceromyopsis*. Mém. Soc. pal. Suisse, vol. xliii.

GUILLAUME, L., 1927. Revision des Posidonomyes jurassiques. Bull. Soc. géol. France [4], vol. xxvii, pp. 217–34.

JOURDY, E., 1924. Histoire naturelle des Exogyres. Ann. Paléont., vol. xiii, pp. 1–104.

LYCETT, J., 1872–83. A Monograph of the British Fossil *Trigoniæ*. Pal. Soc.

MANSEL-PLEYDELL, J. C., 1879. On the Dorset *Trigoniæ*. Proc. Dorset N.F.C., vol. iii, pp. 111–34.

MEEK, F. B., 1864. Remarks on the Family Pteriidæ (= Aviculidæ) with Description of some New Fossil Genera. Amer. Journ. Sci., vol. xxxvii, pp. 212–20.

MOESCH, C., 1875. Monographie der Pholadomyen. Mém. Soc. pal. Suisse, vol. i.

PFANNENSTIEL, M., 1928. Organisation und Entwicklung der Gryphäen. Palaeobiologica, vol. i, pp. 381–418.

PHILIPPI, E., 1898–1900. Morphologie und Phylogenie der Lamellibranchier; Pectinidæ and Limidæ. Zeitschr. Deutsch. Geol. Gesell., vol. l, p. 596, vol. lii, pp. 64, 619.

ROLLIER, L., 1911. Les Faciès du Dogger ou Oolithique dans le Jura et les régions voisines. Mém. publié par la fondation Schuyder von Wartensee à Zurich.

——, 1912–17. Fossiles nouveaux ou peu connus des terrains secondaires du Jura et des contrées environnantes. Mém. Soc. pal. Suisse, vols. xxxviii–xlii. [Catalogues and lists of species ranged according to horizons, with many new names; also Echinoderma, Serpulæ and Brachiopoda in vol. xxxvii, 1911.]

SCHÄFLE, L., 1929. Über Lias und Doggeraustern. Geol. u. Pal. Abhandlungen, Jena, vol. xxi (N.F., vol. xvii), Heft 2, pp. 65–150.

STOLICZKA, A., 1871. The Pelecypoda, with a Review of all Known Genera of this Class, Fossil and Recent. Palaeontologia Indica; Mem. Geol. Surv. India.

WALFORD, E. A., 1885. On the Stratigraphical Position of the *Trigoniæ* of the Lower and Middle Jurassic Beds of North Oxfordshire and Adjacent Districts. Q.J.G.S., vol. xli, pp. 35–47.

GASTROPODA

ALLEN, H. A., 1903–4. Catalogue of Types and Figured Specimens of British Gastropoda from the Rhætic Beds, Lias and Lower, Middle and Upper Oolites, preserved in the Museum of Practical Geology, London. Summ. Prog. Geol. Surv. for 1902, pp. 217–28, and for 1903, pp. 175–87.

COSSMANN, M., 1895–1913. Contributions à la paléontologie française des terrains jurassiques: Études sur les Gastropodes des terrains jurassiques. Mém. Soc. géol. France, vols. v, viii, xix.

COSSMANN, M., 1895–1921. Essais de Paléoconchologie comparée. Livr. i–xii.

HUDLESTON, W. H., & WILSON, E., 1892. A Catalogue of British Jurassic Gasteropoda. London, Dulau.

CEPHALOPODA

AMMONOIDEA

BLAKE, J. F., 1892–3. The Evolution and Classification of the Cephalopoda. P.G.A., vol. xii, pp. 275–95, and vol. xiii, pp. 24–39.

BUCKMAN, S. S., 1909–30. Type Ammonites [and Yorkshire Type Ammonites]. Vols. i–vii. Indices by A. Morley Davies, 1930.

BUCKMAN, S. S., & BATHER, F. A., 1894. Can the Sexes in Ammonites be Distinguished? Nat. Science, vol. iv, pp. 427–32.

CRICK, G. C., 1898. List of Types and Figured Specimens of Fossil Cephalopoda in the British Museum (Natural History). B.M., London.

HYATT, A., 1889. The Genesis of the Arietidæ. Smithsonian Contrib. Knowl., 1889.

LANG, W. D., 1919. The Evolution of Ammonites. P.G.A., vol. xxx, pp. 49–65.

SCHINDEWOLF, O. H., 1925. Entwurf einer Systematik der Perisphincten. Neues Jahrb. für Min., &c., B.-B. liiB, pp. 309–43.

——, 1926. Zur Systematik der Perisphincten. Neues Jahrb. für Min., &c., B.-B., lv B, pp. 497–517.

SCHWARZ, E. H. L., 1894. The Aptychus. Geol. Mag. [4], vol. i, pp. 454–9.

SEITZ, O., 1929. Kritische Bemerkungen zur homöomorphen Ammoniten-Terminologie. Jahrb. Preuss. Geol. Landesanst., vol. l, pp. 148–54.

SIMPSON, M., 1843. A Monograph of the Ammonites of the Yorkshire Lias. London.

SPATH, L. F., 1919. Notes on Ammonites [general], Geol. Mag., vol. lvi, pp. 27–35, 65–71, 115–22, 170–7, 220–5.

——, 1923. A Monograph of the Ammonoidea of the Gault, Part I, pp. 7–13. [Glossary and explanation of modern terms used in ammonite descriptions.]

——, 1923. On Ammonites from New Zealand [Lias–Tithonian ammointes and classification discussed]: Appendix to Trechman's Jurassic Rocks of N.Z. Q.J.G.S., vol. lxxix, pp. 286–312.

——, 1924. On the Blake Collection of Ammonites from Kachh, India. [Review of Families and Genera.] Mem. Geol. Surv. India, vol. ix, Memoir, No. 1, pp. 1–29.

——, 1927–32. Revision of the Jurassic Cephalopod Fauna of Kachh (Cutch). Ammonoidea. Parts i–vi, Mem. Geol. Surv. India, N.S., vol. ix, Memoir, No. 2. [Unfortunately the sixth and final part has not become available at the time of going to press.]

SWINNERTON, H. H., & TRUEMAN, A. E., 1918. The Morphology and Development of the Ammonite Septum. Q.J.G.S., vol. lxxiii, pp. 26–58.

THOMPSON, C., 1913. The Derived Cephalopoda of the Holderness Drift. Q.J.G.S., vol. lxix, pp. 169–83. [Revisions by L. F. Spath, 1925–6, in the Naturalist.]

TRAUTH, F., 1927–31. Aptychenstudien. Annal. Naturhist. Mus. Wien, vol. xli, p. 171; vol. xlii, p. 121; vol. xliv, p. 329; vol. xlv, p. 17.

TRUEMAN, A. E., 1922. Aspects of Ontogeny in Ammonite Evolution. Journal of Geology, Chicago, vol. xxx, pp. 140–3.

WADDINGTON, C. H., 1929. Notes on Graphical Methods of Recording the Dimensions of Ammonites. Geol. Mag., vol. lxvi, pp. 180–6.

WRIGHT, T., 1878–86. Monograph on the Lias Ammonites of the British Islands. Pal. Soc.

——, 1880. On the Modern Classification of the Ammonitidæ. Proc. Cots. N.F.C., vol. vii, pp. 170–221.

NAUTILOIDEA

FOORD, A. H., & CRICK, G. C., 1890. Descriptions of new and imperfectly-defined species of Jurassic *Nautili* contained in the British Museum. Ann. Mag. Nat. Hist. [6], vol. v, pp. 270–90 and 388–409.

SPATH, L. F., 1927. Revision of the Jurassic Cephalopod Fauna of Kachh (Cutch). Nautiloidea. Mem. Geol. Surv. India, N.S., vol. ix, Memoir No. 2, Part I, pp. 19–35. [A Systematic revision of the fossil Nautiloidea, setting forth a general classification.]

WILLEY, A., 1902. Contribution to the Natural History of the Pearly Nautilus. Zool. Results Materials from New Britain, New Guinea, Loyalty Islands, &c.; Cambridge Univ. Press, vol. vi, pp. 691–826. [Contains exhaustive references to bibliography of the Nautiloidea.]

BELEMNOIDEA

HUXLEY, T. H., 1864. On the Structure of Belemnitidæ; with a description of a more complete specimen of a *Belemnites* than any hitherto known, and an account of a new genus of Belemnitidæ: *Xiphoteuthis*. Mem. Geol. Surv., Monograph ii.

LISSAJOUS, M., 1925. Répertoire alphabétique des Bélemnites jurassiques, précédé d'un essai de classification. Travaux du Lab. de Géologie Fac. Sci. Lyon, Fasc. viii, Mém. 7, pp. 1–173.

MILLER, J. S., 1826. Observations on Belemnites. Trans. Geol. Soc. [2], vol. ii, pp. 45–62.

NAEF, A., 1922. Die fossilen Tintenfische. Jena.

PHILLIPS, JOHN, 1865–1909. A Monograph of British Belemnitidæ: Jurassic. Pal. Soc.

THOMPSON, C., 1910. The Belemnites of the Yorkshire Lias. Trans. Hull Geol. Soc., vol. vi, pp. 99–102.

BRACHIOPODA

BUCKMAN, S. S., 1899. List of Types and Figured Specimens of Brachiopoda. Proc. Cots. N.F.C., vol. xiii, pp. 133–40.

——, 1901. Homœomorphy among Jurassic Brachiopoda. Proc. Cots. N.F.C., vol. xiii, pp. 231–90.

——, 1904. Jurassic Brachiopoda. Ann. Mag. Nat. Hist. [7], vol. xiv, pp. 389–97.

——, 1906. Brachiopod Nomenclature. Ann. Mag. Nat. Hist. [7], vol. xviii, pp. 321–7.

——, 1907. Brachiopod Morphology: *Cincta*, *Eudesia*, and the Development of Ribs. Q.J.G.S., vol. lxiii, pp. 338–43.

——, 1916. Terminology for Foraminal Development in Terebratuloids. Trans. New Zealand Inst., vol. xlviii, pp. 130–2. [Brief Anticipation of next item.]

——, 1917. The Brachiopoda of the Namyau Beds, Northern Shan States, Burma. Mem. Geol. Surv. India, N.S., vol. iii, Memoir No. 2. [Numerous British Bathonian and other brachiopods figured.]

DAVIDSON, T., 1851–84. British Fossil Brachiopoda, vols. i–v; Bibliography—vol. vi. 1886. Pal. Soc.

——, 1862. On Scottish Jurassic Brachiopoda. Geologist, vol. v, pp. 443–5 and plate.

DAVIDSON, T., & MORRIS, J., 1847. Descriptions of some Species of Brachiopoda. Ann. Mag. Nat. Hist., vol. xx, pp. 250–7.

HUDLESTON, W. H., & WALKER, J. F., 1877. On the Distribution of the Brachiopoda in the Oolitic Strata of Yorkshire. Ann. Rept. Yorks. Phil. Soc. for 1876, pp. 7–12.

MARSHALL, J. W. D., 1896. Notes on the British Jurassic Brachiopoda. Proc. Bristol Nat. Soc. [3], vol. viii, pp. 17–40.

MOORE, C., 1860–1. On New [Jurassic] Brachiopoda, and on the development of the Loop in *Terebratula*. Geologist, vol. iii, p. 445, vol. iv, pp. 96–9, 190–4.

MORRIS, J., 1846. On the Subdivision of the Genus *Terebratula*. Q.J.G.S., vol. ii, pp. 382–9.

ROLLIER, L., 1915–19. Synopsis des Spirobranches (Brachiopodes) jurassiques Celto-Souabes. Mém. Soc. pal. Suisse, vols. xli–xliv.

SAHNI, M. R., 1928. Morphology and Evolution of certain Jurassic Terebratulids. Ann. Mag. Nat. Hist. [10], vol. ii, pp. 114–38.

TATE, R., 1869. Additions to the List of Brachiopoda of the British Secondary Rocks. Geol. Mag., vol. vi, pp. 550–6.

WALKER, J. F., 1869–70. On Secondary Species of Brachiopoda. Proc. Yorks. Nat. Club, 1869, p. 214; Geol. Mag., vol. vii, 1870, pp. 560–4.

——, 1889. On Oolitic Brachiopoda new to Yorkshire. Ann. Rept. Yorks. Phil. Soc. for 1888, pp. 37–40.

——, 1893. On the Brachiopoda recently discovered in the Yorkshire Oolites. Ann. Rept. Yorks. Phil. Soc. for 1892, pp. 47–51.

POLYZOA

GREGORY, J. W., 1894. Catalogue of the Jurassic Bryozoa in the York Museum [with figures]. Ann. Rept. Yorks. Phil. Soc. for 1893, pp. 58–61.

——, 1895–6. A Revision of the British Jurassic Bryozoa. Ann. Mag. Nat. Hist. [6], vol. xv, pp. 223–8; vol. xvi, pp. 447–51; vol. xvii, pp. 41–9, 151–5, 194–201, 287–95.

——, 1896. Catalogue of the Fossil Bryozoa in the Department of Geology, British Museum. Jurassic Bryozoa. London.

LANG, W. D., 1904. Jurassic forms of the 'Genera' Stomatopora and Proboscina. Geol. Mag. [5], vol. i, pp. 315–22.

——, 1905. On Stomatopora antiqua Haime and its Related Liassic Forms. Geol. Mag. [5], vol. ii, pp. 258–68.

——, 1913. Report of a visit to the exhibits of Polyzoa and corals in the Geological Department of the British Museum. P.G.A., vol. xxiv, pp. 189–93 [with many references].

LONGE, F. D., 1881. On the Relation of the Escharoid Forms of Oolitic Polyzoa to the Cheilostomata and Cyclostomata. Geol. Mag. [2], vol. viii, pp. 23–34.

VINE, G. R., 1880–1. A Review of the Family Diastoporidæ for the Purpose of Classification. Q.J.G.S., vol. xxxvi, pp. 356–61; and Further Notes: Species from the Lias and Oolites, vol. xxxvii, pp. 381–90.

——, 1883. Third Report of the Committee on Fossil Polyzoa. Rept. Brit. Assoc. for 1882, p. 249.

——, 1886–7. Jurassic Polyzoa in the Neighbourhood of Northampton [Inferior Oolite to Oxford Clay]. Journ. Northants Nat. Hist. Soc., vol. iv, pp. 202–11; and in the Gayton Boring, Northants, ibid., 1887, pp. 255–66.

ECHINODERMATA

CARPENTER, P. H., 1882. On some New or little-known Jurassic Crinoids. Q.J.G.S., vol. xxxviii, pp. 29–43.

CRONIES, C., & MCCORMACK, J., 1932. Fossil Holothuroidea [with useful bibliography]. Journ. Palaeont., vol. vi, pp. 111–48.

DEECKE, W., 1928. Über die Jura-Seeigel. Palaeobiologica, vol. i, pp. 419–56.

FORBES, E., 1844. On the Fossil Remains of Starfishes of the Order Ophiuridæ, found in Britain. Proc. Geol. Soc., vol. iv, pp. 232–4.

HAWKINS, H. L., 1912. Classification, Morphology and Evolution of the Echinoidea Holectypoidea. Proc. Zool. Soc., 1912, pp. 440–97.

——, 1917. Morphological Studies in the Echinoidea Holectypoidea and their Allies.
I. Pygaster and Plesiechinus. Geol. Mag. [6], vol. iv, pp. 160–9.
II. Discoides and Conulus. Ibid., pp. 196–205.
III. Holectypus. Ibid., pp. 249–56.
IV. Pygasteridæ. Ibid., pp. 342–50.

——, 1919. The Morphology and Evolution of the Ambulacrum in the Echinoidea Holectypoidea [with bibliography]. Phil. Trans. Roy. Soc., vol. ccix B, pp. 377–480.

M'COY, F., 1848. On some New Mesozoic Radiata. Ann. Mag. Nat. Hist. [2], vol. ii, pp. 397–420.

MOORE, C., 1873. On the Presence of [Holothuria] in the Inferior Oolite and Lias. Rept. Brit. Assoc. for 1872, pp. 117–18.

UPTON, C., 1917. Notes on Chirodota-Spicules [a Holothurian] from the Lias and Inferior Oolite. Proc. Cots. N.F.C., vol. xix, pp. 115–18.

WHITEAVES, J. F., 1861. On the Oolitic Echinodermata of the Neighbourhood of Oxford. Geologist, vol. iv, pp. 174–5.

WRIGHT, T., 1851–3. On the Cidaridæ and Cassidulidæ of the Oolites, with Descriptions of some New Species of those Families. Ann. Mag. Nat. Hist. [2], vol. viii, pp. 241–80, and vol. ix, pp. 81–103; and 1853, Proc. Cots. N.F.C., vol. i, pp. 134–229.

——, 1854. Contributions to the Palaeontology of Gloucestershire. A Description and Figures of some New Species of Echinodermata from the Lias and Oolites. Ann. Mag. Nat. Hist. [2], vol. xiii, pp. 161–73, 312–24, 376–83.

——, 1857–80. Monograph on the British Fossil Echinodermata of the Oolitic Formations. Pal. Soc.

WRIGHT, T., 1860. On a new Genus of Fossil Cidaridæ (*Hemipedina*), with a Synopsis of the Species included therein, and on some New Species from the Oolites. Proc. Cots. N.F.C., vol. ii, pp. 121–30.

——, 1860. A Description of some New Species of Echinodermata from the Lias and Oolites. Proc. Cots. N.F.C., vol. ii, pp. 17–48.

COELENTERA: CORALS

DUERDEN, J. E., 1902–6. The Morphology of the Madreporaria. [Series of papers: for full references see:] Ann. Mag. Nat. Hist. [7], vol. xviii, p. 226 (1906).

FAUROT, L., 1909. Affinités des Tétracoralliaires et des Hexacoralliaires. Ann. de Paléont., vol. iv, pp. 69–107.

GREGORY, J. W., 1900. Jurassic Fauna of Cutch., vol. ii, part ii, The Corals. Mem. Geol. Surv. India, Ser. ix. [Contains modern revision of systematics and nomenclature.]

LANG, W. D., 1917. Homœomorphy in Fossil Corals. P.G.A., vol. xxviii, pp. 85–94.

MILNE-EDWARDS, H., & HAIME, J., 1850–4. A Monograph of the British Fossil Corals. And Supplement by DUNCAN, P. M., 1866.

TOMES, R. F., 1900. Contributions to a History of the Mesozoic Corals of the County of York. Proc. Yorks. Geol. Soc., vol. xiv, pp. 72–85.

WRIGHT, T., 1867. On Coral Reefs Present and Past. Proc. Cots. N.F.C., vol. iv, pp. 97–173.

PORIFERA (SPONGES)

HINDE, G. J., 1883. Catalogue of the Fossil Sponges in the British Museum. 4to. London.

——, 1887–1912. A Monograph of the British Fossil Sponges. Pal. Soc.

PROTOZOA: FORAMINIFERA

CRICK, W. D., & SHERBORN, C. D., 1891. On some Liassic Foraminifera from Northamptonshire. Journ. Northants N.H.S., vol. vi, pp. 208–14.

JONES, T. R., 1882. Catalogue of the Fossil Foraminifera in the British Museum. 8vo. London.

JONES, T. R., & PARKER, W. K., 1871. On Terquem's Researches on the Foraminifera of the Lias and Oolites. Ann. Mag. Nat. Hist. [4], vol. viii, pp. 361–5.

—— ——, 1875. List of some English Jurassic Foraminifera. Geol. Mag. [2], vol. ii, pp. 308–11.

JONES, T. R., & SHERBORN, C. D., 1886. On the Microzoa found in some Jurassic Rocks of England. Geol. Mag. [3], vol. iii, pp. 271–4.

PLANTS

ARBER, E. A. N., & PARKIN, J., 1907–8. On the Origin of Angiosperms. Journ. Linn. Soc., vol. xxxviii, p. 29, 1907. Studies on the Evolution of the Angiosperms. Ann. Bot., vol. xxii, p. 48, 1908.

BROWN, A., 1894. On the Structure and Affinities of the Genus *Solenopora*, with Descriptions of New Species. Geol. Mag. [4], vol. i, pp. 145–51.

CARRUTHERS, W., 1867. On Gymnospermatous Fruits from the Secondary Rocks of Britain. Journ. Botany, vol. v, pp. 1–21.

——, 1867–9. On some Cycadian Fruits from the Secondary Rocks of Britain. Geol. Mag., vol. iv, pp. 101–6, and vol. vi, pp. 97–9.

——, 1869. On some Undescribed Coniferous Fruits from the Secondary Rocks of Britain. Geol. Mag., vol. vi, pp. 1–7.

——, 1870. On the Fossil Cycadean Stems from the Secondary Rocks of Britain. Trans. Linn. Soc., vol. xxvi, pp. 675–708, and Geol. Mag., vol. vii, pp. 573–7.

GARDNER, J. S., 1886. On Mesozoic Angiosperms [*Williamsonia*, &c.]. Geol. Mag. [3], vol. iii, pp. 193–204, 342–5.

GARWOOD, E. J., 1931. Important Additions to our Knowledge of the Fossil Calcareous Algæ since 1913. . . . Q.J.G.S., vol. lxxxvii, Jurassic, pp. xcii–xcvii [with bibliography].

HARRIS, G. F., 1896. The Analysis of Oolitic Structure. P.G.A., vol. xiv, p. 59.

SCOTT, D. H., 1907. The Flowering Plants of the Mesozoic Age, in the Light of Recent Discoveries. Journ. Roy. Micro. Soc., pp. 129–41.

SEWARD, A. C., 1892. Fossil Plants as Tests of Climate. London.

——, 1894. Notes on the Bunbury Collection of Fossil Plants. Proc. Phil. Soc. Cambridge, vol. viii, part 3, p. 187.

——, 1898–1919. Fossil Plants. Cambridge.

——, 1900. Notes on some Jurassic Plants in the Manchester Museum. Mem. Manchester Lit. and Phil. Soc., vol. xiv, part 3, p. 1.

——, 1900–4. Catalogue of the Mesozoic Plants in the Department of Geology, British Museum: vol. i, 1900, Jurassic Flora of Yorkshire Coast; vol. ii, 1904, Jurassic Flora outside Yorkshire.

——, 1901–2. Jurassic Flora of East Yorkshire. Rept. Brit. Assoc. for 1900, p. 756; and for 1901, p. 856; also Geol. Mag. [4], vol. viii, p. 36.

——, 1910. Comparisons of Jurassic Floras. Nature, vol. lxxxv, pp. 258–9.

——, 1911. The Jurassic Flora of Yorkshire. Naturalist, 1911, pp. 1–8, 85–94.

——, 1924. Later Records of Plant Life: Post-Carboniferous Floras. Q.J.G.S., vol. lxxx, pp. lxxxix–xcvii.

——, 1931. Plant Life through the Ages. Cambridge [Jurassic Period, ch. xiv, pp. 335–69].

STOPES, M. C., 1912. Petrifications of the [then] Earliest European Angiosperms. Phil. Trans. Roy. Soc., vol. cciii B, pp. 75–100.

WETHERED, E., 1889. On the Microscopic Structure of the Jurassic Pisolite. Geol. Mag. [3], vol. vi, pp. 196–200.

——, 1895. The Formation of Oolite. Q.J.G.S., vol. li, pp. 196–209.

PALAEOGEOGRAPHY

ANDRÉE, K., 1913, Die paläogeographische Bedeutung sediment-petrographischer Studien. Peterm. Mitteil., Jahrg. lxi, vol. ii, p. 121.

ARLDT, T., 1917–21. Handbuch der Paläogeographie. Berlin (Borntraeger).

BRINKMANN, R., 1924. Der Dogger und Oxford des Südbaltikums. Jahrb. Preuss. Geol. Landesanst., vol. xliv, pp. 477–513.

BUCKMAN, S. S., 1922–3. [British Jurassic] Palaeogeography [with map]. Type Ammonites, vol. iv, pp. 20–5, and 52–4.

CRICKMAY, C. H., 1931. Jurassic History of North America: its Bearing on the Development of Continental Structure. Proc. Amer. Phil. Soc., vol. lxx, pp. 15–102.

DACQUÉ, E., 1913. Paläogeographische Karten und die gegen sie zu erhebenden Einwände. Geol. Rundschau, vol. iv, pp. 186–206.

——, 1915. Grundlagen und Methoden der Paläogeographie. Jena, 1915.

DAVIS, W. M., 1895. The Development of Certain English Rivers. Geograph. Journ., vol. v, pp. 127–46 [peneplains described].

——, 1896. The Peneplain of the Scotch Highlands. Geol. Mag. [4], vol. iii, pp. 525–8.

DUTERTRE, A. P., 1926. Les Aucelles de terrains jurassiques supérieurs du Boulonnais. Bull. Soc. géol. France [4], vol. xxvi, pp. 395–408 [with discussion of boreal connexion].

FREBOLD, H., 1930. Verbreitung und Ausbildung des Mesozoikums in Spitzbergen. Skrifter om Svalbard og Ishavet, No. 31, 1930; also No. 20, 1929. [Numerous palaeogeographical maps and discussion.]

FRECH, F., 1902. Studien über das Klima der geologischen Vergangenheit. Zeitschr. Gesell. für Erdkunde, 1902, pp. 611–93.

GOODCHILD, J. G., 1889. An Outline of the Geological History of the Eden Valley. P.G.A., vol. xi, pp. 277–8, and Trans. Cumb. and Westm. Assoc., no. xiv, pp. 73–90.

GOTHAN, W., 1912. Die Frage der Klimadifferenzierung im Jura und in der Kreideformation im Lichte paläobotanischer Tatsachen. Jahrb. Preuss. Geol. Landesanst., vol. xxix, p. 220.

GREEN, J. F. N., 1920. The Geological Structure of the Lake District. P.G.A., vol. xxxi, pp. 109–26.

GREGORY, J. W., 1907. Climatic Variations, their Extent and Causes. Compte rendu, Congrès géol. int., 1906, Mexico, p. 474.

GROVES, A. W., 1931. The Unroofing of the Dartmoor Granite and the Distribution of its Detritus in the Sediments of Southern England. Q.J.G.S., vol. lxxxvii, pp. 62–96.

HUBBARD, G. D., & WILDER, C. G., 1930. Validity of the Indicators of Ancient Climates. Bull. Geol. Soc. America, vol. xli, pp. 275–92.

JONES, O. T., 1931. Some Episodes in the Geological History of the Bristol Channel Region. Presid. Address, Sect. C; Rept. Brit. Assoc. for 1930, pp. 57–82.

JUKES-BROWNE, A. J., 1911. The Building of the British Isles. 3rd ed.

JÜNGST, H., 1927. Die Meeresverbindung Nord–Süd-Deutschland in der Psiloceraten-Zeit. Neues Jahrb. für Min., &c., B.-B. lviii, Abt. B, pp. 171–214.

KAUENHOWEN, W., 1927. Die Faziesverhältnisse und ihre Beziehungen zur Erdöl-bildung an der Wende Jura-Kreide in Nordwestdeutschland. Neues Jahrb. für Min., &c., B.-B. lviii, Abt. B, pp. 215–72.

LAPPARENT, A. DE, 1906. Traité de Géologie, vol. ii.

MARR, J. E., 1906. The Influence of the Geological Structure of the English Lakeland upon its Present Features. Q.J.G.S., vol. lxii, pp. lxvi–cxxviii.

——, 1929. Deposition of the Sedimentary Rocks.

NEUMAYR, M., 1883. Ueber klimatische Zonen während der Jura- und Kreidezeit. Denkschr. k. Akad. Wiss. Wien, M.–N. Klasse, vol. xviii, pp. 277–310.

——, 1885. Die geographische Verbreitung der Juraformation. Denkschr. k. Akad. Wiss. Wien, vol. l, pp. 57–142.

ORTMANN, A. E., 1896. An Examination of the Arguments given by Neumayr for the Existence of Climatic Zones in Jurassic Times. American Journ. Sci. [4], vol. i, pp. 257–70.

POMPECKJ, J. F., 1908. Die zoogeographischen Beziehungen zwischen den Jurameeren Nordwest- und Süddeutschlands. 1. Jahresber. Niedersächs. Geol. Ver. Hannover, p. 10.

RICHARDSON, L., 1903. Mesozoic Geography of the Mendip Archipelago. Proc. Cots. N.F.C., vol. xiv, pp. 59–72.

SALFELD, H., 1921. Das Problem des borealen Jura und der borealen Unterkreide. Centralblatt für Min., &c., vol. xxii, pp. 169–74.

SAWICKI, L., 1912. Die Einebnungsflächen in Wales und Devon. Sitzungsber. Warschauer Gesell. Wiss., Lief. 2, pp. 123–34.

SCHOTT, W., 1930. Paläogeographische Untersuchungen über den Oberen Braunen und Unteren Weissen Jura Nordwestdeutschlands. Inaugural-Dissertation, Göttingen, 1930; and Abh. Preuss. Geol. Landesanst., N.F., vol. cxxxiii.

SCHUCHERT, C., 1915. Climates of Geologic Time. Ann. Rept. Smithsonian Inst. for 1914, pp. 277–311.

SEMPER, M., 1908. Die Grundlagen paläogeographischer Untersuchungen. Centralblatt für Min., &c., vol. ix, pp. 434–45.

SORBY, H. C., 1852. On the Direction of Drifting of the Sandstone Beds of the Oolitic Rocks of the Yorkshire Coast. Proc. Yorks. Phil. Soc., vol. i, pp. 111–13.

TRAVIS, C. B., 1914. Some Evidences of Peneplanation in the British Isles. Proc. Liverpool Geol. Soc. (Presid. Address), vol. xii, pp. 3–31.

TROTTER, F. M., & HOLLINGWORTH, S. E., 1928. The Alston Block. Geol. Mag., vol. lxv, pp. 433–47.

TROTTER, F. M., 1929. The Tertiary Uplift of the Alston Block. Proc. Yorks. Geol. Soc., N.S., vol. xxi, pp. 161–80.

UHLIG, V., 1911. Die marinen Reiche des Jura und der Unterkreide. Mitteil. Geol. Gesell. Wien, vol. iv, pp. 329–448.

WATTS, W. W., 1911. Geology as Geographical Evolution. Q.J.G.S., vol. lxvii, pp. lxii–xciii.

WILLIS, B., 1910. Principles of Palaeogeography. Science, N.S., vol. xxxi, pp. 241–60.

WILLS, L. J., 1929. The Physiographical Evolution of Britain. London.

WOOD, S. V., 1862. On the Form and Distribution of the Land-Tracts during the Secondary and Tertiary Periods respectively. . . . Phil. Mag. [4], vol. xxiii, pp. 161–71, 269–82, 382–93.

SUBJECT INDEX

Note: To keep the index within reasonable bounds, place-names are not included, but the page headings are intended to act as a geographical as well as a stratigraphical guide. The order of describing each formation is from south to north—from Dorset to Yorkshire and Scotland, and lastly the Weald of Kent.

The only personal names indexed are those of certain pioneers, whose work is specially dealt with in Part I.

INDEX OF STRATA

(Stage-names not indexed, *see* Appendix II, p. 617.)

CORRIGENDA

Pp. 390, 400. *Isognomon* [= *Perna*] *mytiloides* Lamk et auct. The name will have to be changed owing to its having been preoccupied for an Inferior Oolite species by Hermann (1781). Where mentioned the species is probably *I. subplana* (Étallon).

Pp. 468–70. The *Astarte supracorallina* of Blake, Roberts, Woodward, and others, is not *A. supracorallina* d'Orbigny but *A. mysis* d'Orbigny, which is probably synonymous with *A. extensa* Phillips. Both these matters will be dealt with in forthcoming instalments of 'A Monograph of the British Corallian Lamellibranchia'.

P. 613. Explanation of Plate XL, fig. 3. *Provirgatites scythicus* was first named and figured by Vischniakoff in 1882 ('Descr. des Planulati (Perisphinctes) jurass. de Moscou', pl. iii, figs. 1, 2). The types came from Mniovniki near Moscow. Michalski's specimens are at least chorotypes.